Health-Promoting Food Ingredients during Processing

Health-Promoting Food Ingredients during Processing presents a comprehensive science-based approach covering the latest naturally occurring bioactive compounds in seeds, dietary fiber, proteins, fermented bio-compounds, agro-industrial waste by-products, and lactic acid bacteria. A volume in the Food Biotechnology and Engineering Series, the book discusses their identification, characterization, biological activities in terms of their bioavailability, bioaccessibility, and their beneficial effects as inflammatory mediators, probiotics, antioxidants, and hypoglycemic agents, as well as in gastrointestinal digestion and colonic fermentation. Written by an international expert team of food scientists, nutritionists, food biotechnologists, food engineers, and chemists, the book explains how this leads to opportunities in the treatment of diseases such as obesity, diabetes, hypertension, cancer and cognitive disorders.

Key features include:

- Presents original research and relevant peer-reviewed articles written by experts in disciplines such as food science, nutrition, food biotechnology, food engineering, and chemistry.
- Highlights new emerging trends, discoveries, and applications of biologically active compounds from seeds, dietary fiber, proteins, and agro-industrial waste by-products.
- Provides readers with a comprehensive, science-based approach to the identification, characterization, and utilization of food macromolecules, probiotics, lactic acid bacteria, and bioactive compounds.
- Discusses the impact of bioactive compounds in plants, agroindustrial by-products, and fermented biocompounds regarding their bioavailability, bioaccessibility, and potential human health benefits.

Food Biotechnology and Engineering
Series Editor: Octavio Paredes-López

Volatile Compounds Formation in Specialty Beverages
Edited by Felipe Richter Reis and Caroline Mongruel Eleutério dos Santos

Native Crops in Latin America
Biochemical, Processing, and Nutraceutical Aspects
Edited by Ritva Repo-Carrasco-Valencia and Mabel C. Tomás

Starch and Starchy Food Products
Improving Human Health
Edited by Luis Arturo Bello-Perez, Jose Alvarez-Ramirez, and Sushil Dhital

Bioenhancement and Fortification of Foods for a Healthy Diet
Edited by Octavio Paredes-López, Oleksandr Shevchenko, Viktor Stabnikov, and Volodymyr Ivanov

Latin-American Seeds
Agronomic, Processing and Health Aspects
Edited by Claudia Monika Haros, María Reguera, Norma Sammán, and Octavio Paredes-López

Advances in Plant Biotechnology
In Vitro *Production of Secondary Metabolites of Industrial Interest*
Edited by Alma Angélica Del Villar-Martínez, Juan Arturo Ragazzo-Sánchez, and Pablo Emilio Vanegas-Espinoza

Bioconversion of Wastes to Value-added Products
Edited by Olena Stabnikova, Oleksandr Shevchenko, Viktor Stabnikov and Octavio Paredes-López

Health-Promoting Food Ingredients during Processing
Edited by Rosalva Mora-Escobedo, Gloria Dávila-Ortiz, Gustavo F. Gutiérrez López, and José J. Berrios

For more information about this series, please visit: www.routledge.com/Food-Biotechnology-and-Engineering/book-series/CRCFOOBIOENG

Health-Promoting Food Ingredients during Processing

Edited by
Rosalva Mora-Escobedo, Gloria Dávila-Ortiz,
Gustavo F. Gutiérrez López, and José J. Berrios

CRC Press
Taylor & Francis Group
Boca Raton London New York

CRC Press is an imprint of the
Taylor & Francis Group, an **informa** business

Designed cover image: Shutterstock

First edition published 2025
by CRC Press
2385 NW Executive Center Drive, Suite 320, Boca Raton FL 33431

and by CRC Press
4 Park Square, Milton Park, Abingdon, Oxon, OX14 4RN

CRC Press is an imprint of Taylor & Francis Group, LLC

ISBN: 9781032365367 (hbk)
ISBN: 9781032365398 (pbk)
ISBN: 9781003332558 (ebk)

DOI: 10.1201/9781003332558

Typeset in Kepler Std
by Newgen Publishing UK

This book is dedicated to our dearest and most respected mentors, professors:

-Dr. Gloria Dávila-Ortíz

-Dr. Efrén Parada-Arias

-Dr. Lidia Dorantes-Álvarez

-Dr. Octavio Paredes-López

Pioneers of the food science and technology studies in Mexico and founders of the postgraduate programs on this subject at the National Polytechnic Institute of this country. They are a vivid example of vision, and love for our discipline; their legacy has endured and has been a solid reference for many generations of scientists.

Contents

Series Preface

BIOTECHNOLOGY – OUTSTANDING FACTS

The beginning of agriculture started about 12,000 years ago and has since played a key role in food production. We look to farmers to provide the food we need but at the same time, now more than ever, to farm in a manner compatible with the preservation of the essential natural resources of the earth. Additionally, besides the remarkable positive aspects that farming has had throughout history, several undesirable consequences have been generated. The diversity of plants and animal species that inhabit the earth is decreasing. Intensified crop production has had undesirable effects on the environment (i.e., chemical contamination of groundwater, soil erosion, exhaustion of water reserves). If we do not improve the efficiency of crop production in the short term, we are likely to destroy the very resource base on which this production relies. Thus, the role of so-called sustainable agriculture in the developed and underdeveloped world, where farming practices are to be modified so that food production takes place in stable ecosystems, is expected to be of strategic importance in the future; but the future has already arrived.

The biotechnology of plants is a key player in these scenarios of the 21st century. Today, molecular biotechnology in particular is receiving increasing attention because it has the tools of innovation for the agriculture, food, chemical, and pharmaceutical industries. It provides the means to translate an understanding of, and ability to modify, plant development and reproduction into enhanced productivity of traditional and new products. Plant products from seeds, fruits, and plant components and extracts are being produced with better functional properties and longer shelf life; and they need to be assimilated into the commercial agriculture to offer new options to small, industries and finally to consumers. Within these strategies it is imperative to select crops with larger proportions of edible parts as well, thus generating less waste; it is also imperative to consider the selection and development of a more environmentally friendly agriculture.

The development of research innovations for products is progressing, but the constraints of relatively long times to reach the marketplace, intellectual property rights, uncertain profitability of the products, consumer acceptance, and even caution and fear with which the public may view biotechnology are tempering the momentum of all but the most determined efforts. Nevertheless, it appears uncontestable that food biotechnology of plants and microbials is and will emerge as a strategic component to provide food crops and other products required for human well-being.

FOOD BIOTECHNOLOGY AND ENGINEERING SERIES/ OCTAVIO PAREDES-LOPEZ, PH.D., SERIES EDITOR

The Food Biotechnology and Engineering Series aims at addressing a range of topics around the edge of food biotechnology and food microbial world. In the case of foods, it includes agrigenomics, molecular biology, genetic engineering, metabolic aspects, omic sciences, chemistry, nutrition, medical foods and health ingredients, and processing and engineering with traditional- and innovative-based approaches. Environmental aspects and strategies to produce green foods cannot be left aside. At the world level there are foods and beverages produced by different types of microbial technologies to which this series will give attention. It will also consider genetic modifications of microbial cells to produce nutraceutical ingredients and advances in studies on the human microbiome as related to diets.

Preface

Sustainable health of the diverse groups of the population is based on a personal involvement acquired regarding the quality and quantity of the daily food intake and is of paramount importance for the maintenance of a healthy and fulfilling life. For this reason, it is relevant to identify and take advantage of key areas within food science and technology that impact directly on societal wellbeing. And, in this regard, a preventative approach is important concerning proper selection of food and food ingredients. Efforts in favor of this aim include the publication of suitable literature that covers strategic subjects helping to understand and guide students and professionals working in food-related areas.

This book deals with recent advances in health food-related areas with special attention on extraction and delivery of bioactive compounds, as well as process engineering aspects of the production of active compounds, including green processes focused on the key substances such as prebiotic and probiotics, antioxidant capacity, microstructural related aspects, active peptides obtained from complex matrixes, novel dispersed systems as carriers of valuable substances, and colonic physiology including health-promoting agents. It is also considered neoplasia of the gastrointestinal track related to consumption of specific diets and compounds, cancer cell culture, and design of scaffolds for microstructural studies, micro- and nanotechnological aspects of compounds such as chitosan to study DNA recovery, description of health effects of fiber consumption, and relationship with antioxidant capacity of local bioresources at different maturity stages. Local and global sources of healthy food ingredients are discussed with specific emphasis on the description of key healthy ingredients, and in this respect, the overall benefits of such resources are provided as well.

The book is divided in three sections:

- Identification, characterization, use, and effect of bioactive compounds;
- Determination of biological activities in natural products with potential human health benefits; and
- Use of technologies for the scaling up and obtention of bioactive compounds (nutraceuticals) at industrial levels.

The sections are designed to provide a thematic structure to the volume and allow readers to identify particular needs of information and solutions to daily works. Distinguished food scientists and engineers from key institutions have contributed chapters that give a deep analysis of the different subjects.

Readers can find evidence of the multidisciplinary and complex context that influences a healthy diet throughout the different book components. They can also discover how the complex relationships between structure, function, and process influence overall well-being and susceptibility to chronic non-infectious diseases and metabolic syndrome. The common goal is to emphasize the multiple interactions between food intake and health. This book aims to be a valuable addition to the existing literature on healthy food topics and a reference on the matter. Healthy diets, designed according to daily requirements, are the best vaccines for improving the immune system of human beings in the XXIst century.

EDITORS

Rosalva Mora-Escobedo, Gloria Dávila-Ortiz, Gustavo F. Gutiérrez-López, and José J. Berrios

Contributors

Liliana Alamilla-Beltrán, Ph.D.
Instituto Politécnico Nacional, Escuela Nacional de Ciencias Biológicas, Campus Zacatenco, Departamento de Ingeniería Bioquímica, Ciudad de México, México

Guadalupe M. Alvarado-Jasso, Ph.D.
Instituto de Productos Lácteos de Asturias-Consejo Superior de Investigaciones Científicas, Department of Microbiology and Biochemistry of Dairy Products, Villaviciosa Asturias, Spain

Luis G. Bermúdez-Humarán, Ph.D.
Université Paris-Saclay, INRAE, AgroParisTech, Micalis Institute, Jouy-en-Josas, France

Francisco J. Blancas-Benítez, Ph.D.
Tecnológico Nacional de México/Instituto Tecnológico de Tepic, Nayarit, México

José E. Borges-Martínez, Ph.D.
Instituto Politécnico Nacional, Escuela Nacional de Ciencias Biológicas, Campus Zacatenco, Departamento de Ingeniería Bioquímica, Ciudad de México, México

Carolina Buitrago-Arias, Ph.D.
Instituto Politécnico Nacional, Centro de Desarrollo de Productos Bióticos, Yautepec Morelos, México

Gabriela Caez-Ramírez, Ph.D.
Universidad de La Sabana, Facultad de Ingeniería, Campus Puente del Común, Grupo de Investigación en Procesos Agroindustriales, Bogotá, Colombia

Georgina Calderón-Domínguez, Ph.D.
Instituto Politécnico Nacional, Escuela Nacional de Ciencias Biológicas, Campus Zacatenco, Departamento de Ingeniería Bioquímica, Ciudad de México, México

Brenda H. Camacho-Díaz, Ph.D.
Instituto Politécnico Nacional, Centro de Desarrollo de Productos Bióticos, Yautepec Morelos, México

Edén Cano-Sampedro, Ph.D.
Instituto Politécnico Nacional, Escuela Nacional de Ciencias Biológicas, Campus Zacatenco, Departamento de Ingeniería Bioquímica, Ciudad de México, México
Instituto Politécnico Nacional, Unidad Profesional Interdisciplinaria de Ingeniería Campus Tlaxcala, Tlaxcala, México

Maricela Carmona-González, Ph.D.
Instituto Politécnico Nacional, Centro de Desarrollo de Productos Bióticos, Yautepec Morelos, México

Germán A. Chamorro-Cevallos, Ph.D.
Instituto Politécnico Nacional, Escuela Nacional de Ciencias Biológicas Campus Zacatenco, Departamento de Farmacia, Ciudad de México, México

Veera C.S.R. Chittepu, Ph.D.
Instituto Politécnico Nacional, Escuela Nacional de Ciencias Biológicas, Campus Zacatenco, Departamento de Ingeniería Bioquímica, Ciudad de México, México
Harvard Medical School, Department of Medicine, Boston, MA, USA

Stephanie Cid-Gallegos, Ph.D.
Instituto Politécnico Nacional, Unidad Interdisciplinaria de Biotecnología, Ciudad de México, México

Iram Cisneros-García, M.D.
Instituto Politécnico Nacional, Escuela Nacional de Ciencias Biológicas, Campus Zacatenco, Departamento de Ingeniería Bioquímica, Ciudad de México, México

Maribel Cornejo-Mazón, Ph.D.
Instituto Politécnico Nacional, Escuela Nacional de Ciencias Biológicas, Campus Casco de Santo Tomás, Departamento de Biofísica, Ciudad de México, México

Rebeca Dávila-León, M.D.
Instituto Politécnico Nacional, Escuela Nacional de Ciencias Biológicas, Campus Zacatenco, Ciudad de México, México

Gloria Dávila-Ortiz, Ph.D.
Universidad Autónoma Metropolitana Unidad Lerma, Departamento de Ciencias de la Alimentación, División de Ciencias Biológicas y de la Salud, Estado de México, México

Clara G. de los Reyes-Gavilán, Ph.D.
Instituto de Productos Lácteos de Asturias-Consejo Superior de Investigaciones Científicas, Department of Microbiology and Biochemistry of Dairy Products, Villaviciosa Asturias, Spain

Edmundo del Valle-Gallegos, Ph.D.
Instituto Politécnico Nacional, Escuela Superior de Física y Matemáticas, Departamento de Física, Ciudad de México, México

Carlos R. Díaz-Águila, Ph.D.
Universidad de la Habana, Centro de Biomateriales, Cuba

Ruth N. Domínguez-Fernández, Ph.D.
Instituto Politécnico Nacional, Centro Interdisciplinario de Ciencias de la Salud, Ciudad de México, México

Wendoly Domínguez-Rojas, M.D.
Instituto Politécnico Nacional, Escuela Nacional de Ciencias Biológicas, Unidad Zacatenco, Departamento de Ingeniería Bioquímica, Ciudad de México, México

Luisa F. Duque-Buitrago, Ph.D.
Instituto Politécnico Nacional, Escuela Nacional de Ciencias Biológicas, Campus Zacatenco, Ciudad de México, México

María del Rocío Espinosa-Velázquez, Ph.D.
Universidad Anáhuac México Sur, Facultad de Ciencias de la Salud, Ciudad de México, México

Tzayhri Gallardo-Velázquez, Ph.D.
Instituto Politécnico Nacional, Escuela Nacional de Ciencias Biológicas, Campus Casco de Santo Tomás, Departamento de Biofísica, Ciudad de México, México

Evangelina García-Armenta, Ph.D.
Universidad Autónoma de Sinaloa, Ciudad Universitaria, Facultad de Ciencias Químico-Biológicas, Culiacán, México

Ruth M. García-García, Ph.D.
Instituto Politécnico Nacional, Escuela Nacional de Ciencias Biológicas, Campus Zacatenco, Departamento de Ingeniería Bioquímica, Ciudad de México, México

Alitzel B. García-Hernández, M.D.
Universidad Autónoma Metropolitana Unidad Lerma, Departamento de Ciencias de la Alimentación, División de Ciencias Biológicas y de la Salud, Estado de México, México

Vanessa García-Rojas, Ph.D.
Instituto Politécnico Nacional, Escuela Nacional de Ciencias Biológicas, Campus Zacatenco, Departamento de Ingeniería Bioquímica, Ciudad de México, México

Sandra E. García-Solís, M.D.
Instituto Politécnico Nacional, Escuela Nacional de Ciencias Biológicas, Campus Zacatenco, Departamento de Ingeniería Bioquímica, Ciudad de México, México

Cynthia Garfias-Noguez, M.D.
Instituto Politécnico Nacional, Escuela Nacional de Ciencias Biológicas, Campus Zacatenco, Departamento de Ingeniería Bioquímica, Ciudad de México, México

Marcela González-Montoya, Ph.D.
Instituto Politécnico Nacional, Escuela Nacional de Ciencias Biológicas, Campus Zacatenco, Departamento de Ingeniería Bioquímica, Ciudad de México, México
Universidad Michoacana de San Nicolás de Hidalgo, Instituto de Investigaciones Químico-Biológicas, Morelia, Michoacán, México

Marcela González-Vázquez, Ph.D.
Instituto Politécnico Nacional, Escuela Nacional de Ciencias Biológicas, Campus Zacatenco, Departamento de Ingeniería Bioquímica, Ciudad de México, México

Gustavo F. Gutiérrez-López, Ph.D.
Instituto Politécnico Nacional, Escuela Nacional de Ciencias Biológicas, Campus Zacatenco, Departamento de Ingeniería Bioquímica, Ciudad de México, México

Humberto Hernández-Sánchez, Ph.D.
Instituto Politécnico Nacional, Escuela Nacional de Ciencias Biológicas, Campus Zacatenco, Departamento de Ingeniería Bioquímica, Ciudad de México, México

Luis Huerta-González, Ph.D.
Instituto Politécnico Nacional, Centro de Investigación en Biotecnología Aplicada, Tlaxcala, México

Miguel Ángel A. Ibáñez-Hernández, M.D.
Instituto Politécnico Nacional, Escuela Nacional de Ciencias Biológicas, Campus Casco de Santo Tomás, Departamento de Bioquímica, Ciudad de México, México

Incinur Hasbay, Ph.D.
TUBITAK Marmara Research Center Food Institute, Gebze, Kocaeli, Turkey

Judith Jaimez-Ordaz, M.D.
Universidad Autónoma del Estado de Hidalgo, Área Académica de Química, Pachuca, Hidalgo, México

María Eugenia Jaramillo-Flores, Ph.D.
Instituto Politécnico Nacional, Escuela Nacional de Ciencias Biológicas, Campus Zacatenco, Departamento de Ingeniería Bioquímica, Ciudad de México, México

Cristian Jiménez-Martínez, Ph.D.
Instituto Politécnico Nacional, Escuela Nacional de Ciencias Biológicas, Campus Zacatenco, Departamento de Ingeniería Bioquímica, Ciudad de México, México

M. Faridy Juárez-Chairez, Ph.D.
Instituto Politécnico Nacional, Escuela Nacional de Ciencias Biológicas, Campus Zacatenco, Departamento de Ingeniería Bioquímica, Ciudad de México, México
Universidad Autónoma de Zacatecas, Unidad Académica de Ciencias Biológicas, Zacatecas, México

Poonam Kalhotra, Ph.D.
Instituto Politécnico Nacional, Escuela Nacional de Ciencias Biológicas, Campus Casco de Santo Tomás, Departamento de Biofísica, Ciudad de México, México
Harvard Medical School, Dana-Farber Cancer Institute, Department of Medical Oncology, Boston, MA, USA

Diana E. Leyva-Daniel, Ph.D.
Instituto Politécnico Nacional, Escuela Nacional de Ciencias Biológicas, Unidad Zacatenco, Departamento de Ingeniería Bioquímica, Ciudad de México, México

Karen E. Lima-Villegas, Ph.D.
Instituto Politécnico Nacional, Escuela Nacional de Ciencias Biológicas, Campus Zacatenco, Departamento de Ingeniería Bioquímica, Ciudad de México, México

Raúl E. López-Hernández, Ph.D.
Instituto Politécnico Nacional, Escuela Nacional de Ciencias Biológicas, Campus Zacatenco, Departamento de Ingeniería Bioquímica, Ciudad de México, México

Silvia Luna-Suárez, Ph.D.
Instituto Politécnico Nacional, Centro de Investigación en Biotecnología Aplicada, Tlaxcala, México

Deyanira del R. Moguel-Concha, Ph.D.
Instituto Politécnico Nacional, Escuela Nacional de Ciencias Biológicas, Campus Zacatenco, Departamento de Ingeniería Bioquímica, Ciudad de México, México

Ilse Monroy-Rodríguez, Ph.D.
Universidad Autónoma del Estado de Hidalgo, Área Académica de Química, Ciudad del Conocimiento, Hidalgo, México

Rosalva Mora-Escobedo, Ph.D.
Instituto Politécnico Nacional, Escuela Nacional de Ciencias Biológicas, Campus Zacatenco, Departamento de Ingeniería Bioquímica, Ciudad de México, México

J. Ismael Morales-Camacho, Ph.D.
Universidad de las Américas Puebla, Departamento de Ingeniería Química, Alimentos y Ambiental, Puebla, México

Gema Morales-Olán, Ph.D.
Colegio de Postgraduados Campus Córdoba, Veracruz, México

Eduardo Morales-Sánchez, Ph.D.
Instituto Politécnico Nacional, CICATA – Unidad Querétaro, Querétaro, México

Mayra Nicolás-García, Ph.D.
Universidad Autónoma Metropolitana Unidad Lerma, Departamento de Ciencias de la Alimentación, División de Ciencias Biológicas y de la Salud, Estado de México, México
Tecnológico Nacional de México/ITS de Teziutlán, Ingeniería en Industrias Alimentarias, Puebla, México

Juana Angélica Ortega-Cárdenas, Ph.D.
Instituto Politécnico Nacional, Escuela Nacional de Ciencias Biológicas, Campus Zacatenco, Departamento de Ingeniería Bioquímica, Ciudad de México, México

Alicia Ortiz-Moreno, Ph.D.
Instituto Politécnico Nacional, Escuela Nacional de Ciencias Biológicas, Campus Zacatenco, Departamento de Ingeniería Bioquímica, Ciudad de México, México

Guillermo Osorio-Revilla, Ph.D.
Instituto Politécnico Nacional, Escuela Nacional de Ciencias Biológicas, Campus Zacatenco, Departamento de Ingeniería Bioquímica, Ciudad de México, México

Luis E. Palomo-Martínez, M.D.
Instituto Politécnico Nacional, Escuela Nacional de Ciencias Biológicas, Campus Zacatenco, Departamento de Ingeniería Bioquímica, Ciudad de México, México
Instituto Politécnico Nacional, Escuela Nacional de Ciencias Biológicas, Campus Casco de Santo Tomás, Departamento de Bioquímica, Ciudad de México, México

María de Jesús Perea-Flores, Ph.D.
Instituto Politécnico Nacional, Centro de Nanociencias y Micro y Nanotecnologías, Ciudad de México, México

Claudia Pérez-Cruz, Ph.D.
Instituto Politécnico Nacional, Centro de Nanociencias y Micro y Nanotecnologías, Ciudad de México, México

Brenda Pérez-Grijalva, Ph.D.
Universidad Anáhuac Veracruz, Campus Xalapa, Facultad de Ingeniería. Xalapa, Veracruz

Madeleine Perucini-Avendaño, Ph.D.
Universidad Autónoma Metropolitana Unidad Lerma, Departamento de Ciencias de la Alimentación, División de Ciencias Biológicas y de la Salud, Estado de México, México
Universidad Autónoma de Querétaro, Facultad de Química, Programa de Investigación y Posgrado en Ciencia de los Alimentos, Querétaro, México

Lorenzo A. Picos-Corrales, Ph.D.
Universidad Autónoma de Sinaloa, Ciudad Universitaria, Facultad de Ingeniería, Culiacán, México

Morayma Ramírez-Damián, M.D.
Instituto Politécnico Nacional, Escuela Nacional de Ciencias Biológicas, Campus Zacatenco, Departamento de Ingeniería Bioquímica, Ciudad de México, México

Ismael Ramírez-Flores, M.D.
Instituto Politécnico Nacional, Escuela Nacional de Ciencias Biológicas, Campus Zacatenco, Departamento de Ingeniería Bioquímica, Ciudad de México, México

Minerva Rentería-Ortega, Ph.D.
Tecnológico Nacional de México/ TES de San Felipe del Progreso, División de Ingeniería en Industrias Alimentarias, Estado de México, México

Sandra Rivera-Gutiérrez, Ph.D.
Instituto Politécnico Nacional, Escuela Nacional de Ciencias Biológicas, Campus Casco de Santo Tomás, Departamento de Microbiología, Ciudad de México, México

María del Carmen Robles-Ramírez, Ph.D.
Instituto Politécnico Nacional, Escuela Nacional de Ciencias Biológicas, Campus Casco de Santo Tomás, Departamento de Microbiología, Ciudad de México, México

Patricia Rosales-Martínez, Ph.D.
Instituto Politécnico Nacional, Escuela Nacional de Ciencias Biológicas, Campus Casco de Santo Tomás, Departamento de Microbiología, Ciudad de México, México

Nuria Salazar, Ph.D.
Instituto de Productos Lácteos de Asturias-Consejo Superior de Investigaciones Científicas, Department of Microbiology and Biochemistry of Dairy Products, Villaviciosa Asturias, Spain

Ma de la Paz Salgado-Cruz, Ph.D.
Instituto Politécnico Nacional, Escuela Nacional de Ciencias Biológicas, Campus Zacatenco, Departamento de Ingeniería Bioquímica, Ciudad de México, México

María Elena Sánchez-Pardo, Ph.D.
Instituto Politécnico Nacional, Escuela Nacional de Ciencias Biológicas, Campus Zacatenco, Departamento de Ingeniería Bioquímica, Ciudad de México, México

Viviana Sandoval-Flores, Ph.D.
Instituto Politécnico Nacional, Escuela Nacional de Ciencias Biológicas, Campus Zacatenco, Departamento de Ingeniería Bioquímica, Ciudad de México, México

Sonia G. Sáyago-Ayerdi, Ph.D.
Tecnológico Nacional de México/Instituto Tecnológico de Tepic, Nayarit, México

Fátima S. Serrano-Villa, M.D.
Tecnológico Nacional de México/Instituto Tecnológico de Tepic, Nayarit, México

María del Carmen Silva-Lucero, Ph.D.
Universidad Nacional Autónoma de México, Escuela de Medicina, Departamento de Fisiología, Ciudad de México, México

Luz I. Sotelo-Díaz, Ph.D.
Universidad de La Sabana, Facultad de Ingeniería, Campus Puente del Común, Maestría en Diseño y Gestión de Procesos, Bogotá, Colombia

Daniel Tapia-Maruri, Ph.D.
Instituto Politécnico Nacional, Centro de Desarrollo de Productos Bióticos, Morelos, México

Darío I. Téllez-Medina, Ph.D.
Instituto Politécnico Nacional, Escuela Nacional de Ciencias Biológicas, Campus Zacatenco, Departamento de Ingeniería Bioquímica, Ciudad de México, México

Antonio Tornero-Martínez, Ph.D.
Instituto Politécnico Nacional, Escuela Nacional de Ciencias Biológicas, Unidad Zacatenco, Departamento de Ingeniería Bioquímica, Ciudad de México, México

Edgar Torres-Maravilla, Ph.D.
Instituto Politécnico Nacional, Escuela Nacional de Ciencias Biológicas, Campus Zacatenco, Ciudad de México, México

Katherine S. Torres-Ortiz, Ph.D.
Instituto Politécnico Nacional, Escuela Nacional de Ciencias Biológicas, Campus Zacatenco, Ciudad de México, México

Juscelino Tovar, Ph.D.
Instituto Politécnico Nacional, Escuela Nacional de Ciencias Biológicas, Campus Zacatenco, Ciudad de México, México

Arturo A. Velázquez-López, M.D.
Instituto Politécnico Nacional, Escuela Nacional de Ciencias Biológicas Campus Zacatenco, Departamento de Ingeniería Bioquímica, Ciudad de México, México

Julieta del Carmen Villalobos-Espinosa, Ph.D.
Instituto Politécnico Nacional, Escuela Nacional de Ciencias Biológicas Campus Zacatenco, Departamento de Ingeniería Bioquímica, Ciudad de México, México

Hernani Yee-Madeira, Ph.D.
Instituto Politécnico Nacional, Escuela Nacional de Ciencias Biológicas Campus Zacatenco, Departamento de Ingeniería Bioquímica, Ciudad de México, México

Víctor M. Zamora-Gasga, Ph.D.
Tecnológico Nacional de México/Instituto Tecnológico de Tepic, Nayarit, México

Acknowledgements

The editors of *Health-Promoting Food Ingredients during Processing*, a volume in the Food Biotechnology and Engineering series, would like to express their sincere gratitude to Professor Octavio Paredes-López, Series Editor, for considering us as the appropriate editors of a book for this collection.

The editors and authors of the chapters of this book are grateful for the financial support provided by the National Polytechnic Institute (Instituto Politécnico Nacional), Commission for the Operation and Promotion of Academic Activities (COFAA-IPN), and National Council of Humanities, Sciences and Technologies (CONAHCYT).

We would like to thank the authors who contributed chapters to this book, as well as our families for their continuous support. We would also like to express our gratitude to the students who assisted us, particularly Viviana Sandoval-Flores and Vanessa García-Rojas.

Editor Bios

Rosalva Mora-Escobedo is a professor of Nutrition Sciences at the National School of Biological Sciences (ENCB) of the National Polytechnic Institute (IPN) in Mexico. She earned her Ph.D. in Food Science from ENCB-IPN after conducting research at the Centre for Advanced Studies and Research (CINVESTAV) in Irapuato, Mexico. She also holds an MSc in Food Science and Technology from the Autonomous University of Queretaro, Mexico, and a BSc in Chemistry from the Autonomous University of San Luis Potosi, Mexico. Professor Mora-Escobedo received postdoctoral training in the Department of Food Science and Technology at the University of Reading, UK. She is a member of the Mexican Academy of Sciences and an Emeritus Researcher of the National Researcher System of the National Council for Humanities, Science, and Technology in Mexico. Mora-Escobedo has published research articles and book chapters on seeds as a source of bioactive compounds. She coedited a book on the same matter, focusing on amaranth, corn, honey, agave, and prickly pear. Currently, she leads a research group studying seed germination and the influence of seed consumption on diseases such as cancer, cardiovascular conditions, obesity, and diabetes. Her group conducts research on the effects of germination on the anticancer properties of soy protein, as well as the identification and mechanism of action of bioactive compounds responsible for the cytotoxic effect of germinated soybeans.

Gloria Dávila-Ortiz is a professor on the Biochemistry of Vegetable Proteins at the National School of Biological Sciences (ENCB) of the National Polytechnic Institute (IPN), Mexico. She received her Ph.D. in Biochemistry at ENCB and holds a BSc in Biochemical Engineering at the same institution. She is an Emeritus Researcher of the National Researcher System of the National Council for Humanities, Science and Technology, Mexico and is a former director of the Center for the Development of Biotic Products (CEPROBI-IPN). She has received a number of important awards including the *Lázaro Cardenas* medal, given to outstanding members of IPN for their lifetime academic career and received the National Merit Award of the National Council for Humanities, Science and Technology, Mexico, which recognizes the professional and international research trajectory on food science; the *Coatlicue Award*, which is the highest honor granted by the International Coordination of Women, in the Arts to meritorious scientists, artists, journalists and anthropologists; and the Research Award, which is a recognition given by IPN for scientific work. Dávila-Ortiz is an expert in the areas of proteins, protein concentrates, hydrolyzed proteins, bioactive peptides from cereals, legumes (beans, broad beans, chickpea, lupines) and has carried out outstanding research on proteomics, and peptidomics, Her research work has allowed her to coauthored numerous articles and book chapters and, her work on the characterization of bioactive compounds and microstructural changes has impacted the chemical, pharmaceutical, food, and energy industries in Mexico.

Gustavo F. Gutiérrez-López is currently a Professor of Food Engineering at the National School of Biological Sciences (ENCB) of the National Polytechnic Institute (IPN), Mexico and is an Emeritus Researcher of the National Researcher System of the National Council for Humanities, Science and Technology, Mexico. He has successfully accomplished extensive and intensive activities as a leader of the Food Engineering international community by leading academic societies and organizing scientific meetings and was the Chairman of the PhD program in Food Science at ENCB-IPN for 17 consecutive years. Gutiérrez-López received his PhD and MSc in Food Engineering from the University of Reading, UK, a MSc in Food Science and Technology (IPN), and holds a BSc in Biochemical Engineering (IPN). He also carried out postdoctoral training at the University of Reading, UK and, in the Society of Chemical Industry of the United Kingdom as a Seligman-APV Fellow. Gutiérrez-López has consistently worked on structure-function properties of food related materials, focusing in non-linear dynamics and dispersed systems and has authored numerous publications in the fields of food engineering, biological structures including macro, micro and nano-technologies, is co-editor of important books on the subject and an is, currently, associated editor of the prestigious Food Engineering Reviews journal and has been an invited editor of the Journal of Food Engineering (for which he also serves in the Editorial Board) and the Microorganisms journal. He received the *Lázaro Cardenas* medal, given to outstanding members of the IPN community for lifetime academic merits, the IAEF (International Association for Engineering and Food) Lifetime Achievement Award, he is an IAFoST Fellow (IUFoST), an ISOPOW honorary member, and belongs to the ISEKI Food Academy.

José J. Berrios is a Lead Scientist with USDA-ARS at the Western Regional Research Center (WRRC) in California. He is nationally and internationally recognized for his extensive work in applied and basic research studies in Food Science and Agriculture. Leader of the extrusion food program at WRRC, developing processes and formulations to increase legume utilization through the manufacture of new extruded foods and food products with optimized nutritional and functional properties and desirable sensory attributes; pasta products and beverages; high protein and dietary fiber, gluten-free products with low GI and potential cholesterol lowering activity. Author and co-author of more than 100 publications and presentations. External Adviser to Canada´s Pulse Innovation Project, the American Pulse Association, Frito Lay, the U.S.A. Dry Pea and Lentil Council, and several National food companies. Collaborator with National and International organizations: University of California, Davis, CA; Washington State University, Pullman, WA; Kansas State University, Manhattan, KS; University Complutense of Madrid and IATA-CSIC, Valencia, Spain; AgroSup Dijon, France; Nacional University of Panama; National University of Colombia; University of Campinas, EMBRAPA (Rio de Janeiro and Goiania), University of Rio de Janeiro, and University of Pelotas, Brazil; University of Nuevo Leon, National Polytechnic Institute, and CIAD (Guadalajara and Culiacan), Mexico.

Section I

Identification, Characterization, Use, and Effect of Bioactive Compounds

Chapter 1

Soybean

Understanding Benefits of Germination and Obtention of Bioactive Peptides

Viviana Sandoval-Flores
María del Carmen Robles-Ramírez
Vanessa García-Rojas
Edén Cano-Sampedro
Marcela González-Montoya
and Rosalva Mora-Escobedo

CONTENTS

1.1 INTRODUCTION

Soybean has been consumed daily by individuals in East Asian countries, such as China, Japan, and Indochina, for centuries. It is among the most studied food crops worldwide and is known

DOI: 10.1201/9781003332558-2

for its affordability, wide availability, and exceptional nutritional composition (Hu et al. 2020). Owing to their biomolecules and functional groups, soybean consumption has been linked to several health benefits (Agyei et al. 2019). Over time, soybean use has widely spread, making it the most produced legume worldwide. Based on data sources from the United States Department of Agriculture (USDA), Brazil was the primary producer at 40%, followed by the United States at 28%, Argentina at 12%, China at 5%, and India at 3%. Paraguay and Canada each contributed 2%, while the remaining 8% was produced in other countries. Notably, these leading countries possess agricultural systems with advanced technology, a basic requirement for soybean production. Additionally, the dietary guidelines of approximately forty nations, including eight Latin American countries, recommend soybean consumption, according to the Food and Agriculture Organization of the United Nations (FAO). Furthermore, due to environmental awareness, sustainable diets based primarily on plant foods have become a trend; in this context, a diet based on soybean is an excellent choice since the beans contain all the essential amino acids for adults in amounts comparable to animal protein scores (Salinas & López-Sobaler, 2017).

Moreover, this diet is approximately 70% cheaper than a kilogram of beef and requires significantly less water for production, with approximately 4,000 L/kg compared to over 14,000 L for beef production (Hoekstra & Mekonnen, 2012).

In terms of nutritional composition, soybean seeds contain one of the highest protein levels (36–39 g/100 g). In addition, the Protein Digestibility Corrected Amino Acid Score (PDCAAS) of soybean is 1.0, which is close to that of some animal sources, such as meat and dairy products (Qin et al. 2022; Zhao et al. 2021). Soybean also has a high lipid content (16.5–19.9 g/100 g), which are composed of polyunsaturated and monounsaturated fatty acids, especially omega-3 and omega-6 (1.3 and 9.9 g/100 g, respectively). In addition, the dietary fiber content is high (17–25 g/100 g), and more than twenty vitamins and minerals are present. However, it is important to note that the composition of soybean depends on factors, such as variety, location, and climate of cultivation (Robles-Ramírez et al. 2019).

1.2 CHANGES IN THE SOYBEAN BIOACTIVE COMPOUNDS DUE TO GERMINATION

Health benefits attributed to soybean could be enhanced through processing methods such as germination (Waliat et al. 2023). Germination is frequently practiced improving nutritional profiles in numerous crops. The method is gaining increasing attention due to its ability to modify nutritional compositions and generate new bioactive compounds during the process (Hu et al. 2020). The simple act of soaking seeds in water and subsequent imbibition triggers a plethora of intricate biochemical changes in seeds. As seeds develop into seedlings or sprouts, phytochemicals and other metabolites are synthesized, broken down, and chemically modified (Li et al. 2019). This process involves a series of biochemical and morphogenetic mechanisms to transform the embryo into a seedling that can feed itself. Germination begins with seed imbibition, which activates metabolism, protein synthesis, and the mobilization of reserves. The process ends when the embryo elongates and the testa ruptures, allowing the radicle to emerge (Bewley, 1997). As germination mobilizes the nutrients needed for seedling development, the content of certain bioactive compounds is reduced, such as phytates, oligosaccharides, tannins, lectins, and protease inhibitors. These compounds play an essential role in the secondary metabolism of plants, as reserve materials that are used during the germination process, and as defense mechanisms

TABLE 1.1 TOTAL PHENOLICS, ISOFLAVONES, AND SAPONINS CONTENT IN GERMINATED SOYBEAN PROTEIN ISOLATES

Germination Time (days)	Total Phenolics (μg/g)	Saponins (μg/g)	Genistein (μg/g)	Daidzein (μg/g)
0	1342	3942	26	49
2	2225	5464	69	170
3	2295	6165	73	194
4	2576	7066	73	190
5	3145	8070	83	405

Units: μg/g of dry soybean protein isolate.

against microorganisms, insects, and animals (Guzmán-Ortiz et al. 2014; Muzquiz et al. 2012). Mature seeds contain the needed compounds that participate in the mechanisms that are crucial for next-generation plant growth. Germination depends on the regulation of phytohormones, including gibberellic acid (GA), abscisic acid (ABA), ethylene, and auxin. GA and ABA are the two most important hormones that play central and antagonistic roles in the regulation of seed germination, in which GA promotes and ABA inhibits germination. An increase in GA content is always accompanied by a decrease in ABA content during germination (Angelovici et al. 2010). **Table 1.1** shows the germination-induced changes in total phenolics, isoflavones (genistin, genistein, and daidzein), and saponins of soybean protein during the germination process. An increase of almost 200% in the total phenolic content after 5 days of germination was observed. Notably, there were also increases of 100% in saponins, 200% in genistein, and a remarkable 700% in daidzein. The increase in these compounds during germination was reported by different authors (Bueno et al. 2020; Eum et al. 2020; Guzmán-Ortiz et al. 2019).

The role of phenolic compounds against microorganisms, insects, and parasitic plants has been studied, as well as in the reduction of oxidative stress and as signaling molecules to ensure the survival of the seedling (Ndakidemi & Dakora, 2003; Zhu et al. 2016). The content of these compounds in germinated soybean meal is highly variable depending on genetic and environmental factors, such as cultivar, year of cultivation, location, and temperature (Lin & Lai, 2006). Values have been reported from 48 μg/g (Kim et al. 2006) to more than 10,000 μg/g (Dikshit & Ghadle, 2003).

A main metabolic event that occurs during germination is the hydrolysis of storage proteins. This process mobilizes the reserve proteins in the cotyledon protein bodies, producing intermediate molecular weight peptides and amino acids. Also, different studies have demonstrated that germination generates peptides with anticancer, antioxidant, anti-inflammatory, antiobesity, antihypertensive, and antidiabetic effects. These bioactive peptides can interact with the metabolic pathways and signaling cascades that control cell growth, proliferation, and differentiation, cell survival or death, regulating the expression of therapeutic targets and modulating the activity of critical enzymes. Several studies have shown that soybean germination generates peptides with antiproliferative effects against (Caco-2, HT-29, HCT-116) human colon cancer cells, as well as cervical (HeLa, C-33, SiHa, CasKi) and breast (MCF-7 and MDA-MB-231) cancer cell lines (González-Montoya, Hernández-Ledesma, Mora-Escobedo, et al. 2018; González-Montoya, Hernández-Ledesma, Silván, et al. 2018). Additionally, germinated soybean protein hydrolysates exhibited antiadipogenic and lipolytic effects when proven in adipocyte-like cells after (3T3-L) murine fibroblast line differentiation. The positive effects of the peptides derived from germinated soybean protein on tumor growth were also confirmed by an *in vivo assay,* feeding germinated soybean protein to female nude mice inoculated with HeLa cancer cells (González-Espinosa de los Monteros et al. 2011).

Robles-Ramírez, Ramón-Gallegos, Mora-Escobedo, et al. (2012) compared with the casein-fed control, tumor volumes after 5 weeks of treatment were reduced by 44.6%, 98.9%, 97.7%, 94.7%, and 92.7% in animals fed diets with 20% soybean protein: ungerminated, 2-day-germinated, 2-day-germinated without ethanol-soluble phytochemicals (ESPC), 6-day-germinated, and 6-day-germinated without ESPC, respectively. Soybean protein isolates can retain notable amounts of ESPC, such as isoflavones and saponins, which could exhibit antiobesity and anticancer activity (Wang et al. 2017; Boutas et al. 2022); these compounds were removed from the protein isolates to study merely protein activity.

The antiobesity effect of germinated soybean protein was demonstrated by feeding obese C57/BL mice diets with 20% 2-day germination, with and without ESPC. After 4 weeks of treatment, the body weight decreased by 23.5% and 28% in mice fed diets with 2 days of germination and 2 days of germination without ESPC, respectively. The body weight of animals fed diets with casein and ungerminated soybean protein increased by 2.9 and 3.3%, respectively (González-Espinosa de los Monteros et al. 2011). These findings highlight the potential of germination as a process to obtain functional foods or nutraceuticals for cancer and obesity prevention.

1.3 SOYBEAN PROTEIN: A SOURCE OF BIOACTIVE PEPTIDES

The major compound of soybean is protein, at 35–40%. High-molecular-weight reserve proteins, β conglycinin, and glycinin, represent 50 to 70% of total soybean proteins, classified by their solubility in albumins, globulins, prolamins, and glutelins. Albumins represent 10–20% of protein legumes, while globulins represent 70% of the total proteins (González-Montoya et al. 2017). Globulin is a soluble fraction in saline solutions and can be subdivided into two primary components: 11S globulin (glycinin) and 7S globulin (β-conglycinin). Glycinin, with a molecular mass of 360 kDa, is composed of six constituent subunits (A1aB2, A2B1a, AB, A5A4B3, A3B4, and A1bB2), each consisting of an acidic (20 kDa) and a basic (34 kDa) polypeptide linked by a disulfide bond. The β-conglycinin subunit (150 to 200 kDa) is a trimeric glycosylated protein (Osman et al. 2020). Due to the high amount of protein, soybean is an excellent source of peptides from food. Peptides are defined as short chains between 2 and 50 amino acids that are joined together by peptide bonds, derived from dietary proteins, and present in an encrypted form; this form is active once released by specific methods, such as hydrolysis with digestion enzymes *in vivo*, *in vitro*, or *in silico* or by microbial fermentation (Tu et al. 2018).

Bioactive peptides are acknowledged for their roles as therapeutic agents, nutraceuticals, and functional food components that contribute to human health promotion through disease prevention, with a heightened focus on their study, as illustrated in **Figure 1.1**.

Discovering and predicting new bioactive peptides is an increasingly appreciated research area (Mooney et al. 2012). Recently, scientists have begun to isolate, characterize, and identify numerous bioactive peptides from various food proteins, including vegetal or animal proteins such as milk, soybean, amaranth, fish, or flesh. These proteins exert several biological functions through different action mechanisms, such as interaction with cell receptors, hormone activity, enzymatic regulation, and interference with cell cycles (Wang & de Mejia, 2005; Kim et al. 2021).

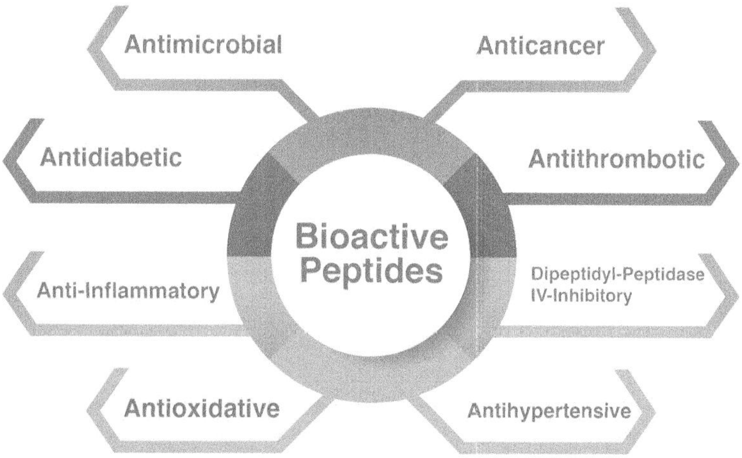

Figure 1.1 Biological activities of peptides. Peptides have a crucial role in some biological processes. The figure illustrates the versatility of their bioactivities, suggesting their potential application as alternative treatments.

González-Montoya, et al. (2018) reported the anti-inflammatory properties of peptides derived from germinated soybean hydrolyzed protein through the inhibition of nitric oxide and pro-inflammatory prostaglandin (PGD2) production in LPS-induced macrophages as well as their cytotoxic effect on human colon cancer cells lines (Caco-2, HT-29, HCT-116). Furthermore, researchers found that some peptides derived from germinated soybean showed antidiabetic potential by inhibiting dipeptidyl peptidase-IV, α-amylase, and α-glucosidase enzymes (González-Montoya, Hernández-Ledesma, Silván, et al. 2018). **Table 1.2** presents examples of peptides derived from plant-based protein sources, such as algae, amaranth, and soybean. Fragments of 4–6 amino acids were obtained from hydrolysis with different digestive enzymes. These peptides were tested in cancer cell models with positive results, as the IC_{50} value, and its mechanism of action were determined (Nwachukwu & Aluko, 2019).

As mentioned above, soybean proteins and their derivate peptides can play an essential role in the prevention and treatment of diseases, exerting a positive effect due to their health-promoting properties (González-Montoya, Hernández-Ledesma, Mora-Escobedo, et al. 2018; Robles-Ramírez, Ramón-Gallegos, Reyes-Duarte, et al. 2012). Peptides with biological activity, such as trypsin inhibitors, have been reported in soybeans. These are low-molecular-weight proteins and perform several functions in seeds, including blocking the use of protein reserves to prevent early germination. However, their function in soybean is to prevent the attack of predators (Grosse-Holz & van der Hoorn, 2016).

The most abundant soybean protease inhibitors are the Kunitz (KSTI) and Bowman-Birk (BBI) peptides, found in the 2S fraction of soybean protein (Wang & de Mejia, 2005). BBI is a 71 amino acid peptide with a molecular weight of approximately 8 kDa and a rigid structure stabilized by seven disulfide bridges that can inhibit the trypsin enzyme and chymotrypsin through two independent, active sites, Lis16-Ser17 and Leu43-Ser44, respectively (Barac et al. 2005). In contrast, the soybean Kunitz Trypsin Inhibitor (KTI) is a large peptide of 181 amino acids (21.5 kDa) that interacts with trypsin through the Arg 63-Ileu 64 junction, which is present in its active site. As

TABLE 1.2 BIOACTIVE PEPTIDES OBTAINED FROM VEGETABLE PROTEIN

Food Source	Peptide	Hydrolytic Enzymes	Mechanism of Action	Cell Model	References
Sarcophyton glaucum	AGAPGG AERQ RDTQ	Papain	Cell Penetration by Amphiphilicity	HeLa Human Cervical Cancer Cells	Quah et al. (2019)
Spirulina platensis	HVLSRAPR	Pepsin, Tripsin and Chymotrypsin	Cell Selectivity	Human Colon Cancer HT-29	Z. Wang & Zhang (2017)
Amaranthus hypochondriacus	SSEDIKE	Alcalase	Negative Regulation of the NF-Kb Pathway	Human Colorectal Cancer Caco-2	Moronta et al. (2016)
Glycine max	VFDGEL IETWNPNNKP FEITPEKNPQ	Pepsin and Pancreatin	Inhibition of Human Topoisomerase II	Lymphocytic Leukemia L 12 10	Huang et al. (2011)
Glycine max	KMLPSYSPY	Thermoase and Ethanol Extraction	Not Investigated	Murine Tumor Cells P388D1	S. E. Kim et al. (2000)
Germinated soybean (Glycine max)	NNDDRDS VVNPDNNEN LSSTEAQQS NAENNQRN IKSQSES QQQQQGGSQSQ QEPQESQQ QQQQQGGSQSQSQKG PETMQQQQQQ	Pepsin and Pancreatin	Inhibition of Nitric Oxide and PGD2 Production (Anti-Inflammatory Activity)	Caco-2, HT-29, HCT-116 Human Colon Cancer Cells	González-Montoya, Hernández-Ledesma, Mora-Escobedo, et al. (2018)

Source: Adapted from Nwachukwu & Aluko (2019).

part of the diet, protease inhibitors decreased protein digestibility, caused pancreas hypertrophy and hyperplasia, and decreased growth in mice. However, the anticancer effects of these peptides have been demonstrated in *in vitro* and *in vivo* studies (Kennedy,1998; Castro-Guillén & Blanco-Labra, 2010; García-Gasca et al. 2012).

Jae et al. (2005) determined that the BBI content in soybean cotyledons gradually increased until six days of germination and totally declined on day nine.

However, other studies report a decrease in the quantity or activity of trypsin inhibitors after 48 h of germination (Donangelo et al. 1995; Bau et al. 1997; Khalil et al. 2006; Kumar et al. 2010). Guzmán-Ortiz et al. (2014) observed a gradual trypsin inhibitor diminution in soybean during germination, reaching a 27% decrease after 6 days. Lectins (hemagglutinins) are another bioactive protein found in all organisms, including plants, bacteria, and viruses, especially in legumes (Muzquiz et al. 2012). Soybean lectin is a tetrameric protein that recognizes D-galactosyl residues. The molecular weight of this tetramer is 120 kDa and constitutes 1–2% of the seed protein. An *in vitro* assay showed that lectins can agglutinate erythrocytes, a recognized physiological effect that results from their high affinity to bind to specific carbohydrates on the surface of cells. Hence, the primary negative impact of lectins, which leads to reduced nutrient absorption and hindered nutrient transfer, arises from their ability to bind with carbohydrates on the epithelial surface of the small intestine (Lajolo & Genovese, 2002; Muzquiz et al. 2012). Some studies indicate that lectins from some plants may exhibit anticarcinogenic and antitumor activity, increasing the susceptibility of tumor cells to macrophage attack or inducing apoptosis (Ganguly & Das, 1994). Germination decreased the lectin content of Brazilian soybean (cultivar BRS 133) by 59%. Germination conditions were determined through response surface methodology with temperatures between 18 and 32°C and germination time between 12 and 72h. Optimal conditions were 25°C for 42 h. In contrast to Brazilian soybean (cultivar BRS 258), an increase in germination time from 12h to 72h at 25°C promoted a decrease in lectin concentration of 72.6% (Paucar-Menacho, Berhow, Mandarino, Chang, et al. 2010; Paucar-Menacho, Berhow, Mandarino, de Mejia, et al. 2010).

Lunasin is a novel peptide with demonstrated anticancer activity. This 43-amino acid peptide isolated from the soybean cotyledon contains nine aspartic residues at its carboxyl end, which is preceded by a cell adhesion motif formed by Arg-Gly-Asp (RGD), in addition to a helix structure homologous to a conserved region of chromatin-bound proteins Glu-Lys-His-Ile-Met-Glu-Lys-Ile-Gln-Gly (EKHIMEKIQG). Studies showed that Lunasin goes inside the cell via its RGD adhesion motif, collocate with hypoacetylated chromatin, and specifically bind to histones H3 and H4, inhibiting their acetylation; this step is crucial in cell proliferation and cancer formation (Galvez et al. 2001). Robles-Ramírez, Ramón-Gallegos, Mora-Escobedo, et al. (2012) reported that pepsin and pancreatin enzymatic hydrolysis of a germinated soybean protein isolate generated peptides with antiproliferative activities. The >5 kDa peptide fraction obtained from the germinated soybean protein hydrolysate was the most active against HeLa cervical cancer cells (IC_{50}=1.83 mg/mL). This fraction downregulated PTTG1 and TOP2A mRNA expression (two genes that are therapeutic targets) and induced apoptosis in cancer cells via the extrinsic and intrinsic pathways, activating the caspase cascade. Further analysis by HPLC-MS/MS of this fraction showed that the peptide Ala-Asp-Cys-Asn-Gly-Ala-Cys-Ser-Pro-Phe-Glu-Val-Pro-Pro-Cys-Arg-Cys-Val-Pro-Ile-Gly-Leu-Phe-Val-Gly-Phe-Cys-Ile-His-Pro-Thr-Gly, (ADCNGACSPFEVPPCRCVPIGLFVGFCIH PTG) could be responsible for its anticancer activity. Likewise, González-Montoya, Hernández-Ledesma, Silván, et al. (2018) found that the 5–10 kDa fraction of a germinated soybean protein hydrolysate was the most active against Caco-2, HT-29, and HCT-116 human colon cancer cells. This peptide fraction was composed of β-conglycine and glycine peptides rich in glutamine.

1.4 OBTENTION OF BIOACTIVE PEPTIDES

In recent years, interest in employing peptides as therapeutic agents has increased, and this area has received significant attention.

The conventional strategy to obtain bioactive peptides involves extracting and isolating proteins from different sources, including plants, microorganisms, and animals, followed by protein fractionation and subsequent hydrolysis. In the next sections, the most applied procedures will be described, including enzymatic, fermentation-based, and *in silico* methods.

1.4.1 Peptides Obtained Through Enzymatic Processes

The use of enzymes is the most commonly employed method for obtaining peptides. After the protein is isolated, various fractions are produced, which comprise a complex of active and nonactive biopeptides. From this protein, the active peptide fractions are further subjected to additional separation and purification protocols, resulting in homogeneous peptides that are suitable for amino acid sequence analysis (Nwachukwu & Aluko, 2019). These process require multiple steps to identify bioactivity, as shown in **Figure 1.2**.

1.4.2 Peptides Obtained by Informatic Tools

The development of informatic tools has grown exponentially, providing alternative methods to search for bioactive peptides and reducing the time and resources needed to research these

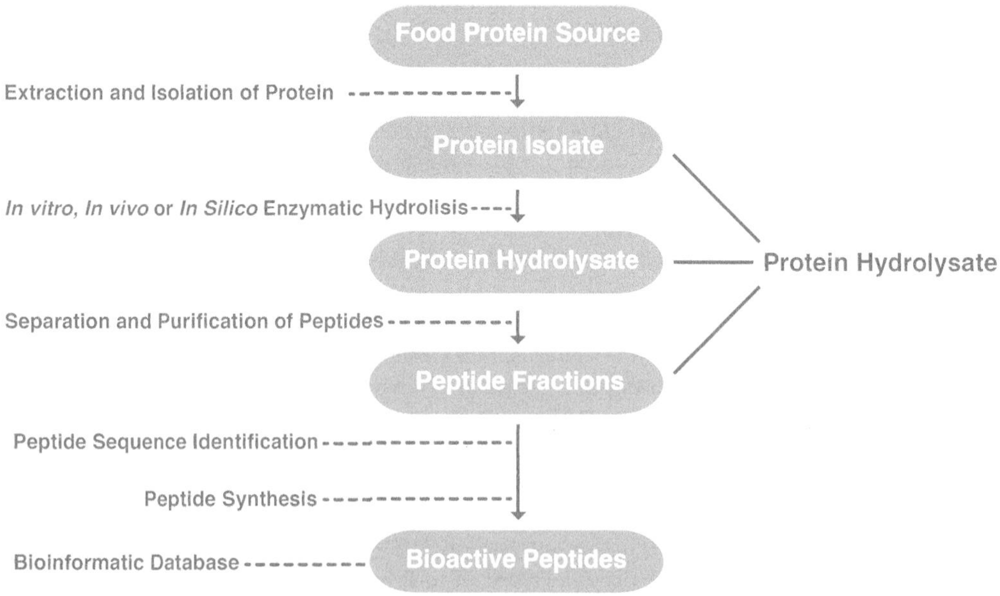

Figure 1.2 Identification of bioactive peptides. This figure outlines the methodologies and techniques employed in the extraction and isolation of bioactive peptides from different sources. The diagram emphasizes the importance of some processes in unlocking the potential of bioactive peptides for applications in nutrition, medicine, and biotechnology.

valuable molecules (Du & Li, 2022). In 2019, the BIOPEP-UWM database (free access) was launched on the website www.uwm.edu.pl/biochemia/index.php/pl/biopep. The purpose of this bioinformatics tool is to help researchers study bioactive peptides derived from food. Of the 4710 peptides identified, 56 are classified as anticancer peptides (accessed in 2023). This database is an invaluable source of information on food-derived peptides that help prevent chronic diseases and a powerful tool that could reduce the time and cost of investigations (Minkiewicz et al. 2019). Through bioinformatic tools, the hydrolysis of a protein can be simulated using powerful learning machines from the databases BIOPEP-UWM or Peptide Cutter of ExPASy (Expert Protein Analysis System). The tools can predict the potential cleavage sites by proteases or chemical reagents in a protein sequence. The website to access this tool is available at www.expasy.com and is free, open to all users, does not require a login, and is available to people who need this information. To summarize, simulating protease digestion can help to assess experiments on a specific portion of a removing a tagged protein when expressing a fusion protein or ensuring that the protein under investigation is not sensitive to endogenous proteases. Additionally, materials can be selected from a list of proteases and chemicals. Using *in silico* methodologies, peptide databases, and software, the experimental results can be predicted from theoretical data, reducing experimentation time and economic resources.

1.4.2.1 Informatic Tools to Predict Anticancer Peptides

Computational tools have been developed to facilitate the prediction of peptide sequences and their bioactivities, such as anticancer activity.

Cancer is a prominent chronic degenerative disease on a global scale and is projected to become a major cause of mortality in the coming decade. Several risk factors contribute to the development of cancer, including a diet high in fats and sugars, a sedentary lifestyle, smoking, and excessive alcohol consumption, among other living habits. However, the consumption of germinated soybean has been associated with a decreased risk of suffering certain types of cancer.

Germinated soybean releases peptides that could possess anticancer activity when subjected to enzymatic hydrolysis, germination, or digestive processes. Anticancer peptides (ACPs) are a subset of antimicrobial peptides that exhibit anticancer activities. These peptides are typically built of short (2–50 amino acids) highly hydrophobic cationic residues that adopt an α-helical structure or fold into a β-sheet while extending into linear forms. Anticancer peptides can interact with anionic cell membrane components of cancer cells and then selectively kill cancer cells (Shoombuatong et al. 2018; Schaduangrat et al. 2019). There are several compelling reasons for these peptides as therapeutic molecules. First, their small size leads to high tissue penetration and efficient cell diffusion, which can potentially reduce the side effects associated with conventional chemotherapy. Additionally, these molecules exhibit low toxicity, further enhancing their therapeutic value (Tyagi et al. 2015).

These peptides are composed of hydrophobic amino acids, including glycine (Gly), leucine (Leu), alanine (Ala), and proline (Pro), along with one or more residues of glutamic acid (Glu), lysine (Lys), arginine (Arg), and threonine (Thr) (Piñeiro et al. 2010; Li et al. 2019). Additionally, Agrawal et al. (2021) reported that lysine was the most abundant amino acid in ACPs. Due to their positively charged amino acids and a considerable proportion of hydrophobic residues, ACPs can interact electrostatically with cancer cell membranes, which typically bear a net negative charge. Consequently, these interactions disrupt the membranes of cancer cells, leading to a cytotoxic effect. Notably, molecular sequences and chemical compositions of peptides influence their ability to interact with cancer cells and exhibit cytotoxicity (Zhang et al. 2023).

TABLE 1.3 ANTICANCER PEPTIDE PREDICTORS WITH BIOINFORMATIC TOOLS

Anticancer Peptide Predictor	Classifier Algorithm	Precision (%)	Website	References
AntiCP 2.0	SVM, RF, KNN, ETree, ANN y Ridge Classifier	90.10	https://webs.iiitd.edu.in/raghava/anticp2/predict.php	Agrawal et al. (2021)
ACPred	SVM RF	95.61	http://codes.bio/acpred/	Schaduangrat et al. (2019)
mACPpred	SVM	91.4	www.thegleelab.org/mACPpred/index.html	Boopathi et al. (2019)
ACPred-Fuse	Feature Learning	89	http://server.malab.cn/ACPred-Fuse/Server.html	Rao et al. (2020)
ACP-DL	ACP-DL	85.42	https://github.com/haichengyi/ACP-DL	Yi et al. (2019)
ACPred-FL	ACPred-FL	90.4	http://server.malab. cn/ACPred-FL	Wei et al. (2018)
MLACP	SVM RF	83	www.thegleelab.org/MLACP/MLACP.html	Manavalan et al. (2017)

ACP: Anticancer Peptides **SVM:** Support Vector Machine **RF:** Random Forest **KNN:** K-Nearest Neighbors **ETree:** Decision Tree **ANN** Artificial Neural Network **Ridge Classifier, Feature Learning, ACP-DL:** Anticancer Peptides-Deep Learning **ACPred-FL:** Anticancer Peptides Features Learning Prediction **MLACP:** Machine Learning Anticancer Peptides

Informatics tools designed for predicting anticancer peptide activity have significantly reduced the time needed for experimental research. **Table 1.3** provides an overview of these tools to assess anticancer activity, including authors and publication year, the classification algorithm employed, the prediction percentage, and the corresponding website to access. These robust resources are built upon classification algorithms with similar characteristics that were derived from comprehensive databases. Moreover, these tools incorporate multiple input features, such as amino acid composition-based models, dipeptide composition-based models, ACP analysis and non-ACP composition, and residue preference.

González-Montoya, Hernández-Ledesma, Mora-Escobedo, et al. (2018) germinated soybean seeds for six days to obtain a protein isolate. Subsequently, this protein isolate underwent enzymatic hydrolysis using pepsin, pancreatin, and peptidase. The resulting protein hydrolysate was fractionated via ultrafiltration, utilizing hydrophilic membranes with cutoff sizes of 5 and 10 kDa, and as a result, the following peptide fractions with different molecular sizes were obtained: one fraction with a molecular weight >10 kDa, another in the range of 5–10 kDa, and a third <5 kDa.

In subsequent stages, a peptide purification process was carried out using semipreparative RP–HPLC of the most active ultrafiltration fraction (5–10 kDa). These peptides were characterized using HPLC-MS/MS.

TABLE 1.4 VALUES OF ANTICANCER PEPTIDES ACTIVITY PREDICTION WITH BIOINFORMATIC TOOLS

Peptide	ACPred	MLACP	AntiCP 2.0	mACPpred	ACPred-Fuse
NNDDRDS	**0.905**	**0.592**	0.400	0.434	**0.624**
VVNPDNNEN	**0.563**	**0.596**	0.350	0.244	**0.788**
LSSTEAQQS	0.232	**0.500**	0.370	0.354	**0.693**
NAENNQRN	**0.707**	**0.630**	**0.550**	**0.600**	**0.769**
IKSQSES	**0.961**	**0.561**	**0.540**	0.083	**0.725**
FEEPQQPQQ	0.339	**0.567**	0.160	0.174	**0.651**
GQSSRPQD	0.050	0.416	0.340	0.376	**0.719**
LAGNQEQE	0.030	0.465	**0.620**	0.157	**0.757**
NLKSQQA	0.026	0.424	0.430	0.455	**0.730**
QEPQESQQ	**0.507**	**0.589**	0.470	0.249	**0.704**
SQKQQLQDSHQ	0.398	0.468	0.270	0.160	**0.657**
QQQQQGGSQSQ	**0.794**	**0.584**	0.350	**0.500**	**0.693**
QQQQQGGSQSQKG	**0.628**	**0.510**	0.190	0.230	**0.678**
PETMQQQQQQ	**0.655**	**0.590**	**0.560**	0.445	**0.666**
SDESTESETEQA	**0.550**	0.364	0.370	0.077	**0.651**

Note: Values greater than 0.5 and closer to 1 are more likely to predict peptides with anticancer activity.

Employing the mention processes, fifteen possible bioactive peptides were identified, and their anticancer activity was assessed by activity predictors, such as ACPred, MLACP, AntiCP 2.0, ACPred-Fuse, and mACPpred; the values obtained are highlighted in bold in **Table 1.4.** According to the criteria of these predictors, values above 0.5 indicated a higher probability of having anticancer activity.

Notably, ACPred-Fuse classifier was not considered in the present research due to its inconsistency with the prediction as shown in **Table 1.4.**

Among the fifteen identified peptides with informatic tools, the peptide Ile-Lys-Ser-Gln-Ser-Glu-Ser (IKSQSES) was selected and synthesized by the authors of this chapter, and its effect on cell viability was evaluated in a HeLa cell line. The results obtained indicated a dose-time-dependent behavior and a reduction of 70% in cell viability. These results showed that employing informatic tools can significantly contribute to the identification of peptides with anticancer activity, which could reduce the resources needed compared to experimental laboratory approaches.

1.4.3 Peptides Obtained by Gut Microbiota

Another widely used method for acquiring bioactive peptides involves the intestinal microbiota produced from a variety of proteases, resulting in the *in vivo* production of various peptides during the digestion of food proteins (S. Wu et al. 2021). Food proteins, peptides, and free amino acids that break out in the digestion and absorption in the small intestine enter the large intestine for subsequent fermentation through the gut microbiota (Fan et al. 2015). Regardless of the pathway through which protein derivatives reach the colon, the gut microbiota is maintained, and protein synthesis is supported. Through this mutual interaction, the homeostasis of the microbial communities and the operation of the intestinal immune response are ensured (L. Wu et al. 2021). In particular, the metabolism of amino acids by bacteria occurs through deamination or decarboxylation reactions and generates short-chain fatty acids (SCFAs) or amines, respectively. These metabolites induce biological functions through various receptors and mechanisms (Fan et al. 2015). For example, soybean peptide Maillard reaction products changed the microbial community structure with beneficial bacteria, such as *Lactobacillus* and *Bifidobacterium* (Zhang et al. 2020). Additionally, the consumption of soybean protein resulted in a greater diversity of microbiota and a decrease in the ketal pH, derived from the production of SCFAs compared to the intake of casein (Tamura et al. 2019). **Table 1.5** summarizes the relationship of some metabolites with respect to the gut microbiota, emphasizing the effect they exert on the colonic flora and how this effect can benefit the health of individuals.

1.5 EFFECT OF GERMINATED SOYBEAN PRODUCTS ON HEALTH

Evidence has confirmed that germination improves the nutritional and functional properties of soybean. However, soybean consumption worldwide is reduced due to the low development of these germinated soybean products. Until now, no updated data on germinated soybean consumption has been provided. Asian countries have reported the highest consumption of soybean worldwide. More information is needed on developing and innovating products that incorporate germinated soybean as ingredients under the hypothesis that this strategy may increase their consumption.

1.5.1 Advancements and Innovations in Products Utilizing Germinated Soybean

An array of products has been meticulously crafted and advanced utilizing germinated soybean as a first ingredient. In these endeavors, researchers have performed comprehensive investigations on the nutraceutical properties of germinated soybean, enhancements in nutritional value, and the refinement of sensory attributes. **Table 1.6** shows the development of assorted products with germinated soybean. As seen, the field of study is wide for researchers in terms of the development of accessible and innovative functional foods for consumers, using soybean as an ingredient. Efforts must be made to offer consumers functional foods based on sprouted soybean.

1.5.1.1 Snack Type Food with Germinated Soybean

Guzmán-Ortiz et al. (2014) and Cruz-Ortiz et al. (2020) developed a snack-type food. This product showed low hardness, a good expansion index, and no increase in the glycemic index. This extruded sample also contained a high proportion of protein compared to similar products available on the market. Extrusion of soybean/corn flour generates low-molecular-weight polypeptides, which were associated with an increased *in vitro* protein digestibility of the extruded material (Guzmán-Ortiz et al. 2014); so therefore, the germinated soybean extruded could be a functional food with high protein and resistant starch.

TABLE 1.5 EFFECT ON GUT MICROBIOTA OF SOYBEAN COMPOUNDS

Compound	Study Conditions	Findings	References
Daidzein Isoflavone	Equol-production and microbiota were analyzed from 1044 subjects.	The intestinal bacteria, which convert daidzein to equol were present in both equol producer and nonproducer subjects. Relative abundance and the prevalence of *Asaccharobacter celatus* and *Slackia isoflavoniconvertens* were significantly higher in equol producer individuals.	Iino et al. (2019)
Defatted Soybean Fermented by *Lactobacillus plantarum* C29 named DW2009	12-week, double-blind, placebo-controlled, randomized clinical trial. Change in gut microbiota, composition of *Bifidobacterium spp., Lactobacillus spp., Clostridium spp.* The cognitive status was evaluated	*Lactobacilli* population was significantly increased in the fermented soybean group (p= 0.045). Cognitive improvement was associated with increased serum brain-derived neurotrophic factor levels after consumption of DW2009.	Hwang et al. (2019)
Equol	Review of current knowledge of equol biosynthesis and summary of clinical trials and meta-analyses of its effects on diverse areas of human health.	The study reported equol producing strains that could be used as probiotics.	Mayo et al. (2019)
Saponins	A randomized crossover trials. Consumption of a soya-rich high-protein weight-loss diet. Metabolite identification in volunteer's plasma after consumption of a soya test meal. Confirmation of the metabolism of Soyasaponin-I by the gut microbiota (*in vitro*).	The proposed mechanism suggests absorption of sapogenols from the colon following metabolism by the gut bacteria. Sapogenol B in the conjugated form is the major metabolite found in plasma.	Neacsu et al. (2020)
Genistein	Assessment of glucose concentration, lipid profile, serum insulin (HOMA) index, oral glucose tolerance test, serum metabolomic analysis, gut microbiota, and a skeletal muscle biopsy.	Genistein treatment reduces intestinal dysbiosis and metabolic endotoxemia, and stimulate *Akkermansia muciniphila* abundance, which may also help improve insulin sensitivity.	Guevara-Cruz et al. (2020)
Soybean-Derived Foods	A cross-sectional subanalysis of data from a prospective hospital-based cohort. Categorized gut microbiome in different enterotypes.	Subjects who consumed soybean and soybean-derived foods had lower fecal concentrations of several gut microbial metabolites. Adherence to a Japanese diet was inversely associated with cognitive decline.	Saji et al. (2022)

TABLE 1.6 PRODUCT DEVELOPMENT UTILIZING GERMINATED SOYBEAN

Product	Conditions	Results	References
Aqueous Extract	Soak time of 8 h at room temperature. Germination was conducted at 21 °C and 12 h light/dark cycle on a Biochemical Oxygen Demand incubator.	Changes in the solubility of some bioactive compounds. This can lead to the develop of soy-based beverages, such as soy milk, with better nutrition and nutraceutical properties	Bueno et al. (2020)
Butter	Soak time of 4 h at room temperature. Germination was conducted at 25 °C, 90% relative humidity for 72 h on a seed germinator.	It contains higher protein and lower fat content, as well as higher organoleptic acceptability than commercial soy butter.	Murugkar et al. (2014)
Soymilk **Tofu** **Kefir**	Soak time of 12 h at room temperature. Germination was conducted at 25 ± 2°C in the dark for 52 h in a growth chamber.	Increase the availability of bioactive peptides and reduce the allergenic effects of soybean.	Oyedeji et al. (2018b)
Soymilk	Soak time of 12 h at room temperature. Germination was conducted at 25 ± 2°C in the dark for 52 h in a growth chamber.	Improved nutritional attributes (higher protein content, phenolic acids, and total amino acids) and reduce the amount of anti-nutrients (trypsin and phytic acid inhibitors) compared to soy milk made with nonsprouted soybean.	Oyedeji et al. (2018a)
Soymilk	Soaked overnight (~12 h) at room temperature. Germination was conducted at room temperature 25 °C until 72 h, in a semi-automatic germination machine.	Significantly higher protein content than traditional soy milk, and a reduced amount of carbohydrates.	Jiang et al. (2013)
Soy Sauce	Soaked at 20 °C for 4 h. Germination was conducted at 20–25 °C for 48 h under a top-irrigation system commercially used for soybean sprout production. Dark and light conditions (1000 lux) were assessed.	Increased organoleptic preferences and levels of isoflavones.	Choi et al. (2011)
Homogenized Lyophilized Product	Soaked overnight. Germination was conducted at 27 °C and 70% RH in an incubator. Different germination times (3, 6, and 9 days) and luminosity (absence and presence of light) were assessed to establish the optimal growth conditions.	Increased the content of Diamine Oxidase, one of the key enzymes involved in the breakdown of dietary histamine, this could be suitable for the formulation of supplements for the treatment of histamine intolerance.	Comas-Basté et al. (2020)

Muffins	Soaked for 8 h. The germination process was conducted for 72 h at room temperature 22 °C with access to natural daylight and darkness at 25 °C and approx. 93% humidity in a humidity chamber.	Enhanced content of free amino acids and concentration of odorous volatile active ingredients providing an intense aroma compared to muffins baked only with wheat.	Kaczmarska et al. (2018)
Fermented Sprouted Soybeans	Soaked for 24 h at 23- 25°C. Germination was conducted in the dark at 25°C for 48 h.	Significantly restored DNFB-induced delayed-type hypersensitivity in C57BL/6N mice. Increased antioxidant activity.	Lee et al. (2021)
Soybeanmilk Okara	Soaked for 2 h at 28°C. Germination was conducted at 28°C in the dark for 48 h in a growth chamber.	Improve antioxidant activity and polyphenol content.	Robles-Ramírez et al. (2019)
Soybean Sauce	Soaked in warm water at 40°C for 2 h. Germination was conducted in a dark room at 25°C for 72 h.	Increased level of GABA, isoflavones, and total polyphenol compared with the control. Moreover, it exhibited a better aroma and taste than the nongerminated soybean.	Zhao et al. (2021)

1.5.1.2 Okara as a Residue from Germinated Soybean

Vong & Liu, (2016) indicated that okara is the solid residue that remains after the soybean milk is extracted. Okara is rich in protein (15.2–37.5%), dietary fiber (14.5–58.1%), fat (8.3–22.3%), and antioxidants (68.4–144.2 mg gallic acid equivalents per 100 g). Okara dietary fibers serve as a prebiotic source for the probiotic *L. rhamnosus* (Castro-Guillén et al., 2010).

Asghar et al. (2022) reported a symbiotic (prebiotics + probiotics) effect of yogurt prepared with 3% okara. The highest total viable count (8.25 log CFU/mL) and probiotic count (8.98 log CFU/mL) were detected. Likewise, the total phenolic content, DPPH (2,2-diphenyl-1-picrylhydrazyl) activity, and ABTS (2,20 -azino-bis (3-ethylbenzothiazoline-6- sulfonic acid) activity was increased due to the addition of okara. These results indicate that okara also shows significant prebiotic potential.

Moreover, Robles-Ramírez et al. (2019) reported that okara obtained from germinated soybean residue retained a good proportion of seed protein after 48 h of germination (27.64 ± 0.59 g/100 g dry okara); as a result, this product is a rich source of protein. Furthermore, the antioxidant capacity of dry okara increased by 646% after 48 h of germination. Consequently, to the germination process, an increase of the hydrolysis of soybean proteins was observed. This may be correlated with an improvement on the digestibility of okara proteins, highlighting the promising potential of germinated soybean-derived okara as a functional food.

Additionally, the health benefits and nutritional quality of okara are often enhanced through fermentation, considering that fermented okara is an inexpensive substrate for the extraction of bioactive compounds. The present research remains at the bench scale, and the main challenges are scaling up, efficiency, and yield. There is a significant opportunity for further exploration into various aspects of okara evaluation, including bioprocessing treatments as a prefermentation step, using combinatorial microbes or enzymes, and evaluating the sensory properties, dietary effects, and potential allergenicity of the fermented products. Research has demonstrated the feasibility and potential for generating a variety of bioactive substances and final consumer products via okara using various techniques, such as enzymatic, chemical, biotransformation, high-pressure microfludization, and fermentation.

In this regard Voss et al. (2021) elaborated four different beverages with okara previously hydrolyzed by *Cynara cardunculus* enzymes and fermented by probiotic bacteria; chickpea proteins contain a well-balanced mix of amino acids, with Glu, Asp, Arg, Leu, Phe, Lys, and Ser being found in the highest proportion. His, Gly, Treo, Ala, Tyr, Val, and ile are present in sufficient quantities to meet the requirements for both adults and children. However, their proteins are deficient in Met and Cys, which is a characteristic of all legume proteins. Furthermore, the beverages showed good angiotensin converting enzyme (ACE) inhibitory activity and an increase in antioxidant activity. After reaching the gastrointestinal tract, all beverages showed an increase in antioxidant and ACE inhibitory activities. The authors concluded that the application of okara to a multifunctional beverage could be a promising strategy for disease prevention and contribute to a zero-waste approach in the food industry.

Due to its prebiotic potential, okara is suitable as a therapeutic agent to prevent a variety of metabolic disorders, such as diabetes, obesity, hypercholesterolemia, and hyperlipidemia.

1.5.1.3 Encapsulation of Hydrolysates of Germinated Soybean

Cano-Sampedro et al. (2021) developed a formulation and encapsulation method for the delivery of protein hydrolysate to the colon. Alginate and xanthan gum were selected as wall materials. Three formulations were used to cover the protein hydrolysate from germinated soybean by ionic gelation followed by freeze-drying. During *in vitro* digestion, the microstructure was monitored by environmental scanning microscopy (ESEM), and the site-specific release of encapsulates was confirmed by fermentation under colonic conditions with human inoculum. These authors observed that encapsulates could retain the protein hydrolysate while passing through the upper gastrointestinal tract; these results are interesting since these peptides can reach the colon intact, be fermented by the colonic microbiota, and exert a beneficial effect against diseases such as cancer (Thomas et al. 2023).

1.5.2 *In vivo* Assays

The MATP/tau protein is hyperphosphorylated in Alzheimer's patients. Therefore, research on the regulation of tau protein phosphorylation is important for understanding Alzheimer's disease. HASPIN is a serine/threonine kinase that is expressed in various cells. To examine whether HASPIN engages in the onset of Alzheimer's disease through tau protein phosphorylation, Tanaka et al. (2023) investigated the effects of a diet including soybean sprouts rich in the HASPIN inhibitor coumestrol in a mouse model of Alzheimer's disease (5xFAD). The results showed that HASPIN was expressed in the hippocampus and phosphorylated tau protein, while the ingestion of soybean sprouts containing coumestrol suppressed the development of spatial cognitive dysfunction in 5xFAD mice; thus, HASPIN may be among the target molecules for the repression of tau phosphorylation in the treatment of Alzheimer's disease. Soybean sprouts show potential to improve human health in diverse ways; however, incorporating them as a portion of food for daily consumption and assessing their effects in clinical trials is a challenge. Therefore, García-Rojas et al. (2021) studied the effect of a snack made from soybean and cornstarch on overweight and obese people, and statistically significant differences were found in waist measurements, glucose levels, and low-density lipoprotein; these results were also found for *Akkermansia muciniphila* in feces (3.36 log DNA/g feces). This is interesting, as *A. muciniphila* has an inverse association with diabetes mellitus, obesity, cardiovascular diseases, immune disorders, pregnancy complications, cancer, tumors, brain disorders, liver diseases, and kidney diseases (Lakshmanan et al. 2022).

1.6 CONCLUSION

Studying the benefits of germination and the obtention of bioactive peptides from soybean presents a promising possibility for enhancing the nutritional and health-promoting properties of this versatile legume. The process of germination not only improves the nutritional composition of soybean but also provides a rich source of bioactive peptides with potential health benefits. These bioactive peptides offer exciting prospects for preventive and therapeutic applications, including anticancer, antidiabetic, and antioxidant properties. Moreover, the development of computational tools for peptide bioactivity prediction has emerged as a valuable resource, streamlining the research process while conserving time and resources themselves. Furthermore, innovations in product development using germinated soybean underscore the potential for creating functional foods that could contribute to better health. Understanding the benefits of germination and the obtention of bioactive peptides from soybean represents a required area of research with significant implications for human well-being and sustainable dietary practices.

REFERENCES

Agrawal, P., Bhagat, D., Mahalwal, M., Sharma, N., & Raghava, G. P. S. (2021). AntiCP 2.0: an updated model for predicting anticancer peptides. *Briefings in Bioinformatics, 22*(3), 1–12. https://doi.org/10.1093/bib/bbaa153

Agyei, D., Bambarandage, E., & Udenigwe, C. C. (2019). The role of bioinformatics in the discovery of bioactive peptides. In P. V. Laurence Melton, & Fereidoon Shahidi (Eds.), *Encyclopedia of Food Chemistry* (pp. 337–344). Elsevier. https://doi.org/10.1016/B978-0-08-100596-5.21863-5

Angelovici, R., Galili, G., Fernie, A. R., & Fait, A. (2010). Seed desiccation: a bridge between maturation and germination. *Trends in Plant Science, 15*(4), 211–218. https://doi.org/10.1016/j.tplants.2010.01.003

Asghar, A., Afzaal, M., Nosheen, F., Saeed, F., Nayik, G. A., AL-Farga, A., Alansari, W. S., Eskandrani, A. A., & Shamlan, G. (2022). Isolation and molecular characterization of processed soybean waste for the development of synbiotic yogurt. *Fermentation, 8*(11), 622. https://doi.org/10.3390/fermentation8110622

Barac, M., Stanojevic, S., & Pesic, M. (2005). Biologically active components of soybeans and soy protein products: a review. *Acta Periodica Technologica, 36*, 155–168. https://doi.org/10.2298/APT0536155B

Bau, H.-M., Villaume, C., Nicolas, J.-P., & Méjean, L. (1997). Effect of germination on chemical composition, biochemical constituents and antinutritional factors of soya bean (*Glycine max*) seeds. *Journal of the Science of Food and Agriculture, 73*(1), 1–9. https://doi.org/10.1002/(SICI)1097-0010(199701)73:1<1::AID-JSFA694>3.0.CO;2-B

Bewley, J. D. (1997). Seed germination and dormancy. *The Plant Cell, 9*(7), 1055–1066. https://doi.org/10.1105/tpc.9.7.1055

Boopathi, V., Subramaniyam, S., Malik, A., Lee, G., Manavalan, B., & Yang, D. C. (2019). MACppred: A support vector machine-based meta-predictor for identification of anticancer peptides. *International Journal of Molecular Sciences*, 20(8). https://doi.org/10.3390/ijms20081964

Boutas, I., Kontogeorgi, A., Dimitrakakis, C., & Kalantaridou, S. N. (2022). Soy isoflavones and breast cancer risk: a meta-analysis. *In Vivo, 36*(2), 556–562. https://doi.org/10.21873/invivo.12737

Bueno, D. B., da Silva Júnior, S. I., Seriani Chiarotto, A. B., Cardoso, T. M., Neto, J. A., Lopes dos Reis, G. C., Glória, M. B. A., & Tavano, O. L. (2020). The germination of soybeans increases the water-soluble components and could generate innovations in soy-based foods. *LWT, 117*, 108599. https://doi.org/10.1016/j.lwt.2019.108599

Cano-Sampedro, E., Pérez-Pérez, V., Osorio-Díaz, P., Camacho-Díaz, B., Maruri, D., Mora-Escobedo, R., & Alamilla-Beltrán, L. (2021). Germinated soybean protein hydrolysate: ionic gelation encapsulation and release under colonic conditions. *Revista Mexicana de Ingeniería Química, 20*, 725–737. https://doi.org/10.24275/rmiq/Alim2319

Castro-Guillén, J. L., García-Gasca, T., & Blanco-Labra, A. (2010). Protease inhibitors as anticancer agents. In C. Mejia Vazquez, et al. (Eds.), *New Approaches in the Treatment of Cancer* (pp. 91–124). Nova Science Publishers.

Choi, U.-K., Jeong, Y.-S., Kwon, O.-J., Park, J.-D., & Kim, Y.-C. (2011). Comparative study of quality characteristics of korean soy sauce made with soybeans germinated under dark and light conditions. *International Journal of Molecular Sciences, 12*(11), 8105–8118. https://doi.org/10.3390/ijms12118105

Comas-Basté, O., Latorre-Moratalla, M. L., Rabell-González, J., Veciana-Nogués, M. T., & Vidal-Carou, M. C. (2020). Lyophilised legume sprouts as a functional ingredient for diamine oxidase enzyme supplementation in histamine intolerance. *LWT, 125*(February), 109201. https://doi.org/10.1016/j.lwt.2020.109201

Cruz-Ortiz, R., Gutiérrez-Dorado, R., Osorio-Díaz, P., Alvarado-Jasso, G. M., Tornero-Martínez, A., Guzmán-Ortiz, F. A., & Mora-Escobedo, R. (2020). Optimized germinated soybean/cornstarch extrudate and its *in vitro* fermentation with human inoculum. *Journal of Food Science and Technology, 57*(3), 848–857. https://doi.org/10.1007/s13197-019-04116-7

Dikshit, M., & Ghadle, M. (2003). Effect of sprouting on nutrients, antinutrients and in vitro digestibility of the macs-13 soybean variety. *Plant Foods for Human Nutrition, 58*(3), 1–11. https://doi.org/10.1023/B:QUAL.0000040357.70606.4c

Donangelo, C. M., Trugo, L. C., Trugo, N. M. F., & Eggum, B. O. (1995). Effect of germination of legume seeds on chemical composition and on protein and energy utilization in rats. *Food Chemistry, 53*(1), 23–27. https://doi.org/10.1016/0308-8146(95)95781-Z

Du, Z., & Li, Y. (2022). Review and perspective on bioactive peptides: a roadmap for research, development, and future opportunities. *Journal of Agriculture and Food Research, 9*, 100353. https://doi.org/10.1016/j.jafr.2022.100353

Eum, H. L., Park, Y., Yi, T. G., Lee, J. W., Ha, K.-S., Choi, I.-Y., & Park, N. Il. (2020). Effect of germination environment on the biochemical compounds and anti-inflammatory properties of soybean cultivars. *PLOS ONE, 15*(4), e0232159. https://doi.org/10.1371/journal.pone.0232159

Fan, P., Li, L., Rezaei, A., Eslamfam, S., Che, D., & Ma, X. (2015). Metabolites of dietary protein and peptides by intestinal microbes and their impacts on gut. *Current Protein & Peptide Science, 16*(7), 646–654. https://doi.org/10.2174/1389203716666150630133657

Galvez, A. F., Chen, N., Macasieb, J., & de Lumen, B. O. (2001). Chemopreventive property of a soybean peptide (lunasin) that binds to deacetylated histones and inhibits acetylation. *Cancer Research, 61*(20), 7473–7478. www.ncbi.nlm.nih.gov/pubmed/11606382

Ganguly, C., & Das, S. (1994). Plant lectins as inhibitors of tumour growth and modulators of host immune response. *Chemotherapy, 40*(4), 272–278. https://doi.org/10.1159/000239204

García-Gasca, T., García-Cruz, M., Hernández-Rivera, E., López-Matínez, J., Castañeda-Cuevas, A. L., Yllescas-Gasca, L., Rodríguez-Méndez, A. J., Mendiola-Olaya, E., Castro-Guillén, J. L., & Blanco-Labra, A. (2012). Effects of tepary bean (*Phaseolus acutifolius*) protease inhibitor and semipure lectin fractions on cancer cells. *Nutrition and Cancer, 64*(8), 1269–1278. https://doi.org/10.1080/01635581.2012.722246

García-Rojas, V., Maldonado-Hernández, J., Mora-Escobedo, R., Izquierdo-Montalvo, G., Verma, M. P., Aragón-Aguilar, A., Ángel-Serrato, D., Encarnación-Fernández, J. S., & San Martín-Martínez, E. (2021). Use of 13c-glucose breath test to assess treatment adherence in nutritional intervention trials. *TIP Revista Especializada En Ciencias Químico-Biológicas, 24*, 1–6. https://doi.org/10.22201/fesz.23958723e.2021.359

González-Espinosa de los Monteros, L. A., Ramón-Gallegos, E., Torres-Torres, N., & Mora-Escobedo, R. (2011). Effect of germinated soybean protein hydrolysates on adipogenesis and adipolysis in 3t3-l1 cells. *Plant Foods for Human Nutrition, 66*(4), 355–362. https://doi.org/10.1007/s11130-011-0263-z

González-Montoya, M., Cano-Sampedro, E., & Mora-Escobedo, R. (2017). Bioactive peptides from legumes as anticancer therapeutic agents. *International Journal of Cancer and Clinical Research, 4*(2). https://doi.org/10.23937/2378-3419/1410081

González-Montoya, M., Hernández-Ledesma, B., Mora-Escobedo, R., & Martínez-Villaluenga, C. (2018). Bioactive peptides from germinated soybean with anti-diabetic potential by inhibition of dipeptidyl peptidase-iv, α-amylase, and α-glucosidase enzymes. *International Journal of Molecular Sciences, 19*(10), 2883. https://doi.org/10.3390/ijms19102883

González-Montoya, M., Hernández-Ledesma, B., Silván, J. M., Mora-Escobedo, R., & Martínez-Villaluenga, C. (2018). Peptides derived from in vitro gastrointestinal digestion of germinated soybean proteins inhibit human colon cancer cells proliferation and inflammation. *Food Chemistry, 242*(June 2017), 75–82. https://doi.org/10.1016/j.foodchem.2017.09.035

Grosse-Holz, F. M., & van der Hoorn, R. A. L. (2016). Juggling jobs: roles and mechanisms of multifunctional protease inhibitors in plants. *New Phytologist, 210*(3), 794–807. https://doi.org/10.1111/nph.13839

Guevara-Cruz, M., Godínez-Salas, E. T., Sánchez-Tapia, M., Torres-Villalobos, G., Pichardo-Ontiveros, E., Guizar-Heredia, R., Arteaga-Sánchez, L., Gamba, G., Mojica-Espinosa, R., Schcolnik-Cabrera, A., Granados, O., López-Barradas, A., Vargas-Castillo, A., Torre-Villalvazo, I., Noriega, L. G., Torres, N., & Tovar, A. R. (2020). Genistein stimulates insulin sensitivity through gut microbiota reshaping and skeletal muscle ampk activation in obese subjects. *BMJ Open Diabetes Research & Care, 8*(1), e000948. https://doi.org/10.1136/bmjdrc-2019-000948

Guzmán-Ortiz, F. A., Castro-Rosas, J., Gómez-Aldapa, C. A., Mora-Escobedo, R., Rojas-León, A., Rodríguez-Marín, M. L., Falfán-Cortés, R. N., & Román-Gutiérrez, A. D. (2019). Enzyme activity during germination of different cereals: a review. *Food Reviews International, 35*(3), 177–200. https://doi.org/10.1080/87559129.2018.1514623

Guzmán-Ortiz, F. A., Robles-Ramírez, M. del C., Sánchez-Pardo, M. E., Berríos, J. D. J., & Mora-Escobedo, R. (2014). Effect of germination on bioactive compounds of soybean (*Glycine max*). In R. Mora Escobedo, J. D. J. Berrios, & G. F. Gutierrez-Lopez (Eds.), *Seeds as Functional Foods and Nutraceuticals: New Frontiers in Food Science* (pp. 23–42). Nova Science Publishers, Inc. https://doi.org/10.1080/01635581.2012.717681

Hoekstra, A. Y., & Mekonnen, M. M. (2012). The water footprint of humanity. *Proceedings of the National Academy of Sciences, 109*(9), 3232–3237. https://doi.org/10.1073/pnas.1109936109

Hu, C., Wong, W.-T., Wu, R., & Lai, W.-F. (2020). Biochemistry and use of soybean isoflavones in functional food development. *Critical Reviews in Food Science and Nutrition, 60*(12), 2098–2112. https://doi.org/10.1080/10408398.2019.1630598

Huang, Y., Wang, X., Wang, H., Liu, Y., & Chen, Y. (2011). Studies on mechanism of action of anticancer peptides by modulation of hydrophobicity within a defined structural framework. *Molecular Cancer Therapeutics, 10*(3), 416–426. https://doi.org/10.1158/1535-7163.MCT-10-0811

Hwang, Y.-H., Park, S., Paik, J.-W., Chae, S.-W., Kim, D.-H., Jeong, D.-G., Ha, E., Kim, M., Hong, G., Park, S.-H., Jung, S.-J., Lee, S.-M., Na, K.-H., Kim, J., & Chung, Y.-C. (2019). Efficacy and safety of lactobacillus plantarum c29-fermented soybean (dw2009) in individuals with mild cognitive impairment: a 12-week, multi-center, randomized, double-blind, placebo-controlled clinical trial. *Nutrients, 11*(2), 305. https://doi.org/10.3390/nu11020305

Iino, C., Shimoyama, T., Iino, K., Yokoyama, Y., Chinda, D., Sakuraba, H., Fukuda, S., & Nakaji, S. (2019). Daidzein intake is associated with equol producing status through an increase in the intestinal bacteria responsible for equol production. *Nutrients, 11*(2), 433. https://doi.org/10.3390/nu11020433

Jae, H.-P., Hyung, J., & De Lumen, B. O. (2005). Contents and bioactivities of lunasin, Bowman–Birk inhibitor, and isoflavones in soybean seed. Journal of Agricultural and Food Chemistry, 53(20), 7686–7690. https://doi.org/10.1021/jf0506481

Jiang, S., Cai, W., & Xu, B. (2013). Food quality improvement of soy milk made from short-time germinated soybeans. *Foods, 2*(2), 198–212. https://doi.org/10.3390/foods2020198

Kaczmarska, Kornelia, T., Chandra-Hioe, M. V., Frank, D., & Arcot, J. (2018). Enhancing wheat muffin aroma through addition of germinated and fermented australian sweet lupin (*Lupinus angustifolius l.*) and soybean (*Glycine max l.*) Flour. *LWT, 96*, 205–214. https://doi.org/10.1016/j.lwt.2018.05.034

Kennedy, A. R. (1998). The Bowman-Birk Inhibitor from soybeans as an anticarcinogenic agent. *The American Journal of Clinical Nutrition, 68*(6), 1406S–1412S. https://doi.org/10.1093/ajcn/68.6.1406S

Khalil, A. A., Mohamed, S. S., Taha, F. S., & Karlsson, E. N. (2006). Production of functional protein hydrolysates from egyptian breeds of soybean and lupin seeds. *African Journal of Biotechnology, 5*(10), 907–916. www.academicjournals.org/AJB

Kim, E. H., Kim, S. H., Chung, J. I., Chi, H. Y., Kim, J. A., Chung, I. M., Kim, S. H., Chung, J. I. M., Chi, H. Y., Kim, J. A., & Chung, J. I. M. (2006). Analysis of phenolic compounds and isoflavones in soybean seeds (*Glycine max (l.) Merill*) and sprouts grown under different conditions. *European Food Research and Technology, 222*(1–2), 201–208. https://doi.org/10.1007/s00217-005-0153-4

Kim, I.-S., Yang, W.-S., & Kim, C.-H. (2021). Beneficial effects of soybean-derived bioactive peptides. *International Journal of Molecular Sciences, 22*(16), 8570. https://doi.org/10.3390/ijms22168570

Kim, S. E., Kim, H. H., Kim, J. Y., Kang, Y. I., Woo, H. J., & Lee, H. J. (2000). Anticancer activity of hydrophobic peptides from soy proteins. *BioFactors, 12*(1–4), 151–155. https://doi.org/10.1002/biof.5520120124

Kumar, V., Sinha, A. K., Makkar, H. P. S., & Becker, K. (2010). Dietary roles of phytate and phytase in human nutrition: a review. *Food Chemistry, 120*(4), 945–959. https://doi.org/10.1016/j.foodchem.2009.11.052

Lajolo, F. M., & Genovese, M. I. (2002). Nutritional significance of lectins and enzyme inhibitors from legumes. *Journal of Agricultural and Food Chemistry, 50*(22), 6592–6598. https://doi.org/10.1021/jf020191k

Lakshmanan, A. P., Murugesan, S., Al Khodor, S., & Terranegra, A. (2022). The potential impact of a probiotic: akkermansia muciniphila in the regulation of blood pressure—the current facts and evidence. *Journal of Translational Medicine, 20*(1), 430. https://doi.org/10.1186/s12967-022-03631-0

Lee, H.-J., Cho, H.-E., & Park, H.-J. (2021). Germinated black soybean fermented with *Lactobacillus pentosus* SC65 alleviates DNFB-induced delayed-type hypersensitivity in C57BL/6N mice. *Journal of Ethnopharmacology, 265*(August 2020), 113236. https://doi.org/10.1016/j.jep.2020.113236

Li, M., Zhang, Y., Xia, S., & Ding, X. (2019). Finding and isolation of novel peptides with anti-proliferation ability of hepatocellular carcinoma cells from mung bean protein hydrolysates. *Journal of Functional Foods, 62*, 103557. https://doi.org/10.1016/j.jff.2019.103557

Lin, P.-Y., & Lai, H.-M. (2006). Bioactive compounds in legumes and their germinated products. *Journal of Agricultural and Food Chemistry, 54*(11), 3807–3814. https://doi.org/10.1021/jf060002o

Martín Salinas, C., & López-Sobaler, A. M. (2017). Benefits of soy in women's health. *Nutricion Hospitalaria, 34*(Suppl 4), 36–40. https://doi.org/10.20960/nh.1569

Manavalan, B., Basith, S., Shin, T. H., Choi, S., Kim, M. O., & Lee, G. (2017). MLACP: Machine-learning-based prediction of anticancer peptides. Oncotarget, 8(44), 77121–77136. https://doi.org/10.18632/oncotarget.20365

Mayo, B., Vázquez, L., & Flórez, A. B. (2019). Equol: a bacterial metabolite from the daidzein isoflavone and its presumed beneficial health effects. *Nutrients, 11*(9), 2231. https://doi.org/10.3390/nu11092231

Minkiewicz, P., Iwaniak, A., & Darewicz, M. (2019). BIOPEP-UWM database of bioactive peptides: current opportunities. *International Journal of Molecular Sciences, 20*(23), 5978. https://doi.org/10.3390/ijms20235978

Mooney, C., Haslam, N. J., Pollastri, G., & Shields, D. C. (2012). Towards the improved discovery and design of functional peptides: common features of diverse classes permit generalized prediction of bioactivity. *PLoS ONE, 7*(10), e45012. https://doi.org/10.1371/journal.pone.0045012

Moronta, J., Smaldini, P. L., Docena, G. H., & Añón, M. C. (2016). Peptides of amaranth were targeted as containing sequences with potential anti-inflammatory properties. *Journal of Functional Foods, 21*, 463–473. https://doi.org/10.1016/j.jff.2015.12.022

Murugkar, D. A., Kotwaliwale, N., Gupta, C., Gulati, P., & Kumar, M. (2014). Optimization of soy-butter from sprouted soybean and its quality parameters. *Journal of Food Quality, 37*(1), 63–72. https://doi.org/10.1111/jfq.12069

Muzquiz, M., Varela, A., Burbano, C., Cuadrado, C., Guillamón, E., & Pedrosa, M. M. (2012). Bioactive compounds in legumes: pronutritive and antinutritive actions. Implications for nutrition and health. *Phytochemistry Reviews, 11*(2–3), 227–244. https://doi.org/10.1007/s11101-012-9233-9

Ndakidemi, P. A., & Dakora, F. D. (2003). Legume seed flavonoids and nitrogenous metabolites as signals and protectants in early seedling development. *Functional Plant Biology, 30*(7), 729. https://doi.org/10.1071/FP03042

Neacsu, M., Raikos, V., Benavides-Paz, Y., Duncan, S. H., Duncan, G. J., Christie, J. S., Johnstone, A. M., & Russell, W. R. (2020). Sapogenol is a major microbial metabolite in human plasma associated with high protein soy-based diets: the relevance for functional food formulations. *Foods, 9*(4), 422. https://doi.org/10.3390/foods9040422

Nwachukwu, I. D., & Aluko, R. E. (2019). Anticancer and antiproliferative properties of food-derived protein hydrolysates and peptides. *Journal of Food Bioactives, 7*, 18–26. https://doi.org/10.31665/jfb.2019.7194

Osman, A., Bin-Jumah, M., Abd El-Hack, M. E., Elaraby, G., Swelum, A. A., Taha, A. E., Sitohy, M., Allam, A. A., & Ashour, E. A. (2020). Dietary supplementation of soybean glycinin can alter the growth, carcass traits, blood biochemical indices, and meat quality of broilers. *Poultry Science, 99*(2), 820–828. https://doi.org/10.1016/j.psj.2019.12.026

Oyedeji, A. B., Mellem, J. J., & Ijabadeniyi, O. A. (2018a). Improvement of some quality attributes of soymilk through optimization of selected soybean sprouting parameters using response surface methodology. *CyTA – Journal of Food, 16*(1), 230–237. https://doi.org/10.1080/19476337.2017.1388292

Oyedeji, A. B., Mellem, J. J., & Ijabadeniyi, O. A. (2018b). Potential for enhanced soy storage protein breakdown and allergen reduction in soy-based foods produced with optimized sprouted soybeans. *LWT, 98*, 540–545. https://doi.org/10.1016/j.lwt.2018.09.019

Paucar-Menacho, L. M., Berhow, M. A., Mandarino, J. M. G., Chang, Y. K., & Mejia, E. G. de. (2010). Effect of time and temperature on bioactive compounds in germinated brazilian soybean cultivar BRS 258. *Food Research International, 43*(7), 1856–1865. https://doi.org/10.1016/j.foodres.2009.09.016

Paucar-Menacho, L. M., Berhow, M. A., Mandarino, J. M. G., de Mejia, E. G., & Chang, Y. K. (2010). Optimisation of germination time and temperature on the concentration of bioactive compounds in brazilian soybean cultivar brs 133 using response surface methodology. *Food Chemistry, 119*(2), 636–642. https://doi.org/10.1016/j.foodchem.2009.07.011

Piñeiro, V., Ortiz-Moreno, A., Mora-Escobedo, R., Hernández-Navarro, M. D., Ceballos-Reyes, G., & Chamorro-Cevallos, G. (2010). Effect of l-arginine oral supplementation on response to myocardial infarction in hypercholesterolemic and hypertensive rats. *Plant Foods for Human Nutrition, 65*(1), 31–37. https://doi.org/10.1007/s11130-009-0143-y

Qin, P., Wang, T., & Luo, Y. (2022). A review on plant-based proteins from soybean: health benefits and soy product development. *Journal of Agriculture and Food Research, 7*, 100265. https://doi.org/10.1016/j.jafr.2021.100265

Quah, Y., Mohd Ismail, N. I., Ooi, J. L. S., Affendi, Y. A., Abd Manan, F., Teh, L.-K., Wong, F.-C., & Chai, T.-T. (2019). Purification and identification of novel cytotoxic oligopeptides from soft coral *Sarcophyton glaucum*. *Journal of Zhejiang University-Science B, 20*(1), 59–70. https://doi.org/10.1631/jzus.B1700586

Rao, B., Zhou, C., Zhang, G., Su, R., & Wei, L. (2020). ACPred-Fuse: Fusing multi-view information improves the prediction of anticancer peptides. Briefings in Bioinformatics, 21(5), 1846–1855. https://doi.org/10.1093/bib/bbz088

Robles-Ramírez, M. del C., Ramón-Gallegos, E., Mora-Escobedo, R., & Torres-Torres, N. (2012). A peptide fraction from germinated soybean protein down-regulates PTTG1 and TOP2A mRNA expression, inducing apoptosis in cervical cancer cells. *Journal of Experimental Therapeutics & Oncology, 9*(4), 255–263. www.ncbi.nlm.nih.gov/pubmed/22545419

Robles-Ramírez, M. del C., Ramón-Gallegos, E., Reyes-Duarte, F. J., & Mora-Escobedo, R. (2012). Effect of germinated soy protein on the growth of hela cervical cancer cells in female athymic mice. *Nutrition and Cancer, 64*(8), 1261–1268. https://doi.org/10.1080/01635581.2012.717681

Robles-Ramírez, M. del C., Rangel-Miranda, S. E., & Mora-Escobedo, R. (2019). Effect of germination on the composition, protein profile and antioxidant capacity of soymilk and okara. In R. Mora-Escobedo, C. Martinez, & R. Reynso (Eds.), *Germination: Types, Process and Effects* (pp. 283–300).

Saji, N., Tsuduki, T., Murotani, K., & Hisada, T. (2022). Relationship between the japanese-style diet, gut microbiota, and dementia: a cross-sectional study. *Nutrition, 94*, 111524. https://doi.org/10.1016/j.nut.2021.111524

Schaduangrat, N., Nantasenamat, C., Prachayasittikul, V., & Shoombuatong, W. (2019). ACPred: a computational tool for the prediction and analysis of anticancer peptides. *Molecules, 24*(10), 1973. https://doi.org/10.3390/molecules24101973

Shoombuatong, W., Schaduangrat, N., & Nantasenamat, C. (2018). Unraveling the bioactivity of anticancer peptides as deduced from machine learning. *EXCLI Journal, 17*, 734–752. https://doi.org/10.17179/excli2018-1447

Tamura, K., Sasaki, H., Shiga, K., Miyakawa, H., & Shibata, S. (2019). The timing effects of soy protein intake on mice gut microbiota. *Nutrients, 12*(1), 87. https://doi.org/10.3390/nu12010087

Tanaka, H., Matsushita, H., Tokuhiro, K., Fukunari, A., & Ando, Y. (2023). Ingestion of soybean sprouts containing a HASPIN inhibitor improves condition in a mouse model of Alzheimer's disease. *Biology, 12*(2), 320. https://doi.org/10.3390/biology12020320

Thomas, A. M., Fidelle, M., Routy, B., Kroemer, G., Wargo, J. A., Segata, N., & Zitvogel, L. (2023). Gut oncomicrobiome signatures (goms) as next-generation biomarkers for cancer immunotherapy. *Nature Reviews Clinical Oncology, 20*(9), 583–603. https://doi.org/10.1038/s41571-023-00785-8

Tu, M., Cheng, S., Lu, W., & Du, M. (2018). Advancement and prospects of bioinformatics analysis for studying bioactive peptides from food-derived protein: sequence, structure, and functions. *TrAC Trends in Analytical Chemistry, 105*, 7–17. https://doi.org/10.1016/j.trac.2018.04.005

Tyagi, A., Tuknait, A., Anand, P., Gupta, S., Sharma, M., Mathur, D., Joshi, A., Singh, S., Gautam, A., & Raghava, G. P. S. (2015). CancerPPD: a database of anticancer peptides and proteins. *Nucleic Acids Research, 43*(D1), D837–D843. https://doi.org/10.1093/nar/gku892

Vong, W. C., & Liu, S.-Q. (2016). Biovalorisation of okara (soybean residue) for food and nutrition. *Trends in Food Science & Technology, 52*, 139–147. https://doi.org/10.1016/j.tifs.2016.04.011

Voss, G. B., Monteiro, M. J. P., Jauregi, P., Valente, L. M. P., & Pintado, M. E. (2021). Functional characterisation and sensory evaluation of a novel synbiotic okara beverage. *Food Chemistry, 340,* 127793. https://doi.org/10.1016/j.foodchem.2020.127793

Waliat, S., Arshad, M. S., Hanif, H., Ejaz, A., Khalid, W., Kauser, S., & Al-Farga, A. (2023). A review on bioactive compounds in sprouts: extraction techniques, food application and health functionality. *International Journal of Food Properties, 26*(1), 647–665. https://doi.org/10.1080/10942912.2023.2176001

Wang, S., Wang, Y., Pan, M.-H., & Ho, C.-T. (2017). Anti-obesity molecular mechanism of soy isoflavones: weaving the way to new therapeutic routes. *Food & Function, 8*(11), 3831–3846. https://doi.org/10.1039/C7FO01094J

Wang, W., & de Mejia, E. G. (2005). A new frontier in soy bioactive peptides that may prevent age-related chronic diseases. *Comprehensive Reviews in Food Science and Food Safety, 4*(4), 63–78. https://doi.org/10.1111/j.1541-4337.2005.tb00075.x

Wang, Z., & Zhang, X. (2017). Isolation and identification of anti-proliferative peptides from *spirulina platensis* using three-step hydrolysis. *Journal of the Science of Food and Agriculture, 97*(3), 918–922. https://doi.org/10.1002/jsfa.7815

Wei, L., Zhou, C., Chen, H., Song, J., & Su, R. (2018). ACPred-FL: A sequence-based predictor using effective feature representation to improve the prediction of anti-cancer peptides. Bioinformatics, 34(23), 4007–4016. https://doi.org/10.1093/bioinformatics/bty451

Wu, L., Tang, Z., Chen, H., Ren, Z., Ding, Q., Liang, K., & Sun, Z. (2021). Mutual interaction between gut microbiota and protein/amino acid metabolism for host mucosal immunity and health. *Animal Nutrition, 7*(1), 11–16. https://doi.org/10.1016/j.aninu.2020.11.003

Wu, S., Bekhit, A. E.-D. A., Wu, Q., Chen, M., Liao, X., Wang, J., & Ding, Y. (2021). Bioactive peptides and gut microbiota: candidates for a novel strategy for reduction and control of neurodegenerative diseases. *Trends in Food Science & Technology, 108,* 164–176. https://doi.org/10.1016/j.tifs.2020.12.019

Zhang, Y., Wang, C., Zhang, W., & Li, X. (2023). Bioactive peptides for anticancer therapies. Biomaterials Translational. Chinese Medical Multimedia Press Co Ltd. *4*(1), 5–17. https://doi.org/http://doi.org/10.12336/ biomatertransl.2023.01.00

Zhang, Z., He, S., Cao, X., Ye, Y., Yang, L., Wang, J., Liu, H., & Sun, H. (2020). Potential prebiotic activities of soybean peptides maillard reaction products on modulating gut microbiota to alleviate aging-related disorders in D-galactose-induced ICR mice. *Journal of Functional Foods, 65,* 103729. https://doi.org/10.1016/j.jff.2019.103729

Zhao, G., Li, J., Zheng, F., & Yao, Y. (2021). The fermentation properties and microbial diversity of soy sauce fermented by germinated soybean. *Journal of the Science of Food and Agriculture, 101*(7), 2920–2929. https://doi.org/10.1002/jsfa.10924

Zhu, L., Deng, X., Zou, L., Wu, J., Zhang, D., & Lin, H. (2016). Proanthocyanidins accelerate the germination of cucumber (*Cucumis sativus L.*) seeds. *Journal of Plant Biology, 59*(2), 143–151. https://doi.org/10.1007/s12374-016-0510-9

Chapter 2

Biological Activities of Chickpea Peptides (*Cicer arietinum* L.)

Deyanira del R. Moguel-Concha
José E. Borges-Martínez
M. Stephanie Cid-Gallegos
M. Faridy Juárez-Chairez
Darío I. Téllez-Medina
and Cristian Jiménez-Martínez

CONTENTS

2.1 INTRODUCTION

Legumes are dicotyledonous seeds, rich in protein, carbohydrates, and dietary fiber; they also have a protein content significantly higher than that of cereals. This makes legumes an important

 DOI: 10.1201/9781003332558-3

source of protein and amino acids for human nutrition. In addition to providing a source of essential amino acids and bioactive peptides, legume proteins greatly influence functional properties, which could help broaden their potential use in developing a wide variety of food products (Juárez-Chairez et al. 2020; Keskin et al. 2021; Singh et al. 2021; Carbas et al. 2023; Xiao et al. 2023). Based on the information provided, legume-based protein products are becoming increasingly popular as a substitute for animal protein due to their high nutritional value, favorable techno-functional characteristics, and low cost (Magalhães et al. 2017; Kaur & Prasad 2021; Keskin et al. 2021). Chickpeas are an essential component of the human diet because of their nutritional and bioactive makeup, and the quality of their protein is superior to that of other legumes (Semba et al. 2021). Chickpeas are the third most widely cultivated legume globally, following soybeans and dry beans, and can be grown in temperate and semi-arid regions such as Europe, Australia, and North America. India is the primary producer of chickpeas, accounting for 68.52% of global production, followed by Pakistan (6.37%), Russia (5.59%), Turkey (5.14%), Australia (4.03%), and Ethiopia (3.21%) (Acevedo-Martínez et al. 2021; FAO 2021; Kaur & Prasad, 2021). Chickpeas are divided into two types: Desi and Kabuli. based on their seed size, color, and seed coat thickness and shape. Desi chickpeas are typically smaller, angular seeds with a thick seed coat that can range from light tan and speckled to solid black. Kabuli chickpeas have larger seeds with thin seed coats with shades of white, pale cream.

Chronic-degenerative diseases like obesity, diabetes, hypertension, hypercholesterolemia, and cancer have become more prevalent due to changes in the environment and diet. Thankfully, there are various treatments that combine nutrition with medication to combat these diseases. While these drugs may be safe, there have been reports of serious side effects (Xue et al. 2017). Studies have shown that amino acids and bioactive peptides found in chickpeas have biological effects on chronic-degenerative diseases, making them a promising alternative treatment option (Juárez-Chairez et al. 2020; Acevedo-Martínez et al. 2021; Kaur & Prasad 2021; Semba et al. 2021; Singh et al. 2021; Wang et al. 2023).

2.2 CHEMICAL COMPOSITION OF CHICKPEA

The quality of the proteins of chickpeas is considered better than that of other legumes since it presents a well-balanced amino acid profile (Paredes-López et al. 1991; Wang et al. 2021). Additionally, they are a rich source of essential minerals, including potassium, calcium, sodium, magnesium, copper, iron, and zinc, albeit in trace amounts. Moreover, chickpeas contain isoflavones like biochanin A, calycosin, formononetin, genistein, trifolirhizin, ononin, and sissotrin (Wang et al. 2021).

2.2.1 Carbohydrates

Chickpea seeds contain a large amount of carbohydrates, making up around 40–70% of their composition, which include monosaccharides, disaccharides, oligosaccharides, and polysaccharides. Notable oligosaccharides found in chickpea seeds are stachyose, ciceritol, raffinose, and verbascose. Of these, ciceritol and stachyose make up a significant portion of the total carbohydrates, accounting for approximately 36–43% and 25%, respectively (Sandhu et al. 2023).

These carbohydrates cannot be absorbed or broken down by the human digestive system, leading to their fermentation by bacteria in the colon, releasing gas formed by methane, CO2, and hydrogen (gastrointestinal flatulence) (Wang et al. 2021). Polysaccharides, on the other hand, are large molecules formed by the condensation and dehydration of monosaccharide molecules.

Polysaccharides are complex molecules that are formed by combining and dehydrating monosaccharide molecules. In chickpea seeds, the predominant polysaccharide is starch, which makes up 41–50% of the seed (35% resistant and 65% available). Starch is composed of two long dextran polymers, amylose (20–30%) and amylopectin (70–80%). The kabuli variety has a higher starch content (38.2–43.9%) than the desi variety (33.1–40.4%) (Juárez-Chairez et al. 2020; Wang et al. 2021). Additionally, dietary fiber represents 18–22% of the seeds total weight, of which 4-8% is soluble and 10–18% is insoluble (Juárez-Chairez et al. 2020).

2.2.2 Proteins

Chickpeas proteins account for 17–23% of their weight, making them the second most abundant compound (Juárez-Chairez et al. 2022). These proteins are highly digestible and have good bioavailability (48–89.01%) (Juárez-Chairez et al. 2020; Boukid, 2021). Chickpea proteins are also well-balanced in terms of amino acids, with Glu, Asp, Arg, Leu, Phe, Lys, and Ser being the most prominent, followed by His, Gly, Treo, Ala, Tyr, Val, and Ile. However, it has a low content of sulfur amino acids such as Met and Cys (Juárez-Chairez et al. 2020). Devi et al. (2020) have reported that chickpeas have higher digestibility of lysine and proline than yellow peas. Additionally, it has also been reported that the hydrolysis of chickpea proteins with different proteases generates a rich variety of chickpea polypeptides with antioxidant, hypoglycemic, and hypolipidemic activities (Shi, Hou, Gou et al. 2019; Grdeń et al. 2023).

According to their solubility chickpea proteins have 8–12% albumin, 56% globulin, 3–7% prolamin, and 19–25% glutelins. Globulins are the most abundant type of protein in all varieties (Juárez-Chairez et al. 2022). The globulins are composed of legumin and vicilin. Legumin, with a molecular weight of 360 kDa, is the main storage protein and represents 97% of all globulins. They have been identified by sodium dodecyl sulfate-polyacrylamide gel electrophoresis (SDS-PAGE): legumin α-subunits (40.6 and 39.6 kDa), legumin β-subunits (23.5 and 22.5 kDa), and vicilin subunits (70.2, 50.7, 35.0, 33.6, 18.9, and 15.5 kDa) (Juárez-Chairez et al. 2020).

2.2.3 Lipids

The lipid content in chickpea seeds is around 4 to 10%; these are made up of saturated fatty acids and unsaturated fatty acids (monounsaturated and polyunsaturated), mainly linoleic (54.7–56.2%), oleic (21.6–22.2%), and linolenic (0.5–0.9%), as well as of saturated fatty acids such as palmitic (18.9–20.4%) and stearic (1.3–1.7%) (Wang et al. 2021). On average, the oleic acid content is more abundant in the kabuli type (27.7–42.46%) than desi type (18.44–28.5%), whereas the linoleic acid content is higher in the desi type (53.1–65.2%) than in the kabuli type (42.2–56.5%) (Juárez-Chairez et al. 2020; Wang et al. 2021).

2.2.4 Vitamins and Minerals

Chickpea contains vitamins including retinol, vitamin C, thiamin, riboflavin, pantothenic acid, pyridoxine, vitamin E, and folic acid. Folic acid, as a coenzyme of carbon transfer enzymatic reactions, plays the role of transporting carbon units. Both varieties of chickpea, desi and kabuli, have a high content of folic acid with concentrations ranging 299.21 μg/100 g and 206.48 μg/100 g, respectively (Wang et al. 2021). In addition to being a good source of folic acid, chickpea seed is also rich in fat-soluble vitamins such as α-tocopherol (vitamin E) and γ-tocopherol and contains small amounts of water-soluble vitamins such as thiamin (B1), riboflavin (B2), niacin (B3), pantothenic acid (B5), and pyridoxine (B6) (Wang et al. 2021). Regarding the main minerals and trace elements of chickpea seed varieties (desi and kabuli), K, Na, Mg, Ca, Fe, Cu, Zn, Mn, and P have been reported (Wang et al. 2021).

2.2.5 Phytochemical Compounds

Chickpea seed has secondary metabolites, some of which are part of the plant's defense system against insects, fungus, bacteria, or diseases. Among these secondary metabolites are phenolic compounds, phytates, saponins, isoflavones, lectins, and others, which exert important biological activities (Cid-Gallegos et al. 2020a) that have increased the interest towards this legume. Domínguez- Arispuro et al. (2018) reported the antioxidant activity of ethanolic extracts from germinated chickpea by absorption capacity of oxygen radicals and scavenging activity of ABTS free radical 15,143 and 14,435 μmol ET/100 g of dry weight, respectively. Likewise, Wahby et al. (2017) found that chickpea extract (mainly with biochanin A and biochanin A-7-Ob-D-glycoside) produced a reduction in the activity of nuclear factor kappa B (NF-κB), cyclooxygenase-2 (COX-2) and TNF-α in Sprague-Dawley rats with neuroinflammation induced by $AlCl_3$.

Chickpeas are rich in carotenoids such as β-carotene, lutein, zeaxanthin, β-cryptoxanthin, lycopene, and α-carotene, as reported by Wang et al. (2021; 2023), with phenolic compound content range of 0.72–1.81 mg/g of seed (Juárez-Chairez et al. (2020). Chickpea seeds also contain 25 different types of isoflavones, including 20 isoflavones, 2 isoflavanones, and 3 phytoalexins. The total isoflavone content in sprouted chickpeas is approximately five times higher than that of sprouted soybeans, as reported by Gao et al. (2015). Megías et al. (2016) discovered seven isoflavones from germinated chickpea seeds, namely BCA, calicosin, formononetin, genistein, trifolirhizin, ononin, and sisotrin.

Chickpeas contain saponins, which are natural compounds with surface-active properties and plasma cholesterol-lowering effects in humans. These compounds have been found in legumes and consist of a complex mixture of pentacyclic triterpenoid glycosides, as reported by Wang et al. in 2021. Saponin B1 may be responsible for the many health benefits of chickpeas, including hypoglycemic, hepatoprotective, anti-inflammatory, and anticancer activities (Cheng et al. 2017).

2.3 BIOACTIVE PEPTIDES

Bioactive peptides are polymers of amino acids produced from protein hydrolysis that can exert health benefits through interactions with biological molecules (Real Hernández & González de Mejía, 2019; Mahgoub et al. 2023). The peptide sequences encoded within the parent protein become active when cleaved, allowing for various physiological activities. Bioactive peptides are released during enzymatic proteolysis, such as gastrointestinal digestion or *in vitro* hydrolysis by proteolytic enzymes like pepsin, pancreatin, alcalase, and flavourzyme. They can also be released during food processing, cooking, fermentation, and ripening for added health benefits (Daliri et al. 2017; Mahgoub et al. 2023).

In enzymatic hydrolysis, the protein material is subjected to hydrolysis at the optimum temperature and pH of the enzyme used. The use of enzymatic hydrolysis to produce bioactive peptides is preferable to microbial fermentation due to short reaction time. More than one proteolytic enzyme can be used to hydrolyze the protein and produce the hydrolysate containing short peptide sequences. However, the addition of the enzymes (whether simultaneously or sequentially) will depend on the optimum pH and temperature of the enzymes. In addition, the enzyme/substrate ratio is important to obtain a good degree of hydrolysis.

Peptide sequences and their biological activities may differ depending on the type of enzyme used. Low-molecular-weight peptides have been found to be more effective as antioxidants and antihypertensives than high-molecular-weight peptides. The use of a buffer has been advised when performing proteolysis because the addition of alkalis often results in high salt concentrations in hydrolysates (Daliri et al. 2017). Protein hydrolysis also requires temperature control, which is generally set at 37°C to mimic average body temperature. Peptides sequenced by pepsin followed by pancreatin have been reported to have mainly a terminal lysine residue (Milán-Noris et al. 2018) but also leucine has been reported as a terminal residue (Torres-Fuentes et al. 2011).

Sánchez-Chino et al. (2019) reported that the consumption of chickpea protein hydrolysates (pepsin-pancreatin) could confer a protective effect against colon carcinogenesis. The peptides resulting from this sequenced enzymatic hydrolysis with pepsin-pancreatin (Sánchez-Chino et al. 2019) and alcalase-flavourzyme (Barbana & Boye, 2010) have shown antioxidant, ACE inhibitor, anti-inflammatory, and anticancer bioactivities (Sánchez-Chino et al. 2019; Tawalbeh et al. 2023). Similarly, alcalase has given rise to highly active renin-inhibiting hydrolysates, and pancreatin and papain have also produced high renin-inhibiting activity (Girgih et al. 2011; He et al. 2013; Fitzgerald et al. 2014).

Additionally, most of the sequenced chickpea peptides are produced from legumin-like globulins, which is the most abundant protein fraction within the seed (Real-Hernández & González de Mejía, 2019). It is important to note that the type of enzyme used, the temperature, and the time for hydrolysis affect the degree of hydrolysis and the type of peptides generated in the hydrolysates (Daliri et al. 2017). Therefore, enzyme selection and degree of hydrolysis require further investigation to ensure stable and high production of bioactive peptides (Boukid, 2021).

2.4 CHICKPEA PEPTIDES WITH BIOLOGICAL ACTIVITIES

Recently, it has been reported that the whole chickpea seed, as well as some of its main components such as proteins, have certain biological activity such as antioxidant, anti- inflammatory, anticancer, antimicrobial, angiotensin-converting enzyme inhibitor (ACE-I), and hypocholesterolemic and hypoglycemic activity (**Figure 2.1**), in addition to improving the integrity of the intestinal barrier (Sánchez-Chino et al. 2017; Monk et al. 2017; Monk et al. 2018; Waili et al. 2018; Xue et al. 2018; Shi et al. 2019; Chandrasekaran et al. 2020; Wang et al. 2023; Mahgoub et al. 2023). The main biological activities of chickpea proteins in human health are described in **Table 2.1**.

2.4.1 Antioxidant Activity (AA)

Oxidative metabolism is an essential part of cell survival and generates free radicals and other reactive oxygen species (ROS) that cause oxidative changes and, when there is an excess of free

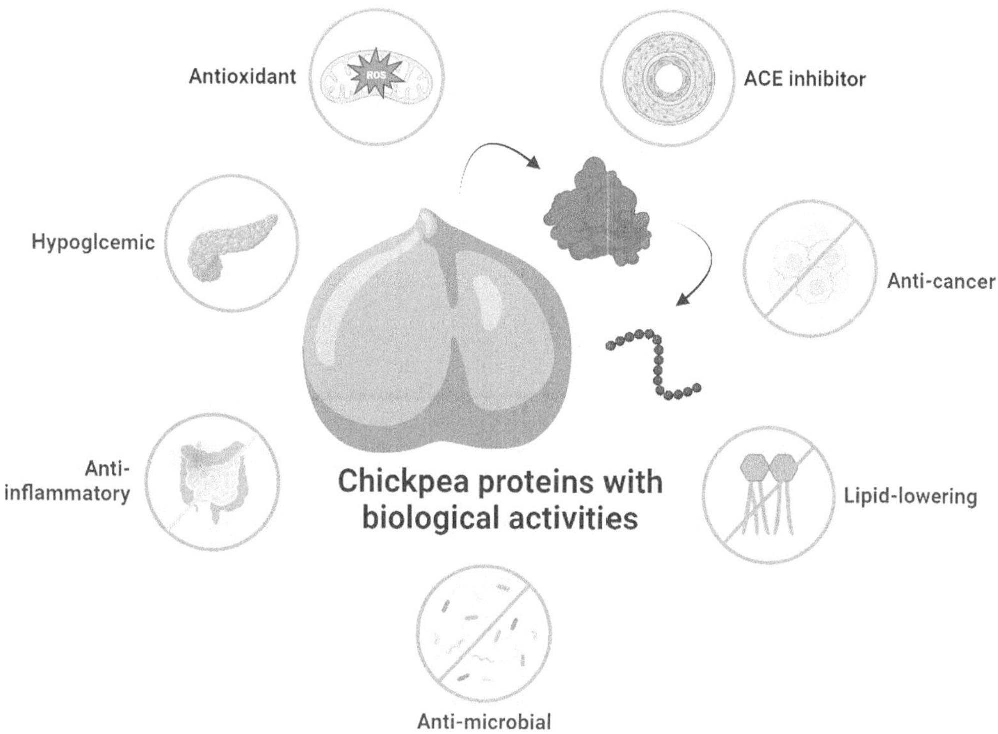

Figure 2.1 Biological activities of chickpea proteins (Kou et al., 2013; Guo et al., 2014; Xue et al., 2015; Xue et al., 2018; Heymich, Nißl, et al., 2021; Quintero-Soto et al., 2021). Created in Biorender.

radicals, protective enzymes are affected. Among the affected enzymes are super oxide dismutase (SOD), catalase (CAT), and peroxidase, which causes destructive and lethal cellular effects on molecules such as lipids, proteins, deoxyribonucleic acid (DNA), and enzymes (Antolovich et al. 2002; Shahidi & Zhong, 2015).

Similarly, the excess of ROS will cause oxidative stress and will contribute to the development of different diseases. For this reason the use of antioxidants may contribute to the prevention of these diseases by counteracting ROS (Torres-Fuentes et al. 2015). Karaś et al. (2015) observed peptide fractions of chickpea seeds that were obtained by *in vitro* gastrointestinal digestion with and without thermal treatment, and they reported that the peptide fraction of 3.5–7.0 kDa of raw chickpea seeds presented a higher antiradical activity against 2,2- diphenyl-1-picrylhydrazyl (DPPH•) and Fe^{2+}-chelating activity with a maximum mean inhibitory concentration value (IC_{50}) of 20.94 and 52.53 μg/mL, respectively.

As for the peptides obtained from the hydrolysate subjected to heat treatment, the highest Cu^{2+} -chelating activity was observed (IC_{50}=56.60 μg/mL), significant activity against 2,2'-azino-bis-3-ethylbenzothiazoline-6-sulfonic acid ($ABTS^{+\bullet}$) with IC50 of 41.01 μg/mL for the peptide fractions of 3.5–7 kDa and for the fraction of <3.5 kDa showed the most significant reducing power (0.362 A700/μg/mL). Kou et al. (2013) analyzed the AA of a chickpea albumin hydrolysate (Alcalase-Flavourzyme) through *in vitro* radical scavenging and reducing power assays and obtained three

TABLE 2.1 BIOLOGICAL ACTIVITIES FROM CHICKPEA PROTEINS

Chickpea Product	Proteolitic Enzymes	Bioactive Compounds	Effect	References
Raw and cooked seed hydrolysates	Pepsin-Pancreatin	Peptide fractions of 3.5–7 kDa	Antioxidant (ABTS, DPPH, metal ion chelating activity [Fe^{2+} and Cu^{2+}])	Karaś et al. (2015)
Albumin protein hydrolysate	Alcalase- Flavourzyme	Peptide fractions, fraction III (RQSHFANAQP)	Antioxidant (ABTS, DPPH, reducing power)	Kou et al. (2013)
Hydrolysates	Pepsin-Pancreatin	Peptide fractions	metal ion chelating activity	Torres-Fuentes, et al. (2011)
Hydrolysates	Pepsin-Pancreatin	3 peptide fractions	Lipid peroxidation inhibition	Torres-Fuentes, et al. (2014)
Peptide sequences		Fractions F2C, F2D, F3D and F3E	Antioxidant (reducing power of Fe^{2+} and DPPH)	Torres-Fuentes, et al. (2015)
Hydrolysates	Alcalase- Flavourzyme	Peptide fractions	Antioxidant (β-carotene bleaching method, reducing power and DPPH) -Hypocolesterolemic activity	Yust et al. (2012)
Hydrolysates	Alcalase	11 Peptide fractions	Antioxidant (DPPH, OH, O2 and metal chelating)	Zhang et al. (2011)
Albumin hydrolisate		CPe-III peptide	Antioxidant (reducing power of Fe^{2+}, •OH radical DPPH and ABTS) -Antiproliferative activity	Xue et al. (2015)
Hydrolisates	Alcalase	2 peptides (DHG VGNI)	Antioxidant (DPPH)	Ghribi et al. (2015)
Hydrolyzed sprouts	Pepsin-Pancreatin	Protein fractions glutelin, albumin and globulin	Antioxidant	Serrano-Sandoval et al. (2019)
Hydrolyzed sprouts	Trypsin, Neutrase, Alcalase and Papain	NF2-4-1 peptide	Antioxidant (DPPH e •OH)	Wali et al. (2020)

Hydrolysates of 45 chickpea seed accessions	Alcalase	BDN-9-3 hydrolysate	Antiproliferative activity against MCF-7 and MDA-MB-231 breast cancer cells	Gupta & Bhagyawant (2019a)
Hydrolisates	Alcalase- Flavourzyme		Antioxidant (DPPH, β-carotene inhibition activity) -Antiproliferative activity in breast cancer cells viz. Ishikawa and MCF-7 -ACE-I inhibition	Gupta & Bhagyawant (2019b)
Germinated (5 days)	Inherent enzymes increased during germination	Sprout Peptides	ACE-I inhibition	Mamilla & Mishra (2017)
Hydrolysates of albumin and globulin fractions of 8 chickpea genotypes	Alcalase	Peptides (FEL, FIE, FKN, FGKG, MEE)	-Antioxidant (ABTS, DPPH) - α-amylase, α-glucosidase and DPP4 inhibition	Quintero-Soto et al. (2021)
Hydrolisates	Alcalase	Pentapeptide NRYHE	Protection of Caco-2 and HT-29 cells through the stimulation of CAT, glutathione reductase and glutathione peroxidase.	Guo et al. (2014)
Sprout concentrate	Pepsin-Pancreatin	peptide fraction	Inhibited NO in LPS-stimulated 264.7 macrophages	Milán-Noris et al. (2018)
Peptide		CPe-III	-Decreased TNF-α expression Attenuates hyperlipidemia	Xue et al. (2018)
Hydrolysates	Pepsin-Pancreatin	peptide fraction	-Decrease aberrant crypts -Decrease triglycerides, LDL, and atherogenic	Sánchez-Chino et al. (2019)
Protease inhibitor concentrates			Antiproliferative activity in breast cancer cell lines MDA-MB-231 and prostate LNCaP and PC-3	Magee et al. (2012)
Protein extracts		protein fraction	- MMP-9 inhibition in HT-29 cells	Lima et al. (2016)

(continued)

TABLE 2.1 (CONTINUED)

Chickpea Product	Proteolitic Enzymes	Bioactive Compounds	Effect	References
Albumin hydrolysate	Alcalase-Flavourzyme		-Decrease in the number of tumors -Decrease in total cholesterol, triglycerides and LDL	Xue et al. (2012)
Isolated protein		C-25	- Anti-proliferative activity in oral cancer -Antifungal activity against: *Candida krusei, Candida tropicalis, and Candida parapsilosis*	Kumar et al. (2014)
Peptides		Peak 8, 9, 10, 11, 12 y 13	Antibacterial activity against *E. coli* and *S.aureus*	Waili et al. (2018)
Hydrolysates		Leg 1 Leg 2	Antibacterial activity against Clostridium perfringens ATCC 13124, *Pseudomonas putida, E. coli, Staphylococcus aureus, Listeria monocytogenes* and *Salmonella typhimurium* (DSM 554)	Heymich, Friedlein, et al. (2021); Heymich, Nißl, et al. (2021)
Hydrolysates	Pepsin		ACE-I inhibition	Boschin et al. (2014)
Fermented with *Cordyceps militaris*		Extracts	ACE-I inhibition	Xiao et al. (2015)
Hydrolysates	Alcalase		Decrease in weight, adipose tissue, total cholesterol, triglycerides, LDL and atherogenic index	Shi, Hou, Liu, et al. (2019)
Hydrolysates	Alcalase	Peptides	decreased total cholesterol, triglycerides and LDL in serum and increased HDL	Shi, Hou, Guo, et al. (2019)
Hydrolysates	Pepsin-Pancreatin and Bromelain		α-amylase inhibition	Chandrasekaran, Luna-Vital, and de Mejia (2020)

Lectin			Antioxidant activity -ACE-I inhibition α-glucosidase and α-amylase inhibition	Bhagyawant et al. (2019)
Hydrolysates of total protein, albumin, and globulin	Alcalase, Papain and Pancreatin	peptide fraction	- Antioxidant activity against DPPH and ABTS -ACE-I inhibition	Medina-Godoy et al. (2012)
Hydrolysates	Alcalase		Antiproliferative activity in MCF7 and MDA-MB-231 cell lines -ACE-I inhibition	Gupta & Bhagyawant (2018)
Hydrolysates	Alcalase/Flavourzyme (1-2 h) and Alcalase/ Flavourzyme (2-1 h),		-Antioxidant activity against DPPH	Xu et al. (2020)
Hydrolysates	Alcalase and Papain		-Antioxidant activity against DPPH	Ahmad Nadzri et al. (2021)
Hydrolysates	Alcalase-Flavourzyme, Pepsin-Pancreatin	Peptide fractions (5–10; 3-5; 1–3 and ≤1 kDa)	Antioxidant activity against DPPH, ABTS, Fe^{3+} reducing power and Cu^{2+} chelating capacity -Anti-inflammatory activity	Júarez-Chairez et al. (2022)

fractions. They observed that fraction III RQSHFANAQP (Arg-Gln-Ser-His-Phe-Ala-Asn-Ala-Gln-Pro [1155 Da]) showed higher antiradical activity against DPPH (41.3%) and ABTS (IC50 0.97 mmol/L) and higher Fe^{2+} reducing power.

Another study reported that three fractions were purified from a hydrolysate (pepsin-pancreatin) of chickpea, which suggests that the composition and distribution of the molecular mass of the His-containing fractions and the low molecular weight of the peptide can influence the chelating activity of metals; in addition, the fraction F3 showed greater chelating activity with 45.3% (Torres-Fuentes et al. 2011).

Likewise, Torres-Fuentes et al. (2014) observed three peptide fractions obtained from a chickpea protein hydrolysate (pepsin-pancreatin) and found that these fractions had the ability to inhibit copper-mediated lipid peroxidation, where subfractions F2C and F2D were able to completely inhibit fatty acid peroxidation at 0.03mg/mL in 24 hours and F2B completely inhibited copper-mediated low-density lipoprotein (LDL) oxidation at a concentration of 0.03 mg/mL. Torres-Fuentes et al. (2015) determined the AA and the peptide sequences included in the studied fractions that were obtained from chickpea protein hydrolysates (pepsin-pancreatin). The F2C and F2D fractions showed significantly higher Fe^{2+} reducing power compared to the butylated hydroxytoluene (BHT) control at 12 μg/mL and 25 μg/mL, respectively; furthermore, F3D and F3E showed similar activity to BHT with an antiradical activity against DPPH around 95%.

Finally, regarding cellular AA, the fractions with the highest activity in Caco-2 cells were F1D (Ala-Leu-Glu-Pro-Asp-His-Arg [836 Da]), F2B (Thr-Glu-Thr-Trp- Asn-Pro-Asn-His-Pro-Glu-Leu [1336 Da]) and F2C (Phe-Val-Pro-His[498Da]). In another, from a chickpea protein hydrolysate (alcalase), 11 peptide fractions were obtained, and of these fractions, fraction seven obtained by means of chromatographic separation techniques (Asn-Arg-Tyr-His-Glu [717.37 Da]) showed the greater AA for DPPH radical scavenging activity (45.33%), hydroxyl (•OH) radical scavenging activity (60.09%), superoxide radical scavenging activity (79.81%), Fe^{2+} (63.08%), and Cu^{2+} (76.92%) chelating activity and against autooxidation caused by linoleic acid (88.81%/8 days) (Zhang et al. 2011).

Yust et al. (2012) evaluated the AA of chickpea protein hydrolysates (alcalase-flavourzyme) by three methods: β-carotene bleaching method, reducing power, and DPPH radical scavenging. They found that oxidation was inhibited by 52 and 77% of β-carotene and DPPH free radical scavenging, respectively, and Fe^{2+} was also reduced. The peptide CPe-III-S (Arg-Gln-Ser-His-Phe-Ala-Asn-Ala-Gln-Pro, 1155 Da) derived from the hydrolysis of chickpea albumin and this peptide showed a high reducing power of Fe^{2+} and •OH radical scavenging activity, as well as great antiradical activity against DPPH and ABTS with an IC_{50} of 8.38 and 3.61 μmol/mL, respectively (Xue et al. 2015).

Ghribi et al. (2015) analyzed two peptides (Asp-His-Gly [327 Da] and Val-Gly-Asn-Ile [402 Da]) isolated from a chickpea protein hydrolysate (alcalase), which showed a great reduction (200 μg/mL) of the capacity of the DPPH radical by 67.32 and 51.66%, respectively. Likewise, Serrano-Sandoval et al. (2019) determined the AA of the protein fractions glutelin, albumin, and globulin that were obtained from hydrolysates (pepsin-pancreatin) of chickpea sprouts treated with selenium in CaCo-2 and observed that the glutelin fraction showed the highest antioxidant capacity. Wali et al. (2021) identified a new peptide, NF2-4-1 (Leu-Thr-Glu-Ile-Ile-Pro [685.41 Da]), from a protein hydrolysate (trypsin, neutrase, alcalase, and papain) of chickpea sprouts and evaluated its AA; they obtained great antiradical activity DPPH and •OH with IC50 0.24±0.06 mg/mL and 0.57±0.04 mg/mL, respectively.

On the other hand, hydrolysates (alcalase and flavourzyme) of two species of chickpea (*C. arietinum* and *C. reticulatum*) were evaluated to determine their AA and it was observed that the protein hydrolysate with alcalase (60 min of hydrolysis) showed great DPPH free radical scavenging activity for *C. arietinum* (55.55 ±6.2%) and *C. reticulatum* (69.12±2.6%). In contrast, the protein hydrolysate with flavourzyme (80 min hydrolysis) from *C. arietinum* (67.4 ± 4.4%) had the highest β-carotene inhibition activity (Gupta & Bhagyawant, 2019b).

Quintero-Soto et al. (2021) analyzed the AA by DPPH and ABTS of hydrolysates (alcalase) of albumin and globulin fractions of eight chickpea genotypes (*C. arietinum* L.) from Mexico and other 10 countries, in which they observed that the albumin hydrolysate showed higher AA against ABTS of 197.62–877.5 μmol ET/L in kabuli genotypes and 459.57–799.4 μmol ET/L in desi type. Likewise, the albumin hydrolysate presented higher DPPH free radical scavenging activity (22.9–64.46%) than that of the globulin hydrolysate (23.07–28.85%). Similarly, albumin hydrolysates of the ICC3671 genotype fractionated by Ultra Performance Liquid Chromatography (UPLC) showed four different peptides with high radical scavenging activity for ABTS and DPPH: FIE (Phe-Ile-Glu), FKN (Phe-Lys-Asn), FGKG (Phe-Gly-Lys-Gly), and MEE (Met-Glu-Glu)

Bhagyawant et al. (2019) evaluated the AA of a purified chickpea lectin by ion exchange chromatography and observed by the β-carotene bleaching inhibitory activity showed AA of lectin chickpea globulin, which showed high AA against DPPH radicals (range 18.78 to 46.76 μmol Trolox equivalent antioxidant capacity [TEAC] – 1) and ABTS (range from 27.73 to 95.61 μmol $TEAC^{-1}$) (Medina-Godoy et al. 2012). Xu et al. (2020) evaluated the AA of a chickpea protein isolate hydrolyzed with alcalase/flavourzyme (1 h:2 h) and alcalase/flavourzyme (2 h:1 h), which presented DPPH· scavenging activity of 44.3 and 48.4%, respectively. Ahmad-Nadzri et al. (2021) analyzed the AA by DPPH of chickpea hydrolysates (alcalase and papain) and observed that the hydrolysate with alcalase (79.57%) showed higher antiradical activity DPPH than the one with papain (61.11%).

On the other hand, Juárez-Chairez et al. (2022) studied protein hydrolysate fractions (alcalase-flavourzyme and pepsin-pancreatin), where AA was evaluated by ABTS, DPPH, Fe^{3+} reducing power, and Cu^{2+}-chelating capacity methods. They observed that the fractions F1-3 kDa showed greater inhibition of ABTS radicals (78%) and DPPH (7.76%); for Cu^{2+}-chelating capacity F≤1 kDa showed greater inhibition (52.60%).

2.4.2 Anti-inflammatory Activity

Inflammation can arise in response to certain infectious processes caused by bacteria, fungi, or viruses in the body, in particular tissues, or in the blood (Isailovic et al. 2015; Pedraza-Alva et al. 2015). Furthermore, in most cases the innate immune response as well as the adaptive immune response are involved in the development of the invasion because the innate immune system is the main defense mechanism against microorganisms and cancer cells and involves the activity of various cells. The adaptive immune system involves activity of more specialized cells such as B and T cells, responsible for eradicating invading pathogens and cancer cells (Fernandes et al. 2015; Rock et al. 2011; Lucas et al. 2006).

Milán-Noris et al. (2018) reported a peptide fraction (24 peptides; 1.1–2.3 kDa) isolated from digested sprouted chickpea protein concentrate (0.5–5 mg/mL) that inhibited nitric oxide (NO) in lipopolysaccharide-stimulated 264.7 macrophages (LPS). In another study with chickpea protein the peptide CPe-III (Arg-Gln-Ser-His-Phe-Ala- Asn-Ala-Gln-Pro, [1155 Da]) showed a decrease in the expression of Tumor Necrosis Factor-Alpha (TNF-α) in an animal model with a high-fat diet (Xue et al. 2018). In addition, peptide fractions of protein hydrolysates (alcalase-flavourzyme and pepsin-pancreatin) were studied to determine their anti-inflammatory activity in LPS and it was observed that the fractions of F5-10 kDa, F3-5 kDa, and F1-3 kDa presented the highest inhibition in the synthesis of NO (76%), TNF-α (93%), and interleukin-1β (90%).

2.4.3 Anticancer Activity

Carcinogenesis is a multistep process that results in the formation of a tumor. It all starts when a normal cell suffers a mutation or modification in the DNA, followed by uncontrolled growth (Cid-Gallegos et al. 2020b). Thus, Sánchez-Chino et al. (2019) conducted a study on the effects of chickpea protein hydrolysates (pepsin-pancreatin) administered at a dosage of 30 mg/kg body weight to ICR strain mice on a high-fat diet. The study demonstrated a reduction in aberrant crypts to baseline levels, indicating a protective effect against colon cancer. Similarly, Guo et al. (2014) analyzed a pentapeptide from chickpea protein hydrolysates (Asn-Arg-Tyr-His-Glu, 717.37 Da) that showed protection to Caco-2 and HT-29 cells from hydrogen peroxide (H_2O_2)-induced injury by means of the stimulation of CAT, glutathione reductase, and glutathione peroxidase. Magee et al. (2012) reported antiproliferative activity of chickpea protease inhibitor concentrates (25–400 µg/mL) in breast cancer cell lines MDA-MB-231 (12–14%) and prostate LNCaP (22–35%) and PC-3 (32–37%).

Lima et al. (2016) showed the anticancer activity of chickpea protein extracts, which inhibited the activity of MMP-9 metalloproteases in HT-29 colon carcinoma cells by 95%, as well as cell migration by 63%. Similarly, another study with the peptide CPe-III showed antiproliferative activity in the cell lines MCF-7 (EC50 2.38 µmol/mL) and MDA-MB-231 (EC50 1.50 µmol/mL) (Xue et al. 2015). On the other hand, Xue et al. (2012) analyzed the anticancer activity of chickpea albumin hydrolysates (alcalase-flavourzyme) at doses of 100mg/kg in mice with tumors induced by inoculation of H-22 tumor cells. As a result, they observed a decrease in tumors of 53.37% during 10 treatment days. For their part, Kumar et al. (2014) isolated an antifungal protein (C-25) from chickpea and tested its antiproliferative potential against human oral cancer cells at a concentration of 37.5 µg/mL by targeting p38 MAP kinase.

In addition, a study with chickpea protein hydrolysates (alcalase) demonstrated antiproliferative activity in MCF-7 and MDA-MB-231 breast cancer cell lines with an IC50 of 0.71 mg/ml and 0.78 mg/ml, respectively (Gupta & Bhagyawant 2018). Likewise, Gupta & Bhagyawant (2019b) reviewed chickpea hydrolysates (alcalase and flavourzyme) from two species *C. arietinum* and *C. reticulatum*, which showed antiproliferative activity with doses of 0.1–1.0 mg in human endometrial adenocarcinoma cell line (Ishikawa) (IC50 0.36 and 0.29 mg/ml) and human breast cancer cell line MCF-7 (IC50 0.47 and 0.41 mg/ml). Lastly, Gupta & Bhagyawant (2019a) studied protein hydrolysates (alcalase) from 45 chickpea seed accessions and found that the BDN-9-3 hydrolysate possessed antiproliferative activity against MCF-7 breast cancer cells (IC50 0.60 mg/ml) and MDA-MB-231 (0.63mg/ml).

2.4.4 Antimicrobial Activity

Antimicrobial peptides are molecules present in legumes such as chickpeas that can exert protective or defense activity against bacteria, yeasts, fungi, and viruses (Zhang & Gallo, 2016). Due to their biological properties, these molecules have been considered useful for the development of drugs, as novel agents in the therapeutics of diseases, applicable to both agriculture and medicine (Rodríguez-Decuadro et al. 2021); they have low molecular weight, containing between 12 and 50 amino acids, of which between two and seven are usually positively charged, and up to 50% are hydrophobic amino acids. Finally, their action is attributed to the lytic action they cause on the membrane and/or their intracellular inhibitory activities (Omidbakhsh Amiri et al. 2021).

Waili et al. (2018) analyzed six peptides from chickpea sprouts (peak 8 [5435.7 Da], peak 9 [6490.1 Da], peak 10 [5078.7 Da], peak 11 [1578.3 Da], peak 12 [1130.3 Da] and peak 13 [1781.6 Da]) that showed antibacterial activity against *Escherichia coli* (peaks 8, 9, 10, and 12) and *Staphylococcus aureus* (peak 8).

On the other hand, a study analyzed the antimicrobial effect of peptides Leg1 (Arg-Ile-Lys-Thr-Val-Thr-Ser-Phe-Asp-Leu-Pro-Ala-Leu-Arg-Phe-Leu-Lys-Leu (2118.59 g/mol) and Leg2 (Arg-Ile-Lys-Thr-Val-Thr-Ser-Phe-Asp-Leu-Pro-Ala-Leu-Arg-Trp-Leu-Lys-Leu 2157.62 g/mol) obtained from chickpea hydrolysates. Both peptides were effective at doses of 125 µmol/L against *Clostridium perfringens* ATCC 13124 and *Pseudomonas putida*. Likewise, the effective minimum inhibitory concentrations (MICs) against *Listeria monocytogenes* (1097/2012) was 250 µmol/L, while for *Enterohemorrhagic Escherichia coli, Staphylococcus aureus*, and *Listeria monocytogenes* (SLCC 5581) it was 500 µmol/L. In addition, the highest concentration applied was 1,000 µmol/L, which had an effect against *Salmonella typhimurium* (DSM 554) (Heymich et al. 2021; Heymich et al. 2021). Likewise, Kumar et al. (2014) reviewed that chickpea protein C-25 exerted antifungal activity against *Candida krusei, Candida tropicalis*, and *Candida parapsilosis* of MIC values 1.56–12.5 µg/L.

2.4.5 Angiotensin Converting Enzyme (ACE) Inhibitory Activity

Hypertension is one of the main cardiovascular diseases that can cause strokes, heart failure, and decrease life expectancy and is therefore considered to be the leading cause of death worldwide (Yu et al. 2014; Kharazmi-Khorassani et al. 2019). The renin-angiotensin system (RAS) is a hormonal system that plays a vital role in the control of blood pressure and fluid balance. In this sense, ACE (EC 3.4.15.1) is an enzyme that catalyzes the conversion of angiotensin I (Ang I) into angiotensin II (Ang II), an active vasoconstrictor (Kharazmi-Khorassani et al. 2019).

Many antihypertensive drugs were developed using the structure of snake venom peptides. For example, captopril, an ACE inhibitor, was the first drug of this kind and has proven to be effective in treating high blood pressure. Studies have shown that ACE inhibition has dual mechanisms of vascular and endothelial protection, leading to increased production of nitric oxide, relaxation of vascular smooth muscles, and improved fibrinolysis (Yu et al. 2014). For example, Boschin et al. (2014) reported a chickpea protein hydrolysate (pepsin) that inhibited ACE-I with an IC50 of 673 µg/mL, and Xiao et al. (2015) carried out a solid-state fermentation of chickpea flour with *Cordyceps militaris* SN-18, from which they subsequently obtained extracts that inhibited ACE-I with an IC50 of 0.668 mg extract/mL. Likewise, Mamilla & Mishra (2017) reported the inhibition of ACE-I of 67.21% by peptides from chickpea sprouts (5 days). In addition, Gupta & Bhagyawant (2019b) achieved ACE-I inhibition with chickpea protein hydrolysates (IC_{50}=0.182 mg/mL with alcalase at 60 min hydrolysis and IC_{50}=0.365 mg/mL with flavourzyme at 100 min hydrolysis) and *Cicer reticulatum* (IC_{50}=0.113 mg/mL with alcalase at 60 min hydrolysis and IC_{50}=0.269 mg/mL with flavourzyme at at 80 min hydrolysis).

Bhagyawant et al. (2019) were able to inhibit ACE-I with a chickpea lectin with an IC50 value of 57.43±1.20 µg/ml compared to captopril. Then, in a study with hydrolysates (alcalase, papain, and pancreatin) of total protein, albumin,\ and chickpea globulin, a great inhibitory activity against ACE-I was found, with IC50 values ranging from 0.101 to 37.33 µg/mL (Medina-Godoy et al. 2012).

Likewise, Gupta & Bhagyawant (2018) reported ACE-I inhibition with an IC50 value of 52.22 μg/ml with chickpea protein hydrolysates (alcalase).

2.4.6 Lipid-Lowering and Hypoglycemic Activity

Hyperlipidemia is defined as an imbalance in blood lipid homeostasis characterized by elevated levels of lipids or lipoproteins (LPL) in serum/plasma, with elevated total cholesterol and high triglycerides being the most common conditions (Ramírez-Jiménez et al. 2015). On hyperglycemia, the range in which the fasting glucose concentration is normally found is 3.98–5.6 mmol/L, and a concentration of 7 mmol/L is considered an elevated concentration, which is defined as such as hyperglycemia. This condition produces a large amount of ROS, through which various processes and signal transduction pathways are activated or inhibited (Tang et al. 2014). This condition is the main biochemical parameter for diagnosing diabetes, chronic hyperglycemia, that in turn induces several conditions such as nephropathy, atherosclerosis, and myocardial infarction. There is also a complication that relates diabetes to atherosclerosis and hypertension called vascular remodeling (El Atab et al. 2022).

Sánchez-Chino et al. (2019) reported that the consumption of protein hydrolysates (pepsin-pancreatin) from chickpeas at doses of 10–30 mg/kg of weight per day in mice with a hypercaloric diet decreased the concentration of triglycerides, LDL, and the atherogenic index. Similarly, Xue et al. (2018) analyzed the chickpea CPe-III peptide and found that it attenuated hyperlipidemia by upregulating genes involved in hepatic fatty acid (FA) synthesis (ACC1 and FAS), cholesterol metabolism (SREBP2, HMGCR, and PCSK9), and upregulating key regulators involved in FA oxidation (PPARα and FABP1), lipolysis (ATGL), LDLR, reverse cholesterol transport (ABCA1, SR-B1, and LXRα), and cholesterol catabolism (CYP7A1).

Furthermore, Xue et al. (2012) reported that chickpea albumin hydrolysates (alcalase-flavourzyme) at a dose of 150 mg/kg in mice decreased cholesterol levels (46.75%), triglycerides (36.55%), and LDL (48.53%), and improved high-density lipoprotein (HDL) (15.34%). Also, in a study they tested chickpea protein hydrolysates (alcalase) in a model of ovariectomized rats and obtained a decrease in weight, adipose tissue size, total cholesterol, triglycerides, LDL, and atherogenic index (Shi, Hou, Liu et al. 2019).

Likewise, a study carried out with peptides obtained from chickpea protein hydrolysates (alcalase) showed a significant decrease in serum total cholesterol, triglycerides, and LDL and an increase in HDL in obese rats fed a high-fat diet (Shi, Hou, Gou, et al. 2019). Yust et al. (2012) studied protein hydrolysates (alcalase-flavourzyme) and reported hypocholesterolemic activity by reducing 50% of the micellar solubilization of cholesterol. On the other hand, Chandrasekaran et al. (2020) evaluated protein hydrolysates (pepsin-pancreatin and bromelain) and observed that they inhibited α-amylase activity in values of 11–38.4% at a concentration of 10 mg/mL.

Quintero-Soto et al. (2021) reported chickpea albumin protein hydrolysates (alcalase) and observed antidiabetic potential by inhibiting the activity of α-amylase (66.6%) and α-glucosidase (57.0%). In addition, they reported the inhibition of α-amylase by the peptides FEI, FEL, and FIE (39.96%) and FGKG (54.12%) and the inhibition of α-glucosidase by the peptides MEE (41.97%) and FGKG (56.01%). Moreover, the peptides inhibited dipeptidyl peptidase-4 (DPP4) with IC50 values of 4.2–12.08 μg/mL (Quintero-Soto et al. 2021). Finally, in a study with a chickpea lectin, they were able to inhibit α-glucosidase and α-amylase activities in a concentration-dependent manner with IC50 values of 85.41±1.21 μg/ml and 65.05±1.2 μg/ml, respectively (Bhagyawant et al. 2019). Therefore, chickpea proteins have shown antidiabetic potential and can be used in antihyperglycemic therapies.

2.5 CONCLUSION

Chickpea seeds are rich in secondary metabolites that function as a defense mechanism against insects, fungi, bacteria, and diseases. These metabolites, which include phenolic compounds, phytates, saponins, isoflavones, and lectins, possess remarkable biological properties. As a result, chickpeas are increasingly being used as a source of functional ingredients, in addition to their nutritional benefits, as they contain an excellent concentration of protein, low digestibility carbohydrates, and dietary fiber. In addition, chickpeas are an excellent source of bioactive peptides with biological activities such as antioxidant, anti-inflammatory, anticancer, antimicrobial, ACE-I inhibitory, antidiabetic, and hypolipidemic, making them a functional food. These peptides can be obtained by methods such as enzymatic hydrolysis using alcalase, flavorzyme, pepsin, and pancreatin, as well as by germination and fermentation. However, their mechanisms of action are not fully understood and further research is needed to learn more about their biological activities and how they function in vivo.

REFERENCES

Acevedo-Martínez, K.A., Yang, M.M., & González de Mejía, E. (2021). Technological properties of chickpea (*Cicer arietinum*): Production of snacks and health benefits related to type-2 diabetes. *Comprehensive Reviews in Food Science and Food Safety*, 20(4): 3762–3787. https://doi.org/10.1111/1541-4337.12762

Ahmad-Nadzri, F.N., Tawalbeh, D., & Sarbon, N.M. (2021). Physicochemical properties and antioxidant activity of enzymatic hydrolysed chickpea (*C. arietinum* L.) protein as influence by alcalase and papain enzyme. *Biocatalysis and Agricultural Biotechnology*, 36: 102131. https://doi.org/10.1016/j.bcab.2021.102131

Antolovich, M., Prenzler, P.D., Patsalides, E., McDonald, S., & Robards, K. (2002). Methods for testing antioxidant activity. *Analyst*, 127(1): 183–189. https://doi.org/10.1039/B009171P

Barbana, C., & Boye, J.I. (2010). Angiotensin I-converting enzyme inhibitory activity of chickpea and pea protein hydrolysates. *Food Research International*, 43(6): 1642–1649. https://doi.org/10.1016/j.foodres.2010.05.003

Bhagyawant, S.S., Narvekar, D.T., Gupta, N., Bhadkaria, A., Gautam, A.K., & Srivastava, N. (2019). Chickpea (*C. arietinum* L.) lectin exhibit inhibition of ACE-I, α-amylase and α-glucosidase activity. *Protein and Peptide Letters*, 26(7). https://doi.org/10.2174/0929866526666190327130037

Boschin, G., Scigliuolo, G.M., Resta, D., & Arnoldi, A. (2014). ACE-inhibitory activity of enzymatic protein hydrolysates from lupin and other legumes. *Food Chemistry*, 145: 495–501 https://doi.org/10.1016/j.foodchem.2013.07.076

Boukid, F. (2021). Chickpea (*Cicer arietinum* L.) protein as a prospective plant- based ingredient: A review. *International Journal of Food Science & Technology*, 56(11): 5435–5444. https://doi.org/10.1111/ijfs.15046

Carbas, B., Machado, N., Pathania, S., Brites, C., Rosa, E.A., & Barros, A.I. (2023). Potential of legumes: Nutritional value, bioactive properties, innovative food products, and application of eco-friendly tools for their assessment. *Food Reviews International*, 39(1): 1–29. https://doi.org/10.1080/87559129.2021.1901292

Chandrasekaran, S., Luna-Vital, D., & de Mejía, E.G. (2020). Identification and comparison of peptides from chickpea protein hydrolysates using either bromelain or gastrointestinal enzymes and their relationship with markers of type 2 diabetes and bitterness. *Nutrients*, 12(12): 3843. https://doi.org/10.3390/nu12123843.

Cheng, K., Gao, H., Wang, R.R., Liu, Y., Hou, Y.X., Liu, X.H., Liu, K. & Wang, W. (2017). Evaluation of extraction and degradation methods to obtain chickpea saponin b1 from chickpea (*C.arietinum* L.). *Molecules*, 22(2): 332. https://doi.org/10.3390/molecules22020332

Cid-Gallegos, M.S., Sánchez-Chino, X.M., Álvarez-González, I., Madrigal-Bujaidar, E., Vásquez-Garzón, V.R., Baltiérrez-Hoyos, R., Villa-Treviño, S., Dávila-Ortíz, G., & Jiménez-Martínez C. (2020a). Modification of *in vitro* and *in vivo* antioxidant activity by consumption of cooked chickpea in a colon cancer model. *Nutrients*, 12(9): 2572. www.mdpi.com/2072-6643/12/9/2572

Cid-Gallegos, M.S., Sánchez-Chino, X.M., Juárez-Chairez, M.F., Álvarez-González, I., Madrigal-Bujaidar, E., & Jiménez-Martínez, C. (2020b). Anticarcinogenic activity of phenolic compounds from sprouted legumes. *Food Reviews International*, 38(1): 18–33 https://doi.org/10.1080/87559129.2020.1840581.

Daliri, E.B., Oh, D.H., & Lee, B.H. (2017). Bioactive peptides. *Foods*, 6(5): 32. https://doi.org/10.3390/foods 6050032

Devi, S., Varkey, A., Dharmar, M., Holt, R.R., Allen, L.H., Sheshshayee, M.S., Preston, T., Keen, L.C., & Kurpad, A.V. (2020). Amino acid digestibility of extruded chickpea and yellow pea protein is high and comparable in moderately stunted south indian children with use of a dual stable isotope tracer method. *Journal of Nutrition*, 150(5): 1178–1185. https://doi.org/10.1093/jn/nxaa004

Domínguez-Arispuro, D.M., Cuevas-Rodríguez, E.O., Milán-Carrillo, J., León-López, L., Gutiérrez-Dorado, R., & Reyes-Moreno, C. (2018). Optimal germination condition impacts on the antioxidant activity and phenolic acids profile in pigmented desi chickpea (*Cicer arietinum* L.) seeds. *Journal of Food Science and Technology*, 55(2): 638–647. https://doi.org/10.1007/s13197-017-2973-1

El Atab, O., Ghantous, C.M., El-Zein, N., Farhat, R., Agouni, A., Korashy, H.M., Djouhri, L., Kamareddine, L., Zibara, K., & Zeidan, A. (2022). Involvement of caveolae in hyperglycemia-induced changes in adiponectin and leptin expressions in vascular smooth muscle cells. *European Journal of Pharmacology*, 919(17):174701. https://doi.org/10.1016/j.ejphar.2021.174701

FAO. (2021). FAOSTAT statistical database. www.fao.org/faostat/

Fernandes, J.V., Cobucci, R.N.O., Jatobá, C.A.N., de Medeiros Fernandes, T.A.A., de Azevedo, J.W.V., & de Araújo, J.M.G. (2015). The role of the mediators of inflammation incancer development. *Pathology & Oncology Research*, 21(3): 527–534. https://doi.org/10.1007/s12253-015-9913-z

Fitzgerald, C., Aluko, R.E., Hossain, M., Rai, D.K., & Hayes, M. (2014). Potential of a renin inhibitory peptide from the red seaweed *Palmaria palmata* as a functional food ingredient following confirmation and characterization of a hypotensive effect in spontaneously hypertensive rats. *Journal of Agricultural and Food Chemistry*, 62(33): 8352–8356. https://doi.org/10.1021/jf500983n

Gao, Y., Yao, Y., Zhu, Y., & Ren, G. (2015). Isoflavone content and composition in chickpea (*C arietinum* L.) sprouts germinated under different conditions. *Journal of Agricultural and Food Chemistry*, 63(10): 2701–2707. https://doi.org/10.1021/jf5057524

Ghribi, A.M., Sila, A., Przybylski, R., Nedjar-Arroume, N., Makhlouf, I., Blecker, C., Attia H., Dhulster, P., Bougatef, A., & Besbes, S. (2015). Purification and identification of novel antioxidant peptides from enzymatic hydrolysate of chickpea (*Cicer arietinum* L.) protein concentrate. *Journal of Functional Foods*, 12: 516–525. https://doi.org/10.1016/j.jff.2014.12.011

Girgih, A.T., Udenigwe, C.C., Li, H., Adebiyi, A.P., & Aluko, R.E. (2011). Kinetics of enzyme inhibition and antihypertensive effects of hemp seed (Cannabis sativa L.) protein hydrolysates. *Journal of the American Oil Chemists' Society*, 88(11): 1767–1774. https://doi.org/10.1007/s11746-011-1841-9

Grdeń, P., & Jakubczyk, A. (2023). Health benefits of legume seeds. *Journal of the Science of Food and Agriculture*. 103(11): 5213-5220. https://doi.org/10.1002/jsfa.12585

Guo, Y., Zhang, T., Jiang, B., Miao, M., & Mu, W. (2014). The effects of an antioxidative pentapeptide derived from chickpea protein hydrolysates on oxidative stress in Caco-2 and HT-29 cell lines. *Journal of Functional Foods*, 7: 719–726. https://doi.org/10.1016/j.jff.2013.12.013

Gupta, N., & Bhagyawant S.S. (2018). Angiotensin-I converting enzyme (ACE-I) inhibitory and antiproliferative potential of chickpea seed protein hydrolysate. *Annals of Plant Science*, 7(3): 2149–2153 https://doi.org/10.21746/aps.2018.7.3.10

Gupta, N., & Bhagyawant, S.S. (2019a). Enzymatic treatment improves ACE-I inhibiton and antiproliferative potential of chickpea. *Vegetos*, 32(3): 2149–2153. https://doi.org/10.1007/s42535-019-00031-6

Gupta, N., & Bhagyawant, S.S. (2019b). Impact of hydrolysis on functional properties, antioxidant, ACE-I inhibitory and antiproliferative activity of *C. arietinum* and *Cicer reticulatum* hydrolysates. *Nutrire*, 44(1): 1–12. https://doi.org/10.1186/s41110-019-0095-4

He, R., Malomo, S.A., Alashi, A., Girgih, A.T., Ju, X., & Aluko, R.E. (2013). Purification and hypotensive activity of rapeseed protein-derived renin and angiotensin converting enzyme inhibitory peptides. *Journal of Functional Foods*, 5(2): 781–789. https://doi.org/10.1016/j.jff.2013.01.024

Heymich, M.L., Friedlein, U., Trollmann, M., Schwaiger, K., Böckmann, R.A., & Pischetsrieder, M. (2021). Generation of antimicrobial peptides Leg1 and Leg2 from chickpea storage protein, active against food spoilage bacteria and foodborne pathogens. *Food Chemistry*, 347: 128917. https://doi.org/10.1016/j.foodchem.2020.128917

Heymich, M.L., Nißl, L., Hahn, D., Noll, M., & Pischetsrieder, M. (2021). Antioxidative, antifungal and additive activity of the antimicrobial peptides Leg1 and Leg2 from chickpea. *Foods*, 10(3): 585. https://doi.org/10.3390/foods10030585

Isailovic, N., Daigo, K., Mantovani, A., & Selmi, C. (2015). Interleukin-17 and innate immunity in infections and chronic inflammation. *Journal of Autoimmunity*, 60: 1–11. https://doi.org/10.1016/j.jaut.2015.04.006

Juárez-Chairez, M.F., Cid-Gallegos, M.S., Meza-Márquez, O.G., & Jiménez-Martínez, C. (2020). Biological activities of Chickpea in human health (*Cicer arietinum* L.). A review. *Plant Foods for Human Nutrition*, 75(2): 142–153. https://doi.org/10.1007/s11130-020-00814-2

Juárez-Chairez, M.F., Meza Márquez, O.G., Márquez-Flores, Y.K., Jiménez-Martínez, C., & Osorio-Revilla, G. (2022). Actividad antiinflamatoria y antioxidante in vitro de fracciones de hidrolizado de proteínas de garbanzo (*Cicer arietinum* L.). *Biotecnia*, 24(2): 59–68. https://doi.org/10.18633/biotecnia.v24i2.1594

Karaś, M., Baraniak, B., Rybczyńska, K., Gmiński, J., Gaweł- Bęben, K., & Jakubczyk, A. (2015). The influence of heat treatment of chickpea seeds on antioxidant and fibroblast growth-stimulating activity of peptide fractions obtained from proteins digested under simulated gastrointestinal conditions. *International Journal of Food Science & Technology*, 50(9): 2097–2103. https://doi.org/10.1111/ijfs.12872

Kaur, R., & Prasad, K. (2021). Technological, processing and nutritional aspects of chickpea (*Cicer arietinum*)- A review.*Trends in Food Science & Technology*, 109: 448–463. https://doi.org/10.1016/j.tifs.2021.01.044

Keskin, S.O., Ali, T.M., Ahmed, J., Shaikh, M., Siddiq, M., & Uebersax, M.A. (2021). Physico-chemical and functional properties of legume protein, starch, and dietary fiber—A review. *Legume Science*, 4(1): e117. https://doi.org/10.1002/leg3.117

Kharazmi-Khorassani, J., Asoodeh, A., & Tanzadehpanah, H. (2019). Antioxidant and angiotensin-converting enzyme (ACE) inhibitory activity of thymosin alpha-1 (Thα1) peptide. *Bioorganic Chemistry*, 87: 743–752. https://doi.org/10.1016/j.bioorg.2019.04.003

Kou, X., Gao, J., Xue, Z., Zhang, Z., Wang, H., & Wang, X. (2013). Purification and identification of antioxidant peptides from chickpea (*Cicer arietinum* L.) albumin hydrolysates. *LWT- Food Science and Technology*, 50(2): 591–598. https://doi.org/10.1016/j.lwt.2012.08.002

Kumar, S., Kapoor, V., Gill, K., Singh, K., Xess, I., Das, S.N., & Dey, S. (2014). Antifungal and antiproliferative protein from *Cicer arietinum*: A bioactive compound against emerging pathogens. *BioMed Research International*, 2014: 387203. https://doi.org/10.1155/2014/387203

Lima, A.I.G., Mota, J., Monteiro, S.A.V S., & Ferreira, R.M.S.B. (2016). Legume seeds and colorectal cancer revisited: Protease inhibitors reduce MMP-9 activity and colon cancer cell migration. *Food Chemistry*, 197: 30–38. https://doi.org/10.1016/j.foodchem.2015.10.063

Lucas, S.M., Rothwell, N.J., & Gibson, R.M. (2006). The role of inflammation in CNS injury and disease. *British Journal of Pharmacology*, 147(S1): S232–S240. https://doi.org/10.1038/sj.bjp.0706400

Magalhães, S.C., Taveira, M., Cabrita, A.R., Fonseca, A.J., Valentão, P., & Andrade, P.B. (2017). European marketable grain legume seeds: Further insight into phenolic compounds profiles. *Food Chemistry*, 215: 177–184 https://doi.org/10.1016/j.foodchem.2016.07.152

Magee, P.J., Owusu-Apenten, R., McCann, M.J., Gill, C.I., & Rowland, I.R. (2012). Chickpea (*Cicer arietinum*) and other plant-derived protease inhibitor concentrates inhibit breast and prostate cancer cell proliferation *in vitro*. *Nutrition and Cancer*, 64(5): 741–748. https://doi.org/10.1080/01635581.2012.688914

Mahgoub, S., Alagawany, M., Nader, M., Omar, S.M., Abd El-Hack, M.E., Swelum, A., & Dhama, K. (2023). Recent development in bioactive peptides from plant and animal products and their impact on the human health. *Food Reviews International*, 39(1): 1–26. https://doi.org/10.1080/87559129.2021.1923027

Mamilla, R.K., & Mishra, V.K. (2017). Effect of germination on antioxidant and ACE inhibitory activities of legumes. *LWT-Food Science and Technology*, 75: 51–58. https://doi.org/https://doi.org/10.1016/j.lwt.2016.08.036

Medina-Godoy, S., Ambriz-Pérez, D.L., Fuentes-Gutiérrez, C.I., Germán-Báez, L.J., Gutiérrez-Dorado, R., Reyes-Moreno, C., & Valdez-Ortiz, A. (2012). Angiotensin-converting enzyme inhibitory and antioxidative activities and functional characterization of protein hydrolysates of hard- to-cook chickpeas. *Journal of the Science of Food and Agriculture*, 92(9): 1974–1981. https://doi.org/10.1002/jsfa.5570

Megías, C., Cortés-Giraldo, I., Alaiz, M., Vioque, J., & Girón-Calle, J. (2016). Isoflavones in chickpea (*Cicer arietinum*) protein concentrates. *Journal of Functional Foods*, 21: 186–192. https://doi.org/10.1016/j.jff.2015.12.012

Milán-Noris, A.K., Gutiérrez-Uribe, J.A., Santacruz, A., Serna-Saldívar, S.O., & Martínez-Villaluenga, C. (2018). Peptides and isoflavones in gastrointestinal digests contribute to the anti-inflammatory potential of cooked or germinated desi and kabuli chickpea (*Cicer arietinum* L.). *Food Chemistry*, 268: 66–76. https://doi.org/10.1016/j.foodchem.2018.06.068

Monk, J.M., Lepp, D., Wu, W., Graf, D., McGillis, L.H., Hussain, A., Carey, C., Robinson, L.E., Liu, R., Tsao, R., Brummer, Y., Tosh, S.M., Power, & K.A. (2017). Chickpea-supplemented di*et al*ters the gut microbiome and enhances gut barrier integrity in C57Bl/6 male mice. *Journal of Functional Foods*, 38:663–674. https://doi.org/10.1016/j.jff.2017.02.002

Monk, J.M., Wu, W., McGillis, L.H., Wellings, H.R., Hutchinson, A.L., Liddle, D.M., Graf, D., Robinson, L.E., & Power, K.A. (2018). Chickpea supplementation prior to colitis onset reduces inflammation in dextran sodium sulfate treated C57Bl/6 male mice. *Applied Physiology, Nutrition, and Metabolism*, 43(9): 893–901. https://doi.org/10.1139/apnm-2017-0689

Omidbakhsh, A.E., Farmani, J., Amiri, Z.R., Dehestani, A., & Mohseni, M. (2021). Antimicrobial activity, environmental sensitivity, mechanism of action, and food application of αs165-181 peptide. *International Journal of Food Microbiology*, 358:109403. https://doi.org/10.1016/j.ijfoodmicro.2021.109403

Paredes-López, O., Ordorica-Falomir, C., & Olivares-Vázquez, M.R. (1991). Chickpea protein isolates: physicochemical, functional and nutritional characterization. *Journal of Food Science*, 56(3), 726–729. https://doi.org/10.1111/j.1365-2621.1991.tb05367.x

Pedraza-Alva, G., Pérez- Martínez, L., Valdez- Hernández, L., Meza- Sosa, K.F., Ando-Kuri, M. (2015). Negative regulation of the inflammasome: Keeping inflammation under control. *Immunological Reviews*, 265(1):231–257. https://doi.org/10.1111/imr.12294

Quintero-Soto, M.F., Chávez- Ontiveros, J., Garzón- Tiznado, J.A., Salazar- Salas, N.Y., Pineda- Hidalgo, K.V., Delgado- Vargas, F., & López- Valenzuela, J.A. (2021). Characterization of peptides with antioxidant activity and antidiabetic potential obtained from chickpea (*Cicer arietinum* L.) protein hydrolyzates. *Journal of Food Science*, 86(7):2962–2977. https://doi.org/10.1111/1750-3841.15778

Ramírez-Jiménez, A.K., Reynoso-Camacho, R., Tejero, M.E., León-Galván, F., & Loarca-Pina, G. (2015). Potential role of bioactive compounds of *Phaseolus vulgaris* L. on lipid-lowering mechanisms. *Food Research International*, 76: 92–104. https://doi.org/10.1016/j.foodres.2015.01.002

Real Hernández, L.M., & González de Mejía, E. (2019). Enzymatic production, bioactivity, and bitterness of chickpea (*Cicer arietinum*) peptides. *Comprehensive Reviews in Food Science and Food Safety*, 18(6): 1913–1946. https://doi.org/10.1111/1541-4337.12504

Rock, K.L., Lai, J.J., & Kono, H. (2011). Innate and adaptive immune responses to cell death. *Immunological Reviews*, 243(1): 191–205. .https://doi.org/10.1111/j.1600-065X.2011.01040.x

Rodríguez-Decuadro, S., da Rosa, G., Radío, S., Barraco-Vega, M., Benko-Iseppon, A.M., Dans, P.D., Smircich, P., & Cecchetto, G. (2021). Antimicrobial peptides in the seedling transcriptome of the tree legume *Peltophorum dubium*. *Biochimie*, 180: 229–242. https://doi.org/10.1016/j.biochi.2020.11.005

Sánchez-Chino, X.M., Jiménez Martínez, C., León-Espinosa, E.B., Garduño-Siciliano, L., Álvarez-González, I., Madrigal-Bujaidar, E., Vásquez-Garzón, V.R., Baltiérrez-Hoyos, R., & Dávila-Ortiz, G. (2019). Protective effect of chickpea protein hydrolysates on colon carcinogenesis associated with a hypercaloric diet. *Journal of the American College of Nutrition,* 38(2): 162–170. https://doi.org/10.1080/07315724.2018.1487809

Sánchez-Chino, X.M., Jiménez-Martínez, C., Vásquez-Garzón, V.R., Álvarez-González, I., Villa-Treviño, S., Madrigal-Bujaidar, E., Dávila-Ortiz, G., & Baltiérrez-Hoyos, R. (2017). Cooked chickpea consumption inhibits colon carcinogenesis in mice induced with azoxymethane and dextran sulfate sodium. *Journal of the American College of Nutrition,* 36 (5): 391–398. https://doi.org/10.1080/07315724.2017.1297744

Sandhu, J.S., Tripathi, S., & Chaturvedi, S.K. (2023). Chickpea Nutritional Status and Value Chain for Sustainable Development. In *Sustainable Food Value Chain Development: Perspectives from Developing and Emerging Economies,* Sapna A Narula & Raj S.P. (Eds). Springer Nature Singapore, 175–183. https://doi.org/10.1007/978-981-19-6454-1_8

Semba, R.D., Ramsing, R., Rahman, N., Kraemer, K., & Bloem, M.W. (2021). Legumes as a sustainable source of protein in human diets. *Global Food Security,* 28: 100520. https://doi.org/10.1016/j.gfs.2021.100520

Serrano-Sandoval, S.N., Guardado-Félix, D., & Gutiérrez-Uribe, J.A. (2019). Changes in digestibility of proteins from chickpeas (*Cicer arietinum* L.) germinated in presence of selenium and antioxidant capacity of hydrolysates. *Food Chemistry,* 285: 290–295. https://doi.org/10.1016/j.foodchem.2019.01.137

Shahidi, F., & Zhong, Y. (2015). Measurement of antioxidant activity. *Journal of Functional Foods,* 18: 757–781. https://doi.org/10.1016/j.jff.2015.01.047

Shi, W., Hou, T., Guo, D., & He, H. (2019). Evaluation of hypolipidemic peptide (Val-Phe-Val-Arg-Asn) virtual screened from chickpea peptides by pharmacophore model in high-fat diet-induced obese rat. *Journal of Functional Foods,* 54: 136–145. https://doi.org/10.1016/j.jff.2019.01.001

Shi, W., Hou, T., Liu, W., Guo, D., & He, H. (2019). The hypolipidemic effects of peptides prepared from Cicer arietinum in ovariectomized rats and HepG2 cells. *Journal of the Science of Food and Agriculture,* 99(2): 576–586. https://doi.org/10.1002/jsfa.9218

Singh, V., Chauhan, Y., Dalal, R., & Schmidt, S. (2021). Chapter 9 – Chickpea. In: *The Beans and the Peas,* edited by Aditya Pratap and Sanjeev Gupta, Woodhead Publishing, 173–215.

Tang, Y., Long, J., & Liu, J. (2014). Hyperglycemia-Associated Oxidative Stress Induces Autophagy: Involvement of the ROS-ERK/JNK-p53 Pathway. In *Autophagy: Cancer, Other Pathologies, Inflammation, Immunity, Infection, and Aging,* Elsevier, 105–115.

Tawalbeh, D., Al-U'datt, M.H., Wan Ahmad, W.A.N., Ahmad, F., & Sarbon, N.M. (2023). Recent advances in *in vitro* and *in vivo* studies of antioxidant, ACE-inhibitory and anti-inflammatory peptides from legume protein hydrolysates. *Molecules,* 28(6): 2423. https://doi.org/10.3390/molecules28062423

Torres-Fuentes, C., Alaiz, M., & Vioque, J. (2011). Affinity purification and characterization of chelating peptides from chickpea protein hydrolysates. *Food Chemistry,* 129(2): 485–490. https://doi.org/10.1016/j.foodchem.2011.04.103

Torres- Fuentes, C., Alaiz, M., & Vioque, J. (2014). Chickpea chelating peptides inhibit copper mediated lipid peroxidation. *Journal of the Science of Food and Agriculture,* 94(15): 3181–3188. https://doi.org/10.1002/jsfa.6668

Torres-Fuentes, C., del Mar Contreras, M., Recio, I., Alaiz, M., & Vioque, J. (2015). Identification and characterization of antioxidant peptides from chickpea protein hydrolysates. *Food Chemistry,* 180: 194–202. https://doi.org/10.1016/j.foodchem.2015.02.046

Wahby, M.M., Mohammed, D.S., Newairy, A.A., Abdou, H.M., & Zaky, A. (2017). Aluminum-induced molecular neurodegeneration: The protective role of genistein and chickpea extract. *Food and Chemical Toxicology,* 107:57–67. https://doi.org/10.1016/j.fct.2017.05.044

Waili, A., Gao, Y.H., Zhang, L.F., Ziyavitdinov, Z.F., Maksimov, V.V., Yili, A., & Aisa, H.A. (2018). Production of antimicrobial peptides from *Cicer arietinum* sprouts and determination of their molecular masses. *Chemistry of Natural Compounds,* 54(6): 1135–1138. https://doi.org/10.1007/s10600-018-2573-y

Wali, A., Mijiti, Y., Yanhua, G., Yili, A., Aisa, H.A., & Kawuli, A. (2021). Isolation and identification of a novel antioxidant peptide from chickpea (*Cicer arietinum* L.) sprout protein hydrolysates. *International Journal of Peptide Research and Therapeutics,* 27(1): 2019–227. https://doi.org/10.1007/s10989-020-10070-2

Wang, J., Li, Y., Li, A., Liu, R.H., Gao, X., Li, D., Kou, X., & Xue, Z. (2021). Nutritional constituent and health benefits of chickpea (*Cicer arietinum* L.): A review. *Food Research International,* 150: 110790. https://doi.org/10.1016/j.foodres.2021.110790

Wang, W., Yang, W., Dai, Y., Liu, J., & Chen, Z.Y. (2023). Production of Food-Derived bioactive peptides with potential application in the management of diabetes and obesity: A review. *Journal of Agricultural and Food Chemistry*, 71(15): 5917–5943. https://doi.org/10.1021/acs.jafc.2c08835

Xiao, S., Li, Z., Zhou, K., & Fu, Y. (2023). Chemical composition of kabuli and desi chickpea (*Cicer arietinum* L.) cultivars grown in Xinjiang, China. *Food Science & Nutrition*, 11: 236–248. https://doi.org/10.1002/fsn3.3056

Xiao, Y., Xing, G., Rui, X., Li, W., Chen, X., Jiang, M., & Dong, M. (2015). Effect of solid-state fermentation with *Cordyceps militaris* SN-18 on physicochemical and functional properties of chickpea (*Cicer arietinum* L.) flour. *LWT-Food Science and Technology*, 63(2): 1317–1324.https://doi.org/10.1016/j.lwt.2015.04.046

Xu, Y., Galanopoulos, M., Sismour, E., Ren, S., Mersha, Z., Lynch, P., & Almutaimi, A. (2020). Effect of enzymatic hydrolysis using endo- and exo-proteases on secondary structure, functional, and antioxidant properties of chickpea protein hydrolysates. *Journal of Food Measurement and Characterization*, 14(1): 343–352. https://doi.org/10.1007/s11694-019-00296-0

Xue, Z., Gao, J., Zhang, Z., Yu, W., Wang, H., & Kou, X. (2012). Antihyperlipidemic and antitumor effects of chickpea albumin hydrolysate. *Plant Foods for Human Nutrition*, 67(4): 393–400. https://doi.org/10.1007/s11130-012-0311-3

Xue, Z., Wang, R., Wen, H., Yu, W., & Kou, X. (2018). Biochanin A and CPe-III peptide improved hepatic inflammation by regulating the hepatic lipid metabolic pathways in diet-induced obese mice. *Transactions of Tianjin University*, 24(3): 234–243. https://doi.org/10.1007/s12209-018-0117-y

Xue, Z., Wen, H., Zhai, L., Yu, Y., Li, Y., Yu, W., Cheng, A., Wang, C., & Kou, X. (2015). Antioxidant activity and anti-proliferative effect of a bioactive peptide from chickpea (*Cicer arietinum* L.). *Food Research International*, 77:75-81. https://doi.org/10.1016/j.foodres.2015.09.027

Xue, Z., Zhang, Q., Yu, W., Wen, H., Hou, X., Li, D., & Kou, X. (2017). Potential lipid-lowering mechanisms of biochanin A. *Journal of Agricultural and Food Chemistry*, 65(19): 3842–3850. https://doi.org/10.1021/acs.jafc.7b00967

Yu, Z., Yin, Y., Zhao, W., Chen, F., & Liu, J. (2014). Antihypertensive effect of angiotensin-converting enzyme inhibitory peptide RVPSL on spontaneously hypertensive rats by regulating gene expression of the Renin–Angiotensin System. *Journal of Agricultural and Food Chemistry*, 62(4): 912–917 https://doi.org/10.1021/jf405189y

Yust, M.D.M., Millán-Linares, M.D.C., Alcaide-Hidalgo, J.M., Millán, F., & Pedroche, J. (2012). Hypocholesterolaemic and antioxidant activities of chickpea (*Cicer arietinum* L.) protein hydrolysates. *Journal of the Science of Food and Agriculture*, 92(9): 1994–2001. https://doi.org/10.1002/jsfa.5573

Zhang, L.J., & Gallo R.L. (2016). *Antimicrobial peptides. Current Biology*, 26 (1): 1543–1575. https://doi.org/https://doi.org/10.1016/j.cub.2015.11.017

Zhang, T., Li, Y., Miao, M., & Jiang, B. (2011). Purification and characterization of a new antioxidant peptide from chickpea (*Cicer arietium* L.) protein hydrolysates. *Food Chemistry*, 128(1): 28–33. https://doi.org/10.1016/j.foodchem.2011.02.072

Chapter 3

Arthrospira and Its Major Components

Effects on Obesity, Inflammation, and Potential Mechanism of Action

Ruth M. García-García
María Eugenia Jaramillo-Flores
and Germán A. Chamorro-Cevallos

CONTENTS

3.1 INTRODUCTION

In the last 30 years, there has been a worldwide increase in the intake of foods high in sugars and fats, a decrease in physical activity, and an increase in a sedentary lifestyle. As a consequence, the incidence and prevalence of overweight and obesity around the world have increased and reached the fifth risk factor responsible for 5.02 million global deaths in 2019 (World Health Organization, 2023; Global Burden of Disease, 2019). As a metabolic disorder, obesity is characterized by the accumulation of excess body fat, manifested physically by chronic excess

DOI: 10.1201/9781003332558-4

energy in adipose tissue. This is associated with a decrease in physical activity, that is, a reduction in energy expenditure (Sharma & Padwal, 2010).

The etiology of this chronic disease is multivariate and complex. Biological factors that predispose to this disease include genetic characteristics, epigenetics, pregnancy, menopause, gut microbiota, and viruses. People who are prone to developing obesity can also be affected by behavioral causes, such as increasing excessive energy intake and portion sizes and practicing a sedentary lifestyle. Genetic predisposition to obesity can be influenced by epigenetic triggers such as high food availability, socioeconomic status, or the presence of chemical contaminants in the environment that could be ingested (González-Casanova et al. 2020). According to the WHO, a person with a body mass index (BMI) greater than or equal to 30 kg/m^2 is considered obese and overweight with a BMI greater than or equal to 25 kg/m^2. Overweight and obesity not only decreases the quality of life and performance of the people who develop them, but also generates high costs for the health system and for families that have a member with these pathologies (World Health Organization, 2023).

Obesity has been closely linked to inflammation, hyperlipidemia, and insulin resistance (IR) (Shah et al. 2008), due to the fact that adipose tissue secretes numerous biologically active substances such as adipokines and chemokines, which play an important role in inflammation (Skrypnik et al. 2018). In addition, chronic excess nutrients lead to metabolic disorders, IR, and dysregulation of glucose, which contribute to increased morbidities such as cancer and chronic diseases of the liver and kidneys (Blasetti-Fantauzzi et al. 2020). While 80% of people diagnosed with type 2 diabetes mellitus (T2DM) are also obese, not all obese people develop this metabolic condition. Family history, hyperglycemia, hypertension, an unhealthy lipidomic profile, abdominal fat, and chronic inflammation all contribute to individual susceptibility to T2DM, suggesting that both genetic and metabolic factors are involved in its onset. However, the reason obesity generates metabolic dysfunction in some cases but not in all remains poorly understood (Di Francesco et al. 2020).

Recent research has focused on understanding the changes in adipose tissue function in central obesity and their association with obesity-related health issues. This has led to an increased search for new drugs and functional food products to combat these diseases. Among the approved medications for obesity are mitochondrial decouplers, sympathomimetics, serotonergic agonists, lipase inhibitors, cannabinoid receptor antagonists, and gastrointestinal peptides. Some approved sympathomimetic drugs include diethylpropion/afepramone, phenmetrazine and phendimetrazine, cathine, sibutramine, and phentermine. However, these medications often cause side effects such as nausea, constipation, insomnia, headaches, seizures, and increased blood pressure (Müller et al. 2018). Orlistat is a lipase inhibitor used for obesity treatment, but it can lead to liver injury and gastrointestinal symptoms. Phentermine/topiramate ER, a sympathomimetic/anticonvulsant drug, has side effects including depression, suicidal ideation, cardiovascular events, memory loss, and birth defects (Garvey et al. 2012). Naltrexone SR/bupropion SR (with titration) (32 mg/360 mg, BID, oral) is a drug that is an opioid receptor antagonist/dopamine and noradrenaline reuptake inhibitor medication that causes seizures, palpitations, and transient elevations of blood pressure as side effects (Greenway et al. 2010). On June 4, 2021, the FDA approved semaglutide for chronic weight management in overweight or obese adults with weight-related conditions (Wilding et al. 2021). This is the second glucagon-like peptide 1 receptor agonist (GLP1R) approved for weight control, following liraglutide's approval in 2014. Both drugs have side effects, including nausea, vomiting, diarrhea, constipation, pancreatitis, and gallstones. Due to the side effects associated with current obesity drugs, natural sources have become attractive for discovering new drugs, nutraceuticals, and functional food ingredients without side effects (Saadi et al. 2015). *Arthrospira*, a complete nutritional source, contains protein, beta-carotene, gamma-linolenic acid (GLA), B vitamins, minerals, chlorophyll, and enzymes. It offers many nutrients that are lacking in typical diets (Soni et al. 2017). This chapter aims to review the effects of *Arthrospira* and its components on obesity and inflammation and discuss their potential mechanisms of action.

3.2 *ARTHROSPIRA MAXIMA* AS A NUTRITIONAL SUPPLEMENT

The two species, *Arthrospira platensis* and *Arthrospira maxima*, belong to the genus *Arthrospira*, phylum *Cyanobacteria*, a class of single-celled organisms formerly known as "blue-green algae." At first, they were included in the genus Spirulina, but several phylogenetic studies concluded that the genera *Spirulina* and *Arthrospira* are different; however, the name *Spirulina* is used to commercially designate the nutritional supplement (Furmaniak et al 2017). The terms cyanobacteria and cyanophyte algae (blue and green) may be compatible, since the first refers to the taxonomic/phylogenetic relationship, while the second represents the ecological/biological correlation (Sánchez et al. 2003). Microalgae are a diverse group of mostly photosynthetic microorganisms that include cyanobacteria (prokaryotic organisms) and, among others, green algae, euglenoids, and diatoms (eukaryotic organisms) (Alvarez et al. 2021).

Arthrospira grows naturally in alkaline reservoirs with high salt content in tropical and subtropical areas of Mexico, Central and South America, Asia, and Central Africa (Dinicolantonio et al. 2020). Ancestral civilizations such as the Aztecs were the first to recognize the nutritional value of these organisms. Europeans had contact with *Arthrospira* during the colonization of America in the sixteenth century, and from that time came the first reports. At present, it is still harvested, dried, and consumed by the Kanembu tribe near Lake Chad in Africa and is called "dihe" (Ciferri, 1983).

Arthrospira contains many functional bioactive ingredients, including vitamins, minerals, polysaccharides, and volatile compounds such as carbonyls, alcohols, aldehydes, esters, and terpenes, among others, which are naturally produced by cyanobacteria and algae and can be used as flavorings. Also, *Arthrospira* contains sterols, which play a fundamental role in the integrity of microalgae membranes, as well as, polyunsaturated fatty acids, phenolic phytochemicals, proteins, and pigments such as phycobiliproteins, C-phycocyanin, and β-carotene (Henrikson, 1994). Phenolic compounds are highly regarded as natural antioxidants and can be used as dietary supplements (Andrade et al. 2018). While phenolic compounds are more commonly found in other microalgae species than cyanobacteria (Jerez-Martel et al. 2017), extracts from various terrestrial cyanobacteria have been found to contain high concentrations of phenols and flavonoids, known for their excellent antioxidant properties (Singh et al. 2020). These extracts contain polyphenols such as gallic, chlorogenic, caffeic, vanillic, and ferulic acids, as well as flavonoids like quercetin and kaempferol. The presence of these polyphenols in terrestrial cyanobacteria contributes to their abilities to scavenge free radicals, chelate metals, and protect against oxidative damage (Singh et al. 2017). In *Arthrospira*, the two most abundant phenolic compounds are acacetin and pinocembrin, while other phenols present include sakuranetine, luteolin, kaempferol, methyl quercetin, quercetin, and apigenin (Guldas et al. 2020).

Other compounds present in *Arthrospira* are long-chain polyunsaturated fatty acids (LCPUFAs). *Arthrospira* contains a significant amount of polyunsaturated fatty acids, such as ω-3 and ω-6. The fatty acids that have been found in greater quantity in *Arthrospira* are palmitic acid as saturated fatty acid, palmitoleic acid as monounsaturated fatty acid, alpha-linolenic acid as ω-3 fatty acid, and gamma linolenic acid as ω-6 fatty acid. However, it has been reported that for Arthrospira, differences in the composition of fatty acids are related to the solvent used for extraction, the climate, or the processing conditions (Guldas et al. 2020). *Arthrospira* contains a high concentration of proteins (60–70% dry weight). *Arthrospira* has been considered beneficial in human nutrition due to the high quality and quantity of its proteins, since it has the highest protein content of all natural foods (Henrikson, 1994) (**Table 3.1**).

Arthrospira contains various pigments, including carotenoids such as beta-carotene, xanthophyll, cryptoxanthin, equinenone, zeaxanthin, and lutein. However, the main pigment found in *Arthrospira* is phycocyanin, which is a protein complex that constitutes a significant portion of the cyanobacteria's dry weight (Henrikson, 1994). The C-phycocyanin crystal structure of the cyanobacterium *A. platensis* has been determined at a resolution of 2.2 Å (Wang et al. 2001). Phycocyanin is highly fluorescent and water-soluble, serving the primary role of transferring excitation energy during photosynthesis. It derives its intense blue color from its tetrapyrrolic linear chromophore called phycocyanobilin (PCB). The crystal structure of C-phycocyanin from *A. platensis* has been determined, revealing the presence of magnesium and iron ions in its molecular ring (Minic et al. 2016). Upon digestion, phycocyanin yields chromopeptides, ranging in size from 2 to 13 amino acid residues, which exhibit antioxidant activities and demonstrate a cytotoxic effect on certain cancer cell lines. These pigments play a vital role in *Arthrospira* by aiding in the synthesis of enzymes involved in metabolic regulation (Minic et al. 2016).

In addition to its nutritional value, *Arthrospira* has also been used in cosmetic applications and has been recognized by the FDA as a safe ingredient (Marles et al. 2011).

3.3 EFFECTS OF *ARTHROSPIRA* AGAINST OBESITY DEVELOPMENT AND ASSOCIATED DISEASES

3.3.1 Anti-obesity and Anti-Inflammatory Effects of *Arthrospira*

White adipose tissue (WAT) is also known as a major secretory organ and a highly active metabolic tissue. It secretes, for example, cholesterol, retinol, steroid hormones, prostaglandins, and proteins known as "adipokines." Some of these molecules may be associated with pathologies such as obesity and IR, as they can increase the risk of metabolic syndrome, cardiovascular diseases, and others (Santillana et al., 2023). WAT stores lipids, and during fasting, it releases fatty acids, which, in the process of β-oxidation, are the source of adenosine triphosphate (ATP). White adipocytes contain a single drop of large lipid that occupies about 90% of the cell volume. The nucleus squeezes towards the cellular periphery, and the cytoplasm forms a very thin edge. The organelles are poorly developed; in particular, mitochondria are small and elongated (Giordano et al. 2014).

Adipogenesis is a controlled process that involves adipocytes maturing and becoming functional. Over 40 years, *in vivo* studies have complemented preadipocyte models like 3T3-L1 and 3T3-F442A. A complex transcriptional cascade controls adipogenesis phases, potentially targeting pharmacological interventions to reduce obesity and metabolic diseases (Ali et al. 2013). Some of the stimulators include the peroxisome proliferator-activated receptor gamma (PPARγ), called the master transcription factor of adipogenesis, as it has been shown that PPARγ2 is necessary for

TABLE 3.1 COMPOUNDS PRESENT IN *ARTHROSPIRA*

Types of Compounds	Compounds	Concentration (ng/g)	Source	References
Phenolic compounds				
	Phloroglucinol	51,000	*Arthrospira platensis*	Gabr et al. (2020) Goiris et al. (2014)
	p-Coumaric acid	920		
	Ferulic acid	0.97		
	Apigenin	6 (μg/100g)		
	Neochlorogenic acid	0.02	*A. platensis*	Gabr et al. (2020) Guldas et al. (2020)
	Chlorogenic	0.01		
	Luteolin	0.68		
	Caffeic acid	0.02		
	Quercetin	0.26		
	Apigenin	0.25		
	Gallic acid	27.23		
	Pinocembrin	0.03		
	Gallic acid	0.08		
	Pinoquercetin	0.13		
	3,3-dimethyl quercetin	0.78		
	Taxifolin	0.11		
	Methylquercetin	0.47		
	Kaempherol	0.53		
	Acacetin	35.37		
Proteins		(%)		
	Proteins	63.35	*A. platensis*	Arslan et al. (2021)
Amino acids		Isotope values		
Essential amino acids	Isoleucine	31.29	*Arthroapira*	Thorp & Bowes (2017)
	Leucine	34.42		
	Phenylalanine	35.17		

(continued)

TABLE 3.1 (CONTINUED)

Types of Compounds	Compounds	Concentration (ng/g)	Source	References
	Valine	35.62		
Non-essential amino acids	Glycine	17.83		
	Proline	16.37		
	Alanine	17.85		
	Aspartic acid	9.20		
	Glutamic acid	19.75		
Pigments		(mg/g dry spirulina)		
	Chlorophyll-1.1	9.26	*A. platensis*	Arslan et al. (2021)
	Chlorophyll-2.2	6.9		
	Phycocyanin-1.2	63.71		
	Phycocyanin-2.1	78.57		
		(mg/mL)		
	Allophycocyanin	0.56	*Arthrospira maxima*	Guzmán-Gómez et al. (2018)
		(mg/g dry spirulina)		
	All-trans-zeaxanthin	1.27	*A. platensis*	Park et al. (2018)
	Diatoxanthin	0.26		
	13-cis-β-Carotene	0.06		
	All-trans-β-carotene	2.30		
	9-cis-β-Carotene	0.38		
	Other carotenoids	0.17		
	Chlorophyll a	10.8		
	C-Phycocyanin	251.2		
Vitamin		mg/100g dry Spirulina		
	Vitamin A	0.34	*Arthrospira*	Andrade et al. (2018)
	Vitamin C	10.1		

TABLE 3.1 (CONTINUED)

Types of Compounds	Compounds	Concentration (ng/g)	Source	References
	Thiamin (vitamin B1)	2.4		
	Riboflavin (vitamin B2)	3.7		
	Niacin (vitamin B3)	12.8		
	Pyridoxine (vitamin B6)	0.4		
	Folate (vitamin B9)	0.94		
	Vitamin E (alpha-tocopherol)	5		
	Vitamin K (phylloquinone)	2.5		
Minerals		mg/g sample		
	Potassium (K)	16	*Arthrospira*	Liestianty et al. (2019)
	Calcium (Ca)	15		
	Phosphorus (P)	10		
	Manganese (Mn)	3		
	Zinc (Zn)	0.07		
	Magnesium (Mg)	3.7		
	Sodium (Na)	2.5		
	Iron (Fe)	1.7		
Polysaccharides		(% mole)		
Immulina	Glucose	21.3	*A. platensis*	Wang et al. (2018)
	Rhamnose	43.6		
	Xylose	2.4		
	Galactose	1.3		
	Arabinose	1.1		
	Glucuronic acid	9.7	*A. platensis*	Pugh et al. (2001)
	Fructose	7.7		
	2-methyl-Rhamnose	5.9		

(continued)

TABLE 3.1 (CONTINUED)

Types of Compounds	Compounds	Concentration (ng/g)	Source	References
	3-methyl-Rhamnose	4.2		
	2-methyl-Xylose	4.2		
	4-methyl-Rhamnose	3.9		
	Mannose	2.4		
	Galacturonic acid	2.0		
	3-methyl-Galactose	2.0		
	Amino sugar	1.5		
	2-3-dimethyl-Fucose	1.2		
	N-acetyl-glucosamine	0.9		
	2-methyl-Glucose	0.5		
	Glycerol	0.4		
Fatty acid		(mg/100 g)		
Saturated Fatty Acids (SFA)	C12:0	87.35	*A. platensis*	Guldas et al. (2020)
	C14:0	97.24		
	C16:0	3380.17		
	C18:0	31.94		
Monounsaturated Fatty Acids (MFA)	C15:1	378.45		
	C16:1	546.14		
	C18:1	69.25		
Polyunsaturated Fatty Acids (PFA)				
ω-3	C16:4	36.14		
	C18:3 (ALA4)	144.76		
	C20:5 (EPA5)	32.65		
	C22:6 (DHA6)	67.49		
ω-6	C18:2 (LA7)	144.81		
	C18:3 (GLA8)	1866.27		
	C20:4 (AA9)	54.86		
	C22:5	46.19		
Volatile compounds		(%)		
	β-cyclocitral	0.65	*Spirulina platensis*	Milovanović et al. (2015)

TABLE 3.1 (CONTINUED)

Types of Compounds	Compounds	Concentration (ng/g)	Source	References
	2-pentylfuran	3.03		
	tetradecane	0.27		
Hydrocarbons	pentadecane	6.48		
	hexadecane	5.12		
	6,9-heptadecadiene	0.62		
	heptadecene	77.67		
Alcohols	2,6-dimetrylcyclohexanol	2.41		
	2-ethyl-1-hexanol	0.38		
	2-methylisoborneol	0.98		
	1,1,3-trimetryl-2-cyclohexanone	0.38		
Ketones	β-ionone	1.59		
Esters	diisobutyric acid 1-tert-buty-l-2-methyl-1,3-propanediyl ester	0.45		

the differentiation of adipocytes and also for the maintenance of this process. Potential binding proteins are C/EBPα, C/EBPβ, and C/EBPδ, which are expressed sequentially during adipocytic differentiation (Farmer, 2006). Other factors involved in the adipogenesis process are the single transducers and transcription activators (STAT) and the transcription factor protein binding to the regulatory element of sterol 1 (SREBP1). SREBPs are a family of transcription factors that regulate cellular lipogenesis and lipid homeostasis by controlling the expression of many enzymes involved in the synthesis of cholesterol, fatty acids, triacylglycerol, and phospholipids (Ali et al. 2013).

WAT secretes products crucial for appetite, energy homeostasis, insulin sensitivity, immune responses, and vascular diseases. Adipocyte-derived factors include leptin, adiponectin, angiotensinogen, prostaglandins, resistin, tumor necrosis factor alpha (TNFα), interleukin 6 (IL-6), interleukin 8 (IL-8), interleukin 10 (IL-10), monocyte chemoattractant protein 1 (MCP1), interferon gamma inducible protein 10, macrophage inflammatory protein 1β, granulocyte colony-stimulating factor, interleukin 1 (IL-1), and alkaline phosphatase. The expression of most (if not all) of these factors is determined by two factors: the size or volume of the adipocytes and the location of the adipocytes (Ali et al. 2013).

In cells 3T3-L1, it has been observed that the 70% ethanol extract of *A. maxima* (SM70EE) has had effects on the lipid droplets reduction; downregulation of adipogenic proteins C/EBPα, PPARγ, and aP2, and lipogenic proteins SREBP1, ACC, FAS, LPAATβ, Lipin1, and DGAT1. In addition, the main components of SM70EE, chlorophyll a and c-phycocyanin, also block the expression of differentiating-inducing proteins (C/EBPα, PPARγ y aP2) during differentiation. However, SM70EE suppresses the expression of regulators involved in the last stage of adipogenesis more forcefully than chlorophyll a and c-phycocyanin in these 3T3-L1 and C3H10T1/2 cells (Seo et al. 2018).

An ethanol extract of *A. platenisis* (SPE) exhibits potent anti-inflammatory effects on macrophages through cellular communication, promoting adipocyte differentiation and maturation, and repressing NF-κB activation in WAT conditions (Pham & Lee, 2016). Another extract of *A. maxima* (SME) also decreases NF-κB expression by inhibiting ERK phosphorylation and attenuating reactive oxygen species (ROS) generation. SME suppresses pro-inflammatory cytokines TNFα, IL-12, IL-1β, and IL-18 upregulation and attenuates inflammasome NLRP3 activation. This suggests SME's effects on innate immunity and inflammasome NLRP3 activation in macrophages (Chei et al. 2020).

A. maxima treatment prevents weight gain and reduces waist circumference, attenuating metabolic parameters and inhibiting adipocyte size. It activates the AMPK pathway and sirtuin 1, suppressing obesity-related mechanisms, including lipid synthesis and oxidation (Heo & Choung, 2018). Similar results were obtained in the experiment when mice on a high-fat diet (HFD) administered with SM70EE (HFD+SM70EE) decreased obesity, reducing adipogenesis and activating the thermogenic program. This effect was also observed in 3T3-L1 cells (Seo et al. 2018). In addition, *A. platensis* also represses the expression and secretion of pro-inflammatory cytokines mentioned in macrophages and splenocytes via inhibition of NF-κB (Ku et al. 2012).

A pepsin-hydrolyzed protein fraction from *A. platensis* protein hydrolysate supplemented to HFD-fed mice reduced body weight, decreased glucose and cholesterol levels, and modulated obesity-associated gene expression. Four peptides inhibited preadipocyte proliferation, with NPVWKRK and CANPHELPNK showing significant decreases in triglyceride accumulation and cell proliferation (Fan et al. 2018). Another study isolated 15 polypeptides from *A. platensis*, showing antiproliferative activities in five cancer cell models, and identified a new peptide with improved inhibitory activity (Wang & Zhang, 2016). Peptide HVLSRAPR, hydrolyzed by trypsin, shows strong inhibition in HT-29 cancer cells and antiproliferative activities, potentially serving as natural anticancer agent (Wang & Zhang, 2017). As cancer is one of the diseases related to obesity, it would be interesting to observe the antiproliferative effect of these peptides in the study with 3T3-L1 cells and thus measure the antiadipogenic effect of these peptides.

Other peptides from *A. maxima* hydrolysate were purified using gastrointestinal endopeptidases (trypsin, α-chymotrypsin, and pepsin), resulting in LDAVNR (P1) and MMLDF (P2) peptides with potent anti-inflammatory benefits, reducing ROS production (Vo et al. 2013). In addition, they showed protective effects against early histamine-induced atherosclerotic responses in EA.hy926 endothelial cells (Vo & Kim, 2013). These peptides significantly inhibited histamine release and production in mast cells RBL-2H3, suppressed IL-8 production in EA.hy926 endothelial cells, and inhibited mast cell degranulation by decreasing histamine release and increasing intracellular Ca^{2+} (Vo et al. 2013). P2 inhibitory activity blocks calcium- and microtubule-dependent signaling pathways, suppresses Cγ-phospholipase activation, ROS production, and IL-4 generation through NFκB translocation depression (Vo et al. 2014). Studies show *Arthrospira*'s peptides as potential antiallergic therapies, potentially regulating TNFα and interleukins in inflammation. These studies could lead to a new nutritional food for preventing and treating chronic inflammatory diseases.

Arthrospira treatment under a high-fat diet prevents weight gain and reduces waist circumference, attenuating changes in adiponectin, leptin, TNFα, glucose, insulin, and lipid profiles. It promotes lipid oxidation and blocks lipid synthesis through AMPK and Sirtuin 1 pathways. *Arthrospira* extracts regulate adipogenic and lipogenic proteins and show anti-inflammatory activity. *Arthrospira* peptides have antiproliferative effects in adipocytes and cancer cell lines.

3.3.2 Effects of *Arthrospira* on Dyslipidemia

Dyslipidemia is the alteration of lipoproteins in the blood, causing increased total cholesterol (TC), triglycerides (TG), low-density lipoproteins (LDL), and the decrease of high-density lipoproteins (HDL), with a prognosis influenced by age and life expectancy (Gómez-Avellaneda & Tarqui-Mamani, 2017).

Li et al. (2018) found that a 95% ethanol extract of *A. platensis* (SPL95) increased HDL, reduced LDL, TG, and TC levels in rats, and significantly decreased appetite, body weight, BMI, and serum TC in obese adults (Zeinalian et al. 2017). *Arthrospira* treatment reduces waist circumference, LDL, and IL-6 in hypertension-treated obese patients, increases insulin sensitivity, total antioxidant status, and reduces TG and C-reactive protein levels (Yousefi et al. 2018). All these results obtained after *Arthrospira* supplementation were recently corroborated in a 2019 meta-analysis, where they evaluated 12 human clinical studies (Moradi et al. 2019).

Similarly, an *in vivo* study showed a cholesterol-lowering effect of soy yogurt with *A. platensis* in hypercholesterolemic mice, reducing total cholesterol and HDL levels (Sengupta et al. 2018). The information obtained shows that *Arthrospira* reduces dyslipidemia in obese individuals by decreasing appetite, body weight, BMI, C-reactive protein, and CT, while polyunsaturated fatty acid extract increases HDL levels.

3.3.3 Effects of *Arthrospira* on Insulin Resistance (IR) and Type 2 Diabetes Mellitus (T2DM)

Chronic overnutrition leads to nutritional stress on adipocytes, causing insulin resistance and adipocyte death. This results in an increase in inflammatory signals, including retinol carrier protein type 4 (RBP-4), which attracts macrophages to white adipose tissue. Inflammatory signaling in macrophages triggers the release of cytokines such as TNFα and IL-1β. These cytokines can directly or indirectly increase adipocyte lipolysis and promote the production of nonesterified fatty acids and glycerol. Inflammation-induced lipolysis has various effects, including increased gluconeogenesis, accumulation of intrahepatic triglycerides, and hepatic insulin resistance. Additionally, it can lead to intramyocellular lipid accumulation and muscle insulin resistance. These processes contribute to elevated hepatic glucose production, worsened hepatic insulin resistance, and impaired glucose removal from muscle, ultimately exacerbating the metabolic dysregulation associated with chronic overnutrition (Petersen & Shulman, 2018). Moreover, rodent studies reveal decreased oxygenation in WAT contributes to obesity-induced IR. A recent study found that decreased oxygen pressure in WAT correlates with hepatic and whole-body insulin sensitivity, while negatively affecting inflammation and extracellular matrix remodeling in obese patients (Cifarelli et al. 2020).

Omega-3 fatty acids (EPA and DHA) regulate glucose uptake and insulin response in adipose tissue and muscle. They convert into anti-inflammatory lipid pro-resolution mediators, including maresin 1, which regulates insulin signaling in human adipocytes and lean, obese mice (Martínez-Fernández et al. 2020). The β-carotene is known to protect against singlet oxygen-mediated lipid peroxidation, interact with intracellular accumulation of ROS, and inhibit inflammatory gene expression (Nasirian et al. 2018). The extract of β-carotene obtained from the biomass of *A. platensis* presented an antihyperglycemic effect in diabetic mice when administered at a dose of 100 mg/kg body weight after 10 days of treatment (Ramos-Romero et al. 2021). Furthermore, the phycocyanin chromophore PCB can inhibit nicotinamide adenine dinucleotide phosphate hydrogen (NADPH) oxidase, a major source of oxidative stress in adipocytes and a key player in inducing insulin resistance, adipokine displacement, and cytokine production in hypertrophied adipocytes. *Arthrospira* may have systemic anti-inflammatory and insulin-sensitizing effects by suppressing oxidative stress in the adipocyte (Prokudina et al. 2017).

Psammomys obesus, a unique animal model of nutritionally induced metabolic syndrome, develops obesity, type 2 diabetes, and insulin resistance when fed a standard laboratory chow diet. However, when treated with *Arthrospira* for 12 weeks at a dose of 2 g/kg, there was a significant reduction in body weight, plasma glucose, insulin, and triglyceride levels. Furthermore, the expression of genes involved in lipogenesis and inflammation, such as SCD1, FAS, ACC1, SREBP-1c, ChREBP, TNF-α, and IL-1β, was significantly reduced in the liver. These findings suggest that *Arthrospira* treatment may improve metabolic parameters and reduce inflammation in this animal model (Saidi et al. 2020).

Postprandial hyperglycemia can be controlled by inhibiting carbohydrate-breaking enzymes such as α-amylase and α-glucosidase (Halimatul et al. 2020). Therefore, in diabetics, inhibition of both enzymes slows down the rate of carbohydrate digestion, and as a result, less glucose is absorbed and transported into the circulation (Mohammadi-Khanaposhtani et al. 2018). *A. platensis* butanol extract exhibited intestinal rat α-glucosidase inhibitory activity and antioxidant properties, with a CI50 value of 37.5 ± 2.8 μg/ml. Its phenol content and flavonoids contribute to its antioxidant, antiproliferative, antihypertensive, α-glucosidase inhibitory, and defense mechanisms against pathogenic microorganisms (Mallikarjun et al., 2015). Another ethanolic extract of *Arthrospira* biomass obtained after hydrolysis showed α-amylase inhibitory activity. The authors showed evidence that chlorogenic acid was found in higher concentrations [648.1 w/(μg/g)] in the extract relative to other phenols, whose antioxidant activity and ability to inhibit mutagenicity promoted by chemical compounds have already been reported (Scaglioni et al. 2018). Metabolic syndrome involves hyperglycemia, hyperlipidemia, and oxidative stress. A phycocyanin-enriched liquid extract of *Arthrospira* (LES) was tested to prevent metabolic abnormalities from hamster diets. Supplementation for 2 weeks showed protective effects against hyperglycemia and lipid accumulation in liver and aorta (Kasbi-Chadli et al. 2021).

A. platensis phycobiliproteins contain 26 peptides with antidiabetic effects. Hydrolyzed using trypsin, the hydrolysate consists of 7–27 amino acids, with seven from C-PC, six from C-PC, and five and eight from APC subunits. The hydrolysate inhibits DPP-IV activity in vitro, with IC50 values of 0.5 and 1.0 mg/ml (Li et al. 2020). DPP-V, α-amylase, and α-glucosidase may target type II diabetes treatment, with DPP-IV lowering postprandial glucose levels and α-amylase converting carbohydrates into monosaccharides, potentially reducing hyperglycemia (Hu et al 2019). *A. platensis* peptide LRSELAAWSR has antidiabetic effects by inhibiting α-amylase, α-glucosidase, and DPP-IV in HepG2 cells, reducing postprandial glycemia, and delaying starch digestion (Hu et al. 2019).

Halimatu et al. (2020) found that phycocyanin inhibits α-amylase and α-glucosidase enzymes by binding to their active sites, disrupting enzyme-substrate binding. This suggests phycocyanin as a potential natural antidiabetic therapeutic agent. In vitro inhibition activity showed phycocyanin inhibiting human salivary amylase by 51.13%. Storage stability tests show that phycocyanin can preserve bioactivity when stored in solid-state conditions.

In this way, *Arthrospira* and its compounds, including β-carotene, phenols, flavonoids, and peptides, have antihyperglycemic effects, reducing oxidative stress and metabolic syndrome. These effects involve inhibiting NADPH, reducing lipogenic gene expression, and inhibiting DPP-IV, making Arthrospira an effective antidiabetic drug or therapeutic agent.

3.3.4 *Arthrospira* Effects on Hypertension

Systemic arterial hypertension is characterized by persistently elevated blood pressure levels and is a major risk factor for cardiovascular diseases (Cifu & Davis, 2017; Villalpando et al. 2020). The vasoconstrictor system becomes predominant in individuals with hypertension, leading to endothelial dysfunction and damage (Martínez-Sámano et al. 2018). Studies have shown that *A. maxima* has potential antihypertensive effects. In a randomized controlled trial, overweight individuals with hypertension who consumed 2 g/day of *A. maxima* for three months experienced a significant reduction in systolic blood pressure. (Miczke, 2016). Additionally, *A. maxima* and its component C-phycocyanin demonstrated beneficial effects in reducing hypertension, left ventricular hypertrophy, renal dysfunction, and oxidative stress in a mouse model of chronic kidney disease (Memije-Lazaro et al. 2018). These findings suggest that *A. maxima* may have beneficial effects in managing hypertension and its associated complications.

High blood pressure is associated with endothelin-1 concentrations and markers of inflammation and endothelial damage such as sICAM-1, sVCAM-1, and soluble forms of E-selectin have been linked to hypertension (Martínez-Sámano et al. 2018). Endothelial dysfunction, a key factor in obesity-related hypertension, is characterized by reduced nitric oxide (NO)-dependent vasodilation and imbalanced coagulation and fibrinolytic systems, which predisposes to a prothrombotic state (Vaněčková et al. 2014). A study on spontaneously hypertensive rats investigated the effect of an aqueous extract of *Arthrospira*, on aortic vasodilatory function. The extract increased NO production, enhanced vasodilator responses, and upregulated proteins associated with vascular protection. Additionally, *Arthrospira*'s phycocyanin component demonstrated antioxidant and anti-inflammatory properties (Villalpando et al. 2020). Furthermore, *A. platensis* phycocyanin has been shown to scavenge free radicals, inhibit NO synthase expression and nitrite production, and inhibit lipid peroxidation (Park et al. 2018). Another study on patients with hypertension receiving angiotensin-converting enzyme inhibitors (ACE) found that *A. maxima* supplementation reduced blood pressure and improved indicators of endothelial damage and oxidative stress. However, further research is needed to fully understand the effects of *Arthrospira* on endothelial damage markers. These findings highlight the potential of *Arthrospira* as a therapeutic agent for hypertension and its associated complications (Martínez-Sámano et al. 2018). *A. platensis* supplementation in soy yogurt increases liver antioxidant enzymes and decreases serum aspartate aminotransferase, demonstrating a new approach to producing functional food products with antiatherogenic, hypolipidemic, and improved liver function (Sengupta et al. 2018).

Obesity disrupts the hemostatic balance and contributes to platelet hyperaggregability and hypercoagulability, leading to an increased risk of hypertension and cardiovascular diseases. Elevated levels of von Willebrand factor, factor VIII, and fibrinogen further promote this process (Russo, 2012). *Arthrospira* extracts, including phycocyanobilin, phycocyanin, polysaccharides, and lipid, were tested for antithrombotic properties in activated rabbit platelets. These extracts significantly inhibited platelet-activating factor (PAF) and thrombin-induced platelet aggregation. The active components responsible for these effects were identified as polar lipid fractions, such as glycolipids and PAF-like phosphatidylcholine residues. Phycocyanobilin inhibited both PAF and thrombin-induced platelet aggregation, as well as had stronger antithrombin activity. *Arthrospira* has the potential to be a long-term source of bioactive compounds with anti-PAF and antithrombotic properties, making it a promising candidate for the development of food supplements and nutraceuticals that target inflammation, thrombosis, and other related disorders. However, additional research is required to investigate the effects of *Arthrospira*-derived ingredients on platelets (Koukouraki et al. 2020).

High concentrations of angiotensin II (AII) have been associated with hypertension, endothelial dysfunction, fibrosis, and cardiovascular problems. The renin-angiotensin-aldosterone system (RAAS) and its action on AII play a crucial role in cardiovascular balance. AII receptor blockers have been introduced to manage hypertension and improve patient longevity. However, studies on hypertensive rats have shown increased sensitivity to blockers, leading to decreased lipolytic activity mediated by AII. RAAS inhibitors had no effect on lipolytic activity in normotensive rat adipocytes but reduced glycerol production in hypertensive rat adipocytes (Borghi et al. 2020). Trypsin-derived protein hydrolysates from *A. platensis*, on the other hand, showed angiotensin I-converting enzyme (ACE-I) inhibitory activity as well as significant antioxidant properties. This could be attributed to the combined action of various peptides and phenols present in *Arthrospira* (Yücetepe et al. 2018). ACE-I inhibitors reduce angiotensin II concentrations, thereby lowering blood pressure. These findings suggest that *Arthrospira*-derived compounds may have potential as natural ACE-I inhibitors and antioxidants (Ramos-Romero et al. 2021).

Peptide fractions of *A. platensis* show antihypertensive effects in hypertensive rats, mainly due to active peptides like IAE, FAL, AEL, IAPG, and VAF (Suetsuna & Chen, 2001). The decapeptide GIVAGDVTPI improves endothelial vasorelaxation and serum nitrite levels, potentially containing pathologies with poor vascular NO regulation (Münzel & Daiber, 2019). Another study found that hyperactivation of the renin-angiotensin system in perivascular adipose tissue (PVAT) contributes to vascular dysfunction in rats with heart failure after myocardial infarction surgery. This leads to reduce NO bioavailability, oxidative stress, and whitening of PVAT, suggesting PVAT as a potential target for treatment (Tavares et al. 2020).

Treatment with *Arthrospira* has been associated with a reduction in systolic blood pressure, as it increases the production of NO and the expression of p-AKt and HO-1, proteins that protect the cell from a pro-oxidant environment. Likewise, it has been observed that phycocyanin decreases free radicals, the expression of NO synthase, and the production of nitrites. In turn, it decreases the levels of markers of endothelial damage such as sVCAM-1, E-selectin, and endothelin 1. In contrast, phycocyanin increases liver antioxidant enzymes such as superoxide dismutase and catalase. On the other hand, lipid fractions and phycocyanobilin showed anti-PAF and antithrombin effects. In addition, some peptides, and phenols of *Arthrospira* decreased the ACE-1 enzyme, thus reducing blood pressure.

3.3.5 *Arthrospira* Effects on Non-Alcoholic Fatty Liver Disease (NAFLD)

Nonalcoholic fatty liver disease (NAFLD) is a metabolic disorder associated with obesity and characterized by the accumulation of liver fat. It ranges from simple steatosis to inflammation, fibrosis, cirrhosis, and cancer. NAFLD is closely linked to other metabolic diseases such as type 2 diabetes, hypertension, and dyslipidemia. Currently, there are no FDA-approved drugs specifically for NAFLD treatment. High-fat diets contribute to increased circulation of free fatty acids, leading to ectopic fat accumulation in the liver. This triggers inflammatory pathways, cellular dysfunction, and lipoapoptosis, contributing to liver injury and dysfunction in NAFLD (Kwon et al. 2020).

The 95% ethanol extract of *A. platensis* (SPL95) has demonstrated the ability to normalize hepatic steatosis and improve liver function tests. This includes reducing transaminases, free fatty acids, and improving the overall lipid profile. These beneficial effects may be attributed to the activation of the AMP-activated protein kinase signaling pathway, which leads to the downregulation of lipid synthesizing genes. This downregulation includes transcription factors involved in sterol-1c binding, 3-hydroxy-3-methyl glutaryl coenzyme A reductase, and acetylCoA carboxylase. As a result, triglyceride levels are reduced, and the synthesis of fatty acids is inhibited (Li et al. 2018). *A. platensis* crude polysaccharides can improve body weight, serum and liver lipid and carbohydrate indices, and liver antioxidant parameters in HFD-fed rats, which were accompanied by regulation of mRNA expression of liver genes involved in lipid and carbohydrate metabolism disorders. In addition, the mRNA expression level of PPAR-α increased significantly with SPLP (*S. platensis* polysaccharides) supplementation compared to the HFD group. Simultaneously, SPLP was able to markedly reduce the expressions of ACC, SREBP-1c, and HMG-CoA (Li et al. 2021).

Zucker fa/fa rats are used as models for obesity and metabolic syndrome due to their sensitivity to hypercaloric and lipoic diets. A recent study evaluated the addition of *Arthrospira* and glucomannan (G) to surimi bars to reduce the adverse effects of high consumption of these bars (i.e., flatulence, abdominal pain, and esophageal or lower gastrointestinal obstruction), used them as functional foods, and tested them on Zucker fa/fa rats. The researchers reported that long-term ingestion of surimi with diets enriched with glucomannan, or glucomannan/*Arthrospira*, affects liver and heart structures and improves NAFLD biomarkers in growing Zucker fa/fa rats, primarily those fed cholesterol-enriched diets (Vázquez-Velasco et al. 2017). According to the above, *Arthrospira* improves NAFLD by decreasing acetyl CoA carboxylase, which has an impact on the decrease in TG levels and fatty acid synthesis in the liver. In addition, polysaccharides present in *A. platensis* influenced the overregulation of the expression of PPARα and a downward regulation of ACC, SREBP1c, and HMG-CoA in the liver.

3.3.6 *Arthrospira* Effects on the Gut Microbiota

The human gut microbiota is crucial for maintaining overall health and is typically stable within individuals (Clemente et al. 2012). However, factors such as diet, supplements, and medications can disrupt the gut microbiota and lead to temporary changes in biological responses and potential disease development (Quigley, 2019). *Arthrospira* has been shown to have effects on the gut microbiota, including lipid-lowering, hypoglycemic, and appetite-control effects. Studies have found that *Arthrospira* supplementation increases the abundance of beneficial bacteria such as *Prevotella, Porphyromonadaceae, Barnesiella, and Paraprevotella.* These bacteria contribute to bile metabolism, short-chain fatty acid production, energy metabolism regulation, improved insulin sensitivity, and reduced body weight and serum LDL levels. Overall, *Arthrospira* supplementation has the potential to positively influence the gut microbiota and improve metabolic disorders and liver health (Li et al. 2018).

In healthy male mice, *Arthrospira* (1.5–3.0 g of spray-dried *A. platensis* powder per Kg of body weight per day for 24 days) modified cecal populations of microbiota at the gender level (*Clostridium* XIVa, *Desulfovibrio, Eubacterium, Barnesiella, Bacteroides,* and *Flavonifractor).* These changes correlated with markers of oxidative stress, blood lipid levels, and serum leptin. The ability of *Arthrospira* to increase leptin levels and reduce lipid metabolism in high doses implies a possible therapeutic application for the clinical management of obesity (Hu et al. 2019). Similarly, water and 55% ethanol extracts of *A. platensis* were found to contribute to hypoglycemic function and intestinal microbiota regulation (SP55). Most of the compounds in SP55 were polyunsaturated fatty acids. After an 8-week high-fat, sucrose diet with *A. platensis* supplements, glucose tolerance improved, and the composition of the gut microbiota was altered. Supplementation

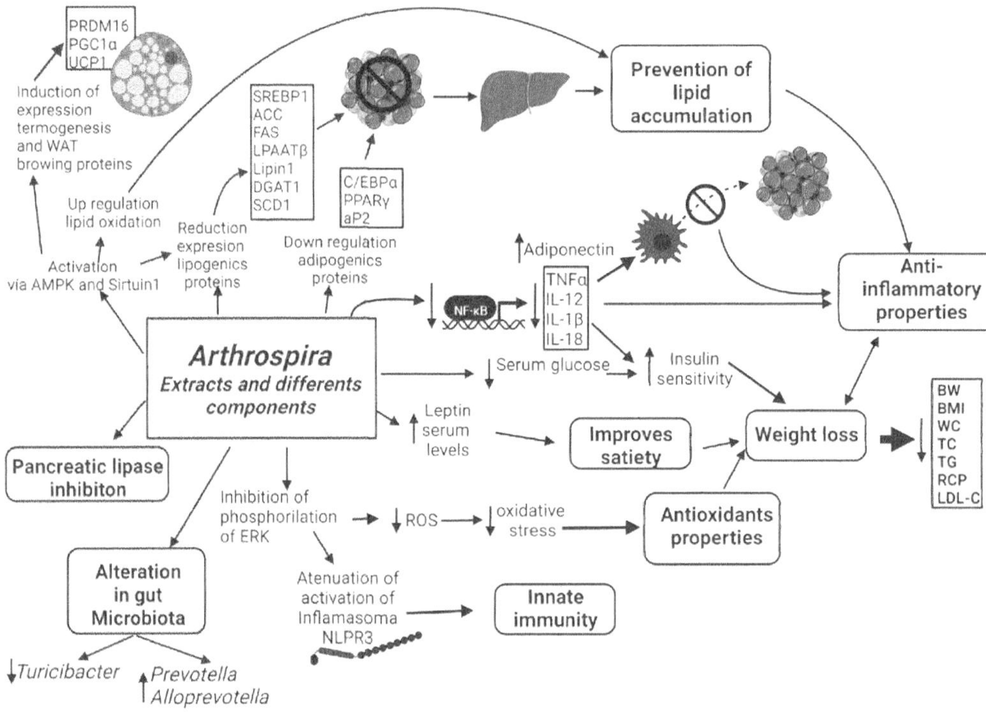

Figure 3.1 Mechanisms of action and reported effects of *Arthrospira* extracts or their various constituents.

with *A. platensis* increased the abundance of *Oscillibacter*, *Parasutterella*, and *Alloprevotella* and decreased the abundance of *Turicibacter*. Therefore, SP55 can be developed as an effective natural food material for preventing diabetes (Wan et al. 2019).

In addition, SPL95 extract, rich in polyunsaturated fatty *acids from A. platensis*, enriched beneficial bacteria, including *Prevotella, Alloprevotella, Porphyromonadaceae, Barnesiella,* and *Paraprevotella*. Treatment with SPL95 led to a decrease in microbes such as *Turicibacter, Romboutsia, Phascolarctobacterium, Olsenella, and Clostridium XVIII*, which were positively correlated with serum levels of triglycerides, total cholesterol, and low-density lipoprotein cholesterol, but were negatively correlated with serum high-density lipoprotein cholesterol levels. These results provide evidence that SPL95 fatty acid can be used as a new adjuvant therapy and functional food to regulate gut microbiota in obese and diabetic individuals (Li et al. 2018). In summary, *Arthrospira* supplementation significantly improves insulin sensitivity, lipid profile, inflammation, and oxidative stress in obesity and hypertension patients, affecting intestinal microbiota composition, increasing beneficial organisms like *Alloprevotella*, and decreasing *Turicibacter* (**Figure 3.1**).

3.4 CONCLUSION

Arthrospira, a plant with numerous bioactive compounds, has been found to have beneficial effects on obesity and its comorbidities. Studies have shown that *Arthrospira* can prevent weight gain, reduce waist circumference, and promote lipid oxidation by activating the AMPK and Sirtuin 1 pathways. It also has anti-inflammatory and antiproliferative effects in adipocytes and cancer cell lines. *Arthrospira* also has antihyperglycemic effects, reducing oxidative stress and the metabolic syndrome by decreasing glucose and plasma insulin. Treatment with *Arthrospira* has been associated with reducing systolic blood pressure and increasing the production of NO and p-AKt and HO-1 proteins. It also decreases free radicals, NO synthase, and nitrites, as well as markers of endothelial damage. *Arthrospira* also increases liver antioxidant enzymes, reducing the ACE-1 enzyme and blood pressure. It also improves NAFLD in the liver by decreasing acetyl-CoA carboxylase, reducing TG levels, and fatty acid synthesis. *Arthrospira*'s polysaccharide can overregulate PPARα expression and decrease ACC, SREBP1c, and HMG-CoA. *Arthrospira* supplementation significantly improves insulin sensitivity, lipid profile, inflammation, and oxidative stress in obesity and hypertension patients, affecting intestinal microbiota composition, increasing beneficial organisms like *Alloprevotella*, and decreasing *Turicibacter*. *Arthrospira*-derived proteins offer bioactive peptides with potential applications in functional foods, including antihypertensive, diabetic, and antioxidant properties.

ACKNOWLEDGMENTS

This work was financially supported through the project SIP-20210177; SIP-20220541.The first author thanks the Mexican National Council for Science and Technology (CONACYT) for the Ph.D. grant provided.

REFERENCES

Ali, A. T., Hochfeld, W. E., Myburgh, R., & Pepper, M. S. (2013). Adipocyte and adipogenesis. *European Journal of Cell Biology*, *92*(6–7), 229–236. https://doi.org/10.1016/j.ejcb.2013.06.001

Alvarez, A. L., Weyers, S. L., Goemann, H. M., Peyton, B. M., & Gardner, R. D. (2021). Microalgae, soil and plants: A critical review of microalgae as renewable resources for agriculture. *Algal Research*, *54*(October 2020), 102200. https://doi.org/10.1016/j.algal.2021.102200

Andrade, L. M., Andrade, C. J., Dias, M., & Nascimento, C. A. O. (2018). *Chlorella* and *Spirulina* microalgae as sources of functional foods, nutraceuticals, and food supplements; an overview growth kinetics of pseudomonas aeruginosa copper stress induced: Intracellular bioaccumulation of copper, proteomic and lipidomic view pro. *MOJ Food Process Technology*, *6*(1), 45-58. https://doi.org/10.15406/mojfpt.2018.06.00144

Arslan, R., Eroglu, E. C., & Aksay, S. (2021). Determination of bioactive properties of protein and pigments obtained from *Spirulina platensis*. *Journal of Food Processing and Preservation*, 45(2), 0–2. https://doi.org/10.1111/jfpp.15150

Blasetti-Fantauzzi, C., Iacobini, C., Menini, S., Vitale, M., Sorice, G. P., Mezza, T., Cinti, S., Giaccari, A., & Pugliese, G. (2020). Galectin - 3 gene deletion results in defective adipose tissue maturation and impaired insulin sensitivity and glucose homeostasis. *Scientific Reports, 0123456789*, 1–15. https://doi.org/10.1038/s41598-020-76952-z

Borghi, F., Morais, C. L., Silva, C., da Silva, P. C., & Grassi-Kassisse, D. M. (2020). The influence of angiotensin II and RAAS blockers on lipolytic and glycolytic activity in isolated adipocytes from wistar, wistar-kyoto and spontaneously hypertensive rats. *Molecular and Cellular Endocrinology*, 520:111086. https://doi.org/10.1016/j.mce.2020.111086

Chei, S., Oh, H. J., Song, J. H., Seo, Y. J., Lee, K., Kim, K. J., & Lee, B. Y. (2020). *Spirulina maxima* extract prevents activation of the NLRP3 inflammasome by inhibiting ERK signaling. *Scientific Reports, 10*(1), 1–10. https://doi.org/10.1038/s41598-020-58896-6

Cifarelli, V., Beeman, S. C., Smith, G. I., Yoshino, J., Morozov, D., Beals, J. W., Kayser, B. D., Watrous, J. D., Jain, M., Patterson, B. W., & Klein, S. (2020). Decreased adipose tissue oxygenation associates with insulin resistance in individuals with obesity. *Journal of Clinical Investigation*, 130(12), 6688-6699. https://doi.org/10.1172/jci141828

Ciferri, O. (1983). *Spirulina*, the edible microorganism. *Microbiological Reviews, 47*(4), 551–578.

Cifu, A., & Davis, A. (2017). Prevention, detection, evaluation, and management of high blood pressure in adults. *JAMA, 318*(21), 2132–2134. https://doi.org/10.1001/JAMA.2017.18706

Clemente, J., Ursell, L., Parfrey, L., & Knight, R. (2012). The impact of the gut microbiota on human health: An integrative view. *Cell, 148*(6), 1258–1270. https://doi.org/10.1016/J.CELL.2012.01.035

Di Francesco, A., Choi, Y., Bernier, M., Zhang, Y., Diaz-Ruiz, A., Aon, M. A., Kalafut, K., Ehrlich, M. R., Murt, K., Ali, A., Pearson, K. J., Levan, S., Preston, J. D., Martin-Montalvo, A., Martindale, J. L., Abdelmohsen, K., Miche, L. C. R, Willmes, D. M., Henke, C., Navas, P., Villalba, J. M., Siegel, D., Gorospe. M., Fritz, K., Biswal, S., Ross, D., & de Cabo, R. (2020). NQO1 protects obese mice through improvements in glucose and lipid metabolism. *NPJ Aging and Mechanisms and Disease* , *6*(1), 13. doi: 10.1038/s41514-020-00051-6.

Dinicolantonio, J. J., Bhat, A. G., & Okeefe, J. (2020). Effects of spirulina on weight loss and blood lipids: A review. *Open Heart, 7*(1), 1–7. https://doi.org/10.1136/openhrt-2018-001003

Fan, X., Cui, Y., Zhang, R., & Zhang, X. (2018). Purification and identification of anti-obesity peptides derived from *Spirulina platensis. Journal of Functional Foods, 47*, 350–360. https://doi.org/10.1016/j.jff.2018.05.066

Farmer, S. R. (2006). Transcriptional control of adipocyte formation. *Cell Metababolism, 4*(4), 263–273. https://doi.org/https://doi.org/10.1016/j.cmet.2006.07.001

Furmaniak, M. A., Misztak, A. E., Franczuk, M. D., Wilmotte, A., Waleron, M., & Waleron, K. F. (2017). Edible cyanobacterial genus *Arthrospira*: Actual state of the art in cultivation methods, genetics, and application in medicine. *Frontiers in Microbiology, 8*(DEC), 1–21. https://doi.org/10.3389/fmicb.2017.02541

Gabr, G. A., El-Sayed, S. M., & Hikal, M. S. (2020). Antioxidant Activities of Phycocyanin: A bioactive compound from *Spirulina platensis. Journal of Pharmaceutical Research International*, 32(2), 73–85. https://doi.org/10.9734/JPRI/2020/V32I230407

Garvey, W. T., Ryan, D. H., Look, M., Gadde, K. M., Allison, D. B., Peterson, C. A., Schwiers, M., Day, W. W., & Bowden, C. H. (2012). Two-year sustained weight loss and metabolic benefits with controlled-release phentermine/topiramate in obese and overweight adults (SEQUEL): A randomized, placebo-controlled, phase 3 extension study. *American Journal of Clinical Nutrition, 95*(2), 297–308. https://doi.org/10.3945/AJCN.111.024927

Giordano, A., Smorlesi, A., Frontini, A., Barbatelli, G., & Cint, S. (2014). White, brown and pink adipocytes: The extraordinary plasticity of the adipose organ. *European Journal of Endocrinology*, 170(5):R159-71. https://doi.org/10.1530/EJE-13-0945

Goiris, K., Muylaert, K., Voorspoels, S., Noten, B., De Paepe, D., E Baart, G. J., & De Cooman, L. (2014). Detection of flavonoids in microalgae from different evolutionary lineages. *Journal of Phycology, 50*(3), 483–492. https://doi.org/10.1111/jpy.12180

Gómez-Avellaneda, G., & Tarqui-Mamani, C. (2017). Vista de prevalencia de sobrepeso, obesidad y dislipidemia en trabajadores de salud del nivel primario. *Duazary, 14*(2), 141–148. https://doi.org/http://dx.doi.org/10.21676/2389783X.1972

González-Casanova, J., Pertuz-Cruz, S., Caicedo-Ortega, N., & Rojas-Gomez, D. (2020). Adipogenesis regulation and endocrine disruptors: Emerging insights in obesity. *BioMed Research International*, 2020:7453786 https://doi.org/10.1155/2020/7453786

Greenway, F. L., Fujioka, K., Plodkowski, R. A., Mudaliar, S., Guttadauria, M., Erickson, J., Kim, D. D., & Dunayevich, E. (2010). Effect of naltrexone plus bupropion on weight loss in overweight and obese adults (COR-I): A multicentre, randomised, double-blind, placebo-controlled, phase 3 trial. *Lancet, 376*(9741), 595–605. https://doi.org/10.1016/S0140-6736(10)60888-4

Guldas, M., Ziyanok-Demirtas, S., Sahan, Y., Yildiz, E., & Gurbuz, O. (2020). Antioxidant and anti-diabetic properties of Spirulina platensis produced in Turkey. *Food Science and Technology, 41*(3), 615–625. https://doi.org/10.1590/FST.23920

Guzmán-Gómez, O., García-Rodríguez, R. V., Quevedo-Corona, L., Pérez-Pastén-Borja, R., Rivero-Ramírez, N. L., Ríos-Castro, E., Pérez-Gutiérrez, S., Pérez-Ramos, J., & Chamorro-Cevallos, G. A. (2018). Amelioration of ethanol-induced gastric ulcers in rats pretreated with phycobiliproteins of Arthrospira (Spirulina) maxima. *Nutrients, 10*(6), 1–15. https://doi.org/10.3390/nu10060763

Halimatu lMunawaroh, H. S., Gumilar, G. G., Nurjanah, F., Yuliani, G., Aisyah, S., Kurnia, D., Wulandari, A. P., Kurniawan, I., Ningrum, A., Koyande, A. K., & Show, P. L. (2020). In-vitro molecular docking analysis of microalgae extracted phycocyanin as an anti-diabetic candidate. *Biochemical Engineering Journal, 161*(June), 107666. https://doi.org/10.1016/j.bej.2020.107666

Henrikson, R. (1994). Microalga *spirulina* (Ediciones). MICROALGA SPIRULINA | ROBERT HENRIKSON | URANO | Casa del Libro México

Heo, M. G., & Choung, S. Y. (2018). Anti-obesity effects of *Spirulina maxima* in high fat diet induced obese rats via the activation of AMPK pathway and SIRT1. *Food and Function, 9*(9), 4906–4915. https://doi.org/10.1039/c8fo00986d

Hu, J., Li, Y., Pakpour, S., Wang, S., Pan, Z., Liu, J., Wei, Q., She, J., Cang, H., & Zhang, R. X. (2019). Dose effects of orally administered *Spirulina* suspension on colonic microbiota in healthy mice. *Frontiers in Cellular and Infection Microbiology, 0,* 243. https://doi.org/10.3389/FCIMB.2019.00243

Jerez-Martel, I., García-Poza, S., Rodríguez-Martel, G., Rico, M., Afonso-Olivares, C., & Gómez-Pinchetti, J. L. (2017). Phenolic profile and antioxidant activity of crude extracts from microalgae and cyanobacteria strains. *Journal of Food Quality, 2017, Volume 2017, Article ID 2924508, 8 page.* https://doi.org/10.1155/2017/2924508

Kasbi-Chadli, F., Coué, M., Aguesse, A., Grit, I., Souque, T., Ferchaud-Roucher, V., & Ouguerram, K. (2021). *Spirulina* liquid extract prevents metabolic disturbances and improves liver sphingolipids profile in hamster fed a high-fat diet. *European Journal of Nutrition, 2021,* 1–12. https://doi.org/10.1007/S00394-021-02589-X

Koukouraki, P., Tsoupras, A., Sotiroudis, G., Demopoulos, C. A., & Sotiroudis, T. G. (2020). Antithrombotic properties of *Spirulina* extracts against platelet-activating factor and thrombin. *Food Bioscience, 37,* 100686. https://doi.org/10.1016/J.FBIO.2020.100686

Ku, C. S., Pham, T. X., Park, Y., Kim, B., Shin, M., Kang, I., & Lee, J. (2012). Edible blue-green algae reduce the production of pro- inflammatory cytokines by inhibiting NF-κB pathway in macrophages and splenocytes. *Bone, 23*(1), 1–7. https://doi.org/10.1038/jid.2014.371

Kwon, D., Kim, S. H., Son, S. W., Seo, J., Jeong, T. Bin, Kim, K. M., Jung, J. C., Jung, M. S., Lee, Y. H., & Jung, Y. S. (2020). Germinated soybean embryo extract ameliorates fatty liver injury in high-fat diet-fed obese mice. *Pharmaceuticals, 13*(11), 1–18. https://doi.org/10.3390/ph13110380

Li, T. T., Huang, Z. R., Jia, R. B., Lv, X. C., Zhao, C., & Liu, B. (2021). *Spirulina platensis* polysaccharides attenuate lipid and carbohydrate metabolism disorder in high-sucrose and high-fat diet-fed rats in association with intestinal microbiota. *Food Research International, 147,* 110530. https://doi.org/10.1016/J.FOODRES.2021.110530

Li, T. T., Liu, Y. Y., Wan, X. Z., Huang, Z. R., Liu, B., & Zhao, C. (2018). Regulatory efficacy of the polyunsaturated fatty acids from microalgae *Spirulina platensis* on lipid metabolism and gut microbiota in high-fat diet rats. *International Journal of Molecular Sciences, 19*(10), 1–14. https://doi.org/10.3390/ijms19103075

Li, Y., Aiello, G., Bollati, C., Bartolomei, M., Arnoldi, A., & Lammi, C. (2020). Phycobiliproteins from *Arthrospira Platensis* (*Spirulina*): A new source of peptides with dipeptidyl peptidase-IV inhibitory activity. *Nutrients,* 12(3):794 1–11.

Liestianty, D., Rodianawati, I., Arfah, R. A., Assa, A., Patimah, S., & Muliadi. (2019). Nutritional analysis of *Spirulina sp* to promote as superfood candidate. *IOP Conference Serires: Materials Science and Engineering, 509,* 012031. https://doi.org/10.1088/1757-899X/509/1/012031

Mallikarjun Gouda, K. G., Kavitha, M. D., & Sarada, R. (2015). Antihyperglycemic, antioxidant and antimicrobial activities of the butanol extract from *Spirulina Platensis*. *Journal of Food Biochemistry*, *39*(5), 594–602. https://doi.org/10.1111/JFBC.12164

Marles, R. J., Barrett, M. L., Barnes, J., Chavez, M. L., Gardiner, P., Ko, R., Mahady, G. B., Dog, T. L., Sarma, N. D., Giancaspro, G. I., Sharaf, M., & Griffiths, J. (2011). United states pharmacopeia safety evaluation of spirulina. *Critical Reviews in Food Science and Nutrition*, *51*(7), 593–604. https://doi.org/10.1080/10408391003721719

Martínez-Fernández, L., González-Muniesa, P., Sáinz, N., Escoté, X., Martínez, J. A., Arbones-Mainar, J. M., & Moreno-Aliaga, M. J. (2020). Maresin 1 regulates insulin signaling in human adipocytes as well as in adipose tissue and muscle of lean and obese mice. *Journal of Physiology and Biochemistry*, **77**, 167–173 (2021)1–7. https://doi.org/10.1007/s13105-020-00775-9

Martínez-Sámano, J., De Oca, A. T. M., O-Bocardo, O. I. L., Torres-Durán, P. V., & Juárez-Oropeza, M. A. (2018). *Spirulina maxima* decreases endothelial damage and oxidative stress indicators in patients with systemic arterial hypertension: Results from exploratory controlled clinical trial. *Marine Drugs*, 16(12):496. https://doi.org/10.3390/md16120496

Memije-Lazaro, I. N., Blas-Valdivia, V., Franco-Colín, M., & Cano-Europa, E. (2018). *Arthrospira maxima* (*Spirulina*) and C-phycocyanin prevent the progression of chronic kidney disease and its cardiovascular complications. *Journal of Functional Foods*, *43*, 37–43. https://doi.org/10.1016/J.JFF.2018.01.013

Miczke, A. (2016). Effects of spirulina consumption on body weight, blood pressure, and endothelial function in overweight hypertensive Caucasians: A double blind, placebo-. *European Review for Medical and Pharmacological Sciences*, 20:150–156. www.europeanreview.org/wp/wp-content/uploads/150-156.pdf

Milovanović, I., Mišan, A., Simeunović, J., Kovač, D., Jambrec, D., & Mandić, A. (2015). Determination of volatile organic compounds in selected strains of cyanobacteria. *Journal of Chemistry*, *2015*. Volume 2015, Article ID 969542, 6 pages https://doi.org/10.1155/2015/969542

Minic, S. L., Stanic-Vucinic, D., Mihailovic, J., Krstic, M., Nikolic, M. R., & Cirkovic Velickovic, T. (2016). Digestion by pepsin releases biologically active chromopeptides from C-phycocyanin, a blue-colored biliprotein of microalga *Spirulina*. *Journal of Proteomics*, *147*, 132–139. https://doi.org/10.1016/j.jprot.2016.03.043

Mohammadi-Khanaposhtani, M., Rezaei, S., Khalifeh, R., Imanparast, S., Faramarzi, M. A., Bahadorikhalili, S., Safavi, M., Bandarian, F., Nasli Esfahani, E., Mahdavi, M., & Larijani, B. (2018). Design, synthesis, docking study, α-glucosidase inhibition, and cytotoxic activities of acridine linked to thioacetamides as novel agents in treatment of type 2 diabetes. *Bioorganic Chemistry*, *80*, 288–295. https://doi.org/10.1016/J.BIOORG.2018.06.035

Moradi, S., Ziaei, R., Foshati, S., Mohammadi, H., Nachvak, S. M., & Rouhani, M. H. (2019). Effects of *Spirulina* supplementation on obesity: A systematic review and meta-analysis of randomized clinical trials. In *Complementary Therapies in Medicine* (Vol. 47, p. 102211). Churchill Livingstone. https://doi.org/10.1016/j.ctim.2019.102211

Müller, T. D., Clemmensen, C., Finan, B., Dimarchi, R. D., & Tschöp, M. H. (2018). Anti-obesity therapy: From Rainbow pills to polyagonists. *Pharmacological Review*, *70*, 712–746. https://doi.org/10.1124/pr.117.014803

Münzel, T., & Daiber, A. (2019). Antihypertensive decapeptide from cyanobacterium *Spirulina platensis*: Superfood with eNOS (Endothelial NO Synthase) enhancer properties. *Hypertension*, *73*(2), 291–293. https://doi.org/10.1161/HYPERTENSIONAHA.118.11893

Nasirian, F., Dadkhah, M., Moradi-Kor, N., & Obeidavi, Z. (2018). Diabetes, metabolic syndrome and obesity: Targets and therapy Dovepress effects of *Spirulina platensis* microalgae on antioxidant and anti-inflammatory factors in diabetic rats. *Diabetes, Metabolic Syndrome and Obesity: Targets and Therapy*, Dovepress.11–375. http://dx.doi.org/10.2147/DMSO.S172104

Park, W. S., Kim, H. J., Li, M., Lim, D. H., Kim, J., Kwak, S. S., Kang, C. M., Ferruzzi, M. G., & Ahn, M. J. (2018). Two classes of pigments, carotenoids and c-phycocyanin, in spirulina powder and their antioxidant activities. *Molecules*, *23*(8), 1–11. https://doi.org/10.3390/molecules23082065

Petersen, M. C., & Shulman, G. I. (2018). Mechanisms of insulin action and insulin resistance. *Physiological Reviews*, *98*(4), 2133–2223. https://doi.org/10.1152/physrev.00063.2017

Pham, T. X., & Lee, J. Y. (2016). Anti-inflammatory effect of *Spirulina platensis* in macrophages is beneficial for adipocyte differentiation and maturation by inhibiting nuclear factor-κB pathway in 3T3-L1 adipocytes. *Journal of Medicinal Food, 19*(6), 535–542. https://doi.org/10.1089/jmf.2015.0156

Prokudina, E., Maslov, L., Ivanov, W., Bespalova, I., Pismennyi, D., & Voronkov, N. (2017). [The Role of reactive oxygen species in the pathogenesis of adipocyte dysfunction in metabolic syndrome. Prospects of pharmacological correction]. *Vestnik Rossiiskoi Akademii Meditsinskikh Nauk, 72*(1), 11–16. https://doi.org/10.15690/VRAMN798

Pugh, N., Ross, S. A., ElSohly, H. N., ElSohly, M. A., & Pasco, D. S. (2001). Isolation of three high molecular weight polysaccharide preparations with potent immunostimulatory activity from *Spirulina platensis*, aphanizomenon flos-aquae and *Chlorella pyrenoidosa*. *Planta Medica, 67*(08), 737–742. https://doi.org/10.1055/S-2001-18358

Quigley, E. (2019). Prebiotics and probiotics in digestive health. *Clinical Gastroenterology and Hepatology: The Official Clinical Practice Journal of the American Gastroenterological Association, 17*(2), 333–344. https://doi.org/10.1016/J.CGH.2018.09.028

Ramos-Romero, S., Torrella, J. R., Pagès, T., Viscor, G., & Torres, J. L. (2021). Edible microalgae and their bioactive compounds in the prevention and treatment of metabolic alterations. *Nutrients, 13*(2), 563. https://doi.org/10.3390/nu13020563

Russo, I. (2012). The prothrombotic tendency in metabolic syndrome: Focus on the potential mechanisms involved in impaired haemostasis and fibrinolytic balance. *Scientifica, 2012*, 1–17. https://doi.org/10.6064/2012/525374

Saadi, S., Saari, N., Anwar, F., Abdul Hamid, A., & Ghazali, H. M. (2015). Recent advances in food biopeptides: Production, biological functionalities and therapeutic applications. *Biotechnology Advances, 33*(1), 80–116. https://doi.org/10.1016/j.biotechadv.2014.12.003

Saidi, H., Bounihi, A., Bouazza, A., Hichami, A., Koceir, E. H. A., & Khan, N. A. (2020). *Spirulina* reduces diet-induced obesity through downregulation of lipogenic genes expression in Psammomys obesus. *Archives of Physiology and Biochemistry*, 128(4):1–9. https://doi.org/10.1080/13813455.2020.1743724

Sánchez, M., Bernal-Castillo, J., Rozo, C., & Rodríguez, I. (2003). *Spirulina* (*Arthrospira*) an edible microorganism a review. *Universitas Scientiarum, 8*(1), 7–24.

Santillana, N., Astudillo-Guerrero, C., D'Espessailles, A., & Cruz, G. (2023). White adipose tissue dysfunction: Pathophysiology and emergent measurements. *Nutrients, 15*(7), 1–15. https://doi.org/10.3390/nu15071722

Scaglioni, P. T., Quadros, L., de Paula, M., Furlong, V. B., Abreu, P. C., & Badiale-Furlong, E. (2018). Inhibition of enzymatic and oxidative processes by phenolic extracts from *Spirulina sp.* and *Nannochloropsis sp. Food Technology and Biotechnology, 56*(3), 344–353. https://doi.org/10.17113/ftb.56.03.18.5495

Sengupta, S., Koley, H., Dutta, S., & Bhowal, J. (2018). Hypocholesterolemic effect of *Spirulina platensis* (SP) fortified functional soy yogurts on diet-induced hypercholesterolemia. *Journal of Functional Foods, 48*, 54–64. https://doi.org/10.1016/J.JFF.2018.07.007

Seo, Y. J., Kim, K. J., Choi, J., Koh, E. J., & Lee, B. Y. (2018). *Spirulina maxima* extract reduces obesity through suppression of adipogenesis and activation of browning in 3T3-L1 cells and high-fat diet-induced obese mice. *Nutrients, 10*(6), 1–15. https://doi.org/10.3390/nu10060712

Shah, A., Mehta, N., & Reilly, M. P. (2008). Adipose inflammation, insulin resistance, and cardiovascular disease. *Journal of Parenteral and Enteral Nutrition, 32*(6), 638–644. https://doi.org/10.1177/0148607108325251

Sharma, A. M., & Padwal, R. (2010). Obesity is a sign – Over-eating is a symptom: An aetiological framework for the assessment and management of obesity. *Obesity Reviews, 11*(5), 362–370. https://doi.org/10.1111/j.1467-789X.2009.00689.x

Singh, D. P., Prabha, R., Verma, S., Meena, K. K., & Yandigeri, M. (2017). Antioxidant properties and polyphenolic content in terrestrial cyanobacteria. *3 Biotech, 7*(2). https://doi.org/10.1007/S13205-017-0786-6

Singh, S., Dwivedi, V., Sanyal, D., & Dasgupta, S. (2020). Therapeutic and nutritional potential of *Spirulina* in combating COVID-19 infection. *International Journal of Creative Research Thoughts (IJCRT), 8*(9), 4051–4063. https://doi.org/10.13140/RG.2.2.16527.41127

Skrypnik, K., Suliburska, J., Skrypnik, D., Pilarski, Ł., Reguła, J., & Bogdański, P. (2018). The genetic basis of obesity complications [pdf]. *Acta Scientiarum Polonorum Technologia Alimentaria, 16*(1), 83–91. https://doi.org/10.17306/j.afs.2017.2017.0442

Soni, R. A., Sudhakar, K., & Rana, R. S. (2017). Spirulina – From growth to nutritional product: A review. *Trends in Food Science and Technology, 69*, 157–171. https://doi.org/10.1016/j.tifs.2017.09.010

Suetsuna, K., & Chen, J. R. (2001). Identification of antihypertensive peptides from peptic digest of two microalgae, *Chlorella vulgaris* and *Spirulina platensis. Marine Biotechnology, 3*(4), 305–309. https://doi.org/10.1007/s10126-001-0012-7

Tavares Fontes, M., Mesquita, S. P., Lino, C. A., Senger, N., Couto, G. K., Luiza, M., Barreto-Chaves, M., Mill, J. G., Rossoni, L. V., & Rossoni, L. V. (2020). Renin-angiotensin system overactivation in perivascular adipose tissue contributes to vascular dysfunction in heart failure. *Clinical Science.* 134(23):3195–3211 https://doi.org/10.1042/CS20201099/897738/cs-2020-1099.pdf

Thorp, J. H., & Bowes, R. E. (2017). Carbon sources in riverine food webs: New evidence from amino acid isotope techniques. *Ecosystems, 20*(5), 1029–1041. https://doi.org/10.1007/s10021-016-0091-y

Vaněčková, I., Maletínská, L., Behuliak, M., Nagelová, V., Zicha, J., & Kuneš, J. (2014). Obesity-related hypertension: Possible pathophysiological mechanisms. *Journal of Endocrinology, 223*(3), R63–R78. https://doi.org/10.1530/JOE-14-0368

Vázquez-Velasco, M., González-Torres, L., García-Fernández, R. A., Méndez, M. T., Bastida, S., Benedí, J., González-Muñoz, M. J., & Sánchez-Muniz, F. J. (2017). Glucomannan or glucomannan plus *Spirulina*-enriched squid-surimi diets reduce histological damage to liver and heart in Zucker fa/fa rats fed a cholesterol-enriched and non-cholesterol-enriched atherogenic diet. *Journal of Medicinal Food, 20*(6), 618–625. https://doi.org/10.1089/jmf.2016.0157

Villalpando, D. M., Verdasco-Martín, C. M., Plaza, I., Gómez-Rivas, J., R De Bethencourt, F., Villarroel, M., García, J. L., Otero, C., & Ferrer, M. (2020). Beneficial effects of *Spirulina* aqueous extract on vasodilator function of arteries from hypertensive rats. *International Journal of Vascular Medicine*, 2020, Article ID 6657077, 9 pages. https://doi.org/10.1155/2020/6657077

Vo, T. S., & Kim, S. K. (2013). Down-regulation of histamine-induced endothelial cell activation as potential anti-atherosclerotic activity of peptides from *Spirulina maxima. European Journal of Pharmaceutical Sciences, 50*(2), 198–207. https://doi.org/10.1016/j.ejps.2013.07.001

Vo, T. S., Ngo, D. H., Kang, K. H., Park, S. J., & Kim, S. K. (2014). The role of peptides derived from *Spirulina maxima* in downregulation of FcεRI-mediated allergic responses. *Molecular Nutrition and Food Research, 58*(11), 2226–2234. https://doi.org/10.1002/mnfr.201400329

Vo, T. S., Ryu, B., & Kim, S. K. (2013). Purification of novel anti-inflammatory peptides from enzymatic hydrolysate of the edible microalgal *Spirulina maxima. Journal of Functional Foods, 5*(3), 1336–1346. https://doi.org/10.1016/j.jff.2013.05.001

Wan, X. Z, Li, T. T., Zhong, R. T., Chen, H. B., Xia, X, Gao, L. Y., Gao, X. X., Liu, B., Zhang, H. Y., & Zhao, C. (2019). Anti-diabetic activity of PUFAs-rich extracts of Chlorella pyrenoidosa and *Spirulina platensis* in rats. *Food and Chemical Toxicology, 128*, 233–239. doi: 10.1016/j.fct.2019.04.017.

Wang, B., Liu, Q., Huang, Y., Yuan, Y., Ma, Q., Du, M., Cai, T., & Cai, Y. (2018). Extraction of polysaccharide from *Spirulina* and evaluation of its activities. Evidence-Based Complementary and Alternative Medicine Volume 2018, Article ID 3425615, 8 pages https://doi.org/10.1155/2018/3425615

Wang, X. Q., Li, L. N., Chang, W. R., Zhang, J. P., Gui, L. L., Guo, B. J., & Liang, D. C. (2001). Structure of C-phycocyanin from *Spirulina platensis* at 2.2 Å resolution: A novel monoclinic crystal form for phycobiliproteins in phycobilisomes. *Acta Crystallographica Section D: Biological Crystallography, 57*(6), 784–792. https://doi.org/10.1107/S0907444901004528

Wang, Z., & Zhang, X. (2016). Inhibitory effects of small molecular peptides from *Spirulina* (*Arthrospira*) *platensis* on cancer cell growth. *Food and Function, 7*(2), 781–788. https://doi.org/10.1039/c5fo01186h

Wang, Z., & Zhang, X. (2017). Isolation and identification of anti-proliferative peptides from *Spirulina platensis* using three-step hydrolysis. *Journal of the Science of Food and Agriculture, 97*(3), 918–922. https://doi.org/10.1002/jsfa.7815

Wilding, J. P. H., Batterham, R. L., Calanna, S., Davies, M., Van Gaal, L. F., Lingvay, I., McGowan, B. M., Rosenstock, J., Tran, M. T. D., Wadden, T. A., Wharton, S., Yokote, K., Zeuthen, N., & Kushner, R. F. (2021). Once-weekly semaglutide in adults with overweight or obesity. *New England Journal of Medicine, 384*(11), 989–1002. https://doi.org/10.1056/NEJMOA2032183/SUPPL_FILE/NEJMOA2032183_DATA-SHARING.PDF

World Health Organization. (2014). *Global Status Report On Noncommunicable Diseases 2014.*

World Health Organization. (2023). *Obesidad y sobrepeso.* www.who.int/es/news-room/fact-sheets/detail/obesity-and-overweight

Yousefi, R., Mottaghi, A., & Saidpour, A. (2018). *Spirulina platensis* effectively ameliorates anthropometric measurements and obesity-related metabolic disorders in obese or overweight healthy individuals: A randomized controlled trial. *Complementary Therapies in Medicine, 40,* 106–112. https://doi.org/10.1016/j.ctim.2018.08.003

Yücetepe, A., Kasapoğlu, K. N., & Özçelik, B. (2018). Angiotensin-I-converting enzyme inhibitory and antioxidant activity of tryptic *Spirulina platensis* Protein hydrolysates: Effect of hydrolysis and *in vitro* gastrointestinal digestion. *E-Journal of New World Sciences Academy, 13*(3), 151–162. https://doi.org/10.12739/NWSA.2018.13.3.5A0104

Zeinalian, R., Farhangi, M. A., Shariat, A., & Saghafi-Asl, M. (2017). The effects of *Spirulina platensis* on anthropometric indices, appetite, lipid profile and serum vascular endothelial growth factor (VEGF) in obese individuals: A randomized double blinded placebo controlled trial. *BMC Complementary and Alternative Medicine, 17*(1), 1–8. https://doi.org/10.1186/s12906-017-1670-y

Chapter 4

Fruit Seeds By-Products as Food Ingredients with High Potential for Human Health Promotion

Silvia Luna-Suárez
Gema Morales-Olán
Luis Huerta-González
and J. Ismael Morales-Camacho

CONTENTS

4.1 INTRODUCTION

Fruits are a vital component of a healthy diet, as they offer essential vitamins, minerals, antioxidants, and most importantly, fiber. They are also used to create a wide variety of food products. Unfortunately, the processing of these fruits generates significant amounts of by-products that are often discarded, causing harm to the environment. One such by-product is seeds, which are often overlooked. However, fruit seeds are rich in valuable nutrients that

 DOI: 10.1201/9781003332558-5

possess functional properties with promising applications in the food and industrial sectors, as well as significant benefits for human health. In this chapter, we will explore the nutritional composition of various fruit seed by-products, their functional properties, their potential as food ingredients, and their health benefits. By doing so, we can begin to appreciate the value of these by-products and make the most of their potential.

4.2 SEEDS AS BY-PRODUCTS OF FRUIT PROCESSING

Fruit processing waste is mainly composed of peels, leaves, residual pulp, stems, and seeds. Seeds constitute more than 50% (w/w) of fruit waste (Mazurek, Chmiel & Górecka, 2017). A small number of discarded seeds is used for animal feed (Rodrigues et al. 2006), or as fertilizer and compost. However, most of the seeds generated from fruit processing are discarded in open-air dumps, causing serious environmental and sanitary problems. Currently, seeds obtained as agro-industrial waste are being revalued, as they are rich in various nutrients, contain beneficial bioactive compounds (Fidelis et al. 2019; de Chaves et al. 2023), and can be a source of techno-functional compounds useful for food processing (Figure 4.1).

Some of the seeds discarded and undervalued are those of apples (*Malus domestica Borkh.*), mangoes (*Mangifera indica* L.), rambutan (*Nephelium lappaceum* L.), guavas (*Psidium guajava* L.), grapes (*Vitis vinifera*), tamarind (*Tamarindus indica* L.), papayas (*Carica papaya* L.), raspberries (*Rubus idaeus* L.), moriche palm (*Mauritia flexuosa*), passion fruit (*Passiflora edulis* Sims f. *flavicarpa* Degener), açai (*Euterpe oleracea*), litchi (*Litchi chinensis*), dates (*Phoenix dactylifera* L.), peaches (*Prunus persica*), pomegranates (*Punica granatum*), and citrus (*Citrus*), among others.

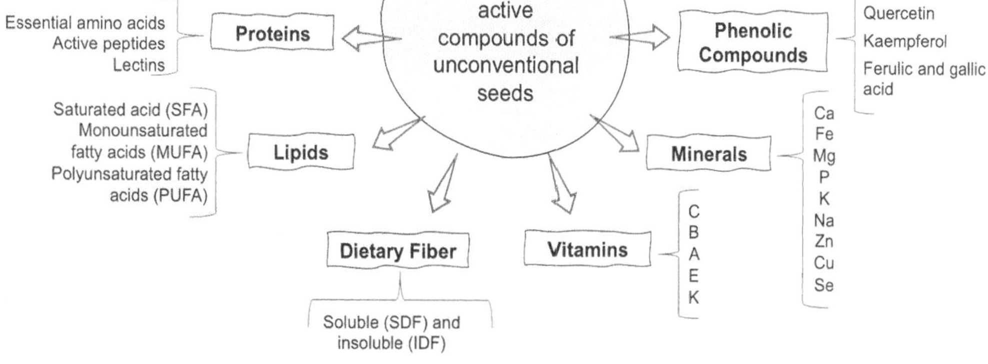

Figure 4.1 Nutrients and bioactive compounds identified in seeds byproducts of fruit processing.

The active compounds and nutrients present in these seeds, their functional properties, and biological activity are described in the following sections.

4.2.1 Composition

4.2.1.1 Simple Carbohydrates, Oligosaccharides, and Polysaccharides

Carbohydrates are one of the macronutrients that can be found in the seeds that are often wasted during fruit processing. Rambutan seeds are a good source of carbohydrates as they contain 46–48.10% (Jahurul et al. 2020), followed by apple seeds with 26.20–29.80% (Kumar et al. 2022a) and mango seeds with 18.2% (Gupta et al. 2022). Guava and orange seeds contain simple sugars such as fructose and glucose (Kumar et al 2022b). In addition, polysaccharides such as xyloglucans, galactoxyloglucans, monosaccharides, octasaccharides, heptasaccharides, and hexasaccharides have been isolated from tamarind seeds. Mango and litchi seeds possess starch (30.0% and 40.7% respectively) (Gupta et al. 2022; Bangar et al. 2021). In a recent study, Bangar et al. (2021) found that litchi seed starch possesses distinctive functional and structural properties owing to its high amylopectin content and low amylose content. These properties make it a promising ingredient for use in a variety of food and nonfood products.

4.2.1.2 Proteins

Proteins are also an important component found in seeds. Peach seed can contain up to 50.69% protein (Ağirbaş, Yavuz-Düzgün and Özçelik, 2022), while apple seed contains between 33–79% (Kumar et al. 2022a) and papaya seed between 24.3% and 31.8% (Dotto & Abihudi, 2021). Citrus seeds such as orange seed contain between 17.9% and 26.5% (Zayed, Badawy, & Farag, 2021). Mango, grape, tamarind, and guava seeds have also significant amounts of protein (6.0–13.0%, 7.8–14.1%, 8.44–25.9%, 12.1%, and 7.6–15.0%, respectively) (Lebaka et al. 2021).

Protein content in seeds depends on the crop, the environmental conditions, and the maturation stage. Their nutritional value lies in their amino acid profile, as they are a good source of essential amino acids. Apple seed proteins contain all essential amino acids except tyrosine (Kumar et al. 2022a). Mango kernel flour contains all the essential amino acids with levels higher than those of the FAO/WHO reference protein (FAO/WHO 2007; Mustafa et al. 2023). In the latter seeds, amino acids such as leucine (6.9 g/100g), valine (5.8 g/100g), isoleucine (4.4 g/100g), lysine (4.3 g/100g), threonine (3.4 g/100g), phenylalanine (3.4 g/100g), tyrosine (2.7 g/100g), and methionine (1.2 g/100g) have been identified (Abdalla et al. 2007). In rambutan seed, the amino acids profile showed lysine, leucine, valine, isoleucine, phenylalanine, histidine, and methionine (Jahurul et al. 2020). Similarly, guava seed proteins showed the presence of aromatic and sulfur-rich amino acids (Kumar et al. 2022b).

4.2.1.3 Lipids

Plant lipids are highly valued in the manufacture of foods. The vegetable's lipid quality is determined by its fatty acid profile and its concentration. **Table 4.1** outlines the key fatty acids, including saturated (SFA), mono-unsaturated (MUFA), and poly-unsaturated (PUFA) that have been identified in seeds that are typically discarded during fruit processing.

Among the fatty acids of food interest, oleic (omega-9), linoleic (omega-6), and linolenic (omega-3) acids stand out. Waste seeds can be considered as attractive unconventional sources of these acids since they have been identified in these important materials.

TABLE 4.1 MAIN FATTY ACIDS IDENTIFIED IN FRUIT SEEDS

Fruit Seeds	Monounsatured Fatty Acids (MUFA)	Polyunsaturated Fatty Acids (PUFA)	Saturated Fatty Acids (SFA)	References
Apple	C16:1 palmitoleic C18:1 oleic	C18:2 linoleic C18:3 linolenic	C12:0 lauric C14:0 myristic C16:0 palmitic C18:0 stearic C20:0 arachidic C22:0 behenic	Yukui et al. (2009); Walia et al. (2014); Górnaś et al. (2014) and Abbas et al. (2019)
Rambutan	C18:1 oleic	C18:2 linoleic	C16:0 palmitic C18:0 stearic C20:0 arachidic C22:0 behenic	Jahurul et al. (2020)
Guava	C18:1 oleic	C18:2 linoleic C18:3 linolenic	C16:0 palmitic C18:0 stearic	Kumar et al. (2022b)
Tamarind	C18:1 oleic	C18:2 linoleic	C16:0 palmitic C20:0 arachidic C22:0 behenic C24:0 lignoceric	Martins et al. (2022)
Papaya	C18:1 oleic	C18:2 linoleic C18:3 linolenic	C16:0 palmitic C18:0 stearic C20:0 arachidic	Dotto & Abihudi, (2021)
Raspberry and Grape	C18:1 oleic	C18:2 linoleic C18:3 linolenic	C16:0 palmitic C18:0 stearic	Marić et al. (2020) Da Porto, Porreto & Decorti, (2013)
Moriche	C16:1 Palmitoleic C18:1 oleic	C18:2 linoleic C18:3 linolenic	C14:0 myristic C16:0 palmitic C18:0 stearic	Quintero-Angel, Martínez-Girón & Orjuela-Salazar, (2022)
Passion fruit and Apricot	C18:1 oleic	C18:2 linoleic	C16:0 palmitic C18:0 stearic	Dos Santos et al. (2019) Priyadarshi et al. (2018)
Açai	C18:1 oleic	C18:2 linoleic C18:3 linolenic	C14:0 myristic C16:0 palmitic C18:0 stearic	Melo et al. (2021)
Litchi, mango, and citrus	C18:1 oleic	C18:2 linoleic C18:3 linolenic	C16:0 palmitic C18:0 stearic C20:0 arachidic C22:0 behenic C24:0 lignoceric	Punia & Kumar, (2021) Zayed, Badawy & Farag, (2021) Gupta et al. (2022); Jahurul et al. (2020; Maldonado-Celis et al. (2019)
Date	C18:1 oleic	C18:2 linoleic	C12:0 lauric C14:0 myristic C16:0 palmitic	Ourradi et al. (2021)

4.2.1.4 Dietary fiber

Açai, raspberry, and guava seeds show the highest content of dietary fiber, with 86.2% (Melo et al. 2021), 65.7% (Marić et al. 2020), and 63.9% (Kumar et al. 2022b), respectively. In litchi seed, dietary fiber accounts for 24.5% of total weight (Punia & Kumar, 2021), in rambutan seed it is 11.6% (Jahurul et al. 2020), and in apple seed it is 3.9–20.6% (Kumar et al. 2022a). The predominant fraction in most of these seeds is insoluble fiber (Pathania & Kaur, 2022).

4.2.1.5 Vitamins and Minerals

Vitamins C, B, A, E, and K have been isolated from many wasted seeds. Tamarind and mango seeds have the highest concentrations with values of 103.2 mg/100g and 257.0 mg/100g (Kumar et al. 2022a; Gupta et al. 2022). Vitamin C is also present in apple, rambutan, and guava seeds. Another important vitamin that has been identified in the lipid fraction of seeds is vitamin E, together with other tocopherols such as γ-tocopherol and δ-tocopherol (Pieszka et al. 2013). Both vitamins are valued for their antioxidant properties. With regards to minerals, Ca, Mg, Na, K, P, Fe, Cu, and Zn have been quantified. For example, papaya seed contains 8435 mg/100 g of Ca and 1635 mg/100 g of K (Dotto & Abihudi, 2021). Apple seed was found to be rich in K (650 mg/100 g) and P (666.5 mg/100 g) (Kumar et al. 2022a).

4.2.1.6 Phenolic Compounds

Quercetin, kaempferol, catechins, gallic and ferulic acid are the most common phenolic compounds identified in unconventional seeds. In apples, the presence of floridzin, which is a hydrochalcone, has attracted attention (Kumar et al. 2022a), and in papaya, coumarins have also been identified (Dotto & Abihudi, 2021). Phenolic compounds are widely studied in these seeds due to their antioxidant properties.

4.2.1.7 Phytosterols and Glucosinolates

With respect to phytosterols, seeds have been shown to contain, in different concentrations, *β*-sitosterol, campesterol, stigmasterol, and avenosterol. For example, 87.0% of total phytosterol content in guava is stigmasterol while only 3.0% is campesterol (Kumar et al. 2022b). In rambutan, passion fruit, dates, and papaya seeds a higher concentration of *β*-sitosterol has been found (Jahurul et al. 2020; Rotta et al. 2018; Maqsood et al. 2020; Dotto & Abihudi, 2021). On the other hand, glucosinolates and sulfur-containing derivatives have been identified in papaya. These compounds have a variety of biological activities (Dotto & Abihudi, 2021).

4.3 USE AS FOOD INGREDIENTS AND FUNCTIONAL PROPERTIES

Seeds that are considered residues, and rather frequently as wastes during the processing of fruits, hold great potential for their functional characteristics. They can be used directly or refined into starch and other macromolecules, flour, or meal, depending on their physical and chemical properties, and their composition (**Figure 4.2**). These versatile seeds can be utilized to enhance the nutritional and nutraceutical value and functional properties of various food preparations or to produce useful ingredients like antimicrobials and colorants, plus bioactive molecules and those components cited before.

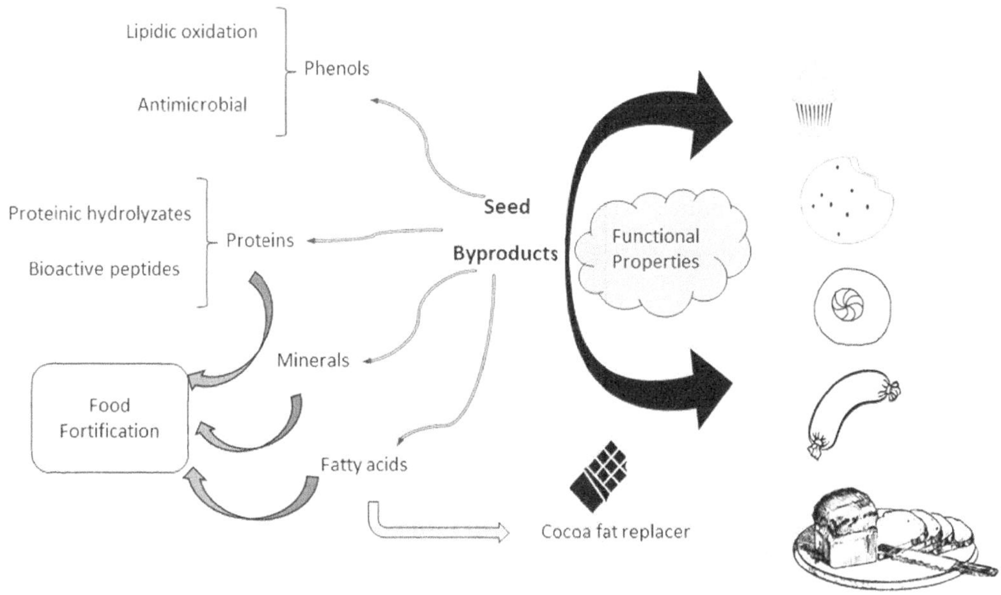

Figure 4.2 Functional properties of seed byproducts.

Tomato

Tomato processed by-products are the skin and seeds (Szabo et al. 2018); the latter component represents 60% in weight, possessing great potential for health improvement. Seeds meal showed good emulsifying activity, good emulsion stability (87.5%) at alkaline pH, an acceptable water holding capacity (3.2 g water/g sample), and good oil absorption capacity (2.0 g oil/g sample) (Maldonado-Torres et al. 2020). These functional properties improved when tomato meal was defatted, as a result of better protein and carbohydrate interactions with the water and fat of the medium. It has been reported that tomato seed protein concentrate showed good oil absorption capacity, as well as foaming, emulsifying, and foaming stability (Shao et al. 2014).

Tomato seed meal can be used in the baking industry, replacing a certain percentage of wheat flour to prepare cookies and bread. Wheat flour replacement increased the protein content in the trial products, although acceptability was compromised by changes in texture and flavor (Sogi et al. 2002).

Apple

Seeds meal has shown good functional properties, high water and oil absorption capacity, as well as emulsifying and foaming properties, indicating that this meal can be used to modify some product formulations like emulsions with high stability (Manzoor et al. 2021).

Defatted apple seed meal has been used to enrich wheat bread using between 5.0% and 20.0% in its formulation. The nutritional profile in terms of protein, ash, fat, and soluble and insoluble fiber increased in the bread made with this meal. They also presented good antioxidant capacity (Purić et al. 2020), although only those added with 5.0% apple meal showed good sensory and texture properties compared to the control. In addition, these seeds contain fructose and glucose, which can be utilized by various industries to produce items such as juices, sweets, soft drinks, and pastries.

Jackfruit

Jackfruit seed and jackfruit seed starch meal present suitable technological properties as an alternative nonconventional starch source (Solanki et al. 2020). This starch can be classified as a meal with a high starch content and used as an ingredient with both nutritional value and gelling properties (Carpiné et al. 2020).

Tamarind and Rambutan

Tamarind seed contains a high content of starch, which can be extracted as recommended by Maninder et al. (2016). This starch has good pasting properties, so it can be used as an emulsifier, stabilizer, and thickening ingredient.

With respect to rambutan seed, its flour has good potential as a thickening agent (Phanthanapratet et al. 2012). This characteristic was increased by treating the flour with dilute alkali (Eiamwat et al. 2016). Additionally, a protein concentrate high in albumins with good solubility at low pH and maximum emulsification and foaming capacity at pH 12 has been obtained (Vuong et al. 2016).

Mango

Kaur et al. (2022) reported the use of mango kernel oil in chocolate processing, being able to replace up to 80.0% of cocoa butter, with improved oxidative stability and increased antioxidant capacity of the final product. This same oil has been used in the processing of gouda cheese, replacing up to 3.5% of milk fat (Khan et al. 2018), increasing the phenolic and flavonoid content in the processed cheeses, and thus improving the oxidative stability and sensory quality. An advantage of using mango kernel oil is that it is not necessary to hydrogenate it before using it (Nadeem et al. 2016). Mango kernel meal has been used for cake processing, where an increase in energy content was observed compared to the control. Panelists preferred cakes with 20% of this meal, and it was observed that cakes could be stored for up to 10 days without the need for preservatives (Das et al. 2019). Likewise, this meal was used for cookie processing, with an improvement in antioxidant capacity (Kaur & Brar, 2017).

Guava

Guava seed meal has been used to make cookies by supplementing wheat flour. An improvement in the volume, diameter, and thickness of the cookies, and a decrease in total carbohydrates, together with an increase in iron content were observed (El-Din & Yassen, 1997). Likewise, this meal has been used to prepare paste. Supplementing semolina with 10.0 and 20.0% of this ingredient, a greater paste volume was observed, increasing the nutritional value by increasing the protein, fat, moisture, ash, and crude fiber content and decreasing the percentage of carbohydrates (Pérez-Rocha et al. 2015).

With respect to guava seed oil, its composition is similar to safflower oil, and it is easily bleached, so it could be used as a substitute for this commercial item (Kumar et al. 2022b).

Grape

Grape seed is the most commonly studied since it is obtained in large quantities during wine production. The ethanol extract has been used in roasted meats and sausages due to its high antioxidant activity that prevents the production of polycyclic aromatic hydrocarbons (Chen et al.

2020b), acrylamide, 5-hydroxymethylfurfural, and other harmful compounds for health (Qi, 2018). Grape seed extract has been reported to prevent fat oxidation and improve the quality of frozen meatballs and tilapia fillets during storage (Zhao et al. 2019). This extract has been used in the preparation of pastries, cakes, and steamed cheeses, improving its structure, specific volume, sensory quality, and antioxidant capacity. The incorporation of grape extract in cheeses improved the inhibitory activity of the angiotensin-converting enzyme (Chen et al. 2020b).

On the other hand, Wu et al. (2013) treated mushrooms with grape seed extract before drying and observed that they maintained their sensory quality with a reduced rate of mass loss and inhibition of bacterial growth. Grape seed extract is used in the formulation and marketing of functional beverages, such as oral liquids and pastes (Wu et al. 2015). Another potential application is as a nitrite substitute in sausage processing. The substitution of nitrates and nitrites in the production of fermented Italian sausages (Pini et al. 2020) and Western-style smoked sausages (Zhou et al. 2020) has been investigated, and inhibition of nitrosamine formation was observed, providing a new method for developing low-nitrate products.

Passion Fruit and Dates

Passion fruit seed is not commonly used, but it possesses high antioxidant activity (Da Costa et al. 2023). It has been used to obtain milk drinks fortified with linseed oil. It was observed that the addition of the polyphenolic extract of passion fruit seed increased the antioxidant activity and strongly reduced lipid oxidation during storage of pasteurized and sterilized beverages (Rotta et al. 2020).

Another nonconventional seed is date kernel, which possesses interesting nutraceutical properties. Date kernel has been used as a functional ingredient in model meat products. The antioxidant extract from this seed improved the oxidative stability of ground meat by reducing the TBARS (Thiobarbituric acid reactive substances) values, like that containing BHT (Butylated Hydroxytoluene) (Amany, Shaker, Abeer, 2012). Date kernel has been also used in bread products such as muffins and pita bread, obtaining products with high fiber content, polyphenols, and antioxidant activity (Mrabet et al. 2016). Using date kernel protein hydrolysate, bread with inhibitory ACE (Angiotensin Converting Enzyme) activity can be obtained (Ambigaipalan & Shahidi, 2015). In addition, date kernel powder is used as a coffee substitute in a commercially available product (Rahman et al. 2007).

Peach, Pomegranate, and Rambutan

Peach and pomegranate seed flours were used to produce cakes. Results showed that pomegranate seed flour produced softer cakes. Adding 10% peach flour produced sensorally acceptable cakes with higher protein content (Ağirbaş, Yavuz-Düzgün & Özçelik, 2022).

With respect to rambutan seed, it is reported that its lipid fraction can be used as a 30% replacement for cocoa fat in chocolate production, as it has a similar oleic acid content (Febrianto et al. 2014).

Pumpkin

Pumpkin seed is commonly considered a waste, but it can be used for oil extraction, with great demand in pharmaceutical applications (Bardaa et al. 2016). Oleic fatty acid (46.9%) and linolenic fatty acid (40.5%) are the main residual oil components. Also, raw pumpkin seedcake is rich in proteins (25.0–51.0%) (Rezig et al. 2016). Cucurbitin is the major storage protein with antifungal activity against seven species (**Table 4.1**) of phytopathogenic molds (Vassiliou et al.1998). In addition, the protein isolated from pumpkin seed that has been hydrolyzed with different proteases individually, or sequentially by combinations of at least two proteases, has shown antioxidant and antihypertensive effects (Ozuna & León-Galván, 2017).

4.4 HEALTH-PROMOTING POTENTIAL

It is interesting to mention that all seeds considered in this report have a great potential for promoting health, as they contain compounds with remarkable biological activities. Some of these materials are summarized in **Table 4.2.**

TABLE 4.2 BIOLOGICAL ACTIVITIES IDENTIFIED IN FRUIT SEEDS

Bioactivity	Fruit Seeds	Compounds Involved	Main Findings and Mechanism	References
Anti-inflammatory	Mango	Phenolic compounds	Balance pro-inflammatory and anti-inflammatory cytokines	Gupta et al. (2022)
	Guava	Nonstarch polysaccharides	Decrease the M1/M2 ratio	Lin & Lin, (2020)
	Grape	Extract	Inhibit COX-2 and inflammatory cytokine	Chen et al. (2020b)
	Date	Powder	Reduces the expression of IL-1b, TGF-b, COX-1 and COX-2	Saryono et al. (2020)
Antioxidant	Apple	Phenolic compounds	ABTS assay, 300 mg Trolox/100 g, and 53 mg Trolox/100 g of DPPH	Kumar et al. (2022a)
	Guava	Phenolic compounds	DPPH assay 81-136 mmol Trolox/100 g	Castro-Vargas et al. (2010)
	Tamarind	Extract	64.5–71.7% of antioxidant activity	Pal & Mukherjee, (2020)
	Papaya	Extract	Reduces radicals DPPH, ABTS, NO_2, and ferric ions	Agada et al. (2021)
Anti-cancerous	Mango	Polyphenols	Effect against breast and colon cancer	Mahmoud & Sharaky, (2021)
	Apple	Coumarins	Cytotoxic potential against breast cancer, mammary gland cancer and cervix cancer	Mustafa et al. (2020)
	Guava	Nonstarch poly- saccharides	Effects on prostate cancer and breast cancer	Kumar et al. (2022b)
	Grape	Proanthocyanidins	Effect on various types of cancer	Chen et al. (2020b)

TABLE 4.2 (CONTINUED)

Bioactivity	Fruit Seeds	Compounds Involved	Main Findings and Mechanism	References
	Tamarind	Extract	Positive effects on renal cell carcinoma	Pal & Mukherjee, (2020)
	Litchi	Extract	Inhibited the growth of cancer cells	Punia & Kumar, (2021)
	Apricot	Extract	Antiproliferation against HepG2 cells	Chen et al. (2020a)
Anti-microbial and anti-fungal bioactivity	Mango	Gallic and ellagic acids, gallates and stable fats rich in saturated fatty acids	Acts against *Candida albicans, E. coli, S. aureus,* and *Vibrio vulnificus*	Gupta et al. (2022)
	Apple	Extract	Activity against *K. pneumonia P. aeruginosa, E. coli,* and *H. influenzae*	Kumar et al. (2022a)
	Rambutan	Monoterpene, Kaempferol	Inhibition against *S. aureus, S. pyogenes, B. subtilis, E. coli.* and *P. aeruginosa*	Bhat & Al-daihan, (2014)
	Grape	Proanthocyanidins	Inhibition against *L. monocytogenes, Salmonella typhimurium,* and *E. coli*	Chen et al. (2020b)
	Passion fruit	Proteins	Fungicidal effects on Aspergillus and Fusarium *spp/*	Ng et al. (2011)
Lipid lowering effect and anti-obesity bioactivity	Apple	Phenolic compounds and dietary fiber	Lower cholesterol levels	Kumar et al. (2022a)
	Grape	Proanthocyanidins	Reduce cholesterol synthesis in the liver	Chen et al. (2020b)
	Citrus	Oil	Reduces elevated levels of triglycerides, total cholesterol, serum LDL cholesterol, and VLDL cholesterol	Chilaka, Ifediba, & Ogamba, (2015)
	Açaí	Extract	In combination with rosuvastatin reduced body weight, improved lipid profile, and improved cardiovascular remodeling	De Moraes Arnoso et al. (2022)

(continued)

TABLE 4.2 (CONTINUED)

Bioactivity	Fruit Seeds	Compounds Involved	Main Findings and Mechanism	References
Anti-diabetic bioactivity	Mango	Dietary fiber, phenolic acid, ellagic acid, and flavonoid	Inhibit the activity of diabetes α-amylase and α-glucosidase	Lasano et al. (2019)
	Apple	Epicatechin and catechin.	Inhibit the activity of α-glucosidase	Ci et al. (2018)
	Guava	Phenolic compounds	Stimulate insulin secretion and reduce glucose uptake	Kumar et al. (2022b)
	Tamarind	Extract	Decreases hyperglycemia	Pal & Mukherjee, (2020)
	Papaya	Extract	Inhibitory potential against a-amylase and intestinal a-glucosidase	Agada et al. (2021)

Reports of the biological activities observed in waste seeds have been evaluated *in vitro* studies; and in some cases, *in vivo* tests, either using experimental animal models or volunteers. The most studied seeds are apple, mango, guava, and grape since many biological properties have been attributed to them. Other seeds, despite their high nutraceutical and health potential, such as rambutan, raspberry, passion fruit, litchi, peach, and açai, have very few reports in the literature.

The main activities evaluated in fruit seeds include anti-inflammatory, antioxidant, anticancer, antiobesity, and hypolipidemic properties, which are mainly attributed to the presence of phenolic compounds, dietary fiber, and polysaccharides, among others. Additionally, and as expected, these seeds exhibit antimicrobial activities against both gram-positive and gram-negative bacteria, as well as antifungal effects.

Although limited information is available so far, isolated compounds from fruit seeds show potential for immunomodulation and neuroprotection against Alzheimer's and Parkinson's and other common diseases.

4.5 CONCLUSION

The agri-food industry is responsible for generating a significant amount of waste, particularly during the processing of fruits where skins, leaves, stems, and mainly seeds are often discarded. However, these wastes in general, but seeds in particular, contain an abundance of bioactive compounds such as proteins, fiber, fatty acids, phenolic compounds, carotenoids, vitamins, and minerals. Discarded wastes possess functional properties such as water-holding capacity, oil-absorption capacity, emulsifying, foaming, and foaming stability. By incorporating these residues into the production of functional food products and nutraceuticals, they can be converted into valuable resources that not only improve economic sustainability but also reduces negative impacts to the environment. Interestingly, and as underlined above, these compounds have been shown to exhibit outstanding bioactive properties against chronic-degenerative diseases such as cancer and diabetes, among others. The high potential benefits of food wastes show in a very clear way, the necessity of intensifying the basic and applied research over the scientific fields described in the previous sections of this chapter; results of such research activities will go directly to the well-being of human societies.

REFERENCES

Abbas, A., Anwar, F., & Ahmad, A. (2019). Apple (*Malus pumila*) seed oil. In *Fruit oils: Chemistry and Functionality*, M. Ramadan, Ed. pp. 495–504. Springer Nature. https://doi.org/10.1007/978-3-030-12473-1_24

Abdalla, A.E., Darwish, S.M., Ayad, E.H., & El-Hamahmy, R.M. (2007). Egyptian mango by-product 1. Compositional quality of mango seed kernel. *Food Chemistry*, 103(4), 1134–1140. https://doi.org/10.1016/j.foodchem.2006.10.017

Agada, R., Thagriki, D., Lydia, D.E., Khusro, A., Alkahtani, J., Shaqha, M.M., Alwahibi, M.S., & Elshikh, M.S. (2021). Antioxidant and anti-diabetic activities of bioactive fractions of Carica papaya seeds extract. *Journal of King Saud University-Science*, 33, 101342. https://doi.org/10.1016/j.jksus.2021.101342

Ağirbaş, H., Yavuz-Düzgün, M., & Özçelik, B. (2022). Valorization of fruit seed flours: Rhelogical characteristics of composite dough and cake quality. *Journal of Food Measurement and Characterization*, 16, 3117–3129. https://doi.org/10.1007/s11694-022-01423-0

Amany, M., Shaker, M.A., & Abeer, A. (2012). Antioxidant activities of date pits in a model meat system. *International Food Research Journal*, 19(1), 223–227.

Ambigaipalan, P., & Shahidi, F. (2015). Date seed flour and hydrolysates affect physicochemical properties of muffin. *Food Bioscience*, 12, 54–60. https://doi.org/10.1016/j.fbio.2015.06.001

Bangar, S., Kumar, M., Whiteside, W., Tomar, M., & Kennedy, J. (2021). Litchi (Litchi chinensis) seed starch: Structure, properties, and applications- A review. *Carbohydrate Polymer Technologies and Applications*, 2, 100080. https://doi.org/10.1016/j.carpta.2021.100080

Bardaa, S., Ben Halima, N., Aloui, F., Ben Mansour, R., Jabeur, H., Bouaziz, M., & Sahnoun, Z. (2016). Oil from pumpkin (*Cucurbita pepo* L.) seeds: Evaluation of its functional properties on wound healing in rats. *Lipids in Health and Disease*, 15(1), 1–12. https://doi.org/10.1186/s12944-016-0237-0

Bhat, R.S., & Al-daihan, S. (2014). Antimicrobial activity of Litchi chinensis and Nephelium lappaceum aqueous seed extracts against some pathogenic bacterial strains. *Journal of King Saud University-Science*, 26(1), 79–82. https://doi.org/10.1016/j.jksus.2013.05.007

Carpiné, D., Dagostin, J.L.A., Mazon, E., Barbi, R.C.T., da Silva, A.F.E., Chaimsohn, F.P., & Hoffmann, R.R. (2020). Valorization of Euterpe edulis Mart. agroindustrial residues (pomace and seeds) as sources of unconventional starch and bioactive compounds. *Journal of Food Science*, 85(1), 94–104. https://doi.org/10.1111/1750-3841.14978

Castro-Vargas, H.I., Rodríguez-Varela, L.I., Ferreira, S.R.S., & Parada-Alfonso, F. (2010). Extraction of phenolic fraction from guava seeds (Psidium guajava L.) using supercritical carbon dioxide and co-solvents. *Journal of Supercritical Fluids*, 51(3), 319–324. https://doi.org/10.1016/j.supflu.2009.10.012

Chaves, J.L.F., Dias, R.E.G., Barboza, T.T., Fitz, V.A. Godoy, C.A., Toci, A.T. Masson, M.L., & Reis, F.R. (2023). Drying, physical, nutritional and bioactive characteristics of apple peel flour subjected to blanching and ultrasound pretreatments. *Plant Foods for Human Nutrition*, 78(4), 704–709.https://doi.org/10.1007/s11130-023-01110-5

Chen, Y., Al-Ghamdi, A.A., Elshikh, M.S., Shah, M.H., Al-Dosary, M.A., & Abbasi, A.M. (2020a). Phytochemical profiling, antioxidant and HepG2 cancer cells' antiproliferation potential in the kernels of apricot cultivars. *Saudi Journal of Biological Sciences*, 27(1), 163–172. https://doi.org/10.1016/j.sjbs.2019.06.013

Chen, Y., Wen, J., Deng, Z., Pan, X., Xie, X., & Peng, C. (2020b). Effective utilization of food waste: Bioactivity of grape seed extraction and its application in food industry. *Journal of Functional Foods*, 73, 104113. https://doi.org/10.1016/j.jff.2020.104113

Chilaka, K.C., Ifediba, E.C., & Ogamba, J.O. (2015). Evaluation of the effects of Citrus sinensis seed oil on blood glucose, lipid profile and liver enzymes in rats injected with alloxan monohydrate. *Journal of Acute Disease*, 4(2), 129–134. https://doi.org/ 10.1016/s2221-6189(15)30022-6

Ci, Z., Kikuchi, K., Hatsuzawa, A., Nakai, A., Jiang, C., Itadani, A., & Kojima, M. (2018). Antioxidant activity, and α-glucosidase, α-amylase and lipase inhibitory activity of polyphenols in flesh, peel, core and seed from mini apple. *Advance Journal of Food Science and Technology*, 6(6), 258–262. https://doi.org/10.12691/ajfst-6-6-5

Da Costa, C.A.R., Machado, G.G.L., Rodrigues, L.J., De Barros, H.E.A, Natarelli, C.V.L., & Boas, E.V.D.B.V. (2023). Phenolic compounds profile and antioxidant activity of purple passion fruit's pulp, peel and seed at different maturation stages. *Scientia Horticulturae*, 321, 112244. https://doi.org/10.1016/j.scienta.2023.112244

Da Porto, C., Porretto, E., & Decorti, D. (2013). Comparison of ultrasound-assisted extraction with conventional extraction methods of oil and polyphenols from grape (*Vitis vinifera* L.) seeds. *Journal of Ultrasonic Sonochemistry*, 20, 1076–1080. http://dx.doi.org/10.1016/j.ultsonch.2012.12.002

Das, P.C., Khan, M.J., Rahman, M.S., Majumder, S., & Islam, M.N. (2019). Comparison of the physicochemical and functional properties of mango kernel flour with wheat flour and development of mango kernel flour based composite cakes. *NFS Journal*, 17, 1–7. https://doi.org/10.1016/j.nfs.2019.10.001

De Moraes Arnoso, B., Magliaccio, F.M., de Araújo, C.A., Soares, R., Barcellos, I., Freitas, G., Fernandes-Santos, C., Teixeira, D., Soares, R., Castro, A., Beltrame, J., & Aguiar da Costa, C. (2022). Açaí seed extract (ASE) rich in proanthocyanidins improves cardiovascular remodeling by increasing antioxidant response in obese high-fat diet-fed mice. *Chemico-Biological Interactions*, 351, 109721. https://doi.org/10.1016/j.cbi.2021.109721

Dos Santos, L.C., Bitencourt, R.G., dos Santos, P., Vieira e Rosa, P., & Martínez, J. (2019). Solubility of passion fruit (*Passiflora edulis* Sims) seed oil in supercritical CO_2. *Journal of Fluid Phase Equilibria*, 493, 174–180. https://doi.org/10.1016/j.fluid.2019.04.002

Dotto, J., & Abihudi, S. (2021). Nutraceutical value of Carica papaya: A review. *Scientific African*, 13, e00933. https://doi.org/10.1016/j.sciaf.2021.e00933

Eiamwat, J., Wanlapa, S., & Kampruengdet, S. (2016). Physicochemical properties of defatted rambutan (Nephelium lappaceum) seed flour after alkaline treatment. *Molecules*, 21(4), 364. https://doi.org/10.3390/molecules21040364

El-Din, M.H.A.S., & Yassen, A.A.E. (1997). Evaluation and utilization of guava seed meal (*Psidium guajava* L.) in cookies preparation as wheat flour substitute. *Nahrung*, 41(6), 344–348. https://doi.org/10.1002/food.19970410605

FAO/WHO. (2007). Protein and amino acid requirements in human nutrition. Report of a joint FAO/WHO/UNU expert consultation (WHO Technical Report Series 935). World Health Organization, Food and Agriculture Organization of the United Nations, United Nations University.

Febrianto, N.A., Issara, U., Yang, T.A., & Wan, A.W.N. (2014). Thermal behavior, microstructure, and texture properties of fermented-roasted rambutan seed fat and cocoa butter mixtures. *Pelita Perkebunan*, 30(1), 65–79.

Fidelis, M., de Moura, C., Junior, T.K., Pap, N., Mattila, P., Makinen, S., Putnik, P., Kovacevic, D.B., Tian, Y., Yang, B., & Granato, D. (2019). Fruit seeds as sources of bioactive compounds: Sustainable production of high value-added ingredients from by-products within circular economy. *Molecules*, 24(21), 3854. https://doi.org/10.3390/molecules24213854

Górnaś, P., Rudzińska, M., & Segliņa, D. (2014). Lipophilic composition of eleven apple seed oils: A promising source of unconventional oil from industry by-products. *Industrial Crops and Products*, 60, 86–91. https://doi.org/10.1016/j.indcrop.2014.06.003

Gupta, A.K., Gurjar, P.S., Beer, K., Pongener, A., Ravi, S.C., Singh, S., Verma, A., Singh, A., Thakur, M., Tripathy, S., & Verma, D.K. (2022). A review on valorization of different byproducts of mango (*Mangifera indica* L.) for functional food and human health. *Food Bioscience*, 48, 101783. https://doi.org/10.1016/j.fbio.2022.101783

Jahurul, M.H.A., Azzatul, F.S., Sharifudin, M.S., Norliza, M.J. Hasmadi, M., Lee, J.S., Patricia, M., Jinap, S., Ramlah, M.R. Firoz, M., & Zaidul, I.S.M. (2020). Functional and nutritional properties of rambutan (*Nephelium lappaceum* L.) seed and its industrial application: A review. *Trends in Food Science & Technology,* 99, 367–374. https://doi.org/10.1016/j.tifs.2020.03.016

Kaur, A., & Brar, J.K. (2017). Use of mango seed kernels for the development of antioxidant rich biscuits. *International Journal of Science and Research,* 78(96), 2319–7064.

Kaur, G., Kaur, D., Kansal, S.K., Garg, M., & Krishania, M. (2022). Potential cocoa butter substitute derived from mango seed kernel. *Food Chemistry,* 372, 131244. https://doi.org/10.1016/j.foodchem.2021.131244

Khan, I.T., Nadeem, M., Imran, M., Ajmal, M., & Ali, S. (2018). Antioxidant activity, fatty acids characterization and oxidative stability of Gouda cheese fortified with mango (Mangifera indica L.) kernel fat. *Journal of Food Science and Technology,* 55(3), 992–1002. https://doi.org/10.1007/s13197-017-3012-y

Kumar, M., Barbhai, M.D., Esatbeyoglu, T., Zhang, B., Sheri, V., Dhumal, S., Rais, N., Radha, et al. (2022a). Apple (*Malus domestica* Borkh.) seed: A review on health promoting bioactivities and its application as functional food ingredient. *Food Bioscience,* 50, 102155. https://doi.org/10.1016/j.fbio.2022.102155

Kumar, M., Kapoor, S., Dhumal, S., Tkaczewska, J., Changan, S., Saurabh, V., Mekhemar, M., Radha Rais, N., Satankar, V., Pandiselvam, R., Sayed, A., Senapathy, M., Anitha, T., Singh, S., Tomar, M., Dey, A., Zengin, G., Amarowicz, R., & Bhyan, D.J. (2022b). Guava (*Psidium guajava* L.) seed: A low-volume, high-value by product for human health and the food industry. *Food Chemistry,* 386, 132694. https://doi.org/10.1016/j.foodchem.2022.132694

Lasano, N.F., Hamid, A.H., Karim, R., Dek, M.S.P., Shukri, R., & Ramli, N.S. (2019). Nutritional composition, anti-diabetic properties and identification of active compounds using UHPLC-ESI-orbitrap-MS/MS in *Mangifera odorata* L. peel and seed kernel. *Molecules*, 24(2), 320. https://doi.org/10.3390/molecules24020320

Lebaka, V., Wee, Y., Ye, W., & Korivi, M. (2021). Nutritional composition and bioactive compounds in three different parts of mango fruit. *International Journal of Environmental Research and Public Health,* 18(741), 1–20. https://doi.org/10.3390/ijerph18020741

Lin, H.C., & Lin, J.Y. (2020). M1 polarization but anti-LPS-induced inflammation and anti-MCF-7 breast cancer cell growth effects of five selected polysaccharides. Evidence-based. *Complementary and Alternative Medicine,* 2020, 1–17. https://doi.org/10.1155/2020/9450246

Mahmoud, M., & Sharaky, M.M. (2021). Compositional quality of mangoes by-product and its anticancer activity in human cancer cell lines: An *in vitro* study. *Egyptian Journal of Food Science,* 49(1), 65–79. https://doi.org/10.21608/ejfs.2021.60913.1094

Maldonado-Celis, M.E., Yahia, E.M., Bedoya, R., Landázuri, P., Loango, N., Aguillón, J., Restrepo, B., & Guerrero-Ospina, J.C. (2019). Chemical composition of mango (*Mangifera indica* L.) fruit: Nutritional and phytochemical compounds. *Frontiers of Plant Science,* 10(1073), 1–21 https://doi.org/10.3389/fpls.2019.01073

Maldonado-Torres, R., Morales-Camacho, J.I., López-Valdez, F., Huerta-González, L., & Luna-Suárez, S. (2020). Assessment of techno-functional and nutraceutical potential of tomato (*Solanum lycopersicum*) seed meal. *Molecules,* 25(18), 4235. https://doi.org/10.3390/molecules25184235

Maninder, K., & Gaganpreet, K.B. (2016). Partial characterization of tamarind (*Tamarindus indica* L.) kernel starch oxidized at different levels of sodium hypochlorite. *International Journal of Food Properties,* 19(3), 605–617. https://doi.org/10.1080/10942912.2015.1038722

Manzoor, M., Singh, J., & Gani, A. (2021). Assessment of physical, microstructural, thermal, techno-functional and rheological characteristics of apple (*Malus domestica*) seeds of Northern Himalayas. *Scientific Reports,* 11(1), 1–10. https://doi.org/ 10.1038/s41598-021-02143-z

Maqsood, S., Adiamo, O., Ahmad, M., & Mudgil, P. (2020). Bioactive compounds from date fruit and seed as potential nutraceutical and functional food ingredients. *Food Chemistry,* 308, 125522. https://doi.org/10.1016/j.foodchem.2019.125522

Marić, B., Abramović, B., Ilić, N., Krulj, J., Kojić, J., Perović, J., Bodroža-Solarov, M., & Teslić, N. (2020). Valorization of red raspberry (*Rubus idaeus* L.) seeds as a source of health beneficial compounds: Extraction by different methods. *Journal of Food Processing and Preservation,* 44(10), ce14744. https://doi.org/10.1111/jfpp.14744

Martins, C.M., Carvalho, J.A., Sousa, E., & Salcador, S.R. (2022). Valorization of tamarind seeds using high-pressure extraction methods to obtain rich fractions in fatty acid and phenolic compounds. *The Journal of Supercritical Fluids*, 183, 1–10. https://doi.org/10.1016/j.supflu.2022.105556

Mazurek, B., Chmiel, M., & Górecka, B. (2017). Fatty acids analysis using gas chromatography-mass spectrometer detector (GC/MSD)-method validation based on berry seed extract samples. *Food Analytical Methods*, 10, 2868–2880. https://doi.org/10.1007/s12161-017-0834-1

Melo, P., Selani, M.M., Gonçalves, R.H., Olivera, J., Massarioli, A., & Matias de Alencar, S. (2021). Açaí seeds: An unexplored agro-industrial residue as a potential source of lipids, fibers, and antioxidant phenolic compounds. *Journal of Industrial Crops & Products*, 161, 113204.

Mrabet, A., Jiménez-Araujo, A., Fernández-Bolaños, J., Rubio-Senent, F., Lama-Muñoz, A., Sindic, M., & Rodríguez-Gutiérrez, G. (2016). Antioxidant phenolic extracts obtained from secondary tunisian date varieties (*Phoenix dactylifera* l.) by hydrothermal treatments. *Food Chemistry*, 196, 917–924. https://doi.org/10.1016/j.foodchem.2015.10.026

Mustafa, M.A.-M., Sorour, M.A.-H., Mehanni, A.-H.E.-S., & Hussien, S.M. (2023). Amino acid profile, physico-chemical properties and fatty acids composition of some fruit seed kernels after detoxification. *Chemical and Biological Technologies in Agriculture*, 10(1), 37. 10.1186/s40538-023-00412-9

Mustafa, Y.F., Khalil, R.R., & Mohammed, E.T. (2020). Antimicrobial activity of aqueous extracts acquired from the seeds of two apples' cultivars. *Systematic Reviews in Pharmacy*, 11(2), 382–387. https://doi.org/10.5530/srp.2020.2.56

Nadeem, M., Imran, M., & Khalique, A. (2016). Promising features of mango (Mangifera indica L.) kernel oil: A review. *Journal of Food Science and Technology*, 53(5), 2185–2195. https://doi.org/10.1007/s13197-015-2166-8

Ng, T., Lam, S., Cheung, R.C.F., Wong, J.H., Wang, H., Ngai, P., Ye, X., Fang, E., & Chan, Y. (2011). Antifungal protein from Passion Fruit (Passiflora edulis) seeds. In *Nuts and Seeds in Health*, V.R. Preedy, R.R. Watson & V.B. Patel, Ed. 865–871. Academic Press. https://doi.org/10.1016/B978-0-12-375688-6.10102-1

Ourradi, H., Ennahli, S., Viuda, M., Hernadez, F., Dilorenzo, C., Hssaini, L., Elantari, A., & Hanine, H. (2021). Proximate composition of polyphenolic, phytochemical, antioxidant activity content and lipid profiles of date palm seeds oils (*Phoenix dactylifera* L.). *Journal of Agriculture and Food Research*, 6, 100217. https://doi.org/10.1016/j.jafr.2021.100217

Ozuna, C., & León-Galván, M. (2017). Cucurbitaceae seed protein hydrolysates as a potential source of bioactive peptides with functional properties. *BioMed Research International*, 2017, 1–16. https://doi.org/10.1155/2017/2121878

Pal, D., & Mukherjee, S. (2020). Tamarind (*Tamarindus indica*) seeds in health and nutrition. In *Nuts and Seeds in Health and Disease Prevention*, V.R. Preedy & R.R. Watson, Ed. 171–182. Academic Press. https://doi.org/10.1016/B978-0-12-818553-7.00014-0

Pathania, S., & Kaur, N. (2022). Utilization of fruits and vegetable by-products for isolation of dietary fibres and its potential application as functional ingredients. *Bioactive Carbohydrates and Dietary Fibre* 27, 100295. https://doi.org/10.1016/j.bcdf.2021.100295

Perez-Rocha, K.A., Guemes-Vera, N., Bernardino-Nicanor, A., Gonzalez-Cruz, L., Hernandez-Uribe, J.P., & Sanchez, A.T. (2015). Fortification of white bread with guava seed protein isolate. *Pakistan Journal of Nutrition*, 14(11),828–833. https://doi.org/10.3923/pjn.2015.828.833

Phanthanapratet, W., Trangwacharakul, S., Wanlapa, S., Eiamwat, J., Boonkosum, P., & Lauhakul, K. (2012). Application of rambutan seed flour for low calorie salad dressing product. *Agricultural Sciences*, 43(2), S517–S520.

Pieszka, M., Tombarkiewicz, B., Roman, A., Migdal, W., & Niedziolka, J. (2013). Effect of bioactive substances found in rapeseed, raspberry and strawberry seed oils on blood lipid profile and selected parameters of oxidative status in rats. *Environmental Toxicology and Pharmacology*, 36(3), 1055–1062. https://doi.org/10.1016/j.etap.2013.09.007

Pini, F., Aquilani, C., Giovannetti, L., Viti, C., & Pugliese, C. (2020). Characterization of the microbial community composition in Italian Cinta Senese sausages dry-fermented with natural extracts as alternatives to sodium nitrite. *Food Microbiology*, 89, 103417. https://doi.org/10.1016/j.fm.2020.103417

Priyadarshi, R., Sauraj., Kumar, B., Deeba, F., Kulshreshtha, A., & Negi, Y.S. (2018). Chitosan films incorporated with Apricot (*Prunus armeniaca*) kernel essential oil as active food packaging material. *Journal of Food Hydrocolloids*, 85, 158–166. https://doi.org/10.1016/j.foodhyd.2018.07.00393

Punia, S., & Kumar, M. (2021). Litchi (*Litchi chinenis*) seed: Nutritional profile, bioactivities, and its industrial applications. *Trends in Food Science & Technology*, 108, 58–70. https://doi.org/10.1016/j.tifs.2020.12.005

Purić, M., Rabrenović, B., Rac, V., Pezo, L., Tomašević, I., & Demin, M. (2020). Application of defatted apple seed cakes as a by-product for the enrichment of wheat bread. *LWT–Food Science and Technology*, 130, 109391. https://doi.org/10.1016/j.lwt.2020.109391

Qi, Y., Zhang, H., Wu, G., Zhang, H., Gu, L., Wang, L., Qian, H., & Qi, X. (2018). Mitigation effects of proanthocyanidins with different structures on acrylamide formation in chemical and fried potato crisp models. *Food Chemistry*, 250(1), 98–104. https://doi.org/10.1016/j.foodchem.2018.01.012

Quintero-Angel, M., Martínez-Girón, J., & Orjuela-Salazar, S. (2022). Agroindustrial valorization of the pulp and peel, seed, flour, and oil of moriche (*Mauritia flexuosa*) from the Bita River, Colombia: a potential source of essential fatty acids. *Biomass Conversion and Biorefinery*, 12, 1–9. https://doi.org/10.1007/s13399-022-02298-3

Rahman, M., Kasapis, S., Al-Kharusi, N., Al-Marhubi, I., & Khan, A. (2007). Composition characterisation and thermal transition of date pits powders. *Journal of Food Engineering*, 80(1), 1–10. https://doi.org/10.1016/j.jfoodeng.2006.04.030

Rezig, L., Riaublanc, A., Chouaibi, M., Guéguen, J., & Hamdi, S. (2016). Functional properties of protein fractions obtained from pumpkin (*Cucurbita maxima*) seed. *International Journal of Food Properties*, 19(1), 172–186. https://doi.org/10.1080/10942912.2015.1020433

Rodrigues, R.B., Lichtenthaler, R., Zimmermann, B.F., Papagiannopoulos, M., Fabricius, H., & Marx, F. (2006). Total oxidant scavenging capacity of *Euterpe oleracea* Mart. (Açaí) seeds and identification of their polyphenolic compounds. *Journal of Agricultural and Food Chemistry*, 54(12), 4162–4167. https://doi.org/10.1021/jf058169p

Rotta, E.M., da Silva, M.C., & Visentainer, J.V. (2018). Ultrasound-assisted saponification coupled with gas chromatography-flame ionization detection for the determination of phytosterols from passion fruit seed oil. *Journal of the Brazilian Chemical Society*, 29, 579–586. https://doi.org/10.21577/0103-5053.20170170

Rotta, E.M., Giroux, H.J., Lamothe, S., Bélanger, D., Sabik, H., Visentainer, J.V., & Britten, M. (2020). Use of passion fruit seed extract (Passiflora edulis Sims) to prevent lipid oxidation in dairy beverages during storage and simulated digestion. *LWT–Food Science and Technology* 123, 109088. https://doi.org/10.1016/j.lwt.2020.109088

Saryono, W., Isworo, A., & Sarmoko. (2020). Anti-inflammatory activity of date palm seed by downregulating interleukin-1β, TGF-β, cyclooxygenase-1 and -2: A study among middle age women. *Saudi Pharmaceutical Journal*, 28(8), 1014–1018. https://doi.org/10.1016/j.jsps.2020.06.024

Shao, D., Atungulu, G.G., Pan, Z., Yue, T., Zhang, A., & Fan, Z. (2014). Characteristics of isolation and functionality of protein from tomato pomace produced with different industrial processing methods. *Food and Bioprocess Technology*, 7(2), 532–541. https://doi.org/10.1007/s11947-013-1057-0

Sogi, D.S., Sidhu, J.S., Arora, M.S., Garg, S.K., & Bawa, A.S. (2002). Effect of tomato seed meal supplementation on the dough and bread characteristics of wheat (PBW 343) flour. *International Journal of Food Properties*, 5(3), 563–571. https://doi.org/10.1081/JFP-120015492

Solanki, S., Bhardwaj, R., Vasudeva, R., Chourey, S., & Archak, S. (2020). Biochemical composition of pulp and seed of wild jack (*Artocarpus hirsutus* Lam.) fruit. *Plant Foods for Human Nutrition*, 75, 659–660. https://doi.org/10.1007/s11130-020-00849-5

Szabo, K., Cătoi, A.F., & Vodnar, D.C. (2018). Bioactive compounds extracted from tomato processing by-products as a source of valuable nutrients. *Plant Foods for Human Nutrition*, 73, 268–277. https://doi.org/10.1007/s11130-018-0691-0

Vassiliou, A.G., Neumann, G.M., Condron, R., & Polya, G.M. (1998). Purification and mass spectrometry-assisted sequencing of basic antifungal proteins from seeds of pumpkin (*Cucurbita maxima*). *Plant Science*, 134(2), 141–162. https://doi.org/10.1016/S0168-9452(98)00052-1

Vuong, H.T.H., Tran, N.M.C., Tran, T.T.T., Ton, N.M.N., & Le, V.V.M. (2016). Effects of pH and salt concentration on functional properties of rambutan (*Nephelium lappaceum* L.) seed albumin concentrate. *International Journal of Food Science and Technology*, 51, 1212–1219. https://doi.org/10.1111/ijfs.13087

Walia, M., Rawat, K., Bhushan, S., Padwad, Y.S., & Singh, B. (2014). Fatty acid composition, physicochemical properties, antioxidant, and cytotoxic activity of apple seed oil obtained from apple pomace. *Journal of the Science of Food and Agriculture*, 94(5), 929–934. https://doi.org/10.1002/jsfa.6337

Wu, N., Jin, C., Huang, J., Song, J., Zhang, L., Jiang, T., & Li, J. (2013). Effects of grape seed extract on the preservation effect of lentinus edodes. *Food Science and Technology*, 34(08), 299–302. https://doi.org/10.3390/foods9050611

Wu, Z.M., Zheng, Y., Chen, L., & Liu, W. (2015). Development of grape seed polyphenol cake. *Food Research and Development*, 36(16), 93–96.

Yukui, R., Wenya, W., Rashid, F., & Qing, L. (2009). Fatty acids composition of apple and pear seed oils. *International Journal of Food Properties*, 12, 774–779. https://doi.org/10.1080/10942910802054320

Zayed, A., Badawy, M., & Farag, M. (2021). Valorization and extraction optimization of citrus seeds for food and functional food applications. *Food Chemistry*, 355, 129609. https://doi.org/10.1016/j.foodchem.2021.129609

Zhao, X., Zhou, Y.G., Zhao, L., Chen, L., He, Y., & Yang, H.S. (2019). Vacuum impregnation of fish gelatin combined with grape seed extract inhibits protein oxidation & degradation of chilled tilapia fillets. *Food Chemistry*, 294, 316–325. https://doi.org/10.1016/j.foodchem.2019.05.054

Zhou, Y., Wang, Q., & Wang, S. (2020). Effects of rosemary extract, grape seed extract and green tea polyphenol on the formation of N-nitrosamines and quality of western-style smoked sausage. *Journal of Food Processing and Preservation*, 44(6), e14459. https://doi.org/10.1111/jfpp.14459

Chapter 5

Magnetic Nanoparticles Coencapsulated to Peptides in Poly Lactic-*co*-Glycolic Acid (PLGA) as a Delivery System for Cancer Treatment

Carlos R. Díaz-Águila
Juana Angélica Ortega-Cárdenas
Viviana Sandoval-Flores
Rosalva Mora-Escobedo
Edmundo del Valle-Gallegos
and Hernani Yee-Madeira

CONTENTS

DOI: 10.1201/9781003332558-6

5.1 INTRODUCTION

Iron oxide nanoparticles (IONPs) are well known for their ability to be oriented under an external magnetic field and have been investigated for potential biomedical applications (Maity et al. 2010). This chapter evaluates degradable poly lactic-*co*-glycolic acid (PLGA) capsules loaded with magnetic nanoparticles and bioactive peptides (BP) from germinated soybean as potential peptide delivery systems for cancer treatment.

Cancer is a generic term for a large group of diseases that can affect any part of the body. One defining feature of cancer is the rapid creation of abnormal cells that grow beyond their usual boundaries, which can then invade adjoining parts of the body and spread out to other organs; the latter process is referred to as metastasis. Widespread metastases are the primary cause of death from cancer. Currently, cancer treatment relies on chemotherapy, endotherapy, immunotherapy, radiotherapy, and surgical interventions depending on the type of cancer (Ghaly et al. 2023). Despite the treatment options mentioned above, which aid significantly in extending the lifespan of individuals, they result in unbearable side effects such as alopecia, anemia, dementia, nausea, peripheral neuropathy, etc. (Farzanegan & Tahmasbi, 2023). Furthermore, these traditional chemotherapeutic regimens are becoming nugatory against cancer cells because of the development of resistance to chemotherapeutic agents in cancer (Bhutta & Choi, 2023). Therefore, there is an urgent need to search for alternatives to these traditional chemotherapeutic agents.

With this aim, different products from natural sources have been evaluated as anticancer components with potential clinical use. One example is germinated soybean protein isolates, whose effects are mainly attributed to their protein and peptide composition. These bioactive molecules can interfere with various processes critical to tumor cell development, including growth, proliferation, and cell survival (Mora-Escobedo et al. 2009; Subramaniam et al. 2019). In addition, González-Montoya et al. (2018) have shown that simulated gastrointestinal digestion increases the antiproliferative activity of germinated soybean protein isolates against colorectal, cervical, and breast cancer. The peptide composition of these cytotoxic isolates has also been identified and evaluated (Robles-Ramírez et al. 2012; González-Montoya et al. 2016).

The use of peptide-loaded capsules has emerged as an effective strategy to optimize drug and active molecule administration. This alternative is based on the unique ability of PLGA capsules to facilitate controlled and sustained release of loaded compounds. By enabling this gradual release, the effectiveness of medical treatments is enhanced, while mitigating potential side effects that could arise from a more abrupt release.

The use of peptide and protein delivery systems based on degradable PLGA capsules represents a promising solution to various challenges associated with conventional drug administration. These capsules offer the ability to precisely control the release of active ingredients, thereby allowing a more refined therapeutic response and minimizing potential adverse effects on patients. In this context, the versatility and efficacy of PLGA capsules stand out as a key tool in improving the quality and safety of modern medical treatments (Tabatabaei Mirakabad et al. 2014).

This explores the synergistic benefits derived from employing BP extracted from germinated soybean in conjunction with magnetic IONs encapsulated within PLGA spheres. This innovative combination serves as a potent delivery system for cancer treatment. Additionally, the chapter delves into the role of ferroptosis within this context, providing a comprehensive examination of its implications in the proposed therapeutic approach.

5.2 PROTEINS, PEPTIDES, AND ANTICANCER ACTIVITY

Proteins and peptides are the most abundant biomolecules within living organisms. They are involved in a wide variety of processes in the cell and present an enormous diversity of functions directly related to their structures.

A peptide is defined as a natural polymer of amino acids covalently bonded by a substituted amide bond, commonly known as a peptide bond. This bond is formed by dehydration of the α -carboxyl group of one amino acid and the amino group of another (Jakubczyk et al. 2020).

Peptide molecules with a molecular mass lower than 10 kDa are classified as oligo- and polypeptides, while above this molecular mass are called proteins.

The structure of proteins is a determinant of the function they exert and is studied by analyzing their levels of complexity, and structure order, according to an established hierarchy (Nelson & Cox, 2001) .

Tumor cells develop by escaping the immuno-vigilance mechanisms of the immune system; that is, failures must occur in the process that allows the immune system to keep malignant cells under control (Arum et al. 2010; Beg et al. 2013). This capacity in tumor cells is determined by their ability to change the properties of the cell membrane by varying its peptide composition and establishing immunosuppressive networks that operate locally and systemically (Motz & Coukos, 2011; Olson & McNeel, 2012). Neoplastic cells can express a set of polypeptides that differ from those corresponding to healthy cells and thus lead to a specific favorable immune response to their continuing development.

The new peptide profile on the cell surface allows the transformation of the tumor and the synthesis of other malignant proteins and influences gene expression at the cellular level (Overwijk et al. 2013; Wick et al. 2014), followed by an increase in the number of immunogenic polypeptides (Negrini et al. 2010).

The conjunction of these factors favors the adaptive responses of neoplastic cells. It triggers the expression of associated tumor antigens, proteins that differ qualitatively or quantitatively from those expressed in healthy cells of the same type (Coulie et al. 2014).

In recent decades, numerous immunological cancer treatments have been developed by utilizing tumor-associated antigens to trigger a beneficial adaptive immune response (Bonifaz et al. 2004; Kreutz et al. 2013). To achieve this, peptides and proteins, obtained either through recombinant methods or purification processes, are administered along with immune-stimulatory compounds known as adjuvants. (Hailemichael & Overwijk, 2013; Ricupito et al. 2013).

5.2.1 Soybean with Anticancer Activity

Soybean (*Glycine max*) is a dicotyledonous angiosperm plant of the legume family. The seeds are an important source of proteins and lipids for humans. Soybean protein is one of the least expensive sources of dietary proteins and is considered a good substitute for animal protein as soybean protein contains all the essential amino acids, including lysine, which is especially low in other staple crops (Agyenim-Boateng et al. 2023).

There are several nutrients and micronutrients present in the soybean with known anticancer properties; among them, many are proteins and peptides. The enzymatic digestion of soybean proteins releases biologically active peptides against different types of cancer due to the functions exerted by these proteins when they interact with different cell receptors. Antitumor peptides include protease inhibitor peptides, lectins, lunasin, and other low molecular-weight peptides (Barac et al. 2005; Lath et al. 2023).

One of the best-known protease inhibitors is the Bowman-Birk inhibitor (BBI) which is a low molecular weight protein of 8 kDa. Its primary sequence consists of 71 amino acids and 7 disulfide bridges that provide great structural rigidity. Its main action is the inhibition of the protease's trypsin and chymotrypsin through the active sites Lys 16-Ser 17 and Leu 43-Ser 44, which operate independently (Gitlin-Domagalska et al. 2020). The action of BBI as a suppressor of carcinogenesis, the process by which cancer occurs and develops, has been demonstrated (González de Mejía & Prisecaru, 2005) in multiple systems. Its anticancer activity is associated with the selective inhibitory action of serine proteases that are crucial in the growth processes of malignant cells (Sarkar & Bhaskara Rao, 2023). Protease enzymes play a crucial role in the mechanism of life-threatening diseases like cancer, malaria, and AIDS. Hence, blocking these enzymes with potential inhibitors can be an efficient approach in drug therapy for these diseases.

Kunitz trypsin inhibitor (KTI), which appears in soybean, is a peptide with a molecular weight of 21.5 kDa. It consists of 181 amino acids, whose spatial orientation causes the formation of two disulfide bonds (Gillman et al. 2015). Its inhibition activity is determined by the binding of the active site Arg 63-Ile 64 with trypsin. This peptide supresses cancer cell invasion *in vitro* and peritoneal disseminated metastasis *in vivo* by blocking the upregulation of the urokinase-like plasminogen activator (Kobayashi et al. 2004).

Lectins are bioactive proteins present in a wide variety of organisms that can recognize polysaccharide molecules. Specifically, soybean lectin is a tetramer with a molecular weight of 120 kDa and an affinity for recognizing D-galactosyl-type residues. Lectins constitute 1–2% of seed proteins, and can bind erythrocytes due to affinity for the carbohydrates present on the surface of these cell types. Research has shown the anticancer cytotoxicity of lectins (i.e., the inhibition of cell growth or even apoptosis of malignant cells) (Wen et al. 2021); their activity is determined by the binding mechanisms of lectins to cell membranes or neoplastic receptors. In this way, they induce cell death or increase the susceptibility of tumor cells to macrophage attack. The inhibition of angiogenesis, a process by which the tumor creates new blood vessels from existing vessels and thus maintains a constant supply of nutrients for its development, has been studied by Sheih et al. (2010).

Finally, lunasin is a 43-amino acid peptide in soybean seeds with proven anticancer activity. The presence of arg-gly-asp (RGD) in its amino acid sequence gives the peptide the capability to penetrate the cell membrane, and once inside the tumor cell, the peptide inhibits histone H3 and H4 acetylation, a primary process in carcinogenesis (Galvez et al 2001; Hernández-Ledesma et al. 2009 Gonzalez de Mejia et al. 2021). RGD dermal application has been shown to reduce skin tumor incidence by 70% in studies exposed to mice with agents that generate carcinomas (Chatterjee et al. 2018).

5.2.1.1 Peptides of Germinated Soybean

The germination of soybeans is a simple and inexpensive process that eliminates or inactivates some non-nutritional factors present in the seed, facilitating food digestion. Furthermore, gastrointestinal enzymatic digestion can generate anticancer peptides. The germination process deploys the protein reserve stored in the cotyledons of soybean and forms low-molecular-weight peptides (Eum et al. 2020).

Several research groups have highlighted the potential of the soybean germination process to generate peptides and amino acids with possible biological activity. The protein isolates of soybeans, germinated for 72 h, inhibited the development of HeLa cancer cells by 50% (Vernaza et al. 2012). Additionally, it has been concluded that protein hydrolysates of germinated soybeans in simulated gastrointestinal fluids considerably increase the cytotoxicity of soybean to leukemia L1210 cells (Torres y Torres & Tovar-Palacio, 2009).

However, the therapeutic use of these molecules is not simple. Proteins and polypeptides are subject to enzymatic degradation in the gastrointestinal tract, which gives them a short blood half-life and reduces the efficiency of therapies (Ma et al. 2019). One of the approaches to overcoming this problem is the encapsulation of BP to obtain delivery systems (McClements, 2018; Chronopoulou et al. 2019). The most reported systems include polymeric nano and microcapsules, liposomes, and micelles (Campardelli et al. 2016; Fang et al. 2017; Tao et al. 2020). The primary function of these carriers is to protect the integrity of the encapsulated molecule against the hostile environment. Furthermore, they promote the slow release of the active molecules for a sustained period, which increases the circulatory half-life and the bioavailability of the protein or peptide, with a consequent reduction in side effects compared to the conventional treatment (Martini & Tagliazucchi, 2023; Monroy et al. 2023).

5.3 IRON OXIDE MAGNETIC NANOPARTICLES

The application of IONPs as a delivery system for therapeutic interventions is of fundamental importance due to their unique magnetic properties, enabling remote manipulation and precise targeting. To understand those special characteristics, it is convenient to analyze the properties of common magnetic materials.

Two types of iron oxide, magnetite and maghemite, play a crucial role in this context. Magnetite is a ferrimagnetic iron oxide with a cubic inverse spinel structure. Its magnetic arrangement involves an interpenetration of Fe-ion sublattices comprises Fe_{+3} cations, where spins align antiparallel to each other along the crystalline plane direction.

The cubic inverse spinel structure comprises cations in tetrahedral positions and a combination of Fe^{+2} and Fe^{+3}, in the octahedral ones.

The term "inverse" distinguishes it from normal spinel, where each type of site contains only one type of cation. Maghemite shares the same structure as magnetite, with all iron cations having an oxidation state of Fe^{3+}. These compounds can be described as members of a solid solution phase $\left(Fe_1^{3+}\right)_A\left[Fe_{1-3\delta}^{2+}Fe_{1+2\delta\square\delta}^{3+}\right]_B O_4$, where □ represents an atom vacancy. For magnetite Fe_3O_4 δ is zero, and for maghemite $Fe_{2.67}^{3+}O_4$ or γFe_2O_3 δ is 1/3 (Issa et al. 2013).

Across this range of compounds (from $\delta=0$ to $\delta=1$) all are ferrimagnetic. Magnetic interaction is controlled by super exchange through the oxygen anions (Fe-O-Fe) due to the large distances between iron atoms, preventing direct exchange. The alignment of spins in octahedral sites is opposite to that in tetrahedral sites, resulting in a net ferrimagnetic order. In the case of magnetite, most octahedral iron ions are twice the tetrahedral ones (Bianchetti & Di Valentin, 2022).

In summary, the distinct magnetic properties of magnetite and maghemite, compounds of IONPs, highlight their potential as delivery systems. The ferrimagnetic nature and controlled magnetic interaction underscore the versatility of IONPs for targeted applications. These nanoparticles stand as promising branches in the landscape of advanced therapeutic strategies.

5.3.1 Magnetic Properties of Iron Oxide Nanoparticles

When a magnetic field with intensity H is applied to material, individual atomic moments contribute to its magnetic response. The magnetic induction B is given by (Eq. 5.1):

$$B=\mu_0(H+M) \quad \text{(Eq. 5.1)}$$

where μ_0 is the vacuum magnetic permeability and M is the magnetic moment per unit of the volume material (Laurent et al. 2008).

Magnetic materials are classified according to their magnetic susceptibility values per unit volume χ, where (Eq. 5.2):

$$\chi=M/H. \quad \text{(Eq. 5.2)}$$

If the X is between $10^{-6}-10^{-1}$ the material is considered paramagnetic and if the χ is negative the material is considered diamagnetic.

On the other hand, magnetic properties of some materials can be present without any external magnetic field (**Figure 5.1**). Depending on the distances between the atoms, the ions spins are stably aligned in these substances, which contain paramagnetic ions. This alignment is not due to the interaction between the magnetic moments of the atoms but to electrical and quantum mechanical exchange forces. The structure adopted by the alignment will be the one that achieves the minimum electrical interaction energy: with all the spins parallel (ferromagnetic materials) or with two networks of spins with different directions (antiferromagnetic or ferrimagnetic, where in the latter case, there is an imbalance between the subnets) (Morrish, 2001).

The domain structures of materials classified as ferromagnetic and ferrimagnetic are shown in **Figure 5.1**.

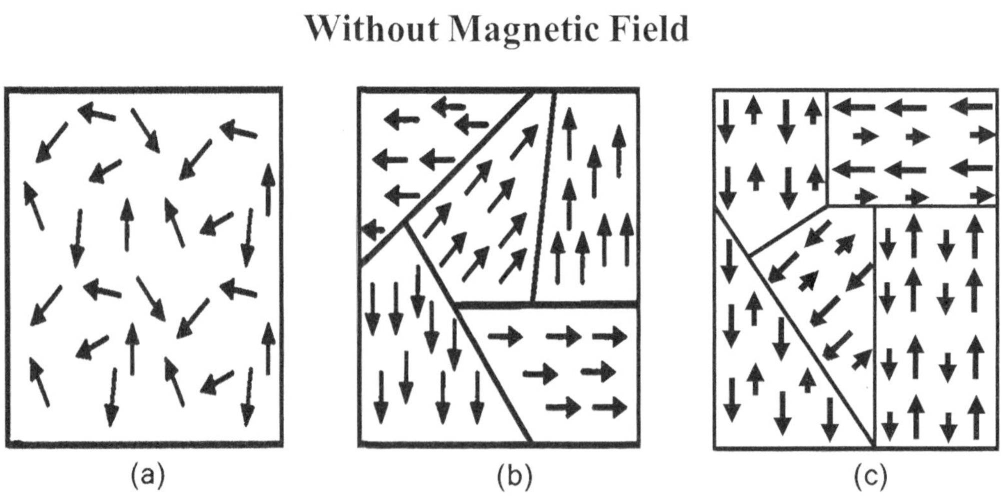

Figure 5.1 The behavior of materials according to their magnetic properties: (a) paramagnetic material (b) ferromagnetic material (c) ferrimagnetic material.

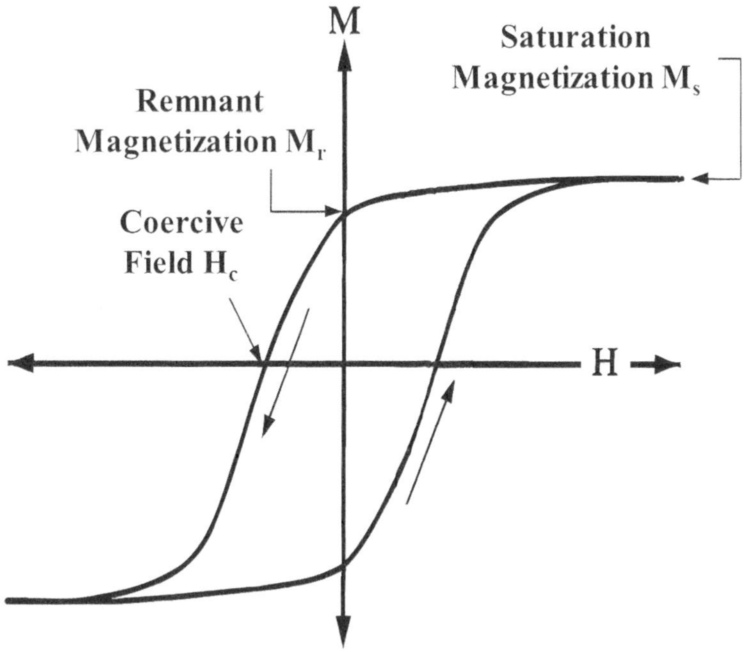

Figure 5.2 Magnetization curve obtained of magnetization against applied magnetic field strength.

The main characterization method for these materials is to obtain the magnetization curve (magnetization vs. applied magnetic field strength) as shown in **Figure 5.2.** The application of a sufficiently intense magnetic field H generates a maximum value of magnetization M, called saturation magnetization M_s .When the magnitude of the applied field is diminished, the atomic spins cease to align with the field, which causes a decrease in the total magnetization. Magnetization rapidly decays to zero when the magnetic field is completely nullified in the paramagnetic materials. Nonetheless, for ferromagnetic and ferrimagnetic materials, a certain magnetization value will be maintained, called the remnant magnetization M_r. The remannce ratio is presented as the M_r/M_s fraction and will take values between 0 and 1. Additionally, the coercive field value H_c is the magnitude of the magnetic field that must be applied in the opposite direction to completely nullify the material's magnetization (Leslie-Pelecky & Rieke, 1996).

Frenkel & Doefman (1930), regarding energy considerations, reported that small enough magnetic particles could present a single group of spins oriented in the same direction; that is, a single magnetic domain (monodomain particles). When the particle size decreases to a critical diameter value D_c the formation of domain walls becomes energetically unfavorable and the multidomain structure changes to monodomain with a very robust, almost square hysteresis curve, a property that makes these particles very suitable for magnetic recording. Interesting properties of nanoscale magnetic materials manifest at sizes smaller than D_c.

The phenomenon of superparamagnetism, which was first observed in the mid-twentieth century, represents one of the most remarkable properties of magnetic nanoparticles. In this state, the thermal energy exceeds the magnetic anisotropic energy per particle, which normally maintains the magnetic moments aligned in a specific direction. Consequently, the system behaves like a paramagnet, with magnetic moments changing direction spontaneously. Nevertheless, in contrast to paramagnetism, where one deals with individual atomic magnetic moments, each superparamagnetic nanoparticle exhibits a significantly enhanced net magnetic moment (Knobel et al. 2008).

This case corresponds to the superparamagnetic regime, in which the values of ***M*** are comparable in magnitude with those of ferromagnetic materials; nevertheless, the relationship of remnant M_r and M_s is nullified, and the values of the coercive field H_c tend to zero. These properties give rise to magnetization curves where no hysteresis loops are present.

The practical consequence is that if an external magnetic field is applied, the magnetic moment will stop fluctuating and orient itself with the applied field; if it is removed, the nanoparticles will return to the superparamagnetic state.

This characteristic, in addition to their large surface/size ratio, the ability to bind molecules of interest to their surface, and the fact that their small size enables their interaction with cellular structures, make them very attractive for various biomedical and biotechnological applications (Mahmoudi et al. 2010; Montiel Schneider et al. 2022).

5.3.2 Iron Oxide Magnetite Nanoparticles Synthesis by Coprecipitation

There are different types of synthesis for obtaining IONPs, including physical, microbial, or chemical methods. Specifically, chemical synthesis in solution offers the possibility of obtaining nanoparticles with uniform morphological characteristics, controlling the thermodynamics of the crystallization stages of the solid. This process includes stages of nucleation, growth, maturation, and agglomeration of the nuclei (Mahmoudi et al. 2010; Arias et al. 2018).

Coprecipitation reactions are the simplest and most efficient chemical processes in synthesizing nanoparticles with superparamagnetic properties, with average diameters below 20 nm. This methodology consists of placing a base in contact with an aqueous solution of Fe^{2+} and Fe^{3+} salts in stoichiometric quantities (by dropping the ion solution over the base or vice versa or by a one-pot reaction), resulting in the precipitation of Fe_3O_4, as shown in (Eq. 5.3):

$$Fe^{2+}(ac)+2Fe^{3+}(ac)+8OH^{-}(ac)\rightarrow Fe_3O_4(s)+4H_2O \quad \text{(Eq. 5.3)}$$

As Fe^{2+} ions in solution are extremely sensitive to oxidation in the presence of oxygen and basic medium, the reaction must be accomplished in an inert atmosphere of argon or nitrogen. There are many parameters involved in the coprecipitation reactions, such as pH, ionic strength, and temperature values; the nature of the salts used (perchlorates, chlorides, sulfates, and nitrates); and the stabilizing adsorbents or proportions of Fe^{2+} and Fe^{3+}. The size and shape of the particles can be controlled by adjusting the parameters mentioned earlier to obtain nanoparticles with physical and chemical properties compatible with the human body (Corchero & Villaverde, 2009).

5.3.2.1 Toxicity of Iron Oxide Nanoparticles

Magnetite stands out as a preferred material for biomedical and bioengineering applications due to its remarkable characteristics. As mentioned this preference is founded on its outstanding physicochemical properties, encompassing high magnetization values, nano sizes, and a consistently narrow particle size distribution. This uniformity ensures stable and reproducible physical and chemical properties over time, complemented by its inherent biocompatibility and low cytotoxicity (Arriortua et al. 2018; Fernández-Acosta et al. 2022).

Also, recent investigations have indicated that the level of cytotoxicity induced by IONPs depends on the size, concentration, surface charge, coating type, and functional groups of the nanoparticles (Rascol et al. 2017; Yarjanli et al. 2017; Marzano et al. 2021).

Calero et al. in 2015 discovered that superparamagnetic IONs, coated with dimercaptosuccinic acid, exhibit remarkable efficiency and biocompatibility when targeting breast cancer cells (MCF-7). To substantiate these findings, cancer cells were exposed to these nanoparticles for different time intervals, ranging from 0.5 to 72 h. Results from time-dependent uptake studies revealed the highest accumulation of dimercaptosuccinic acid-coated superparamagnetic IONs after 24 h of incubation. Afterwards a gradual remove of nanoparticles from cells was observed.

The internalization occurred by energy-dependent endocytosis and localized in endosomes. Transmission electron microscopy studies showed macropinocytosis uptake and clathrin-mediated internalization depending on the nanoparticles aggregate size (clathrin a protein that is essential for clathrin-mediated endocytosis (CME) in mammalian cells). Remarkably, MCF-7 cells accumulated these nanoparticles without inducing significant effect on cell morphology, cytoskeleton organization, cell cycle distribution, ROS generation, or cell viability, showing a similar behavior to untreated control cells (Calero et al. 2015).

Cellular internalization predominantly occurs through endocytosis. For smaller nanoparticles, this process is primarily mediated by the cellular membrane protein clathrin and fossa-mediated endocytosis. In contrast, the internalization of bigger nanoparticles and aggregates is mainly facilitated through macrocytosis and phagocytosis (Peynshaert et al., 2014).

Conversely, the potential distribution and internalization mechanisms of IONPs in the body and cellular metabolism have been studied. It has been observed that, within the intricate circulatory system, these nanoparticles undergo initial uptake in the liver and spleen.

In therapeutic applications, low cytotoxicity has been observed. Upon cellular absorption, the nuclear body degrades IONPs, releasing iron ions. This process further regulates molecular expression in a therapy based on mesenchymal stem cells (MSCs). MSCs have the capacity to differentiate into various cell types, including osteoblasts, chondrocytes, myocytes, and adipocytes (Chen & Hou, 2023).

Li et al. (2019) found that the incorporation of Fe_3O_4 @PDA nanoparticles into MSCs upregulate the expression of the fusin receptor (CXCR4 has been shown to enhance tumor immune evasion), thereby enhancing the migration of MSCs to the site of laser burn injury. Additionally, the labeled MSC group exhibited heightened cytokine levels and reduced proinflammatory factors. Moreover, there was an increase in the expression of healing-related genes. It has been confirmed that IONPs could trigger the production of ROS. Furthermore, this induction of ROS has been demonstrated to subsequently increase the expression of CXCR4 in MSC derived from the bone marrow (Li et al. 2019).

On the other hand, it has been studied that exposure to IONPs enhances the quantity, permeability, and alkalinity of lysosomes. Furthermore, IONPs are susceptible to degradation by enzymes within lysosomes (pH~4.5), resulting in the liberation of free iron from nanoparticles and impacting iron homeostasis (Ortega & Reguera, 2019).

It is also known that the storage of free iron occurs in the form of proteins, such as ferritin and hemosiderin, which serve for future utilization by the host. The biocompatibility of IONPs is attributed to their effective modulation of cellular iron metabolism, facilitating iron circulation, and enabling the exocytosis of excess iron.

Hence, by controlling the specific parameters of IONPs, adverse effects on cell viability, metabolic activity, oxidative stress, proliferation, and differentiation can be mitigated (Mulens-Arias et al. 2020).

Another extensively studied mechanism involves the production of ROS. It is well-documented that iron overload induces the generation of ROS. The primary cause of IONPs cytotoxicity is the excessive production of this molecules, which can induce oxidative stress in cells by activating pro-inflammatory mediators, ultimately leading to cell necrosis or apoptosis (Yarjanli et al. 2017).

IONPs are sensitive to weak acids. When entering cancer cells, they can release abundant Fe^{2+} ions in response to the acidic tumor microenvironment and internal body conditions to participate in the Fenton reaction and catalyze H_2O_2 to form highly toxic $\cdot OH$, which will increase cell oxidative stress, damage mitochondria and the cell membrane and induce tumor cell ferroptosis, autophagy, apoptosis and necrosis. The adsorption of plasma proteins onto the IONPs surface depends on the size, hydrophilicity, shape, and surface charge of the particles. Generally, smaller nanoparticles absorb a lower number of proteins and enable an increase in the plasma half-life of the IONPs. In addition, IONPs smaller than 10 nm are rapidly removed by the renal clearance process, while IONPs with diameters of approximately 50–100 nm accumulate in the hepatic parenchyma, whereas particles with sizes higher than 200 nm result in spleen uptake, and the larger particles (more than 4 μm diameter) are mainly captured and retained in the lungs, raising the risk of embolism (Chen & Hou, 2023).

Considering the factors mentioned afore, it is crucial to determine optimal particle sizes, synthesis and coating material, dosage, magnetization values, and other physical and chemical factors. This comprehensive evaluation is essential to ensure the safe and effective utilization of IONPs.

5.3.2.2 Ferroptosis and Lipid Peroxidation

5.3.2.2.1 Ferroptosis

Although, as explained above, iron homeostasis is well controlled when using IONPs, ferroptosis (a newly identified type of cell death related to the production of ROS, the accumulation of intracellular iron, and lipid peroxidation) must be considered.

Ferroptosis is an iron-dependent form of cell death caused by excessive oxidation of polyunsaturated fatty acids. It can be activated by iron-based nanoparticles for cancer treatment (Huang et al. 2023).

Ferroptosis was proposed as a newly identified mode of regulated cell death (Li et al. 2020) characterized by the production of ROS with iron dependence. Molecular oxygenO_2 (a biradical with two unpaired electrons that arise during cellular respiration) and hydrogen peroxide H_2O_2 (with two endogenous forms of generation: enzymatic and nonenzymatic) are the first ROS implicated in the reactions with Fe^{2+} (Andrés et al. 2022):

$$Fe^{2+} + O_2 \rightarrow Fe^{3+} + O_2^- \quad \text{(Eq. 5.4)}$$

$$Fe^{2+} + H_2O_2 \rightarrow Fe^{3+} + OH^- + \cdot OH \quad \text{(Eq. 5.5)}$$

$$Fe^{3+} + H_2O_2 \rightarrow Fe^{2+} + HO_2^{\cdot} + H^+ \quad \text{(Eq. 5.6)}$$

$$Fe^{3+} + HO_2^{\cdot} \rightarrow Fe^{2+} + O_2 + H^+ \quad \text{(Eq. 5.7)}$$

The last three equations are the Fenton reactions (Braughler et al. 1986). Eq. 5.5 represents the Fenton reaction itself. In Eq. 5.6, the Fe^{3+} ion is reduced by reaction with H_2O_2, forming Fe^{2+} ion again and more hydroxyl radicals. This process, called the Fenton-like reaction, occurs more slowly than the first process and allows the regeneration of Fe^{2+}. The biologically pertinent free radicals derived from oxygen for lipid reactions are $O_2^{\cdot-}$ (the superoxide anion) and $^{\cdot}OH$ (the hydroxyl radical).

5.3.2.2.2 Lipid Peroxidation

Lipid peroxidation is a process under which free radicals attack carbon-carbon double bond(s), especially polyunsaturated fatty acids that involve hydrogen abstraction from carbon, with oxygen insertion resulting in lipid peroxyl radicals and hydroperoxides. Lipids are essential components of cell membranes that maintain the structure and control the function of cells, so this attack strongly affects the cell (Rochette et al. 2023).

The cell controls lipid peroxidation by controlling the production of ROS, for which the crucial elements are a catalyst and its substrate: glutathione peroxidase 4 (GPX4) and glutathione (GSH). GPX4, an enzyme based on selenocysteine (a compound like cysteine with selenium instead of sulfur), is a component of the cellular antioxidant defense system, in this case with the function of catalyzing the reaction of reduction of H_2O_2 to water and hydroperoxides to the corresponding alcohols, by glutathione (thiol), which is in turn oxidized to glutathione disulfide in the process. Glutathione is a tripeptide of glutamic acid, cysteine, and glycine (Sonkusare & Chaturvedi, 2023).

The so-called Xc- system is important in the cellular synthesis of glutathione. The Xc- system, composed of members of the subunit solute transporter family (SLC7A11 and SLC3A2), mediates the exchange of cystine and glutamate across the plasma membrane (Rochette et al. 2023).

Given the role of iron in producing oxygen-free radicals, iron levels in tissues must be tightly regulated. At the cellular level, the iron regulatory proteins IRP1 and IRP2 record cytosolic iron concentrations and regulate the expression of iron metabolism genes. Iron is imported via endocytosis of transferrin (Tf) loaded with Fe^{3+}, which interacts with its TfR1 and TfR2 receptors in tightly regulated feedback control. On the other hand, accumulating evidence indicates that ferroptosis may require autophagy machinery for its execution. The term ferritinophagy describes the removal of the major iron storage protein, ferritin, by the autophagy machinery, thereby releasing stored iron and making it available for biosynthetic pathways (Rochette et al. 2023).

Lipid peroxidation and intracellular iron accumulation are two central biochemical events leading to ferroptosis. The characteristics (morphological, biochemical, and genetic) of the substances that induce and inhibit ferroptosis are unique, as detailed below (Li et al. 2020).

Certain changes occur in cells during ferroptosis, such as mitochondrial volume reduction, bilayer membrane density increase, and mitochondrial cristae reduction. There are biochemical changes, such as intracellular GSH depletion, decreased GPX4 activity, the presence of lipid peroxides that cannot be metabolized by the GPX4-catalyzed reduction reaction, and lipid oxidation. The interrelationship between ferroptosis and cancer progression has been validated using ferroptotic agents. The mechanisms underlying the susceptibility and resistance to ferroptosis in cancer have been an intense field of research in recent years (Lee & Roh, 2023). Furthermore, recent studies have shown that ferroptosis plays an important regulatory role in the occurrence and development of many diseases and has become a focus of research on the treatment and prognosis improvement of related diseases (Li et al. 2020).

Contrastingly, within the realm of magnetic nanoparticle applications, the concept of ferroptosis has not received thorough exploration. Hence, it is imperative to investigate this potential risk and concurrently assess its opportunities in the context of cancer.

5.4 MICRO- AND NANOENCAPSULATION

The terminology used to describe microparticles can be confusing for those unfamiliar with the field. The term microparticle refers to a particle with a diameter of 1–1000 μm; while "microspheres" refers to those that are spherical, and the subgroup of "microcapsules" encompasses those that have a core surrounded by a different material (Singh et al. 2010). Microencapsulation is the coating of solid, liquid, or gaseous materials with a layer of polymeric or fatty material, which allows formation of capsules at the micro- or nanometer scale (Dubey et al. 2009).

When microcapsules have sizes below 1 μm, they are considered to have nano- or submicrometric dimensions. Micro- and nanoencapsulation can be applied in various fields, such as agriculture (fertilizer and pesticide encapsulation), the cosmetic industry (manufacture of deodorants, nebulizers), and the pharmaceutical industry. However, one of the branches with the most significant potential is biomolecule-controlled release systems (Kumari et al. 2010; Bhujbal et al. 2014). Micro- and nanoencapsulation are used to decrease the frequency of medicine administration, increase therapeutic efficiency due to the elimination of fluctuations in drug serum levels, and decrease the total dose required for treatment. These applications reduce adverse effects since they significantly lower the doses of medicine released into the body (Freiberg & Zhu, 2004).

The materials that can be used to manufacture micro- and nanocapsules are classified into three categories: fats, proteins, and polymers (Saez et al. 2007). In the specific field of controlled drug release, the most commonly used covering materials are natural and synthetic polymers; the most commonly studied are homopolymers and copolymers derived from lactic acid and glycolic acid, commonly referred to as PLGA (Vlachopoulos et al. 2022).

PLGA are polyesters obtained by linear polycondensation of hydroxy acids or by opening the ring of the corresponding lactones (Gomzyak et al. 2021). These polymers are soluble in organic solvents such as dichloromethane, chloroform, ethyl acetate, acetone, and tetrahydrofuran. However, they are not soluble in water. Some are semicrystalline, and others are amorphous depending on the ratio of lactic acid and glycolic acid, with amorphous formulations being the most commonly used in the design of controlled release formulations. PLGA differs, besides the proportion of the monomers that compose them, in their molecular mass and the presence of a carboxyl-terminal group (COOH) or an ester group (COOR). These three parameters determine, to a large extent, the hydrophobicity, degradation kinetics, encapsulation efficiency, and release rate of the encapsulated substance (Tamber et al. 2005).

Concurrently, the biodegradability of these polymers refers to the hydrolysis of their ester bonds by contact with biological or artificial fluids. This reaction produces lactic acid and glycolic acid, which are metabolized in the Krebs cycle to form water and CO_2, when these materials are used as matrices to obtain micro- and nanospheres. Their degradation occurs in two stages: the first involves the hydrolytic cleavage of the ester bonds, properly called degradation, in which oligomers and monomers are generated with a consequent decrease in the molecular mass of the original polymer. In the second stage, called erosion, the loss of mass of the matrix occurs, and the cleavage rate of the bonds can increase because of the autocatalytic action of acidic degradation products. It is precisely the fact that these polymers are not soluble in water but can be degraded by it, which makes them extraordinarily useful in developing controlled drug release systems.

On the other hand, PLGA are biocompatible since they do not present toxicity when introduced into the body by different routes (Schoenhammer et al. 2010). In fact, for many years, these polymers have been used in producing biodegradable suture threads and in manufacturing orthopedic fixing devices such as discs, screws, and bolts. For dissolving PLGA, ethyl acetate, chloroform, and dichloromethane have been used, the latter most commonly for its high volatility, low boiling point, and immiscibility with water. Moreover, it has been proven that the micro- and nanospheres obtained with dichloromethane as a solvent have greater encapsulation efficiency and better morphological characteristics (Li et al. 2008).

There are various methods for obtaining micro- and nanoparticles of PLGA, and they are classified into three groups: physical, chemical, and mechanical. Physical, chemical, and simple and complex coacervation methods have been widely studied on a laboratory scale yet have not reached a great application at the industrial level (Vlachopoulos et al. 2022). Physical and mechanical methods have been extensively used in laboratories and industry, resulting in their feasibility for application. Among them are solvent extraction, evaporation, and spray drying. Chemical methods such as interfacial and heterogeneous polymerization are usually accomplished under conditions that are well tolerated by fragile molecules.

5.5 NANOENCAPSULATION IN POLY LACTIC-*CO*-GLYCOLIC ACID

One of the objectives of encapsulating bioactive compounds within PLGA microspheres or nanoparticles is to attain prolonged delivery through sustained release (Pei et al. 2023). PLGA is a biodegradable polymer easily administered and has been approved in the form of microparticles, nanoparticles, solid implants, and gel-forming implants (Park et al. 2019). Active molecule release occurs through diffusion and capsule erosion processes as PLGA undergoes hydrolysis in aqueous medium through ester bond cleavage. As a result, the formed carboxylic groups act as catalysts for further hydrolysis (Tabatabaei Mirakabad et al. 2014).

One of the essential advantages of using PLGA structures is the possibility of tailoring their physicochemical properties by chemical modification, which allows better association between the polymer and the active molecule, leading to higher encapsulation efficiency. However, these polymer particles cannot guarantee that bioactive compound release will occur only at the target site, and off-target release decreases the treatment's effectiveness. To overcome this drawback, various researchers have presented targeted delivery approaches.

The size (<200 nm) of IONPs makes them extremely advantageous for tumor-targeted drug delivery since angiogenic tumors tend to generate permeable vasculature and can passively accumulate circulating NPs (enhanced permeability and retention effect). However, they are also versatile in conjugating with a variety of micro- and macrobiomolecules that can target the drug-loaded NPs to a specific site of action (Sharma et al. 2023)

Distally controlled magnetic PLGA nanocapsules are promising systems for achieving the specific delivery and release of antitumor molecules. They can be obtained by encapsulating superparamagnetic iron oxide nanoparticles (SPIONs) into the polymeric matrix through conventional encapsulation methodologies.

SPIONs show several advantages as materials for biomedical applications, including low toxicity, ease of fabrication, and characteristic magnetic properties (McCarthy & Weissleder, 2008; Huang et al. 2017).

5.6 COENCAPSULATION OF IRON OXIDE NANOPARTICLES AND BIOACTIVE PEPTIDES FROM GERMINATED SOYBEAN IN MATRICES

An attempt at the encapsulation of magnetite nanoparticles and peptides from germinated soybean protein in PLGA matrices was performed by our group (Rodriguez-Rivero, 2018), to the best of our knowledge for the first time.

IONPs were coated with silica using tetraethyl orthosilicate and the Stöber methodology (Stöber & Fink, 1968). After synthesis, the $Fe_3O_4@SiO_2$ IONPs were characterized by FT-IR, XRD, and TEM; a mean size of 22 nm was observed.

PLGA capsules were loaded with $Fe_3O_4@SiO_2$ IONPs and BP (IONPs/BP–PLGA) through a double emulsion/solvent evaporation methodology (Rosca et al. 2004). A peptidic fraction (> 10 kDa) from the enzymatic hydrolysis of germinated soybean was used as BP. The cytotoxic effect of this isolate has been evaluated on HeLa and cervical cancer C33 cells, and it was found that the growth of both cell lines was inhibited by 87% (González-Montoya et al. 2016).

The size and morphology of the composites were analyzed through SEM. Recorded images show that this methodology obtained spheroidal-shaped structures of 70 ± 30 μm.

Determination of the encapsulation content and encapsulation efficiency corresponding to IONPs/BP–PLGA capsules is considered a critical step in the rigorous characterization of these composite materials. These parameters were determined by spectrophotometric analysis of the capsules hydrolyzed in basic medium. The obtained results were similar to those reported for encapsulating polar molecules in PLGA matrices (Sun et al. 2003), which were convenient for modified release systems and analogous to those reported by Frounchi & Shamshiri (2015).

They indicated that *an in vitro* release study with IONPs/BP-PLGA capsules was performed by incubating a known mass of microcapsules, in a phosphate regulatory solution medium at pH 7.4. At different times, aliquots of the suspension were centrifuged, and the total mass of BP in the supernatant was determined. For this purpose, Bradford's protein assay was utilized. The peptide release curve obtained is shown in **Figure 5.3.**

The observation reveals that, during the initial measurement, the solution contained approximately 7% of the total BP. This occurrence is linked to the degradation of polymer capsules when stored under inadequate preservation conditions. In general, the greatest change in peptide concentration in the supernatant, from 7% to 30%, was observed during the first 10 h of incubation. This stage in the system's behavior is associated with the presence of peptide molecules on the surface of the microspheres and in the pores or channels that are part of the internal structure of the capsules. In the first hours of the test, the liquid medium penetrates these morphological changes and allows for the rapid release of the analytes (Rodriguez-Rivero, 2018).

In the subsequent hours of the study, the release of the peptides into the solution was determined mainly by erosion of PLGA, in the aqueous phase, leading to the appearance and increasing size of the pores in the microcapsules (**Figure 5.4**).

The peptide concentration in the solution continues to increase until approximately 50 h, at which time 60% of total peptides was released. These released levels were maintained with slight

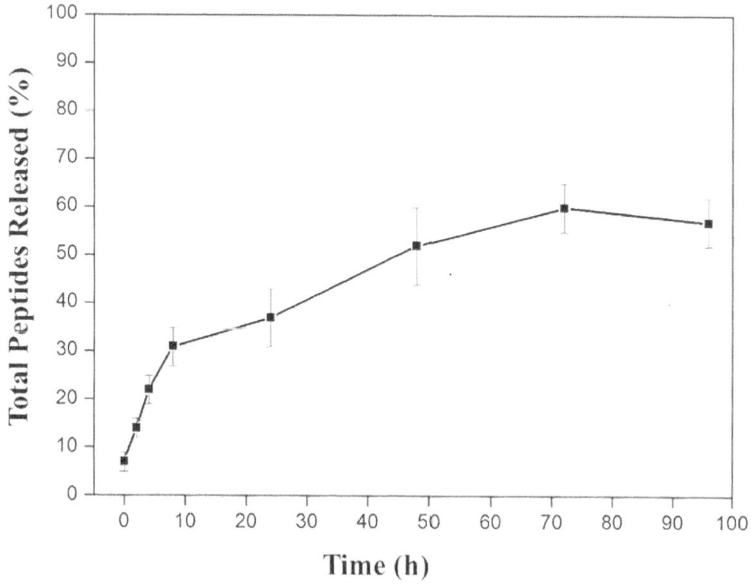

Figure 5.3 *In Vitro* release curve of peptides obtained for PLGA capsules loaded with Fe_3O_4 nanoparticles and protein hydrolysates from germinated soybean.

Figure 5.4 SEM images of capsules loaded with nanoparticles.

variation until the end of the test at 96 h. The curve adopted an asymptotic shape resembling the loose curves reported for PLGA capsules loaded with polar proteins (Gouveia et al. 2023).

Consequently, the obtained charged microcapsules satisfy the desired characteristics for controlled drug release systems, as they modify the liberation of the active molecules compared to those supplied conventionally.

5.7 CONCLUSION

The coencapsulation IONPs/BP-PLGA has proven to be a promising and effective strategy as a release system for cancer treatment. The combination of the magnetic properties of IONPs with the loading capacity of peptides in the PLGA biomaterial offers a multifaceted approach to address challenges associated with the delivery of therapeutic agents in the context of cancer treatment.

This delivery system presents significant advantages, including the magnetic targeting capability for specific localization at the tumor site, controlled and sustained release of therapeutic peptides, and the biocompatibility of the PLGA vehicle. These combined attributes suggest considerable potential for improving therapeutic efficacy while minimizing unwanted side effects.

As we progress in the research and development of drug delivery systems, the application of IONPs coencapsulated with peptides in PLGA for cancer treatment emerges as an innovative and promising strategy. However, continuous and rigorous research is required to focus on specific challenges, such as optimizing drug loading, understanding detailed release kinetics, and conducting comprehensive toxicity assays. Ultimately, this research contributes to the growing body of knowledge supporting the development of more effective and specific therapies for cancer treatment.

REFERENCES

Agyenim-Boateng, K. G., Zhang, S., Zhang, S., Khattak, A. N., Shaibu, A., Abdelghany, A. M., Qi, J., Azam, M., Ma, C., Feng, Y., Feng, H., Liu, Y., Li, J., Li, B., & Sun, J. (2023). The nutritional composition of the vegetable soybean (maodou) and its potential in combatting malnutrition. *Frontiers in Nutrition, 9*, 1–21. https://doi.org/10.3389/fnut.2022.1034115

Andrés, C. M. C., Pérez de la Lastra, J. M., Juan, C. A., Plou, F. J., & Pérez-Lebeña, E. (2022). Chemistry of hydrogen peroxide formation and elimination in mammalian cells, and its role in various pathologies. *Stresses, 2*(3), 256–274. https://doi.org/10.3390/stresses2030019

Arias, L., Pessan, J., Vieira, A., Lima, T., Delbem, A., & Monteiro, D. (2018). Iron oxide nanoparticles for biomedical applications: A perspective on synthesis, drugs, antimicrobial activity, and toxicity. *Antibiotics, 7*(2), 46. https://doi.org/10.3390/antibiotics7020046

Arriortua, O. K., Insausti, M., Lezama, L., Gil de Muro, I., Garaio, E., de la Fuente, J. M., Fratila, R. M., Morales, M. P., Costa, R., Eceiza, M., Sagartzazu-Aizpurua, M., & Aizpurua, J. M. (2018). RGD-functionalized Fe3O4 nanoparticles for magnetic hyperthermia. *Colloids and Surfaces B: Biointerfaces, 165*, 315–324. https://doi.org/10.1016/j.colsurfb.2018.02.031

Arum, C.-J., Anderssen, E., Viset, T., Kodama, Y., Lundgren, S., Chen, D., & Zhao, C.-M. (2010). Cancer immunoediting from immunosurveillance to tumor escape in microvillus-formed Niche: A study of syngeneic orthotopic rat bladder cancer model in comparison with human bladder cancer. *Neoplasia, 12*(6), 434–442. https://doi.org/10.1593/neo.91824

Barac, M., Stanojevic, S., & Pesic, M. (2005). Biologically active components of soybeans and soy protein products: A review. *Acta Periodica Technologica*, (36), 155–168. https://doi.org/10.2298/apt0536155b

Beg, A. A., Khan, T., & Antonia, S. J. (2013). A new role for NFκB in immunosurveillance and its implications for cancer immunotherapy. *OncoImmunology, 2*(10), e25963. https://doi.org/10.4161/onci.25963

Bhujbal, S. V., de Vos, P., & Niclou, S. P. (2014). Drug and cell encapsulation: Alternative delivery options for the treatment of malignant brain tumors. *Advanced Drug Delivery Reviews, 67–68*, 142–153. https://doi.org/10.1016/j.addr.2014.01.010

Bhutta, Z. A., & Choi, K.-C. (2023). Antitumor effects of dietary compounds. *Nutrients, 15*(10), 2349. https://doi.org/10.3390/nu15102349

Bianchetti, E., & Di Valentin, C. (2022). Mechanism of spin ordering in Fe3O4 nanoparticles by surface coating with organic acids. *Materials Today Nano, 17*, 100169. https://doi.org/10.1016/j.mtnano.2021.100169

Bonifaz, L. C., Bonnyay, D. P., Charalambous, A., Darguste, D. I., Fujii, S.-I., Soares, H., Brimnes, M. K., Moltedo, B., Moran, T. M., & Steinman, R. M. (2004). In vivo targeting of antigens to maturing dendritic cells via the DEC-205 receptor improves T Cell vaccination. *Journal of Experimental Medicine, 199*(6), 815–824. https://doi.org/10.1084/jem.20032220

Braughler, J. M., Duncan, L. A., & Chase, R. L. (1986). The involvement of iron in lipid peroxidation. Importance of ferric to ferrous ratios in initiation. *The Journal of Biological Chemistry, 261*(22), 10282–10289. https://doi.org/10.1016/S0021-9258(18)67521-0

Calero, M., Chiappi, M., Lazaro-Carrillo, A., Rodríguez, M. J., Chichón, F. J., Crosbie-Staunton, K., Prina-Mello, A., Volkov, Y., Villanueva, A., & Carrascosa, J. L. (2015). Characterization of interaction of magnetic nanoparticles with breast cancer cells. *Journal of Nanobiotechnology, 13*(1), 1–15. https://doi.org/10.1186/s12951-015-0073-9

Campardelli, R., Espirito Santo, I., Albuquerque, E. C., de Melo, S. V., Della Porta, G., & Reverchon, E. (2016). Efficient encapsulation of proteins in submicro liposomes using a supercritical fluid assisted continuous process. *The Journal of Supercritical Fluids, 107*, 163–169. https://doi.org/10.1016/j.supflu.2015.09.007

Chatterjee, C., Gleddie, S., & Xiao, C.-W. (2018). Soybean bioactive peptides and their functional properties. *Nutrients, 10*(9), 1211. https://doi.org/10.3390/nu10091211

Chen, Y., & Hou, S. (2023). Recent progress in the effect of magnetic iron oxide nanoparticles on cells and extracellular vesicles. *Cell Death Discovery, 9*(1), 195. https://doi.org/10.1038/s41420-023-01490-2

Chronopoulou, L., Domenici, F., Giantulli, S., Brasili, F., D'Errico, C., Tsaouli, G., Tortorella, E., Bordi, F., Morrone, S., Palocci, C., & Silvestri, I. (2019). PLGA based particles as "drug reservoir" for antitumor drug delivery: characterization and cytotoxicity studies. *Colloids and Surfaces B: Biointerfaces, 180*(February), 495–502. https://doi.org/10.1016/j.colsurfb.2019.05.006

Corchero, J. L., & Villaverde, A. (2009). Biomedical applications of distally controlled magnetic nanoparticles. *Trends in Biotechnology, 27*(8), 468–476. https://doi.org/10.1016/j.tibtech.2009.04.003

Coulie, P. G., Van den Eynde, B. J., van der Bruggen, P., & Boon, T. (2014). Tumour antigens recognized by T lymphocytes: at the core of cancer immunotherapy. *Nature Reviews Cancer, 14*(2), 135–146. https://doi.org/10.1038/nrc3670

Dubey, R., Shami, T. C., & Bhasker, R. (2009). Microencapsulation technology and applications. *Defence Science Journal, 59*(1), 82–95. https://doi.org/10.14429/dsj.59.1489

Eum, H. L., Park, Y., Yi, T. G., Lee, J. W., Ha, K.-S., Choi, I.-Y., & Park, N. Il. (2020). Effect of germination environment on the biochemical compounds and anti-inflammatory properties of soybean cultivars. *PLOS ONE, 15*(4), e0232159. https://doi.org/10.1371/journal.pone.0232159

Fang, X., Xie, H., Duan, H., Li, P., Yousaf, M., Xu, H., Yang, Y., & Wang, C. (2017). Anti-tumor activity of nanomicelles encapsulating CXCR4 peptide antagonist E5. *PLOS ONE, 12*(8), 1–18. https://doi.org/10.1371/journal.pone.0182697

Farzanegan, Z., & Tahmasbi, M. (2023). Evaluating the applications and effectiveness of magnetic nanoparticle-based hyperthermia for cancer treatment: A systematic review. *Applied Radiation and Isotopes, 198*, 110873. https://doi.org/10.1016/j.apradiso.2023.110873

Fernández-Acosta, R., Iriarte-Mesa, C., Alvarez-Alminaque, D., Hassannia, B., Wiernicki, B., Díaz-García, A. M., Vandenabeele, P., Vanden Berghe, T., & Pardo Andreu, G. L. (2022). Novel iron oxide nanoparticles induce ferroptosis in a panel of cancer cell lines. *Molecules, 27*(13), 3970. https://doi.org/10.3390/molecules27133970

Freiberg, S., & Zhu, X. X. (2004). Polymer microspheres for controlled drug release. *International Journal of Pharmaceutics, 282*(1–2), 1–18. https://doi.org/10.1016/j.ijpharm.2004.04.013

Frenkel, J., & Doefman, J. (1930). Spontaneous and Induced Magnetisation in Ferromagnetic Bodies. *Nature, 126*(3173), 274–275. https://doi.org/10.1038/126274a0

Frounchi, M., & Shamshiri, S. (2015). Magnetic nanoparticles-loaded PLA/PEG microspheres as drug carriers. *Journal of Biomedical Materials Research Part A, 103*(5), 1893–1898. https://doi.org/10.1002/jbm.a.35317

Galvez, A. F., Chen, N., Macasieb, J., & de Lumen, B. O. (2001). Chemopreventive property of a soybean peptide (lunasin) that binds to deacetylated histones and inhibits acetylation. *Cancer Research, 61*(20), 7473–7478. www.ncbi.nlm.nih.gov/pubmed/11606382

Ghaly, G., Tallima, H., Dabbish, E., Badr ElDin, N., Abd El-Rahman, M. K., Ibrahim, M. A. A., & Shoeib, T. (2023). Anti-Cancer Peptides: Status and Future Prospects. *Molecules, 28*(3), 1148. https://doi.org/10.3390/molecules28031148

Gillman, J. D., Kim, W.-S., & Krishnan, H. B. (2015). Identification of a new soybean Kunitz trypsin inhibitor mutation and its effect on Bowman–Birk protease inhibitor content in soybean seed. *Journal of Agricultural and Food Chemistry, 63*(5), 1352–1359. https://doi.org/10.1021/jf505220p

Gitlin-Domagalska, A., Maciejewska, A., & Dębowski, D. (2020). Bowman-Birk inhibitors: Insights into family of multifunctional proteins and peptides with potential therapeutical applications. *Pharmaceuticals, 13*(12), 421. https://doi.org/10.3390/ph13120421

Gomzyak, V. I., Sedush, N. G., Puchkov, A. A., Polyakov, D. K., & Chvalun, S. N. (2021). Linear and branched lactide polymers for targeted drug delivery systems. *Polymer Science, Series B, 63*(3), 257–271. https://doi.org/10.1134/S1560090421030064

González-Montoya, M., Hernández-Ledesma, B., Silván, J. M., Mora-Escobedo, R., & Martínez-Villaluenga, C. (2018). Peptides derived from *in vitro* gastrointestinal digestion of germinated soybean proteins inhibit human colon cancer cells proliferation and inflammation. *Food Chemistry, 242*(June 2017), 75–82. https://doi.org/10.1016/j.foodchem.2017.09.035

González-Montoya, M., Ramón-Gallegos, E., Robles-Ramírez, M. del C., & Mora-Escobedo, R. (2016). Evaluation of the antioxidant and antiproliferative effects of three peptide fractions of germinated soybeans on breast and cervical cancer cell lines. *Plant Foods for Human Nutrition, 71*(4), 368–374. https://doi.org/10.1007/s11130-016-0568-z

Gonzalez de Mejia, E., Castañeda-Reyes, E. D., Mojica, L., Dia, V., Wang, H., Wang, T., & Johnson, L. A. (2021). Potential health benefits associated with lunasin concentration in dietary supplements and lunasin-enriched soy extract. *Nutrients, 13*(5), 1618. https://doi.org/10.3390/nu13051618

González de Mejía, E., & Prisecaru, V. I. (2005). Lectins as bioactive plant proteins: A potential in cancer treatment. *Critical Reviews in Food Science and Nutrition, 45*(6), 425–445. https://doi.org/10.1080/10408390591034445

Gouveia, M. G., Wesseler, J. P., Ramaekers, J., Weder, C., Scholten, P. B. V., & Bruns, N. (2023). Polymersome-based protein drug delivery – quo vadis? *Chemical Society Reviews, 52*(2), 728–778. https://doi.org/10.1039/D2CS00106C

Hailemichael, Y., & Overwijk, W. W. (2013). Peptide-based anticancer vaccines. *OncoImmunology, 2*(7), e24743. https://doi.org/10.4161/onci.24743

Hernández-Ledesma, B., Hsieh, C.-C., & de Lumen, B. O. (2009). Lunasin, a novel seed peptide for cancer prevention. *Peptides, 30*(2), 426–430. https://doi.org/10.1016/j.peptides.2008.11.002

Huang, Q., Hu, Q., Wen, Z., & Li, Y. (2023). Iron oxide nanoparticles inhibit tumor growth by ferroptosis in diffuse large B-cell lymphoma. *American Journal of Cancer Research, 13*(2), 498–508.

Huang, Y., Mao, K., Zhang, B., & Zhao, Y. (2017). Superparamagnetic iron oxide nanoparticles conjugated with folic acid for dual target-specific drug delivery and MRI in cancer theranostics. *Materials Science and Engineering: C, 70*, 763–771. https://doi.org/10.1016/j.msec.2016.09.052

Issa, B., Obaidat, I., Albiss, B., & Haik, Y. (2013). Magnetic nanoparticles: Surface effects and properties related to biomedicine applications. *International Journal of Molecular Sciences, 14*(11), 21266–21305. https://doi.org/10.3390/ijms141121266

Jakubczyk, A., Karas, M., Rybczynska-Tkaczyk, K., Zielinska, E., & Zielinski, D. (2020). Current trends of bioactive peptides – New sources and therapeutic effect. *Foods*, 9(7),846. https://doi.org/10.3390/foods9070846

Knobel, M., Nunes, W. C., Socolovsky, L. M., De Biasi, E., Vargas, J. M., & Denardin, J. C. (2008). Superparamagnetism and other magnetic features in granular materials: A review on ideal and real systems. *Journal of Nanoscience and Nanotechnology, 8*(6), 2836–2857. https://doi.org/10.1166/jnn.2008.15348

Kobayashi, H., Suzuki, M., Kanayama, N., & Terao, T. (2004). A soybean Kunitz trypsin inhibitor suppresses ovarian cancer cell invasion by blocking urokinase upregulation. *Clinical & Experimental Metastasis, 21*(2), 159–166. https://doi.org/10.1023/B:CLIN.0000024751.73174.c2

Kreutz, M., Tacken, P. J., & Figdor, C. G. (2013). Targeting dendritic cells—why bother? *Blood, 121*(15), 2836–2844. https://doi.org/10.1182/blood-2012-09-452078

Kumari, A., Yadav, S. K., & Yadav, S. C. (2010). Biodegradable polymeric nanoparticles based drug delivery systems. *Colloids and Surfaces B: Biointerfaces, 75*(1), 1–18. https://doi.org/10.1016/j.colsurfb.2009.09.001

Lath, A., Santal, A. R., Kaur, N., Kumari, P., & Singh, N. P. (2023). Anti-cancer peptides: their current trends in the development of peptide-based therapy and anti-tumor drugs. *Biotechnology and Genetic Engineering Reviews, 39*(1), 45–84. https://doi.org/10.1080/02648725.2022.2082157

Laurent, S., Forge, D., Port, M., Roch, A., Robic, C., Vander Elst, L., & Muller, R. N. (2008). Magnetic iron oxide nanoparticles: Synthesis, stabilization, vectorization, physicochemical characterizations, and biological applications. *Chemical Reviews, 108*(6), 2064–2110. https://doi.org/10.1021/cr068445e

Lee, J., & Roh, J.-L. (2023). Epigenetic modulation of ferroptosis in cancer: Identifying epigenetic targets for novel anticancer therapy. *Cellular Oncology*, 46,1605–1623. https://doi.org/10.1007/s13402-023-00840-7

Leslie-Pelecky, D. L., & Rieke, R. D. (1996). Magnetic properties of nanostructured materials. *Chemistry of Materials, 8*(8), 1770–1783. https://doi.org/10.1021/cm960077f

Li, J., Cao, F., Yin, H., Huang, Z., Lin, Z., Mao, N., Sun, B., & Wang, G. (2020). Ferroptosis: past, present and future. *Cell Death & Disease, 11*(2), 88. https://doi.org/10.1038/s41419-020-2298-2

Li, M., Rouaud, O., & Poncelet, D. (2008). Microencapsulation by solvent evaporation: State of the art for process engineering approaches. *International Journal of Pharmaceutics, 363*(1–2), 26–39. https://doi.org/10.1016/j.ijpharm.2008.07.018

Li, X., Wei, Z., Li, B., Li, J., Lv, H., Wu, L., Zhang, H., Yang, B., Zhu, M., & Jiang, J. (2019). In vivo migration of Fe 3 O 4 @polydopamine nanoparticle-labeled mesenchymal stem cells to burn injury sites and their therapeutic effects in a rat model. *Biomaterials Science, 7*(7), 2861–2872. https://doi.org/10.1039/C9BM00242A

Ma, B., Niu, F., Qu, X., He, W., Feng, C., Wang, S., Ouyang, Z., Yan, J., Wen, Y., Xu, D., Shao, Y., Ma, P. X., & Lu, W. (2019). A tetrameric protein scaffold as a nano-carrier of antitumor peptides for cancer therapy. *Biomaterials, 204*(January), 1–12. https://doi.org/10.1016/j.biomaterials.2019.03.004

Mahmoudi, M., Simchi, A., & Imani, M. (2010). Recent advances in surface engineering of superparamagnetic iron oxide nanoparticles for biomedical applications. *Journal of the Iranian Chemical Society, 7*(S2), S1–S27. https://doi.org/10.1007/BF03246181

Maity, D., Chandrasekharan, P., Feng, S. S., & Jun, D. (2010). Synthesis and studies of APTES functionalized magnetite nanoparticles. *ICONN 2010 – Proceedings of the 2010 International Conference on Nanoscience and Nanotechnology, February*, 94–97. https://doi.org/10.1109/ICONN.2010.6045190

Martini, S., & Tagliazucchi, D. (2023). Bioactive peptides in human health and disease. *International Journal of Molecular Sciences, 24*(6), 5837. https://doi.org/10.3390/ijms24065837

Marzano, M., Bou-Dargham, M. J., Cone, A. S., York, S., Helsper, S., Grant, S. C., Meckes, D. G., Sang, Q.-X. A., & Li, Y. (2021). Biogenesis of extracellular vesicles produced from human-stem-cell-derived cortical spheroids exposed to iron oxides. *ACS Biomaterials Science & Engineering, 7*(3), 1111–1122. https://doi.org/10.1021/acsbiomaterials.0c01286

McCarthy, J. R., & Weissleder, R. (2008). Multifunctional magnetic nanoparticles for targeted imaging and therapy. *Advanced Drug Delivery Reviews, 60*(11), 1241–1251. https://doi.org/10.1016/j.addr.2008.03.014

McClements, D. J. (2018). Encapsulation, protection, and delivery of bioactive proteins and peptides using nanoparticle and microparticle systems: A review. *Advances in Colloid and Interface Science, 253*, 1–22. https://doi.org/10.1016/j.cis.2018.02.002

Monroy, G. R., Murguiondo Pérez, R., Weintraub Ben Zión, E., Vidal Alcántar-Garibay, O., Loza-López, E. C., Tejerina Marion, E., Blancarte Hernández, E., Navarro-Torres, L., & Ibarra, A. (2023). Immunization with neural-derived peptides in neurodegenerative diseases: A narrative review. *Biomedicines, 11*(3), 919. https://doi.org/10.3390/biomedicines11030919

Montiel Schneider, M. G., Martín, M. J., Otarola, J., Vakarelska, E., Simeonov, V., Lassalle, V., & Nedyalkova, M. (2022). Biomedical applications of iron oxide nanoparticles: Current insights progress and perspectives. *Pharmaceutics, 14*(1), 204. https://doi.org/10.3390/pharmaceutics14010204

Mora-Escobedo, R., Robles-Ramírez, M. del C., Ramón-Gallegos, E., & Reza-Alemán, R. (2009). Effect of protein hydrolysates from germinated soybean on cancerous cells of the human cervix: An *in vitro* study. *Plant Foods for Human Nutrition, 64*(4), 271–278. https://doi.org/10.1007/s11130-009-0131-2

Morrish, A. H. (2001). *The Physical Principles of Magnetism* (1 st ed., p. 700) Wiley-IEEE Press.

Motz, G. T., & Coukos, G. (2011). The parallel lives of angiogenesis and immunosuppression: cancer and other tales. *Nature Reviews Immunology, 11*(10), 702–711. https://doi.org/10.1038/nri3064

Mulens-Arias, V., Rojas, J. M., & Barber, D. F. (2020). The intrinsic biological identities of iron oxide nanoparticles and their coatings: Unexplored territory for combinatorial therapies. *Nanomaterials, 10*(5), 837. https://doi.org/10.3390/nano10050837

Negrini, S., Gorgoulis, V. G., & Halazonetis, T. D. (2010). Genomic instability — An evolving hallmark of cancer. *Nature Reviews Molecular Cell Biology, 11*(3), 220–228. https://doi.org/10.1038/nrm2858

Nelson, D. L., & Cox, M. M. (2001). *Lehninger Biochemie.* Springer Berlin Heidelberg. https://doi.org/10.1007/978-3-662-08289-8

Olson, B. M., & McNeel, D. G. (2012). Antigen loss and tumor-mediated immunosuppression facilitate tumor recurrence. *Expert Review of Vaccines, 11*(11), 1315–1317. https://doi.org/10.1586/erv.12.107

Ortega, G., & Reguera, E. (2019). Biomedical applications of magnetite nanoparticles. In *Materials for Biomedical Engineering* (pp. 397–434). Elsevier. https://doi.org/10.1016/B978-0-12-816913-1.00013-1

Overwijk, W. W., Wang, E., Marincola, F. M., Rammensee, H.-G., & Restifo, N. P. (2013). Mining the mutanome: developing highly personalized Immunotherapies based on mutational analysis of tumors. *Journal for ImmunoTherapy of Cancer, 1*(1), 11. https://doi.org/10.1186/2051-1426-1-11

Park, K., Skidmore, S., Hadar, J., Garner, J., Park, H., Otte, A., Soh, B. K., Yoon, G., Yu, D., Yun, Y., Lee, B. K., Jiang, X., & Wang, Y. (2019). Injectable, long-acting PLGA formulations: Analyzing PLGA and understanding microparticle formation. *Journal of Controlled Release, 304*(May), 125–134. https://doi.org/10.1016/j.jconrel.2019.05.003

Pei, Y., Wang, J., Khaliq, N. U., Meng, F., Oucherif, K. A., Xue, J., Horava, S. D., Cox, A. L., Richard, C. A., Swinney, M. R., Park, K., & Yeo, Y. (2023). Development of poly(lactide-co-glycolide) microparticles for sustained delivery of meloxicam. *Journal of Controlled Release, 353*, 823–831. https://doi.org/10.1016/j.jconrel.2022.12.019

Peynshaert, K., Manshian, B. B., Joris, F., Braeckmans, K., De Smedt, S. C., Demeester, J., & Soenen, S. J. (2014). Exploiting intrinsic nanoparticle toxicity: The Pros and Cons of nanoparticle-induced autophagy in biomedical research. *Chemical Reviews, 114*(15), 7581–7609. https://doi.org/10.1021/cr400372p

Rascol, E., Daurat, M., Da Silva, A., Maynadier, M., Dorandeu, C., Charnay, C., Garcia, M., Lai-Kee-Him, J., Bron, P., Auffan, M., Liu, W., Angeletti, B., Devoisselle, J. M., Guari, Y., Gary-Bobo, M., & Chopineau, J. (2017). Biological fate of Fe3O4 core-shell mesoporous silica nanoparticles depending on particle surface chemistry. *Nanomaterials, 7*(7), 162. https://doi.org/10.3390/nano7070162

Ricupito, A., Grioni, M., Calcinotto, A., & Bellone, M. (2013). Boosting anticancer vaccines. *OncoImmunology, 2*(7), e25032. https://doi.org/10.4161/onci.25032

Robles-Ramírez, M. del C., Ramón-Gallegos, E., Reyes-Duarte, F. J., & Mora-Escobedo, R. (2012). Effect of germinated soy protein on the growth of HeLa cervical cancer cells in female athymic mice. *Nutrition and Cancer, 64*(8), 1261–1268. https://doi.org/10.1080/01635581.2012.717681

Rochette, L., Dogon, G., Rigal, E., Zeller, M., Cottin, Y., & Vergely, C. (2023). Lipid peroxidation and iron metabolism: Two corner stones in the homeostasis control of ferroptosis. *International Journal of Molecular Sciences, 24*(1), 449. https://doi.org/10.3390/ijms24010449

Rodriguez Rivero, Y. G. (2018). Coencapsulación de nanopartículas de Fe3O4 y péptidos antitumorales en matrices de ácido láctico-cóglicólico. Master degree thesis. Escuela Superior de Físico Matemáticas, Instituto Politécnico Nacional.

Rosca, I. D., Watari, F., & Uo, M. (2004). Microparticle formation and its mechanism in single and double emulsion solvent evaporation. *Journal of Controlled Release, 99*(2), 271–280. https://doi.org/10.1016/j.jconrel.2004.07.007

Saez, V., Hernández, J. R., & Peniche, C. (2007). Microspheres as delivery systems for the controlled release of peptides and proteins. *Biotecnologia Aplicada, 24*(2), 108–116.

Sarkar, A., & Bhaskara Rao, K. V. (2023). Unraveling anticancer potential of a novel serine protease inhibitor from marine yeast Candida parapsilosis ABS1 against colorectal and breast cancer cells. *World Journal of Microbiology and Biotechnology, 39*(8), 225. https://doi.org/10.1007/s11274-023-03670-9

Schoenhammer, K., Boisclair, J., Schuetz, H., Petersen, H., & Goepferich, A. (2010). Biocompatibility of an injectable in situ forming depot for peptide delivery. *Journal of Pharmaceutical Sciences, 99*(10), 4390–4399. https://doi.org/10.1002/jps.22149

Sharma, P., Kaur, N., & Shanavas, A. (2023). Targeting strategies using PLGA nanoparticles for efficient drug delivery. In *Poly(lactic-co-glycolic acid) (PLGA) Nanoparticles for Drug Delivery* (pp. 123–151). Elsevier. https://doi.org/10.1016/B978-0-323-91215-0.00016-9

Sheih, I. C., Fang, T. J., Wu, T. K., & Lin, P. H. (2010). Anticancer and antioxidant activities of the peptide fraction from algae protein waste. *Journal of Agricultural and Food Chemistry, 58*(2), 1202–1207. https://doi.org/10.1021/jf903089m

Singh, M., Hemant, K. S., Ram, M., & Shivakumar, H. G. (2010). Microencapsulation: A promising technique for controlling drug delivery. *International Journal For Multidisciplinary Research, 4*(6), 65–77. https://doi.org/10.36948/ijfmr.2022.v04i06.1106

Sonkusare, S., & Chaturvedi, A. (2023). Glutathione: The master antioxidant and its role in health and disease. *The USA Journals, 05*(05), 1–4.

Stöber, W., & Fink, A. (1968). Controlled growth of monodisperse silica spheres in the micron size range. *Journal of Colloid and Interface Science, 6*(2), 1–18.

Subramaniam, S., Selvaduray, K. R., & Radhakrishnan, A. K. (2019). Bioactive Compounds: Natural defense against cancer? *Biomolecules, 9*(12), 1–15. https://doi.org/10.3390/biom9120758

Sun, S.-W., Jeong, Y.-I., Jung, S.-W., & Kim, S.-H. (2003). Characterization of FITC-albumin encapsulated poly(DL-lactide-co-glycolide) microspheres and its release characteristics. *Journal of Microencapsulation, 20*(4), 479–488. https://doi.org/10.1080/0265204031000093014

Tabatabaei Mirakabad, F. S., Nejati-Koshki, K., Akbarzadeh, A., Yamchi, M. R., Milani, M., Zarghami, N., Zeighamian, V., Rahimzadeh, A., Alimohammadi, S., Hanifehpour, Y., & Joo, S. W. (2014). PLGA-based nanoparticles as cancer drug delivery systems. *Asian Pacific Journal of Cancer Prevention, 15*(2), 517–535. https://doi.org/10.7314/APJCP.2014.15.2.517

Tamber, H., Johansen, P., Merkle, H., & Gander, B. (2005). Formulation aspects of biodegradable polymeric microspheres for antigen delivery. *Advanced Drug Delivery Reviews, 57*(3), 357–376. https://doi.org/10.1016/j.addr.2004.09.002

Tao, A., Huang, G. Lo, Igarashi, K., Hong, T., Liao, S., Stellacci, F., Matsumoto, Y., Yamasoba, T., Kataoka, K., & Cabral, H. (2020). Polymeric micelles loading proteins through concurrent ion complexation and pH-cleavable covalent bonding for in vivo delivery. *Macromolecular Bioscience, 20*(1), 1900161. https://doi.org/10.1002/mabi.201900161

Torres y Torres, N., & Tovar-Palacio, A. R. (2009). La historia del uso de la soya en México, su valor nutricional y su efecto en la salud (The importance of soy in Mexico, its nutritional value and effect on health). *Salud Pública de México, 51*(3), 246–254. https://doi.org/10.1590/S0036-36342009000300016

Vernaza, M. G., Dia, V. P., Gonzalez de Mejia, E., & Chang, Y. K. (2012). Antioxidant and antiinflammatory properties of germinated and hydrolysed Brazilian soybean flours. *Food Chemistry, 134*(4), 2217–2225. https://doi.org/10.1016/j.foodchem.2012.04.037

Vlachopoulos, A., Karlioti, G., Balla, E., Daniilidis, V., Kalamas, T., Stefanidou, M., Bikiaris, N. D., Christodoulou, E., Koumentakou, I., Karavas, E., & Bikiaris, D. N. (2022). Poly(Lactic Acid)-based microparticles for drug delivery applications: An overview of recent advances. *Pharmaceutics, 14*(2), 359. https://doi.org/10.3390/pharmaceutics14020359

Wen, Y., Liu, A., Meng, C., Li, Z., & He, P. (2021). Quantification of lectin in soybeans and soy products by liquid chromatography-tandem mass spectrometry. *Journal of Chromatography B, 1185*, 122987. https://doi.org/10.1016/j.jchromb.2021.122987

Wick, D. A., Webb, J. R., Nielsen, J. S., Martin, S. D., Kroeger, D. R., Milne, K., Castellarin, M., Twumasi-Boateng, K., Watson, P. H., Holt, R. A., & Nelson, B. H. (2014). Surveillance of the tumor mutanome by T cells during progression from primary to recurrent ovarian cancer. *Clinical Cancer Research, 20*(5), 1125–1134. https://doi.org/10.1158/1078-0432.CCR-13-2147

Yarjanli, Z., Ghaedi, K., Esmaeili, A., Rahgozar, S., & Zarrabi, A. (2017). Iron oxide nanoparticles may damage to the neural tissue through iron accumulation, oxidative stress, and protein aggregation. *BMC Neuroscience, 18*(1), 51. https://doi.org/10.1186/s12868-017-0369-9

Section II

Determination of Biological Activities in Natural Products with Potential Human Health Benefits

Chapter 6

Computational and Technological Tools to Develop Therapeutic Foods To Be Used in Combination with Existing Drugs to Improve Efficacy in Diabetes Mellitus Treatment

Poonam Kalhotra
Veera C.S.R. Chittepu
Tzayhri Gallardo-Velázquez
and Guillermo Osorio-Revilla

CONTENTS

6.1 FOOD AS A SOURCE OF BIOACTIVE MOLECULES

Bioactive molecules found in food have already proven beneficial in reducing the risk of chronic diseases like cancer (Subramaniam et al. 2019), diabetes (Tran et al. 2020), and infectious diseases (Tomas et al. 2022). Varied crops, plants, and animal species are essential components of the human diet and, when consumed, play a significant role in different metabolic processes (Schaal, 2019; Dahiya et al. 2023). Plants produce primary metabolites that are directly involved

DOI: 10.1201/9781003332558-8

in plant growth and the metabolism of critical plant components (Pang et al. 2021). Secondary metabolism is linked to primary metabolism by creating a wide range of specialized bioactive molecules that benefit various therapeutic applications. Bioactive substances present in plants and animal sources involve nutrients and non-nutrients with physiological effects and nutritional qualities (Rein et al. 2013).

Fungi, algae, and other marine organisms are additional sources of these chemicals. Numerous publications today provide qualitative and quantitative information on bioactive compounds in foods. Both *in vitro* and *in vivo* reports show the beneficial effects of these bioactive compounds. Even though the health benefits of foods have been recognized since early times, their role as a source of valuable bioactive compounds has gained significant interest in recent decades. Recent studies show that a close relationship between health status and food consumption coexists (Hollman et al. 1999).

Bioactive compounds found in foods include steroids, alkaloids, terpenoids, flavonoids, sterols, lignins, tannins, curcumins, saponins, glucosides, phenolic compounds, essential oils, vitamins, minerals, and fibers, among others (Tungmunnithum et al. 2018; Dahiya et al. 2023); within these, phenolic compounds are the predominant phytochemicals in plants and are generally recognized as natural health-promoting compounds (Shahidi et al. 2005). Phenolic compounds are characterized by containing one or more aromatic rings with one or more hydroxyl groups in their chemical structure (Rahman et al. 2022); among these are flavonoids, stilbenes, coumarins, and tannins (Mutha et al. 2021). Adding phenolic compounds into our diet may provide additional health benefits (Pandey et al. 2009) associated with a reduced risk of developing chronic diseases (Sreenivasulu et al. 2022). Based on clinical study investigations, consuming these compounds, specifically, flavonoids like quercetin, in doses of 50–650 mg/day provides health benefits (Ożarowski et al. 2018).

Wild blueberry and blackberry have the highest total phenolic contents among more commonly consumed fruits, followed by pomegranate, cranberry, plum, raspberry, strawberry, red grape, and apple. So far, more than 5000 distinct phytochemicals have been extracted and identified in fruits, vegetables, and grains (Liu et al. 2013). Yet, a considerable fraction of these compounds still needs to be discovered despite advanced technologies in the characterization of chemicals in a complex mixture (Atanasov et al. 2021).

6.2 PREVALENCE OF DIABETES MELLITUS

Based on the report of Saeedi et al. (2019), globally, 578 million individuals will have diabetes by 2030, which is predicted to rise to 700 million by 2045. The prevalence of this disease will be 10.8% in urban and 7.2% in rural areas, 10.4% in high-income and 4.0% in low-income countries.

In diabetic people, free radicals scavenging and antioxidant capacity function are diminished due to an imbalance between the production of oxygen-derived radicals and the antioxidant system (Bajaj et al. 2012). Chronic hyperglycemia is a progressive situation linked with oxidative stress illness (Ji et al. 2020), and it is usually a significant challenge to provide effective treatment. People with a deficiency in the production of insulin are categorized as type 1 diabetes mellitus (T1DM) (Paschou et al. 2017), and when the function of insulin is impaired despite being produced by the pancreas, it is categorized as type 2 diabetes mellitus (T2DM) (Cohrs et al. 2020). T2DM is inherited in families as idiopathic diabetes mellitus (IDM) (Lonescu-Tirgoviste et al. 2018). Additionally, when it is present in women with pregnancy, it is categorized as gestational diabetes (Dall et al. 2014).

Exogenous insulin is commonly used to raise insulin levels in T1DM and T2DM therapies. However, insulin injections are inadequate to stop the course of diabetes and are related to the onset of long-term complications. Besides exogenous insulin, several treatments used to control glucose levels include glucagon-like peptide 1 (GLP1) analogs (Rowlands et al. 2018), dipeptidyl peptidase-4 (DPP-4) inhibitors (Gomez-Peralta et al. 2018), sulfonylureas (Sola et al. 2015), and repaglinide (Hezarkhani et al. 2013). Additionally, the World Health Organization estimated in 2019 that 80–99% of diabetes patients already use herbal medicine for their treatment, alone or in conjunction with existing antidiabetic medications (www.who.int/ttraditionalcomplementary integrativemedicine/WhoGlobalReportOnTraditionalAndComplementaryMediciene2019.pdf).

6.3 DIPEPTIDYL PEPTIDASE-4 AS A THERAPEUTIC DRUG TARGET

The serine protease DPP-4 enzyme is an integral cell membrane protein present in most tissue cells, including endocrine tissues, respiratory system, proximal digestive tract, gastrointestinal tract, liver and gall bladder, pancreas, kidney, urinary bladder, skin, bone marrow, and lymphoid tissues. It has a large extracellular domain C-terminal region and a short intracellular domain N-terminus, as illustrated in **Figure 6.1 (top)**. The catalytic site residues are present in the C-terminal region of the extracellular part of the protein (Trzaskalski et al. 2020). When food enters the duodenum, enteroendocrine L cells of the gut produce GLP-1 (the glucagon-like peptide 1). The incretin hormone GLP-1 reduces blood glucose levels by boosting insulin production and blocking glucagon release, but it is inactivated by the DPP-4 enzyme producing the GLP-1 inactive form disrupting glucose maintenance homeostasis (**Figure 6.1 bottom**). Hence, any mechanism to protect and maintain levels of GLP-1 is a way to control glucose homeostasis; thus, DPP-4 inhibitors like sitagliptin are approved for clinical use (Gallwitz, 2007).

DPP-4 inhibitors not only control glucose levels but also reduce hemoglobin A1c (HbA1c), a glycated hemoglobin. This is another biomarker for diabetes present in red blood cells that carry oxygen via hemoglobin (Hb). Studies have revealed that protease inhibitors effectively reduce HbA1c levels by 0.5–0.6% (Aschner et al. 2006, Defronzo et al. 2008). Notably, DPP-4 inhibitors are proven to reduce A1c more effectively in combination with other drugs like metformin, sulfonylureas, and thiazolidinediones (Ahrén et al. 2008). DPP-4 inhibitors work well, lowering both fasting and post-prandial glucose levels, and contributing to improving overall blood glucose control. Nevertheless, these agents tend to have a more significant effect on postprandial glucose than fasting glucose (Hinnen et al. 2015). The function of pancreatic beta cells and the development of diabetes go hand in hand because the role of beta cells is essential to maintain insulin levels when blood sugar levels rise. Therefore, therapies that can improve cell function may help slow the development of insulin resistance. DPP-4 inhibitors improve beta-cell function. Sitagliptin was the only DPP-4 inhibitor that improved insulin resistance when used alone (Fonseca et al. 2009; Hinnen et al. 2015).

There are other potential benefits associated with DPP-4 inhibitors such as reduced blood pressure to prevent chronic kidney disease (CKD) (Hattori et al. 2011), minimal risk of hypoglycemia (Hinnen et al. 2015), and renoprotective effect by reducing the urine albumin/creatinine ratio (ACR) (Fujita et al. 2014). In contrast to sulfonylureas and thiazolidinediones, DPP-4 inhibitors do not affect body weight (Goldstein et al. 2007). In addition to these benefits, DPP-4 inhibitors prevent retinopathy by increasing vasodilation (Forst et al. 2016) and improve wound healing ability in diabetic patients (Marfella et al. 2012). Despite the potential benefits, existing

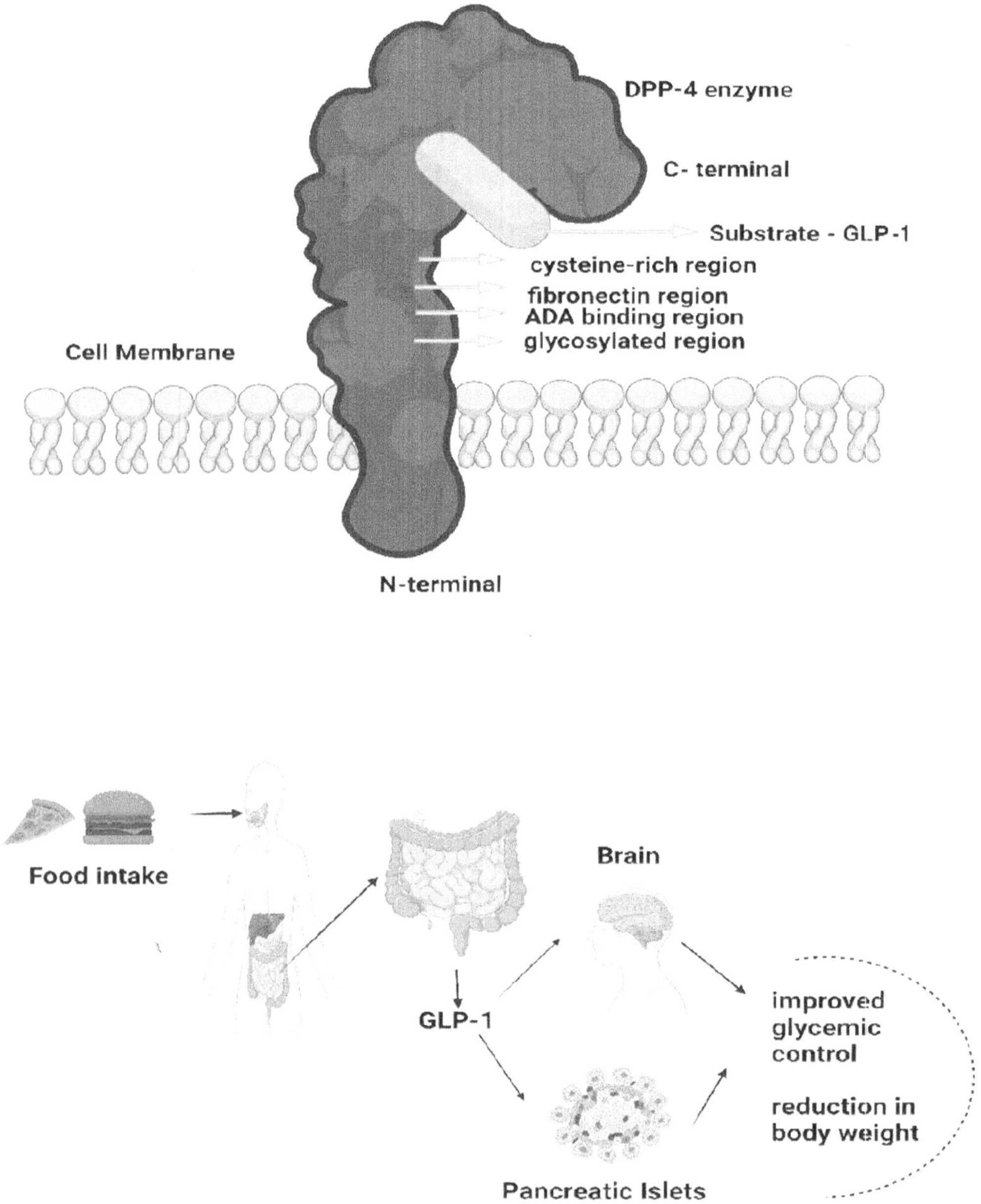

Figure 6.1 **(Top)** Dipeptidyl peptidase-4 (DDP-4) showing the large extracellular domain C-terminal region and the short intracellular domain N-terminal. In the figure the substrate incretin hormone GLP-1 is interacting with the extracellular C-terminal domain. **(Bottom)** Role of glucagon-like peptide-1 (GLP-1) as an antidiabetic agent. This image was created with BioRender.com.

DPP-4 inhibitors have some adverse effects, occurring in less than 5% of the population. The most commonly observed impacts are upper respiratory tract infection (Marfella et al. 2012), headache (Pathak et al. 2010), pancreatitis (Lee et al. 2019), arthralgia (Men et al. 2017), and nasopharyngitis (Willemen et al. 2011).

In recent years, it has been revealed that plant-based systems are abundant sources of bioactive compounds and have been employed by numerous ancient societies, particularly in North America. Today, plants continue to be a source of novel leads and new chemical entities in drug development (Atanasov et al. 2021). Furthermore, natural substances derived from plants in general possess less toxicity than some of the drugs approved for clinical use. An example of this is quercetin, a flavonoid that besides its ability to inhibit DPP-4, also shows anticancer and anti-inflammatory properties (Vafadar et al. 2020). It has an acute toxicity dosage (dose that cause 50% mortality) of 160 mg/kg (Salehi et al. 2020) compared with cisplatin, an FDA-approved drug to treat cancer, whose acute toxicity dosage is 32.7 mg/kg (Perse et al. 2018). This is one example to demonstrate that natural molecules provide several advantages over synthetic molecules, including acute toxicity and fewer side effects.

6.4 COMBINATION OF BIOACTIVE PEPTIDES AND POLYPHENOLS WITH INSULIN TO BENEFIT OUTCOME IN PATIENTS WITH DIABETES MELLITUS

Using computational technologies, critical features of DPP-4 inhibition activity in specific segments were studied by Kalhotra et al. (2019), developing the Activity Atlas, Average of Actives and Activity Cliff models for several DPP-4 inhibitors both in clinical use and in preclinical stage. **Figure 6.2 (top)** illustrates the resultant model of 107 inhibitors that summarizes the characteristics of chemical space as a function of Actives and Activity Cliff summary. Based on this model, a field template based on proven DPP-4 inhibitors was applied to get an idea of the Activity summary of four natural products: flavone, luteolin, resveratrol, and sitagliptin, a DPP-4 inhibitor approved by the FDA shown in **Figure 6.2 (bottom)**. This template was used to align already-proven DPP-4 inhibitors, leading us to develop the Activity Atlas Model. The generated model was then applied to screen or identify new polyphenols that could inhibit DPP-4. Following this, we used the molecular docking algorithm Flex X to obtain binding energies. *In silico* molecular docking calculations revealed that the flavonoids chrysin and galangin possess -21.4 kJ/ mol, and -24.162 kJ/mol free energy, respectively. The DPP-4 amino acids TRP 629, Val 546, TRP627, LYS 554, TYR 547, SER 630, and GLY 628 of DPP-4 interact with chrysin, and in the case of galangin, the interacting residues are with LYS 554, TYR 547, ASP 545, VAL 546, LY 632, TYR 631, SER 630, GLY 633, and GLY 628 shown in **Figure 6.3**. Both polyphenols discovered in our lab as DPP-4 inhibitors possess interacting residues SER 630 and TYR 547, which belong to the DPP-4 druggable region. The actual polyphenols' ability to inhibit DPP-4 as a therapeutic drug target was evaluated using *in vitro* assays and it was observed that they inhibited DPP-4, and that the effect was increased by the -OH group presence in galangin (whose IC50 is 40.13 μM) (Kalhotra et al. 2019).

Efficacy studies in patients have revealed that combining insulin with sitagliptin (Sato et al. 2015; Vilsbøll et al. 2010) improved glycemic response and was well tolerated by patients with diabetes mellitus. Also, cell-based model studies revealed that sitagliptin (a DPP-4 inhibitor), in combination with insulin, increased insulin function (increased glucose uptake), improving the efficacy of the use of insulin for the treatment of diabetes mellitus.

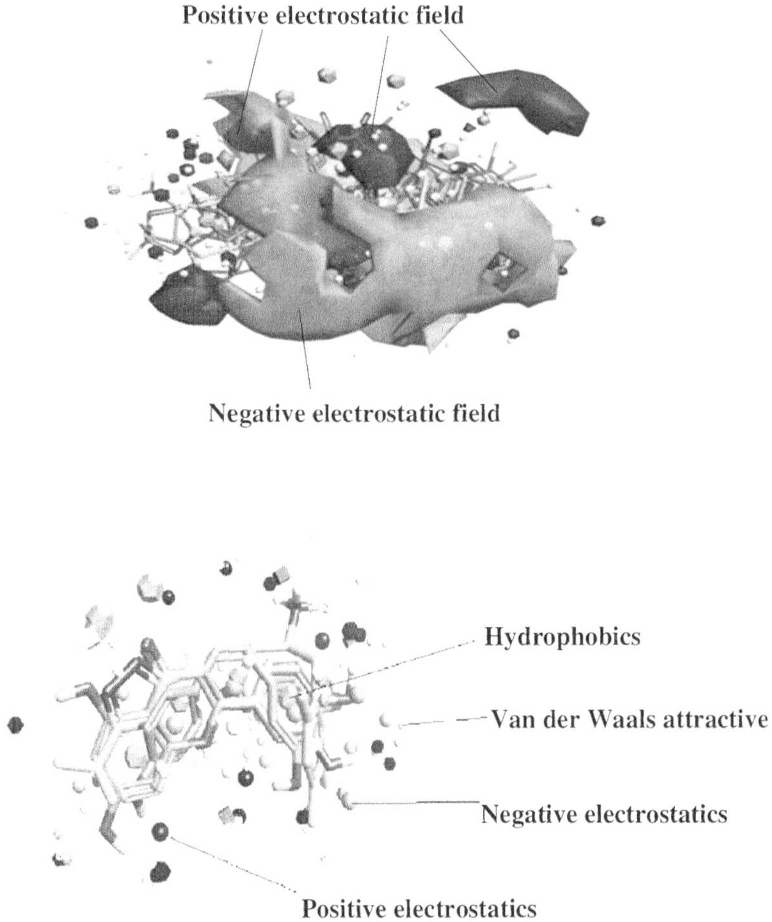

Figure 6.2 **(Top)** Inhibition of dipeptidyl peptidase-4 (DPP-4) enzyme activity by 107 chemicals as a function of positive and negative electrostatic fields. **(Bottom)** Field template tool of four bioactive compounds: resveratrol, flavone, luteolin, and sitagliptin. The figures were obtained from previous publications (Kalhotra et al. 2018; Chittepu et al. 2019).

Polyphenols like galangin are not toxic to skeletal muscle cells either alone or in combination with insulin, and indeed, it has been proven that polyphenols reduce the toxicity induced by insulin on muscle cells and benefit the uptake of glucose levels (Kalhotra et al. 2019). This finding adds to potential use of polyphenols in diabetes management strategies, offering dual benefits in terms of both glycemic control and cellular health.

Peptides from natural sources have also been proven to inhibit DPP-4. It was demonstrated that oxytocin, a cyclic natural peptide, can inhibit DPP-4 in a concentration-dependent manner (**Figure 6.4 bottom**). Infinity molecular docking simulations revealed that oxytocin binding is stable in the presence of many water molecules, and interactions are favorable. Upon binding oxytocin to DPP-4, the key interacting residues are ARG 356, PHE 355, TYR 663, GLU 204, GLU 203, HIS 124, PHE 355, GLU 203, GLU 204, and ASP 664. The inhibitory concentration (IC50) of natural

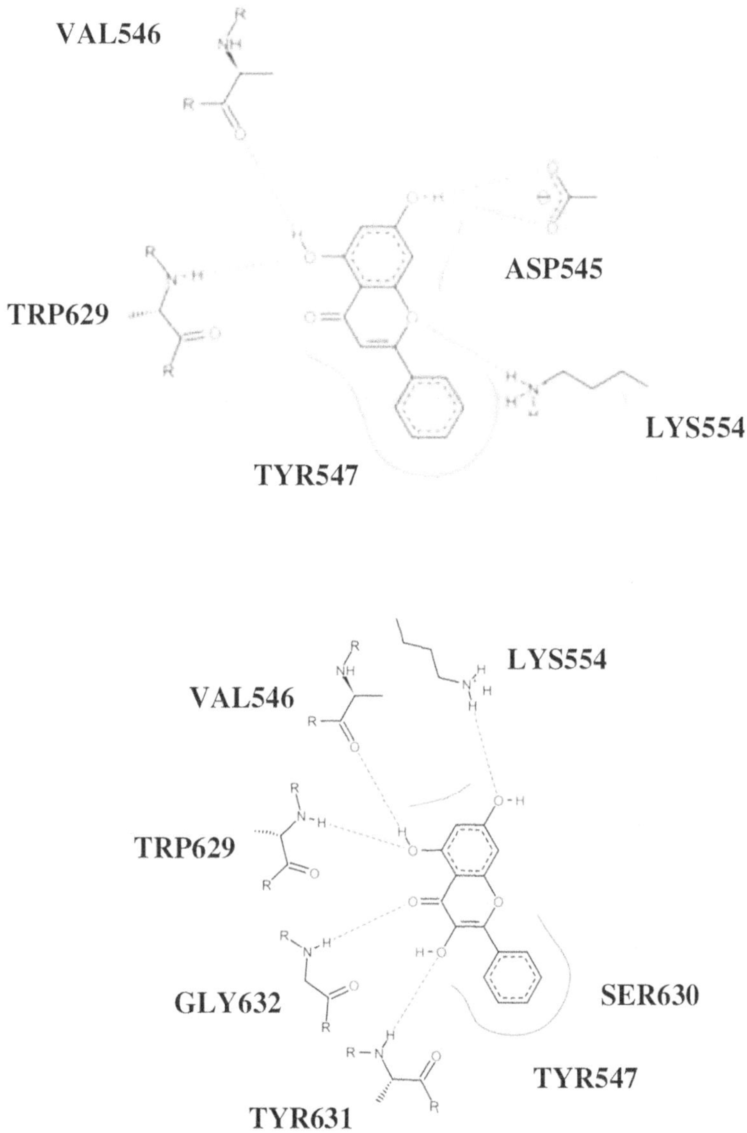

Figure 6.3 **(Top)** Visualization of a two-dimensional view of chrysin and galangin. **(Bottom)** Interacting residues with dipeptidyl peptidase-4 (DPP-4). Image obtained using the poseview tool, after molecular docking calculations.

peptide oxytocin is 110.7 nM (Chittepu et al. 2019), which is significantly effective in comparison with peptides from barley 3.91 mg/mL, buckwheat 1.98 mg/mL (Wang et al. 2015), whey protein 1.08 mg/mL(Nongonierman et al. 2013), amaranth flour 1.1 mg/mL (Velarde-Salcedo et al. 2013), and small peptides generated by dietary hydrolysates, which range from 3.5 to 2240 μM (Jao et al. 2015). Thus, the procedure and models previously described act as a guide for identifying the roles of biomolecules as DPP-4 inhibitors.

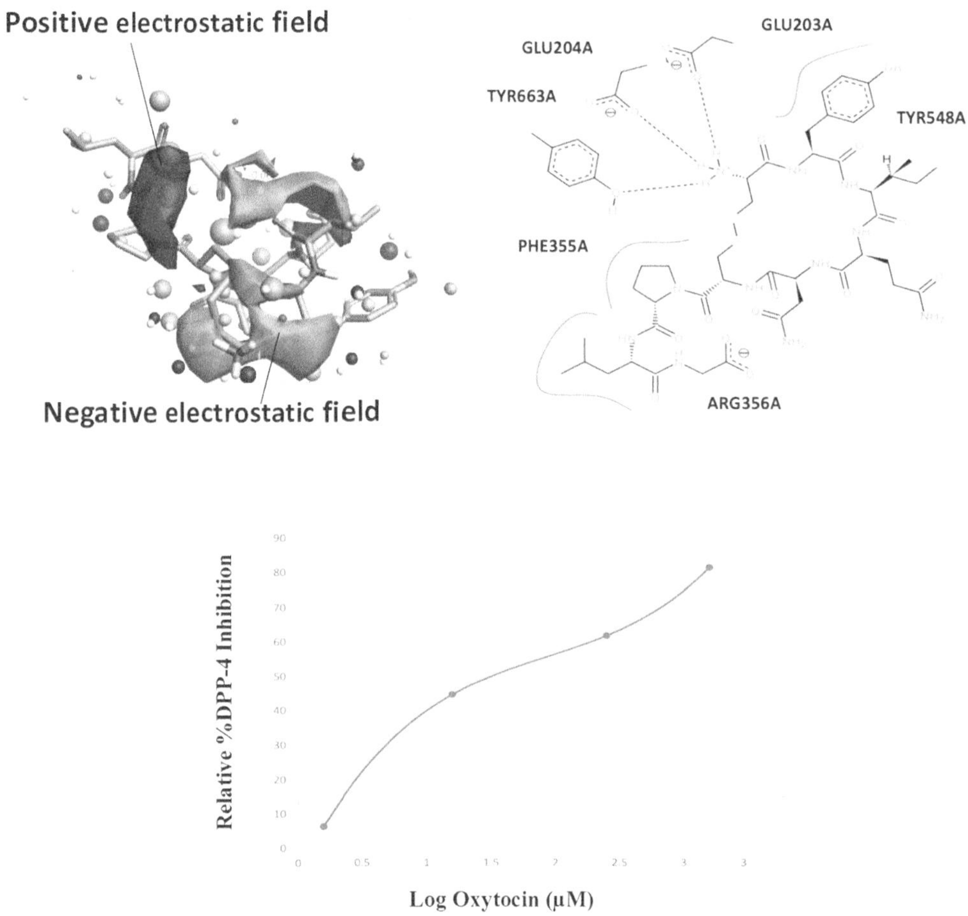

Figure 6.4 **(Top)** Visualization of positive and negative electrostatic fields and a two-dimensional view of oxytocin interacting residues with dipeptidyl peptidase-4 (DPP4). **(Bottom)** DPP4 inhibition curve. Interacting enzyme residues were obtained using the poseview tool. The figure related to half maximal inhibitory concentration (IC_{50}) were adopted from previous publication (Chittepu et al. 2019).

6.5 APPLICATION OF ULTRASOUND-ASSISTED EXTRACTION METHODS TO RETRIEVE BIOACTIVE COMPOUNDS AND BIOLOGICS FROM FOOD SOURCES AND FOOD BY-PRODUCTS

A study was conducted (Kalhotra et al. 2020) to reveal the possibility of producing or developing therapeutic food preparations containing polyphenols that can inhibit the DPP-4 enzyme. In this study, an ultrasonic-assisted garlic bulb extract was prepared, obtaining a half-maximal inhibitory concentration (IC50) value of 70.9 μg/mL. The obtained extract showed a higher DDP-4 inhibition capacity than previously reported plant extracts, such as black tea, green tea, and white tea (Ekayanti et al. 2018). Furthermore, the ultrasonic-assisted garlic bulb extract contained the polyphenols: malonylgenistin, catechin, caffeic acid 4-O-glucoside, and calenduloside E (determined using HPLC MS/MS), which proved to be responsible for the observed DPP-4 inhibition activity.

The prepared extract also induces skeletal muscle cell proliferation (Kalhotra et al. 2020) demonstrating the antidiabetic effect and its possible mechanism of action.

This methodology can also be applied for the successful extraction of bioactive compounds from food-byproducts, adding value to these food waste materials. In the following section we address the possibility of using microencapsulation to improve the stability and bioavailability of biomolecules from food, for which encapsulation technology is described herewith.

6.6 MICROENCAPSULATION TECHNOLOGIES TO IMPROVE THE BIOAVAILABILITY OF THERAPEUTIC FOOD COMPOUNDS

In the food and pharmaceutical industry, spray drying and freeze drying are the two microencapsulation processes predominantly utilized in developing microcapsules in powder form. However, hydrophilic bioactive compounds can be better protected using the combination of Water in Oil in Water (W/O/W) double emulsions and complex coacervation, protecting in this way the double emulsion globules with a double layer of protein and a polysaccharide. This coacervated double emulsion can then be spray dried to obtain a fluid powder (Cardenas-Bailon et al. 2015; Rojas-Moreno et al. 2018; García-Miguel et al. 2019). Spray-dried complex coacervates can be used to improve the functional properties of bioactive molecules in foods such as peptides and polyphenols, protecting them from the digestive tract, increasing the storage lifetime, and controlling the release of encapsulated compounds. **Figure 6.5** illustrates the practical application

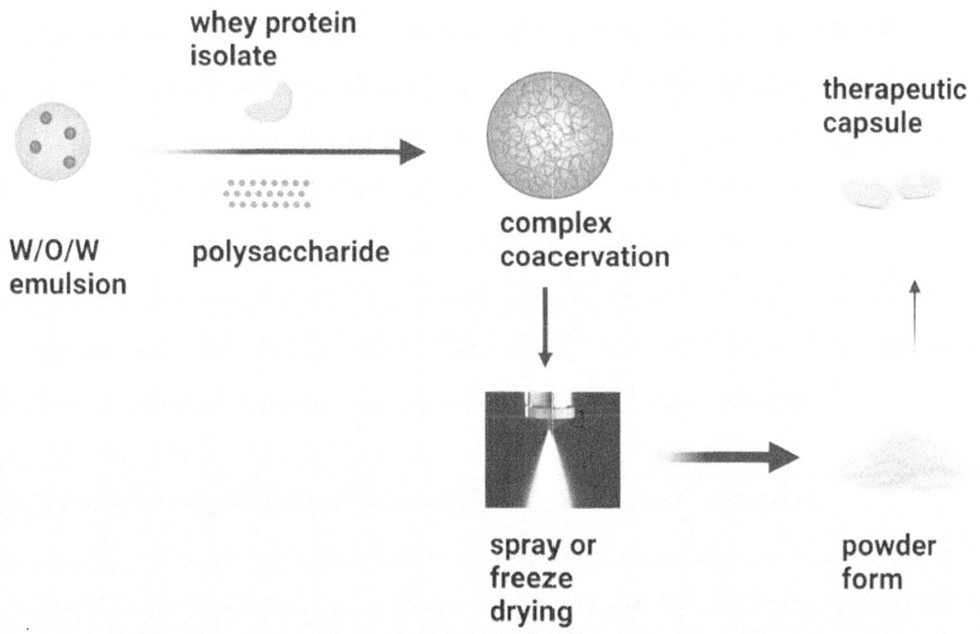

Figure 6.5 Technology to prepare water in oil in water (W/O/W) coacervated double emulsion microcapsules using spray drying method to benefit oral delivery of therapeutic food extracts with dipeptidyl peptidase-4 (DPP4) inhibition capacity. This image was created with BioRender.com.

of W/O/W double emulsions-complex coacervation-spray drying technology developed in our lab for the oral delivery of therapeutic bioactive compounds from foods. This technology enables the encapsulation of bioactive compounds possessing DPP-4 inhibition capacity that can be delivered as therapeutic foods, protecting them from the digestive tract.

This technology involves, initially, the formation of a W/O/W double emulsion, in which the bioactive compound of interest is present in the inner water globules. This double emulsion is prepared with soybean or other food oil as the oily phase, using polyglycerol polyricinoleate as the lipophilic surfactant and whey protein isolate as the hydrophilic surfactant. Later, these double emulsions are subjected to complex coacervation by coating with polysaccharides such as sodium alginate or carboxymethyl cellulose. These polysaccharides have proven to be the best for encapsulating bioactive molecules from food products, like biomolecules from essential oils, cactus pear fruit (*Opuntia ficus indica*), and proteins like insulin (Cardenas-Bailon et al. 2015; Rojas-Moreno et al. 2018; Toledo-Madrid et al. 2018; Boyano-Orozco et al. 2020; García-Miguel et al. 2019).

As a result, the technology of coating double emulsions with proteins such as WPI and a polysaccharide followed by spray drying yields the microcapsules. When these microcapsules are subjected to *in vitro* simulated digestion, the coat of polysaccharide protects the microcapsule from the gastric phase, and in the intestinal phase, they release small peptides from WPI that *per se* can inhibit DPP-4 in addition to the bioactive compound encapsulated within the double emulsion (García-Miguel et al. 2019). This evidence supports the idea that microcapsule complex with WPI alone hold the potential to be therapeutic. Consequently, this technology could be used as a vehicle for oral delivery of bioactive molecules from food. In the future, food capsules can be developed for treating chronic diseases to be used alone or in combination with existing treatments.

6.7 CONCLUSION

Computational methods like Activity Atlas Models and Molecular Docking Simulations were discussed in this chapter to assist in identifying mechanisms of DPP-4 inhibition for new and existing bioactive molecules in food. These methods can be used to develop and improve the efficacy of existing drugs in clinical use when applied in combination with bioactive compounds from foods to improve healthcare treatment outcomes, particularly to treat diabetes mellitus and other DPP-4-related diseases. Moreover, the use of ultrasound-assisted technology to extract bioactive molecules from food and food by-products offers an opportunistic method to develop plant extracts to enhance therapeutic outcomes.

Additionally, W/O/W double emulsions-complex coacervation-spray drying technology to assist in synthesizing therapeutic microcapsules was considered as a possible option for the protection of the gastrointestinal tract and delivery of polyphenols and peptides with DPP-4 inhibitory capacity. These microcapsules with custom DPP-4 inhibitory activity *per se* could be used for oral delivery of bioactive molecules from food to treat chronic diseases. It is expected that patient-tailored therapeutic food products will be available in the future. As a result, this chapter could potentially serve as a basic starting point for researchers worldwide interested in exploring the development of food as a therapeutic product that may have the potential to enhance the current treatment outcomes of patients with chronic diseases.

REFERENCES

Ahrén, B. (2008). Novel combination treatment of type 2 Diabetes DPP-4 inhibition + metformin. *Vascular Health and Risk Management* 4 (2), 383–394. https://doi.org/10.2147/vhrm. s1944

Aschner, P., Kipnes, M.S., Lunceford, J.K., Sanchez, M., Mickel, C., & Williams-Herman, D.E. (2006). Effect of the dipeptidyl peptidase-4 inhibitor sitagliptin as monotherapy on glycemic control in patients with type 2 diabetes. *Diabetes Care* 29 (12), 2632–2637. https://doi.org/10.2337/DC06-0703

Atanasov, A.G., Zotchev, S.B., Dirsch, V.M., Orhan, I.E., Banach, M., Rollinger, J.M., & Barreca, D. (2021). Natural products in drug discovery: Advances and opportunities. *Nature Reviews Drug Discovery* 20 (3), 200–216. https:// doi:10.1038/s41573-020-00114-z

Bajaj, S., & Khan, A. (2012). Antioxidants and diabetes decreased antioxidant defenses. *Indian Journal Endocrinology Metabolism* 16 (Suppl 2): S267–S271. https://doi.org/10.4103/2230-8210.104057

Boyano-Orozco, L., Gallardo-Velázquez, T., Meza-Márquez, O.G., & Osorio-Revilla, G. (2020). Microencapsulation of rambutan peel extract by spray drying. *Foods,* 9 (7), 899. https://dx.doi.org/ 10.3390/FOODS9070899

Cardenas-Bailon, F. Osorio-Revilla, G., & Gallardo-Velazque, T. (2015). Microencapsulation of insulin using a W/O/W double emulsion followed by complex coacervation to provide protection in the gastro-intestinal tract. *Journal of Microencapsulation* 32 (3), 308–316. https://dx.doi.org/10.3109/02652 048.2015.1017619

Chittepu, V.C.S.R., Kalhotra, P., Osorio-Gallardo, T., Jiménez-Martínez, C., Torre, R.R.R., Gallardo-Velazquez, T., & Osorio-Revilla, G. (2019). New molecular insights into the inhibition of dipeptidyl peptidase-4 by natural cyclic peptide oxytocin. *Molecules* 24 (21), 3887. https://dx.doi.org/10.3390/molecules24213887

Cohrs, C.M., Panzer, J.K., Drotar, D.M., Enos, S.J., Kipke, N., Chen, C., & Bozsak, R. (2020). Dysfunction of persisting β cells Is a key feature of early type 2 diabetes Pathogenesis *Cell Reports* 31(1), 107469. https:// dx.doi.org/10.1016/j.celrep.2020.03.033

Dahiya, D., Terpou, A., Dasenaki, M., & Nigam, P.S. (2023). Current status and future prospects of bioactive molecules delivered through sustainable encapsulation techniques for food fortification. *Sustainable Food Technology,* 1, 500-510. https://dx.doi.org/10.1039/d3fb00015j

Dall, T.M., Yang, W., Halder, P., Pang, B., Massoudi, M., Wintfeld, N., Semilla, A.P., Franz, J., & Hogan, P.F. (2014). The economic burden of elevated blood glucose levels in 2012: Diagnosed and undiagnosed diabetes, gestational diabetes mellitus, and prediabetes. *Diabetes Care* 37(12), 3172–3179. https://dx.doi.org/ 10.2337/dc14-1036

Defronzo, R.A, Fleck, P.R., Wilson, C.A., & Mekki, Q. (2008). Efficacy and safety of the dipeptidyl peptidase-4 Inhibitor alogliptin in patients with type 2 diabetes and inadequate glycemic control: A randomized, double-blind, placebo-controlled study. *Diabetes Care* 31 (12), 2315–2317. https://doi.org/10.2337/ dc08-1035

Ekayanti, M., Sauriasari, R., & Elya, B. (2018). Dipeptidyl peptidase IV inhibitory activity of fraction from white tea ethanolic extract (*Camellia Sinensis* (L.) Kuntze) ex vivo. *Pharmacognosy Journal.* https://doi. org/10.5530/pj.2018.1.32

Fonseca, V.A. (2009). Defining and characterizing the progression of type 2 diabetes. *Diabetes Care* 32 (suppl_ 2), S151–S156. https://doi.org/10.2337/DC09-S301.

Forst, T., Michelson, G., Diessel, S., Jahnke, J., & Kapitza, C. (2016). Microvascular effects of the inhibition of dipeptidylpeptidase IV by linagliptin in nondiabetic hypertensive patients. *Journal of Hypertension, 34*(2), 345–350. https://doi.org/10.1097/HJH.0000000000000776

Fujita, H., Taniai, H., Murayama, H., Ohshiro, H., Hayashi, H., Sato, S., Kikuchi, N., Komatsu, T., Komatsu, K., Komatsu, K., Narita, T., & Yamada, Y. (2014). DPP-4 inhibition with alogliptin on top of angiotensin II type 1 receptor blockade ameliorates albuminuria via up-regulation of SDF-1α in type 2 diabetic patients with incipient nephropathy. *Endocrine journal, 61*(2), 159–166. https://doi.org/10.1507/endo crj.ej13-0305

Gallwitz, B. (2007). Sitagliptin: Profile of a novel DPP-4 inhibitor for the treatment of type 2 diabetes. *Drugs Today* (Barc) 43 (1), 13–25. https://doi.org/10.1358/dot.2007.43.1.1043909

García-Miguel, E., Veera, C.S.R., Chittepu, P.K., Proal-Nájera, J., Osorio-Revilla, G., & Gallardo-Velázquez, T. (2019). In vitro digestion of microcapsule carriers for oral delivery of bioactive compounds for diabetes treatment and their inhibitory effect on the DPP-4 enzyme. *Applied Sciences* 9 (23), 5041. https://doi.org/10.3390/app9235041

Goldstein, B. J., Feinglos, M. N., Lunceford, J. K., Johnson, J., Williams-Herman, D. E., & Sitagliptin 036 Study Group (2007). Effect of initial combination therapy with sitagliptin, a dipeptidyl peptidase-4 inhibitor, and metformin on glycemic control in patients with type 2 diabetes. *Diabetes Care, 30*(8), 1979–1987. https://doi.org/10.2337/dc07-0627

Gomez-Peralta, F., Abreu, C., Gomez-Rodriguez, S., Barranco, R.J., & Umpierrez, G.E. (2018). Safety and efficacy of DPP-4 inhibitor and basal insulin in type 2 diabetes: An updated review and challenging clinical scenarios. *Diabetes Therapy* 9 (5), 1775–1789. https://doi.org/10.1007/s13300-018-0488-z

Hattori S. (2011). Sitagliptin reduces albuminuria in patients with type 2 diabetes. *Endocrine Journal*, 58(1), 69–73. https://doi.org/10.1507/endocrj.k10e-38

Hezarkhani, S., Bonakdaran, S., Rajabian, R., Shahini, N., & Marjani, A. (2013). Comparison of glycemic excursion in patients with new onset type 2 diabetes mellitus before and after treatment with repaglinide. *The Open Biochemistry Journal* 7, 19–23. https://doi.org/10.2174/1874091x01307010019

Hinnen, D.A. (2015). Therapeutic options for the management of postprandial glucose in patients with type 2 diabetes on basal insulin. *Clinical Diabetes: A Publication of the American Diabetes Association* 33 (4), 175–180. https://doi.org/10.2337/diaclin.33.4.175

Hollman, P.C., & Katan, M.B. (1999). Dietary flavonoids: Intake, health effects and bioavailability. *Food and Chemical Toxicology* 3 (9-10), 937–942. https://doi.org/10.1016/S0278-6915(99)00079-4

Ionescu-Tirgoviste, C., Gagniuc, P., & Guja, C. (2018). A commentary on classification of diabetes: Latent autoimmune diabetes in adults (LADA) or intermediary diabetes mellitus (IDM). *Acta Endocrinológica* 14 (4), 520–524. https://doi.org/10.4183/AEB.2018.520

Jao, C.L., Hung, C.C., Tung, Y.S., Lin, P.Y., Chen, M.C., & Hsu, K.C. (2015). The development of bioactive peptides from dietary proteins as a dipeptidyl peptidase IV Inhibitor for the management of Type 2 Diabetes. *BioMedicine* 5 (3), 14. https://doi.org/10.7603/S40681-015-0014-9

Ji, G., Rong, S., Haiyan, H., Fengsen, X., Xinjuan, Y., Veeraraghavan, V.P., Surapaneni, K.M., & Xiaoyan, C. (2020). Vannilic acid ameliorates hyperglycemia-induced oxidative stress and inflammation in streptozotocin-induced diabetic rats. *Journal of King Saud University – Science* 32 (7), 2905–2911. https://doi.org/10.1016/j.jksus.2020.04.010

Kalhotra, P., Chittepu, V. C. S. R., Osorio-Revilla, G., & Gallardo-Velázquez, T. (2018). Structure–activity relationship and molecular docking of natural product library reveal chrysin as a novel dipeptidyl peptidase-4 (DPP-4) inhibitor: An integrated in silico and in vitro study. *Molecules/Molecules Online/Molecules Annual, 23*(6), 1368. https://doi.org/10.3390/molecules23061368

Kalhotra, P, Chittepu, V.C.S.R., Osorio-Revilla, G., & Gallardo-Velázquez, T. (2019). Discovery of galangin as a potential DPP-4 inhibitor that improves insulin-stimulated skeletal muscle glucose uptake: A combinational therapy for diabetes. *International Journal of Molecular Sciences* 20 (5), 12–28 https://doi.org/10.3390/ijms20051228

Kalhotra, P, Chittepu, V.C.S.R., Osorio-Revilla, G., & Gallardo-Velázquez, T. (2020). Phytochemicals in garlic extract inhibit therapeutic enzyme DPP-4 and induce skeletal muscle cell proliferation: A possible mechanism of action to benefit the treatment of diabetes mellitus. *Biomolecules* 10 (2), 305. https://doi.org/10.3390/biom10020305

Lee M, Sun J, Han M, Cho Y, Lee J.Y, Nam C.M, Kang E.S. Nationwide Trends in Pancreatitis and Pancreatic Cancer Risk Among Patients With Newly Diagnosed Type 2 Diabetes Receiving Dipeptidyl Peptidase 4 Inhibitors. *Diabetes Care.* 2019 42(11):2057–2064. doi: 10.2337/dc18-2195. Epub 2019 Aug 20. PMID: 31431452.

Liu, R.H. (2013). Health-promoting components of fruits and vegetables in the diet. *Advances in Nutrition* 4 (3), 384S–392S. https://doi.org/10.3945/AN.112.003517

Marfella, R., Sasso, F. C., Rizzo, M. R., Paolisso, P., Barbieri, M., Padovano, V., Carbonara, O., Gualdiero, P., Petronella, P., Ferraraccio, F., Petrella, A., Canonico, R., Campitiello, F., Della Corte, A., Paolisso, G., & Canonico, S. (2012). Dipeptidyl peptidase 4 inhibition may facilitate healing of chronic foot ulcers in patients with type 2 diabetes. *Experimental diabetes research, 2012*, 892706. https://doi.org/10.1155/2012/892706

Men, P., He, N., Song, C., & Zhai, S. (2017). Dipeptidyl peptidase-4 inhibitors and risk of arthralgia: A systematic review and meta-analysis. *Diabetes & metabolism, 43*(6), 493–500. https://doi.org/10.1016/j.diabet.2017.05.013

Mutha, R.E., Tatiya, A.U.S., & Surana, J. (2021). Flavonoids as natural phenolic compounds and their role in therapeutics: An overview future. *Journal of Pharmaceutical Sciences* 7 (1), 25. https://doi.org/10.1186/S43094-020-00161-8

Nongonierma, A.B., & FitzGerald, R.J. (2013). Dipeptidyl peptidase IV inhibitory properties of a whey protein hydrolysate: Influence of fractionation, stability to simulated gastrointestinal digestion and food–drug interaction. *International Dairy Journal* 32 (1), 33–39. https://doi.org/10.1016/J.IDAIRYJ.2013.03.005

Ożarowski, M., Mikołajczak, P., Kujawski, R., Wielgus, K., Klejewski, A., Wolski, H., & Seremak-Mrozikiewicz, A. (2018). Pharmacological effect of quercetin in hypertension and its potential application in pregnancy-induced hypertension: Review of in vitro, in vivo, and clinical studies. *Evidence-Based Complementary and Alternative Medicine*, 2018,1–19. https://doi.org/10.1155/2018/7421489

Pandey, K.B., & Rizvi, S.I. (2009). Plant polyphenols as dietary antioxidants in human health and disease. *Oxidative Medicine and Cellular Longevity* 2 (5): 270–278. https://doi.org/10.4161/OXIM.2.5.9498.

Pang, Z., Chen, J., Wang, T., Gao, C., Li, Z., Guo, L., Xu, J., & Cheng, Y. (2021). Linking plant secondary metabolites and plant microbiomes: A review. *Frontiers in Plant Science* 12, 621276. https://doi.org/10.3389/FPLS.2021.621276/BIBTEX

Paschou, S.A., Papadopoulou-Marketou, N., Chrousos, G.P., & Kanaka-Gantenbein, C. (2017). On type 1 diabetes mellitus pathogenesis. *Endocrine Connections* 7 (1), R38–R46. https://doi.org/10.1530/ec-17-0347

Pathak R, Bridgeman MB. (2010). Dipeptidyl Peptidase-4 (DPP-4) Inhibitors In the Management of Diabetes. P T. 35(9):509–13. PMID: 20975810; PMCID: PMC2957740.

Perše, M., & Večerić-Haler, Ž. (2018). Cisplatin-induced rodent model of kidney injury: characteristics and challenges. *BioMed Research International*. 12, 1462802. https://doi.org/10.1155/2018/1462802

Rahman, M.M., Rahaman, M.S., Islam, M.R., Rahman, F., Mithi, F.M., Alqahtani, T., Almikhlafi, M.A., Alghamdi, S.Q., Alruwaili, A.S., Hossain, M.S., Ahmed, M., Das, R., Emran, T.B., & Uddin, M.S. (2022). Role of phenolic compounds in human disease: Current knowledge and future prospects. *Molecules* 27 (1), 233. https://doi.org/10.3390/MOLECULES27010233

Rein, M.J., Renouf, M., Cruz-Hernandez, C., Actis-Goretta, L., Thakkar, S.K., & da Silva-Pinto, M. (2013). Bioavailability of bioactive food compounds: A challenging journey to bioefficacy. *British Journal of Clinical Pharmacology* 75 (3), 588–602. https://doi.org/10.1111/J.1365-2125.2012.04425.X

Rojas-Moreno, S., Osorio-Revilla, O., Gallardo-Velázquez, M., Cárdenas-Bailón, M., & Meza-Márquez, G. (2018). Effect of the cross-linking agent and drying method on encapsulation efficiency of orange essential oil by complex coacervation using whey protein isolate with different polysaccharides. *J Microencapsul*. 35(2),165–180. https://doi.org/10.1080/02652048.2018.1449910

Rowlands, J., Heng, J., Newsholme, P., & Carlessi, R. (2018). Pleiotropic effects of GLP-1 and analogs on cell signaling, metabolism, and function. *Frontiers in Endocrinology* 9, 672. https://doi.org/10.3389/fendo.2018.00672

Saeedi, P., Petersohn, I., Salpea, P., Malanda, B., Karuranga, S., Unwin, N., Colagiuri, S., Guariguata, L., Motala, A.A., Ogurtsova, K., & Shaw, J.E.; IDF Diabetes Atlas Committee. (2019). Global and regional diabetes prevalence estimates for 2019 and projections for 2030 and 2045: Results from the international diabetes federation diabetes atlas, 9th Edition. *Diabetes Research and Clinical Practice* 157, 107843. https://doi.org/10.1016/j.diabres.2019.107843

Salehi, B., Machin, L., Monzote, L., Sharifi-Rad, J., Ezzat, S.M., Salem, M.A., Merghany, R.M., El Mahdy, N.M., Kılıç, C.S., Sytar, O., Sharifi-Rad, M., Sharopov, F., Martins, N., Martorell, M., & Cho, W.C. (2020). Therapeutic potential of quercetin: New insights and perspectives for human health. *ACS Omega* 5(20), 11849–11872. https://doi.org/10.1021/ACSOMEGA.0C01818/ASSET/IMAGES/LARGE/AO0C01818_0002.JPEG

Sato, S., Saisho, Y., Kou, K., Meguro, S., Tanaka, M., Irie, J., Kawai, T., & Itoh, H. (2015). Efficacy and safety of sitagliptin added to insulin in Japanese patients with type 2 diabetes: The EDIT randomized trial. *PLOS ONE* 10 (3), e0121988. https://doi.org/10.1371/JOURNAL.PONE.0121988

Schaal, B. (2019). Plants and people: Our shared history and future. *Plants, People, Planet* 1 (1), 14–19. https://doi.org/10.1002/PPP3.12

Shahidi, F., & Ho, Chi-Tang. (2005). Phenolics in food and natural health products: An overview. In: *Phenolic Compounds in Foods and Natural Health Products*. ACS Symposium Series. pp.1–8. https://pubs.acs.org/sharingguidelines.

Sola, D., Rossi, L., Schianca, G.P., Maffioli, P., Bigliocca, M., Mella, R., Corlianò, F., Fra, E.G.P., & Bartoli, D.G. (2015). Sulfonylureas and their use in clinical practice. *Archives of Medical Science* 11 (4), 840–848. https://doi.org/10.5114/aoms.2015.53304

Sreenivasulu, N., & Fernie, A.R. (2022). Diversity: Current and prospective secondary metabolites for nutrition and medicine. *Current Opinion in Biotechnology* 74, 164–170. https://doi.org/10.1016/J.COPBIO.2021.11.010

Subramaniam, S., Selvaduray, K.R., & Radhakrishnan, A.K. (2019). Bioactive compounds: Natural defense against cancer? *Biomolecules* 9 (12), 75. https://doi.org/10.3390/BIOM9120758

Toledo-Madrid, K., Gallardo-Velázquez, T., & Osorio-Revilla, G. (2018). Microencapsulation of purple cactus pear fruit (*Opuntia Ficus* Indica) extract by the combined method W/O/W double emulsion-spray drying and conventional spray drying: A comparative study. *Processes*, 6 (10), 189. https://doi.org/10.3390/pr6100189

Tomas, M., Capanoglu, E., Bahrami, A., Hosseini, H., Akbari-Alavijeh, S., Shaddel, R., Rehman, A., Rezaei, A., Rashidinejad, A., Garavand, F., Goudarzi, M., & Jafari, S.M. (2022). The direct and indirect effects of bioactive compounds against coronavirus. *Food Frontiers* 3 (1), 96–123. https://doi.org/10.1002/FFT2.119

Tran, N.B, Pham, B., & Le, L. (2020). Bioactive compounds in anti-diabetic plants: From herbal medicine to modern drug discovery. *Biology* 9 (9), 1–31. https://doi.org/10.3390/BIOLOGY9090252

Trzaskalski, N.A., Fadzeyeva, E., & Mulvihill, E.E. (2020). Dipeptidyl peptidase-4 at the interface between inflammation and metabolism. *Clinical Medicine Insights: Endocrinology and Diabetes* 13, 1179551420912972. https://doi.org/10.1177/1179551420912972

Tungmunnithum, D., Thongboonyou, A., Pholboon, A., & Yangsabai, A. (2018). Flavonoids and other phenolic compounds from medicinal plants for pharmaceutical and medical aspects: An overview. *Medicines* 5 (3), 93. https://doi.org/10.3390/MEDICINES5030093

Vafadar, A., Shabaninejad, Z., Movahedpour, A., Fallahi, F., Taghavipour, M., Ghasemi, Y., Akbari, M., Shafiee, A., Hajighadimi, S., Moradizarmehri, S., Razi, E., Savardashtaki, A., & Mirzaei, H. (2020). Quercetin and cancer: new insights into its therapeutic effects on ovarian cancer cells. *Cell Bioscience* 10, 32. https//:doi: 10.1186/s13578-020-00397-0s

Velarde-Salcedo, A.J., Barrera-Pacheco, A., Lara-González, S., Montero-Morán, G.M., Díaz-Gois, A., González De Mejia, E., & Ana P. Barba De La Rosa, A.P. (2013). *In vitro* inhibition of dipeptidyl peptidase IV by peptides derived from the hydrolysis of amaranth (Amaranthus Hypochondriacus L.) proteins. *Food Chemistry* 136 (2), 758–764. https://doi.org/10.1016/J.FOODCHEM.2012.08.032

Vilsbøll, T., Rosenstock, J., Yki-Järvinen, H., Cefalu, W.T., Chen, Y., Luo, E., Musser, J., Andryuk, Y., Ling, K., Kaufman, D., Amatruda, J.M., Engel, S.S., & Katz, L. (2010). Efficacy and safety of sitagliptin when added to insulin therapy in patients with type 2 diabetes. *Diabetes, Obesity and Metabolism* 12 (2), 167–177. https://doi.org/10.1111/J.1463-1326.2009.01173.X

Wang, F., Yu, G., Zhang, Y., Zhang, B., & Fan, J. (2015). Dipeptidyl peptidase IV inhibitory peptides derived from oat (*Avena Sativa* L.), buckwheat (*Fagopyrum Esculentum*), and Highland Barley (*Hordeum Vulgare Trifurcatum (L.) Trofim*) proteins. *Journal of Agricultural and Food Chemistry* 63(43), 9543–9549. https://doi.org/10.1021/ACS.JAFC.5B04016

Willemen, M. J., Mantel-Teeuwisse, A. K., Straus, S. M., Meyboom, R. H., Egberts, T. C., & Leufkens, H. G. (2011). Use of dipeptidyl peptidase-4 inhibitors and the reporting of infections: a disproportionality analysis in the World Health Organization VigiBase. *Diabetes care*, *34*(2), 369–374. https://doi.org/10.2337/dc10-1771

Chapter 7

Anti-Inflammatory Effect of *Aloe barbadensis* on Cognitive Disorders

Arturo A. Velázquez-López
María del Carmen Silva-Lucero
Claudia Pérez-Cruz
Brenda Pérez-Grijalva
Rosalva Mora-Escobedo
and Antonio Tornero-Martínez

CONTENTS

7.1 INTRODUCTION

Aloe barbadensis is a plant native to sub-Saharan Africa from the Saudi Arabian Peninsula. *Aloe barbadensis* Miller was classified by the botanist Philip Miller, who reported the plant as native to the island of Barbados and introduced it to the world as a product of maritime trade in the Caribbean (Sánchez-Machado et al. 2017; Majumder et al. 2019).

Taxonomically, the Aloe plant belongs to the Aloeaceae family, related to the Liliaceae family (garlic and onion). Some of the best-known species are *Aloe arborescens, Aloe chinensis, Aloe socotrino*, and *Aloe ferox* (Domínguez-Fernández et al. 2012).

DOI: 10.1201/9781003332558-9

Aloe barbadesis is a perennial, drought-resistant plant with thick, conical, lance-shaped green, juicy, basal, pointed leaves with jagged edges. The leaves form wrapped to the stem forming a rosette. The leaves contain latex and gel; the juice or sap is the latex that is present below the epidermis layer, which occupies approximately 20 to 30% of the weight of the entire leaf (Chelu et la. 2023).

The leaves are constituted of bark (skin), which is a thick epidermis covered with a cuticle that surrounds the mesophyll, which represents 20 to 30% of the total weight of the plant; this can be differentiated into chlorenchyma cells and thinner wall cells that form the parenchyma (65 to 80% of total weight). The leaves secrete two exudates, a reddish-yellow juice called latex or acibar, produced by the pericyclic cells under the cutinized epidermis of leaves. The other exudate, a transparent and slippery mucilage or gel called Aloe gel, is produced by the thin-walled tubular cells in the leaf's inner central area (parenchyma) (Sánchez-Machado et al. 2017).

Aloe barbadensis contains more than 70 active components, including anthraquinones, vitamins, non-essential amino acids, essential amino acids, and inorganic compounds; the wide variety of components allows these compounds to penetrate the skin tissue and provide essential nutrients (Farrugia et al. 2019). These compounds are related to anti-inflammatory and neuroprotective effects.

7.1.1 Chemical Composition of *Aloe barbadensis*

The *Aloe barbadensis* chemical composition depends on various factors: the type and conditions of cultivation, harvest time, climate, the position of the leaves on the stem, and the method for harvesting. The optimal time for harvesting the leaves is after three years of plant development because it has the highest polysaccharides and flavonoids. On a dry matter basis, aloe gel consists of polysaccharides (55%), sugars (17%), minerals (16%), proteins (7%), lipids (4%), and phenolic compounds (~1%); the latter is attributed to the potential healing present in the inner parenchymal tissue of the leaf (Kumar et al. 2019).

Aloe barbadensis is characterized by the presence of potentially active components (**Table 7.1**); polysaccharides (pectin (0.02%), hemicellulose (0.3%), polymannan (0.7%), flavonoids (0.047%), sterols (0.0095%), enzymes (0.013%), and minerals (calcium, chromium, copper, iron, magnesium, manganese, potassium, phosphorus, sodium, and zinc) (0.25%). In addition to traces of terpenoids, lecithins, saponins, fatty acids, anthraquinones, tannins, salicylic acid, and vitamins (A, C, E, β-carotene, B1, B2, B3, choline, B12, folic acid) (Borges-Argáez et al. 2019; Hęś et al. 2019).

The Aloe phenolic compounds are classified into two groups: the first group comprises chromones, such as aloesin, aloesol, and isoaloeresin, and the second group denominated anthraquinones (free and glycosylated), including barbaloin, isobarbaloin, and aloe-emodin (AE), located in the inner layer of epidermal cells (Majumder et al. 2019).

Latex is composed mainly of secondary metabolites, principally aloin A, aloin B, aloesin, aloeresin, and AE. The aloins are used as a laxative due to the presence of substitution with hydroxyls at carbon eight (C8) (Rodríguez et al. 2010).

7.1.2 Polysaccharides of *Aloe barbadensis*

Polysaccharides are the most abundant compounds in aloe gel, the most common being glucose-mannose polysaccharides, mainly polymannans (~ 0.7%) consisting of straight chains having a higher amount of mannose with a lower amount of glucose. Among the polymannans,

TABLE 7.1 CHEMICAL COMPOSITION OF ALOE GEL

Group	Chemical Compounds	References
Polysaccharides	Mannans, acemannans, glucogalactomannans, galactans, galactogalacturans, arabinogalactans, galactoglucoarabinomannans, pectic substances, xylans, cellulose	Reynolds & Dweck (1999); Bajpai, (2018)
Polyphenols (Anthraquinones, anthrones, and chromones)	AE, aloetic acid, anthranol, aloin A and B (known as barbaloin), isobarbaloin, emodin, and cinnamic acid ester 8-C-glucosyl-(2'-O-cinnamoyl)-7-O-methylalloediol A, 8-C-glucosyl-(S) aloesol, 8-C-glucosyl-7-O-methyl-(S)-aloesol, 8 - C-glucosyl-7-O methylalloediol, 8-C-glucosyl-noreugenin, isoaloeresin D, isorabaicchromone, neoaloesin A.	Kumar et al. (2019)
Enzymes	Alkaline phosphatase, amylase, carboxypeptidase, catalase, cyclo-oxidase, cyclooxygenase, lipase, oxidase, phosphoenolpyruvate carboxylase, superoxide dismutase	Gullón et al. (2015)
Inorganic compounds	Calcium, chlorine, chromium, copper, iron, magnesium, manganese, potassium, phosphorus, sodium, zinc	Domínguez - Fernández et al. (2012)
Organic compounds and lipids	Arachidonic acid, γ-linolenic acid, steroids (campesterol, cholesterol, β-sitosterol), triglycerides, triterpenes, gibberellins, lignins, potassium sorbate, salicylic acid, uric acid	Bajpai (2018)
Essential and non-essential amino acids	Alanine, arginine, aspartic acid, glutamic acid, glycine, histidine, hydroxyproline, isoleucine, leucine, lysine, methionine, phenylalanine, proline, threonine, tyrosine, valine	Nair et al. (2016)
Vitamins	B1, B2, B6, C, β-carotene, choline, folic acid, α-tocopherol	Rodríguez et al. (2010)

acemannan is the main polysaccharide composed of one or more polymers of different lengths (Kumar et al. 2019). There are a variety of aloe polysaccharides, such as galactans, arabinans, pectins, hemicellulose, acid galactoarabinans, and arabinorhamno-galactan, that form complex mucopolysaccharides (Gupta & Rawat, 2017), which could have a synergistic action with the other active compounds (Shi et al. 2018).

Conjugated phenolic compounds bind to soluble molecules such as carbohydrates, proteins, and some lipids by esterification at the carboxylic moiety or esterification at the hydroxyl group. Insoluble phenolic compounds are usually covalently attached to polymers, such as polysaccharides and lignin, through an ester bond and released through acid, alkaline, or enzymatic hydrolysis (Hęś et al. 2019).

Acemannans in aloe gel contain mainly (1→4)-β-mannopyranosyl units with small amounts of (1→4)-β-glucopyranosyl units. The amount of mannose varies from 2.8:1 to 22:1. The branches present units of galactose, mannose, and arabinose linked along the leading chains. Acetyl groups attached to the main chain are essential functional groups; the content varies from 1.1 to 25%, and the acetylated positions could be O-2, O-3, O-6, and O-2,3. Structurally, acetyl groups are the only non-sugar functional groups present in acemannan (Minjares-Fuentes & Femenia, 2019). The molecular weight of acemannan varies from 1.2 MDa to >200 kDa or 1.5 kDa; these differ due to the plant's geographical locations of development and growth, seasonal variations, culture conditions, extraction methods, and preparation of polysaccharides (Shi et al. 2018).

7.1.3 Polyphenols

Aloe barbadensis is utiliced as a medicinal plant due to the content of bioactive compounds: the high percentage of polysaccharides with biological activities, anthraquinones, anthrones, and chromones. Most phenolic compounds are found in the exudate or latex, which is yellow-brown (Minjares-Fuentes & Femenia, 2019). Anthraquinones represent around 0.1% of the phenolic compound composition in aloe gel, including aloin A, aloin B, AE, chrysophanol, aloenin, and aloesaponol I, II, III, IV (Gao et al. 2019). Anthraquinones are compounds derived from the anthracene of the quinone group; they are known for their solid and mild laxative activities and exhibit antioxidant, antiviral and cytotoxic effects (Rodríguez et al. 2010). AE and aloin are essential anthraquinones. AE is a compound with a primary alcohol group found in Aloe gel. Anthrones are considered typical constituents of aloe leaf exudate, most of which are represented by aloin (Minjares-Fuentes & Femenia, 2019). Aloin is an AE derivative with glycosylation.

Generally, aloin occurs naturally as a mixture of two diastereoisomers (aloin A and B) that are unstable in aqueous solutions (Rodríguez et al. 2010). Most of the chromones are derivatives of the chemical compound C-glucosyl-7-hydroxy-5-methyl-2-propyl-4-chromone. These compounds vary depending on the degree of oxidation in the propyl side chain, the methylation of the hydroxyl group in C-7, and the esterification of the glucose moiety (Minjares-Fuentes & Femenia, 2019).

7.2 BIOMEDICAL APPLICATION OF *ALOE BARBADENSIS*

Aloe barbadensis has medicinal uses; for example, it is used for constipation, gastrointestinal disorders, and immunological deficiencies, and it has antioxidant, antimicrobial antitumor, hypolipidemic, and antidiabetic activity (Tariq et al. 2019; Altinkaynak et al. 2023). *Aloe barbadensis* also reduced the severity of mucositis and stomatitis because anthraquinones are analogous structures of tetracycline that inhibit protein synthesis in bacteria and reduce the viability of cancer cell lines (breast, cervical, and melanoma) (de Carvalho et al. 2022).

Some research indicates that the consumption of aloe gel is beneficial to the human body because it reduces blood glucose levels, increases insulin levels, improves pancreatic islets and the well-being of diabetics, and also reduces the inflammatory process, myocardial oxidative stress, and antiviral activity by acemannan (Kumar et al. 2019; Sánchez et al. 2020).

7.2.1 Anti-Inflammatory Activity

Inflammation is a central and required process in defense of animal cells against injuries or infections; it regularly develops into an acute or chronic infection, causing various diseases, including neurodegenerative disorders. The neuroinflammation pathway involves modification of the physical location of white blood cells (monocytes, basophils, eosinophils, and neutrophils), plasma, and fluids at the inflamed site, and the secretion of mediators and other signaling molecules released by cells of immune defense that contribute to the inflammatory process.

Inflammatory processes can be start with a state of physical tension and release reactive oxygen species (ROS) and nitrogen species (NO) that activate the nuclear factor-kb chain and enhancer of activated B cells (NF-kB). This activation leads to an increase in the production of the tumor necrosis factor-alpha (TNF-α), interleukin-6 (IL-6), interleukin 1-B (IL-1b), and interleukin-2 (IL-2) (**Figure 7.1**). Chronic inflammation is distinguished by mononuclear cell infiltration, fibroblast proliferation, connective tissue formation, and the persistence of inflammatory stimuli,

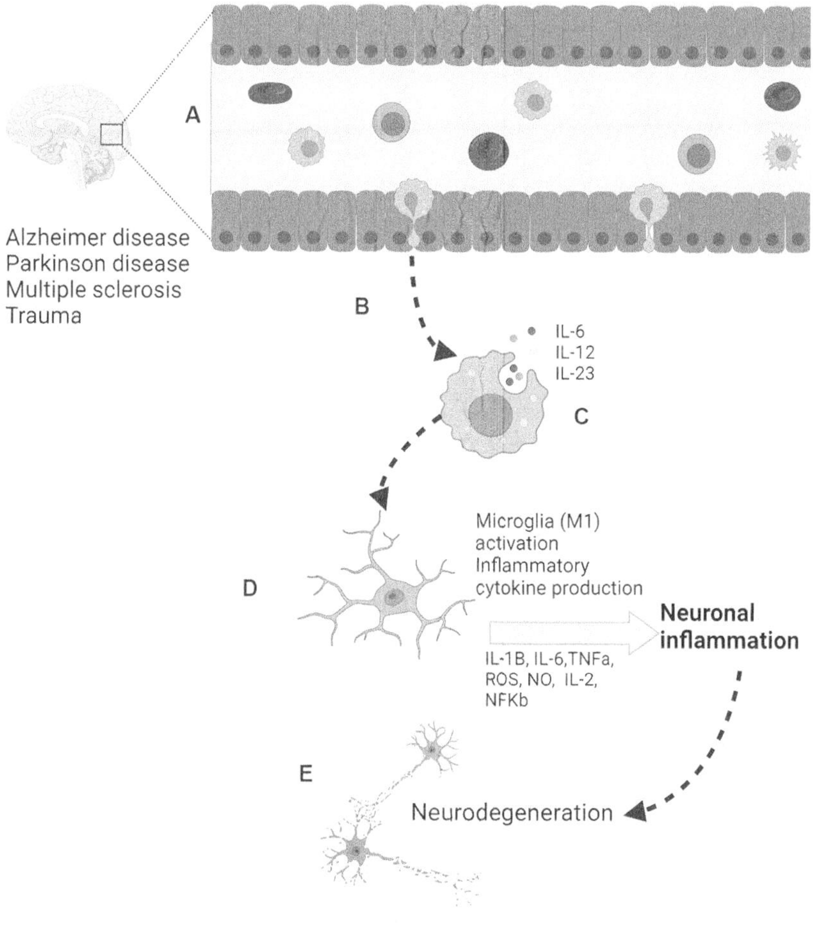

Figure 7.1 Description of neuroinflammation caused by chronic degenerative diseases such as Alzheimer's, Parkinson's, and Multiple Sclerosis, among others. (A) Increase in proinflammatory cytokines such as IL-6, IL-8, IL-23, and TNFa in neurons. (B) These cytokines generated in the brain cross the brain blood barrier, functioning as signals from the nervous system. (C) Stimulation of the vagus nerve increases the release of proinflammatory cytokines and chemokines in the central nervous system. (D) Microglia and astrocytes activation, triggering neuronal inflammation. (E) Prolonged neuronal inflammation by activation of astrocytes and microglia promote degeneration and development of cognitive disorders.

contributing to the development of complex diseases (neurological, metabolic, and cardiovascular) (Lewis et al. 2021).

Aloe barbadensis has anti-inflammatory properties, decreasing the levels of TNF-α in plasma and colon and also reducing the synthesis of interleukin-1β. Aloesin decreases the synthesis of NF-kB in adipose tissue, the production of proinflammatory cytokines and leukocyte adhesion, and increases IL-10 levels. AE decreases the expression of metalloproteins via glycolysis and NF-kB, and inhibits the expression of nitric oxide synthase, the levels of cyclooxygenase, and the production of prostaglandins (Chandraa et al. 2023). The aloe gel reduces the production of prostaglandin E2 and IL-8 in epithelial cells of the colon. AE and barbaloin can be hydrolyzed into salicylic acid through the Kolbe reaction, which has anti-inflammatory activity.

7.2.2 Cognitive Disorders

Many neuroinflammatory diseases are linked to stressors, depression, anxiety, multiple sclerosis, epilepsy, Parkinson's disease (PD), Alzheimer's (AD), and other dementias (Visser et al. 2022). These diseases are called neurodegenerative or cognitive diseases, and are defined as medical conditions that affect the nervous system, capable of modifying the structural, biochemical, or electrical functions of the brain, spinal cord, or nerves. Epilepsy, schizophrenia, AD, PD, multiple sclerosis, and migraine are classified as neuropsychiatric and neurological sequelae, such as cerebrovascular diseases, poliomyelitis, tetanus, meningitis, and encephalitis (Jeromin & Bowser, 2017).

The pathophysiology of cognitive diseases has essential characteristics, such as accumulating certain compounds, generally proteins that affect the nervous system. For instance, the neuropathological characteristics of AD are extracellular Aβ plaques and intracellular neurofibrillary tangles. Another main characteristic is the development of gliosis, activated astrocytes, and microglia found in abundance near neurons and plaques. Once activated, astrocytes and microglia synthesize many proinflammatory signaling molecules (Ball et al. 2020). A balanced diet is related to general health, preventing diseases, and improving quality of life. Aloe polysaccharides (0.04 g/L) can affect neurobiology and cognitive function (Chandraa et al. 2023).

Aloe polysaccharides combined with proteins and lipids accomplish synaptogenesis and produce neurotransmitters, contributing to cognitive enhancement. Polymannans, at a dose of 20 g/day, present a significant relationship between mature brain-derived neurotrophic factor (BNDF) and its precursor (proBNDF), which is crucial in neuro synaptic function, apoptosis, plasticity, and learning memory processes, improving symptoms of cognitive diseases (Lewis et al. 2021).

Another important compound found in aloe gel is AE, with neuroprotective activity (20 μM/day); it decreases malondialdehyde, increases neuronal cell intensity and production of acetylcholine, decreases the apoptosis of ganglion cells, and forms the shrunken and irregular shape of a neuronal cell in the hippocampus brain region improving cognitive deficit in neurodegenerative diseases (Ma et al. 2014). Furthermore, AE offers protection against reperfusion of bilateral carotid arteries (MO/RCA) by decreasing the levels of ROS and malondialdehyde (MDA) and increasing glutathione (GSH) and catalase (CAT) levels. Moreover, AE inhibits the effect on TNF-α, caspase-3 (CASP3), nitric oxide synthase (NOS), and glutamate receptor (GR), suggesting its substantial protective effect against the cerebral injury (Pasala et al. 2022). AE showed neuroprotector activity by increasing the activities of superoxide dismutase, glutathione peroxidase, and acetylcholine content, modulating oxidative stress in the brain and improving cognitive disease symptoms (Dong et al. 2020).

In the colon, various undigested and unabsorbed compounds are fermented, improving the production of short chain fatty acids (SCFA) and promoting their absorption. SCFA, principally acetate, propionate, and butyrate, are produced by the fermentation of indigestible polysaccharides (lignin, cellulose, and some hemicelluloses). Butyrate, the most important SCFA, allows stimulating colonocytes, conferring multiple anti-inflammatory properties on gut epithelial, immune cells and thus neuroprotective properties (Shandilya et al. 2021).

7.3 MOLECULAR BIOMARKERS ON COGNITIVE DISORDERS

Biomarker is an quantitative indicator of a definable biological or pathologic process, observed outside the patient or person, that may be used in diagnosis or to monitor therapy. The quantification of biomarkers supports the diagnosis of lesions in cognitive diseases, monitors the progression of the disease, and predicts the appearance of symptoms (Visser et al. 2022). Microglia and astrocytes are the prominent participants in the inflammatory response in the central nervous system. Microglia are the resident macrophages in the central nervous system; in an active state, they secrete numerous proinflammatory cytokines and chemokines such as IL-1, IL-6, IL-12, and TNF-α (Enache et al. 2019). Older people have microglia, which respond to more stimuli and produce higher amounts of proinflammatory cytokines that can cause persistent neuroinflammation and develop neuropathological diseases. Astrocytes are the most numerous glial cells, and are activated by proinflammatory cytokines produced by microglia; they have phagocytic properties and secretion of proinflammatory cytokines (Enache et al. 2019).

Some treatments with phytochemicals decrease the concentration of proinflammatory cytokines in various types of cells (leukemia, lungs, hepatoma, neuroectodermal, glial cells, and neurons) (Acevedo et al. 2004, Almeida et al. 2020, Saleh et al. 2021).

AE applicated (40 μM) in glial cell decreases the concentration and viability of activated glial cell; likewise, it decreases the activity of PKC, inhibits SVG and U-373 MG cells, and induces apoptosis (Almeida et al. 2020). In addition, the treatment in rats with AE (10 mg/Kg) decreases the brain damage previously caused with controlled explosions, decreasing the concentration of nitrate (NO) related to neuroinflammation (Ma et al. 2014).

The application of AE (20–80 mg/kg) in a mouse model reverses the effects of depression, improving motor skills and corticosterone levels, working as an antidepressant (Li et al. 2021). Zhong et al. (2019) report that rats with brain damage induced improved remarkably with aloin treatment (30 mg/kg) for eight weeks, observing amelioration in motor deficiencies, decreased lipid peroxidation, and increased superoxide activity dismutase in the hippocampus.

Colonic fermentation of glucomannans and acemannans increases the production of SFCA (acetate and butyrate) and variability of concentration of phenolic compounds, allowing different effects in the brain, such as reducing the activation of astrocytes (Gupta & Rawat, 2017). Rats with hippocampal damage improve cognitive behavior with methanolic extracts of aloe vera (32 mg/kg), decreasing lipoperoxidation and carbonyl content in the hippocampus (Farrugia et al. 2019).

Tornero-Martinez et al. (2022) evaluated aloe gel, polysaccharide extract, and colonic fermentation of both. Human astrocytoma cell line U373 MG was used. Cells treated with colonic ferments of aloe gel (40mg/mL) for 4 and 24 hours are reduced by 69% and 9.69%, respectively. The first hours of polysaccharide fermentation present high cell reduction due to the synergistic effect of the phenolic compounds and a high concentration of butyric acid produced, blocking the mitogen-activated protein kinases (MAPK) pathway, inhibiting of interleukin-6 (IL-6) production.

Phenolic compounds can pass through the digestive system, through the blood, to reach the blood-brain barrier, and cross it to the brain. In longer colonic fermentation, the phenolic compounds can be degraded or transformed over time, reducing their toxicity and efficacy. Colonic fermentation of indigestible polysaccharides from *Aloe barbadensis* allows the activation of several factors such as NF-κB, c-Jun N-terminal kinase (JNK), and protein extracellular signal-regulated kinase (ERK) pathway, blocking inflammatory neurokinin-1 receptor (NK-1R), and p38 phosphorylation, causing a decrease in the viability of astrocytoma (~75%) and reduction in NF-κB phosphorylation (Tariq et al. 2019).

7.4 RELATIONSHIP BETWEEN INTESTINAL MICROBIOTA AND NEURODEGENERATIVE DISEASES

Multiple factors can contribute to the development of neurodegenerative diseases, mainly aging, degradation of anatomical pathways, environmental factors, mitochondrial and immune system dysfunction, genetic factors, and changes in gut microbiota. Gut microbiota is an ecological community of microorganisms that live in a symbiotic and pathogenic way in the human body; it plays essential roles in regulating health and disease (Megur et al. 2020).

The microbiota consists of more than 1500 species, distributed in more than 50 phyla, where Bacteroidetes and Firmicutes are the dominant genera, followed by *Proteobacteria, Fusobacteria, Tenericutes, Actinobacteria,* and *Verucomicrobia* (Gooma, 2020). Besides acting as a barrier against pathogenic microorganisms, the microbiota is involved in the absorption of nutrients, fermentation of carbohydrates, and synthesis of vitamins and other metabolites (Shandilya et al. 2021).

Many investigations indicate that the gut microbiota is modified as the person develops and grows. In the infant period, *Lactobacillus* and *Prevotella* spp. are overcome, whereas, in older people, a decrease in the microbiota attributed to a limited diet is seen. Changes in the gut microbiota directly affect the membranes of the colonocytes, making them permeable (Shandilya et al. 2021). These alterations, called dysbiosis, increase intestinal permeability that can increase obesity, colorectal cancer, inflammatory bowel disease, heart failure, type 2 diabetes, and neurodegenerative disorders (AD, PD, and multiple sclerosis) (Megur et al. 2020). The gut microbiota of older adults represents a proinflammatory phenotype, favoring the entry of neurotoxins into the systemic circulation, which can be a precursor to neuroinflammatory processes, developing pathogenic pathways of communication between in the immune system (Wu et al. 2021). Likewise, the microbiota influences the synthesis of several neurotransmitters and neuromodulators (Megur et al. 2020).

Aloe polysaccharides are considered bioactive macromolecules, exhibiting antioxidant, neuroprotective, and hepatoprotective activities and having obesity, cancer, and immune regulation effects (Liu et al. 2021). They are considered as indigestible because they resist the hydrolytic enzymes of the stomach and small intestine, while the gut microbiota, through glycosidases and proteins, can partially or entirely hydrolyze them. In this sense, polysaccharides allow to nourish and regulate the variety of intestinal microorganisms, changing their diversity and number (Liu et al. 2021). Hydrolyzed polysaccharides can be metabolized by the host's gut microbiota, producing various metabolites, mainly SCFA. These compounds regulate the metabolism of lipids in the liver and function as protection of the intestinal barrier. Acetic acid, propionic acid, and butyric acid are the main SCFA synthesized by the colonic microbiota in people with a diet rich in indigestible polysaccharides such as acemannan (Liu et al. 2015). Acetic acid affects the biosynthesis of endogenous fatty acids that modulate leukocyte function, and propionic acid stimulates the expression of fatty acid oxidation genes, reducing total cholesterol in the liver (Tornero-Martínez et al. 2019). An increase in propionic acid synthesis and a decrease in the concentration of acetic acid are positive indicators of the lipid-lowering effect (Guo et al. 2021). Butyric acid increases insulin sensitivity and affects the transcription of specific genes, acting as an inhibitor of histone deacetylases (Kumar et al. 2019). These metabolites can pass the blood-brain barrier; exhibiting neuroprotective and anti-inflammatory activity in the brain by inhibiting microglia activation (Jiang et al. 2021).

7.5 BRAIN-GUT-MICROBIOTA AXIS

Previously, it was considered that neurodegenerative diseases were caused by damage to the nervous system, ignoring the relation of microorganisms in the small and large intestines with the development of diseases (Liu et al. 2015; Tang et al. 2021).

Currently, intestinal dysbiosis and neuroinflammation are consistent factors in the pathophysiology of neurological disorders (**Figure 7.2**). The specific relationship between the gastrointestinal tract, the nervous system, and the gut microbiota is called the *gut-microbiota-brain axis*, which consists of a bidirectional exchange between the two parts (Gomaa, 2020). The gut microbiota can regulate the homeostasis of the host's central nervous system (Wu et al. 2021) directly or indirectly through various chemical signals, such as SCFA, serotonin, bile acids, and gamma-aminobutyric acid (GABA). Bacteria such as *Escherichia*, *Lactobacillus*, *Saccharomyces*, and *Bacillus* can synthesize various chemicals such as amino acids, 5-hydroxytryptamine, dopamine, butyrate, histamine, and serotonin, which play essential roles in human brain activity (Murciano-Brea et al. 2021).

The SCFA produced by the gut microbiota **(Figure 7.2)** can cross the intestinal barrier and affect neurological processes through an extracellular interaction as endogenous ligands of G protein-coupled receptors. Likewise, they interact with neuroimmune pathways by reducing systemic inflammation and regulating the activation of microglia associated with neuroinflammation. SCFA reduces inflammation in two ways: a) by improving the intestinal barrier and reducing the number of pathogenic bacteria and b) through metabolites passing into the peripheral blood and reaching the brain.

Besides the production of SCFA, the gut microbiota can synthesize neurotransmitters. For example, bacteroides such as *Lactobacillus*, *Bifidobacterium*, and *Parabacteroides* produce GABA in the large intestine; this neurotransmitter participates in gut-brain connections through enteroendocrine and neuroimmune pathways (Gomaa, 2020). Neurotransmitters and metabolites produced by the gut microbiota can be associated with the gut-brain axis through spinal and vagal fibers, while the brain sends efferent sympathetic and parasympathetic innervations to the gastrointestinal tract (Naveed et al. 2021; Liu et al. 2021).

Some polysaccharides have beneficial bioactivities such as antioxidants and anti-inflammatory and intestinal protectors. Specifically, mannans can be fermented by *Lactobacillus*, increasing

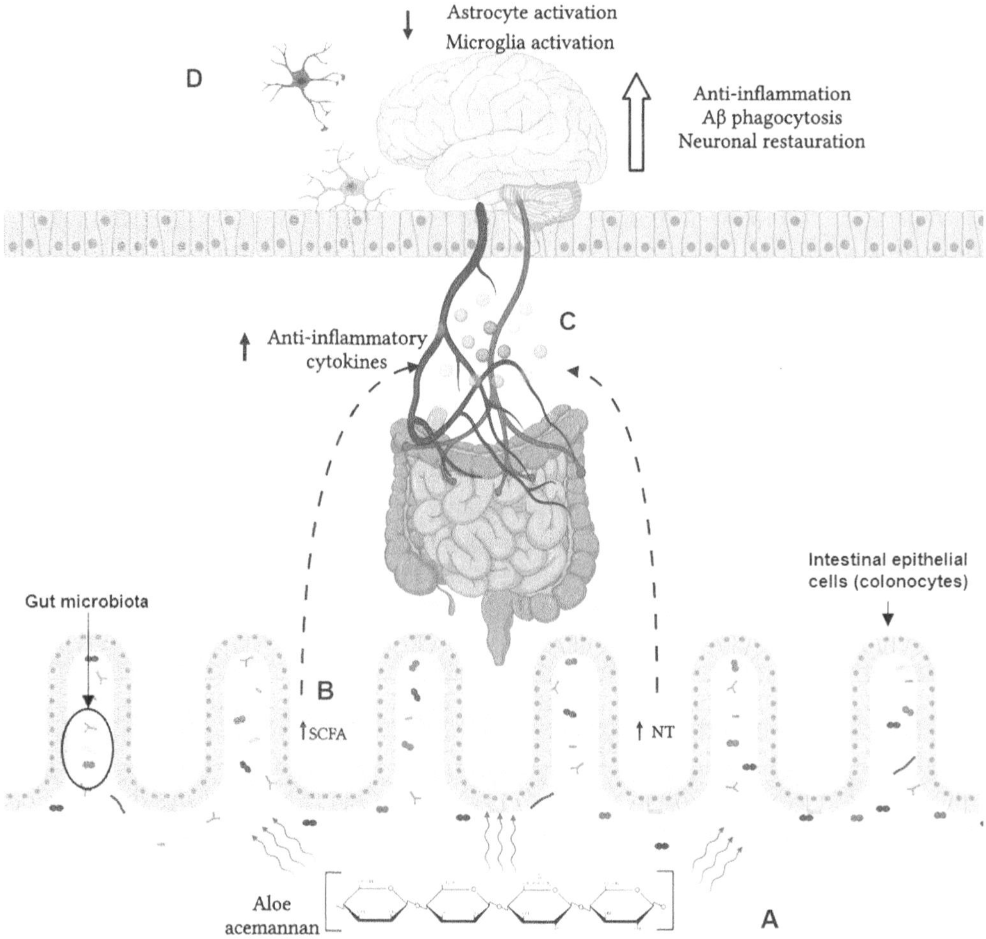

Figure 7.2 Communication between the gut microbiome and the brain (Gut-Microbiota-Brain Axis). (A) Dietary carbohydrates (*Aloe acemannan*) are metabolized as SCFA (Short Chain Fatty Acids) by colonic microbiota. (B) SCFA are absorbed in the colon, a rich source for the host. (C) SCFA, mainly butyrate, function as signal molecules for the increase in the production of anti-inflammatory cytokines, which cross the blood-brain barrier. (D) This signaling allows the increase of anti-inflammatory cytokines in the brain, decreasing the activation of astrocytes and microglia and improving neuronal restoration.

the production of SCFA and modifying the composition of the gut microbiota. The ingesting of polysaccharides from aloe gel increases the abundance of *Lactobacillus*, increasing the production of butyrate and propionate, improves cognitive function, increasing interactions with the mineralocorticoid receptors in the brain and decreasing the concentration of lipopolysaccharides generated by pathogenic bacteria, related to neuroinflammation present in cognitive diseases (Liu et al. 2021).

7.6 CONCLUSION

Aloe barbadensis is a source of anthraquinones, anthrones, chromones, and polysaccharides with beneficial activities to human health. The presence of indigestible polysaccharides in aloe gel allows the modulation of the intestinal microbiota, mainly colon bacteria. Bacteria can synthesize neurotransmitters that cross the intestinal blood barrier, regulating some brain processes and reducing the production of acetylcholinesterase and proinflammatory factors. The gut microbiota can regulate the homeostasis of the host's central nervous system directly or indirectly through chemical signals due to the production of SCFA, serotonin, bile acids, and GABA. The consumption of aloe gel and its compounds improves some cognitive diseases by producing compounds with neurotransmitter activity.

REFERENCES

Almeida, P.G.C., Nani, J.V., Oses, J.P., Brietzke, E., & Hayashi, M.A.F. (2020). Neuroinflammation and glial cell activation in mental disorders. *Brain, Behavior and Immunity-Health*, 2, 100034. https://doi.org/10.1016/j.bbih.2019.100034

Altinkaynak, C., Haciosmanoglu, E., Ekremoglu, M., Hacioglu, M., & Özdemir, N. (2023). Anti-microbial, anti-oxidant and wound healing capabilities of Aloe vera-incorporated hybrid nanoflowers. *Journal of Bioscience and Bioengineering*, 135(4), 321–330 https://doi.org/10.1016/j.jbiosc.2023.01.004.

Bajpai, S. (2018). Biological importance of Aloe vera and its active constituents. *Synthesis of Medicinal Agents from Plants*, 177–203. https://doi.org/10.1016/B978-0-08-102071-5.00008-8

Ball, H.A., McWhirter, L., Ballard, C., Bhome, R., Blackburn, D.J., Edwards, M.J., Fleming, S.M., Fox, N.C., Howard, R., Huntley, J., Isaacs, J.D., Larner, A.J., Nicholson, T.R., Pennington, C.M., Poole, N., Price, G., Price, J.P., Reuber, M., Ritchie, C., Rossor, M.N., Schott, J.M., Teodoro, T., Venneri, A., Stone, J., & Carson, A.J. (2020). Functional cognitive disorder: Dementia's blind spot. *Brain: A Journal of Neurology*, 143(10), 2895–2903. https://doi.org/10.1093/brain/awaa224

Borges-Argáez, R., Chan-Balan, R., Cetina-Montejo, L., Ayora-Talavera, G., Sansores-Peraza, P., Gómez-Carballo, J., & Cáceres-Farfán, M. (2019). In vitro evaluation of anthraquinones from *Aloe vera* (*Aloe barbadensis* Miller) roots and several derivatives against strains of influenza virus. *Industrial Crops and Products*, 132, 468–475. https://doi.org/10.1016/j.indcrop.2019.02.056

Chandraa, S., Gahlot, M., Choudhary, A., Palai, S., de Almeida, R., de Vasconselos, J., Vieira, F., Maia, P., & Melo, H. (2023). Scientific evidences of anticancer potential of medicinal plants. *Food Chemistry Advances*, 2, 100239. https://doi.org/10.1016/j.focha.2023.100239

Chelu, M., Musuc, A.M., Aricov, L., Ozon, E.A., Iosageanu, A., Stefan, L.M., Prelipcean, A.M., Popa, M., & Moreno, J.C. (2023). Antibacterial Aloe vera based biocompatible hydrogel for use in dermatological applications. *International Journal of Molecular Sciences*, 24(4), 3893. https://doi.org/10.3390/ijms24043893.

de Carvalho, E., Barros, G., Sobhie, R., Schetter, M., Castilho, J., & Justo, J. (2022). Effects of carbon nanomaterials and Aloe vera on melanomas- Whe are we? Recent updates. *Pharmaceutics*, 14 (10), 1–34. https://doi.org/10.3390/pharmaceutics14102004

Domínguez-Fernández, R.N., Arzate-Vázquez, I., Chanona-Pérez, J.J., Welti-Chanes, J.S., Alvarado-González, J.S., Calderón-Domínguez, G., Garibay-Febles, V., & Gutiérrez-López, G.F. (2012). El gel de Aloe vera: estructura, composición química, procesamiento, actividad biológica e importancia en la industria farmacéutica y alimentaria. *Revista Mexicana de Ingeniería Química*, 11, 23–43. www.scielo.org.mx/scielo.php?script=sci_arttext&pid=S1665-27382012000100003&lng=es&nrm=iso>

Dong, X., Zeng, Y., Liu, Y., You, L., Yin, X., Fu, J., & Ni, J. (2020). Aloe-emodin: A review of its pharmacology, toxicity, and pharmacokinetics. *Phytotherapy Research*, 34(2), 270–281. https://doi.org/10.1002/ptr.6532.

Enache, D., Pariante, C.M., & Mondelli, V. (2019). Markers of central inflammation in major depressive disorder: A systematic review and meta-analysis of studies examining cerebrospinal fluid, positron emission tomography and post-mortem brain tissue. *Brain, Behavior, and Immunity*, 81, 24–40. https://doi.org/10.1016/j.bbi.2019.06.015

Farrugia, C., Burke, E.S., Haley, M.E., Bedi, K.T., & Gandhi, M.A. (2019).The use of Aloe vera in cancer radiation: An updated comprehensive review. *Complementary Therapies in Clinical Practice*, 35, 126–130. https://doi.org/10.1016/j.ctcp.2019.01.013

Gao, Y., Kuok, K.I., Jin, Y., & Wang, R. (2019). Biomedical applications of Aloe vera. *Critical Reviews in Food Science and Nutrition*, 59, S244–S256. https://doi.org/10.1080/10408398.2018.1496320

Gomaa E.Z. (2020). Human gut microbiota/microbiome in health and diseases: A review. *Antonie Van Leeuwenhoek*, 113(12), 2019–2040. https://doi.org/10.1007/s10482-020-01474-7

Gullón, B., Gullón, P., Tavaria, F., Alonso, J.L., & Pintado, M. (2015). In vitro assessment of the prebiotic potential of Aloe vera mucilage and its impact on the human microbiota. *Food Function*, 6, 525–531. https://doi.org/10.1039/C4FO00857J

Guo, C., Wang, Y., Zhang, S., Zhang, X., Du, Z., Li, M., & Ding, K. (2021). Crataegus pinnatifida polysaccharide alleviates colitis via modulation of gut microbiota and SCFAs metabolism. *International Journal of Biological Macromolecules*, 181, 357–368. https://doi.org/10.1016/j.ijbiomac.2021.03.137

Gupta, A., & Rawat, S. (2017). Clinical importance of Aloe vera. *Research Journal of Topical and Cosmetic Sciences*, 8, 30–39. http://dx.doi.org/10.5958/2321-5844.2017.00004.8

Hęś, M., Dziedzic, K., Górecka, D., Jędrusek-Golińska, A., & Gujska, E. (2019). Aloe vera (L.) Webb: Natural sources of antioxidants – A review. *Plant Foods for Human Nutrition*, 74(3), 255–265. https://doi.org/10.1007/s11130-019-00747-5

Jeromin, A., & Bowser, R. (2017). Biomarkers in neurodegenerative diseases. *Advances in Neurobiology*, 15, 491–528. https://doi.org/10.1007/978-3-319-57193-5_20

Jiang, Y., Li, K., Li, X., Xu, L., & Yang, Z. (2021). Sodium butyrate ameliorates the impairment of synaptic plasticity by inhibiting the neuroinflammation in 5XFAD mice. *Chemico-Biological Interactions*, 341, 109452. https://doi.org/10.1016/j.cbi.2021.109452

Kumar, R., Singh, A.K., Gupta, A., Bishayee, A., & Pandey, A.K. (2019). Therapeutic potential of Aloe vera -A miracle gift of nature. *Phytomedicine*, 60, 152996. https://doi.org/10.1016/j.phymed.2019.152996

Lewis, J.E., Poles, J., Shaw, D.P., Karhu, E., Khan, S.A., Lyons, A.E., Sacco, S.B., & McDaniel, H.R. (2021). The effects of twenty-one nutrients and phytonutrients on cognitive function: A narrative review. *Journal of Clinical and Translational Research*, 7(4), 575–620.

Liu, Q., Xi, Y., Wang, Q., Liu, J., Li, P., Meng, X., Liu, K., Chen, W., Liu, X., & Liu, Z. (2021). Mannan oligosaccharide attenuates cognitive and behavioral disorders in the 5xFAD Alzheimer's disease mouse model via regulating the gut microbiota-brain axis. *Brain, Behavior, and Immunity*, 95, 330–343. https://doi.org/10.1016/j.bbi.2021.04.005

Ma, Y., Xia, X., Cheng, J., Kuang, Y., Yang, T., Yang, L., Fan, K., & Gu, J. (2014). Emodin inhibits inducible nitric oxide synthase in a rat model of craniocerebral explosive injury. *Neurochemical Research*, 39, 1809–1816.

Majumder, R., Das, C.K., & Mandal, M. (2019). Lead bioactive compounds of Aloe vera as potential anticancer agent. *Pharmacological Research*, 148, 104416. https://doi.org/10.1016/j.phrs.2019.104416

Megur, A., Baltriukienė, D., Bukelskienė, V., & Burokas, A. (2020). The microbiota-gut-brain axis and Alzheimer's disease: Neuroinflammation is to blame?. *Nutrients*,13(1), 37. https://doi.org/10.3390/nu13010037

Minjares-Fuentes, R., & Femenia, A. (2019). *Nonvitamin and Nonmineral Nutritional Supplements*. Academic Press. https://doi.org/10.1016/B978-0-12-812491-8.00070-9

Murciano-Brea, J., Garcia-Montes, M., Geuna, S., & Herrera-Rincon, C. (2021). Gut microbiota and neuroplasticity. *Cells*, 10(8), 2084. https://doi.org/10.3390/cells10082084

Nair, G. R., Naidu, G. S., Jain, S., Nagi, R., Makkad, R. S., & Jha, A. (2016). Clinical effectiveness of aloe vera in the management of oral mucosal diseases – A systematic review. *Journal of Clinical and Diagnostic Research*, 10(8), ZE01–ZE7. https://doi.org/10.7860/JCDR/2016/18142.8222

Naveed, M., Zhou, Q.G., Xu, C., Taleb, A., Meng, F., Ahmed, B., Zhang, Y., Fukunaga, K., & Han, F. (2021). Gut-brain axis: A matter of concern in neuropsychiatric disorders. *Progress in Neuro-Psychopharmacology & Biological Psychiatry*, 104, 110051. https://doi.org/10.1016/j.pnpbp.2020.110051

Pasala, P.K., Abbas Shaik, R., Rudrapal, M., Khan, J., Alaidarous, M.A., Jagdish Khairnar, S., Bendale, A.R., Naphade, V.D., Kumar Sahoo, R., Zothantluanga, J.H., & Walode, S.G. (2022). Cerebroprotective effect of Aloe Emodin: In silico and in vivo studies. *Saudi Journal of Biological Sciences*, 29, 998–1005. https://doi.org/10.1016/j.sjbs.2021.09.077

Reynolds,T., & Dweck, A.C. (1999). Aloe vera leaf gel: A review update. *Journal of Ethnopharmacology*, 68 (1), 3–37. https://doi.org/10.1016/s0378-8741(99)00085-9

Rodríguez, E., Darias Martin, J., & Diaz Romero, C. (2010). Aloe vera as a functional ingredient in foods. *Critical Reviews in Food Science and Nutrition*, 50, 305–326. https://doi.org/10.1080/10408390802544454

Saleh, H.A., Yousef, M.H., & Abdelnaser, A. (2021). The anti-inflammatory properties of phytochemicals and their effects on epigenetic mechanisms involved in TLR4/NF-κB-mediated inflammation. *Frontiers in Immunology*, 12, 606069. https://doi.org/10.3389/fimmu.2021.606069

Sánchez, M., González Burgos, E., Iglesias, I., & Gómez-Serranillos, M.P. (2020). Pharmacological update properties of Aloe Vera and its major active constituents. *Molecules*, 25(6), 1324. https://doi.org/10.3390/molecules25061324

Sánchez-Machado, D.I., López-Cervantes, J., Sendón, R., & Sanches-Silva, A. (2017). Aloe vera: Ancient knowledge with new frontiers. *Trends in Food Science & Technology*, 61, 94–102. https://doi.org/10.1016/j.tifs.2016.12.005

Shandilya, S., Kumar, S., Kumar Jha, N., Kumar Kesari, K., & Ruokolainen, J. (2021). Interplay of gut microbiota and oxidative stress: Perspective on neurodegeneration and neuroprotection. *Journal of Advanced Research*, 38, 223–244. https://doi.org/10.1016/j.jare.2021.09.005

Shi, X.D., Yin, J.Y., Huang, X.J., Que, Z.Q., & Nie, S.P. (2018). Structural and conformational characterization of linear O-acetyl-glucomannan purified from gel of Aloe barbadensis Miller. *International Journal of Biological Macromolecules*, 120, 2373–2380. https://doi.org/10.1016/j.ijbiomac.2018.09.005

Tang, Y., Wang, Q., & Liu, J. (2021). Microbiota-gut-brain axis: A novel potential target of ketogenic diet for epilepsy. *Current Opinion in Pharmacology*, 61, 36–41. https://doi.org/10.1016/j.coph.2021.08.018

Tariq, H., Zia, M., Ihsan ul, H., Muhammad, S.A., Khan, S.A., Fatima, N., Mannan, A., Abbasi, A.M., & Zhang, M. (2019). Antioxidant, antimicrobial, cytotoxic, and protein kinase inhibition potential in Aloe vera L. *BioMedical Research International*, 2019, 6478187. https://doi.org/10.1155/2019/6478187

Tornero-Martínez, A., Cruz-Ortiz, R., Jaramillo-Flores, M.E., Osorio-Díaz, P., Ávila-Reyes, S.V., Alvarado-Jasso, G.M., & Mora-Escobedo, R. (2019). *In vitro* fermentation of polysaccharides from Aloe vera and the evaluation of antioxidant activity and production of short chain fatty acids. *Molecules*, 24(19), 3605. https://doi.org/10.3390/molecules24193605

Tornero-Martínez, A., Del Carmen Silva-Lucero, M., Sampedro, E.C., Ramón-Gallegos, E., Pérez-Cruz, C., Pérez-Grijalva, B., & Mora-Escobedo, R. (2022). Aloe vera and fermented extracts exhibit an anti-inflammatory effect on human glioblastoma/astrocytoma U373 MG Cells. *Plant Foods for Human Nutrition*, 77(1), 37– 43. https://doi.org/10.1007/s11130-022-00957-4

Visser, K., Koggel, M., Blaauw, J., van der Horn, H.J., Jacobs, B., & van der Naalt, J. (2022). Blood-based biomarkers of inflammation in mild traumatic brain injury: A systematic review. *Neuroscience and Biobehavioral Reviews*, 132, 154–168. https://doi.org/10.1016/j.neubiorev.2021.11.036

Wu, M.L., Yang, X.Q., Xue, L., Duan, W., & Du, J.R. (2021). Age-related cognitive decline is associated with microbiota-gut-brain axis disorders and neuroinflammation in mice. *Behavioural Brain Research*, 402, 113125. https://doi.org/10.1016/j.bbr.2021.113125

Zhong, J., Wang, F., Shen, C., Zheng, Y., Ma, F., Zhu, T., Chen, L., Tang, Q., & Zhu, J. (2019). Aloin attenuates cognitive impairment and inflammation induced by D-galactose via down-regulation ERK, p38 and NF-kb signaling pathway. *International Immunopharmacology*, 72, 48–54.

Chapter 8

Antioxidant Activity During Gastrointestinal *In Vitro* Digestion of Agro-Industrial Waste By-Products

Karen E. Lima-Villegas
Georgina Calderón-Domínguez
Ruth N. Domínguez-Fernández
Marcela González-Vázquez
Rebeca Dávila-León
and Ma de la Paz Salgado-Cruz

CONTENTS

8.1 INTRODUCTION

A significant percentage of worldwide food production (20 to 30%) goes to waste (Kaza et al. 2018). This percentage changes depending on the geographical area and type of food produced (FAO, 2016). During food processing, the non-edible part of foods generates additional waste, and this proportion can vary from low to high levels, depending on the type of food (Muñoz de Chávez et al. 1996). It is reported that over twenty billion dollars in income could be gained from the twenty million tons of agro-industrial waste generated globally if it were to be reused (FAO, 2018).

 DOI: 10.1201/9781003332558-10

It has also been reported that this waste, particularly the non-edible part of foods, is a significant source of vitamins, minerals, and dietary fiber, as well as of antioxidant bioactive compounds (Palafox-Carlos et al. 2011, Gonçalves et al 2023), and can represent a source of raw materials to produce functional health products.

The importance of these materials has led to the development of a large and potential market for antioxidant, bioactive, and fiber products (Rodríguez et al. 2006), which aim to slow the generation of oxidative damage in the tissues or DNA of cells, by deactivating reactive oxygen species (ROS) (Palafox-Carlos et al. 2011). However, when these fiber products are consumed, the interaction with the gastric enzymes and pH conditions of human digestive tract affect the availability of these components. These changes are followed by *in vitro* gastrointestinal studies.

This chapter discusses scientific information on the functional properties of dietary fiber, with emphasis on the antioxidant activity of agro-industrial residues during *in vitro* gastrointestinal digestion, seeking to highlight the importance of these materials in human health.

8.2 DIETARY FIBER

Most agro-industrial waste is composed of cell walls that are rich in fiber. This fiber is defined as plant substances or similar carbohydrates that cannot be digested and absorbed by the small intestine but can be partially or fully fermented in the large intestine (Mu et al. 2017). Fiber components include polysaccharides, oligosaccharides, lignin, and some other plant materials. Fiber has been classified based on different criteria (Mu et al. 2017). For example: 1) by its function in the plant, 2) by the components of the fiber, 3) by the type of polysaccharides, 4) by the digestion site, 5) by its *in vitro* gastrointestinal solubility, 6) by the digestion products, and 7) by its physiological function. However, none of these classifications is considered completely satisfactory. Of these classification systems, one of the most accepted is that based on the concept of gastrointestinal solubility using a buffer at a defined pH and classifying dietary fiber as insoluble and soluble (Mudgil, 2017), having both different components with different nutraceutical properties as shown in **Table 8.1**.

8.2.1 Insoluble Dietary Fiber

Insoluble dietary fiber refers to those components of plant food residue that are resistant to hydrolysis by human digestive enzymes, as well as being water insoluble (Capuano, 2017). It is composed of cellulose, hemicelluloses, and lignin, presenting high porosity and low density, and the capacity to retain water, forming low-viscosity mixtures that help reduce intestinal transit time during gastric digestion (Topping, 2007; Codex Alimentarius, 2010).

Cellulose is a polysaccharide formed by glucose units linked by $\beta(1\text{-}4)$ bonds. It is not hydrolyzed by enzymes during the digestive process when consumed; however, it is an ideal substrate for the intestinal microbiota (Nsor-Atindana et al. 2020), and also helps peristaltic movement, since it interacts with the intestinal mucosa increasing motility and protecting it (McRorie & McKeown, 2017).

Lignin is a polymer made up of polyphenols (trans-coniferyl, trans-sinapyl and trans p-cumaril alcohol), with high antioxidant and antimicrobial activities (Mahmood et al. 2018), but is not yet fully defined and understood. Like cellulose, it is not hydrolyzed by digestive enzymes but by the intestinal microbiota, resulting in a prebiotic effect (Harini et al. 2018). In addition, it has been reported that lignin interacts with bile salts and cholesterol preventing them from being absorbed

TABLE 8.1 POLYSACCHARIDES, STRUCTURE AND FUNCTIONAL ACTIVITY

Fiber Type	Polysaccharides	Functional Activity	Base Components and Chemical Bond Type	Reference
Insoluble fiber	Cellulose	Increased satiety, increased motility, antioxidant, antimicrobial and prebiotic activity	Glucopyrans bound with ß 1-4 bonds	McRorie & McKeown (2017); Jongaroontaprangsee et al (2018); Robles et al. (2018); Won Ko et al. (2018), Mugwagwa et al. (2020), Nsor-Atindana et al. (2020)
	Lignin	Antioxidant, antimicrobial activity, prebiotic activity	Alcohol trans-coniferyl Alcohol trans-sinapyl Trans-p-cumaril alcohol, ortho-OH groups	Jiang et al. (2018); Alzagameem et al. (2019)
	Hemicellulose	Hypocholesterolemic, antithrombotic activity, anticoagulant, antioxidant increased motility, prebiotic activity	Pentoses, hexoses, uronic acid, with ß 1-4 bonds	Doliska et al. (2012) Mudgil & Barak, (2013); Bouaziz et al. (2016); Banerjee et al. (2019);
	Starch resistant (R1, R2, R3, R4, R5)	Prebiotic activity, hypoglycemic effect	Glucose units with α(1-4) and α (1-6) bonds	Thompson, (2007); Mudgil & Barak, (2013); Stewart & Zimmer (2018); Gourini et al. (2020); Bojarczuk et al. (2022)
Soluble fiber	Pectin	Delays gastric emptying, hypocholesterolemic, anti-inflammatory at the intestinal level, hypoglycemic, antioxidant activity, prebiotic activity	D-galacturonic acid with α-D-1-4 bonds	Escudero & González, (2006); Faith et al. (2017); Pereira et al. (2017); Coman et al. (2019); Jiang et al (2020); Singh et al. (2020); Sun Chen & Zhu, (2020)
	Gums and mucilages	Delays gastric, hypocholesterolemic, hypoglycemic emptying, increased satiety, antioxidant and prebiotic activity	Ionic, neutral, linear, or branched heteropolysaccharides (hexoses) and uronic acids	Viuda-Martos et al., (2010, 2017); Mohsin et al. (2018); Punia et al., (2019); Ferreira et al., (2020); Li et al. (2020)
	FOS	Increased motility, decreased intestinal colic, antioxidant activity, antidepressant, prebiotic activity	ß 1-2 fructosyl-fructose. Link between glucose and fructose	Russo et al. (2017); Capriles & Areas, (2013); Crispin et al. (2015); Muller et al. (2016); Ortiz-Basurto, (2017); Coman et al. (2019).

in the small intestine (Alzagameem et al. 2019). Lignin and cellulose have also been linked to phenolic compounds in their structure (Jiang et al. 2018), which are then released from the polymeric matrix during digestion (Piña et al. 2011).

Hemicellulose is a polysaccharide composed of different sugars (xylose, arabinose, glucose, mannose, galactose, rhamnose) that can also contain uronic acids such as glucuronic and galacturonic acids. Hemicellulose has been reported to have anticoagulant and antithrombotic activities (Doliska et al. 2012), due to its hydrophobic nature and its high content of carboxymethyl sulfate groups binding fatty acids and triglycerides, preventing these from being absorbed in the small intestine.

Resistant starch (RS) has been classified as insoluble fiber by some authors (Thompson, 2007) and can be divided into five types: 1) RS1, physically inaccessible starch; 2) RS2, raw starch with indigestible crystal structure; 3) RS3, retrograded starch; 4) RS4, chemically modified starch; and 5) RS5, starch linked by free fatty acids (Mudgil & Barak, 2013; Bojarczuk et al. 2022). Its structure is made up of amylose (D-glucose linked by α (1-4) bonds) and amylopectin (D-glucose linked by α (1-4) and α (1-6) bonds), which gives it its branched structure. The most outstanding nutritional property of resistant starch is its capacity to be fermented by the intestinal microbiota generating energy and short-chain fatty acids, including acetate, butyrate, and propionate. Additionally, it has been shown to have a hypoglycemic effect, helping reduce postprandial blood glucose levels (Stewart & Zimmer, 2018, Gourineni et al. 2020).

8.2.2 Soluble Dietary Fiber

Soluble dietary fiber refers to those components of plant food residue that are water soluble (Capuano, 2017), do not undergo modification during their transit through the small intestine, and are fermented by the microbiota in the large intestine. It includes pectins, gums, mucilages, and fructooligosaccharides (FOS). Some of them (pectins, gums, and mucilages) have the ability to form gels when in contact with water, and due to their viscosity, they have a healthy effect on metabolism, reducing the speed at which the stomach empties, stimulating motility and decreasing inflammation in the small intestine (Escudero & González, 2006). Also, a hypocholesterolemic effect is attributed to these components, due to the binding between intestine bile salts and uronic acids, which prevents them from being reabsorbed (Escudero & González, 2006).

Pectin is a polysaccharide formed by D-galacturonic acid linked via α-D-1-4 bonds, forming homo and hetero galacturonic chains. It can be acetylated in O-2 or O-3 positions and methylated in the C-6. The hetero galacturonic forms of pectin can be attached to rhamnose or arabinogalactans (Singh et al. 2020). Pectin has antioxidant activity, but which chemical structure causes this effect has not been specified (Jiang et al. 2020). It is also reported that pectin presents prebiotic activity due to the presence of rhamnose in its composition (Singh et al. 2020). Some authors explain that the antioxidant activity could be related to the high concentration of hydroxyl radicals present in galacturonic acid (Sun et al.2020).

Regarding gums and mucilages, they are considered plant hydrocolloids that yield uronic acids and sugars on hydrolysis. They are formed by linear neutral or branched polysaccharides consisting of β D-xylose, α D-glucose, and uronic acids (Câmara et al. 2020); they been reported to have prebiotic effects (Li et al. 2020).

Fructooligosaccharides (FOS) are a type of oligosaccharide formed by fructose units linked by β (2-1) fructosyl fructose glycosidic bonds with a glucose residue in their chain. They are divided into oligofructans and inulin according to their size. Oligofructans are generally composed of 2 to 7 fructose units, whereas inulin is composed of up to 60 (Crispin et al. 2015; Kaprosob et al. 2018). These polysaccharides are resistant to enzymatic hydrolysis by the gastrointestinal system due to the bonds and ramifications of their structure, so they are highly fermentable (prebiotic) by the intestinal microbiota, which increases gastrointestinal motility and therefore decreases intestinal inflammation (Russo et al. 2017; Capriles & Areas, 2013). It is also reported that FOS present antioxidant activity (Ortiz-Basurto et al. 2017) because of the high affinity they have with water, which limits the mobility of molecules, avoiding chemical reactions of oxidizing agents.

All the dietary fiber components mentioned in this section present nutraceutical activity (**Table 8.1**), mainly antioxidant properties that, when combined, give rise to the concept of Antioxidant Dietary Fiber (ADF). This concept was first mentioned by Saura-Calixto et al. (1998) and defined as the dietary fiber associated with polyphenolic compounds in a single material (Herrera-Balandrano et al. 2020).

8.3 PHENOLIC COMPOUNDS

Phenolic compounds (PC) are one type of secondary metabolites, characterized by having at least one aromatic ring with one or more hydroxyl groups, ranging from single aromatic-ringed compounds to highly polymerized substances such as complex tannins (hydrolyzable and condensed). They are contained in almost all edible parts of plant tissues such as fruits, seeds, leaves, stems, roots, etc., and compartmentalized in the vacuoles of the plant cell wall, in the tonoplast or in the cytoplasmic lipid membranes, depending specifically on the type of PC (Acosta-Estrada et al. 2014; González-Aguilar et al. 2017). They are classified as: 1) non-flavonoids and 2) flavonoids (Crozier et al. 2006; Morandi-Vuola et al. 2019).

The main non-flavonoid compounds are phenolic acids, which are derived from benzoic acid or cinnamic acid. Hydroxybenzoic acids (HBAs) are formed via the shikimate pathway, giving rise to the formation, among others, of gallic acid, which is a precursor of hydrolyzable tannins. (Morandi-Vuola et al. 2019). On the other hand, cinnamic acid, caffeic, ferulic, or p-coumaric acids are synthesized from HCAs (hydroxycinnamic acids) via the phenylpropanoid pathway and hence are also referred to as phenylpropanoids (Crozier et al. 2006). Phenolic acids are typically insoluble or bound in nature, and in the case of HCA and HBA, can be associated with lignin through ether linkages with hydroxyl groups from the aromatic ring, and structural carbohydrates and proteins via ester linkages with carboxylic groups (Acosta-Estrada et al. 2014).

Flavonoids constitute the largest group of PC, including flavones, flavonols, flavanones, isoflavonoids, flavanols, and anthocyanins, having the phenylbenzopyrane skeleton as their basic structure. Among these, flavanols, and specifically flavan-3-ols, are reported as the most abundant flavonoids in nature, having good antioxidant activity when containing galloyl moieties (De la Rosa et al. 2018). **Table 8.2** shows some examples of these compounds, where the main food sources are fruits, vegetables, herbs, teas, and cereals (Rothwell et al. 2013, Morandi-Vuolo et al. 2019), along with their nutraceutical properties, such as antioxidant, antimicrobial, anti-inflammatory, and antitumoral activities. It is important to mention that not all phenolic components are contained in all the mentioned types of foods, but some are dominant while others are absent. However, the published information and the free web databases regarding food phenolic composition present in dietary fiber are still scarce, as well as reports about their bioavailability and bio-accessibility during their consumption, even though more than 8000 polyphenol components from fruits and vegetables have been studied (Bhagwat 2011; Rothwell et al. 2013).

TABLE 8.2 PHENOLIC COMPOUNDS CLASSIFICATION

Classification	Skeleton	Data Base Reported Compounds	Examples	Food Sources	Reported Nutraceutical Effects
Non flavonoids					
Phenolic acids (HBA)	C6-C1	40	Benzoic, Ellagic, Gallic, Gallagic, Gentisic, Hippuric, Syringic, Salicylic and Vanillic acids	Blackberries, strawberries, apricot, blueberry, citrus, grapes, tomato, eggplant, pepper, melon, cucumber, green and black teas	Antioxidant
Phenolic acids (HCA)	C6-C3	88	Caffeic, Chicoric, Cinnamic, Coumaric, Ferulic, Isoferulic, Rosamarinic, Sinapic, Sinapine, Verbacoside acids.	Apples, cherries, berries, peaches, cranberry plums, citrus fruits, coffee, grape, pear, wheat seeds, peanuts, cranberry, apple, and orange	Antioxidant
Polyphenolic stilbenes	C6-C2-C6	27	Palladiol, Resveratrol, Viniferin, Pallidol, Dihydroresveratrol	Grapes, red wine, berries, peanuts, pistachio	Antioxidant, Chemopreventive, cardioprotective
Flavonoids					
Flavonols	C6-C3-C6	82	Galangin, Kaemferol Morin, Isorhamnetin, Myricetin, Quercetin, Rahmanetin	Herbs, kale, common beans, berries, grapes, cherries, apples, artichoke, Chinese cabbage, hot peppers, lettuce, onion, walnut	Antioxidant, antimicrobial, antiviral, hepatoprotective, anti-inflammatory, antidiabetic
Flavones		49	Apigenin, Baicalein, Chrysin, Cirsilineol, Cirsimaritin, Luteolin, Tangeretin	Celery, parsley, thyme, cantaloupe, watermelon, hot peppers, Chinese cabbage, artichoke, citrus peel, citrus juices.	Antioxidant, anti-inflammatory, antibacterial, antifungal, antiviral, anticarcinogenic
Flavanones		32	Didymin, Eriocitrin, Hesperetin, Hesperidin, Naringenin, Naringin, Narirutin	Citrus fruit, citrus juices, Mexican oregano, pepper mint, wines	Antioxidant, anti-inflammatory, antitumoral, antiviral antimicrobial

(Continued)

TABLE 8.2 (CONTINUED)

Classification	Skeleton	Data Base Reported Compounds	Examples	Food Sources	Reported Nutraceutical Effects
Isoflavonids		92	Angolesin, Butin, Daidzein, Daidzin, Genistin, Glycetein, Koparin,	Lucerne, clovers, soybean, and other legumes	Antioxidant, anti-inflammatory, antitumoral, hepatoprotective
Flavanols		45	Epicatechin, Catechin, Gallocatechin, Cinnamtannin A2, Theoflavin	Chocolate, red wine, nuts, grapes, strawberries, blackberry, peach, nectarine, apple, plum, beans	Antioxidant, anticarcinogenic, cardiopreventive, antimicrobial, antiviral, neuroprotective
Anthocyanins		89	Cyanidin, Delphinidin, Malvidin, Vitisin	Berries, grapes, plum, peach, black beans, red lettuce, red onion, red wine.	Antioxidant, increase sight acuteness, anticarcinogenic

Sources: Adapted from: Crozier et al. (2006); Rothwell et al. (2013); Acosta-Estrada et al. (2014); De la Rosa et al.(2018); Kumar & Goel (2019); Morandi-Vuolo (2019); Carpena et al.(2021); Vidal-Casanella et al. (2021).

8.4 ANTIOXIDANT DIETARY FIBER

Antioxidant dietary fiber, as mentioned, contains large amounts of antioxidant compounds (equivalent to 200mg of Vitamin E or more) and is associated intermolecularly with the fiber of the food matrix (Saura-Calixto, 1998). It has beneficial effects for human health as it stimulates colonic fermentation, helps regulates blood glucose levels and insulin response, and helps to decrease obesity and cardiovascular diseases, in addition to other types of bioactivities dependent of fiber source (Costa et al. 2019). The phenolic compounds of this antioxidant dietary fiber reduce the oxidative stress generated by free radicals and are involved in anti-inflammatory, modulation, and signaling of cell growth processes (Comert & Gokmen, 2020).

In this context, and according to Celik et al. (2015), Comert & Gokmen, (2020), and Maurer et al. (2020), the beneficial effects could be due to the synergy between bound and unbound phenolic compounds to dietary fiber, deactivating the free radicals of the medium by hydrogen donation, and regenerating the antioxidant activity. Once this fiber reaches the colon, the gut microbiota starts to hydrolyze the prebiotic components of the fiber, slowly and continuously releasing the phenolic compounds, playing an essential role in the intestinal health in anti-proliferative and anti-inflammatory activities (Vázquez- Sánchez et al. 2018).

Many of these biological properties vary depending on the matrix in which the antioxidant fiber is found and the way it was processed before consumption (Rao et al. 2020), modifying how it responds to the gastric digestion process. All these factors could result in different bioaccessibility and bioavailability activities of these nutraceutical compounds (Blancas-Benitez et al. 2015).

8.4.1 *In Vitro* Digestion Process

The study of the changes on bioaccessibility and bioavailability caused by the digestive process on the different components of dietary fiber when consumed can be carried out in various ways; with in vitro digestion being one of the most commonly used methods. This technique helps to analyze the alterations of food structure, predict the bioavailability of food components, and observe the interaction that is created in the food matrices in an economical, easily accessible way and without ethical implications (Ting et al. 2015).

The *in vitro* digestion methodologies can be divided into four phases mimicking the digestive process that occurs in the human body: oral, gastric, duodenal-intestinal and, in some cases, fermentative (Bender & Mayes, 2012).

The oral phase is the first step of the *in vitro* digestive process. In this step, α-amylase, salivary fluid, and water are added to transform the food into the bolus by hydrolyzing the α bonds of starch, glycogen, and dextrins (Minekus et al. 2014). However, other carbohydrates present in the agro-industrial residues suffer minimal changes, since the ß bonds in their composition cannot be hydrolyzed by the enzymes present in salivary fluid (Hall & Hall, 2021; Cárabez, Julián, & Riveros 2018; Bender & Mayes, 2012). In the stomach or gastric phase, α-amylase is deactivated when the bolus is mixed with the gastric fluids (pH=3) and pepsin is added (Minekus et al. 2014). The high acidity and the pepsin start the hydrolysis of lipids and proteins present in the bolus, which are broken down into monoglycerides and chains of amino acids. Food rich in carbohydrates present in the residues remains in the stomach for a shorter time, compared to foods with a higher lipid content. Finally, the food bolus changes to chyme (Bender & Mayes, 2012; Cárabez, Julián & Riveros, 2018; Hall & Hall, 2021). The intestinal phase is the most important since it is here where all the nutrients are absorbed and digested. In this step, the chyme is mixed with synthetic intestinal fluid, water, pancreatin and bile salts, and for the simulation of the intestinal nutrient absorption, a dialysis bag filled with a sodium bicarbonate solution is used (Minekus et al. 2014). The last phase of the *in vitro* digestion technique is the fermentative phase, where the colon microbiota hydrolyzes the ß-linked carbohydrates to monosaccharides such as glucose, fructose, and galactose are absorbed through specific transporters (Louis et al. 2016), along with the production of short chain fatty acids.

8.4.2 Antioxidant Availability of Fiber during *In Vitro* Gastric Digestion

In vitro gastric digestion analysis has been widely employed to study the bioaccesibility and biodisponibility of many different types of products, including agricultural and agro-industrial residues, following the change after digestion of their main chemical components as well as some secondary metabolites. From these reports, antioxidants became one of the most studied components over the last five years, followed by carbohydrates and proteins at a very similar level, and lipids coming in last (**Figure 8.1**). It is important to mention that some reports include the study of food components that can be related to several of the macronutrients and present antioxidant activity at the same time, depending on the nature of the food.

Table 8.3 summarizes some of these published works where several *in vitro* digestion studies with different food products and subproducts are analyzed, following the antioxidant changes, via DPPH or ABTS activities and total phenolic compounds (TPC). From this data, the bioaccessibility of TPC is calculated. This parameter is defined as the digested material that is available for absorption after being released during the gastrointestinal process (Gómez-Maqueo et al. 2020).

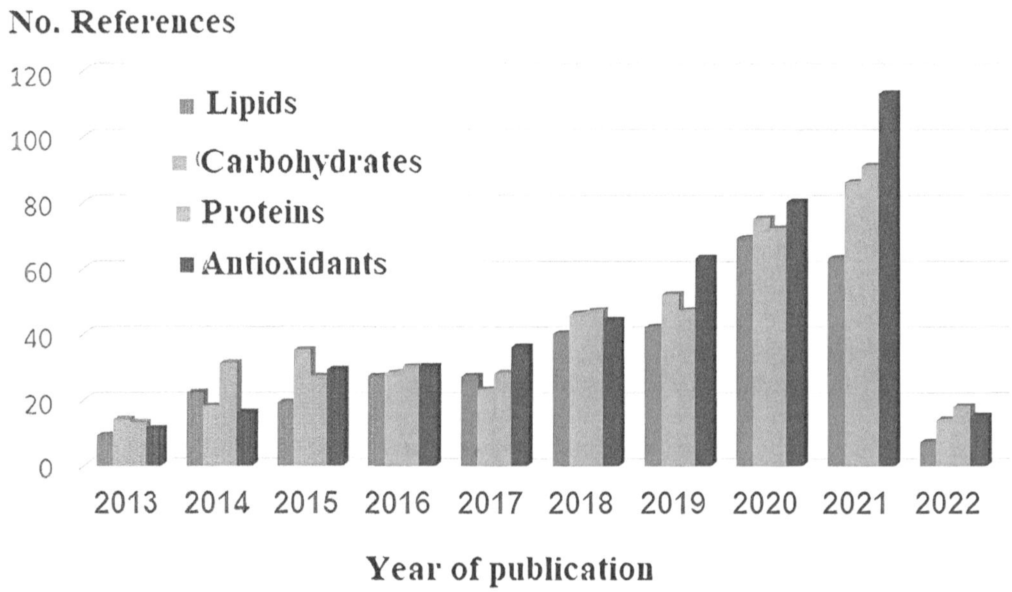

Figure 8.1 *In vitro* digestion studies. Trends for different food components. Data extracted from Scopus database.

It is seen in **Table 8.3** that most products present a higher proportion of insoluble than soluble fiber, except for mango pulp (Blancas-Benitez et al. 2015), wheat noodles (Wang et al. 2021), and grape pomace extract (Beres et al. 2019), which show insoluble fiber contents of 8.0%, 3.9%, and 20.4% respectively. Spent coffee grounds have the highest concentration (68.8%), while the lowest values were observed in food, mainly papaya (4.4±0.2%) and pineapple (2.4±0.2%) pulps and wheat noodles (2.6±0.1%).

The reported antioxidant activities of these materials before being digested fluctuate from 100.4±0.17 mgGAE/g to 0.02±0.01 mgGAE/g measured as TPC for grape pomace extract (Beres et al. 2019) and grape pomace (Solari-Godiño et al. 2017), respectively, but most of the products in **Table 8.3** have much lower values (1.0<TPC<6.1 mgGAE/g). As expected, the TPC content did not statistically correlate with DPPH and ABTS data, as blueberry bagasse powder showed the highest DPPH and ABTS activities as compared to grape pomace extract (Bas-Bellver et al. 2020).

When analyzing the changes in the antioxidant activity during the *in vitro* digestion, with respect to the products fiber content, they did not follow a homogeneous behavior, as some of them showed increased bioaccesibility, while others decreased, and no statistical correlation was found between the type of fiber or the ratio of soluble-insoluble fiber with antioxidants or bioaccesibility.

This behavior has already been reported in other products, as shown by Su et al. (2017) and Ng and See, (2019). They analyzed 12 functional products including fruits, herbs, and roots such as Chinese yam (*Dioscorea polystachya*) and licorice (*Glycyrrhiza glabra*). Both groups reported that the content of total phenols was higher in the gastric phase compared to the intestinal phase. However, the complete opposite results have also been published, showing the release of antioxidants compounds during *in vitro* digestion.

Antioxidants from different foods are commonly immersed in various macromolecules such as carbohydrates, lipids, and proteins to form the food matrix. In plant tissue carbohydrates are the main compounds found in free and conjugated forms (Manach et al. 2004), affecting the bioaccessibility and bioavailability of antioxidants (Palafox et al. 2011). Some of these results show that during gastric digestion different antioxidant metabolites, such as kaempferol 3-O-glucoside, luteolin 4'-O-glucoside, apigenin O-rhamnosylglucoside, and luteolin O-hexoside, are released from seeds (Czubinski et al. 2019), while in other cases, some antioxidants decreased their concentration (Gullon et al. 2015). In all these cases, the relationship between the type of fiber is unclear, and the change in the rheological properties of the fibers during digestion have not been included. These parameters may be influencing biodisponibility of antioxidant by preventing the phenolic compounds from leaving the food matrix during oral and gastric *in vitro* digestion, but releasing these compounds in the intestinal phase, where the fiber can be metabolized. The evaluation of the digestibility of the produced prebiotics confirms that they reach the colon without being hydrolyzed, while fermentability assays allow verifying their selectivity to promote the growth of beneficial microorganisms and to follow its conversion into bioactive metabolites, such as short chain fatty acids (SCFA), which play very important roles in human physiology and metabolism (Gonçalves, et al., 2023).

8.5 CONCLUSION

Residues in roots or husks generate fiber that may have many different applications; for example, in biogas production, as clarifying agents, or as raw materials to develop new products enriched

TABLE 8.3 CHANGES IN TOTAL PHENOLIC CONTENT DURING *IN VITRO* GASTROINTESTINAL STUDIES OF DIFFERENT PRODUCTS

Source	Soluble fiber (SF) (%)	Insoluble Fiber (IF) (%)	TPC Bio-accessibility %	Antioxidant activity DPPH content x 10^2 mg TE/g		Antioxidant activity ABTS content 10^1 mg TE/g		Reference
				Undigested sample	*Intestinal digested	Undigested sample	*Intestinal digested	
Food Products								
Papaya Pulp	2.9±0.1	4.4±0.2	38.1	4.0±0.5	2.1±0.2	9.0±0.1	4.9±0.5	Velderrain-Rodríguez et al. (2016)
Pineapple Pulp	1.8±0.1	2.4±0.2	54.5	3.0±0.1	16±0.1	8.6±0.01	4.0±0.4	
Mango ulp (Mangifera indica L.)	8.0±0.5	7.0±1.3	5400	-	-	-	-	Blancas-Benítez et al. (2015)
Jícama	6.5 ± 0.2	16.6 ± 0.1	49.4	-	-	-	-	González-Vázquez et al. (2022)
Wheat noodles	3.9±0.4	2.6±0.1	111	-	-	13±1.0	20±1.0	Wang et al. (2021)
Wheat noodles/ White mushroom	3.9±0.1	7.0±0.1	280	-	-	40±1.0	50±1.0	
Seeds *Amaranthus viridis*	0.7±0.1	29.9±0.1	200	200±10	3±1	180±1.0	20±0.9	Silva et al. (2021)
Root *R.pimpinellifolia*	-------	-------	62.9	148.1±19	26.6±3.6	81.6±13	26.2±2.5	Macit et al. (2023)
Traditional nixtamalized corn tortillas	------	------	960.7	6500	4000	10000	30000	Menchaca-Armental et al. (2023)
Extruded nixtamalized blue corn tortillas	-------	-------	162.83	21250	6250	2000	35000	

Food by-products or agricultural residues								
Grape pomace extract	20.4	3.0	4.2	-	-	390±10	480±80	Beres et al. (2019)
Grape pomace flour	2.4	14.3	11.7	-	-	200±60	430±80	
Wheat bran	3.1±0.1	38.3±0.1	3.8	300±10	600±10	15±0.1	20±0.1	Zhang et al. (2022)
Grape pomace	3.4	65.4	150	-	-	-	-	Solari-Godiño et al. (2017)
Blueberry bagasse powder	9.4±0.1	21.0±0.2	18.9	6680±80	1390±60	5760±20	893±7	Bas-Bellver et al. (2020)
Persimmon waste powder	7.6±0.1	10.4±0.1	36.4	900±20	270±12	313±11	152±2	
Mango peel (*Mangifera indica* L.)	18.6±1.3	22.8 ± 2.3	5000	-	-	-	-	Blancas-Benítez, et al. (2015)
Olive pomace powder	16.0	52.2 ± 0.01	40	200±10	400±90	30± 2	20±1.2	Ribeiro, et al. (2021)
Spent coffee (*Coffea arabica L.*) grounds	0.3 ± 0.1	68.8 ± 0.3	89	-	-	-	-	Vázquez-Sánchez, et al. (2018)
Melon peel flour	4.5 ± 0.5	37.5 ± 2.5	122.2	25±2	128±10	-	-	Gómez-García, et al. (2022)
Tomato bagasse flour	13.0 ± 0.6	46.0 ± 0.13	128.6	-	-	-	-	Coelho, et al. (2021)
Aloe barbadensis Miller peel	5.2 ±	27.5 ±	560	-	-	-	-	Lima-Villegas et al. (2023, own data

with antioxidant dietary fiber. It is important to point out that the biological activity of potential prebiotics must be evaluated before commercialization and before making any health claim on these products. This will be an important contribution, in addition to giving added value to these waste materials.

ACKNOWLEDGMENTS

Karen E. Lima-Villegas, thanks Instituto Politécnico Nacional (BEIFI-IPN) and CONACYT for the scholarships provided. This research was funded through projects: SIP-20221471; 20221510; 20230985; 20230993 from the Instituto Politécnico Nacional (IPN, Mexico).

REFERENCES

Acosta-Estrada, B. A., Gutiérrez-Uribe, J. A., & Serna-Saldívar, S. O. (2014). Bound phenolics in foods, A review. *Food Chemistry*, 152, 46–55. https://doi.org/10.1016/j.foodchem.2013.11.093

Alzagameem, A., Klein, S. E., Bergs, M., Tung, D. X., Korte, I., Dohlen, S., Hüwe, C., Kreyenschmidt, J., Kamm, B., Larkins, M., & Schulze, M. (2019). Antimicrobial activity of lignin and lignin-derived cellulose and chitosan composites against selected pathogenic and spoilage microorganisms. *Polymers*, 11, 670. https://doi.org/10.3390/polym11040670

Banerjee, S., Patti, A., Ranganathan, V., & Arora, A. (2019). Hemicellulose based biorefinery from pineapple peel waste: Xylan extraction and its conversion into xylooligosaccharides. *Food and Bioproducts Processing*, 117, 38–50. https://doi.org/10.1016/j.fbp.2019.06.012

Bas-Bellver, C., Andrés, C., Segui, L., Barrera, C., Jimenez-Hernandéz, N., Artacho, A., Betoret, N., & Gosalbes, M. J. (2020). Valorisation of persimmon and blueberry by-products to obtain functional powders: *In vitro* digestion and fermentation by gut microbiota. *Journal of Agricultural and Food Chemistry*, 68, 8080–8090. https://doi.org/10.1021/acs.jafc.0c02088

Bender, D. A., & Mayes, P. A. (2012). Nutrición, digestión y absorción. In R. K. Murray, D. A. Bender, K. M. Botham., P. J. Kennely, V. W. Rodwell, & P. A. Weil (Eds.), *Harper Bioquímica Ilustrada* (8th ed.) (pp. 517–524). Mc Graw Hill Lange. ISBN: 978-607-15-0914-7

Beres, C., Pereira-Freitas, S., de Oliveira-Godoy, R. L., Rodrigues-de Oliveira, D. C., Deliza, R., Lacomini, M., Mellinger-Silva, C., & Correa-Cabral, L. M. (2019). Antioxidant dietary fibre from grape pomace flour or extract: Does it make any difference on the nutritional and functional value? *Journal of Functional Foods*, 56, 276–285. https://doi.org/10.1016/j.jff.2019.03.014

Bhagwat, S., Haytowitz, D. B., & Holden, J. M. (2011). *USDA database for the flavonoid content of selected foods, release 3*. US Department of Agriculture: Beltsville, MD, USA, 159.

Blancas-Benitez, F., Mercado-Mercado, G., Quirós-Sauceda, A., Montalvo-González, E., González-Aguilar, & Sáyago-Ayerdi, G. (2015). Bioaccesibility of polyphenols associated with dietary fiber and *in vitro* kinetics release of polyphenols in Mexican "Ataulfo mango" (Mangifera indica L.) by-products. *Royal Society of Chemistry*, 6, 859–868. https://doi.org/10.1039/c4fo00982g

Bojarczuk, A., Skapska, S., Mousavi, K. A., & Marszalek, K. (2022). Health benefits of resistan starch: A review of the literature. *Journal of Functional Foods*, 93, 105094. https://doi.org/10.1016/j.jff.2022.105094

Bouaziz, F., Koubaa, M., Ben, K., Kallel, F., Boisset, C., Khelfa, A., & Ghorbel, R. (2016). Water-soluble polysaccharides and hemicelluloses from almond gum: Functional and prebiotic properties. *International Journal of Biological Macromolecules*, 93, 359-368. https://doi.org/10.1016/j.ijbiomac.2016.08.032

Câmara, A. K. F. I., Okuro, P. K., Lopes-da Cunha, R., Herrero, A. M., Ruíz-Capillas, C., & Rodrigues-Pollonio, M. A. (2020). Chia (*Salvia hispanica* L.) mucilage as a new fat substitute in emulsified meat products: Technological, physicochemical, and rheological characterization. *LWT Food Science and Technology*, 125, 109193. https://doi.org/10.1016/j.lwt.2020.109193

Capriles, V. D., & Arêas, J. A. G. (2013). Effects of prebiotic inulin-type fructans on structure, quality, sensory acceptance and glycemic response of gluten-free breads. *Food & Function*, 4, 104–110. https://doi.org/10.1039/c2fo10283h

Capuano, E. (2017). The behavior of dietary fiber in the gastrointestinal tract determines its physiological effect. *Critical Reviews in Food Science and Nutrition*, 57, 3543–3564. https://doi.org/10.1080/10408398.2016.1180501

Cárabez, T. A., Julián, S. A., & Riveros, R. H. (2018). Metabolismo de los carbohidratos. In F. M. Martínez, J. P. V. Pardo & H. R. Riveros (Eds.), *Bioquímica de Laguna y Piña* (29th ed.). (pp. 591–659). Mc Graw Hill Lange. ISBN: 978-607-15-0914-7

Carpena, M., Caleja, C., Nuñez-Estevez, B., Pereira, E., Fraga-Corral, M., Reis, F. S., Simal-Gandara, J., Ferreira. I, Prieto, M. A., & Barros, L. (2021). Flavonoids: A group of potential food additives with beneficial health effects. In M. A. Prieto & P. Otero (Eds.), *Natural Food Additives*. IntechOpen. https://doi.org/10.5772/intechopen.91548

Celik, E. E., Gokmen, V., & Skibsted, L. H. (2015). Synergism between soluble and dietary fiber bound antioxidants. *Journal of Agriculture and Food Chemistry*, 63, 2338–2343. https://doi.org/10.1021/acs.jafc.5b00009

Codex Alimentarius. (2010). Guidelines on Nutrition Labelling Cac/Gl 2-1985. Roma.

Coelho, M. C., Ribeiro, T. B., Oliveira, C., Batista, P., Castro, P., Monforte, A. R., Rodrigues, A. S., Teixeira, J., & Pintado, M. (2021). *In vitro* gastrointestinal digestion impact on the bioaccessibility and antioxidant capacity of bioactive compounds from tomato flours obtained after conventional and ohmic heating extraction. *Foods*, 10, 554. https://doi.org/10.3390/foods10030554

Coman, V., Teleky, B. E., Mitrea, L., Martau, G. A., Szabo, K., Calinoiu, L. F., & Vodnar, D. C. (2019). Bioactive potential of fruit and vegetable wastes. In F. Toldrá (Ed.), *Advances in Food and Nutrition Research* (pp. 157–225). Academic Press. https://doi.org/10.1016/bs.afnr.2019.07.001

Comert, E. D., & Gokmen, V. (2020). Physiological relevance of food antioxidants. In F. Toldrá (Eds.), *Advances in Food and Nutrition Research* (Vol. 93, pp. 205–250). Academic Press. https://doi.org/10.1016/bs.afnr.2020.03.002

Costa, J. R., Amorim, M., Vilas-Boas, A., Tonon, R. V., Cabral, L. M., Pastrana, L., & Pintado, M. (2019). Impact of *in vitro* gastrointestinal digestion on the chemical composition, bioactive properties, and cytotoxicity of *Vitis vinifera* L. cv. Syrah grape pomace extract. *Food & Function*, 10(4), 1856–1869. https://doi.org/10.1039/c8fo02534g

Crispin-Isidro, J., Lobato-Cabellos, C., Espinosa-Andrews, H., Alvarez-Jimenez, J., & Vernon-Carter, E. J. (2015). Effect of inulin and agave fructans addition on the rheological, microstructural and sensory properties of reduced-fat stirred yogurt. *LWT- Food Science and Technology*, 62, 438-444. https://doi.org/10.1016/j.lwt.2014.06.042

Crozier, A., Jaganath, I. B., & Clifford, M. N. (2006). Phenols, polyphenols and tannins: an overview, In A. Crozier, M. N. Clifford, & H. Ashihara (Eds.), *Plant Secondary Metabolites: Occurrence, Structure, and Role in the Human Diet* (Vol. 1, pp. 1–25). Blackwell Publishing. https://doi.org/10.1002/9780470988558.ch1

Czubinski, J., Wroblewska, K., Czyzniejewski, M., Górnas, P., Kachlicki, P., & Siger, A. (2019). Bioaccessibility of defatted lupin seed phenolic compounds in a standardized static *in vitro* digestion system. *Food Research International*, 116, 1126–1134. https://doi.org/10.1016/j.foodres.2018.09.057

De la Rosa, L. A., Moreno-Escamilla, J. O., Rodrigo-García, J., & Alvarez-Parrilla, E. (2018). Phenolic compounds. Postharvest physiology and biochemistry of fruits and vegetables. In E. Yahia & A. Carrillo-Lopez (Eds.). *Postharvest Physiology and Biochemistry of Fruits and Vegetables* (pp. 253–271). Elsevier-WP Woodhead Publishing. https://doi.org/10.1016/b978-0-12-813278-4.00012-9

Doliska, A., Willfor, S., Strnad, S., Ribitsch, V., Stana, K., Eklund, P., & Xu, C. (2012). Antithrombotic properties of sulfated wood-derived galactoglucomannas. *Wood Research and Technology Holzforschung*, 66, 149–154. https://doi.org/10.1515/HF.2011.136

Escudero, A. E., & González, S. P. (2006). La fibra dietética. *Nutrición Hospitalaria*, 21, 61–72. ISSN 0212-1611

Faith, C. W. S., Meijerink, M., Zeuner, B., Holck, J., Louis P., Meyer, A. S., Wells, J. M., Flint, H. J., & Duncan, S. H. (2017). Prebiotic potential of pectin and pectic oligosaccharides to promote anti-inflammatory commensal bacteria in the human colon. *FEMS Microbiology Ecology*, 93, 11. https://doi.org/10.1093/femsec/fix127

FAO. (2016). Save Food. Perdidas y desperdicios de alimentos en América Latina y el Caribe.

FAO. (2018). Agriculture News from Latin America and the Caribbean. México wastes 20 million tons of food per year. www.fao.org/in-action/agronoticias/detail/en/c/1129783/

Ferreira, I., Vilar, G., Kiyomi, O., Maróstica, J., Lopés, da C., & Rodrigues P. (2020). Satiety and *in vitro* digestibility of low saturated fat Bologna sausages added of chia mucilage powder and chia mucilage-based emulsion gel. *Journal of Functional Foods*, 65, 103753. https://doi.org/10.1016/j.jff.2019.103753

Gómez-García, R., Sánchez-Gutiérrez, M., Freitas-Costa, C., Vilas-Boas, A. A., Campos, D. A., Aguilar, C. N., Madureira, A. R., & Pintado, M. (2022). Prebiotic effect, bioactive compounds and antioxidant capacity of melon peel (*Cucumis melo* L. inodorus) flour subjected to *in vitro* gastrointestinal digestion and human faecal fermentation. *Food Research International*, 154, 111045. https://doi.org/10.1016/j.foodres.2022.111045

Gómez-Maqueo, A., Antunes-Ricardo, M., Welti-Chanes, J., & Cano, M. P. (2020). Digestive stability and bioaccessibility of antioxidants in prickly pear fruits from the Canary Islands: Healthy foods and ingredients. *Antioxidants*, 9, 164. https://doi.org/10.3390/antiox9020164

Gonçalves, D. A., González, A., Roupar, D., Teixeira, J. A., & Nobre, C. (2023). How prebiotics have been produced from agro-industrial waste: An overview of the enzymatic technologies applied and the models used to validate their health claims. *Trends in Food Science & Technology*, 135, 74–92. https://doi.org/10.1016/j.tifs.2023.03.016

González-Aguilar, G. A., Blancas-Benítez, F. J., & Sáyago-Ayerdi, S. G. (2017). Polyphenols associated with dietary fibers in plant foods: Molecular interactions and bioaccessibility. *Current Opinion in Food Science*, 13, 84–88. https://doi.org/10.1016/j.cofs.2017.03.004

González-Vázquez, M., Calderón- Domínguez, G., Mora-Escobedo, R., Salgado-Cruz, M. P., Arreguín-Centeno, J. H., & Monterrubio-López, R. (2022). Polysaccharides of nutritional interest in jicama (*Pachyrhizus erosus*) during root development. *Food Science & Nutrition*, 00, 1–13. https://doi.org/10.1002/fsn3.2746

Gourineni, V., Stewart, M. L., Wilcox, M. L., & Maki, K. C. (2020). Nutritional bar with Potato-based resistant starch attenuated post-prandial glucose and insulin response in healthy adults. *Foods*, 9, 1679. https://doi.org/10.3390/foods9111679

Gullon, B., Pintado, M., Barber, X., Fernández-López, J., Pérez-Álvarez, J., & Viuda-Martos, M. (2015). Bioaccessibility, changes in the antioxidant potential and colonic fermentation of date pits and apple bagasse flours obtained from co-products during simulated *in vitro* gastrointestinal digestion. *Food Research International*, 78, 169–176. https://doi.org/10.1016/j.foodres.2015.10.021

Hall, J. E., & Hall, M. E. (2021). General principles of gastrointestinal function-motility, nervous control, and blood circulation. In J. E. Hall & M. E. Hall (Eds.), *Text of Medical Physiology* (14th ed.) (pp 787–839). Elsevier. ISBN: 978-0-323-59712-8

Harini, K., Chandra, C., Ramya, K., Karthikeyan, S., & Sukumar, M. (2018). Effect of Punica granatum peel extracts on antimicrobial properties in Walnut shell cellulose reinforced Bio-thermoplastic starch films from cashew nut shells. *Carbohydrate Polymers*, 184, 231–242. https://doi.org/10.1016/j.carbpol.2017.12.072

Herrera-Balandrano, D. D., Báez-González, J. G., Carvajal-Millán, E., Muy-Rangel, D., Urías-Orona, V., Martínez-López, A. L., Marquez-Escalante, J. A., Heredia, J. B., Beta, T., & Niño-Medina, G. (2020). Alkali-extracted feruloylated arabinoxylans from nixtamalized maize bran byproduct: A synonymous with soluble antioxidant dietary fiber. *Waste and Biomass Valorization*, 11(2), 403–409. https://doi.org/10.1007/s12649-018-0462-z

Jiang, B., Zhang Y., Gu, L., Wu, W., Zhao, H., & Jin, Y. (2018). Structural elucidation and antioxidant activity of lignin isolated from rice straw and alkali oxygen black liquor. *International Journal of Biological Macromolecules*, 116, 513–519. https://doi.org/10.1016/j.ijbiomac.2018.05.063

Jiang, Y., Xu, Y., Li, F., Li, D., & Huang, Q. (2020). Pectin extracted from persimmon peel: A physicochemical characterization and emulsifying properties evaluation. *Food Hydrocolloids*, 101, 105561. https://doi.org/10.1016/j.foodhyd.2019.105561

Jongaroontaprangsee, S., Chiewchan, N., & Devahastin S. (2018). Production of nanofibrillated cellulose with superior water redispersibility from lime residues via a chemical-free process. *Carbohydrate Polymers*, 193, 249–258. https://doi.org/10.1016/j.carbpol.2018.04.008

Kaprosob, R., Kerdchoechuen, O., Laohakunjit N., & Sonboonpanyalkul, P. (2018). B vitamins and prebiotic fructooligosaccharides of cashew apple fermented with probiotic strains *Lactobacillus* spp., *Leuconostoc mesenteroides* and *Bifidobacterium longum. Process Biochemistry*, 70, 9–19. https://doi.org/10.1016/j.procbio.2018.04.009

Kaza, S., Yao, L., Bhada-Tata, P., & Van Woerden, F. (2018). *What a Waste 2.0: A global snapshot of solid waste management to 2050. Urban Development Series*. Washington, DC: World Bank. License: Creative Commons Attribution CC BY 3.0 IGO.

Kumar, N., & Goel, N. (2019). Phenolic acids: Natural versatile molecules with promising therapeutic applications. *Biotechnology Reports*, 24, e00370. https://doi.org/10.1016/j.btre.2019.e00370

Li, Y. X., Liu, H. J., Shi, Y. Q., Yan, Q. J., You, X., & Jiang, Z. Q. (2020). Preparation, characterization, and prebiotic activity of manno-oligosaccharides produced from cassia gum by a glycoside hydrolase family 134 β-mannanase. *Food Chemistry*, 309, 125709. https://doi.org/10.1016/j.foodchem.2019.125709

Lima-Villegas, K. E. (2023) Disponibilidad de compuestos bioactivos de la corteza de Aloe *barbadensis* Miller durante la digestión gastorointestinal *in vitro* [Ph.D. thesis]. Escuela Nacional de Ciencias Biológicas, Instituto Politecnico Nacional. México City, México.

Louis, P., Flint, H. J. & Michel, C. (2016). How to manipulate the Microbiota: Prebiotics. In A. Schwiertz (Ed.), *Microbiota of the Human Body, Implications in Health and Disease* (Vol. 902, pp. 119–142). Board Springer. ISBN: 978-3-319-31248-4

Macit, M., Aras, A., Capanoglu Guven, E., & Bakir, S. (2023). Investigating the content and bioaccessibility of phenolic compounds in roots of rosa canina l. and *Rosa pimpinellifolia L. Yuzuncu Yil University Journal of Agricultural Sciences*, 33(2), 163–173. https://doi.org/10.29133/yyutbd.1231881

Mahmood, Z., Yameen, M., Jahangeer, M., Riaz, G. A., & Javid, I. (2018). Lignin as natural antioxidant capacity. In M. Poletto (Ed.), *Lignin. Trends and applications.* 181–206. techOpen. https://doi.org 10.5772/intechopen.68464

Manach, C., Scalbert, A., Morand, C., Rémésy, C. & Jímenez L. (2004). Polyphenols: food sources and bioavailability. *The American Journal of Clinical Nutrition*, 79(5), 727–742. https://doi.org 10.1093/ajcn/79.5.727

Maurer, L. H., Betim, C. B. B., Quatrin, A., Marafiga, N. S., Machado, M. N., Ferreira, T. C., Manica, C. I. B., Maróstica J. M. R., & Emanuelli, T. (2020). Dietary fiber and fiber-bound polyphenols of grape peel powder promote GSH recycling and prevent apoptosis in the colon of rats with TNBS-induced colitis. *Journal o Functional Foods*, 64. https://doi.org/10.1016/j.jff.2019.103644

McRorie, W. J., & McKeown, M. N. (2017). Understanding the physics of functional fibers in the gastrointestinal tract: An evidence-based approach to resolving enduring misconceptions about Insoluble and soluble fiber. *Journal of the Academy of Nutrition and Dietetics*, 117(2), 251–264. https://doi.org/10.1016/j.jand.2016.09.021

Menchaca-Armenta, M., Frutos, M. J., Ramírez-Wong, B., Valero-Cases, E., Muelas-Domingo, R., Quintero-Ramos, A., Torres-Chávez P. I., Carbonell-Barrachina, A. A., Ledesma-Osuna, A. I., & Campas-Baypoli, O. N. (2023). Changes in phytochemical content, bioaccesibility and antioxidant capacity of corn tortillas during simulated in vitro gastrointestinal digestion. *Food Chemistry, 405*, 134223. https://doi.org/10.1016/j.foodchem.2022.134223

Minekus, M., Alminger, M., Alvito, P., Balance, S., Bohn, T., Bourlieu, C., Carriere, F., Boutrou, R., Corredig, M., Dupont, D., Dufour, C., Egger, L., Golding, M., Karakaya, S., Kirkhus, B., Le Feunteun, S., Lesmes, U., Marcierzanka, A., Mackie, A., Marze, S., McClements, D. J., Ménard, O., Recio, I., Santos, C. N., Singh, R. P., Vegarud, G. E., Wickham, M. S. J., Weitschies, W., & Brodkorb, A. (2014). A ulplcaed static *in vitro* digestion method suitable for food – an international consensus. *Royal Society of Chemistry*, 5, 1113–1124. https://doi.org/10.1039/c3fo60702j.

Mohsin, A., Zhang, K., Hu J., Salim-ur-Rehman, Tariq, M., Qamar, Z., Mahmood, K., Zhuang Y., & Guo, M. (2018). Optimized biosynthesis of xanthan via effective valorization of orange peels using response surface methodology: A kinetic model approach. *Carbohydrates Polymers*, 181, 793–800. https://doi.org/10.1016/j.carbpol.2017.11.076

Morandi, V. M., Silva, L. V., & Maróstica, J. M. R. (2019). Phenolic compounds: structure, classification, and antioxidant power. In M. R. Segura (Ed.), *Bioactive Compounds* (pp. 33–50). Elsevier Woodhead Publishing. https://doi.org/10.1016/C2017-0-02265-6

Mu, T., Sun, H., Zhang, M., & Wang, C. (2017). Sweet potato dietary fiber. In T. Mu, H. Sun, M. Zhang & C. Wang (Eds.) *Sweet Potato Processing Technology* (pp. 121–181). Academic Elsevier. ISBN: 978-0-12-812871-8.

Mudgil, D. (2017). The Interaction between insoluble and soluble fiber. In R. A. Sadmaan (Ed.) *Dietary Fiber for the Prevention of Cardiovascular Disease* (pp. 35–59) Academic Press. ISBN: 978-0-12-805130-6

Mudgil, D., & Barak, S. (2013). Composition, properties, and health benefits of indigestible carbohydrate polymers as dietary fiber: A review. *International Journal of Biological Macromolecules*, 61, 1–6. https://doi.org/10.1016/j.ijbiomac.2013.06.044

Mugwagwa, L., & Chimphango, A. (2020). Enhancing the functional properties of acetylated hemicellulose films for active food packaging using acetylated nanoellulose reinforcement and polycaprolactone coating. *Food Packaging and Shelf Life*, 24, 100481. https://doi.org/10.1016/j.fpsl.2020.100481

Muller, M., Reiner, J., Fleischhacker, L., Viernstein, H., Loeppert, R., & Praznik, W. (2016). Growth of selected probiotic strains with fructans from ulplca sources relating to degree of polymerization and structure. *Journal of Functional Food*, 24, 264–275. https://doi.org/10.1016/j.jff.2016.04.010

Muñoz de Chávez, M., Chávez, V .A., Roldán, A. J. A., Pérez-Gil, R. F., Ledezma, S. J. A., Hernández, C. S. L., Mendoza, M. E., & Chaparro F. A. G. (1996). Tabla de valor nutritivo de los alimentos de mayor consumo en México. México: Instituto Nacional de Nutrición Salvador Zubirán. Editorial Pax, 330

Ng, Z. X., & See, A. N. (2019). Effect of *in vitro* digestion on the total polyphenol and flavonoid, antioxidant activity and carbohydrate hydrolyzing enzymes inhibitory potential of selected functional plant-based foods. *Journal of Food Processing and Preservation*, 43 (4),13903. https://doi.org/10.1111/jfpp.13903

Nsor-Atindana. J., Xing. Z., Nazmus. S., Chen. M., Douglas. G., Ma. J., & Zhong. F. (2020). Enhancing the prebiotic effect of cellulose biopolymer in the gut by physical structuring via particle size manipulation. *Food Research International*, 131, 108935. https://doi.org/10.1016/j.foodres.2019.108935

Ortiz-Basurto, R. I., Rubio-Ibarra, M. E., Ragazzo-Sanchez, J. A., Beristain, C. I., & Jiménez-Fernández, M. (2017). Microencapsulation of Eugenia uniflora L. juice by spray drying using fructans with different degrees of polymerization. *Carbohydrate Polymers*, 175, 603–609. https://doi.org/10.1016/j.carbpol.2017.08.030

Palafox-Carlos, H., Ayala-Zavala, J. F., & Gonzalez-Aguilar, G. A. (2011). The role of dietary fiber in the bioaccessibility and bioavailability of fruit and vegetables antioxidants. *Journal of Food Science*. 76, R7–R13. https://doi.org/10.1111/j.1750-3841.2010.01957.x

Pereira, S. A. C., Gurak, P. D., & Ferreira, M. L. D. (2017). Maltodextrin, pectin and soy protein isolate as carrier agents in the encapsulation of anthocyanins-rich extract from jaboticaba pomace. *Food and Bioproducts Processing*, 102, 186–194. https://doi.org/10.1016/j.fbp.2016.12.012

Piña, I., Ysambertt, F., Arias, M., Chirinos, J. & Castillo, M. (2011). Capacidad antioxidante de la lignina extraída del licor negro en polietileno de baja densidad. *Revista iberoamericana de Polímeros*. 124(4):216-226. https://reviberpol.files.wordpress.com/2019/07/2011-pina.pdf

Punia, S., & Bala, D. S. (2019). Chia seed (Salvia ulplca L.) mucilage (a heteropolysaccharide): Functional, ulpl, rheological behaviour and its utilization. *International Journal of Biological Macromolecules*, 140, 1084–1090. https://doi.org/10.1016/j.ijbiomac.2019.08.205

Rao, H., Baricevic, I., Bernard, H., Smith, F., Sayers, R., Balasundaram, A., Costello, C. A., Padfield P., Semic-Jusufagic, A., Simpson, A., Adel-Patient, K., Xue, W., & Mills, C. (2020). The effect of the food matrix on the *in vitro* bio-accessibility and IgE reactivity of peanuts allergens. *Molecular Nutrition & Food Research*, 64, 1901093. https://doi.org/10.1002/mnfr.201901093

Ribeiro, T. A., Campos, D., Oliveira, A., Nunes, J., Vicente, A. A., & Pintado, M. (2021). Study of olive pomace antioxidant dietary fiber powder throughout gastrointestinal tract as multisource of phenolics, fatty acids and dietary fiber. *Food Research International*, 142, 110032. https://doi.org/10.1016/j.foodres.2020.110032

Robles, E., Fernández-Rodríguez, J., Barbosa, A., Gordobil, O., Carreño, N., & Labidi, J. (2018). Production of cellulose nanoparticles from blue agave waste treated with environmentally friendly processes. *Carbohydrate Polymers*, 183, 294–302. https://doi.org/10.1016/j.carbpol.2018.01.015

Rodríguez, R., Jimenez, A., Fernández-Bolanos, J., Guillen, R., & Heredia, A. (2006). Dietary fibre from vegetable products as source of functional ingredients. *Trends in Food Science & Technology*, 17(1), 3–15. https://doi.org/10.1016/j.tifs.2005.10.002

Rothwell, J. A., Pérez-Jiménez, J., Neveu, V., Medina-Ramon, A., M'Hiri, N., Garcia Lobato, P., Manachj, C., Knox, K., Eisner, R., Wishart, D., & Scalbert A. (2013). Phenol-Explorer 3.0: A major update of the Phenol-Explorer database to incorporate data on the effects of food processing on polyphenol content. Database, 10.1093/database/bat070. Full text (free access)

Russo, R., Cristiano, C., Avagliano, C., De Caro, C., La Rana, G., Mattace, R. G., Berni, C. R., Meli, R., & Calignano, A. (2017). Gut-brain axis: Role of lipids in the regulation of inflammation, pain and CNS diseases. *Current Medicinal Chemistry*, 24, 1–22. https://doi,org/10.2174/0929867324666170216113756

Saura-Calixto, F. (1998). Antioxidant dietary fiber product: A new concept and a potential food ingredient. *Journal of Agricultural and Food Chemistry*, 46(10), 4303–4306. Https://doi.org/10.1021/jf9803841

Silva, A. D., Ávila, S., Tulio, K. R., Padovan, D. S. M., Tadeu, G. M., Pereira, P. C., de, Q., Gomes, M. O. & Rodrigues, F. S. M. (2021). *In vitro* bioaccessibility of proteins, phenolics, flavonoids and antioxidant activity of amaranthus viridis. *Plant Foods for Human Nutrition*, 76, 478–486. https://doi.org/10.1007/s11130-021-00924-5

Singh, R., Prakash, S., Bhatia, R., Negi, M., Singh, J., Bishnoi, M., & Kondepudi, K. (2020). Generation of structurally diverse pectin oligosaccharides having prebiotic attributes. *Food Hydrocolloids*, 108,. https://doi.org/10.1016/j.foodhyd.2020.105988

Solari-Godiño, A., Pérez-Jiménez, J., Saura-Calixto, F., Borderías, A. J., & Moreno, H. M. (2017). Anchovy mince (Engraulis ringens) enriched with polyphenol-rich grape pomace dietary fibre: *In vitro* polyphenols bioaccessibility, antioxidant and physico-chemical properties. *Food Research International*, 102, 639–646. https://doi.org/10.1016/j.foodres.2017.09.044

Stewart, M. L., & Zimmer, P. (2018). Postprandial glucose and insulin response to a high-fiber muffin top containing resistant starch type 4 in healthy adults: A double-blind, randomized, controlled trial. *Nutrition*. 53, 59–63. https://doi.org/10.1016/j.nut.2018.01.002

Su, D., Liu, H., Zeng, Q., Qi, X., Yao, X., & Zhang, J. (2017). Changes in the phenolic contents and antioxidant activities of citrus peels from different cultivars after *in vitro* digestion. *International Journal of Food Science and Technology*, 52(11), 1–8. https://doi.org/10.1111/ijfs.13532

Sun, D., Chen, X., & Chu, C. (2020). Physicochemical properties and antioxidant activity of pectin from Hawthorn wine pomace: A comparison of different extraction methods. *International Journal of Biological Macromolecules*, 158, 1239–1247. https://doi.org/10.1016/j.ijbiomac.2020.05.052

Thompson D. B. (2007). Resistant starch. In C. G. Biliaderis & M. S. Izydorczyk (Eds). *Functional Food Carbohydrates, Functional Food and Nutraceuticals Series* (pp. 73–96). CRC Press Taylor & Francis Group. ISBN: 978-0-8493-1822-1

Ting, Y., Zhao, Q., Xia, C., & Huang, Q. (2015). Using *in vitro* and in vivo models to evaluate the oral bioavailability of nutraceuticals. *Journal of Agricultural and Food Chemistry*, 63 (5), 1332–1338. https://doi.org/10.1021/jf5047464

Topping, D. (2007). Cereal complex carbohydrates and their contribution to human health. *Journal of Cereal Science*, 46(3), 220–229. https://doi.org/10.1016/J.JCS.2007.06.004

Vázquez-Sánchez, K., Martinez-Saenz, N., Rebollo-Hernanz, R., Del Castillo, M. D., Gaytán-Martínez, M., & Campos-Vega, R. (2018). *In vitro* health promoting properties of antioxidant dietary fiber extracted from spent coffee (Coffee arabica L.) grounds. *Food Chemistry*, 261 253–259. https://doi.org/10.1016/j.foodchem.2018.04.064

Velderrain-Rodríguez, G., Quirós-Sauceda, A., Mercado-Mercado, G., Ayala-Zavala, J. F., Astiazarán-García, H., Robles-Sánchez, R. M., Wall-Medrano, A., Sayago-Ayerdi, S., & González-Aguilar, G. A. (2016). Effect of dietary fiber on the bioaccessibility of phenolic compounds of mango, papaya and pineapple fruits by an *in vitro* digestion model. *Food Science and Technology* (Campinas), 36, 188-194.~~0101–2061~~. https://doi.org/10.1590/1678-457X.6729

Vidal-Casanella, O., Nuñez, O., Granados, M., Saurina, J., & Sentellas, S. (2021). Analytical methods for exploring nutraceuticals based on phenolic acids and polyphenols. *Applied Sciences*, 11, 8276. https://doi.org/10.3390/app11188276

Viuda-Martos M., Férnandez-López, J., & Pérez-Álvarez, J. A. (2010). Pomegranate and its many functional components as related to human health: A review. *Comprehensive reviews in Food Science and Food Safety*, 9, 635–654. https://doi.org/10.1111/j.1541-4337.2010.00131.x

Viuda-Martos, M., Lucas-Gonzalez, R., Ballester-Costa, C., Pérez-Álvarez, J. A., Muñoz, A., & Fernández-López, J. (2017). Evaluation of protective effect of different dietary fibers on polyphenolic profile stability of maqui berry (Aristotelia chilens (Molina) Stuntz) during *in vitro* gastrointestinal digestion. *Royal Society of Chemistry*, 9, 573–584. https://doi.org/10.1039/C7FO01671A

Wang, L., Brennan, M. A., Guan, W., Liu, J., Zhao, H., & Brennan, C. S. (2021). Edible mushrooms dietary fibre and antioxidants: Effects on glycaemic load manipulation and their correlations pre-and post-simulated *in vitro* digestion. Food Chemistry, 351, 129320. https://doi.org/10.1016/j.foodchem.2021.129320

Won, K. S., Soriano, J. P., Rajan, U. A., Yeon, L. J., Hee, P. C., & Sang, K. C. (2018). Development of bioactive cellulose nanocrystals derived from dominant cellulose polymorphs I and II from Capsosiphon Fulvescens for biomedical applications. *International Journal of Biological Macromolecules*, 110, 531–539. https://doi.org/10.1016/j.ijbiomac.2017.11.047

Zhang, L., Wu, T., Zhang, Y., Chen, Y., Ge, X., Sui, W., Zhu, Q., Geng, J., & Zhang, M. (2022). Release of bound polyphenols from wheat bran soluble dietary fiber during simulated gastrointestinal digestion and colonic fermentation *in vitro*. *Food Chemistry*, 402, 134111. https://doi.org/10.1016/j.foodchem.2022.134111

Chapter 9

Gastrointestinal Digestion and Colonic Fermentation

Beyond Classic Food Science

Francisco J. Blancas-Benítez
Víctor M. Zamora-Gasga
Juscelino Tovar
and Sonia G. Sáyago-Ayerdi

CONTENTS

9.1 INTRODUCTION

Today, consumers are demanding food with clean labels driven by climate change concerns, well-being, animal welfare, and the production systems involved in developing new foods. Consumers are looking for safe foods with more significant health benefits. The changes in the awareness of the relationship between food and health result in a market for foods with health-enhancing properties. At the same time, people are changing their dietary preferences and introducing options such as vegan, vegetarian, organic, or flexitarian habits, with preferential use of ingredients of natural origin over synthetic ones (Rosenthal et al. 2021, Teixé-Roig, et al.

DOI: 10.1201/9781003332558-11

2023). Thus, introducing emerging technologies to improve nutritional properties and adding bioactive compounds (BCs) as natural ingredients with potential functional properties must be considered when developing new products. BCs have been widely studied, but they are frequently only characterized by analytical methods without considering the impact of the food digestion process that follows their consumption. In this sense, the study of food digestion has gained new relevance not only for macro- and micronutrients (e.g., starch and protein) (Colombo et al. 2021), whose digestibility has been extensively studied in the past, but also in recent years consideration has been given to the evaluation of their bioaccessibility (BA), bioavailability, and colonic fermentation-linked bioconversion when present in the whole food matrix, and not only with the isolated compounds. These studies should ideally be performed *in vivo* in humans, but this is only sometimes technically feasible due to financial or ethical constraints. In addition, the interindividual variation that may be involved in this type of study can be large. For this reason, *in vitro* digestion systems have been widely applied in the last decade (Zhoe, Than, & MacClements, 2023). They closely mimic the processes in the human GI and the colonic fermentation of the non-digestible fraction of the food matrix, allowing the detailed evaluation of the changes of these metabolites during their passage through the large intestine (Li et al. 2020). *In vitro* studies also consider the BC functional properties when absorbed, or their impact on the growth of specific genera of the gut microbiota, raising new research questions on the BC field, including the gut-brain axis and food development aspects of these compounds. The clinical trials to evaluate the bioavailability of BC have become a valuable tool to follow the bioconversion of BC metabolites to phase I or II compounds, which show the bioactive property. In this chapter, some static, dynamic, and semi-dynamic models of gastrointestinal digestion are described, in addition to an approach to clinical trials aiming to investigate the bioavailability and colonic fermentation systems and metabolites, highlighting the relevance of the studies on simulated GI digestion of food samples.

9.2 RELEVANCE OF PROTOCOLS THAT EVALUATE GASTROINTESTINAL DIGESTION AND BIOACCESSIBILITY

The relevance of evaluating GI digestion while developing new food products goes beyond assessing the nutritional composition of macro- and micronutrients. Today, including natural food compounds or BCs in new-product formulas makes it necessary to evaluate, beyond the amount added, the BA of these compounds. In this sense, *in vitro* digestion methods are widely used to study the behavior of foods or pharmaceuticals along the GI tract. Although human intervention studies are still considered the "gold standard" for addressing dietary questions, *in vitro* methods have the advantage of being faster, less expensive, less complex, and free from ethical restrictions. This allows a relatively large number of samples to be measured in parallel. The reproducibility, the choice of controlled conditions, and the easy sampling at the different stages of the simulated gastrointestinal process make *in vitro* models very suitable for elucidating mechanisms of action of food components and hypothesizing their possible effects (Minekus et al. 2014). The existence of different methods to evaluate the *in vitro* GI makes this topic an important tool to consider when development and innovation departments and researchers undertake the development of a new food product.

9.3 STATIC MODELS

In the last decades, different alternatives have been explored to simulate the GI system, including the physiological conditions and all the stages of the digestive process. Among the different methods, the static ones (biochemical methods) are the most straightforward and easily reproducible. They usually comprise three digestion stages, oral, gastric, and intestinal, all occurring in a single static bioreactor (Etcheverry et al. 2012). In each phase, the food product is incubated for a specific time and at a defined temperature with artificial saliva, followed by gastric and small intestinal digestive fluids. The pH is generally kept at a fixed value using a buffer. While this may seem simple, the need for more consensus regarding the simulation of the physiological conditions applied has led to different models, and thus results from different research teams cannot be compared (Minekus et al. 2014).

These methods simulate a limited number of physiological parameters and do not mimic physical processes such as particle size reduction, mixing, changes in conditions over time, or peristalsis (Li et al. 2020). An alternative methodological approach is the use of a dialysis bag after gastric digestion of the food sample, where the soluble components pass through a semi-permeable membrane without removing the dialyzed compounds; this stage also allows discrimination between components of high and low molecular weight, managing to simulate the processes of passive absorption that can occur in the small intestine (Blancas-Benitez et al. 2015; Hernández-Maldonado et al. 2019). As a first approximation, this method is applied in homogenized foods or isolated compounds in a closed system, followed by determining the amount of soluble compound present in the supernatant obtained by centrifugation or filtration. The quantity of solubilized compounds can be used as a measure of the BA of a BC or food constituent in general.

The main differences between the existing *in vitro* digestion models relate to the chemical composition of the digestive solutions used in each phase,(e.g., type and concentrations of enzymes, the salts and buffers used, and the use of tensoactive components) (Brodkorb et al. 2019). The COST-INFOGEST network has proposed a standardized and practical general static digestion method based on relevant conditions that can be applied for various purposes. The goal was to harmonize static *in vitro* systems that simulate digestive processes by defining key parameters and conditions (Brodkorb et al. 2019).

This harmonized static *in vitro* digestion method for foods should contribute to producing more comparable data in the future. However, it has been reported that the harmonized method has some limitations, which must be considered for each study case, depending on the nature of the food to be analyzed, such as in the case of almonds, which are rich in lipophilic compounds (carotenoids and plant sterols, among others) (Rodríguez-Roque et al. 2016). Despite all these limitations of static *in vitro* digestion models, a review of the correlation between *in vivo* and *in vitro* data on food digestion concluded that, although *in vitro* static models are too simplistic and do not reproduce all the dynamic aspects of the GI tract, they sometimes help predict *in vivo* digestion (Bohn et al. 2017).

9.4 DYNAMIC MODELS

In vitro dynamic models have emerged in recent decades as an innovative strategy that provides an improved and more sophisticated digestion simulation method. They can vary from single to multi-compartment systems and incorporate the dynamic aspects of the digestion process. Among them are shearing, mixing, hydration, or peristalsis phenomena that occur throughout the

GI tract (Cardoso et al. 2015). The goal is to improve the ability of the models to simulate variations in enzyme concentration, viscosity, pH, particle size and partitioning of nutrients, transport of digested products, and continuous secretions of digestive fluids observed in the GI tract of the human body (Cardoso et al. 2015). This methodology makes it possible to approximate better the mechanisms and processes that occur during digestion, the structural changes suffered by food products, and a more precise evaluation of BA.

The three most important factors to be considered in the design of dynamic models are the geometry (vertical alignment, horizontal alignment, or beaker) (Minekus et al. 2014); the biochemistry (the different digestive secretions are added to the model compartments over time); and the physical forces (simulated with Teflon rollers, flexible discs, or water jackets) (Brodkorb et al. 2019). Some more advanced dynamic digestion models have geometry designed to represent the fundus and antrum of the stomach and/or duodenum (Gonçalves et al. 2021). These designs allow for the simulation of the physical forces exerted on the digesta during transit through the gastrointestinal tract, which in turn allows for the simulation of the non-homogeneous nature of the digesta and localized biochemical environments (Brodkorb et al. 2021; Gonçalves et al. 2021).

Colombo, Paolillo & Papeli (2023) proposed a new dynamic model to evaluate the intestinal absorption and toxicity on different vegetables food secondary metabolites by using cell-based bioreactors (dynamic bioreactor). This new dynamic platform represented a promising *in vitro* methodology that could provide more information than static models. They reported experiments under dynamic conditions, plated in the millifluidic chambers at a density of 5000 cells/cm^2 in the presence of 5% fetal bovine Caco-2 cells (Colombo et al. 2023a).

9.5 EFFECT OF *IN VITRO* GI DIGESTION ON BIOACCESIBILITY OF BIOACTIVE COMPOUNDS (BC)

In recent years, several BCs with potential health benefits, such as phenolic compounds (PCs), carotenoids, phytoestrogens, etc., have been explored in developing functional foods, nutraceuticals, pharmaceuticals, and other applications (Sotomayor-Gerding et al. 2016). However, it should be noted that the health beneficial effects of the compounds mentioned above are determined by their BA and bioavailability due to their stability and interactions with the food matrix (Cilla et al. 2018; Miedes et al. 2023). Furthermore, their lipophilic nature and insolubility challenge their delivery and absorption (Alminger et al. 2014). For these reasons, when analyzing the potential functionality of BC, their BA and bioavailability are more relevant than the concentration of the BC present in the food.

It is well known that fruits and vegetables are essential BC (mainly PC) sources that may exert numerous beneficial effects *in vivo*, mainly related to their high antioxidant potential (Gowe, 2015). For this reason, the bioavailability of antioxidant compounds after digestion is essential, and several studies have indicated that the bioavailability of individual antioxidant compounds is not complete (Dufour et al. 2018). The impact of *in vitro* GI digestion processes on the stability of BC and, therefore, on their antioxidant properties has been one of the most studied topics during the last decade. Different studies have analyzed the effects of GI conditions *in vitro*, observing variability in the antioxidant activity of different plant-based foods that are sources of antioxidant PC (e.g., chili, tomato, guava) (Zhoe, Than, MacClements, 2023). In most of these studies, the stability of PC was evaluated by determining the total PC content, including the hydrolyzable polyphenols or condensed tannins, which does not provide information on the specific type of PC, and by determination of antioxidant capacity by methods such as DPPH, FRAP, and ABTS (Cárdenas-Castro et al. 2021; Blancas-Benitez et al. 2022). The results are controversial; in some cases, the apparent content of total polyphenols increased, and therefore the antioxidant activities after the gastric and intestinal phase of *in vitro* digestion were not reduced (compared to the gastric passage) (Cárdenas-Castro et al. 2019; Blancas-Benitez et al. 2022) while in other cases, a slight decrease was observed in the antioxidant capacities analyzed (Wootton-Beard & Ryan, 2011). This indicates that digestion can alter the antioxidant properties of foods, depending not only on the content of antioxidant compounds but also on the profile of compounds present. Further analyses of specific PC, especially their possible degradation products, should be performed to clarify these conflicting results (Sayago-Ayerdi et al. 2021; Macedo-Dantas et al. 2023).

9.6 *IN VITRO* VERSUS *IN VIVO* DIGESTION MODELS

Few studies have evaluated the *in vivo-in vitro* correlation of food results. Therefore, it takes work to assess current *in vitro* assays accuracy adequately. Some reviews consider that there is indeed a correlation in the case of carotenoids and that a measure of BA could be sufficient to estimate how bioavailable a carotenoid is from the food in question (Etcheverry et al. 2012). Therefore, when using *in vitro* methods, it must be considered how to adapt the digestion conditions according to the composition of the sample and/or the components of the food under study, seeking a balance between technical simplification and precision, and always maintaining the *in vivo* situation as a reference to reproduce the physiological situation as much as possible (Li et al. 2020). Thus, there is still an urgent need for harmonization and standardization of *in vitro* techniques, particularly static methods. Furthermore, these methods must be validated with adequate assessments of human GI physiology to allow for better study designs.

9.7 CLINICAL TRIALS TO EVALUATE THE BIOAVAILABILITY OF BC

The *in vitro* models bring a first valuable approximation to the transformations and BA of BC during their GI transit before reaching the target tissues. Both human and animal models have been used to study the bioavailability and metabolism of BC through *in vivo* assays. Each model has its advantages and constraints, such as ethical aspects (i.e., the Helsinki Declaration must be considered), in addition to possible additional ethical requirements of each institution and country. Human studies should be registered online in a recognized registry, such as www.clinicalTrials.gov, and require an informed consent from the volunteers. In this section, we discuss only clinical trials to briefly describe the pharmacokinetics of the BC.

The first-pass metabolism occurs when the BC crosses the intestinal barrier before reaching systemic circulation. It is crucial to consider if the BC is free from the food matrix, if it is encapsulated, or in an emulsion since it must cross the epithelium barrier. The second step is bioavailability; much research is focused on the evaluation of the bioavailability of BC. This means the proportion of a BC administrated that can reach the systemic circulation, which is a function of the opposing processes of absorption and clearance. This can be calculated by the concentration among the time and area under the curve (AUC) of the initial concentration of the BC divided by the amount found at the venous level (AUC). The bioavailability of BC, mainly PC, is highly variable and has been reported; for example, it was reported that flavonoids are poorly absorbed glycosides that require hydrolysis by b-glucosidases in the small intestine before being absorbed as aglycones (Ávila-Román et al.2021).

The bioavailability studies aim to determine the formation of active metabolites from BC after absorption. It is essential to evaluate if the metabolic transformations affect the BC biological activities instead of focusing on those of the original BC. Once absorbed, BCs are subjected to different bioconversion reactions in the liver, producing Phase I and II metabolites. The methylation, sulfation, and glucuronidation facilitate their biliary and urinary elimination. The methylation reactions are catalyzed by catechol-methyl-transferase with S-adenosyl-L-methionine as the methyl group donor (Lee et al. 2002). Sulfation is catalyzed by sulfotransferases where the 3'-phosphoadenosine-5'-phosphosulfate is the sulfate group donor (Falany, 1997). These metabolites may be transferred back into the intestine through the enterohepatic cycle (**Figure 9.1**).

The bioavailability of a BC can be assessed by evaluating its plasma and urine concentration after ingestion of the compound either in pure form, included in a shot, or as a nutraceutical. It can be evaluated indirectly by the antioxidant capacity of plasma after the BC consumption. These parameters can be evaluated by a single dose in acute trials, where the phenol concentration increases in plasma transitionally and reflects the ability of the organism to take these compounds from the food matrix. However, it is highly recommended to evaluate the whole diet and the use of repeated doses (Pandareesh et al., 2015). **Table 9.1** lists some studies on the bioavailability of polyphenols.

The most common techniques used in the analysis of these types of samples are mass spectrometry (MS), nuclear magnetic resonance (NMR), and chromatographic techniques, predominantly liquid chromatography (LC) or gas chromatography (GC). These are valuable techniques for metabolomics. However, for bioavailability and metabolism studies it is highly recommended to use liquid chromatography coupled with mass spectrometry (Schön et al. 2018). The triple quadruple MS analyzer (QQQ) and quadrupole ion trap (Q-IT) are commonly used, as well as the tandem mass analyzers (MS/MS), which allow the selection of specific metabolites (Agulló et al., 2021). The untargeted methods, which search for a global analysis without knowing the identity of the metabolites in advance, have the advantage of discovering novel metabolites, but it is necessary to use high-resolution MS analyzers, such as time-of-flight (TOF) or Orbitrap, which can provide very accurate mass values (Schrimpe-Rutledge et al. 2016). Currently, bioavailability assessment methods have a highly accurate focus on the potential bioconversion of the original BC, bringing a new approach to studying functional properties in foods naturally rich in – or added with – the compound of interest.

9.8 MODELING THE HUMAN GUT MICROBIOME *IN VITRO* AND *IN VIVO*

Scientific evidence suggests that microbial metabolites participate in the communication of the microbiome-gut-brain axis. However, the exact mechanisms by which these intestinal metabolites access the brain are still unknown. Species differences, ethical constraints, high costs, and the laborious nature of animal studies (e.g., germ-free and gnotobiotic mice) have given way to the design of *in vitro* models that allow simplified, well-controlled, and reproducible study of interactions between gut and health. *In vitro* colonic fermentation models allow for exploring the influence of dietary compounds, pharmaceuticals, microbial pathogens, bioactive, and toxic substances on the composition and metabolism of the gut microbiota. Nissen et al. (2020) described the main characteristics of different *in vitro* systems that emulate the conditions prevailing in the

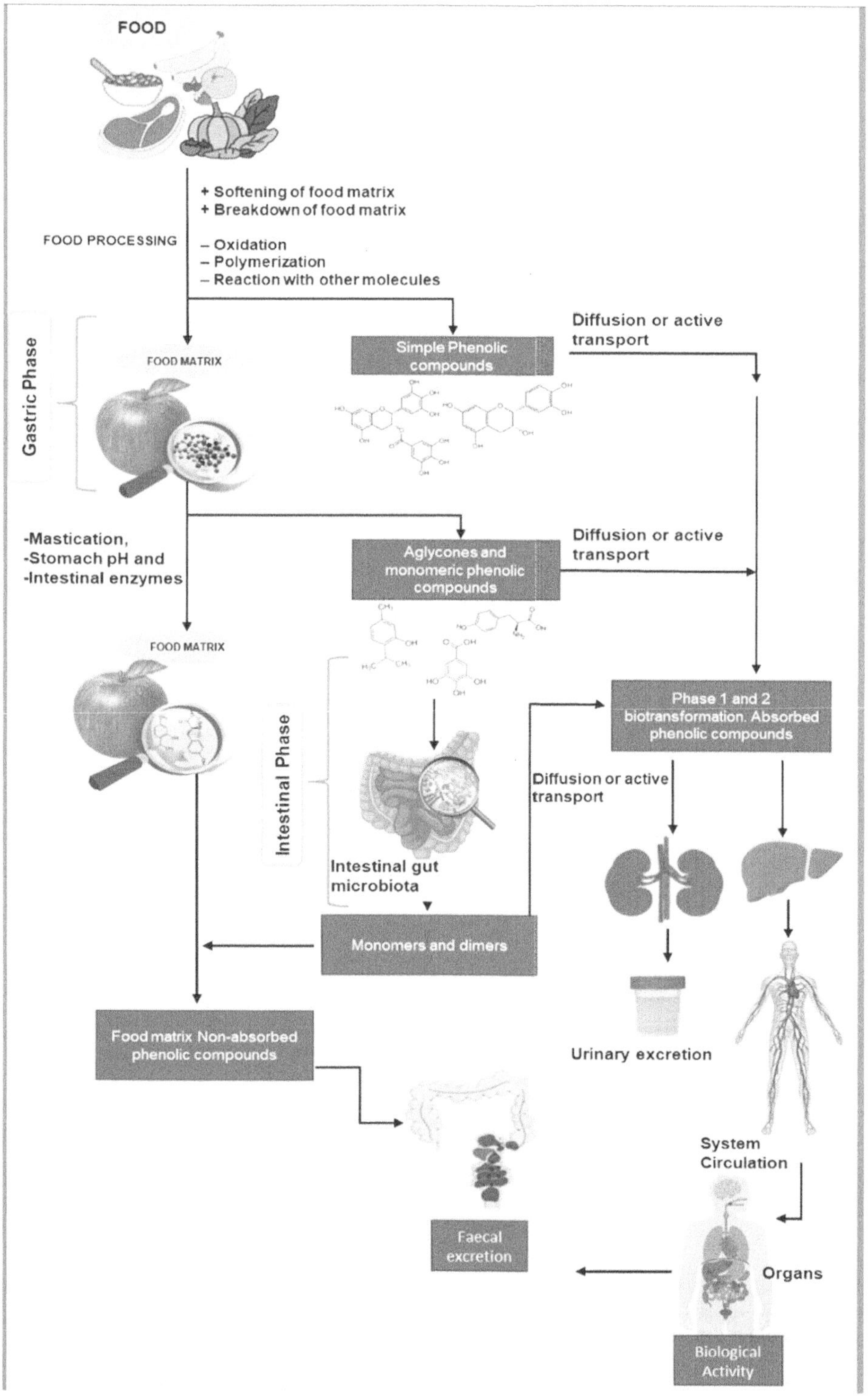

Figure 9.1 Digestion and absorption of phenolic compounds.

TABLE 9.1 BENEFICIAL EFFECTS AND BIOAVAILABILITY IN HUMAN SUBJECTS IN PLASMA AND URINE BY BIOACTIVE COMPOUNDS

Functional Matrix	Human Model	Relevant Resolts	References
Phenolic acids, flavonoids.			
Rosemary tea	HHV (n = 12)	Phase II metabolites bioavailables. Metabolism by colonic microbiota	Achour et al. (2021)
Yerba mate	HHV (n = 12)	Sulfated conjugates of caffeic and ferulic/isoferulic acids.	Gómez-Juaristi et al. (2018a)
Coffee Beans	HHV (n = 12)	↑ Metabolites sulfate and methyl derivatives.	Gómez-Juaristi et al. (2018b)
green/roasted		↓ Glucuronides metabolites	
Hydroxytyrosol (HT) enriched biscuits	HHV (n = 12)	↑ Fast rate of excretion of metabolites in urine ↑ HT-sulphate and 3,4- DOPAC-sulphate,	Mateos et al. (2016)
Anthocyanins, hydroxycinnamic acids and proanthocyanidins			
A blackcurrant juice drink	HHV (n = 20)	No effect on vascular reactivity. ↓ Anthocyanins concentrations in urine	Jin et al. (2011)
Aronia melanocarpa extract	HAFS (n=49)	↓ 8% fasting plasma total colesterol	Xie et al. (2017)
Flavonols			
Cocoa chocolate	HHV (n=18)	↑ Epicatechin metabolites concentration in plasma	Montagnana et al. (2018)
EMIQ® maltodextrin, and Cottee's Raspberry flavored cordial	Subjects with CVD risk (n=25)	↑ Quercetin metabolites concentration in plasma	Bondonno et al. (2020)
Phenolic acid and phenolic metabolites			
Coffee	HHV (n = 15)	↑ Total plasma chlorogenic acids metabolites (CgAM) ↑ FMD	Mills et al. (2017)
Grape pomace	HHV (n = 10)	Phenolic metabolites; 35 in plasm and 28 in urine ↑ Phenyl-γ-valerolactones, hydroxybenzoic acids and phenols ↑ Hydroxyphenylpropionic acids, hydroxycinnamates,	Castello et al. (2018)

CVD, Cardiovascular Diseases; DOPAC, Dihydroxyphenylacetic acid; FMD, Flow-mediated Dilatation; HAFS, Healthy Adult Former Smokers; HHV, Healthy Human Volunteers

colon. Batch or static fermentation models consist of controlled reactors (time, temperature, pH, anaerobiosis, in basal chemical medium) and represent the simplest and most accessible way to assess the accumulation of microbial metabolites dependent on substrate depletion, using microbial inoculum controls or fecal suspensions. These systems have been used, for instance, in the study of the interactions occurring between indigestible compounds and microbiota (Zamora-Gasga et al. 2015; Del Juncal-Guzmán et al. 2022).

On the other hand, the *in vitro* dynamic colonic fermentation model consists of several stages that mimic the ecological niches of the proximal, transverse, and distal colon in microaerophilic and anoxic conditions with N_2 or CO_2, modifying the pH conditions through a constant feeding of nutrients by peristaltic pumps. Among the most commonly used models are the TNO *in vitro* Model of the Colon (TIM-2); the simulator of the human intestinal microbial ecosystem (SHIME), M-SHIME; the SIMulator gastrointestinal (SIMGI); polyfermentor intestinal model (PolyFermS); small-scale reactors (MiniBio); and the most diminutive intestine *in vitro* model (TSI) (Nissen et al., 2020). In particular, the TIM-2 model is a dynamic GI model, and is used for in vitro research. The TIM-2 model can be used to study changes in the microbiota composition and the production of beneficial molecules in the gut. It resembles the human large intestine, including peristalsis, dialysis, and microorganisms. It is a good model that provides short experiment durations and the studies done with it are highly reproducible. This particular dynamic model has been used to evaluate the bioconversion of polyphenols in mango (*Mangifera indica* L) peel, jamaica (*Hibiscus sabdariffa* L) calyces, and agave (*Agave tequilana* Weber) fructans (Sáyago-Ayerdi et al. 2020, 2021).

In vitro, human cell cultures mimic the *in vivo* environment and are essential tools for toxicology and nutrition studies. Among the most commonly used intestinal cell lines are Caco-2, HT-29, and T84. In particular, the human epithelial cell line Caco-2 is characterized as an enterocyte model expressing hydrolase enzymes and different membrane transporters, making it useful in absorption, transport, and bioavailability studies. However, one of the limitations is its inability to produce mucus. This cell line has been used for the evaluation of the stability, transport, and metabolism of different BC from plant sources and their intestinal metabolites. The human colon adenocarcinoma cell line HT29 is another pluripotent intestinal cell line since its differentiation is not spontaneous and depends on the nutritional conditions of the culture (typically in a glucose-free medium, in a similar way as the Caco-2 line). However, it can differentiate into goblet cells and produce mucus. Colonic metabolites from digested guava derivatives of polyphenols have shown antiproliferative effects or death-promoting action on HT-29 cells via apoptosis, necrosis, and autophagy, suggesting a potential benefit for intestinal health (Blancas-Benitez et al. 2022). A third cell line frequently used in intestinal studies is T84, characterized by its differentiation into crypt-like cells induced by recombinant transforming human growth factor (TGFβ1) or other soluble cellular factors of mesenchymal cells. It has been used in studies of signaling pathways related to the inflammatory response of the intestinal epithelium and the interaction between bacteria and enterocytes (Ponce de León-Rodríguez et al. 2019).

The 3D printer revolution and biotechnological advances have allowed the development of more complex intestinal models, such as organoids and organs on a chip. Organoids are derived from primary tissues, exhibit varying levels of cellular complexity, and have physiological similarity to native organs and are capable of long-term growth. Various protocols exist for establishing GI and brain organoids derived from adult tissue biopsies and induced pluripotent stem cells (Moysidou & Owens, 2021). Gastrointestinal organoid cultures have been applied in studies of host-microbe and host-pathogen interactions to understand apical-out vs. basal-out enteroid tissue and the induction of pluripotent stem cells and intestinal organoids derived from patients with inflammatory bowel disease. Organoids have been derived from intestinal stem cells and colonies using bacterial incorporation methods such as dissociation–human intestinal monolayers, dissociation–Transwell monolayers, and microinjection (Moysidou & Owens, 2021). Through organoid cultures, Fellows et al. (2018) proposed that SCFA, particularly butyrate, might induce central nervous system histone crotonylation, thereby influencing brain functions.

An "organ on a chip" contains cells of interest with physiological, structural, and functional characteristics close to that of the original tissue contained in a microfluidic device and have been used for modeling aspects of the microbiota-gut-brain axis. Gut-on-a-chip allows altering luminal and basal conditions to monitor in real-time key parameters such as barrier integrity, oxygen, and growth factors with a collection of bacteria and the study of composition and metabolism in the intestinal lumen. In this sense, Jalili-Firoozinezhad et al. (2019) found that a transluminal hypoxia gradient increased intestinal barrier function and maintained a physiologically relevant level of microbial diversity (>200 unique operational taxonomic units from 11 different genera, obligate anaerobic bacteria) compared to aerobic coculture conditions. These models provide increased knowledge about the interactions between the gut microbiota and human health. Technological advances in the coming years will allow a deeper understanding of the role of the microbiota and its metabolism in human health.

9.9 INTESTINAL METABOLITES AS REGULATORS IN THE GUT-BRAIN AXIS

The microbiota of the GI tract represents the highest density and abundance of microorganisms that inhabit the human body. Throughout life, bacterial communities adapt and modify their composition and metabolism in response to different host-related factors, such as age and genetics, diet, and use of drugs. Thus, the microbiota can influence human health, regulating chemical signals that circulate through the body. Current scientific evidence indicates that the development of neurological diseases involves multiple systems (immune system, chemical transmitters, and neuronal pathways) that allow bidirectional communication between the brain and the gut microbiota, called the gut-microbiota-brain axis (Martin et al., 2018). Studies in digestion models contribute greatly to the understanding of these phenomena. Intestinal metabolites regulate chemical signals between the intestine and the brain. Short-chain fatty acids are produced during colonic fermentation of dietary fiber (digestion-resistant carbohydrates) and act as endogenous ligands for orphan G protein-coupled receptors (GPR43 and GPR41, renamed as free fatty acid receptors FFAR2 and FFAR3, respectively) (Dalile et al., 2019), and affect gene expression by inhibiting histone deacetylases involved in Alzheimer disease, depression, schizophrenia, and addictions (Volmar & Wahlestedt, 2015).

Propionate acts via FFAR3 on the periportal afferent neutral system, which subsequently signals to peripheral and central nervous system areas to induce intestinal gluconeogenesis (De Vadder et al., 2014). In addition to SCFA, neuroactive molecules such as serotonin (5-HT), kynurenines (quinolinic acid and kynurenic acid), tryptamine, and indole compounds are products of tryptophan metabolism by the microbiota and are of importance in the control of signals in the gut-brain axis. Due to its agonist properties in the glutamatergic ionotropic N-methyl-D-aspartate (NMDA) receptors, quinolinic acid can act as an excitotoxin both at the peripheral level and in the central nervous system, causing a series of biochemical events: activation of protein kinases, phospholipases, oxidase nitric synthase, protease, mitochondrial alterations, formation of ROS and reactive nitrogen species, lipid peroxidation, and cytoskeletal destabilization, which can lead to apoptotic or necrotic cell death (Bosi et al. 2020).

Kynurenic acid acts as an antagonist of NMDA receptors and α-nicotinic receptors and as agonist of the G protein-coupled receptor GPR35, which gives it neuroprotective properties (Bosi et al. 2020). For its part, enteric 5-HT, through the activation of serotonergic receptors (5-HT1 to 5HT7), modulates mood, behavior, and cognitive functions and can be converted into melatonin after successive enzymatic transformations that include conversion to 5-hydroxy-3-indole acetaldehyde by monoamine oxidase (Tordjman et al., 2017).

Indole and its derivatives (indole-3-aldehyde, indole-3-acetic acid, indole-3-propionic acid, and indole-3-acetaldehyde) are ligands of the aryl hydrocarbon receptor AhR, which are involved in encoding enzymes involved in xenobiotic metabolism, in the detoxification of carcinogenic polycyclic aromatic hydrocarbons (Hubbard et al. 2015), in the modulation of epithelial cell turnover, and in the immune response at the intestinal level (Stockinger et al. 2014).

The vagus nerve innervates the intestine and the enteric nervous system, the main neuronal pathway physically connecting the intestine and the brain. The intestinal microbiota and its metabolism indirectly stimulate vagal afferents affecting brain function. *Lactobacillus johnsonii* BS15 protected restraint stress-induced mice from the impaired gut barrier by enhancing mRNA levels of tight junction proteins and exerted beneficial effects on resistant starch-reduced levels of anti-inflammatory cytokines (Wang et al. 2021). Muller et al. (2020) suggest that microbial metabolites can modulate the neuronal pathways of the gut-brain axis since the administration of SCFA-producing intestinal microorganisms suppresses neuronal pathways related to motility. Although these findings open the possibility of treating diseases by activating the vagus nerve by modifying the intestinal environment, the challenge is to demonstrate the mechanism in animal models and, even more, to translate these findings to humans.

The microbiota also affects the development and function of innate immune cells in the brain, known as microglia. Administration of xylo-oligossaccharide and *Lactobacillus paracasei* HII01 for 13 weeks, individually or in combination, reduced hippocampal oxidative stress and restored cognitive function through decreased microglial activation in male obese-insulin-resistant rats induced by a high-fat diet (Chunchai et al. 2018).

Another important link between the gut microbiota and the brain is determined by changes in systemic immunity caused by cytokines and chemokines that access the brain through direct transport through the blood–brain barrier. For example, lower levels of brain-gut peptides (such as vasoactive intestinal polypeptide, cholecystokinin, and somatostatin) may be associated with the pathogenesis of Moyamoya disease and act as clinically useful biomarkers along with the levels of proinflammatory cytokines (Han et al. 2020). Changes in inflammatory markers may be related to neurological disorders such as autism spectrum disorder, depression, anxiety, or Parkinson's. **Table 9.2** shows the effects of probiotic supplementation on neurological disorders and highlights the connections between the microbiota and its metabolism with systemic immunity and effects in the brain.

9.10 CONCLUSION

The relevance of assessing GI digestion is real. Research on the digestibility of macronutrients (carbohydrates, protein, and fat) and micronutrients (vitamins and minerals) has dealt with these

TABLE 9.2 CLINICAL EVIDENCE OF PROBIOTIC SUPPLEMENTATION EFFECTS ON NEUROLOGICAL DISEASES PROVIDED BY RANDOMIZED, DOUBLE-BLIND, CONTROLLED TRIALS

Study Population	Intervention	Intervention Results	References
Autism Preschoolers with or without gastrointestinal symptoms (n=85)	Probiotics for six months	↓ Total Autism Diagnostic	Santocchi et al. (2020)
		↓ Severity Score	
		↑ Adaptive function and sensory profile	
Depressive Subjects with MID underwent PCI (n=44)	Lactobacillus rhamnosus capsules for 12 weeks.	↓ Total Beck Depression Inventory score and malondialdehyde	Moludi et al. (2019)
		↑ Quality of life and antioxidant capacity	
Hemodialysis patients with Anxiety (n=75)	Symbiotic group for 12 weeks	↓ Hospital Anxiety and Depression Scale score	Haghighat et al. (2021)
		↑ SBNF	
Geriatric patients with Parkinson´s disease (n=60)	Probiotic group (8 × 109 CFU/day) for 12 weeks.	↓ UPDRS	Tamtaji, Heidari-Soureshjani, et al. (2019)
		↓ High-sensitivity C-reactive protein and malondialdehyde	
		↑ glutathione levels.	
		↑ insulin sensitivity	
Adults with chronic schizophrenia (n=60)	8 × 109 CFU/day probiotic plus 200 μg/day selenium for 12 weeks	↓ high-sensitivity C-reactive protein levels, fasting glucose, insulin levels	Jamilian & Ghaderi, (2021)
		↑ Total antioxidant capacity and total glutathione	
Geriatric patients with Alzheimer's disease (n=79)	Selenium (200 μg/day) plus probiotic (2 × 109 CFU/day each) for 12 weeks	↓ Serum high sensitivity C reactive protein, insulin levels	Tamtaji, Taghizadeh, et al. (2019)
		↑ MMSES	
		↑ total antioxidant capacity	

CFU, Colony Forming Units; MID, Myocardial Infarction Diagnosis; MMSES, mini-mental state examination score; PCI, Percutaneous Coronary Intervention; SBNF Serum Brain-derived Neurotrophic Factor; UPDRS, Unified Parkinson's Disease Rating Scale.

aspects for years. Developing new food products related to clean labels, functional foods, use of emerging technologies, etc., has forced an updated vision of the GI and the digestive processes. Bioaccessibility was introduced and is being more intensively used as the volume of research on food digestion increases. It is clear that *in vitro* digestion models are a helpful tool for evaluating bioactive compounds since, with bioavailability trials on the rise, researchers realize that derived metabolites have more functional properties than the original bioactive compounds. This is the case for the gut-brain axis, a field that provides the most novel reasons for incorporating inputs from food digestion studies into comprehensive food research.

Gastrointestinal digestion models are evolving towards more physiologically relevant designs, aiming to replicate the conditions and processes of the human GI tract with greater precision. This includes simulating pH, intestinal motility, and digestive enzymes in the human body. Furthermore, these models are expected to increasingly incorporate the presence and interaction with the intestinal microbiota, which is crucial for understanding how food and bioactive compounds affect its composition and activity and how this influences health. Additionally, GI digestion models can expand their focus to evaluate the absorption and metabolism of nutrients and bioactive compounds in the human body, allowing for a more comprehensive understanding of the interactions between food and the body and their effects on health. Integrating these models with other techniques and approaches, such as genomics, metabolomics, and transcriptomics, provides a complete picture of the impact of food on health and enables a personalized approach to diet and nutrition. To ensure the reliability and comparability of results, it is important to establish validation standards and protocols for using these GI digestion models, promoting confidence in their utility and applicability in both research and the food industry.

REFERENCES

Achour, M., Bravo, L., Sarriá, B., Fredj, M. Ben, Nouira, M., Mtiraoui, A., Saguem, S., & Mateos, R. (2021). Bioavailability and nutrikinetics of rosemary tea phenolic compounds in humans. *Food Research International, 139*, 109815. https://doi.org/10.1016/j.foodres.2020.109815

Agulló, V., García-Viguera, C., & Domínguez-Perles, R. (2021). Beverages based on second quality citrus fruits and maqui berry, a source of bioactive (poly) phenols: Sorting out urine metabolites upon a longitudinal study. *Nutrients, 13*(2), 312. https://doi.org/10.3390/nu13020312

Alminger, M., Aura, A. M., Bohn, T., Dufour, C., El, S. N., Gomes, A., Karakaya, S., Martínez-Cuesta, M. C., Mcdougall, G. J., Requena, T., & Santos, C. N. (2014). *In vitro* models for studying secondary plant metabolite digestion and bioaccessibility. *Comprehensive Reviews in Food Science and Food Safety, 13*(4), 413–436. https://doi.org/10.1111/1541-4337.12081

Ávila-Román, J., Soliz-Rueda, J. R., Bravo, F. I., Aragonès, G., Suárez, M., Arola-Arnal, A., Mulero, M., Salvadó, M .J., Arola, L., & Torres-Fuentes, C. (2021). Phenolic compounds and biological rhythms: Who takes the lead? *Trends in Food Science & Technology, 113*, 77–85. https://doi.org/10.1016/j.tifs.2021.04.050

Blancas-Benitez, F. J., Mercado-Mercado, G., Quirós-Sauceda, A. E., Montalvo-González, E., González-Aguilar, G. A., & Sáyago-Ayerdi, S. G. (2015). Bioaccessibility of polyphenols associated with dietary fiber and *in vitro* kinetics release of polyphenols in Mexican 'Ataulfo' mango (*Mangifera indica* L.) by-products. *Food & Function, 6*(3), 859–868. https://doi.org/10.1039/C4FO00982G

Blancas-Benitez, F. J., Pérez-Jiménez, J., Sañudo-Barajas, J. A., Rocha-Guzmán, N. E., González-Aguilar, G. A., Tovar, J., & Sáyago-Ayerdi, S. G. (2022). Indigestible fraction of guava fruit: Phenolic profile, colonic fermentation and effect on HT-29 cells. *Food Bioscience, 2021*, 101566. https://doi.org/10.1016/j.fbio.2022.101566

Bohn, T., Carriere, F., Day, L., Deglaire, A., Egger, L., Freitas, D., Golding, M., Le Feunteun, S., Macierzanka, A., & Menard, O. (2017). Correlation between *in vitro* and *in vivo* data on food digestion. What can we predict with static *in vitro* digestion models? *Critical Reviews in Food Science and Nutrition*, 58(13), 2239–2261. https://doi.org/10.1080/10408398.2017.1315362

Bondonno, N. P., Bondonno, C. P., Ward, N. C., Woodman, R. J., Hodgson, J. M., & Croft, K. D. (2020). Enzymatically modified isoquercitrin improves endothelial function in volunteers at risk of cardiovascular disease. *British Journal of Nutrition, 123*(2), 182–189. https://doi.org/10.1017/S0007114519002137

Bosi, A., Banfi, D., Bistoletti, M., Giaroni, C., & Baj, A. (2020). Tryptophan metabolites along the microbiota-gut-brain axis: An interkingdom communication system influencing the gut in health and disease. *International Journal of Tryptophan Research, 13*, 1178646920928984. https://doi.org/10.1177/1178646920928984

Brodkorb, A., Egger, L., Alminger, M., Alvito, P., Assunção, R., Ballance, S., Bohn, T., Bourlieu-Lacanal, C., Boutrou, R., Carrière, F., Clemente, A., Corredig, M., Dupont, D., Dufour, C., Edwards, C., Golding, M., Karakaya, S., Kirkhus, B., Le Feunteun, S., ... Recio, I. (2019). INFOGEST static *in vitro* simulation of gastrointestinal food digestion. *Nature Protocols, 14*(4), 991–1014. https://doi.org/10.1038/s41596-018-0119-1

Brodkorb, A., Egger, L., Alminger, M., Alvito, P., Assunção, R., Ballance, S., Bohn, T., Bourlieu-Lacanal, C., Boutrou, R., Carrière, F., Clemente, A., Corredig, M., Dupont, D., Dufour, C., Edwards, C., Golding, M., Karakaya, S., Kirkhus, B., Le Feunteun, S., ... Sips, A. J. A. M. (2021). Comparison of five *in vitro* digestion models to study the bioaccessibility of soil contaminants. *Nature Protocols, 0*(4), 428–438. https://doi.org/10.1080/10408398.2021.1884841

Cárdenas-Castro, A. P., Perales-Vázquez, G. del C., De la Rosa, L. A., Zamora-Gasga, V. M., Ruiz-Valdiviezo, V. M., Alvarez-Parrilla, E., & Sáyago-Ayerdi, S. G. (2019). Sauces: An undiscovered healthy complement in Mexican cuisine. *International Journal of Gastronomy and Food Science, 17*(February), 100154. https://doi.org/10.1016/j.ijgfs.2019.100154

Cárdenas-Castro, A. P., Zamora-Gasga, V. M., Alvarez-Parrilla, E., Ruíz-Valdiviezo, V. M., Venema, K., & Sáyago-Ayerdi, S. G. (2021). *In vitro* gastrointestinal digestion and colonic fermentation of tomato (*Solanum lycopersicum* L.) and husk tomato (Physalis ixocarpa Brot.): Phenolic compounds released and bioconverted by gut microbiota. *Food Chemistry, 360*(May). https://doi.org/10.1016/j.foodchem.2021.130051

Cardoso, C., Afonso, C., Lourenço, H., Costa, S., & Nunes, M. L. (2015). Bioaccessibility assessment methodologies and their consequences for the risk–benefit evaluation of food. *Trends in Food Science & Technology, 41*(1), 5–23. https://doi.org/10.1016/j.tifs.2014.08.008

Castello, F., Costabile, G., Bresciani, L., Tassotti, M., Naviglio, D., Luongo, D., Ciciola, P., Vitale, M., Vetrani, C., & Galaverna, G. (2018). Bioavailability and pharmacokinetic profile of grape pomace phenolic compounds in humans. *Archives of Biochemistry and Biophysics, 646*, 1–9. https://doi.org/10.1016/j.abb.2018.03.021

Chunchai, T., Thunapong, W., Yasom, S., Wanchai, K., Eaimworawuthikul, S., Metzler, G., Lungkaphin, A., Pongchaidecha, A., Sirilun, S., & Chaiyasut, C. (2018). Decreased microglial activation through gut-brain axis by prebiotics, probiotics, or synbiotics effectively restored cognitive function in obese-insulin resistant rats. *Journal of Neuroinflammation, 15*(1), 1–15. https://doi.org/10.1186/s12974-018-1055-2

Cilla, A., Bosch, L., Barberá, R., & Alegría, A. (2018). Effect of processing on the bioaccessibility of bioactive compounds–A review focusing on carotenoids, minerals, ascorbic acid, tocopherols and polyphenols. *Journal of Food Composition and Analysis, 68*, 3–15. https://doi.org/10.1016/j.jfca.2017.01.009

Colombo, R., Ferron, L., Frosi, I., & Papetti, A. (2021). Advances in static *in vitro* digestion models after the COST action Infogest consensus protocol. *Food & Function, 12*(17), 7619–7636. https://doi.org/10.1039/D1FO01089A

Colombo, R., Paolillo, M., Frosi, I., Ferron, L., & Papetti, A. (2023). Effect of the simulated digestion process on the chlorogenic acid trapping activity against methylglyoxal. *Food Functional, 14*, 541–549.

Colombo, R., Paolillo, M., & Papetti, M. (2023a). A dynamic *in vitro* model for testing intestinal absorption of different vegetable food secondary metabolites. *Applied Science, 13*(8), 5033. https://doi.org/10.3390/app13085033

Dalile, B., Van Oudenhove, L., Vervliet, B., & Verbeke, K. (2019). The role of short-chain fatty acids in microbiota–gut–brain communication. *Nature Reviews Gastroenterology & Hepatology, 16*(8), 461–478. https://doi.org/10.1038/s41575-019-0157-3

Dantas, A. M., Fernandes, F. G., Magnani, M., da Silva C., & Borges, G. (2023). Gastrointestinal digestion assays for evaluating the bioaccessibility of phenolic compounds in fruits and their derivates: An overview. *Food Research International,* 170, 112920. https://doi.org/10.1016/j.foodres.2023.112920

De Vadder, F., Kovatcheva-Datchary, P., Goncalves, D., Vinera, J., Zitoun, C., Duchampt, A., Bäckhed, F., & Mithieux, G. (2014). Microbiota-generated metabolites promote metabolic benefits via gut-brain neural circuits. *Cell, 156*(1–2), 84–96. https://doi.org/10.1016/j.cell.2013.12.016

Del Juncal-Guzmán, D., Antunes-Ricardo, M., Sánchez-Burgos, J. A., Sáyago-Ayerdi, S. G., & Gutiérrez-Uribe, J. A. (2022). Immunomodulatory effect of metabolites from digested and fermented fractions from irradiated pineapple (Annanas comosus L.) snack-bars. *Food Chemistry, 373, 131375* (August 2021). https://doi.org/10.1016/j.foodchem.2021.131375

Dufour, C., Loonis, M., Delosière, M., Buffière, C., Hafnaoui, N., Santé-Lhoutellier, V., & Rémond, D. (2018). The matrix of fruit & vegetables modulates the gastrointestinal bioaccessibility of polyphenols and their impact on dietary protein digestibility. *Food Chemistry, 240*(June 2017), 314–322. https://doi.org/10.1016/j.foodchem.2017.07.104

Etcheverry, P., Grusak, M. A., & Fleige, L. E. (2012). Application of *in vitro* bioaccessibility and bioavailability methods for calcium, carotenoids, folate, iron, magnesium, polyphenols, zinc, and vitamins B6, B12, D, and E. *Frontiers in Physiology, 3,* 317. https://doi.org/10.3389/fphys.2012.00317

Falany, C. N. (1997). Enzymology of human cytosolic sulfotransferases. *The FASEB Journal, 11*(4), 206–216. https://doi.org/10.1096/fasebj.11.4.9068609

Fellows, R., Denizot, J., Stellato, C., Cuomo, A., Jain, P., Stoyanova, E., Balázsi, S., Hajnády, Z., Liebert, A., & Kazakevych, J. (2018). Microbiota derived short chain fatty acids promote histone crotonylation in the colon through histone deacetylases. *Nature Communications, 9*(1), 1–15. https://doi.org/10.1038/s41467-017-02651-5

Gómez-Juaristi, M., Martínez-López, S., Sarria, B., Bravo, L., & Mateos, R. (2018a). Absorption and metabolism of yerba mate phenolic compounds in humans. *Food Chemistry, 240,* 1028–1038. https://doi.org/10.1016/j.foodchem.2017.08.003.

Gómez-Juaristi, M., Martínez-López, S., Sarria, B., Bravo, L., & Mateos, R. (2018b). Bioavailability of hydroxycinnamates in an instant green/roasted coffee blend in humans. Identification of novel colonic metabolites. *Food & Function, 9*(1), 331–343. https://doi.org/10.1039/c7fo01553d

Gonçalves, A., Estevinho, B. N., & Rocha, F. (2021). Methodologies for simulation of gastrointestinal digestion of different controlled delivery systems and further uptake of encapsulated bioactive compounds. *Trends in Food Science and Technology, 114*(October 2020), 510–520. https://doi.org/10.1016/j.tifs.2021.06.007

Gowe, C. (2015). Review on potential use of fruit and vegetables by-products as a valuable source of natural food additives. *Food Science and Quality Management, 45*(5), 47–61.

Haghighat, N., Rajabi, S., & Mohammadshahi, M. (2021). Effect of synbiotic and probiotic supplementation on serum brain-derived neurotrophic factor level, depression and anxiety symptoms in hemodialysis patients: A randomized, double-blinded, clinical trial. *Nutritional Neuroscience, 24*(6), 490–499. https://doi.org/10.1080/1028415X.2019.1646975

Han, W., Jin, F., Zhang, H., Yang, M., Cui, C., Wang, C., & Jiang, P. (2020). Association of brain-gut peptides with inflammatory cytokines in Moyamoya disease. *Mediators of Inflammation, 2020.* https://doi.org/10.1155/2020/5847478.

Hernández-Maldonado, L. M., Blancas-Benítez, F. J., Zamora-Gasga, V. M., Cárdenas-Castro, A. P., Tovar, J., & Sáyago-Ayerdi, S. G. (2019). *In vitro* gastrointestinal digestion and colonic fermentation of high dietary fiber and antioxidant-rich mango (Mangifera indica L.)"Ataulfo"-based fruit bars. *Nutrients, 11*(7), 1564. https://doi.org/10.3390/nu11071564

Hubbard, T. D., Murray, I. A., & Perdew, G. H. (2015). Indole and tryptophan metabolism: Endogenous and dietary routes to Ah receptor activation. *Drug Metabolism and Disposition, 43*(10), 1522–1535. https://doi.org/10.1124/dmd.115.064246

Jalili-Firoozinezhad, S., Gazzaniga, F. S., Calamari, E. L., Camacho, D. M., Fadel, C. W., Bein, A., Swenor, B., Nestor, B., Cronce, M. J., & Tovaglieri, A. (2019). A complex human gut microbiome cultured in an anaerobic intestine-on-a-chip. *Nature Biomedical Engineering, 3*(7), 520–531. https://doi.org/10.1038/s41551-019-0397-0

Jamilian, H., & Ghaderi, A. (2021). The effects of probiotic and selenium co-supplementation on clinical and metabolic scales in chronic schizophrenia: A randomized, double-blind, placebo-controlled trial. *Biological Trace Element Research, 199*(12), 4430–4438. https://doi.org/10.1007/s12011-020-02572-3

Jin, Y., Alimbetov, D., George, T., Gordon, M. H., & Lovegrove, J. A. (2011). A randomised trial to investigate the effects of acute consumption of a blackcurrant juice drink on markers of vascular reactivity and bioavailability of anthocyanins in human subjects. *European Journal of Clinical Nutrition, 65*(7), 849–856. https://doi.org/10.1038/ejcn.2011.55

Lee, M.-J., Maliakal, P., Chen, L., Meng, X., Bondoc, F. Y., Prabhu, S., Lambert, G., Mohr, S., & Yang, C. S. (2002). Pharmacokinetics of tea catechins after ingestion of green tea and (−)-epigallocatechin-3-gallate by humans: Formation of different metabolites and individual variability. *Cancer Epidemiology and Prevention Biomarkers, 11*(10), 1025–1032.

Li, C., Yu, W., Wu, P., & Chen, X. D. (2020). Current *in vitro* digestion systems for understanding food digestion in human upper gastrointestinal tract. *Trends in Food Science & Technology, 96*, 114–126. https://doi.org/10.1016/j.tifs.2019.12.015

Martin, C. R., Osadchiy, V., Kalani, A., & Mayer, E. A. (2018). The brain-gut-microbiome axis. *Cellular and Molecular Gastroenterology and Hepatology, 6*(2), 133–148. https://doi.org/10.1016/j.jcmgh.2018.04.003

Mateos, R., Martínez-López, S., Arévalo, G. B., Amigo-Benavent, M., Sarriá, B., & Bravo-Clemente, L. (2016). Hydroxytyrosol in functional hydroxytyrosol-enriched biscuits is highly bioavailable and decreases oxidised low density lipoprotein levels in humans. *Food Chemistry, 205*, 248–256. https://doi.org/10.1016/j.foodchem.2016.03.011

Miedes, D., Makran, M., Cilla, A., Barberá, R. Garcia-Llatas, G., & Alegría, A. (2023). Function aging-related gastrointestinal conditions decrease the bioaccessibility of plant sterols in enriched wholemeal rye bread: *In vitro* static digestion. *Food & Function, 14*, 602. https://doi.org/10.1039/d3fo00710c

Mills, C. E., Flury, A., Marmet, C., Poquet, L., Rimoldi, S. F., Sartori, C., Rexhaj, E., Brenner, R., Allemann, Y., & Zimmermann, D. (2017). Mediation of coffee-induced improvements in human vascular function by chlorogenic acids and its metabolites: Two randomized, controlled, crossover intervention trials. *Clinical Nutrition, 36*(6), 1520–1529. https://doi.org/10.1016/j.clnu.2016.11.013

Minekus, M., Alminger, M., Alvito, P., Ballance, S., Bohn, T., Bourlieu, C., Carrière, F., Boutrou, R., Corredig, M., Dupont, D., Dufour, C., Egger, L., Golding, M., Karakaya, S., Kirkhus, B., Le Feunteun, S., Lesmes, U., Macierzanka, A., Mackie, A., & Brodkorb, A. (2014). A standardised static *in vitro* digestion method suitable for food – An international consensus. *Food & Function, 5*(6), 1113–1124. https://doi.org/10.1039/c3fo60702j

Moludi, J., Alizadeh, M., Mohammadzad, M. H. S., & Davari, M. (2019). The effect of probiotic supplementation on depressive symptoms and quality of life in patients after myocardial infarction: Results of a preliminary double-blind clinical trial. *Psychosomatic Medicine, 81*(9), 770–777. https://doi.org/10.1097/PSY.0000000000000749

Montagnana, M., Danese, E., Angelino, D., Mena, P., Rosi, A., Benati, M., Gelati, M., Salvagno, G. L., Favaloro, E. J., & Del Rio, D. (2018). Dark chocolate modulates platelet function with a mechanism mediated by flavan-3-ol metabolites. *Medicine, 97*(49). e13432 https://doi.org/10.1097/MD.0000000000013432

Moysidou, C.-M., & Owens, R. M. (2021). Advances in modelling the human microbiome–gut–brain axis in vitro. *Biochemical Society Transactions, 49*(1), 187–201. https://doi.org/10.1042/BST20200338

Muller, P. A., Schneeberger, M., Matheis, F., Wang, P., Kerner, Z., Ilanges, A., Pellegrino, K., Del Mármol, J., Castro, T. B. R., & Furuichi, M. (2020). Microbiota modulate sympathetic neurons via a gut–brain circuit. *Nature, 583*(7816), 441–446. https://doi.org/10.1038/s41586-020-2474-7

Nissen, L., Casciano, F., & Gianotti, A. (2020). Intestinal fermentation *in vitro* models to study food-induced gut microbiota shift: An updated review. *FEMS Microbiology Letters, 367*(12), fnaa097. https://doi.org/10.1093/femsle/fnaa097

Pandareesh, M. D., Mythri, R. B., & Bharath, M. M. S. (2015). Bioavailability of dietary polyphenols: Factors contributing to their clinical application in CNS diseases. *Neurochemistry International, 89*, 198–208. https://doi.org/10.1016/j.neuint.2015.07.003

Ponce de León-Rodríguez, M. del C., Guyot, J. P., & Laurent-Babot, C. (2019). Intestinal *in vitro* cell culture models and their potential to study the effect of food components on intestinal inflammation. *Critical Reviews in Food Science and Nutrition, 59*(22), 3648–3666. https://doi.org/10.1080/10408398.2018.1506734

Rodríguez-Roque, M. J., de Ancos, B., Sánchez-Vega, R., Sánchez-Moreno, C., Cano, M. P., Elez-Martínez, P., & Martín-Belloso, O. (2016). Food matrix and processing influence on carotenoid bioaccessibility and lipophilic antioxidant activity of fruit juice-based beverages. *Food Function., 7*(1), 380–389. https://doi.org/10.1039/C5FO01060H

Santocchi, E., Guiducci, L., Prosperi, M., Calderoni, S., Gaggini, M., Apicella, F., Tancredi, R., Billeci, L., Mastromarino, P., & Grossi, E. (2020). Effects of probiotic supplementation on gastrointestinal, sensory and core symptoms in autism spectrum disorders: A randomized controlled trial. *Frontiers in Psychiatry, 944.* 550593 https://doi.org/10.3389/fpsyt.2020.550593

Sayago-Ayerdi, S., García-Martínez, D. L., Ramírez-Castillo, A. C., Ramírez-Concepción, H. R., & Viuda-Martos, M. (2021). Tropical fruits and their co-products as bioactive compounds and their health effects: A review. *Foods, 10*(8), 1–25. https://doi.org/10.3390/foods10081952

Sáyago-Ayerdi, S. G., Venema, K., Tabernero, M., Sarriá, B., Bravo, L. L., & Mateos, R. (2021). Bioconversion by gut microbiota of predigested mango (Mangifera indica L) 'Ataulfo' peel polyphenols assessed in a dynamic (TIM-2) *in vitro* model of the human colon. *Food Research International, 139* 109963 (November 2020). https://doi.org/10.1016/j.foodres.2020.109963

Sáyago-Ayerdi, S. G., Zamora-Gasga, V. M., & Venema, K. (2020). Changes in gut microbiota in predigested Hibiscus sabdariffa L calyces and Agave (Agave tequilana weber) fructans assessed in a dynamic *in vitro* model (TIM-2) of the human colon. *Food Research International, 132*(January), 109036. https://doi.org/10.1016/j.foodres.2020.109036

Schön, C., Wacker, R., Micka, A., Steudle, J., Lang, S., & Bonnländer, B. (2018). Bioavailability study of maqui berry extract in healthy subjects. *Nutrients, 10*(11), 1720. https://doi.org/10.3390/nu10111720

Schrimpe-Rutledge, A. C., Codreanu, S. G., Sherrod, S. D., & McLean, J. A. (2016). Untargeted metabolomics strategies—challenges and emerging directions. *Journal of the American Society for Mass Spectrometry, 27*(12), 1897–1905. https://doi.org/10.1007/s13361-016-1469-y

Sotomayor-Gerding, D., Oomah, B. D., Acevedo, F., Morales, E., Bustamante, M., Shene, C., & Rubilar, M. (2016). High carotenoid bioaccessibility through linseed oil nanoemulsions with enhanced physical and oxidative stability. *Food Chemistry, 199*, 463–470. https://doi.org/10.1016/j.foodchem.2015.12.004

Stockinger, B., Meglio, P. Di, Gialitakis, M., & Duarte, J. H. (2014). The aryl hydrocarbon receptor: Multitasking in the immune system. *Annual Review of Immunology, 32*, 403–432. https://doi.org/10.1146/annurev-immunol-032713-120245

Tamtaji, O. R., Heidari-Soureshjani, R., Mirhosseini, N., Kouchaki, E., Bahmani, F., Aghadavod, E., Tajabadi-Ebrahimi, M., & Asemi, Z. (2019). Probiotic and selenium co-supplementation, and the effects on clinical, metabolic and genetic status in Alzheimer's disease: A randomized, double-blind, controlled trial. *Clinical Nutrition, 38*(6), 2569–2575. https://doi.org/10.1016/j.clnu.2018.11.034

Tamtaji, O. R., Taghizadeh, M., Kakhaki, R. D., Kouchaki, E., Bahmani, F., Borzabadi, S., Oryan, S., Mafi, A., & Asemi, Z. (2019). Clinical and metabolic response to probiotic administration in people with Parkinson's disease: A randomized, double-blind, placebo-controlled trial. *Clinical Nutrition, 38*(3), 1031–1035. https://doi.org/10.1016/j.clnu.2018.05.018

Teixé-Roig, J, Oms-Oliu, G., Odriozola-Serrano, I., & Martín-Belloso, O. (2023). Emulsion-based delivery systems to enhance the functionality of bioactive compounds: Towards the use of ingredients from natural, sustainable sources. *Foods, 12*(7), 1502. https://doi.org/10.3390/foods12071502

Tordjman, S., Chokron, S., Delorme, R., Charrier, A., Bellissant, E., Jaafari, N., & Fougerou, C. (2017). Melatonin: Pharmacology, functions and therapeutic benefits. *Current Neuropharmacology, 15*(3), 434–443. https://doi.org/10.2174/1570159X14666161228122115

Volmar, C.-H., & Wahlestedt, C. (2015). Histone deacetylases (HDACs) and brain function. *Neuroepigenetics, 1*, 20–27. https://doi.org/10.1016/j.nepig.2014.10.002

Wang, H., He, S., Xin, J., Zhang, T., Sun, N., Li, L., Ni, X., Zeng, D., Ma, H., & Bai, Y. (2021). Psychoactive effects of Lactobacillus johnsonii against restraint stress-induced memory dysfunction in mice through modulating intestinal inflammation and permeability—A study based on the gut–brain axis hypothesis. *Frontiers in Pharmacology, 12*, 1299. https://doi.org/10.3389/fphar.2021.662148

Wootton-Beard, P. C., & Ryan, L. (2011). Improving public health?: The role of antioxidant-rich fruit and vegetable beverages. *Food Research International, 44*(10), 3135–3148. https://doi.org/10.1016/j.foodres.2011.09.015

Xie, L., Vance, T., Kim, B., Lee, S. G., Caceres, C., Wang, Y., Hubert, P. A., Lee, J.-Y., Chun, O. K., & Bolling, B. W. (2017). Aronia berry polyphenol consumption reduces plasma total and low-density lipoprotein cholesterol in former smokers without lowering biomarkers of inflammation and oxidative stress: A randomized controlled trial. *Nutrition Research, 37*, 67–77. https://doi.org/10.1016/j.nutres.2016.12.007

Zamora-Gasga, V. M., Loarca-Piña, G., Vázquez-Landaverde, P. A., Ortiz-Basurto, R. I., Tovar, J., & Sáyago-Ayerdi, S. G. (2015). *In vitro* colonic fermentation of food ingredients isolated from Agave tequilana Weber var. azul applied on granola bars. *LWT-Food Science and Technology, 60*(2), 766–772. https://doi.org/10.1016/j.lwt.2014.10.032

Zhou, H., Tan, Y., & McClements, D. J. (2023). Applications of the INFOGEST *in vitro* gigestion model to foods: A review. *Annual Review of Food Science and Technology, 14*, 135–156. https://doi.org/10.1146/annurev-food-060721- 01223

Chapter 10

The Potential of Agavins as a Prebiotic

Guadalupe M. Alvarado-Jasso
Luisa F. Duque-Buitrago
Carolina Buitrago-Arias
Brenda H. Camacho-Díaz
Maricela Carmona-González
Clara G. de los Reyes-Gavilán
and Nuria Salazar

CONTENTS

DOI: 10.1201/9781003332558-12

10.1 INTRODUCTION

The prebiotic concept was first introduced in 1995 by researchers Glenn Gibson and Marcel Roberfroid. It was described as a non-digestible food ingredient that beneficially affects the host by selectively stimulating the growth or activity of one or a limited number of bacteria in the colon and thus improving host health (Gibson & Roberfroid, 1995). According to this definition, only a few carbohydrate-based compounds, such as short- and long-chain fructooligosaccharides (FOS) and inulin, lactulose, and galactooligosaccharides (GOS), can be considered prebiotics. In 2004, the definition of prebiotics was altered to selectively fermented ingredients that allow specific changes in the composition or activity in the gastrointestinal microflora that confers benefits upon host well-being and health (Gibson et al. 2004). Later in 2008, the 6th Meeting of the International Scientific Association of Probiotics and Prebiotics (ISAPP) defined dietary prebiotics as a selectively fermented ingredient that results in specific changes in the composition or activity of the gastrointestinal microbiota, thus conferring benefit(s) upon host health (Gomez et al. 2010). Subsequently, with the improved knowledge of gut microbiota composition and mechanisms of prebiotics, the definition was updated in 2017 as a substrate that is selectively utilized by host microorganisms conferring a health benefit by an international and consensus panel from ISAPP (Gibson et al. 2017). This definition expands the concept of prebiotics to include non-carbohydrate substances (polyphenols and conjugated polyunsaturated fatty acids), applications to body sites other than the gastrointestinal tract, to have the potential to improve not only human but also animal health, and to reduce the risk of diseases mediated by microbiota. Prebiotics are usually administered orally, but with this updated definition, they can be administered directly to other microbially colonized body sites. However, prebiotics are not the only substances that can affect the microbiota, and for that reason, currently, there are some criteria to classify a substrate as prebiotic: I) Must not be degraded by the target host enzymes; II) must be selectively used by live host microorganisms conferring a health benefit; and III) its efficacy and safety should be demonstrated for the specific target.

Dietary fibers such as FOS, GOS, inulin, resistant starch, and some fructans, galactans, and mannans are considered prebiotics as they have been related to healthy physiological functions linked to the composition and activity of the intestinal microbiota. These soluble fibers, mostly non-digested, reach the colon, and the gut microbiota ferments them to produce short-chain fatty acids (SCFA), which can benefit the host (Ríos-Covián et al. 2016).

Agavins are complex natural fructose-based polymers (fructans), highly branched, that contain an internal glucose moiety linked to inulin-type β (2→1) or to levan-type bonds β (2→6). One of the most important fructan-producing plant families is the *Asparagaceae,* which includes asparagus, onion, and agave species. They present a particular chemical structure only found and described in *Agave* species, which is why the established name of "Agavins." Agavins constitute the main storage carbohydrate form in *Agave* species and are located in all plant tissues, including roots, rhizomes, and flowers, accumulating predominantly in the oversized stems (Pérez-López et al. 2021).

Agavins have high economic value in countries such as Mexico, providing the raw material for popular alcoholic beverages such as Tequila and Mezcal. However, they also have the potential to be utilized as renewable bioenergy sources and as dietary prebiotic supplements (Pérez-López et al. 2021). The interest in agavins as prebiotics has increased recently based on studies proving their ability to beneficially modulate the intestinal microbiota. The benefits of agavins intake include promoting a selective growth of beneficial bacteria such as bifidobacteria and increased production of SCFA while moderately reducing the colonic pH and ammonium release.

Nevertheless, few studies have been carried out on the prebiotic capacity of agavins, most of them performed using in vitro models of colonic fermentation (Gomez et al. 2010; Ramnani et al. 2015; Allsopp et al. 2013; Kilua et al. 2021).

10.2 AGAVINS

10.2.1 Chemical Structure

Fructans are naturally synthesized by various plant species to be used as reserve carbohydrates and allow them to tolerate extreme climates (Barros et al. 2022). Fructans are the primary photosynthetic product of plants from *Asterales, Buxales, Poales,* and *Asparagales* orders, which are composed of fructose polymers derived from a sucrose molecule (glucose, fructose) (Mellado-Mojica et al. 2017). Fructans in the agave plant are synthesized and stored in vacuoles. Their primary function in crassulacean acid metabolism using plants is energy storage, and as osmo-protectors during drought and allow physiological adaptation in arid environments (Mancilla-Margalli & López, 2006; Michel-Cuello et al. 2012).

Fructans are fructose polymers derived from the sucrose molecule, a disaccharide of fructose and glucose. The type of fructan is established according to its chemical structure, the presence of branches, the type of bond between the adjacent fructose unit, and the position of the glucose residues, as well as the degree of polymerization (DP) ranging from three to hundreds of fructose units, with a wide variety of bonds and fructosyl residues (Franco-Robles and López, 2015). One type of fructans is the mixed ones, which have two different bonds β (2→1) and β (2→6) between fructose and contain one internal glucose, which is typical of this group. These fructans can be found in several species of agave and are known as agavins (Nava-Cruz et al. 2015).

Agavins of various species have a mixture of branched and linear fructans; they also have many complex structures with a DP from 3 to greater than 40 fructose units, depending on the phenological stage of the plant (Pérez-López & Simpson, 2020). However, most studies of agavins focus on the species *Agave tequilana* Weber due to the denomination of origin and the impact of this commercialization certification. Currently, there is a great interest in knowing and taking advantage of the similarities and differences between the agavins obtained from different species, their possible applications, and their compartmentalization along the stem (Aldrate-Herrera et al. 2019).

Due to the beta configuration of the bonds between fructose monomers, specifically by alpha-glycosidic bonds, all fructans are resistant to hydrolysis by human digestive enzymes, so they pass undigested through the upper part of the gastrointestinal tract to the colon. For this reason, fructans are classified as indigestible oligosaccharides (López Pérez & Urías-Silvas, 2007; Roberfroid, 2007).

10.2.2 Sources of Fructans and Synthesis

In nature, fructans are synthesized from sucrose by a restricted number of plant species and many microorganisms, including archaea, fungi, and a wide range of bacteria. Plant fructans have a low DP, and bacterial inulin and levans reach much higher molecular masses, offering a broader spectrum of biotechnological applications.

In the case of fructans that are biosynthesized by various microorganisms, eubacteria are synthesized as oligosaccharides and extracellular polysaccharides. The main mechanism of action involves the transfer of a fructosyl unit from a sucrose molecule to an acceptor. The inulobiose and the levanbious are the simplest structures of fructans existing in nature. Unlike fructans that biosynthesize plants that catalyze the cleavage and transfer of the fructosyl unit to an acceptor by at least two distinct enzymes, microorganisms only require the enzymatic activity of a single enzyme to synthesize fructan polymers: fructosyltransferase, levansacarases, or inulosacarasas (Banguela & Hernández, 2006; Velázquez-Hernández et al. 2009).

While levans are synthesized by many gram-positive and gram-negative eubacteria, microbial inulin synthesis has only been observed in some gram-positive species, mainly lactic acid bacteria (Kralj et al. 2018; Zannini et al. 2016). Unlike Eubacteria, in plants, fructans accumulate intracellularly in addition to the fructans type levan and inulin, the highly branched neo-fructans.

The synthesis of fructans in plants occurs within vacuoles, where sucrose accumulation induces the synthesis of the responsible enzymes. The action of the different fructosyltransferases in the metabolism of plants generates a wide variety of structures of fructans, the six types mentioned above. To synthesize agavins and the other types of fructans in plants four fructosyltransferase activities are needed (1-SST, 1-FFT, 6-SFT, and 6G-FFT), while the degradation of fructans is carried out by fructan exohydrolases that can be specific to β (1→2) or β (1→6) bonds or act on both. These enzymes in common with vacuolar and cell wall invertases are members of the family of vegetable glycoside hydrolases (Pérez-López and Simpson, 2020).

10.2.3 Patents

By searching for related patent documents with the term *agavins*, four applications were found. This is compared to sixty-four applications with the term *agave fructans*. In comparison with those involving methods for obtaining agave polysaccharides in the national database administered by the Mexican Institute of Industrial Property (IMPI), eighteen results of patent documents were found to be registered in Mexico. The institutions with the highest number of patent applications concerning the methods of obtaining agave polysaccharides the Center for Research and Advanced Studies of the National Polytechnic Institute (CINVESTAV) with six registered documents, and the Center for Research and Assistance in Technology and Design of the State of Jalisco (CIATEJ) with five registered documents.

The most relevant and most similar applications are shown in **Table 10.1**, which shows the principal applicants concerning the compositions and methods in terms of agave fructans as search keyword according to its technological classifier, the most cited or requested patent documents are in Mexico and allow to see that there are two Mexican research centers. They are also part of direct competition between the different methods of extraction and obtaining fructans from agave.

The technology developed is from educational institutions that have made these requests, such as the CIATEJ, the CINVESTAV, and the Center for the Development of Biotic Products (CEPROBI) of the National Polytechnic Institute, which describe similar processes. However, the latter involves the intellectual protection of systems (machines, devices, devices) to extract agave polysaccharides. Therefore, another common factor found among patent registrations is that they describe the method for obtaining agave polysaccharides and complement the application by describing the compositions obtained from fructans and their health benefits. The benefits composition allows, when used in the elaboration of a product, to emphasize the sectors in which it can be used (health, food, and pharmaceutical).

TABLE 10.1 SUMMARY OF PATENTS OF PROCESSES OF OBTAINING FRUCTANS OF AGAVE IN MEXICO

Registration Number and Title	Summary
MX/a/2015/016512 Modular system and process for obtaining various products of agave fructans.	The difference with respect to other technologies is that it is not carried out by enzymatic methods, and it is not previously dehydrated or cooked material at high pressures and using flocculation by cold as clarification method.
MX/a/2015/014523 Agave fractionated fructans and their process of obtaining them at pilot and industrial level.	The process uses two stages of ultrafiltration and one of nanofiltration separated by molecular chain size or DP into two types: a) DP between 3 and 10 monomeric units. b) Average polymerization grade greater than 10.
MX 2013004901 A Process for obtaining fructans from young agave and its use as a functional ingredient.	This process is performed with young agave, consists of washing the core, crushing, and tearing it to leach it with hot water and obtain an aqueous extract containing the soluble solids of the agave. The extract is clarified by coagulation, sedimentation, and ultrafiltration to separate the insoluble solids present in it. The extract can be concentrated or dehydrated.
MX 2011003876 A 20110411 Process for producing and applying fructanase enzymes resulting from *Kluyveromyces marxianus* yeasts in the hydrolysis of branched fructans of *agavaceae*, grasses and/or grass plants.	This invention replaces the traditional methods of acid-thermal hydrolysis currently used in fructose-rich syrup and agave fructans. The products were obtained by applying fructhanases from *Kluyveromyces marxianus* yeasts.
MX PA04011739 A 20041126 Extraction of agavins (fructans of agave) and use of them as prebiotics	The invention further describes the general structure of agavins. The bifidogenic (*Bifidobacteria*) and prebiotic (*Lactobacilli)* effect of some agavins on the *in vitro* growth of *Bifidobacterium breve, Lactobacillus casei,* and *L. paracasei* was better than that shown in most commercial inulins

10.3 PHYSIOLOGICAL BENEFITS OF AGAVINS

The physiological benefits of agavins are attributed to their soluble dietary fiber characteristic and fermentability. The main physiological effects achieved by the consumption of agavins and their possible mechanisms of action are shown in **Figure 10.1**. Soluble dietary fiber stimulates the absorption of water and electrolytes from the colonic lumen, increasing fecal viscosity and decreasing gut transit time (Bauer et al. 2006; Mudgil and Barak, 2013). Furthermore, dietary fiber promotes healthier lipoprotein profiles for improved endothelial health and can attenuate elevated systemic inflammation (Santisteban et al. 2017). In addition, it promotes colonic microbiota and their fermentation by-products like SCFA, leading to potentially improved systemic health (Hu et al. 2018). This section summarizes the results of some investigations obtained from *in vitro* and *in vivo* studies using agavinas on the physiological effect that they can have.

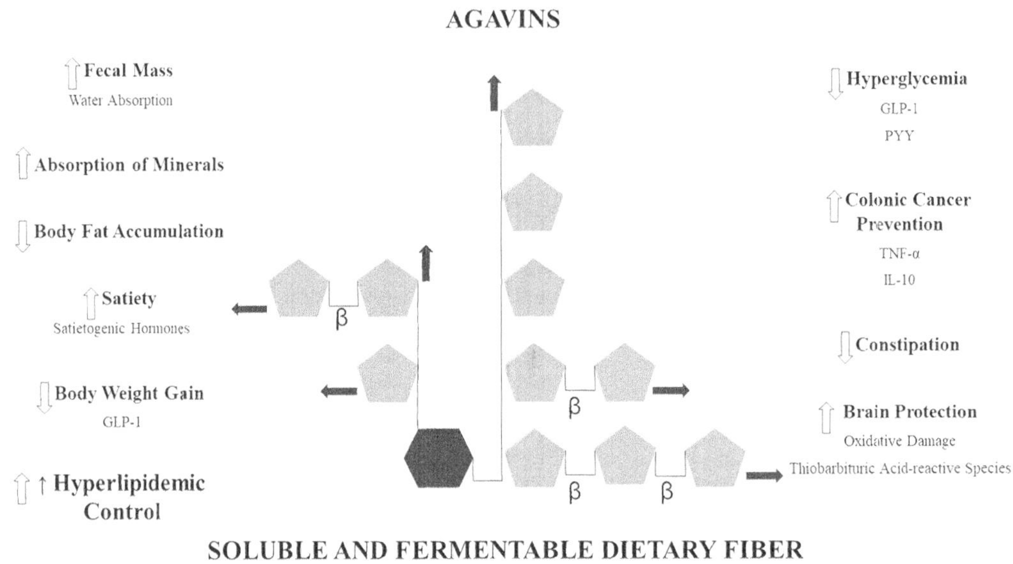

Figure 10.1 Possible physiological effects of the consumption of Agavins in *in vitro* and *in vivo* tests.

10.3.1 Fecal Mass Increment and Constipation

Agavins improved fecal excretion in animals and humans. For example, in a model of healthy rats supplemented with agavins of *A. salmiana* for 35 days, the fecal output was significantly higher with higher doses, and the feces were softer compared to the control group (Castillo Andrade et al. 2018); these effects were accompanied by the corresponding modulation in SCFA production (Castillo Andrade et al. 2019). The increment in fecal excretion with other sources of agavins has also been proved with 10% supplementation with agavins of *A. tequilana* in healthy mice, with 17% more feces collected than in the control group (Urías-Silvas et al. 2008).

The first clinical trial with agavins involved healthy adults consuming daily doses of 0.5 g and 7.5 g for 21 days each, with 7 days of washout between the two dosing periods (Holscher et al. 2014). In this clinical trial, laxation improved with the administration of up to 7.5 g/day without increasing diarrhea symptoms (Holscher et al. 2015). A later clinical trial confirmed the laxative effect in children with chronic constipation, where 4 g/day of agavins improved the stool consistency and decreased the reports of painful defecation with symbiotic treatment with the agavins and *Lactobacillus reuteri* DSM 17938 (García-Contreras et al. 2020).

10.3.2 Satiety, Body Weight Gain, and Body Fat Accumulation

Body weight gain may be modulated by the same effects as those reported by the consumption of dietary fiber by increasing satiety, increasing stomach distention, reducing gastric emptying, and modulating the production of intestinal hormones (Chambers et al. 2015). In addition, agavins of different *Agave* species have been described as regulators of appetite, food intake, and subsequent weight loss in different animal models.

Castillo Andrade *et al.* (2018) reported that healthy rats supplemented with agavins of *A. salmiana* decreased weight by up to 32% compared to the control group. Agavins of *A. tequilana* with small

DP (<10) reduced the final body weight by 30% and fat accumulation by 51% (Márquez-Aguirre et al. 2016). Mice fed with a high-fat diet and then switched to a standard diet with agavins supplementation showed that agavins reversed metabolic disorders that originated from the high-fat diet. Only the animals with the diet change and agavins supplementation significantly decreased body weight by about 9% compared with a 4% body weight decrement in mice with a standard diet with no supplementation. Also, the energy intake was significantly lower with agavins supplementation (Huazano-García & López, 2015). Later reported, agavins supplement reduced the adipose tissue by around 32%, the size of the adipocytes, and showed a tendency to reduce membrane thickness in relation to the high-fat diet mice group (Huazano-García et al. 2020; Huazano-García & López, 2015). Another study supplementing healthy rats with 2.5 g of agavins of *A. tequilana* and *A. salmiana* for 20 weeks showed a slight reduction in abdominal adipose tissue compared with the group without supplementation with no differences between the agavins origin (Regalado-Rentería et al. 2020).

Satiety modulation by agavins has been associated with satietogenic hormones in mice. Healthy mice fed with a standard diet and supplemented with agavins of *A. tequilana* presented increased levels of caecum proglucagon mRNA by over 30% versus the standard diet alone. The increase in the medial colon was 20% and in the proximal colon 12%. When measured in blood, glucagon-like peptide-1 (GLP-1) concentrations were 55% higher than in the standard group (Urías-Silvas et al. 2008). Healthy mice raised in a high-fat diet and then changed to a standard one and supplemented with agavins of *A. tequilana* presented higher concentrations of GLP-1 and leptin, showing that the supplementation with agavins reversed the adverse effects of a hypercaloric and fatty diet, more than the diet change alone (Huazano-García and López, 2015). With no changes in diet, mice fed with a high-fat diet and agavins of *A. tequilana* also exhibited the highest concentration of GLP-1 with an increment of over two-fold regarding the standard diet and high-fat diet group (Huazano-García et al. 2020).

Other sources of agavins also have satisfactory results. Compared with the standard diet-fed mice, supplementation with agavins of *A. potatorum* and *A. angustifolia* showed higher GLP-1 concentrations of 40% and 93% more, respectively. The agavins also stimulated the GLP-1 concentration in the cecum, proximal, and medial colon more than in the mice fed with the standard diet. This increase was statistically significant in the cecum of *A. angustifolia*-agavins-fed mice and the proximal colon for the *A. potatorum*-agavins in the medial colon, where both concentrations increased two-fold compared with the standard diet. Also, the ghrelin concentration was lower in the group fed with agavins of *A. angustifolia* (12%) or *A. potatorum* (38%) (Santiago-García & López 2014).

In a clinical trial with obese subjects following a low-calorie diet (1500 kcal/day), the ingestion of agavins (96 mg/kg twice a day for one week and daily for 11 weeks) decreased body mass index and total body fat. The average body mass index decreased from 35.3 to 33.0 kg/m^2, while the control low-calorie group without agavins remained unchanged. Total body fat reduction was 10% in the group supplemented with agavins and 3% without agavins (Padilla-Camberos et al. 2018). In another clinical study performed with subjects with hypercholesterolemia, thirty people took simvastatin plus ezetimibe, and another thirty took simvastatin and agavins. All patients from the two groups reduced weight, body mass index, and waist circumference (Martínez-Abundis et al. 2016). These studies showed that agavins of different botanical sources help regulate satiety by hormone control.

10.3.3 Hyperlipidemic Control

In healthy rats supplemented with commercial and experimental agavins of *A. tequilana*, a 21% decrease in total cholesterol with both agavins and an 18% decrease in triglycerides with commercial agavins and 30% with experimental ones were identified (Regalado-Rentería et al. 2020). Agavins of *A. salmiana* reduced cholesterol by 21%, and the reduction in triglycerides was 10%. Jasso-Padilla et al. proposed that symbiotic supplementation could enhance various effects of agavins administration. Healthy Wistar rats fed with commercial diets and supplemented with *A. salmiana* agavins (20 g/kg) and *Bacillus animalis* subsp. *Lactis* for 12 weeks presented a 19% reduction in triglyceride concentration (Jasso-Padilla et al. 2017).

Mice fed high-fat diets and supplemented with agavins of *A. tequilana* (10%) had significantly reduced triglyceride levels (Franco-Robles & López, 2016). Regalado-Rentería *et al.* (2020) fed a dose of 0.38 g/day ay in their drinking water showed a decrease of 57% in triglycerides in relation to high-fat group mice without agavins supplementation (Huazano-García et al. 2020). This effect was also seen with supplementation with agavins of *A. angustifolia* at doses of 15%, where its administration significantly reduced total cholesterol levels in obese diabetic rats (Rendón-Huerta et al. 2012). Regarding the different degrees of polymerization that agavins can present, a study in obese mice fed with *A. tequilana* agavins with different degrees of polymerization showed that agavins with a higher degree of polymerization and the mixture show statistical differences in triglycerides in comparison with the control group with a reduction of 52.7% and 51.5%, respectively. On the contrary, the short-chain agavins did not improve triglyceride or cholesterol (Márquez-Aguirre et al. 2016).

Regarding human studies, a remarkable study performed with healthy infants in apparent good health supplemented a standardized infant formula with agavins of high (>10) and low (<10) degrees of polymerization plus probiotics *ad libitum* for three months. The total cholesterol, triglycerides, and lipoproteins were significantly lower in the group supplemented with both types of agavins compared to the infants with no supplemented formula, with the lower values being for the infants fed with a mix of both agavins and probiotics (López-Velázquez et al. 2015). In addition, people with hypercholesterolemia receiving simvastatin plus ezetimibe or simvastatin plus agavins presented reduced cholesterol, LDL-C, and triglyceride levels with both treatments.

Treatment with simvastatin plus inulin achieved recommended levels of total cholesterol below 200 mg/day and triglycerides below 150 mg/day in 60% and 30% of patients, respectively, compared to 76.9% and 34.6% with simvastatin plus ezetimibe.. These results showed a non-inferiority effect with agavins treatment (Martínez-Abundis et al. 2016). In another study with overweight and obese subjects who received 15 g/day of agavins of *A. tequilana*, the cholesterol concentration decreased by 37%, LDL by 23%, and total triglycerides by 53%. Regarding the clinical diagnoses, 6.7% of subjects reversed their hypercholesterolemia diagnosis, 13.3% their hypertriglyceridemia, 6.6% changed their dyslipidemia, and 23.3% modified their pre-hypertensive diagnosis (Hernández-González et al. 2016). Obese subjects ingesting agavins (96 mg/kg twice a day for one week and daily for 11 weeks) decreased triglyceride levels by 33% with agavins treatment compared with the placebo group (Padilla-Camberos et al. 2018).

Regarding blood pressure, mice fed a high-fat diet and agavins reduced their blood pressure levels, similar to those fed the standard diet (systolic blood pressure of the high-fat diet was 137.36 mm Hg against 115.34 high-fat diet and agavins) (Alvarado-Jasso et al. 2020).

10.3.4 Control of Glycemia

The glucose metabolism can be altered through the physical interference with the digestion or absorption of macronutrients (Mudgil & Barak, 2013), reduction of food intake (Chater et al. 2015; Slavin & Green, 2007), or by hormone regulation such as the release of GLP-1 and PYY through intestinal L cells followed by SCFA stimulation (Murphy & Bloom, 2006; Guyenet & Schwartz, 2012).

Marquez-Aguirre et al. (2016) reported that in mice with induced obesity by a high-fat diet and supplemented with agavins of high and low DP, the addition of agavins of lower DP decreased glucose and oral glucose tolerance test compared with the obese mice fed only with the high-fat diet. The supplementation treatment with the mixture of agavins of both DP decreased hyperglycemia by 20%, and agavins of lower DP decreased hyperglycemia by 25%. Overweight mice fed high-fat diets and agavins lowered their glucose concentrations, similar to those fed the standard diet (Huazano-García & López 2015). Furthermore, glucose reduction was associated with higher SCFA production and satiety hormones such as GLP-1 and leptin, resulting from agavins fermentation (Huazano-García et al. 2017; Jasso-Padilla et al. 2017). This effect was also seen in a study with mice receiving 0.38 g/day of agavins of *A. tequilana* as a supplement, with a decrease of 6% in glucose levels in relation to the high-fat diet group. This effect was accompanied by a reduction of about 32% in insulin levels (Huazano-García et al. 2020).

Diabetic mice supplemented with *A. angustifolia* agavins (15%) presented reduced blood glucose levels by 23% compared with the diabetic group without supplementation. In obese non-diabetic rats, the supplementation significantly reduced blood glucose levels by 28% (Rendón-Huerta et al. 2012). Also, in healthy Wistar rats supplemented with agavins of *A. salmiana* (5, 10, 12.5, 15, and 20%), the glucose concentrations were maintained constant and within the minimum reference interval during the 35 days of treatment with all the doses (Castillo Andrade et al. 2018). García-Curbelo et al. (2015) found that glucose concentrations in mice fed with a diet supplemented with agavins of *A. fourcroydes* (10%) with DP <10 were significantly lower than those in mice fed a standard diet.

Obese individuals on a low-calorie diet (1500 kcal/day) supplemented with agavins (96 mg/kg twice a day for one week and daily for 11 weeks) presented glucose values within the normal range and did not change during treatment. Insulin showed statistically significant differences with the agavins supplementation compared to the group baseline with an increment of 78%. In contrast, the placebo group with only the low-calorie diet presented a 36% increment (Padilla-Camberos et al. 2018). In overweight and dyslipidemic patients receiving 15 g/day of agavins of *A. tequilana* in the morning for 60 days, the fasting glucose reduced by 7.9% with statistical significance. In addition, increased insulin concentrations by 23.5% were also found (Hernández-González et al. 2016).

10.3.5 Colonic Cancer Prevention

The administration of agavins of *A. tequilana* of lower DP inhibited the development of colitis and colon cancer in mice. Specifically, the agavins treatment reduced the concentration of tumor necrosis factor-alpha and interleukin-10 and inhibited the development of intestinal polyps, villous atrophy, and lymphoid hyperplasia in the colon and jejunum (Rivera-Huerta et al. 2017).

Castillo Andrade et al. (2019) supplemented healthy rats with agavins of *A. salmiana* at different doses (5, 10, 12.5, 15, and 20%) for 35 days and evaluated mucosal integrity, bacterial proliferation, and inflammatory response. Rats on doses of 5, 10, and 12.5% showed a thickening in the epithelial tissue, mucosa, and submucosa layers. Cellular proliferation was present in the mucous layer in rat groups fed all doses. However, there was greater cell proliferation at the 12.5% dose, with a strong relationship to the production of SCFA, as this dose presented the highest concentration of butyric acid and total SCFA. The animals administered doses of 15 and 20% suffered cellular lysis on the muscle membrane and epithelial tissue and the disappearance of the mucosa and submucosa layers.

Another study performed with healthy rats supplemented with 2.5 g of agavins of *A. tequilana* and *A. salmiana* for 20 weeks showed a trend to reduce the length of the small intestine because of a reduction of retention time in the intestine of rats administered with agavins, especially of *A. salmiana*. The cecal content in the agavins groups was similar, and all were higher than the control group with an increase of over 50%, and the colon content agavins group had 50% less. These results showed that agavins supplementation increases feces volume, helps defecation, and reduces retention time, preventing constipation. Another significant result from the study was that colon grave crypt width and size were similar for all treatments, and the morphology was intact. The calciform cells of the crypts corresponding to agavins treatments presented more vacuole structures with intracellular mucin, which is beneficial for colonic health and microbiota modulation (Regalado-Rentería et al. 2020).

10.3.6 Brain Protection

A high-fat diet has been associated with inflammation, overproduction of reactive oxygen species, and damage of macromolecules in different organs. In addition, lipid peroxidation produces toxic compounds that cause neuronal apoptosis (Park et al. 2010; Popa-Wagner et al. 2013; Franco-Robles et al. 2019). Furthermore, elevated intake of saturated fat and simple sugars increases the incidence of neurodegenerative diseases and impairs cognitive abilities (Stranahan et al. 2008; Zhang et al. 2009; Kanoski & Davidson, 2011).

Mice fed a high-fat diet supplemented with agavins of *A. tequilana* (5 and 10%) preserved normal neurotrophic factors levels in the frontal cortex and cerebellum, especially for the high-fat diet group supplemented with 10% agavins, helping to support neuronal interconnections. Agavins also kept low oxidative brain damage levels in the hippocampus, frontal cortex, and cerebellum of the obese mice in comparison with the control group of obese mice without supplementation (Franco-Robles & López, 2016). Later reported by the same authors, agavins with small DP (<10) also decreased oxidative stress markers in the brain of overweight mice. In addition, lipid peroxidation decreased by agavins supplementation in the hippocampus, frontal cortex, and cerebellum. Protein oxidation measured as carbonyl levels decreased in the hippocampus 52%, and cerebellum 46% showing that agavins may have anti-oxidative potential and can be used as an alternative treatment to prevent the consequences of this pathology (Franco-Robles et al. 2019).

10.3.7 Mineral Absorption

Deoxypyridinoline is a bone resorption marker released in children during bone growth. In a study with healthy infants, supplementation of formula with agavins of high (>10) and low (<10) DP besides probiotics resulted in a tendency of increasing deoxypyridinoline comparable to the reference group fed only with breast milk. This shows preliminarily that agavins supplementation could help equate formula-fed infants with breast milk-fed infants regarding bone turnover (López-Velázquez et al. 2015).

10.4 AGAVINS AS A PREBIOTIC: FROM *IN VITRO* TO HUMAN STUDIES

It is well known that dietary fibers can regulate the yield and molar ratios of SCFA. The fermentation of aldehydes produces acetate and butyrate (e.g., glucose, galactose, mannose, xylose), whereas propionate is produced by ketones fermentation (e.g., fructose, arabinose) (Yang et al. 2013; Plongbunjong et al. 2017). The chemical composition of agavins increases propionate concentration through their colonic fermentation. This effect has been shown in *in vitro* fermentation assays with human fecal samples (Gomez et al. 2010; Allsopp et al. 2013) and in animal models fed with standardized normal and high-fat diets (**Table 10.2**). In addition, the increase in SCFA antagonizes the growth of pathogenic bacterial strains (Blaut, 2002) and favors mucin production in the colon (Barcelo et al. 2000). Moreover, butyrate decreases the glutamine requirement for epithelial cells and alters epithelial cell gene expression (Jenkins et al. 1999; Sanderson, 2007). In diet-induced obesity, mice fed with high-fat diets were supplemented with agavins of *A. tequilana* Weber and oligofructose (DP= 5) (Huazano-García et al. 2017). Also, agavins of *A. angustifolia* Haw (Alvarado-Jasso et al. 2020) increased the SCFA contents.

TABLE 10.2 EXAMPLES OF *IN VITRO* AND *IN VIVO* STUDIES DESIGNED TO DETERMINE THE EFFICACY POTENTIAL PREBIOTIC EFFECT OF AGAVINS

Agavins Source	Study Type	Prebiotic Effect	References
A. tequilana Weber, *A. angustifolia*, *A. potatorum*, *A. cantala*, *A. fourcroides*.	*In vitro.*	↑Bifidobacteria and lactobacilli.	Urías-Silvas et al. (2008)
A. angustifolia ssp, *A. tequilana* Haw.	*In vivo* (rats)	↑Bifidobacteria and lactobacilli.	Rendón-Huerta et al. (2012)
A. tequilana Weber.	*In vitro* and *in vivo* (mice).	↑Bifidobacteria.	Márquez-Aguirre et al. (2013)
A. *tequilana* Weber.	*In vitro* and *in vivo* (humans).	↑Bifidobacteria and lactobacilli.	Ramnani et al. (2015)
A. fourcroydes.	*In vivo* (mice).	↑Bacterial diversity.	García-Curbelo et al. (2015)
A. tequilana.	*In vitro.*	↑Bifidobacteria.	Mueller et al. (2016)
A. salmiana.	*In vivo* (rats).	↓Fecal pH. ↑Lactic acid bacteria.	Jasso-Padilla et al. (2017)
A. tequilana Weber.	*In vivo* (mice).	↑Bacterial diversity. ↑SCFA.	Huazano-García et al. (2017)
A. salmiana var. liso, *A. salmiana* var. chino, *A. salmiana* spp. *crassipina*, *A. atrovirens*, *A. tequilana* var. cenizo and *A. tequilana.*	*In vitro.*	↓Growth pathogens. ↑*Lactobacillus* species.	García Gamboa et al. (2018)

(continued)

TABLE 10.2 (CONTINUED)

Agavins Source	Study Type	Prebiotic Effect	References
A. tequilana.	*In vivo* (mice).	↑SCFA.	González-Herrera et al. (2019)
A. fourcroydes.	*In vivo* (poultry).	↑ Acetic and butyric acids in cecum. ↓Cecal pH. ↑ Bursa of fabricius weight. ↑ Hemagglutination inhibition.	García Curbelo et al. (2020)
A. angustifolia Haw, commercial and acetylated	*In vitro.*	↑Growth of *Saccharomyces boulardii.*	Londoño-Moreno et al. (2020)
A. tequilana.	*In vivo* (mice).	↓ Lipopolysaccharide and pro-inflammatory cytokines. ↓ Lipid droplet in the liver. ↓ Adipocyte size.	Huazano-García et al. (2020)
A. tequilana agavins: Hibiscus sabdariffa 70:30.	*In vitro.*	↑Bacterial diversity.	Sáyago-Ayerdi et al. (2020)
A. tequilana Weber/ Probiotics: *L. rhamnosus and B. lactis.*	*In vivo* (humans).	↓Bacterial load. ↑Bifidobacteria.	González-Hernández et al. (2012)
A. angustifolia Haw.	*In vitro.*	↑Bifidobacteria. ↑SCFA. ↓Ammonia production.	Allsopp et al.(2013)
A. angustifolia Haw.	*In vitro.*	↑Bifidobacteria.	Velázquez-Martínez et al. (2014)
A. salmiana.	*In vitro.*	↑Bifidobacteria. ↑SCFA. ↓pH.	Martínez-Gutiérrez et al. (2017)

A = Agave; SCFA = short-chain fatty acids

In vitro fermentation assay with human fecal samples showed the prebiotic potential of agavins of *A. angustifolia*, promoting the growth of fecal *Bifidobacterium* and *Lactobacilli* samples (Allsopp et al. 2013), and with *A. tequilana* agavins contributing to the suppression of potentially pathogenic bacteria (Gomez et al. 2010). *In vitro* growth of *Bifidobacterium breve* and *Lacticaseibacillus casei* (before *Lactobacillus casei)* with A. *angustifolia* agavins produced higher yields than other commercial inulin (López Pérez & Urías-Silvas, 2007); and *A. angustifolia* Haw agavins fractions with DP of 3–60 fructose units, 2–40, and 2–22 DP selectively promoted the growth of a distinct species of *Bifidobacterium* and *Lactobacillus*, especially the lower DP were the preferred carbon source for probiotic bacteria. The bifidogenic activity of agavins has been shown in HIV patients in a double-blind study with *A. tequilana* Weber agavins as prebiotics and *Lactobacillus rhamnosus* and *Bifidobacterium lactis* as probiotics. In addition, the treatment with symbiotics decreased the bacterial load in feces (González-Hernández et al. 2012). The prebiotic effect of agavins also depends on DP and the ramification. The agavins with a smaller DP produce more rapid growth of lactobacilli and bifidobacteria. In contrast, the greater length chains have a more prolonged effect over the time reached for the bacteria of the distal and proximal colon, leading to higher production of SCFA (González-Ávila et al. 2014; Mueller et al. 2016).

10.5 APPLICATION OF PHYSICOCHEMICAL AND FUNCTIONAL PROPERTIES OF NATIVE AND MODIFIED AGAVINS

In recent years, agavins have been used as prebiotics in the food industry. The branched structure containing bonds β (2→1) and β (2→6) makes agavins different from other types of fructans and gives them particular technological properties such as texture and viscosity modifiers and carriers for microencapsulation (Espinosa-Andrews & Rodríguez-Rodríguez, 2018; García-García et al. 2019). Agavins have also been studied as wall materials for encapsulating bioactive compounds and adjuvants in the pharmaceutical industry (Delgadillo et al. 2015; Santiago-García et al. 2017; Sánchez-Madrigal et al. 2019). Low-molecular-weight agavins contain branched bonds and are highly soluble, an important property in the design of functional products (Olvera-Carranza et al. 2015). The properties that agavins provide, such as viscosity and gel formation capacity, are directly related to the DP and molecular branching (Tungland, 2018).

10.5.1 Physicochemical Studies and Technological Properties of Native and Modified Agavins

The structural analysis of the agavins has been carried out using different instrumental techniques such as high-performance anion exchange chromatography with pulsed amperometric detection (HPAEC-PAD), size exclusion chromatography (SEC), nuclear magnetic resonance (RMN), matrix-assisted laser desorption/ionization time-of-flight mass spectrometry (MALDI-TOF MS) (López et al. 2003; Arrizon et al. 2010; Ávila-Fernández et al. 2011), and infrared spectroscopy (FTIR) (Apolinário et al. 2017; Ignot-Gutiérrez et al. 2020; Miramontes-Corona et al. 2020). These techniques helped the elucidation of agavins molecules, corroborating what other authors had described of its branched structure.

The oligosaccharide profile of *A. tequilana* changes depending on the age of each plant (Arrizon et al. 2010). Espinosa-Andrews and Urías-Silvas, (2012), through differential scanning calorimetry studies, reported that the agavins showed an endothermic relaxation of the amorphous polymer chains with a degradation temperature less than 200 °C. These studies help to identify the storage and manipulation conditions when using agavins in food matrices (Santiago-García et al. 2021). The distribution of chain length or DP of agavins is important as influences human tolerance to dietetic interventions. Short-chain fractions, which ferment more easily, have a lower tolerance than longer-chain fructan fractions (Olvera-Carranza et al. 2015; Espinosa-Andrews et al. 2021).

10.5.2 Actual and Potential Market of Agavins Products as Ingredients

The use of agavins in the food industry has increased in recent years due to their functional attributes (**Table 10.3**). Due to their solubility and water properties, agavins have been used as ingredients in food products such as baked goods, beverages, and some dairy products (Palatnik et al. 2017; Santiago-García et al. 2021). For example, high DP agavins as a fat substitute in cookies have been found to enhance the functional properties due to the added prebiotic activity.

An attractive option that has been investigated is the addition of agavins to ice creams low in fat and sugar. Jardines et al. (2020) reported that adding a concentration of 3% of agavins improved sensory properties such as a soft and creamy texture, making these agavins a feasible alternative as sugar and fat substitutes.

Agavins have been used as a wall material to encapsulate different bioactive compounds (**Table 10.4**). Ortiz-Basurto et al. (2017) encapsulated pitanga (*Eugenia uniflora L.*) juice using agavin and maltodextrin as wall material and determined its antioxidant properties and anthocyanin retention. They found that the highest yield and concentration of the pigments depended on the proportion of the wall material, with 1:6 being the best proportion. Agavins were more effective than maltodextrin for the protection of antioxidants during drying and their respective storage. Ramos-Hernández et al. (2018) found that the encapsulation of β-carotene with agavins by electrospraying protects it from the photosensitivity of the molecule up to 50 h of exposure, concluding that this material is a practical choice as an encapsulating material. However, some authors suggest that due to their hydrophilic nature, it is necessary to modify the agavins to extend their applications (Delgadillo et al. 2015).

Chemically, agavins have hydroxyl groups in their terminal fructose, which makes them ideal to be modified by different techniques such as oxidation, sulfation, esterification, and amidation, among others (Miramontes-Corona et al. 2020; Mizrahy & Peer, 2012). These modifications include esterification with acetic anhydride and cross-linking with polyfunctional reagents such as phosphoryl chloride ($POCl_3$). These modifications improved the encapsulation efficiency of bioactive

TABLE 10.3 TECHNOLOGICAL FUNCTIONALITY OF AGAVINS IN FOOD MATRICES

Type	Food Application	References
Agave nectar.	Sports drink.	Phard & Dince, (2009)
Agave nectar.	Ice cream.	Zwarg & Ralphs, (2015)
A. tequilana.	Granola bars.	Zamora-Gasga et al. (2014)
A. tequilana.	Enhanced sensory properties in yogurt.	Crispín-Isidro et al. (2015)
A. tequilana Weber.	Improved the mechanical properties of hardness and stickiness in dehydrated apple matrix.	González-Herrera et al. (2016)
Agavins.	Improves nutritional composition in guava puree.	García-López et al. (2017)
A. tequilana.	Fat replacer.	Palatnik et al. (2017)
Agavins.	Fat and sugar replacer.	Pintor-Jardines et al. (2018)
Agave syrup.	Natural sweetener in gummy bear.	Čižauskaite et al. (2019)
A. tequilana.	Increased the stability and springiness during the kneading.	Morales-Hernández et al. (2019)
Agave syrup.	Sucrose replacement in muffins.	Ozuna et al.(2020)

TABLE 10.4 TECHNOLOGICAL USES OF NATIVE AND MODIFIED AGAVINS

Agavin	Source	Functionality	References
Native.	*A. tequilana* Weber var. Azul.	Encapsulation of spray-dried mango and pineapple juices with maltodextrin and agavins.	Jimenez-Sánchez et al. (2018)
Native.	*A.tequilana* Weber var. Azul.	Encapsulation of blue maize extract using agavins as a carrier.	Sánchez-Madrigal et al. (2019)
Native.	*A.tequilana* Weber var. Azul.	Encapsulation using agavins and inulin as wall materials.	Bermúdez-Quiñones et al. (2021)
Esterified.	*A. tequilana* Weber var. Azul.	Vehicle in the encapsulation of drugs.	Miramontes-Corona et al. (2020)
Acylated.	*A. tequilana* Weber var. Azul.	Agavins modified showed better reduction in solubility and a greater retention of water and oil.	Ignot-Gutiérrez et al. (2020)
Esterified.	*A. Angustifolia* Haw.	Prebiotic to *S. boulardii.*	Buitrago-Arias et al. (2021)

compounds (Delgadillo et al. 2015; Gracia et al. 2013). Recently Ignot-Gutiérrez *et al.* (2020) modified agavins by acylation with lauric acid improving their functional properties such as a reduction in solubility and greater water and oil retention compared to native agavins.

Another modification performed was the esterification of agavins with lauroyl, acetyl, and palmitoyl groups to encapsulate ibuprofen and evaluate the release of the drug under simulated colon conditions (Miramontes-Corona et al. 2020). These authors reported that the encapsulation efficiency for non-esterified agavins was 0.8%, while acetylation-modified agavins and palmitoylate had efficiencies of 17% and 21.5%, respectively. In contrast, the unmodified encapsulates released a large amount of the drug (85%) under gastric conditions, while the acetylated and palmitoylated agavins capsules had a release of 57.5% and 28%, respectively. This study showed that it is important to modify agavins to change their polarity and use them as vehicles for drug delivery.

Due to their technological and physiological properties previously discussed in this chapter, the use of agavins as food ingredients is attracting much attention from both research and industry. Several scientific articles have been published in the last few years pointing out the potential benefits of these products. However, in addition to these, in a highly competitive sector such as the one on sugars/prebiotics, the raw material availability, manufacturing costs, and the interest and acceptability of consumers are of utmost importance in terms of market success. In this context, as discussed below, agavins offer an exciting opportunity. It is also important to underline that, although not extensively discussed in this chapter, in addition to their use as food ingredients agavins may also offer biotechnological opportunities such as their use as matrixes for the microencapsulation of other bioactive compounds or probiotic microorganisms (Bermúdez-Quiñones et al. 2021).

The source of agavins, the agave plant, is produced in arid climates with countries such as Mexico or Colombia as the leading producers. There, the agave is used for producing distilled spirituous drinks. Just taking Mexico as an example, the yearly production reaches 1500 metric tons; therefore, after distillation hundreds of tons of agavins by-products from the liquor industry are available. In this context, using these products for agavins extraction offers opportunities for increasing the sustainability of the sector and its economic turnover. Furthermore, as indicated in section 10.2 of this chapter, different cost-effective extraction methods are available for obtaining agavin-rich fractions from agave and agave industry by-products. Therefore, two critical criteria are fulfilled for developing agavins products for food applications: the availability of raw materials and the existence of cost-effective manufacturing processes for obtaining enriched or purified fractions.

Nevertheless, another critical requirement must also be achieved for market success, the acceptance and interest of consumers in the product. The accumulation of evidence on the potential physiological benefits of agavins consumption and the data on their prebiotic potential becoming available in the last years (previously discussed in sections 10.3 and 10.4 of this chapter) point out these compounds as interesting functional ingredients. In a context where the global prebiotics market is expected to reach 8.54 billion US dollars by 2025 (Statista, 2019) the future for novel functional prebiotics results is promising. The continuously increasing interest of consumers and authorities in health-promoting foods paves the way for the continuous growth of this market sector in the next few years. Moreover, consumers already know and well accept agave, as underlined by the extensive use of agave-derived products as sugar substitutes. After stevia, sucralose, saccharine, and, at the same level as aspartame, agave is among the top seller sugar replacers, making agave a product and ingredient source already well recognized by consumers. In addition to the consumer's awareness of agave, their perception and willingness to accept it are also of great relevance. In this regard, a recent market study found agave-based products as the fourth sweetener more likely to be consumed, just after honey, brown sugar, and sucrose, and ahead of other sweeteners, including stevia or polyols, among others (Statista, 2021, 2022).

With consumer interest in health-promoting foods, availability of affordable raw materials, cost-effective manufacturing processes, and consumer awareness and acceptance, there is good current and future potential market for developing agavins as food ingredients. However, as is always the case for functional foods, given the restrictive legislative framework existing in some countries (e.g., in Europe) the future success of these products will highly likely be affected by the confirmation of the promising preliminary results available so far in well designed and conducted human intervention studies. Also, as is always the case for foods, including the ingredients in products with good organoleptic properties and adapted to the needs of consumers will constitute a primary target for the food industry interested in such ingredients.

10.6 CONCLUSION

The prebiotic potential of agavins is supported daily as studies are conducted on their influence on gut microbiota. The interrelation between agavins and host gut microbiota is being explained. However, the mechanisms behind microbial interactions still require significant understanding. Several in vitro and in vivo studies have shown that the regular consumption of agavins improves the growth of beneficial colon bacteria and reduces several problems related to glucose metabolism, obesity, metabolic syndrome, calcium absorption, inflammation, oxidative stress, cancer, and mental disorders. Despite these efforts, there is a need to increase the number of clinical studies in populations with different diseases that include agavins in their daily diet or with a therapeutic dose. In addition, it is necessary to conduct a more significant number of nutritional intervention studies related to the effects of agavins addition that involve their nutritional influence in specific populations.

Another significant issue that needs to be addressed is the establishment of standard procedures to demonstrate the prebiotic potential of different agavins via in vitro or in vivo tests. In this sense, there is enough information to warrant the use and safety of agavins as a prebiotic and functional ingredient. However, it is still to be seen how these functional properties of the native molecule are altered by chemically modifying it by replacing its functional groups, such as acetylation. Another essential aspect of agavins is the physicochemical and nutritional characteristics that they contribute to food products. According to this chapter, its biological and techno-functional properties depend on the source, average DP, concentration, processing, storage conditions, and interaction with other ingredients in some food matrices. Therefore, in-depth knowledge about the composition profile, distribution, and type of branching structures the different Agave species present and the relationship between the chemical structure of the agavins and their biological and technological approaches is fundamental.

This knowledge will allow generating strategies in the growing industry of agavins to optimize the standardization of extraction processes to obtain fractions that guarantee the benefits of the consumer by including them as part of their diet. Additionally, these efforts will let it open niches of opportunity for innovation in the design of more straightforward, cheaper, and more environmentally sustainable extraction processes. Finally, with the expansion of the prebiotics market, especially in the United States and Japan, the appearance of agavins could facilitate this demand.

REFERENCES

Aldrate-Herrera, P.I., López, M.G., Medina-Torres, L., Ragazzo-Sánchez, J.A., Calderón-Santoyo, M., González-Ávila, M., & Ortiz Basurto, R.I. (2019). Physicochemical composition and apparent degree of polymerization of fructans in five wild agave varieties: potential industrial use. *Foods, 8*(9), 404. https://doi.org/10.3390/foods8090404

Allsopp, P., Possemiers, S., Campbell, D., Oyarzábal, I.S., Gill, C., & Rowland, I. (2013). An exploratory study into the putative prebiotic activity of fructans isolated from *Agave angustifolia* and the associated anticancer activity. *Anaerobe, 22*, 38–44. https://doi.org/10.1016/j.anaerobe.2013.05.006

Alvarado-Jasso, G.M., Camacho-Díaz, B.H., Arenas-Ocampo, M.L., Jiménez-Ferrer, J.E., Mora-Escobedo, R., & Osorio-Díaz, P. (2020). Prebiotic effects of a mixture of agavins and green banana flour in a mouse model of obesity. *Journal of Functional Foods, 64*,103685. https://doi.org/10.1016/j.jff.2019.103685

Apolinário, A.C., de Carvalho, E.M., de Lima-Damasceno, B.P.G., da Silva, P.C.D., Converti, A., Pessoa, A., & da Silva, J.A. (2017). Extraction, isolation and characterization of inulin from *Agave sisalana* boles. *Industrial Crops and Products, 108*(July), 355–362. https://doi.org/10.1016/j.indcrop.2017.06.045

Arrizon, J., Morel, S., Gschaedler, A., & Monsan, P. (2010). Comparison of the water-soluble carbohydrate composition and fructan structures of *Agave tequilana* plants of different ages. *Food Chemistry, 122*(1), 123–130. https://doi.org/10.1016/j.foodchem.2010.02.028

Ávila-Fernández, Á., Galicia-Lagunas, N., Rodríguez-Alegría, M.E., Olvera, C., & López-Munguía, A. (2011). Production of functional oligosaccharides through limited acid hydrolysis of agave fructans. *Food Chemistry, 129*(2), 380–386. https://doi.org/10.1016/j.foodchem.2011.04.088

Banguela, A., & Hernández, L. (2006). Fructans: from natural sources to transgenic plants. *Biotecnologia Aplicada, 23*, 202–210. https://doi.org/10.1024/0301-1526.32.1.54

Barcelo, A., Claustre, J., Moro, F., Chayvialle, J.A., Cuber, J.C., & Plaisancié, P. (2000). Mucin secretion is modulated by luminal factors in the isolated vascularly perfused rat colon. *Gut, 46*(2), 218–224. https://doi.org/10.1136/gut.46.2.218

Barros, K.A., Inaba, M., Martins, A.O., & Sulpice, R. (2022). High-throughput extraction and enzymatic determination of sugars and fructans in fructan-accumulating plants. *Methods in Molecular Biology (Clifton, N.J.), 2398*, 107–119. https://doi.org/10.1007/978-1-0716-1912-4_10

Bauer, E., Williams, B.A., Verstegen, M.W.A., & Mosenthin, R. (2006). Fermentable carbohydrates: potential dietary modulators of intestinal physiology, microbiology and immunity in pigs. In Mosenthin, R., Zentek, J., & Żebrowska, T. (Eds.), *Biology of Growing Animals* (Vol. 4, pp. 33–63). Elsevier.

Bermúdez-Quiñones, G., Ochoa-Martínez, L.A., Gallegos-Infante, J.A., Rutiaga-Quiñones, O.M., Lara-Ceniceros, T.E., Delgado-Licon, E., & González-Herrera, S.M. (2021). Synbiotic microcapsules using agavins and inulin as wall materials for *Lactobacillus casei* and *Bifidobacterium breve*: viability, physicochemical properties, and resistance to *in vitro* oro-gastrointestinal transit. *Journal of Food Processing and Preservation, 45*(12), 1–11. https://doi.org/10.1111/jfpp.16106

Blaut, M. (2002). Relationship of prebiotics and food to intestinal microflora. *European Journal of Nutrition 41*(SUPPL 1), 11–16. https://doi.org/10.1007/s00394-002-1102-7

Buitrago-Arias, C., Londoño-Moreno, A., Avila-Reyes, S.V., Arenas-Ocampo, M.L., Alamilla-Beltran, L., Jimenez-Aparicio, A.R., & Camacho-Diaz, B.H. (2021). Evaluation of the fermentation of acetylated agave fructans (agavins), with *Saccharomyces boulardii* as a probiotic. *Revista Mexicana de Ingeniería Química, 20*(3), 1–13. https://doi.org/10.24275/rmiq/poly2533

Castillo-Andrade, A.I., Rivera-Bautista, C., Godínez-Hernández, C., Ruiz-Cabrera, M.A., Fuentes-Ahumada, C., García-Chávez, E., & Grajales-Lagunes, A. (2018). Physiometabolic effects of *Agave salmiana* fructans evaluated in Wistar rats. *International Journal of Biological Macromolecules, 108*, 1300–1309. https://doi.org/10.1016/j.ijbiomac.2017.11.043

Castillo-Andrade, A.I., Rivera-Bautista, C., Ruiz-Cabrera, M.A., Soria-Guerra, R.E., García-Chávez, E., Fuentes-Ahumada, C., & Grajales-Lagunes, A. (2019). *Agave salmiana* fructans as gut health promoters: prebiotic activity and inflammatory response in Wistar healthy rats. *International Journal of Biological Macromolecules, 136*(9), 785–795. https://doi.org/10.1016/j.ijbiomac.2019.06.045

Chambers, L., McCrickerd, K., & Yeomans, M.R. (2015). Optimising foods for satiety. *Trends in Food Science and Technology, 41*(2), 149–160. https://doi.org/10.1016/j.tifs.2014.10.007

Chater, P.I., Wilcox, M.D., Pearson, J.P., & Brownlee, I.A. (2015). The impact of dietary fibres on the physiological processes governing small intestinal digestive processes. *Bioactive Carbohydrates and Dietary Fibre, 6*(2), 117–132. https://doi.org/10.1016/j.bcdf.2015.09.002

Čižauskaite, U., Jakubaityte, G., Žitkevičius, V., & Kasparavičiene, G. (2019). Natural ingredients-based gummy bear composition designed according to texture analysis and sensory evaluation *in vivo*. *Molecules, 24*(7). https://doi.org/10.3390/molecules24071442

Crispín-Isidro, G., Lobato-Calleros, C., Espinosa-Andrews, H., Alvarez-Ramirez, J., & Vernon-Carter, E.J. (2015). Effect of inulin and agave fructans addition on the rheological, microstructural and sensory properties of reduced-fat stirred yogurt. *LWT – Food Science and Technology, 62*(1), 438–444. https://doi.org/10.1016/j.lwt.2014.06.042

Delgadillo, E., Corona-González, R.I., Toriz, G., Contreras, H.J., Sadeghifar, H., Baobing, W., Yang, G., Lucia, L.A., & Delgado, E. (2015). Coacervated liposoluble fructan-based host-guest microspheres as unique drug delivery materials. *RSC Advances, 5*(83), 67759–67766. https://doi.org/10.1039/c5ra10164f

Espinosa-Andrews, H., & Rodríguez-Rodríguez, R. (2018). Water state diagram and thermal properties of fructans powders. *Journal of Thermal Analysis and Calorimetry, 132*(1), 197–204. https://doi.org/10.1007/s10973-017-6868-1

Espinosa-Andrews, H., & Urías-Silvas, J.E. (2012). Thermal properties of agave fructans (*Agave tequilana* Weber var. Azul). *Carbohydrate Polymers, 87*(4), 2671–2676. https://doi.org/10.1016/j.carbpol.2011.11.053

Espinosa-Andrews, H., Urías-Silvas, J.E., & Morales-Hernández, N. (2021). The role of agave fructans in health and food applications: a review. *Trends in Food Science and Technology, 114*, 585–598. https://doi.org/10.1016/j.tifs.2021.06.022

Franco-Robles, E., & López, M.G. (2015). Implication of fructans in health: immunomodulatory and antioxidant mechanisms. *Scientific World Journal, 2015*, ID 289267, 585-598. https://doi.org/10.1155/2015/289267

Franco-Robles, E., & López, M.G. (2016). Agavins increase neurotrophic factors and decrease oxidative stress in the brains of high-fat diet-induced obese mice. *Molecules, 21*(8), 998. https://doi.org/10.3390/molecules21080998

Franco-Robles, E., Ramírez-Emiliano, J., & López, M.G. (2019). Agave fructans and oligofructose decrease oxidative stress in brain regions involved in learning and memory of overweight mice. *Natural Product Research, 33*(10), 1527–1530. https://doi.org/10.1080/14786419.2017.1423297

García-Contreras, A.A., Vásquez-Garibay, E.M., Sánchez-Ramírez, C.A., Fafutis-Morris, M., & Delgado-Rizo, V. (2020). *Lactobacillus reuteri* dsm 17938 and agave inulin in children with cerebral palsy and chronic constipation: a double-blind randomized placebo controlled clinical trial. *Nutrients, 12*(10), 1–12. https://doi.org/10.3390/nu12102971

García-Curbelo, Y., Bocourt-Salabarría, R., Rodríguez-Sanchez, B., Albelo-Dorta, N., Albelo-González, A., Herrera-Villafranca, M., & López-Pérez, M.G.(2020). Prebiotic effects of agavins in poultry. *Livestock Research for Rural Development, 32*(8).

García-Curbelo, Y., López, M.G., Bocourt, R., Collado, E., Albelo, N., & Nuñez, O. (2015). Structural characterization of fructans from *Agave fourcroydes* (Lem.) with potential as prebiotic. *Cuban Journal of Agricultural Science, 49*(1), 75–80.

García-Gamboa, R., Ortiz-Basurto, R.I., Calderón-Santoyo, M., Bravo-Madrigal, J., Ruiz-Álvarez, B.E., & González-Avila, M. (2018). *In vitro* evaluation of prebiotic activity, pathogen inhibition and enzymatic metabolism of intestinal bacteria in the presence of fructans extracted from agave: a comparison based on polymerization degree. *LWT – Food Science and Technology, 92*(February), 380–387. https://doi.org/10.1016/j.lwt.2018.02.051

García-García, A.B., Ochoa-Martínez, L.A., Lara-Ceniceros, T.E., Rutiaga-Quiñones, O.M., Rosas-Flores, W., & González-Herrera, S.M. (2019). Changes in the microstructural, textural, thermal and sensory properties of apple leathers containing added agavins and inulin. *Food Chemistry, 30*, 124590. https://doi.org/10.1016/j.foodchem.2019.03.143

García-López, E.G., Sáyago-Ayerdi, S.G., Vivar-Vera, M.D.L.A., García-Magaña, M.D.L., Aldrate-Herrera, P.I., & Montalvo-González, E. (2017). Guava purees with addition of agave fructans and natural sweeteners as potential functional products. *Journal of Food and Nutrition Research, 56*(3), 265–276.

Gibson, G.R., Hutkins, R., Sanders, M.E., Prescott, S.L., Reimer, R.A., Salminen, S.J., Scott, K., Stanton, C., Swanson, K.S., Cani, P.D., Verbeke, K., & Reid, G. (2017). Expert consensus document: the International Scientific Association for Probiotics and Prebiotics (ISAPP) consensus statement on the definition and scope of prebiotics. *Nature Reviews Gastroenterology & Hepatology, 14*(8), 491–502. https://doi.org/10.1038/nrgastro.2017.75

Gibson, G.R., Probert, H.M., Loo, J. Van, Rastall, R.A., & Roberfroid, M.B. (2004). Dietary modulation of the human colonic microbiota: updating the concept of prebiotics. *Nutrition Research Reviews, 17*(2), 259–275. https://doi.org/10.1079/nrr200479

Gibson, G.R., & Roberfroid, M.B. (1995). Dietary modulation of the human colonic microbiota: introducing the concept of prebiotics. *Journal of Nutrition, 125*(6), 1401–1412. https://doi.org/10.1093/jn/125.6.1401

Gomez, E., Tuohy, K.M., Gibson, G.R., Klinder, A., & Costabile, A. (2010). *In vitro* evaluation of the fermentation properties and potential prebiotic activity of agave fructans. *Journal of Applied Microbiology, 108*(6), 2114–2121. https://doi.org/10.1111/j.1365-2672.2009.04617.x

González-Ávila, M., Prado-Ramirez, R., Flores-Montano, J.L., Perez-Martinez, F.J., Gaspar-Ramirez, O., & Alonso-Segura, D. (2014). Evaluation of prebiotic potential of agave fructans from different regions of Colima and Zacatecas, Mexico. *International Journal of Probiotics and Prebiotics, 9*(3), 93–100.

González-Hernández, L.A., Jave-Suarez, L.F., Fafutis-Morris, M., Montes-Salcedo, K.E., Valle-Gutierrez, L.G., Campos-Loza, A.E., Enciso-Gómez, L.F., & Andrade-Villanueva, J.F. (2012). Synbiotic therapy decreases microbial translocation and inflammation and improves immunological status in HIV-infected patients: a double-blind randomized controlled pilot trial. *Nutrition Journal, 11*(90). https://doi.org/10.1186/1475-2891-11-90

González-Herrera, S.M., Rutiaga-Quiñones, O.M., Aguilar, C.N., Ochoa-Martínez, L.A., Contreras-Esquivel, J.C., López, M.G., & Rodríguez-Herrera, R. (2016). Dehydrated apple matrix supplemented with agave fructans, inulin, and oligofructose. *LWT – Food Science and Technology, 65*, 1059–1065. https://doi.org/10.1016/j.lwt.2015.09.037

González-Herrera, S.M., Simental-Mendía, L.E., López, M.G., Rocha-Guzmán, N.E., Rutiaga-Quiñones, O.M., Rodríguez-Herrera, R., & Gamboa-Gómez, C.I. (2019). Effect of agave fructans on the production of short chain fatty acid in mice. *Food Science and Biotechnology*, *28*(5), 1493–1498. https://doi.org/10.1007/s10068-019-00572-1

Gracia, M.I., Tinoco, M.M., Rivera, H.M., Sanchez, B.F., Tapia, P.G., Altamirano, L.M., Romero, R.L., & García, O.L. (2013). Acute toxicity and genotoxic evaluation of Metlin® and Metlos® (organic agave fructans). *Food and Nutrition Sciences*, *04*(07), 106–112. https://doi.org/10.4236/fns.2013.47a013

Guyenet, S.J., & Schwartz, M.W. (2012). Regulation of food intake, energy balance, and body fat mass: implications for the pathogenesis and treatment of obesity. *Journal of Clinical Endocrinology and Metabolism*, *97*(3), pp. 745–755). https://doi.org/10.1210/jc.2011-2525

Hernández-González, S.O., Bricio-Ramírez, R.I., & Ramos-Zavala, M.G. (2016). Effect of Inulin from *Agave tequilana* Weber blue variety on the metabolic profile of overweight and obese dyslipidemic patients. *Journal of Clinical Trials*, *06*(254), 2167–0870. https://doi.org/10.4172/2167-0870.1000254

Holscher, H.D., Bauer, L.L., Gourineni, V., Pelkman, C.L., Fahey, G.C., & Swanson, K.S. (2015). Agave inulin supplementation affects the fecal microbiota of healthy adults participating in a randomized, double-blind, placebo-controlled, crossover trial. *Journal of Nutrition*, *145*(9), 2025–2032. https://doi.org/10.3945/jn.115.217331

Holscher, H.D., Doligale, J.L., Bauer, L.L., Gourineni, V., Pelkman, C.L., Fahey, G.C., & Swanson, K.S. (2014). Gastrointestinal tolerance and utilization of agave inulin by healthy adults. *Food and Function*, *5*(6), 1142–1149. https://doi.org/10.1039/c3fo60666j

Hu, J., Lin, S., Zheng, B., & Cheung, P.C.K. (2018). Short-chain fatty acids in control of energy metabolism. *Critical Reviews in Food Science and Nutrition*, *58*(8), 1243–1249. https://doi.org/10.1080/10408398.2016.1245650

Huazano-García, A., & López, M.G. (2015). Agavins reverse the metabolic disorders in overweight mice through the increment of short chain fatty acids and hormones. *Food and Function*, *6*(12), 3720–3727. https://doi.org/10.1039/c5fo00830a

Huazano-García, A., Shin, H., & López, M.G. (2017). Modulation of gut microbiota of overweight mice by agavins and their association with body weight loss. *Nutrients*, *9*(9), 821. https://doi.org/10.3390/nu9090821

Huazano-García, A., Silva-Adame, M.B., Vázquez-Martínez, J., Gastelum-Arellanez, A., Sánchez-Segura, L., & López, M.G. (2020). Highly branched neo-fructans (agavins) attenuate metabolic endotoxemia and low-grade inflammation in association with gut microbiota modulation on high-fat diet-fed mice. *Foods*, *9*(12), 1792. https://doi.org/10.3390/foods9121792

Ignot-Gutiérrez, A., Ortíz-Basurto, R.I., García-Barradas, O., Díaz-Ramos, D.I., & Jiménez-Fernández, M. (2020). Physicochemical and functional properties of native and modified agave fructans by acylation. *Carbohydrate Polymers*, *245*(June), 116529. https://doi.org/10.1016/j.carbpol.2020.116529

Jardines, A.P., Arjona-Román, J.L., Severiano-Pérez, P., Totosaus-Sánchez, A., Fiszman, S., & Escalona-Buendía, H.B. (2020). Agave fructans as fat and sugar replacers in ice cream: sensory, thermal and texture properties. *Food Hydrocolloids*, *108*, *106032*.. https://doi.org/10.1016/j.foodhyd.2020.106032

Jasso-Padilla, I., Juárez-Flores, B., Alvarez-Fuentes, G., De la Cruz-Martínez, A., González-Ramírez, J., Moscosa-Santillán, M., González-Chávez, M., Oros-Ovalle, A.C., Prell, F., Czermak, P., & Martínez-Gutiérrez, F. (2017). Effect of prebiotics of *Agave salmiana* fed to healthy Wistar rats. *Journal of the Science of Food and Agriculture*, *97*(2), 556–563. https://doi.org/10.1002/jsfa.7764

Jenkins, D.J.A., Kendall, C.W.C., & Vuksan, V. (1999). Inulin, oligofructose and intestinal function. *Journal of Nutrition*, *129*(7 SUPPL.), 1431S–1433S. https://doi.org/10.1093/jn/129.7.1431s

Jimenez-Sánchez, D.E., Calderón-Santoyo, M., Ortíz-Basurto, R.I., Bautista-Rosales, P.U., & Ragazzo-Sánchez, J.A. (2018). Effect of maltodextrin reduction and native agave fructans addition on the physicochemical properties of spray-dried mango and pineapple juices. *Food Science and Technology International*, *24*(6), 519–532. https://doi.org/10.1177/1082013218769168

Kanoski, S.E., & Davidson, T.L. (2011). Western diet consumption and cognitive impairment: links to hippocampal dysfunction and obesity. *Physiology and Behavior*, *103*(1), 59–68. https://doi.org/10.1016/j.physbeh.2010.12.003

Kilua, A., Pelpolage, S., Goto, A., Nakayama, Y., Kitazono, E., Toyohara, K., Nagata, R., Fukuma, N., Han, K.H., & Fukushima, M. (2021). Deciphering the colonic fermentation characteristics of agavin and digestion-resistant maltodextrin in a simulated batch fermentation system. *International Journal of Biological Macromolecules, 189*(August), 151–159. https://doi.org/10.1016/j.ijbiomac.2021.08.063

Kralj, S., Leeflang, C., Sierra, E.I., Kempiński, B., Alkan, V., & Kolkman, M. (2018). Synthesis of fructooligosaccharides (FosA) and inulin (InuO) by GH68 fructosyltransferases from *Bacillus agaradhaerens* strain WDG185. *Carbohydrate Polymers, 179*, 350–359. https://doi.org/10.1016/j.carbpol.2017.09.069

Londoño-Moreno, A., Buitrago-Arias, C., Ávila-Reyes, S., Jiménez-Aparicio, A., Arenas-Ocampo, M., & Camacho-Díaz, B.H. (2020). Crecimiento de *Saccharomyces boulardii* con agavinas acetiladas como fuente de carbono. *Revista Tendencias En Docencia e Investigación En Química, 6*(6), 621–626.

López, M.G., Mancilla-Margalli, N.A., & Mendoza-Diaz, G. (2003). Molecular structures of fructans from *Agave tequilana* weber var. azul. *Journal of Agricultural and Food Chemistry, 51*(27), 7835–7840. https://doi.org/10.1021/jf030383v

López Pérez, M.G., & Urías-Silvas, J.E. (2007). Prebiotic effect of fructans from Agave, Dasylirion and Nopal. *Acta Horticulturae, 744*, 397–404. https://doi.org/10.17660/ActaHortic.2007.744.45

López-Velázquez, G., Parra-Ortiz, M., De la Mora, I., García-Torres, I., Enríquez-Flores, S., Alcántara-Ortigoza, M.A., González-del Angel, A., Velázquez-Aragón, J., Ortiz-Hernández, R., Cruz-Rubio, J.M., Villa-Barragán, P., Jiménez-Gutiérrez, C., & Gutiérrez-Castrellón, P. (2015). Effects of fructans from Mexican agave in newborns fed with infant formula: a randomized controlled trial. *Nutrients, 7*(11), 8939–8951. https://doi.org/10.3390/nu7115442

Mancilla-Margalli, N.A., & López, M.G. (2006). Water-soluble carbohydrates and fructan structure patterns from Agave and Dasylirion species. *Journal of Agricultural and Food Chemistry, 54*(20), 7832–7839. https://doi.org/10.1021/jf060354v

Márquez-Aguirre, A.L., Camacho-Ruiz, R.M., Arriaga-Alba, M., Padilla-Camberos, E., Kirchmayr, M.R., Blasco, J.L., & González-Ávila, M. (2013). Effects of *Agave tequilana* fructans with different degree of polymerization profiles on the body weight, blood lipids and count of fecal *Lactobacilli/Bifidobacteria* in obese mice. *Food and Function, 4*(8), 1237–1244. https://doi.org/10.1039/c3fo60083a

Márquez-Aguirre, A.L., Camacho-Ruiz, R.M., Gutiérrez-Mercado, Y.K., Padilla-Camberos, E., González-Ávila, M., Gálvez-Gastélum, F.J., Díaz-Martínez, N.E., & Ortuño-Sahagún, D. (2016). Fructans from *Agave tequilana* with a lower degree of polymerization prevent weight gain, hyperglycemia and liver steatosis in high-fat diet-induced obese mice. *Plant Foods for Human Nutrition, 71*(4), 416–421. https://doi.org/10.1007/s11130-016-0578-x

Martínez-Abundis, E., Barrera-Durán, C., González-Ortiz, M., & Hernández-Salazar, E. (2016). Efecto de simvastatina más inulina en comparación con simvastatina más ezetimiba en el tratamiento de la dislipidemia mixta. *Gaceta Médica de México.*

Martínez-Gutiérrez, F., Ratering, S., Juárez-Flores, B., Godinez-Hernandez, C., Geissler-Plaum, R., Prell, F., Zorn, H., Czermak, P., & Schnell, S. (2017). Potential use of *Agave salmiana* as a prebiotic that stimulates the growth of probiotic bacteria. *LWT, 84*, 151–159. https://doi.org/10.1016/j.lwt.2017.05.044

Mellado-Mojica, E., González de la Vara, L.E., & López, M.G. (2017). Fructan active enzymes (FAZY) activities and biosynthesis of fructooligosaccharides in the vacuoles of *Agave tequilana* Weber Blue variety plants of different age. *Planta, 245*(2), 265–281. https://doi.org/10.1007/s00425-016-2602-7

Michel-Cuello, C., Ortiz-Cerda, I., Moreno-Vilet, L., Grajales-Lagunes, A., Moscosa-Santillán, M., Bonnin, J., González-Chávez, M.M., & Ruiz-Cabrera, M. (2012). Study of enzymatic hydrolysis of fructans from *Agave salmiana* characterization and kinetic assessment. *The Scientific World Journal, 2012.* https://doi.org/10.1100/2012/863432

Miramontes-Corona, C., Escalante, A., Delgado, E., Corona-González, R.I., Vázquez-Torres, H., & Toriz, G. (2020). Hydrophobic agave fructans for sustained drug delivery to the human colon. *Reactive and Functional Polymers, 146*(September 2019). https://doi.org/10.1016/j.reactfunctpolym.2019.104396

Mizrahy, S., & Peer, D. (2012). Polysaccharides as building blocks for nanotherapeutics. *Chemical Society Reviews, 41*(7), 2623–2640. https://doi.org/10.1039/c1cs15239d

Morales-Hernández, J.A., Chanona-Pérez, J.J., Villanueva-Rodríguez, S.J., Perea-Flores, M.J., & Urías-Silvas, J.E. (2019). Technological and structural properties of oat cookies incorporated with fructans (*Agave tequilana* Weber). *Food Biophysics, 14*(4), 415–424. https://doi.org/10.1007/s11483-019-09589-9

Mudgil, D., & Barak, S. (2013). Composition, properties and health benefits of indigestible carbohydrate polymers as dietary fiber: A review. *International Journal of Biological Macromolecules, 61*(3), 1–6. https://doi.org/10.1016/j.ijbiomac.2013.06.044

Mueller, M., Reiner, J., Fleischhacker, L., Viernstein, H., Loeppert, R., & Praznik, W. (2016). Growth of selected probiotic strains with fructans from different sources relating to degree of polymerization and structure. *Journal of Functional Foods, 24*, 264–275. https://doi.org/10.1016/j.jff.2016.04.010

Murphy, K.G., & Bloom, S.R. (2006). Gut hormones and the regulation of energy homeostasis. *Nature, 444*(7121), 854–859. https://doi.org/10.1038/nature05484

Nava-Cruz, N.Y., Medina-Morales, M.A., Martinez, J.L., Rodriguez, R., Aguilar, C.N., Nava-Cruza, N.Y., Medina-Moralesa, M.A., Martineza, J.L., & Rodrigueza, R. (2015). Agave biotechnology: an overview. *Critical Reviews in Biotechnology, 35*(4), 546–559. https://doi.org/10.3109/07388551.2014.923813

Olvera-Carranza, C., Ávila-Fernandez, A., Bustillo-Armendáriz, G.R., & López-Munguía, A. (2015). Processing of fructans and oligosaccharides from agave plants. In Preedy, V., (Eds.), *Processing and impact on active components in food* (pp. 121–129). Academic Press. https://doi.org/10.1016/B978-0-12-404699-3.00015-9

Ortíz-Basurto, R.I., Rubio-Ibarra, M.E., Ragazzo-Sanchez, J.A., Beristain, C.I., & Jiménez-Fernández, M. (2017). Microencapsulation of *Eugenia uniflora* L. juice by spray drying using fructans with different degrees of polymerisation. *Carbohydrate Polymers, 175*, 603–609. https://doi.org/10.1016/j.carbpol.2017.08.030

Ozuna, C., Trueba-Vásquez, E., Moraga, G., Llorca, E., & Hernando, I. (2020). Agave syrup as an alternative to sucrose in muffins: impacts on rheological, microstructural, physical and sensorial properties. *Foods, 9*(7), 895.

Padilla-Camberos, E., Barragán-Álvarez, C.P., Diaz-Martinez, N.E., Rathod, V., & Flores-Fernández, J.M. (2018). Effects of agave fructans (*Agave tequilana* weber var. azul) on body fat and serum lipids in obesity. *Plant Foods for Human Nutrition, 73*(1), 34–39. https://doi.org/10.1007/s11130-018-0654-5

Palatnik, D.R., Aldrate-Herrera, P.I., Rinaldoni, A.N., Ortíz-Basurto, R.I., & Campderrós, M.E. (2017). Development of reduced-fat cheeses with the addition of agave fructans. *International Journal of Dairy Technology, 70*(2), 212–219. https://doi.org/10.1111/1471-0307.12334

Park, H.R., Park, M., Choi, J., Park, K.Y., Chung, H.Y., & Lee, J. (2010). A high-fat diet impairs neurogenesis: involvement of lipid peroxidation and brain-derived neurotrophic factor. *Neuroscience Letters, 482*(3), 235–239. https://doi.org/10.1016/j.neulet.2010.07.046

Pérez-López, A.V., & Simpson, J. (2020). The sweet taste of adapting to the desert: fructan metabolism in agave species. *Frontiers in Plant Science, 11*. https://doi.org/10.3389/fpls.2020.00324

Pérez-López, A.V., Simpson, J., Clench, M.R., Gomez-Vargas, A.D., & Ordaz-Ortiz, J.J. (2021). Localization and composition of fructans in stem and rhizome of *Agave tequilana* Weber var. azul. *Frontiers in Plant Science, 11*(January), 1–16. https://doi.org/10.3389/fpls.2020.608850

Phard, D.S., & Dince, N. (2009). *Sports drink containing rice syrup and agave nectar US 2009 /0220638A1.*

Pintor-Jardines, A., Arjona-Román, J.L., Totosaus-Sánchez, A., Severiano-Pérez, P., González-González, L.R., & Escalona-Buendia, H.B. (2018). The influence of agave fructans on thermal properties of low-fat, and low-fat and sugar ice cream. *LWT – Food Science and Technology, 93*(February), 679–685. https://doi.org/10.1016/j.lwt.2018.03.060

Plongbunjong, V., Graidist, P., Knudsen, K.E.B., & Wichienchot, S. (2017). Starch-based carbohydrates display the bifidogenic and butyrogenic properties in pH-controlled faecal fermentation. *International Journal of Food Science and Technology, 52*(12), 2647–2653. https://doi.org/10.1111/ijfs.13553

Popa-Wagner, A., Mitran, S., Sivanesan, S., Chang, E., & Buga, A.M. (2013). ROS and brain diseases: the good, the bad, and the ugly. *Oxidative Medicine and Cellular Longevity*, 2013(1), 963520 https://doi.org/10.1155/2013/963520

Ramnani, P., Costabile, A., Bustillo, A.G.R., & Gibson, G.R. (2015). A randomised, double- blind, cross-over study investigating the prebiotic effect of agave fructans in healthy human subjects. *Journal of Nutritional Science, 4, e10*. https://doi.org/10.1017/jns.2014.68

Ramos-Hernández, J.A., Ragazzo-Sánchez, J.A., Calderón-Santoyo, M., Ortíz-Basurto, R.I., Prieto, C., & Lagaron, J.M. (2018). Use of electrosprayed agave fructans as nanoencapsulating hydrocolloids for bioactives. *Nanomaterials, 8*(11), 1–11. https://doi.org/10.3390/nano8110868

Regalado-Rentería, E., Aguirre-Rivera, J.R., Godínez-Hernández, C.I., García-López, J.C., Oros-Ovalle, A.C., Martínez-Gutiérrez, F., Martinez-Martinez, M., Ratering, S., Schnell, S., Ruíz-Cabrera, M.Á., & Juárez-Flores, B. (2020). Effects of agave fructans, inulin, and starch on metabolic syndrome aspects in healthy Wistar rats. *ACS Omega, 5*(19), 10740–10749. https://doi.org/10.1021/acsomega.0c00272

Rendón-Huerta, J.A., Juárez-Flores, B., Pinos-Rodríguez, J.M., Aguirre-Rivera, J.R., & Delgado-Portales, R.E. (2012). Effects of different sources of fructans on body weight, blood metabolites and fecal bacteria in normal and obese non-diabetic and diabetic rats. *Plant Foods for Human Nutrition, 67*(1), 64–70. https://doi.org/10.1007/s11130-011-0266-9

Ríos Covián, D., Ruas-Madiedo, P., Margolles, A., Gueimonde, M., De los Reyes-Gavilán, C.G., & Salazar, N. (2016). Intestinal short chain fatty acids and their link with diet and human health. *Frontiers in Microbiology, 7*(FEB), 1–9. https://doi.org/10.3389/fmicb.2016.00185

Rivera-Huerta, M., Lizárraga-Grimes, V.L., Castro-Torres, I.G., Tinoco-Méndez, M., Macías-Rosales, L., Sánchez-Bartéz, F., Tapia-Pérez, G.G., Romero-Romero, L., & Gracia-Mora, M.I. (2017). Functional effects of prebiotic fructans in colon cancer and calcium metabolism in animal models. *BioMed Research International, 2017, 9758982.* https://doi.org/10.1155/2017/9758982

Roberfroid, M.B. (2007). Inulin-type fructans: Functional food ingredients. *The Journal of Nutrition. 137*(11), 2493–2502.

Sánchez-Madrigal, M., Quintero-Ramos, A., Amaya-Guerra, C.A., Meléndez-Pizarro, C.O., Castillo-Hernández, S.L., & Aguilera-González, C.J. (2019). Effect of agave fructans as carrier on the encapsulation of blue corn anthocyanins by spray drying. *Foods, 8*(7), 268. https://doi.org/10.3390/foods8070268

Sanderson, I.R. (2007). Dietary modulation of GALT. *Journal of Nutrition, 137*(11). https://doi.org/10.1093/jn/137.11.2557s

Santiago-García, P.A., & López, M.G. (2014). Agavins from *Agave angustifolia* and *Agave potatorum* affect food intake, body weight gain and satiety-related hormones (GLP-1 and ghrelin) in mice. *Food and Function, 5*(12), 3311–3319. https://doi.org/10.1039/c4fo00561a

Santiago-García, P.A., Mellado-Mojica, E., León-Martínez, F.M., Dzul-Cauich, J.G., López, M.G., & García-Vieyra, M.I. (2021). Fructans (agavins) from *Agave angustifolia* and *Agave potatorum* as fat replacement in yogurt: Effects on physicochemical, rheological, and sensory properties. *LWT – Food Science and Technology, 140*(April), 110846. https://doi.org/10.1016/j.lwt.2020.110846

Santiago-García, P.A., Mellado-Mojica, E., León-Martínez, F.M., & López, M.G. (2017). Evaluation of *Agave angustifolia* fructans as fat replacer in the cookies manufacture. *LWT – Food Science and Technology, 77(April)*, 100–109. https://doi.org/10.1016/j.lwt.2016.11.028

Santisteban, M.M., Qi, Y., Zubcevic, J., Kim, S., Yang, T., Shenoy, V., Cole-Jeffrey, C.T., Lobaton, G.O., Stewart, D.C., Rubiano, A., Simmons, C.S., Garcia-Pereira, F., Johnson, R.D., Pepine, C.J., & Raizada, M.K. (2017). Hypertension-linked pathophysiological alterations in the gut. *Circulation Research, 120*(2), 312–323. https://doi.org/10.1161/CIRCRESAHA.116.309006

Sáyago-Ayerdi, S.G., Zamora-Gasga, V.M., & Venema, K. (2020). Changes in gut microbiota in predigested *Hibiscus sabdariffa* L calyces and Agave (*Agave tequilana* weber) fructans assessed in a dynamic *in vitro* model (TIM-2) of the human colon. *Food Research International, 132*(January), 109036. https://doi.org/10.1016/j.foodres.2020.109036

Slavin, J., & Green, H. (2007). Dietary fibre and satiety. *Nutrition Bulletin, 32*(SUPPL.1), 32–42. https://doi.org/10.1111/j.1467-3010.2007.00603.x

Statista. (2019). *Prebiotics market value worldwide 2014-2025.* www.statista.com/statistics/1053702/prebiotics-market-value-worldwide/

Statista. (2021). *Sweeteners use of consumers by type U.S. 2021.* www.statista.com/statistics/1287641/sweeteners-consumption-by-type-us/

Statista. (2022). *Agave production volume in Mexico 2019.* www.statista.com/statistics/884792/mexico-agave-production-volume/

Stranahan, A.M., Norman, E.D., Lee, K., Cutler, R.G., Telljohann, R.S., Egan, J.M., & Mattson, M.P. (2008). Diet-induced insulin resistance impairs hippocampal synaptic plasticity and cognition in middle-aged rats. *Hippocampus, 18*(11), 1085–1088. https://doi.org/10.1002/hipo.20470

Tungland, B. (2018). Nondigestible Fructans as Prebiotics. In Tungland. B. (Eds.), *Human Microbiota in Health and Disease: From Pathogenesis to Therapy.* (pp. 349–379). Elsevier. https://doi.org/10.1016/C2017-0-01893-1

Urías-Silvas, J.E., Cani, P.D., Delmée, E., Neyrinck, A., López, M.G., & Delzenne, N.M. (2008). Physiological effects of dietary fructans extracted from *Agave tequilana* Gto. and Dasylirion spp. *British Journal of Nutrition, 99*(2), 254–261. https://doi.org/10.1017/S0007114507795338

Velázquez-Martínez, J.R., González-Cervantes, R.M., Hernández-Gallegos, M.A., Campos Mendiola, R., Jiménez Aparicio, A.R., & Arenas Ocampo, M.L. (2014). Prebiotic potential of *Agave angustifolia* haw fructans with different degrees of polymerization. *Molecules, 19*(8), 12660–12675. https://doi.org/10.3390/molecules190812660

Velázquez-Hernández, M.L., Baizabal-Aguirre, V.M., Bravo-Patiño, A., Cajero-Juárez, M., Chávez-Moctezuma, M.P., & Valdez-Alarcón, J.J. (2009). Microbial fructosyltransferases and the role of fructans. *Journal of Applied Microbiology, 106*(6), 1763–1778. https://doi.org/10.1111/j.1365-2672.2008.04120.x

Yang, J., Martínez, I., Walter, J., Keshavarzian, A., & Rose, D.J. (2013). *In vitro* characterization of the impact of selected dietary fibers on fecal microbiota composition and short chain fatty acid production. *Anaerobe, 23*, 74–81. https://doi.org/10.1016/j.anaerobe.2013.06.012

Zamora-Gasga, V.M., Bello-Pérez, L.A., Ortíz-Basurto, R.I., Tovar, J., & Sáyago-Ayerdi, S.G. (2014). Granola bars prepared with *Agave tequilana* ingredients: chemical composition and *in vitro* starch hydrolysis. *LWT – Food Science and Technology, 56*(2), 309–314. https://doi.org/10.1016/j.lwt.2013.12.016

Zannini, E., Waters, D.M., Coffey, A., & Arendt, E.K. (2016). Production, properties, and industrial food application of lactic acid bacteria-derived exopolysaccharides. *Applied Microbiology and Biotechnology, 100*(3), 1121–1135. https://doi.org/10.1007/s00253-015-7172-2

Zhang, L., Bruce-Keller, A.J., Dasuri, K., Nguyen, A.T., Liu, Y., & Keller, J.N. (2009). Diet-induced metabolic disturbances as modulators of brain homeostasis. *Biochimica et Biophysica Acta – Molecular Basis of Disease, 1792*(5), 417–422 https://doi.org/10.1016/j.bbadis.2008.09.006

Zwarg, J., & Ralphs, S. (2015). *Shelf-stable ice cream and ice cream mix and process for making same with agave nectar. No.: US 2015 /0334785 A1.*

Chapter 11

Probiotic Activity of Lactic Acid Bacteria Isolated from Traditional Mexican Fermented Beverages

An Update

María Elena Sánchez-Pardo
Morayma Ramírez-Damián
Cynthia Garfias-Noguez
Alicia Ortiz-Moreno
Edgar Torres-Maravilla
and Luis G. Bermúdez-Humarán

CONTENTS

11.1 INTRODUCTION

The word probiotic is derived from a Greek word that means for life and is used to define viable nonpathogenic microbes and their beneficial effects on hosts. More than a century ago, Elie Metchnikoff (a Russian scientist, Nobel laureate, and professor at the Pasteur Institute in Paris) postulated that lactic acid bacteria (LAB) were beneficial to health and capable of promoting

DOI: 10.1201/9781003332558-13

longevity; today, probiotics are known as live microorganisms that confer a health benefit when administered in adequate amounts (Guarner et al. 2017). Among emerging health concepts in the probiotics field, it has been proposed that inactivated or nonviable probiotic cells, as well as their metabolic byproducts herein, referred to as postbiotics and paraprobiotics, can also provide benefits to the host's health (Martín & Langella, 2019). Two aspects of these new probiotic-related concepts allow us to establish final statements: the biological responses of postbiotics and paraprobiotics have been observed in cell cultures in animal models and have been confirmed by several human trials, and postbiotic and paraprobiotic applications in food, animal feed, and pharmaceutical industries are a promising and acceptable proposal to achieve beneficial effects.

Functional foods are part of the essential diet of the host since it has been confirmed that they provide an adequate nutritional contribution. In addition to being metabolized, they contain active components that convey essential benefits for controlling and regulating host homeostases, such as omega-3 polyunsaturated fatty acids, phytosterols, dietary fiber, and green tea catechin in the treatment of hyperlipidemia (Chen et al. 2014). Even concerning chronic or acute pathology or disease, these active components intervene beneficially; therefore, it is necessary to study how these active compounds help reduce the mechanisms of damage due to the accumulation of reactive oxygen species or the release of free radicals that accumulate in the cell and cause damage. Moreover, their application in functional foods is attractive, such as proposing a green tea with a high percentage of catechin (46%) in the diet of hyperlipidemia patients (Peluso & Serafini, 2016).

In many regions of Mexico, pulque, a natural beverage of artisanal elaboration with low alcohol content, can be a functional food with probiotics as an active compound extracted from its sediment (xaxtle). Probiotic bacteria such as *Lactobacillus plantarum, L. paracasei, L. brevis,* and *L. composti* have been isolated from some regions, including Nanacamilpa, Tlaxcala and Milpa Alta, Mexico City. The pulque contains LAB that can reduce weight loss, as in the case of *L. sanfranciscensis* LBH1068, which attenuated weight loss in a model of chronic colitis in mice. This study demonstrates the probiotic potential of LAB isolated from traditional Mexican beverages (Torres-Maravilla et al. 2016). However, a considerable proportion of the general population does not accept sensory characteristics (appearance, smell, aroma, and taste). A small population consumes pulque, a highly beneficial product prepared artisanal. The sensory characteristics are not widely liked due to their consistency, and, in some cases, the flavour is only accepted and consumed by a small population; this leads to most people preferring a refined and processed product with a high amount of sugar and artificial antioxidants that harm the intestinal microbiota (Ounnas et al. 2017).

Most products of traditional origins, such as pozol, mezcal, tepache, and sweets made with different regional fruits, such as nanchi, mango, zapote, yaca, guamúchil, coyul, and chicozapote, in the region of Guerrero (Mexico) are made in an artisanal way by using raw materials from the region. Bello et al. (1999) determined the correct conservation method and standardization of a suitable procedure for the large-scale processing of functional foods in the presentation of beverages, nectars, or foods such as bread, preserves, cookies, and even sweets.

11.2 FRESH AND TRADITIONAL MEXICAN FERMENTED MILK BEVERAGES

Currently, the dairy industry is a sector in constant growth, having as raw material the milk coming primarily from cows; this has led to the study of its beneficial properties and identification of the beneficial components that make it so nutritious, focusing attention on the probiotics content. In addition, understanding how the feeding of cows affects the intestinal microbiota of the animal to produce the milk of high quality is important, as is the case in a study performed in the municipality of Aculco in the central highlands of Mexico (Celis-Alvarez et al. 2016). The effect of feeding whole barley and black oats to cows and its impacts on milk productivity and composition were studied. Modifying the content of saturated fatty acids and linoleic acid can have effects, as these acids are essential for probiotic bacteria; they could have a beneficial effect on the colonization of the gastric tract and the productivity and increase of milk nutrients (Gómez-Miranda et al. 2021).

The previous findings led to the formulation of a diet reinforced with cereals such as whole barley and black oats in the diet of dairy cows to strengthen their intestinal microbiota, including probiotic bacteria, to increase the nutritional and beneficial capacity of milk. Additionally, combining milk with products such as cocoa or processed chocolate has shown remarkable capacity as a functional food due to probiotic bacteria. The preparation of milk supplemented with chocolate prepared using ingredients provided by the Mexican School of Confectionery and Chocolate in San Luis Potosí, Mexico, has been proposed. The probiotic *L. plantarum* 299v demonstrates a highly beneficial effect and increases omega-3 polyunsaturated fatty acids, which shows those products of Mexican origin and their addition to dairy products bring out great benefits with probiotics the primary ingredient (Gómez-Fernández et al. 2021). Hurtado-Romero et al. (2021) reported the isolation of LAB from different foods of Mexican origins, such as *Lactococcus, Lactobacillus, Leuconostoc, and Kluyveromyces*, from artisanal milk kefir from different locations in Guadalajara, in the State of Jalisco, Mexico.

11.3 PERSPECTIVE: A GREATER CONSUMPTION OF TRADITIONAL MEXICAN FERMENTED BEVERAGES

One perspective is to propose a greater consumption of drinks of traditional origin from regions with great cultural richness. It has been shown that many of these drinks contain LAB such as *Lactobacillus brevis* (Torres-Maravilla et al. 2016) that, by colonizing the gastrointestinal tract, improve the absorption of macronutrients and micronutrients by intestinal microvilli, which aids in digestibility and weight gain in the individual. In addition, increasing research into the bioactive compounds of these beverages, as in the case of pulque, standardizes the ingredients that make it up and identifies their beneficial effects. Pulque is a traditional Mexican alcoholic beverage obtained from the fermentation of the sap of various agaves, which contains a great diversity of LAB, particularly *Lactobacillus* and *Leuconostoc* (Escalante et al. 2008). As already mentioned, probiotics have a wide application in therapeutic areas, the veterinary field, the industrial area of food, and the technology sector to produce nutraceuticals and pharmaceuticals; however, it is still necessary to study various models of isolation of probiotic bacteria as well as different sources of isolation.

Additionally, we propose the isolation of probiotic bacteria from artisanal beverages, as is the case for the isolation of *Weissella confusa* from pozol (a drink based on cocoa and nixtamalized corn) of freshly ground nixtamal mass from two producers in the Pino Suárez market in Tabasco, Mexico (Hernández-Oaxaca et al. 2021).

The isolation of *Lactobacillus pentosus* ABHEAU-05 from pineapple tepache (a traditional Mexican fermented beverage) has the capacity to produce peptides that are beneficial for the gastric tract (Escobar-Ramírez et al. 2020) and the use of *Saccharomyces boulardii* CNCM I-745, and its possible use as a probiotic bacterium in craft brewing, has been proposed (Ramírez-Cota et al. 2021).

This study has led to the proposal of improving the sensory characteristics of yoghurt, primarily those of natural origin. The people do not accept many due to their acidity, color, and refreshing smell, characteristic of lactic acid fermentation and flavor, suggesting a mixture of fermented yoghurt with LAB and its combination with natural ingredients from Mexican zones.

As done previously, they enhanced sensory and biological potential from hydrolysates of the bean protein by encapsulating and incorporating it into a Greek-style yoghurt matrix to improve sensory characteristics. The improvement was accomplished using ingredients from native regions, such as raw black beans (*Phaseolus vulgaris* L.) grown in the region of Sombrerete, Zacatecas, Mexico (23 37053.9“ N 103 38029.5” W), concentrated whey protein obtained from the region of Jalisco, Mexico and canola oil acquired from a local market in Jalisco, Mexico (Free-Manjarrez et al. 2022). This study showed how using natural ingredients could improve the sensory characteristics of yoghurt and increase its probiotic potential and consumption.

Moreover, supplementation with food helps people with cholesterolemia, as is the case of *Enterococcus faecium* isolated from rhizospheric soils, which have a proven ability to eliminate cholesterol *in vitro* and possess essential probiotic attributes (Singhal et al. 2019).

11.4 STRAINS COMMONLY RECOGNIZED AS PROBIOTICS

Agave plants are of great economic importance and have numerous biotechnological advantages. The high fructose, sugar, and fiber composition of agave make it valuable for food, beverage, and biofuel production. Mezcal is a distilled alcoholic beverage fermented from agave, and its production has increased in recent years. Agave species are a source of various products for human use, such as food, fiber, and beverages, including mezcal. Agave is an excellent source of LAB. *Lactobacillus plantarum, Lactobacillus rhamnosus, Enterococcus faecium,* and *Lactococcus lactis* strains with probiotic attributes were isolated and characterized for their resistance under gastrointestinal conditions, such as lysozyme, acidic pH, and bile salts. In addition, the adherence of these LAB to human intestinal epithelial cells (Caco-2 and HT-29 cells) was tested *in vitro*, and their antioxidant and immunomodulatory profiles were determined using cellular models (Hernández-Delgado et al. 2021).

The most common genera to which the probiotic microorganisms belong include *Lactobacillus, Bifidobacterium, Lactococcus, Streptococcus, Bacillus,* and *Enterococcus.* In Mexico, novel candidate probiotic strains were isolated and characterized from pulque (xaxtle) (Torres-Maravilla et al. 2016). Fourteen strains were obtained from xaxtle samples isolated from three different Mexican regions. The isolates belonged to the *Lactobacillus plantarum, Lactobacillus paracasei, Lactobacillus brevis,* and *Lactobacillus composti* phylogenetic groups, with *Lactobacillus brevis* being the most dominant group (**Table 11.1**).

Currently, the probiotic potential of LAB isolated from traditional fermented products is attracting significant interest, particularly in developing countries where access to probiotics is limited and where the development of new products and food supplements could help fight against certain diseases (Vinderola et al. 2008).

Lactobacillus brevis LBH1073 from Nanacamilpa, Tlaxcala, Mexico, was patented. The invention (Patent République Française N° de publication: 3 082 207) relates to a new probiotic strain of *Lactobacillus brevis* isolated from pulque, endowed with anticancer properties, as well as compositions comprising this strain. This strain of *Lactobacillus brevis* was deposited according to the Budapest Treaty with the CNCM (National Collection of Cultures of Microorganisms, 25 rue du Docteur Roux, 75724 Paris Cedex 15) on May 30, 2018, under the number 1-5321.

TABLE 11.1 STRAINS OF *LACTOBACILLUS* FROM XAXTLE (SEDIMENT OF PULQUE) OF THREE DIFFERENT MEXICAN REGIONS, IDENTIFIED BY THE API TEST AND 16S RNA ANALYSES

	ENCB Number	Microorganisms	Region	API 50CHL (%)	16S rRNA (%)
1.	ENCB1062	*Lactobacillus plantarum* LBH1062	Santiago Villalta	71.1	99
2.	ENCB1063	*Lactobacillus plantarum* LBH1063	Santiago Villalta	99.6	99
3.	ENCB1064	*Lactobacillus plantarum* LBH1064	Santiago Villalta	99.9	99
4.	ENCB1065	*Lactobacillus paracasei ssp. paracasei* LBH1065	Santiago Villalta	96.6	99
5.	ENCB1066	*Lactobacillus paracasei ssp. paracasei* LBH1066	Santiago Villalta	99.4	99
6.	ENCB1067	*Lactobacillus paracasei ssp. paracasei* LBH1067	Santiago Villalta	84.3	99
7.	ENCB1068	*Lactobacillus sanfranciscensis* LBH1068	Santiago Villalta	99.5	99
8.	ENCB1069	*Lactobacillus brevis* LBH1069	Milpa Alta	96.1	99
9.	ENCB1070	*Lactobacillus brevis* LBH1070	Milpa Alta	96.1	99
10.	ENCB1071	*Lactobacillus brevis* LBH1071	Milpa Alta	96.1	99
11.	ENCB1072	*Lactobacillus composti* LBH1072	Nanacamilpa	Not detected	99
12.	ENCB1073	*Lactobacillus brevis* LBH1073	Nanacamilpa	98.6	99
13.	ENCB1074	*Lactobacillus composti* LBH1074	Nanacamilpa	Not detected	99
14.	ENCB1075	*Lactobacillus plantarum* LBH1075	Milpa Alta	99.9	99

Strain CNCM 1-5321 was identified as belonging to the *L. brevis* species using an API 50 CHL test and 16S RNA sequencing. Partial sequence of the 16S RNA of the CNCM 1-5321 strain (SEQ ID NO: 1) AGCCAAGTCTGATGGAGCATGCCGCGTGAGTGAAGAAGGGTTTCGGCTCGTAA AACTCTGTTGTTAAAGAAGAACACCTTTGAGAGTAACTGTTCAAGGGTTGACGGTATTTA ACCAGAAAGCCACGGCTAACTACGTGCCAGCAGCCGCGGTAATACGTAGGTGGCAAGCGT TGTCCGGATTTATTGGGCGTAAAGCGAGCGCAGGCGGTTTTTTAAGTCTGATGTGAAAGC CTTCGGCTTAACCGGAGAAGTGCATCGGAAACTGGGAGACTTGAGTGCAGAAGAGGACA GTGGAACTCCATGTGTAGCGGTGGAATTA.

It also has anticancer properties, inhibiting the proliferation of tumor cell lines and modifying the expression of specific pro-apoptotic genes or certain genes involved in tumor development. As shown in the examples, the anticancer properties of the CNCM I-5321 strain are independent of its state of viability. Consequently, the present invention encompasses the CNCM I-5321 strain regardless of its viability, not only in live form but also in dead form, inactivated form, or in the form of a bacterial lysate.

There is great interest in isolating new probiotic LAB strains from unconventional sources. New bacterial strains were isolated and characterized from agave in mezcal production, and their antioxidant (*in vitro*) and anti-inflammatory (*in vivo*) properties were determined to establish their probiotic potential. Agave samples were collected from mezcal production in the Santiago Matatlán, Tlacolula de Matamoros, and Macuilxóchilt de Ártigas Carranza localities (in the state of Oaxaca, Mexico). Twenty Gram-positive and catalase-negative bacterial strains were isolated from the mezcal batch samples according to 16S gene amplification (**Table 11. 2**).

TABLE 11.2 LACTIC ACID BACTERIA ISOLATED FROM AGAVE-FERMENTED SAMPLES OF THREE DIFFERENT REGIONS OF OAXACA, MEXICO

Code	Microorganism	Region	16S rRNA(%)
LM01	*Lactobacillus plantarum*	Macuilxóchilt de Ártigas	98.8
LM02	*Lactobacillus plantarum*	Macuilxóchilt de Ártigas	99
LM03	*Lactobacillus plantarum*	Macuilxóchilt de Ártigas	95.7
LM04	*Lactococcus lactis ssp. lactis*	Tlacolula de Matamoros	99
LM05	*Lactobacillus plantarum*	Tlacolula de Matamoros	97.4
LM06	*Enterococcus faecium*	Santiago Matatlán	99
LM07	*Lactobacillus rhamnosus*	Tlacolula de Matamoros	100
LM08	*Lactobacillus plantarum*	Tlacolula de Matamoros	98.1
LM09	*Lactobacillus plantarum*	Macuilxóchilt de Ártigas	97.5
LM10	*Lactobacillus plantarum*	Macuilxóchilt de Ártigas	97.4
LM11	*Lactobacillus plantarum*	Tlacolula de Matamoros	96.7
LM12	*Lactobacillus plantarum*	Santiago Matatlán	96
LM13	*Lactobacillus plantarum*	Tlacolula de Matamoros	97
LM14	*Lactobacillus plantarum*	Tlacolula de Matamoros	97.6
LM15	*Enterococcus faecium*	Macuilxóchilt de Ártigas	99
LM16	*Lactobacillus rhamnosus*	Tlacolula de Matamoros	99
LM17	*Lactobacillus plantarum*	Tlacolula de Matamoros	99.9
LM18	*Lactobacillus plantarum*	Tlacolula de Matamoros	97
LM19	*Lactobacillus plantarum*	Santiago Matatlán	99
LM20	*Lactobacillus plantarum*	Macuilxóchilt de Ártigas	99
LM14	*Lactobacillus plantarum*	Tlacolula de Matamoros	97.6
LM15	*Enterococcus faecium*	Macuilxóchilt de Ártigas	99
LM16	*Lactobacillus rhamnosus*	Tlacolula de Matamoros	99
LM17	*Lactobacillus plantarum*	Tlacolula de Matamoros	99.9
LM18	*Lactobacillus plantarum*	Tlacolula de Matamoros	97
LM19	*Lactobacillus plantarum*	Santiago Matatlán	99
LM20	*Lactobacillus plantarum*	Macuilxóchilt de Ártigas	99

Lactobacillus plantarum (15 strains) was the predominant isolated species (from the three different regions in the samples), followed by *Lactobacillus rhamnosus* (2 strains), *Enterococcus faecium* (2 strains), and *Lactococcus lactis* (1 strain). Tlacolula de Matamoros region samples displayed the greatest isolated strains (Hernández et al. 2021). *Lactobacillus rhamnosus* LM07 and *Lactobacillus plantarum* LM17 have the probiotic potential for use in the context of inflammatory bowel disease. The importance of agave spp. as a source of microorganisms with beneficial effects on intestinal health has been noted, as *Lactobacillus rhamnosus* LM07 and *Lactobacillus plantarum* LM17 improved some colitis symptoms in the DNBS-induced colitis murine model.

11.5 PREBIOTICS

Prebiotics are nondigestible components of food that help improve host health conditions by stimulating the growth of microbes inhabiting the gastrointestinal region of the host. All prebiotics are fibers, but not all dietary fibers have prebiotic effects. The association between dietary fibers and the prevention of cardiovascular and metabolic diseases, such as diabetes, obesity, and cancer, is well known (Slavin, 2013).

Research has focused on prebiotics since many of these polysaccharides can be metabolized by the intestinal microbiota, producing short-chain fatty acids, including acetate, butyrate, and propionate (Holscher, 2017).

Prebiotics were reported to be selectively fermented, allowing specific changes in the composition or activity of microbes inhabiting the human body's gastrointestinal tract (GIT) and resulting in improved health benefits (Gibson et al. 2004).

For probiotics, the ability to tolerate gastrointestinal fluids and colonize the intestinal tract are key factors affecting their proposed health effects, and probiotics can effectively use prebiotics to promote their growth; prebiotics can be used in place of probiotics or as an additional supplement, stimulating the growth and activity of beneficial bacteria in the gastrointestinal tract (Li et al. 2020).

11.6 POSTBIOTICS

The concept of postbiotics is relatively new and currently refers to bioactive compounds (Szydłowska & Sionek 2022). Therefore, postbiotics may be defined as bioactive soluble factors (products or metabolic byproducts) produced by some food-grade microorganisms during growth and fermentation. Many LAB are considered probiotic, and their postbiotic compounds exhibit similar or additional health benefits to the consumer (Moradi et al. 2020). In addition, the mechanisms of action at the molecular level and application in a functional postbiotic food are being actively studied. A postbiotic is a complex mixture of compounds produced during the metabolic activity of probiotic bacteria within the host.

Short-chain fatty acids, vitamins, and proteins of low molecular weight are postbiotics-parabiotics (inactive or nonviable microbial cells, which, when administered or ingested in sufficient quantities, confer benefits to the host). Probioceuticals/probiotaceuticals (factors derived from probiotics, for example, the return of *Lactobacillus reuteri* with both beneficial and harmful effects on the host (Nataraj et al. 2020, Guglielmetti, 2023).

The isolation of *Bacillus coagulans* MA-13 with proven production activity of galactoside hydrolyzing enzymes isolated from canned beans has been reported (Aulitto et al. 2021). This use of prebiotics and postbiotics leads to a reclassification of probiotic foods since the evidence shows the isolation of probiotic bacteria.

Ensuring the food goes through a pasteurization process is important since these processes modify a food's primary characteristics, such as sensory, physical, chemical, and nutritional characteristics. The survival of probiotic bacteria through these processes is possible, or it is also proposed to include foods that, after the technological process to reduce their microbial load (such as pasteurization), are subjected to added probiotic bacteria to raise their level of benefit to the host. In this way, a classification is proposed that includes probiotic foods of natural origin (dairy and nondairy) and processed foods (with probiotic bacteria active after the process and with foods that, after the industrial process, are supplemented with probiotics) and their application in the food industry considering current technologies, equipment, and tools.

11.7 SYNBIOTICS

The synbiotic concept was first described in 1995 by Gibson and Roberfroid, and referred to the combination of probiotics and prebiotics that, in said association, induce beneficial effects for the host. The word synbiotic was formed from the Greek prefix syn, which means together, and the suffix biotic, which means belonging to life (Kolida and Gibson, 2011; Rovinaru and Pasarin, 2020). The synergistic combination of probiotics and prebiotics in products such as foods, drugs, and supplements are known as synbiotics (Mohanty et al. 2018).

The definition of synbiotics has evolved based on changes made in the definitions of probiotics and prebiotics; in 2019, a panel of academic scientific experts convened by the International Scientific Association of Probiotics and Prebiotics (ISAPP). Updated the definition of synbiotic to a mixture comprising living microorganisms and substrate(s) selectively used by the host's autochthonous (resident) or allochthonous (external) microorganisms that confer a health benefit on the host. It is essential to clarify that in the etymology of the term synbiotic the prefix syn meaning together does not strictly imply a synergy, since the panel recognizes that the current use of the term includes the combination of probiotics and prebiotics that may not have evidence of codependent function (Kolida and Gibson, 2011; Swanson et al. 2020). Considering the above, two approaches (complementary and synergistic) for the design of synbiotics are described (**Figure 11.1**).

Complementary synbiotics combine a probiotic and a prebiotic that work independently to produce one or more health benefits; the probiotic is chosen based on the specific desired beneficial effects for the host, and the prebiotic is chosen independently to modulate and favor the growth of beneficial autochthonous microorganisms to the host. In contrast, synergistic synbiotics combine a probiotic and a prebiotic that work together to generate beneficial effects for health. In this approach, the probiotic is also selected based on the desired beneficial effects for the host, but the prebiotic is chosen to be used selectively by the coadministered alien microorganism (probiotic), favoring its growth, survival, and functionality in the host (Kolida & Gibson, 2011; Krumbeck et al. 2018; Swanson et al. 2020). Determining the living microorganism's safety, identity, purity, and potency is necessary. As well as the substrate to be used under suitable methods that comply with the required

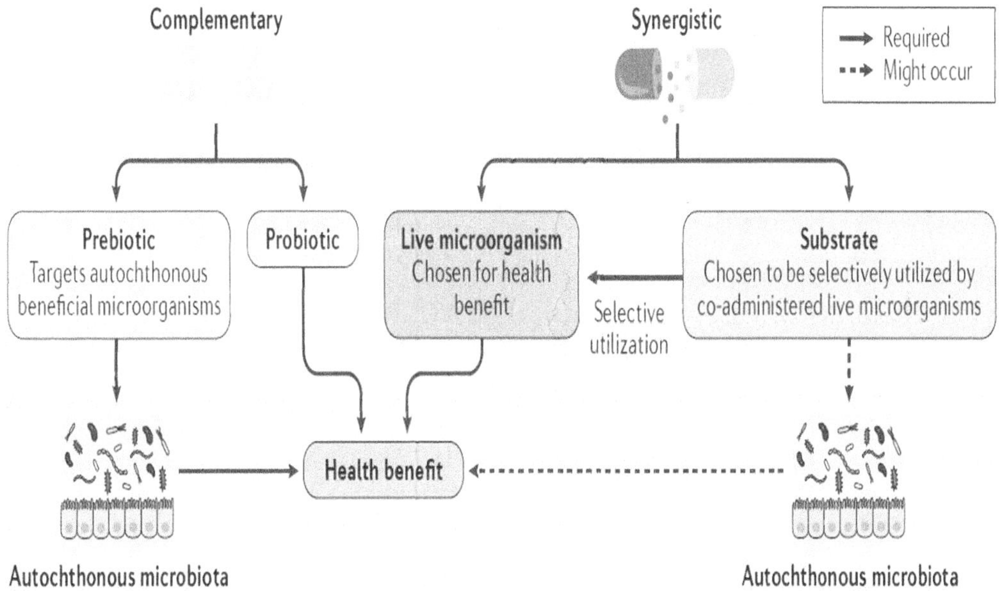

Figure 11.1 Design and mechanism of action of complementary and synergistic symbiotic.

regulatory standards in the same way, scientific evidence of the beneficial effect of the synbiotic must be generated through experimental trials designed and applied correctly in the target host (Swanson et al. 2020).

Synbiotics have better probiotic functions than either probiotics or prebiotics alone, and dairy products are excellent probiotic carriers. Our study demonstrated the benefit of probiotics and prebiotics in healthy foods, especially dairy products, providing a theoretical basis for developing functional foods (Li et al. 2020).

11.8 MECHANISM OF ACTION OF PROBIOTICS

Probiotic bacteria attach to epithelial cells, thereby blocking pathogens. Moreover, a signaling cascade is initiated as soon as the probiotic strain attaches to epithelial cells, resulting in modification of the immunological response. On the other hand, the release of specific soluble components might result in indirect or direct activation of immune system cells; this might be beneficial in treating and preventing various communicable diseases and help resolve chronic inflammation of the digestive tract (Isolauri et al. 2002).

Several pieces of evidence favour different benefits associated with probiotics, including enhancing the immune system, improving intestinal health, preventing cancer, and reducing serum cholesterol. Li et al. (2020) demonstrated that the use of probiotics (*L. bulgaricus* and *L. rhamnosus*) or emerging prebiotics, xylooligosaccharide (XOS) and red ginseng extracts, accelerated the recovery of intestinal microbiota (damaged by antibiotics) by allowing the proliferation of beneficial bacteria and inhibiting that of harmful bacteria. This study used the probiotic biomass's targeted approach and anti-inflammatory action to lessen the side effects of therapeutic agents of ulcerative colitis. Another study used a spray-drying method to prepare mesalamine-loaded eudragit S-100 with probiotic microparticles (Thakur et al. 2019).

The inclusion of probiotics in the diet helps to maintain the gut microbial balance since it prevents the colonization of pathogens, in addition to other beneficial effects such as attenuating lactose intolerance, diarrhoea, and constipation, and acts as an aid for immune regulation and cancer prevention and protects mucosal barrier integrity in both animals and humans (Gu et al. 2019).

11.9 HEALTH BENEFITS: PROBIOTIC ACTIVITY TO ATTENUATE DIFFERENT DISEASES

In 2007, knowledge regarding the role of probiotics in peptic ulcer disease was limited; researchers studied the effect of the probiotic strain *Lactobacillus rhamnosus* GG on gastric ulcers and elucidated the mechanisms involved. In rats, acetic acid-induced gastric kissing ulcers (60% v/v) were induced. *L. rhamnosus* GG was administered intragastrically at 8-log10 CFU/day or 9-log10 CFU/day for three consecutive days after ulcer induction. *L. rhamnosus* GG successfully colonized the gastric mucosa, especially at the ulcer margin. It also significantly and dose-dependently reduced the gastric ulcer area (Lam et al. 2007).

Bioactive compounds and probiotics fight the symptomatic complications associated with COVID-19: they exert an immunomodulatory role by regulating innate and acquired immunity. Therefore, it is pertinent to examine/probe their efficacy against COVID-19 infection in detail (Bhushana et al. 2021).

The intestinal microbiota affects the development of the central nervous system through the gut-brain axis, and modifications in the gut microbiome can lead to disorders in the central nervous system immune response, which may result in related diseases such as depression, Parkinson's disease, or schizophrenia (Zhu et al. 2018).

Probiotics are live microorganisms that confer health benefits to the host when administered adequately. Their use is increasingly widespread, preventing and treating diseases, including traveller's diarrhoea and inflammatory bowel diseases. *Lactobacillus sanfranciscensis* LBH1068 improved mouse health, as observed by attenuated weight loss, significant decreases in gut permeability, and cytokine modulation. Altogether, these results highlight the potential of lactobacilli isolated from pulque, particularly the strain *L. sanfranciscensis* LBH1068 as a novel probiotic to treat inflammatory bowel diseases. It has also been experimentally demonstrated that certain strains of lactobacilli (in particular strains of *Lactobacillus sanfranciscensis, Lactobacillus plantarum,* and *Lactobacillus composti*) isolated from pulque have anti-inflammatory properties that are potentially beneficial for the treatment of inflammatory bowel diseases (Torres-Maravilla et al. 2016). Recent studies suggest that certain pre-Hispanic cultures in Mexico administer enemas using pulque to combat diseases and disorders of the digestive system (Lemus-Fuentes 2006).

The anti-*Helicobacter pylori* effect of the *Lactobacillus fermentum* UCO-979C strain in a simulated gastric system confirmed the probiotic activity of this strain previously shown in *in vitro* and animal model studies. The anti-*Helicobacter pylori* effect was enhanced when the probiotic was encapsulated in carrageenan, demonstrating collaboration between the probiotic strain and carrageenan. These results also revealed that consuming the probiotic under fasting conditions provided better anti-*Helicobacter pylori* efficacy than other foods (Gutiérrez-Zamorano et al. 2019).

It is important to control the optimal intake of the bacteria of this genus through dose-effect studies to understand and use their existing mechanisms and then to design the clinical study with the aim of improving the lives of patients, promoting the development of specific therapeutic products for inflammatory bowel diseases (Chen et al. 2023).

11.10 FOOD APPLICATIONS

Currently, there is great interest in the isolation of new probiotic LAB strains from unconventional sources, such as koumiss (a traditional fermented food produced from mare's milk kefir in China); kimchi (a traditional fermented and functional plant food from Korea); and mezcal (a traditional fermented beverage from Agave spp. in Mexico) (Hernández-Delgado et al. 2021).

Pickled fermented vegetables are traditional foods from China; its fermentation process is the product of the interaction of the same bacteria with the vegetables under certain specific conditions that are generated in the pickles, which present different characteristics in their chemical composition, aromatic substances, and the bacterial diversity present. Bacteria of the genus Lactobacillus resulted in decreased pH, increased total acid content, nitrite degradation, and organic acid (lactic acid and malic acid) production of the fermented pickles. The main strains are *L. plantarum* R5, *L. pentosus* R8, and *L. plantarum* R5 plus *L. pentosus* R8. At the end of the study, 80 flavors were detected in the pickles fermented for 14 days, which were the product of a negative correlation with pH and a strong positive correlation with malic acid and lactic acid, demonstrating how microorganisms could acclimate to the changing fermentation environment of the pickles (Lin et al. 2023).

The most traditional places across different regions propose the isolation of yeasts and probiotic bacteria from locally produced food. In the case of Mexico, in many regions, such as in the state of Guerrero, inhabitants are dedicated to the planting of a fruit called white zapote, containing zapotin (5,6,2',6'-tetramethoxyflavone), a phytochemical with antioxidant activity (Murillo et al. 2007), leading to the proposal of fruit-based foods such as zapote and its addition to probiotic bacteria for a healthy diet.

This study arises because milk and its derivatives are consumed food products around the world with a constant increase in their production and consumption, setting a trend in the market due to their nutritional contribution. Cheese being one of the derivatives with great demand, demonstrating that the presence, quantity, and availability of the bioactive compounds present varies, due to factors such as the type of milk used for its preparation, the type of microorganism used, as well as the technological process used (López-Mendoza et al. 2023).

In addition to bioactive compounds, probiotic microorganisms can confer enormous health benefits. The most broadly studied probiotic bacteria are *Lactobacillus* and *Bifidobacterium* (Yan & Polk 2011).

Studies have shown that functional foods, such as vegetables, like tomatoes (*Lycopersicon esculemtum*), contain lycopene (phytochemical carotenoid). Lycopene acts as an antioxidant due to the absence of the chemical structure of the γ-ionic ring, allowing interaction between ions and aromatic chemical compounds. This results in its ability to trap reactive oxygen species effectively to reduce damage at the cellular level. In addition, tomatoes and other vegetables have been confirmed to have bioactive compounds such as cinnamic acid, chlorogenic acid, and gallic acid molecularly coupled with insulin receptors to combat type II diabetes (Roy et al. 2020).

Feeding based on probiotic supplementation leads to the proposal of beverages from juices based on fruits rich in carotenes and lycopene, such as orange, mango, and even beets, as a perspective in the future. Foods such as cookies, whole grain loaves of bread, and even tortillas or bread based on basic low-fat formulations supplemented with probiotic bacteria such as *Lactobacillus rhamnosus* represent a functional food that increases antioxidant levels and neutralizes the effects of reactive oxygen species caused by practicing sports at high levels (Martarelli et al. 2011).

In addition to proposing dairy sera, products such as cheeses, yoghurt, or food gels supplemented with strains of *Lactobacillus* could be integrated into a diet aimed at young children (Giovannini et al. 2007). Foods ingested continuously manage to colonize the gastric tract and, in this way, prevent enteric diseases, gastrointestinal diseases, and pathologies associated with excessive amounts of free radicals due to oxidative stress and metabolic syndromes due to poor diet in middle age (Ramakrishna et al. 2009).

The subjection of vegetables, meats, and fruits to a cold plasma atmosphere eliminates *Salmonella*, *Escherichia coli* O157:H7, *Listeria monocytogenes*, and *Staphylococcus aureus* (Niemira 2012). Electric pulse fields for developing texture and flavor in milk and yoghurt improve the physicochemical changes produced by fermentation by probiotic bacteria to the product (Sfakianakis & Tzia 2014). Isolation of probiotic bacteria from milk that has gone through a pasteurization process has been reported due to the addition of LAB to the product after the technological process (Aljutaily et al. 2020). The preparation of bread using sourdough with *Lactobacillus plantarum* SPC-SNU 72-2 isolated from kimchi and its resistance to the cooking process have been reported, as isolation of the bacteria was possible even after heating the dough (Park et al. 2019).

Probiotics uphold defense mechanisms, including adaptive and innate immune responses, by modulating the function of macrophages, dendritic cells, B-lymphocytes, and T-lymphocytes. One such immune-modulatory mechanism occurs by activating Toll-like receptors. The two primary elemental mechanisms of the action of probiotics are either the signaling pathways in host cells or regulating gene expression. Functional foods have been developed to manage the intestinal microecological environment using probiotic and prebiotic supplements (Li et al. 2020).

Research has focused on the use of dietary fibers and prebiotics, since many of these polysaccharides can be metabolized by intestinal microbiota, produces short-chain fatty acids (Luca-Guarino et al. 2020). Furthermore, various plants, such as *Curcuma longa*, *Ocimum sanctum*, *Azadirachta indica*, *Elettaria cardamom*, *Syzygium aromaticum*, *Viola odorata*, and *Momordica charantia* possess enormous medicinal and therapeutic properties (Tufarelli & Laudadio 2016). Probiotic-enriched foods carry enough viable microorganisms (at least 9-log10 CFU per serving portion) that can change the microbiota to create beneficial effects in the host (Tufarelli & Laudadio 2016).

The viability of probiotics in yoghurt can be affected by food components: sugars, proteins, fat, vitamins, minerals, flavoring agents, antioxidants, and amino acids; processing-related factors: heat treatments, homogenization, and fermentation temperature; and microbiological factors: type of strains and inoculum level (Meybodi et al. 2020).

Sourdough, a leavening agent in bread making, is a method that can be traced back to ancient times. Sourdoughs harbor a complex microbiota affected by multiple factors, including those related to cereal plants, grains, and sourdough processing techniques. *Lactobacillus sanfranciscensis* is the key autochthonous bacterium of the traditional sourdough microbiota, and it is said to be a sourdough-adapted species. Despite the significant dominance of this bacterium in sourdoughs, the origin of this species remains unclear. *Lactobacillus sanfranciscensis* positively influences all aspects of sourdough and fermented foods. However, the positive influence of this species on sourdough is a strain-dependent characteristic. The first purpose of this review was to discuss factors affecting the microbiota of sourdoughs, with particular emphasis on the reasons underlying the remarkable prevalence of *L. sanfranciscensis* in this ecological niche. The second objective was to discuss the genotypic and phenotypic classification of *L. sanfranciscensis* strains and the influence of this species on the technological and functional characteristics of sourdough, including its influence on the rheological properties of dough and bread characteristics, texture, aroma, and shelf life through the inhibition of fungal growth (Zhang et al. 2019).

Many mechanisms demonstrate how carotenoids and the identification of more bioactive compounds, including polyunsaturated acids and phytochemicals in vegetables such as red pepper, tomato, carrots, pumpkin, and some fruits such as orange and mango, have a beneficial effect on the host. However, much remains to be understood, including the oxidative prevention mechanisms and the application of functional foods that contain an antioxidant as the primary active compound, such as tomato (*Lycopersicon esculemtum*), a tomato sauce rich in lycopene that helps the host reduce the probability of developing colon cancer by oxidative stress at the cellular level and combatting type II diabetes can be proposed. In the same way, dressings, jellies, and condiment formulations have been proposed with each of the vegetables containing lycopene increased by some molecular genetic engineering technique, proposing the use of probiotics as a reinforcement of the formulation (Heider et al. 2014).

It has been confirmed that colonizing the intestinal microbiome in the small and large intestines by probiotic bacteria has implications for preventing the oxidative stress mechanisms of intestinal cells. The antioxidant capacity of various probiotic strains has been studied, as is the case for *Lactobacillus plantarum* and *Streptococcus thermophilus*, and their antioxidant capacity in HaCaT cells results in capturing the radicals ABTS+, DPPH, OH- and O_2-, exerting a protective effect against oxidative damage and cellular cytotoxicity (Liu et al. 2021).

Another probiotic that has been studied is *Lactococcus lactis* NZ9000, which was genetically modified to produce selenium nanoparticles coated with polysaccharides. Tests revealed that the modification significantly reduced the increase in the concentration of malondialdehyde (produced during lipid peroxidation), the decrease in the activity of the enzyme glutathione peroxidase (GPx), and total superoxide dismutase (T-SOD) in a study using porcine intestinal epithelial cells (IPEC-J2) inducing simulated oxidative stress with hydrogen peroxide (H_2O_2) (Xu et al. 2019).

Additionally, studies with *Lactobacillus gasseri* 505 in combination with *Cudrania tricuspidata* leaf extracts prevent liver toxicity in rat-induced colorectal cancer, decreasing inflammatory activity due to low levels of inflammation markers such as cyclooxygenase-2, myeloperoxidase, tumor necrosis factor-α, and the kappa light chain enhancer of activated B-cell nuclear factor (Oh et al. 2020).

It has been proposed that soymilk supplemented with *Lactobacillus fermentum* CQPCo4 increases the activity of the enzymes superoxide dismutase (SOD), glutathione peroxidase (GSH-Px), total superoxide dismutase (T-SOD), and catalase (CAT). Catalyzing the decrease of 02 – in H_2O_2 hydrogen peroxide is an intermediate in the signaling of programmed cell death produced in lipid peroxidation of the cell membrane (Zhou et al. 2021).

A new strain, *Lactobacillus brevis* LBH1073, shows anticancer properties isolated from pulque, a traditional Mexican alcoholic beverage obtained from the fermentation of the sap of various agaves; other new bacterial strains isolated and characterized from agave used by obtaining mezcal, *Lactobacillus rhamnosus* LM07 and *Lactobacillus plantarum* LM17, have the probiotic potential for use in the context of inflammatory bowel disease. In addition, functional foods have been developed to manage the intestinal microecological environment using probiotic and prebiotic supplements.

11.11 CONCLUSION

The microorganisms most used as probiotics belong to the LAB group, primarily including species of the genera *Lactobacillus* and *Bifidobacterium,* which are commonly consumed in fermented foods. Herein, 14 probiotic strains were isolated from the sediment (xaxtle) of an artisanal beverage named pulque, belonging to the phylogenetic group Lactobacillus. *Lactobacillus sanfranciscensis* strain LBH1068 tended to block the secretion of cytokines that promote the expression of proinflammatory genes involved in the pathogenesis of inflammatory bowel disease. The fermentation of agave in mezcal production allowed it to isolate 20 new LAB strains with potential probiotic properties. *Lactobacillus plantarum* LM17 and *L. rhamnosus* LM07 strains displayed antioxidant properties, they were evaluated in CaCo-2 and HT-29 cell lines with hydrogen peroxide demonstrated anti-inflammatory properties in TNF-α-stimulated HT-29 cells. These studies showed the probiotic activity of endemic LAB in Mexico and encourage us to continue evaluating and characterizing traditional Mexican fermented beverages.

REFERENCES

Aljutaily, T., Huarte, E., Martinez-Monteagudo, S., Gónzalez-Hernandez, J., Rovai, M., & Sergeev, I. (2020). Probiotic-enriched milk and dairy products increase gut microbiota diversity: A comparative study. *Nutrition Research.* 82: 25–33. https://doi: 10.1016/j.nutres.2020.06.017

Aulitto, M., Strazzulli, A., Sansone, F., Cozzolino, F., Monti, M., Moracci, M., Fiorentino, G., Limauro, D., Bartolucci, S., & Contursi, P. (2021). Prebiotic properties of *Bacillus coagulans* MA-13: Production of galactoside hydrolyzing enzymes and characterization of the transglycosylation properties of a GH42 β-galactosidase. *Microbial Cell Factories.* 20(1): 71. https://doi: 10.1186/s12934-021-01553-y

Bello, L., Agama-Acevedo, E., Sánchez-Hernández, L., & Paredes-López, O. (1999). Isolation and partial characterization of banana starches. *Journal Agriculture Food Chemistry.* 47(3): 854–857. http://doi.org/10.1021/jf980828t

Bhushan, I., Sharmaa, M., Mehta, M., Badyal, S., Sharma, V., Sharma, I., Singh, H., & Sistla S. (2021). Bioactive compounds and probiotics-a ray of hope in COVID-19 management. *Food Science and Human Wellness.* 10: 131–140. http://doi.org/10.1016/j.fshw.2021.02.001

Celis-Álvarez, M., López-González, F., Galdino, C., Martínez-García, J., Estrada-Flores G., & Arriaga-Jordán, C. (2016). Oat and ryegrass silage for small-scale dairy systems in the highlands of central Mexico. *Tropical Animal Health and Production.* 48(6): 1129–1134. https://doi: 10.1007/s11250-016-1063-0

Chen, G., Hong, Xu, Z., & Shang-Tian Y. (2014). Nutraceuticals and functional foods in the management of hyperlipidemia. *Critical Reviews in Food Science and Nutrition.* 54(9): 1180–1201. https://doi: 10.1080/10408398.2011.629354

Chen, Y., Gao, H., Zhao, J., Paul, R., Stanton, C., Zhan, H., Chen, W., & Yang, B. (2023). Exploiting lactic acid bacteria for inflammatory bowel disease: A recent update. *Trends in Food Science & Technology.* 138: 126–140. https://doi.org/10.1016/j.tifs.2023.06.007

Escalante, A., Giles-Gómez, M., Hernández, G., Córdova-Aguilar, M., López-Munguía, A., Gosset, G., & Bolívar, F. (2008). Analysis of bacterial community during the fermentation of pulque, a traditional Mexican alcoholic beverage, using a polyphasic approach. *International Journal of Food Microbiology.* 124(2): 126–34. https://doi: 10.1016/j.ijfoodmicro.2008.03.003

Escobar-Ramírez, M., Jaimez-Ordaz, J., Escorza-Iglesias, V., Rodríguez-Serrano, G., Contreras-López, E., Ramírez-Godínez, J., Castañeda-Ovando, A., Morales-Estrada, A., Félix-Reyes, N., & González-Olivares, L. (2020). *Lactobacillus pentosus* ABHEAU-05: An *in vitro* digestion resistant lactic acid bacterium isolated from a traditional fermented Mexican beverage. *Revista Argentina de Microbiología.* 52(4): 305–314. https://doi:10.1016/j.ram.2019.10.005

Free-Manjarrez, S., Mojica, L., Espinosa-Andrews, H., & Morales-Hernández, N. (2022). Sensory and biological potential of encapsulated common bean protein hydrolysates incorporated in a Greek-style yogurt matrix. *Polymers.* 14(5): 854. https://doi.org/10.3390/polym14050854

Gibson, G. R., Probert, H. M., Loo, J. V., Rastall, R. A., & Roberfroid, M. B. (2004). Dietary modulation of the human colonic microbiota: updating the concept of prebiotics. *Nutrition Research Reviews.* 17(2): 259–275. https://doi.org/10.1079/NRR200479

Giovannini, M., Agostoni, C., Riva, E., Salvini, F., Ruscito, A., Vincenzo, G., & Radaelli, G. (2007). Felicita Study Group. A randomized prospective double blind controlled trial on effects of long-term consumption of fermented milk containing *Lactobacillus casei* in pre-school children with allergic asthma and/or rhinitis. *Pediatric Respiratory.* 62(2): 215–20. https://doi: 10.1203/PDR.0b013e3180a76d94

Gómez-Fernández, A., Faccinetto-Beltrán, P., Orozco-Sánchez, N., Pérez-Carrillo, E., Marín-Obispo, L., Hernández-Brenes, C., Santacruz, A., & Jacobo-Velázquez, D. (2021). Sugar-free milk chocolate as a carrier of omega-3 polyunsaturated fatty acids and probiotics: A potential functional food for the diabetic population. *Foods.* 10(8): 1866. https://doi.org/10.3390/foods10081866

Gómez-Miranda, A., Colín-Navarro, V., López-González, F., Morales-Almaráz, E., Estrada-Flores, J., & Arriaga-Jordán, C. (2021). The effect of feeding whole-crop barley or black oat silage in the fatty acid content of milk from cows in small-scale dairy systems in the highlands of Mexico. *Tropical Animal Health and Production.* 53(3): 399. https://doi:10.1007/s11250-021-02805-0

Gu, J., & Roberts, K. (2019). Probiotics and prebiotics, in: Corrigan, M. L., K. Roberts, & E. Steiger (Eds.). *Adult Short Bowel. Syndrome Nutritional. Medical and Surgical Management.* Academic Press. pp. 67–80.

Guarner, F., Sanders, M. E., Eliakim, R., Fedorak, R., Gangl, A., Garisch, J., Kaufmann, P., Karakan, T., Khan, A. G., Kim, N., De Paula, J. A., Ramakrishna, B., Shanahan, F., Szajewska, H., Thomson, A., Le Mair, A., Merenstein, D., & Salminen, S. (2017). *Guías Mundiales de la Organización Mundial de Gastroenterología. Probióticos y prebióticos. World Gastroenterology Organisation.* pp. 1–35.

Guglielmetti, S. (2023). Safety considerations in the use of nonviable microbial cells as health-promoting agents in food and dietary supplements. *Current Opinion in Food Science.* 54: 101105. https://doi.org/10.1016/j.cofs.2023.101105

Gutiérrez-Zamorano, C., González-Ávila, M., Díaz-Blas, G., Smith, C., González-Correa, C., & García-Cancino, A. (2019). Increased anti-*Helicobacter pylori* effect of the probiotic *Lactobacillus fermentum* UCO-979C strain encapsulated in carrageenan evaluated in gastric simulations under fasting conditions. *Food Research International.* 121: 812–816. https://doi:10.1016/j.foodres.2018.12.064

Heider, S., Peters-Wendisch, P., Wendisch, V., Beekwilder, J., & Brautaset, T. (2014). Metabolic engineering for the microbial production of carotenoids and related products with a focus on the rare C50 carotenoids. *Applied Microbiology and Biotechnology.* 98(10): 4355–4368. https://doi: 10.1007/s00253-014-5693-8.

Hernández-Delgado, N., Torres-Maravilla, E., Mayorga-Reyes, L., Martín, R., Langella, F., Pérez-Pastén-Borja, R., Sánchez-Pardo. M., & Bermúdez- Humarán, L. G. (2021). Antioxidant and anti-inflammatory properties of probiotic candidate strains isolated during fermentation of agave (*Agave angustifolia* Haw). *Microorganisms.* 9(5): 1063. https://doi.org/10.3390/microorganisms9051063

Hernández-Oaxaca, D., López-Sánchez, R., Lozano, L., Wacher-Rodarte, C., Lorenzo-Segovia, & López-Munguía, A. (2021). Diversity of *Weissella confuse* in *pozol* and its carbohydrate metabolism. *Frontiers in Microbiology.* 12: 629449. https://doi:10.3389/fmicb.2021.629449

Holscher, H. (2017). Dietary fiber and prebiotics and the gastrointestinal microbiota. *Gut Microbes.* 8(2): 172–184.

Hurtado-Romero, A., Del Toro-Barbosa, M., Gradilla-Hernández, M., García-Amezquita, L., & García-Cayuela, T. (2021). Probiotic properties, prebiotic fermentability, and GABA-producing capacity of microorganisms isolated from Mexican milk kefir grains: A clustering evaluation for functional dairy food applications. *Foods.* 10(10): 2275. https://doi:10.3390/foods10102275

Isolauri, E., Kirjavainen, P., & Salminen, S. (2002). Probiotics: A role in the treatment of intestinal infection and inflammation?. *Gut.* 50 (Suppl 3): III54–III59. https://doi.org/10.1136/gut.50.suppl_3.iii54.

Kolida, S., & R. Gibson, G. (2011). Synbiotics in health and disease. *Annual Review of Food Science and Technology.* 2: 373–393. https://doi.org/10.1146/annurev-food-022510-133739

Krumbeck, J., Walter J., & Hutkins, W. R. (2018). Synbiotics for improved human health: Recent developments, challenges, and opportunities. *Annual Review of Food Science and Technology.* 9: 451–479. https://doi.org/10.1146/annurev-food-030117-012757

Lam, E., Lyu, H., KK-Wu, G., Shin, V., KK-Tai, E., So, W., Woo, P., & Cho, C. (2007). Probiotic *Lactobacillus rhamnosus* GG enhances gastric ulcer healing in rats. *European Journal of Pharmacology.* 565: 171–179. https://doi:10.1016/jejphar.2007.02.050

Lemus-Fuentes, E. (2006). Los enemas prehispánicos como instrumentos para aplicar probióticos. *Temas de Ciencia y Tecnología.* 10(29): 17–20. Universidad Tecnológica de la Mixteca, Instituto de Agroindustrias, Huajuapan de León, Oaxaca. México. ISSN: 2007-0977

Li, C., Zhihua, N., Meijuan, Z., Liu, M., Wang, X., Gu, X., Lu, H., Hongtao T., & Rajesh J. (2020). Probiotics, prebiotics, and synbiotics regulate the intestinal microbiota differentially and restore the relative abundance of specific gut microorganisms. *Journal of Dairy Science.* 103: 5816–5829. https://doi.org/10.3168/jds.2019-18003

Lin, X., Bakyrbay, S., Liu, L., Tang, X., & Liu, Y. (2023). Microbiota succession and chemical composition involved in lactic acid bacteria-fermented pickles. *Quality and Sensory Analysis of Fermented Products.* 9(4): 330. https://doi.org/10.3390/fermentation9040330

Liu, H., Peng, F., Ming, Y., Sheng, N., Qing, L., & Wei, W. (2021). Optimization of fermentation process of pomegranate peel and *Schisandra chinensis* and the biological activities of fermentation broth: Antioxidant activity and protective effect against H_2O_2-induced oxidative damage in HaCaT cells. *Molecules.* 26(11): 3432. https://doi:10.3390/molecules26113432

López-Mendoza, J., Adriano, L., Gálvez, D., & Vázquez, A. (2023). Bioactive compounds in cheeses: Biosynthesis, biological activity, and contribution of lactic acid bacteria. *Mesoamerican Agronomy*, 34(2), 22. https://doi.org/10.15517/am.v34i2.51432

Luca-Guarino, M., Annamaria, S., Claudia di Rosa, M., Paola, P., Lovino, P., Giulia, R., & Michelle, C. (2020). Mechanisms of action of prebiotics and their effects on gastro-intestinal disorders in adults. *Nutrients.* 12: 1037.

Martarelli, D., Verdenelli, M., Scuri, S., Cocchioni, M., Silvi, S., Cecchini C., & Pompei, P. (2011). Effect of a probiotic intake on oxidant and antioxidant parameters in plasma of athletes during intense exercise training. *Current Microbiology.* 62(6): 1689–1696. https://doi: 10.1007/s00284-011-9915-3

Martin, R., & Langella, P. (2019). Emerging health concepts in the probiotics field: streamlining the definitions. Frontiers in Microbiology. 10: 1047. doi: 10.3389/fmicb.2019.01047

Meybodi, N., Mortazavian, A., Masoumeh, A., & Amene, N. (2020). Probiotic viability in yoghurt: A review of influential factors. *International Dairy Journal.* 109: 104793. https://doi.org/10.1016/j.idairyj.2020.104793

Mohanty, D., Misra, S., Mohapatra, S., & Soumyaranjan-Sahu, P. (2018). Prebiotics and synbiotics: Recent concepts in nutrition. *Food Bioscience.* 26: 152–160. https://doi.org/10.1016/j.fbio.2018.10.008

Moradi, M., Seyedeh, A., Hadi, A., Arash, A., Jonas, T., Guimarães, N., & Lotfi, A. (2020). Postbiotics produced by lactic acid bacteria: The next frontier in food safety. *Reviews in Food Science and Food Safety.* 19: 3390–3415.

Murillo, G., Hirschelman, W. H., Ito, A., Moriarty, R. M., Kinghorn, A. D., Pezzuto, J. M., & Mehta, R. G. (2007). Zapotin, a phytochemical present in a Mexican fruit, prevents colon carcinogenesis. *Nutrition and Cancer.* 57(1): 28–37. https://doi: 10.1080/01635580701268097

Nataraj, B., Syed, A., V. Behare, P., & Hariom, Y., (2020). Postbiotics-parabiotics: The new horizons in microbial biotherapy and functional foods. *Microbial Cell Factories.* 19(1): 168. https://doi: 10.1186/s12934-020-01426-w

Niemira, B. (2012). Cold plasma decontamination of foods. *Annual Review of Food Science and Technology.* 3: 125–42. https://doi:10.1146/annurev-food-022811-101132

Oh, N. S., Lee, J. Y., Kim, Y. T., Kim, S. H., & Lee, J. H. (2020). Cancer-protective effect of a synbiotic combination between Lactobacillus gasseri 505 and a Cudrania tricuspidata leaf extract on colitis-associated colorectal cancer. *Gut Microbes*, 12(1). https://doi.org/10.1080/19490976.2020.1785803

Ounnas, F., Michel de Lorgeril, P., Laporte, F., Calani, L., Mena, P., Brighenti F., & Del Rio C. (2017). Rye polyphenols and the metabolism of n-3 fatty acids in rats: A dose dependent fatty fish-like effect. *Scientific Reports.* 7: 40162. https://doi: 10.1038/srep40162

Park, D., Jae-Han, B., Min, K., Hyeontae, K., Shin, S., Deukbuhm, L., Jin-Ho, S., Hee, K., & Nam, S. (2019). Suitability of *Lactobacillus plantarum* SPC-SNU 72-2 as a probiotic starter for sourdough Fermentation. *Journal of Microbiology and Biotechnology.* 29(11): 1729–1738. https://doi:10.4014/jmb.1907.07039

Patent République Française N° de publication: 3 082 207. Institut National de la Propriété Industrielle (à n'utiliser que pour les commandes de reproduction) .21) N° d'enregistrement national: 18 54912 COURBEVOIE © Int Cl8: C 12 N 1/20 (2018.01), A 23 L 33/135, A 61 K35/747, A 61 P 35/00, C 12 R 1/24 Date de mise à la disposition du public de la demande: 13.12.19 Bulletin 19/50. 71) Demandeur(s): Institut National de la Recherche Agronomique Etablissement public — FR et Institute Politecnico Nacional — MX. (72) Inventeur(s): Bermudez, L., Torres, E., Langella Philippe Et., Sánchez-Pardo, M.

Peluso, I., & Serafini, M. (2016). Antioxidants from black and green tea: From dietary modulation of oxidative stress to pharmacological mechanisms: Antioxidant and pharmacological properties of tea. *British Journal of Pharmacology.* 174(11): 1195–1208 https://doi.10.1111/bph.13649

Ramakrishna, B. (2009). Probiotic-induced changes in the intestinal epithelium: Implications in gastrointestinal disease. *Tropical Gastroenterology.* 30(2): 76–85. PMID: 19760989.

Ramírez-Cota, G., Yedid, E., López-Villegas, O., Jiménez-Aparicio, A., & Hernández-Sánchez, H. (2021). Modeling the ethanol tolerance of the probiotic yeast *Saccharomyces cerevisiae* var. Boulardii cncm i-745 for its possible use in a functional beer. *Probiotics and Antimicrobial Proteins.* 13(1): 187–194. https://doi:10.1007/s12602-020-09680-5

Rovinaru, C., & Pasarin, D. (2020). Application of microencapsulated synbiotics in fruit-based beverages. *Probiotics and Antimicrobial Proteins.* 12(2): 764–773. https://doi: 10.1007/s12602-019-09579-w

Roy, A., Ponnulakshmi, R., & Lakshmi, T. (2020). Molecular docking analysis of compounds from *Lycopersicon esculentum* with the insulin receptor to combat type 2 diabetes. *Bioinformation.* 16(10): 748–752. https://doi: 10.6026/97320630016748

Sfakianakis, P., & Tzia, C. (2014). Conventional and innovative processing of milk for yogurt manufacture; development of texture and flavor: A review. *Foods.* 3(1): 176–193. https://doi: 10.3390/foods 3010176

Singhal, N., Anay, M., Shilpa, M., Manish, K., & Jugsharan, V. (2019). Evaluation of bile salt hydrolases, cholesterol-lowering capabilities, and probiotic potential of *Enterococcus faecium* isolated from rhizosphere. *Frontiers in Microbiology.* 10: 1567. https://doi:10.3389/fmicb.2019.01567

Slavin, J. (2013). Fiber and prebiotics: Mechanisms and health benefits. *Nutrients* 5: 1417–1435. https://doi: 10.3390/nu5041417

Swanson, K., Gibson, G. R., Hutkins, R., Reimer, R. A., Reid, G., Verbeke, K., ..., & Sanders, M. E. (2020). The International Scientific Association for Probiotics and Prebiotics (ISAPP) consensus statement on the definition and scope of synbiotics. *Nature Reviews Gastroenterology and Hepatology.* 17(11): 687–701. https://doi.org/10.1038/nrgastro.2017.75

Szydłowska, A., & Sionek, B. (2022). Probiotics and postbiotics as the functional food components affecting the immune response. *Microorganisms.* 11(1),104 doi: 10.3390/microorganisms11010104

Thakur, V., Singh, A., Nabin, J., & Neeraj, M. (2019). Spray dried formulation of mesalamine embedded with probiotic biomass for the treatment of ulcerative colitis: In-vitro and in-vivo studies. *Drug Development and Industrial Pharmacy.* 45(11): 1807–1820. https://doi: 10.1080/03639045.2019.1665059

Torres-Maravilla, E., Lenoir, M., Mayorga-Reyes, L., Thibault, A., Sokol, H., Langella, P., Sánchez-Pardo M., & Bermúdez-Humarán, L. (2016). Identification of novel anti-inflammatory probiotic strains isolated from pulque. *Applied Microbiology and Biotechnology.* 100(1): 385–396. https://doi: 10.1007/s00 253-015-7049-4.E

Tufarelli, V., & Vito L. (2016). An overview on the functional food concept: Prospectives and applied researches in probiotics, prebiotics and synbiotics. *Journal of Experimental Biology and Agricultural Sciences.* 4(3S): 273–278. https://dx.doi.org/10.18006/2016.4(3S).273.278

Vinderola, G., Capellini, B., Villarreal, F., Suárez, V., Quiberoni, A., & Reinheimer, J. (2008). Usefulness of a set of simple in vitro tests for the screening and identification of probiotic candidate strains for dairy use. *LWT – Food Science and Technology.* 41: 1678e1688. https://doi: 10.1016/j.lwt.2007.10.008

Xu, C., Lei, Q., Li, M., Shuqi, Y., Yu, G., Xina, D., Baohua, Z., & Alexandra, R. (2019). Biosynthesis of polysaccharides-capped selenium nanoparticles using *Lactococcus lactis* NZ9000 and their antioxidant and anti-inflammatory activities. *Frontier Microbiology.* 10: 1632. https://doi: 10.3389/fmicb.2019.01632

Yan, F., & D. B. Polk. (2011). Probiotics and immune health. *Current Opinion in Gastroenterology.* 27(6): 496–501. https://doi.org/10.1097/MOG.0b013e32834baa4d

Zhang, G, Jian, T., Faizan A., Weizhen, Z., & Wei, W. (2019). Prevalence, genetic diversity, and technological functions of the *Lactobacillus sanfranciscensis* in sourdough: A review. *Comprehensive Reviews in Food Science and Food Safety.* 18(4): 1209–1226. https://doi.org/10.1111/1541-4337.12459

Zhou, X., Du, H., Jiang, M., Zhou, Ch., Deng, Y., Long, X., & Zhao, X. (2021). Antioxidant Effect of Lactobacillus fermentum CQPC04-Fermented Soy Milk on D-Galactose-Induced Oxidative Aging Mice. *Frontiers in Nutrition.* 8. 10.3389/fnut.2021.727467

Zhu, C., Ramesh, G., Patterson, T., & Nicholson, S. E. (2018). A review of traumatic brain injury and the gut microbiome: Insights into novel mechanisms of secondary brain injury and promising targets for neuroprotection. *Brain Science.* 8: 113. https://doi.org/10.3390/brainsci8060113

Chapter 12

Probiotics, Gut Microbiota, and Health

Wendoly Domínguez-Rojas
Antonio Tornero-Martínez
Diana E. Leyva-Daniel
Rosalva Mora-Escobedo
and Brenda Pérez-Grijalva

CONTENTS

12.1 INTRODUCTION

Throughout history, the importance of microorganisms for human health has been gradually discovered (Rastogi et al. 2022). It is known that some pathogenic bacteria can be transmitted to humans through foods, and they can cause diseases and even death; therefore, several techniques have been investigated to eliminate or prevent food contamination (Zhang et al. 2021). Additionally, certain bacteria can be harnessed to produce foods and provide them with

DOI: 10.1201/9781003332558-14

unique sensorial attributes (Rastogi et al. 2022). Still, little is known on the influence that the bacteria in the human diet exert on metabolism and health in general (Agrizzi et al. 2022). Current studies show the existence of a bilateral communication between the gut and the brain (Li et al. 2021). The latter can affect the intestinal environment while gut microbiota can alter the brain microenvironment through neuronal, endocrine, and immune pathways. Understanding the mechanisms through which probiotics regulate different functions in the organism, improving health, is key to developing potential therapeutic interventions aimed at preventing diseases (Abdhul et al. 2014; Zhang & Mo, 2023).

12.2 PROBIOTICS THROUGH THE HISTORY

The beneficial role of milk fermented with microorganisms has been known for a very long time. In 1907, Ilya Metchnikoff, a Russian microbiologist at the Pasteur Institute, observed that people in an underprivileged area of Bulgaria lived longer. Most of the population in the area drank fermented milk and their average life expectancy reached 100 years. Although Louis Pasteur had already discovered the process of milk fermentation and inhibition through pasteurization, Metchnikoff managed to isolate the microorganisms (*Lactobacillus bulgaricus*) to study their properties. He attributed the beneficial effects of consuming yogurt microorganisms on the host's health to a reduction in the bacteria-producing toxins in the digestive tract (Metchnikoff, 1907). In 1965, Lilly & Stillwell named as *probiotics* the substances produced by microorganisms that promote the growth of other microorganisms (Lilly & Stillwell, 1965). In 1989, Fuller defined probiotics as "A live microbial feed supplement that has a beneficial effect on the host animal by improving the intestinal microbial balance" (Fuller, 1989). Later, another definition considers "oral probiotics are living micro-organisms, which upon ingestion in certain numbers exert health benefits beyond inherent basic nutrition" (Guarner & Schaafsma, 1998). Therefore, they can be ingested as food components or non-food preparations. Then, functional foods with added probiotics are defined as "a product containing viable organisms of human origin, that when given in high concentrations improve balance of the endogenous gut microflora, inhibit growth of pathogenic organisms and produce health benefits beyond inherent general nutrition." It must be noted that bacteria can be considered probiotic when they survive the highly acidic environment in the stomach and colonize the small and large intestines. As a consequence, there is a scientific debate on whether the consumption of foods added with probiotics allows enough of these microorganisms to reach the intestine, since most of them are often destroyed by the hydrochloric acid in the stomach. This evidenced the need to legislate the functional statements surrounding the products manufactured using probiotics. In the 1980s, a legislation aimed to regulate probiotics and prebiotics, understanding that a prebiotic is "a non-digestible food ingredient that beneficially affects the host by selectively stimulating the growth and/or activity of one or a limited number of bacteria in the colon, and thus improves host health" (Gibson & Roberfroid, 1995). After a post-war government study, the Japanese FOSHU (Foods for Special Health Use) determined the functions of daily foods beyond nutrition. The program also established the importance of probiotics and their effect on life expectancy. A decade later, interest in functional foods, especially probiotics, spread across Europe and the United States, giving rise to adequate regulations to protect consumers and guarantee products that establish a dose–response relationship with recommended minimum and maximum effective amounts. In 1998, probiotics were defined as living microorganisms that promote the host's health when administered in adequate doses, participating in several mechanisms and functions of the organism (Guarner & Schaafsma, 1998). In 2001, a group of experts from the United Nations' Food and Agriculture Organization (FAO) and the World Health Organization (WHO) defined probiotics as "live microorganisms which when administered in adequate amounts confer a health benefit on the host." The following year, the working group FAO/WHO published the guidelines to help regulators, scientists, and industry interpret the original document (FAO/WHO, 2002). Although both documents establish the protocol to evaluate the efficacy of probiotics delivered as food, there are inconsistencies when considering scientific developments and applications. For instance, several probiotics are not administered as foods, and the definition does not provide thresholds on the type of effects on health and the target population, leading to the misuse of the term probiotic in products that do not meet the required standards. Therefore, the International Scientific Association for Probiotics and Prebiotics (ISAPP) sought to protect the consumer from deceptive claims and, in 2013, it organized an expert meeting to re-examine the concept of probiotic. The panel of experts concluded that the definition proposed by FAO/WHO in 2001, "live microorganisms that, when administered in adequate amounts, confer a health benefit on the host" is comprehensive enough to include a wide range of products developed and sufficiently restrictive to establish core requirements (Hill et al. 2014).

The following must be considered when using the term probiotic: attachment to intestinal cells, colonization in the gastrointestinal tract, resistance to acid and bile, production of antimicrobial substances, and exertion of beneficial effects on the host's health. However, when probiotics are used to improve health, the strain must be considered, along with the minimum daily amount recommended and the period of consumption. In addition, it must be noted that probiotics can have different means of administration, target populations, and target sites, and they must be safe to use. These data must be the result of *in vitro* and *in vivo* studies and, if possible, their effects must be tested in humans.

The panel of ISSAP experts (Hill et al. 2014) classifies probiotics into four categories:

- Probiotic in food or supplements without health claim: It contains strains that could have an effect similar to that of other taxonomic groups, the benefits of which have been proven in humans.
- Probiotic in food or supplements with a specific health claim: It contains defined probiotic strains with proven activity in the specified health indication and, at the end of shelf-life, has viable strains at the effective dose.
- Probiotic drug: It contains defined strains of live microorganisms with specific indication for the treatment or prevention of a disease and can deliver viable strains at the effective dose at the end of shelf-life.
- Not probiotic: It contains live and active cultures, including any food fermentation microbes; the metabolites it excretes and substances constituting it are not considered probiotics.

These specifications promote the continuous progress in probiotics research and awareness of their beneficial effects in patients and consumers.

The most commercial probiotics available in the foods market are Gram-positive bacteria of the genera *Lactobacillus* and *Bifidobacterium,* used as probiotics since they are often considered to promote health. Additionally, they are classified as "Generally Recognized as Safe" (GRAS) by the United States Food and Drug Administration (FDA) (Min et al., 2019) and as QPS (Qualified Presumtion of Safety) by the European Food Safety Authority (EFSA).

One of the most commonly studied groups are lactic acid bacteria (LAB), which include the genera *Enterococcus, Lactobacillus, Lactococcus, Leuconostoc, Aerococcus, Alloiococcus, Bifidobacterium, Carnobacterium, Dolosigranulum, Lactosphaera, Melissococcus, Oenococcus, Tetragenococcus, Vagococcus, Weissella, Pediococcus,* and *Streptococcus* (Mitsuoka, 2009).

The genus *Lactobacillus* is highly important due to its role in food fermentation and human and animal health. Still, since 1901, around 250 species have been assigned to this genus, and modern tools for DNA study have allowed revealing that many of these species are faintly related. Therefore, experts agreed to reclassify these bacteria in 25 genera, 23 of which are new, and arranged them in functionally important groups: *Lactobacillus, L. delbrueckii, Paralactobacillus, Acetilactobacillus, Agrilactobacillus, Amylolactobacillus, Apilactobacillus, Bombilactobacillus, Companilactobacillus, Dellaglioa, Fructilactobacillus, Furfurilactobacillus, Holzapfelia, Lacticaseibacillus, Lactiplantibacillus, Lapidilactobacillus, Latilactobacillus, Lentilactobacillus, Levilactobacillus, Ligilactobacillus, Limosilactobacillus, Liquorilactobacillus, Loigolactobacillus, Paucilactobacillus, Schleiferilactobacillus,* and *Secundilactobacillus* (Zheng et al. 2020).
Products fortified with probiotic bacteria must meet certain criteria:

a) Probiotic bacteria must be prepared in a viable manner at large scale, b) they must remain viable during their use and storage, c) they must survive the conditions in the gastrointestinal tract, d) they must provide beneficial effects to the host, and e) they must be proven to be safe for consumption. Thus, good manufacturing practices (GMP) must be applied in the making of such foods to guarantee product quality and viability of microorganisms during their shelf-life.

12.3 GUT MICROBIOTA

Gut microbiota is a diverse community of bacterial species metabolically active that have complex interactions to keep a mutualistic relationship between beneficial symbionts and the host; these interactions are defined as intestinal homeostasis or eubiosis. Gut microbiota consists of two major domains: a) Bacteria and b) Archaea. The first is mostly constituted (90%) by the phyla Bacteroidetes that includes gram-negative bacteria, Firmicutes that includes gram-positive bacteria, and Actinobacteria (Rinninella et al. 2019). A smaller domain, Archaea, consists of methanogenic archaea that produce methane. The most common methanogen in the gut microbiota is *Methanobrevibacter smithii*. Archaea overgrow in patients with inflammatory bowel disease, periodontal disease, obesity, cancer, and diverticulosis when compared to healthy individuals. Still, there is no evidence indicating they ever act as pathogens (Mafra et al. 2022).

Given the acidity conditions and content of digestive enzymes, the stomach contains around 10^3 colony-forming units per milliliter (CFU/mL) of gastric juice. The predominant bacteria here are acid-tolerant, such as *Lactobacillus* and *Enterococcus*. In the ileum and large intestine, there is a final concentration of 10^9–10^{12} bacteria/g content (Ghavami et al. 2018), and most are forced anaerobic bacteria. Still, this bacterial content can vary depending on age, genetic factors, drug treatments, and interaction with diet components (Montassier et al. 2021), among others. Alterations in the composition of gut microbiota lead to an unbalance or intestinal dysbiosis and the clinical expression of several diseases. There must be a healthy balance between the host and microorganisms so that metabolic and immune functions are carried out optimally and the development of diseases is prevented (Rinninella et al. 2019). Microorganisms take part in functions such as vitamin and SCFA synthesis and fermentation of non-digestible nutrients while acting as an energy source for intestinal epithelial cells and strengthening the intestinal barrier. In addition, the microbiota takes part in the regulation of the immune and endocrine systems and metabolism. It modulates the signaling pathways involved in the homeostasis of the intestinal mucosa and brain physiology modulation (McCann et al. 2006).

12.3.1 Interaction between Probiotics and Diet

Gut microbiota is characterized by an inter-individual variability due to genetic and environmental factors (Bibbò et al. 2016). Diet is one of the main regulators of gut microbiota composition. Given the preference of microorganisms in the microbiota for a certain substrate and the existence of an intense competition for resources, alterations in diet components lead to changes in the growth of specific bacterial strains and direct effects on health (Gareau et al. 2010; Bibbò et al. 2016).

Microbial colonization starts after birth when several microbes fast colonizes the gut. In adult age, the microbiota is more resistant to changes because of its lower plasticity but can be altered by environmental factors throughout life (Laitinen et al. 2022).

The formation of human gut microbiota starts at birth and is established after the first year of life; the breast milk promotes the healthy development of the gut microbial community, the optimal maturation of the immune system, and the defense against pathogens. Infants fed with breast milk show a large proportion of Bifidobacterium, a bacterial species that ferments oligosaccharides in breast milk. This is not observed in those infants fed with formula added with prebiotics (as oligosaccharides from breast milk) that imitate the functions of the unique components of breast milk. Those infants show significantly higher proportions of bacteria from the phyla Firmicutes and Bacteroidetes, linked to overweight and obesity. Then, during the stage of transition to solid foods, they contribute to the maturation of the infant's gut microbiome, reducing the proportion of Bifidobacterium and increasing the presence of Firmicutes (Westfall et al. 2018).

The western diet characterized by foods low in fiber and phytochemicals but rich in refined carbohydrates and saturated fatty acids modifies the gut microbiota (Bibbò et al. 2016). The western diet has been linked to the risk of developing chronic degenerative disorders such as obesity, diabetes mellitus, cancer, and cardiovascular diseases. This diet is correlated to intestinal disbiosis, which triggers an alteration in the intestinal barrier functions and increases gut permeability leading to an inflammatory state (Westfall et al. 2018).

A healthy diet including abundant fruits, vegetables, whole-grain cereals, fermented foods, and legumes (Bibbò et al. 2016) helps to prevent chronic degenerative diseases and others. Compounds that are accessible to the gut microbiota, such as polyphenols and fiber, act as prebiotics. For example, anthocyanins and gallic acid promote the growth of genus *Bifidobacterium* and *Lactobacillus.* Additionally, some polyphenols also inhibit the growth of pathogenic bacteria. For example, quercetin suppresses the growth of *Erysipelotrichaceae*, *Bacillus*, and *Eubacterium cylindroides* associated with obesity (Li et al. 2021). Shi et al. (2022) suggested the possibility of a synergistic activity between fiber and polyphenols.

In the colon, SCFAs are the main metabolite in the fermentation of dietary fiber; they are easily absorbed as nutrients by colonocytes and other tissues. In addition, SCFAs can reduce the pH value in the colon (Ye et al. 2022) to prevent the excessive growth of pathogenic bacteria and modulate inflammatory processes and show positive effects on health, including better tolerance to glucose by regulating the ratio of *Bacteroides*/*Prevotella* and reducing the risk of hepatocellular carcinoma, among others (Wu et al. 2020).

12.4 METABOLITES RELEASED DURING PROBIOTIC METABOLISM

The term postbiotic was introduced recently to describe the metabolites that are liberated during probiotic metabolism and play a key role in the host's health. The metabolites produced can be classified according to their molecular nature. Some of the major metabolites are SCFA, amino acids, proteins, bacteriocins, vitamins, and neurotransmitters.

It is known that SCFA are carboxylic acids of no more than six carbons. They are produced in the proximal colon by probiotics and commensal bacteria, such as *Faecalibacterium prausnitzii, Eubacterium rectale, Eubacterium hallii,* and *Ruminococcus bromii* that can convert prebiotic polysaccharides or oligosaccharides into acetate, butyrate, and propionate (Koh et al. 2016). Acetate and butyrate are commonly produced by *Bifidobacterium* species, while *Lactobacillus* species (e.g., *Lentilactobacillus buchneri, Lactobacillus diolivorans, Limosilactobacillus reuteri*) produce propionate. A combination of probiotics and prebiotics (synbiotics) is known to selectively stimulate predominant bacteria and SCFA production. For instance, *Lacticaseibacillus rhamnosus* strain GG (LGG) produces SCFA when mixed with prebiotics (e.g., fructo-oligosaccharides (FOS) and pectin) (Slavin, 2013). *Bifidobacteria* such as as *Bifidobacterium longum* SP 07/03 and *Bifidobacterium bifidum* MF 20/5 produce and release propionate and acetate (LeBlanc et al. 2017). The mechanisms through which SCFA exert their beneficial effect are varied. Butyrate induced apoptosis due to its capacity to suppress histone deacetylases (Flint et al. 2012). Anti-inflammatory actions are the result of SCFA such as butyrate, propionate, and acetate reducing the signaling of the kappa light chain enhancer of the nuclear factor kappa B (NF- κB), nitric oxide production, and reactive oxygen species (ROS) production in renal tubular epithelial cells (Park et al. 2007; Andrade-Oliveira et al. 2015).

12.4.1 Amino Acids

Probiotics can synthesize some essential amino acids in response to several adverse conditions (Chávarri et al. 2021). Probiotics such as *Lactobacillus plantarum* or *Lactococcus lactis* can produce glutamic acid, aspartic acid, proline, and alanine against osmotic stress (Papadimitriou et al. 2016). The production of these amino acids entails multiple benefits since they participate in the synthesis of hormones, neurotransmitters, nucleic acids, and melanins (Aliu & Arnold, 2018). Some amino acids can be used by probiotics to produce other metabolites. The effects of indolic acid derivatives produced from tryptophan (e.g., indole-3-acetic acid, indole-3-aldehyde, indolyacryloylglycine, indole lactic acid, indole acrylic acid, indole propionic acid) can affect intestinal homeostasis (Liu et al. 2018). It has been reported that *Lactobacillus reuteri* produces indole-3-aldehyde that contributes to antifungal resistance and the protection of mucosa against inflammation (Zelante et al. 2013). In addition, there are reports on the production of extracellular proteins by *Lactobacillus plantarum* BMCM12, which attenuates the attachment to pathogens and the protection of the intestinal barrier (Sánchez & Urdaci, 2012).

Bacteriocins are peptides with antimicrobial activity produced by *Lactobacillus, Enterococcus,* and *Brevibacillus* spp. (Goderska et al. 2018). It has been reported that the bacteriocin produced by *Lactobacillus acidophilus* KS400 can potentially inhibit the growth of vaginal pathogens such as *Gardnerella vaginalis,* Streptococcus agalactiae, and *Pseudomonas eruginos* (Gaspar et al. 2018).

12.4.2 Vitamins

Vitamins are organic molecules that play an essential role in biological processes in the body. LAB produce abundant quantities of group B vitamins. Some species of lactobacillus, such as *Lactococcus lactis, Lactobacillus gasseri, Lactobacillus reuteri*, and *Bifidobacterium*, produce vitamins (Thakur, et al. 2016). Folate or vitamin B11 participates in DNA repair, methylation, and replication, as well as DNA decrease in Alzheimer's disease, cancer, or coronary disease (Ibrahim et al. 2015).

12.4.3 Neurotransmitters

Neurotransmitters are chemical compounds that regulate neuronal signaling to ensure brain–body homeostasis (Mora et al. 2012). Recent studies indicate the existing relationship between gut microbiota and the gut–brain axis. This communication takes place through neuroactive and neuroendocrine molecules, like serotonin, γ-aminobutyric acid (GABA), histamine, noradrenaline, and adrenaline. GABA is a nonprotein amino acid classified as inhibitory neurotransmitter of the central nervous system (CNS), acting through three receptors: GABAA, GABAB, and GABAC. Through these receptors, GABA regulates mood, sleep disorders, and short-term and spatial memory (Sigel & Steinmann, 2012). Lactobacilli can convert glutamate into GABA, which inhibits pain. Serotonin can be produced by *Lactococcus lactis, Lactobacillus plantarum*, or *Streptococcus thermophilus* (Liang et al. 2018) using tryptophan as precursor metabolite. Furthermore, *Aspergillus oryzae* is capable of synthesizing 3,4-dihydroxyphenyl Ala (L-dopa) having positive effects on Parkinson's disease symptoms. The enzymatic oxidation of tyrosine (Try) promotes L-dopa production (Ali a & Haq, 2010), while some probiotics as *Lactococcus lactis* can even use L-dopa as a precursor of dopamine, another essential neurotransmitter (Vodolazov et al. 2018).

Among metabolites produced by probiotics are organic acids, reuterin, ribosomal proteins, linoleic acid, and secondary biliary acids, which have antimicrobial activity. The increase in organic acids, including lactic acid and SCFAs, acidifies the micro-environment to fight pathogenic microorganisms that prefer a slightly alkaline environment (Kim & White, 2009).

12.5 PROBIOTICS, MICROBIOTA AND THE INTESTINAL BARRIER

The correct functioning of intestinal mucosa largely depends on the integrity of the intestinal barrier. The integrity of the epithelial barrier provides better absorption as it keeps an effective defense against pathogens and harmful chemical elements. An increasing number of studies have proven the relationship between the altered integrity of the epithelial barrier and several gastrointestinal diseases (Mu et al. 2022). The intestinal epithelium creates a protective barrier against luminal content, such as microorganisms and food antigens. The function of this barrier is mediated through the tight junctions of epithelial cells in which paracellular proteins such as zonula occludens 1 (ZO-1) take part (Donato et al. 2010).

As evidence of the relationship between epithelial integrity and prevention of intestinal inflammation, Hermiston & Gordon, (1995) studied cell adhesion molecules (cadherins) in some regions of the small intestine in a murine model, where only the expression of N-cadherin and not of E-cadherin was shown. This disrupts the interactions of E-cadherin that help to maintain junction integrity. The intestinal regions expressing N-cadherin were more permeable than those expressing E-cadherin and developed local lesions due to inflammation. Several studies using mice models have proven that the anomalies in the pathways maintaining the epithelial integrity, or the regulation of mucosal immune responses, lead to the destruction of the barrier and colitis. The latter is triggered by bacterial antigens and microbe-associated molecular patterns (MAMPs) (Hermiston & Gordon, 1995).

The presence of probiotics benefits intestinal barrier integrity through NF-κB pathway inhibition, extracellular-signal regulated kinase (ERK) 1/2, and the activation of toll-like receptor-2 (TLR2) (Arseneault-Bréard et al. 2012). The activation of NF-κB is induced by the tumor necrosis factor-α (TNF-α), and it plays a modulating role in the expression and localization of ZO-1. Although its action is not fully defined, there are three possibilities: (1) activated NF-κB is bound to DNA in the ZO-1 promoting region or around it, creating a reduction in the expression of the protein, (2) NF-κB might induce the transcription and translation of certain genes that inhibit ZO-1 expression, and (3) NF-κB could accelerate ZO-1 degradation, likely through proteasome activation. On the other hand, NF-κB acts as an antiapoptotic factor given that it represses the apoptotic action caused by TNF-α (Ma et al. 2004). Then, NF-κB mediates both disruptive and protective effects.

It has been observed that the inoculation of *Lacticaseibacillus Rhamnosus* GG (LGG) on a human colon cell monolayer (Caco-2bbe) inhibits the nuclear translocation of NF-κB subunit p65 and, therefore, its activation. In addition, LGG can partially inhibit NF-κB activation since it could exacerbate interferon γ (IFN-γ) and TNF-α levels, leading to the disruption of the epithelial barrier. Additionally, it must be noted that there is no positive effect of the probiotic on such pathway if ERK-1/2 is blocked. These kinases are not necessary to maintain ZO-1 apical localization (Donato et al. 2010). The maintenance of the intestinal barrier can also be achieved through the microbial recognition of TLR2, so probiotics could be beneficial (Arseneault-Bréard et al. 2012). A study shows that TLR2 stimulation preserves ZO-1 association, which maintains intestinal barrier integrity against damage by induced stress; then, the activation of PI3K/Akt pathway by TLR2 is essential (Cario et al. 2007). When intestinal cells (Caco-2) are treated with the probiotic *Lacticaseibacillus casei*, there is an increase in TLR2 and Akt protein in contrast to untreated cells. This increase leads to enhanced trans epithelial resistance as an integrity measure of the intestinal barrier. Therefore, *L. casei* can revert the dysfunction in the intestinal barrier through TLR2 (Eun et al. 2011). There are also studies using an established model of porcine intestinal epithelial cells (IPEC-J2) that show a protective effect of *Lactobacillus plantarum* against the epithelial rupture caused by enterotoxigenic *Escherichia coli*. It was confirmed that *L. plantarum* reduced the genetic expression of interleukin-8 (IL-8) and TNF-α in differentiated IPEC-J2 cells. These cytokines are overexpressed in the inflamed intestine and are involved in increased gut permeability. It has been proven that TNF-α-stimulated intestinal epithelial cells reduce the expression of claudin-1 and occludin and ZO-1 localization and expression (Ye & Sun, 2017; Andrews et al. 2018). This is how the bacteria protect against damage of the epithelial barrier, maintaining the intestinal homeostasis and intestinal tight junctions. Furthermore, it was proven that *L. plantarum* partially suppressed the expression of inflammatory cytokines by modulating NF-κB and MAPK (Adenosine monophosphate-activated protein kinase) pathways (Wu et al. 2016).

Common changes in the intestinal barrier take place through pathogens, especially bacteria and viruses. Several studies proved that the permeability of the intestinal barrier produced by pathogens could improve with probiotics. The action mechanism is a competition between bacteria, like *Lactobacillus plantarum*, that rivals the enteropathogenic bacteria *Escherichia coli* (ETEC), for attachment to the intestinal mucosa (Pretzer et al. 2005). The competitive exclusion of ETEC by recognition of the same adherence sites on the intestinal epithelial surface is a potential action mechanism. It is currently unclear whether these mechanisms are enough to antagonize pathogens since they cannot be exactly measured *in vivo*. Additionally, a large amount of microbiota resides in the mucosal lumen; still, there is always a thick, firm mucus layer that offers a protective barrier (Jakobsson et al. 2015). Other mechanisms by which probiotics act against pathogens are competition for macronutrients and by the production of secondary metabolites that can be bactericides (Derrien and van Hylckama Vlieg, 2015). One of them is class II bacteriocin Abp118 produced by *Lactobacillus salivarius* UCC118 that effectively inhibits the infection by *Listeria monocytogenes* and breaks the barrier in vivo in a mouse model (Corr et al. 2007).

For the reasons above, we can conclude that an adequate treatment with probiotics, such as as *Lactobacillus plantarum* (Karczewski et al. 2010), might help maintain and recover the intestinal barrier. This would prevent endotoxins and lipopolysaccharide (LPS) from entering the bloodstream and could lead to hyperactivation of the immune system.

12.6 PROBIOTICS AND MICROBIOTA-GUT-BRAIN AXIS

In recent years, there has been an increasing interest in the influence of the microbiota at brain level since it can modulate emotional behavior and brain physiology. It regulates neurotransmitter production, neurotrophic factors such as the brain-derived neurotrophic factor (BDNF), and synaptogenesis (Heijtz et al. 2011). However, the communication mechanisms are still under investigation.

There is a bidirectional communication between gut microbiota and the brain (microbiota–gut–brain axis) through neuroendocrine, neuroimmune, autonomous nervous system, and enteric nervous system pathways (Mayer, 2011). As mentioned before, when there is an unbalance in the composition of the gut microbiota, the integrity of the epithelial barrier is compromised. The metabolites produced by these bacteria, such as LPS, toxins, and functional amyloid proteins, enter the bloodstream and could reach the CNS.

Nevertheless, the brain and the spinal cord are protected by a specialized vascular barrier known as the blood–brain barrier (BBB). It works as an immune and metabolic barrier that restricts the access of substances and molecules in the bloodstream to the nervous parenchyma, thus maintaining the homeostasis of the CNS chemical microenvironment. The BBB is constituted by endothelial cells (ECs) with tight junctions created by the interaction of transmembrane proteins, making the barrier highly selective when allowing substances access to the brain (Michel & Prat, 2016). These tight junctions are constituted by claudin, occludin, junction adhesion molecules (JAM proteins), and accessory cytoplasmic proteins like ZO (Vorbrodt & Dobrogowska, 2003) and cingulin (**Figure 12.1**). Additionally, the BBB is also created by pericytes, perivascular astrocytes, neurons, extracellular matrix, and microglia, which provide functional and structural support (Parker et al. 2020).

Several pathological conditions modify the functioning and normal development of the BBB. Microbiota can affect the integrity of the BBB since it is involved in the organization of the tight junctions through the regulation of zonulin (Shandilya et al., 2022), occludin, and claudin-5 production. The colonization of germ-free mice with pathogen-free gut flora has been linked to the enhanced integrity of the BBB and a higher expression of occludin and claudin-5 (Braniste et al. 2014). This is related to the presence of SCFA-producing bacteria in the healthy flora, which are involved in angiogenesis, neurogenesis, and the consolidation of long-term memory. Although the mechanisms through which SCFAs act remain unclear, studies show they might act in two ways: binding to G-protein-coupled receptors (GPCR41 and GPCR43) (Brown et al. 2003) or entering the cells to act as histone deacetylase (HDAC) inhibitors to modulate epigenetic processes (Eun et al. 2011).

There is an increasing interest in methylamines produced by the microbial metabolism of some dietary components such as choline and L-carnitine. The bacterial degradation of these dietary components in fish, red meat, and eggs produced trimethylamine (TMA), which is transformed into trimethylamine N-oxide (TMAO) in the liver (Hoyles et al., 2018). Both methylamines have been related to vascular diseases and integrity alterations of the BBB. However, a protective effect of TMAO has also been reported in hypertension and atherosclerosis animal models. In 2017, Dumas reported that TMAO reduces lipogenesis in adipocytes and improves insulin secretion and homeostasis of glucose in mice fed a high-fat diet (Dumas et al. 2017). Recently, studies reported that TMAO improves BBB integrity by protecting the barrier from inflammatory challenge through annexin A1 regulation. In contrast, TMAO damages the integrity of tight junctions, affecting the function of the BBB. Additionally, it has shown a protective effect of the cognitive function against an inflammatory challenge in a murine model, restricting astrocyte and microglia reactivity (Hoyles et al. 2021).

Other studies have focused on the relation between gut microbiota, tryptophan metabolism, and its effect in the brain. Diets with an adequate amount of tryptophan help to maintain a healthy microbiota (Dai et al. 2014; Wikoff et al. 2009). In turn, a healthy microbiota ensures that tryptophan results in a larger production of serotonin (which reduces risk of depression), melanin (which improves sleep), and BDNF (Musumeci et al., 2015). A study conducted in aged mice that received a diet low in tryptophan for 8 weeks revealed harmful changes in gut microbiota bacteria and higher systemic inflammation. The researchers reported that animals with lower tryptophan also showed reduced levels of *Clostridium* sp. (metabolizes tryptophan to produce metabolites beneficial to the host, as serotonin) and Mucispirillum (associated to production factors participating in the reduction of reactive oxygen species) and a significant increase in Acetatifactor (a bacterial genus from the family of *Lachnospiraceae*), linked to intestinal inflammation. These harmful alterations in microbiota composition promoted the release of higher amounts of proinflammatory cytokines (IL-6, IL-17A, IL-1a) and a considerable reduction in IL-27, which suppresses inflammation (Yusufu et al. 2021). However, diets rich in tryptophan both affect intestinal homeostasis and neurogenesis. Supplementing a diet with 0.4% tryptophan in aged mice significantly decreased oxidative stress and inflammation by regulating AMPK and NF-kB. In addition, the diet increased indole levels in the colon and the relative abundance of *Akkermansia muciniphila* a probiotic that potentially prevents colitis, diabetes, and aging. Supplementation with tryptophan was observed to inhibit microglial and astrocyte activation, promoted cAMP-response element-binding protein (CREB), and upregulated the expression of BDNF and postsynaptic density protein 95 (PSD95), involved in synapse formation and maintenance (Yin et al., 2021). The data obtained proved the positive modulation of the diet using 4% tryptophan on aging-induced neurodegeneration.

12.7 PROBIOTICS AND THEIR EFFECT ON HEALTH

As stated throughout this chapter, a dysbiotic gut microbiota has been associated with several diseases and disorders, such as obesity, diabetes mellitus (Wang et al., 2021), colon cancer,

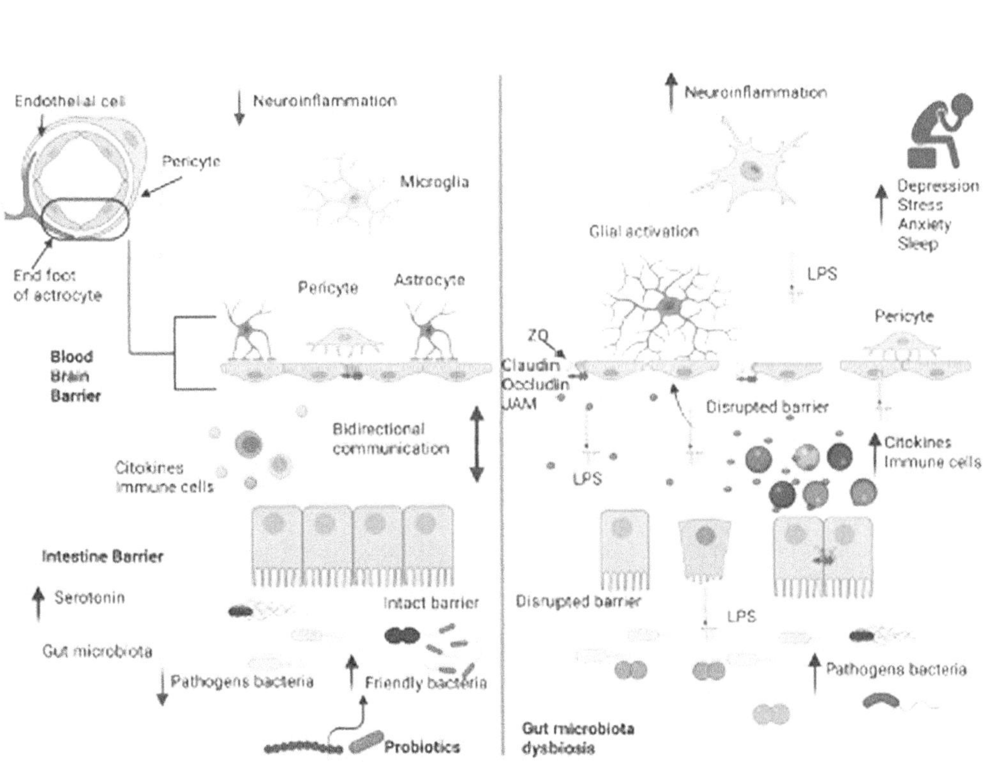

Figure 12.1 Effect of gut microbiota dysbiosis and probiotics on the integrity of the intestinal barrier and the BBB. (a) Intestinal mucosa composed of monolayer of epithelial cells and a subepithelial region containing the own lamela, enteric nervous system, connective tissue, and muscle layers. Compounds presents in the intestinal lumen can cross the epithelium through transcellular or paracellular routes. (b) BBB composed of brain microvascular endothelial cells, pericytes, the continuous basement membrane, and the perivascular feet of astrocytes. BBB interacts with other components such as extracellular matrices, neurons, and microglia. (c) The gut microbiota secretes metabolites, endotoxins, exopolysaccharides, neurotransmitters, and amilyde peptides that can affect the structure and function of the BBB. BBB= Blood Brain Barrier *LPS= lipopolysaccharides, zo= zonula ocludens, JAM= Junctional Adhesion Molecules.

food allergies, sleep disorders, stress, anxiety (Ma et al. 2021), depression, Alzheimer's disease (Shamsipour et al. 2021), Parkinson's disease, Multiple Sclerosis, and the gut retina-axis (Shang & Mo, 2023) (**Figure 12.2**).

Intestinal dysbiosis caused by antibiotics or certain pathologies increases intestinal permeability, allowing toxins such as LPS and bacterial amyloid proteins to enter the brain, causing neuroinflammation. Probiotics could restore the microbiota, promoting health and maintaining the integrity of the intestine and mucosa. They act as a pathogen suppressor agent, increasing the diversity of "good" bacteria, due to the secretion of metabolites such as SCFA that acidify the environment, regulate the immune response, and maintain the integrity of the BBB.

Probiotics produce neurotransmitters such as dopamine, serotonin, GABA, and neurotrophic factors such as BDNF, promoting neurogenesis in the brain. In addition, they intervene in the

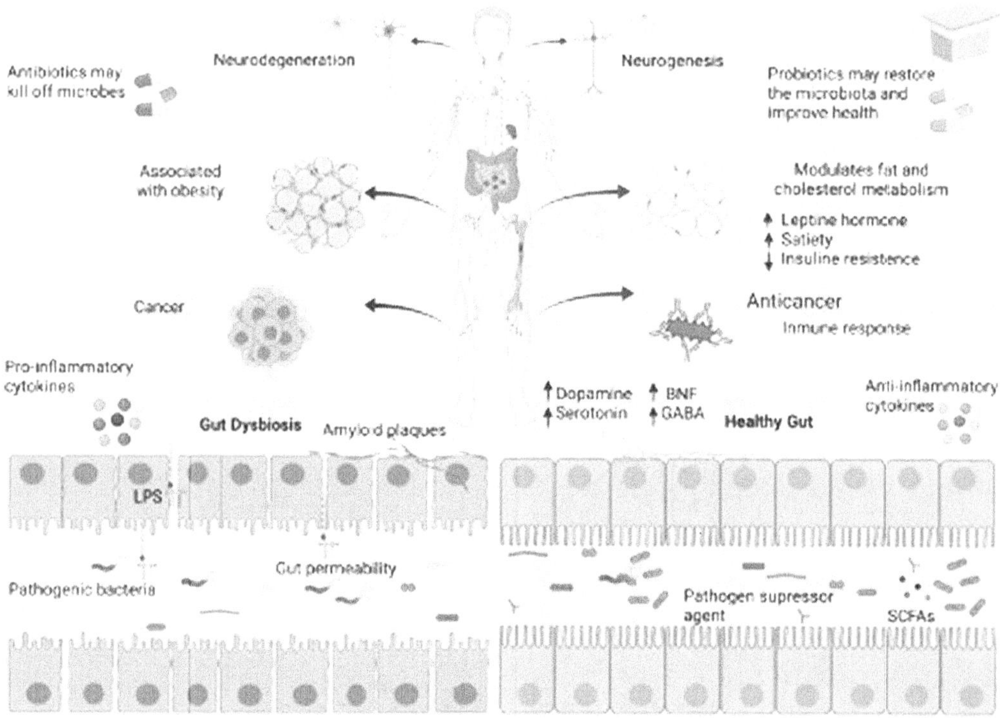

Figure 12.2 The effect of probiotics on the microbiota and the impact on health. Intestinal dysbiosis caused by antibiotics or certain pathologies increases intestinal permeability allowing the migration of toxins to the brain such as LPS and bacterial amyloid proteins, causing neuroinflammation, related to the development of pathologies including cancer, obesity, and Alzheimer's disease, among others. Probiotics restore the microbiota, promoting health, maintaining the integrity of the intestine and mucosa. They act as a pathogen suppressor agent, increasing the diversity of "good" bacteria, due to the secretion of metabolites such as SCFA that acidify the environment, regulate the immune response, and maintain the integrity of the BBB. Probiotics produce neurotransmitters such as dopamine, serotonin, GABA, and neutrophils such as Brain-Derived Neurotrophic Factor (BDNF), promoting neurogenesis in the brain. They intervene in the metabolism of lipids, cholesterol, and in the regulation of leptin, increasing satiety and reducing insulin resistance. LPS= lipopolysaccharides.

metabolism of lipids and cholesterol and in the regulation of leptin, increasing satiety and reducing insulin resistance, for which they are attributed an anti-obesity effect.

12.7.1 Modulation of Gut Microbiota Composition

Human health is related to the diversity of the intestinal microbiota, and its modulation through probiotics, which is key to the prevention and treatment of some diseases (Alberda et al., 2018). A high gut microbiota diversity is assumed to be important for maintaining a healthy physiological state, and some diseases have been associated with reduced gut microbial diversity, and gut dysbiosis could even partially perturb normal energy metabolism (Ma et al. 2021). The main mechanisms through which probiotics modulate the composition of the intestinal microbiota are the competition against pathogens for the substrate, production of growth substrate for other bacteria (vitamins and exopolysaccharides), and competition for binding sites on the enteric wall (O'Toole & Cooney, 2008).

Several works have proven the efficacy of probiotics. Gao et al. (2015) demonstrated that people with colorectal cancer (CRC) who were administered an oral mix of the probiotics *Bifidobacterium longum, Lactobacillus acidophilus,* and *Enterococcus faecalis* (1:1:1) showed a significant decrease in pathogenic bacteria (*Peptostreptococcus, Comamonas, Fusobacterium*) and an increase in *Enterococcus* and Proteobacteria in the mucosally adherent microbiota.

Other studies show that probiotics such as *Bifidobacterium breve* UCC2003 produces a exocellular polysaccharide, which helps probiotic bacteria to survive under adverse conditions (high acid and bile concentrations) in the intestine also protects them from the host's immune response and reduces colonization levels of the gut pathogens (Fanning et al. 2012). Probiotic *Lactobacillus salivarius* acts as suppressor agent of *Helicobacter pylori* by secreting lactic acid (Aiba et al. 1998). Since *H. pylori* is associated with gastric adenocarcinoma, probiotics could be involved in the prevention of gastric cancer development (Bultman, 2014).

12.7.2 Immune System Activation

It has been proven that probiotics positively affect the immune system and systemic inflammation through several mechanisms (Taverniti et al. 2022). Immunostimulant probiotics are probiotic microorganisms that induce responses in the production of proinflammatory cytokines (IL-6, IL-8, IL-10, IL-17, and TNF-α) (Zeinali et al. 2022). Immunoregulatory probiotics are bacterial strains that tend to activate regulatory T cells (Tregs).

There is evidence of SCFA anti-inflammatory activity through the regulation of T and dendritic cell proliferation or the suppression of proinflammatory cytokine production by antigen-presenting cells (Chang et al. 2014). Furthermore, the production of other metabolites such as polysaccharide A (PSA) produced by Bacteroides fragilis, lipid 654 produced by several Bacteroidetes species (Maslowski et al. 2009), and peptidoglycan (PGN) (Visser et al. 2005) directly affect the modulation of the immune system. *Bifidobacterium breve* prevents intestinal inflammation by inducing interleukin 10 production in the colon (Jeon et al. 2012). Supplementation with *Lactobacillus acidophilus* and *Bifidobacterium* spp. increases the immune response against cholera toxin (Tejada-Simon et al. 1999). Probiotics also affect diseases, such as obesity. In this disease, the hypertrophy of fat tissue creates hypoxia, which leads to cell apoptosis, which triggers an increase in macrophages and proinflammatory cytokines (Yazdani et al. 2018). Supplementation with *Lactobacillus* and *Bifidobacterium* increases anti-inflammatory cytokine levels.

12.7.3 Metabolism Regulation through Microbial Metabolite Production

Metabolites released by probiotics can promote body weight and adiposity loss, while controlling the metabolism and glucose (De Vadder et al. 2014). This effect is widely attributed to specific bacterial strains, such as *Lactobacillus rhamnosus* GG (Patterson et al. 2019), *Lactobacillus spp., Bifidobacterium spp., Lactobacillus gasseri, Lactobacillus casei,* and *Bifidobacterium pseudocatenulatum* (Núñez et al. 2014). This supplementation has a direct effect on the profile of the gut microbiota, reverting intestinal permeability. There is an increase in the proportion of firmicutes vs. bacteroidetes, reducing pathogenic bacteria and promoting bacterial diversity. There are also effects on leptin hormone, increasing satiety, reducing insulin resistance, and modulating fat and cholesterol metabolism (Wang et al. 2021).

12.7.4 Brain Physiology Modulation

Psychobiotics are defined as live microorganisms (probiotics) that confer mental health benefits to the host through interactions with commensal gut bacteria when administered in the right amounts (Ma et al. 2021). These interactions with gut microbiota can regulate the homeostasis of the host's CNS directly through vagus nerve signaling and indirectly via modulating the host's metabolism by producing neuroactive compounds such as SCFAs, serotonin (5-hydroxytryptamine [5-HT]), bile acids, and γ-aminobutyric acid (GABA) (Yu & Zhao, 2021). Specifically, serotonin is a key neurotransmitter in the brain-gut axis: It regulates the signal transmission of enteric glial cells and plays an essential role in intestinal motility, liquid secretion, and gastrointestinal sensation.

Psychobiotics such as *Lactiplantibacillus plantarum* P8 (Shamsipour et al. 2021), L. spp., *Bifidobacterium* spp., and *Lacticaseibacillus rhamnosus* JB-1 (Tette et al. 2022) are able to synthesize dopamine, serotonin, and GABA (neurotransmitters and neuromodulators), which can reduce stress-induced behaviors (Shamsipour et al., 2021), depressive states, and anxiety (Amirani et al. 2020; Yu & Zhao, 2021; Tette et al. 2022). *Bacillus* spp. produces noradrenaline and dopamine, while *Lactobacillus* spp. produces acetylcholine (Lyte, 2011); they promote neurogenesis in the hippocampus; *L. helveticus* is used in elderly people to prevent and improve cognitive functions.

12.8 CONCLUSION

Today, it is well-known that a balanced gut microbiota is key to maintaining our health and preventing a number of diseases. The different metabolites generated by probiotics and their deficiency, SCFA (as acetic, propionic, and butyric acids), amino acids (as glutamic acid, aspartic acid, proline, and alanine), and neurotransmitters (GABA, noradrenaline, and adrenaline), among others, have been associated with diseases and disorders (type-2 diabetes, obesity, colon cancer, sleep disorders, stress, anxiety, depression, Alzheimer's disease, and Parkinson's disease). Scientific studies prove the efficiency of using probiotics to modulate the gut microbiota and the production of such metabolites to prevent and treat these diseases. Nevertheless, one of the main challenges is to find the mechanisms responsible for the effect of probiotics since the efficacy of these therapies depends on factors like strain, minimum effective dose, target population, and interaction with other food components, among others. Finally, it is key to have adequate biomarker selection to evaluate the efficacy of these products.

REFERENCES

Abdhul, K., Ganesh, M., Shanmughapriya, S., Kanagavel, M., Anbarasu, K., & Natarajaseenivasan, K. (2014). Antioxidant activity of exopolysaccharide from probiotic strain Enterococcus faecium (BDU7) from Ngari. *International Journal of Biological Macromolecules, 70*, 450–454. https://doi.org/10.1016/j.ijbiomac.2014.07.026

Agrizzi, T., Sant, C., Grancieri, M., Tako, E., Cristina, M., Paes, D., Stampini, H., & Martino, D. (2022). Black corn (*Zea mays* L.) flour has the potential to improve the gut microbiota composition and goblet cell proliferation in mice fed a high-fat diet. *Nutrition Research, 108*, 60–72. https://doi.org/10.1016/j.nutres.2022.10.005

Aiba, Y., Suzuki, N., Kabir, A.M.A., Takagi, A., & Koga, Y. (1998). Lactic acid-mediated suppression of *Helicobacter pylori* by the oral administration of *Lactobacillus salivarius* as a probiotic in a gnotobiotic murine model. *American Journal of Gastroenterology, 93*(11), 2097–2101. https://doi.org/10.1111/j.1572-0241.1998.00600.x

Alberda, C., Marcushamer, S., Hewer, T., Journault, N., & Kutsogiannis, D. (2018). Feasibility of a lactobacillus casei drink in the intensive care unit for prevention of antibiotic associated diarrhea and clostridium difficile. *Nutrients, 10*(5): 539. https://doi.org/10.3390/nu10050539

Ali, S., & Haq, I. (2010). Production of 3, 4-dihydroxy L-phenylalanine by a newly isolated Aspergillus niger and parameter significance analysis by Plackett-Burman design. *BMC Biotechnology,* 10:86. https://doi.org/10.1186/1472-6750-10-86

Aliu, E., Kanungo, S., & Arnold, G.L. (2018). Amino acids disorders. *Annals of Translational Medicine, 6*(24), 471. https://doi.org/10.21037/atm.2018.12.12

Amirani, E., Milajerdi, A., Mirzaei, H., Jamilian, H., Mansournia, M.A., Hallajzadeh, J., & Ghaderi, A. (2020). The effects of probiotic supplementation on mental health, biomarkers of inflammation and oxidative stress in patients with psychiatric disorders: A systematic review and meta-analysis of randomized controlled trials. *Complementary Therapies in Medicine, 49,* 102361. https://doi.org/10.1016/j.ctim.2020.102361

Andrade-Oliveira, V., Amano, M.T., Correa-Costa, M., Castoldi, A., Felizardo, R.J.F., de Almeida, D.C., Bassi, E., Moraes-Vieira, P.M., Hiyane, M.I., Rodas, A.C.D., Peron, J.P.S., Aguiar, C.F., Reis, M.A., Ribeiro, W.R., Valduga, C.J., Curi, R., Vinolo, M.A.R., Ferreira, C.M., & Câmara, N.O.S. (2015). Gut Bacteria products prevent AKI induced by ischemia-reperfusion. *Journal of the American Society of Nephrology, 26*(8), 1877–1888. https://doi.org/10.1681/ASN.2014030288

Andrews, C., McLean, M.H., & Durum, S.K. (2018). Cytokine tuning of intestinal epithelial function. *Frontiers in Immunology, 9.* https://doi.org/10.3389/fimmu.2018.01270

Arseneault-Bréard, J., Rondeau, I., Gilbert, K., Girard, S.A., Tompkins, T.A., Godbout, R., & Rousseau, G. (2012). Combination of Lactobacillus helveticus R0052 and Bifidobacterium longum R0175 reduces post-myocardial infarction depression symptoms and restores intestinal permeability in a rat model. *British Journal of Nutrition, 107*(12), 1793–1799. https://doi.org/10.1017/S0007114511005137

Bibbò, S., Ianiro, G., Giorgio, V., Scaldaferri, F., Masucci, L., Gasbarrini, A., & Cammarota, G. (2016). The role of diet on gut microbiota composition. *European Review for Medical and Pharmacological Sciences, 20*(22), 4742–4749. PMID: 27906427

Braniste, V., Al-Asmakh, M., Kowal, C., Anuar, F., Abbaspour, A., Tóth, M., Korecka, A., Bakocevic, N., Ng, L.G., Kundu, P., Gulyás, B., Halldin, C., Hultenby, K., Nilsson, H., Hebert, H., Volpe, B.T., Diamond, B., & Pettersson, S. (2014). The gut microbiota influences blood-brain barrier permeability in mice. *Science Translational Medicine, 6*(263), 1–12. https://doi.org/10.1126/scitranslmed.3009759

Brown, A.J., Goldsworthy, S.M., Barnes, A.A., Eilert, M.M., Tcheang, L., Daniels, D., Muir, A.I., Wigglesworth, M.J., Kinghorn, I., Fraser, N.J., Pike, N.B., Strum, J.C., Steplewski, K.M., Murdock, P.R., Holder, J.C., Marshall, F.H., Szekeres, P.G., Wilson, S., Ignar, D.M., ...& Dowell, S.J. (2003). The Orphan G protein-coupled receptors GPR41 and GPR43 are activated by propionate and other short chain carboxylic acids. *Journal of Biological Chemistry, 278*(13), 11312–11319. https://doi.org/10.1074/jbc.M211609200

Bultman, S.J. (2014). Emerging roles of the microbiome in cancer. *Carcinogenesis, 35*(2), 249–255. https://doi.org/10.1093/carcin/bgt392

Cario, E., Gerken, G., & Podolsky, D.K. (2007). Toll-like receptor 2 controls mucosal inflammation by regulating epithelial barrier function. *Gastroenterology, 132*(4), 1359–1374. https://doi.org/10.1053/j.gastro.2007.02.056

Chang, P.V., Hao, L., Offermanns, S., & Medzhitov, R. (2014). The microbial metabolite butyrate regulates intestinal macrophage function via histone deacetylase inhibition. *Proceedings of the National Academy of Sciences, 111*(6), 2247–2252. https://doi.org/10.1073/pnas.1322269111

Chávarri, M., Diez-Gutiérrez, L., Marañón, I., & Barron, L.J. (2021). Secondary metabolites from probiotic metabolism. *Advances in Probiotics Microorganisms in Food and Health, 1,* 260–261. https://doi.org/10.1016/B978-0-12-822909-5.00017-4

Corr, S.C., Li, Y., Riedel, C.U., O'Toole, P.W., Hill, C., & Gahan, C.G.M. (2007). Bacteriocin production as a mechanism for the antiinfective activity of Lactobacillus salivarius UCC118. *Proceedings of the National Academy of Sciences, 104*(18), 7617–7621. https://doi.org/10.1073/pnas.0700440104

Dai, Z., Wu, Z., Hang, S., Zhu, W., & Wu, G. (2014). Amino acid metabolism in intestinal bacteria and its potential implications for mammalian reproduction. *Molecular Human Reproduction, 21*(5), 389–409. https://doi.org/10.1093/molehr/gav003

De Vadder, F., Kovatcheva-Datchary, P., Goncalves, D., Vinera, J., Zitoun, C., Duchampt, A., Bäckhed, F., & Mithieux, G. (2014). Microbiota-generated metabolites promote metabolic benefits via gut-brain neural circuits. *Cell, 156*(1–2), 84–96. https://doi.org/10.1016/j.cell.2013.12.016

Derrien, M., & van Hylckama Vlieg, J.E.T. (2015). Fate, activity, and impact of ingested bacteria within the human gut microbiota. *Trends in Microbiology, 23*(6), 354–366. https://doi.org/10.1016/j.tim.2015.03.002

Donato, K.A., Gareau, M.G., Wang, Y.J.J., & Sherman, P.M. (2010). *Lactobacillus rhamnosus* GG attenuates interferon-γ and tumour necrosis factor-α-induced barrier dysfunction and pro-inflammatory signalling. *Microbiology, 156(*11), 3288–3297. https://doi.org/10.1099/mic.0.040139-0

Dumas, M.E., Rothwell, A.R., Hoyles, L., Aranias, T., Chilloux, J., Calderari, S., Noll, E.M., Péan, N., Boulangé, C.L., Blancher, C., Barton, R.H., Gu, Q., Fearnside, J.F., Deshayes, C., Hue, C., Scott, J., Nicholson, J.K., & Gauguier, D. (2017). Microbial-host co-metabolites are prodromal markers predicting phenotypic heterogeneity in behavior, obesity, and impaired glucose tolerance. *Cell Reports, 20*(1), 136–148. https://doi.org/10.1016/j.celrep.2017.06.039

Eun, C.S., Kim, Y.S., Han, D.S., Choi, J.H., Lee, A.R., & Park, Y.K. (2011). *Lactobacillus casei* prevents impaired barrier function in intestinal epithelial cells. *APMIS, 119*(1), 49–56. https://doi.org/10.1111/j.1600-0463.2010.02691.x

Fanning, S., Hall, L.J., Cronin, M., Zomer, A., MacSharry, J., Goulding, D., Motherway, M.O.C., Shanahan, F., Nally, K., Dougan, G., & Van Sinderen, D. (2012). Bifidobacterial surface-exopolysaccharide facilitates commensal-host interaction through immune modulation and pathogen protection. *Proceedings of the National Academy of Sciences of the United States of America, 109*(6), 2108–2113. https://doi.org/10.1073/pnas.1115621109

FAO & WHO. (2002). *Guidelines for the evaluation of probiotics in food.* Food and Agriculture Organization of the United Nations, 1–11.

Flint, H.J., Scott, K., Loius, P., & Duncan, S. (2012). The Role of the gut microbiota in nutrition and health. *Nature Reviews Gastroenterology and Hepatology, 9*, 577–589. https://doi.org/10.1038/nrgastro.2012.156

Fuller, R. (1989). Probiotics in man and animals. *Journal of Applied Bacteriology, 66*(5), 365–378. https://doi.org/10.1111/j.1365-2672.1989.tb05105.x

Gao, Z., Guo, B., Gao, R., Zhu, Q., Wu, W., & Qin, H. (2015). Probiotics modify human intestinal mucosa-associated microbiota in patients with colorectal cancer. *Molecular Medicine Reports, 12*(4), 6119–6127. https://doi.org/10.3892/mmr.2015.4124

Gareau, M.G., Sherman, P.M., & Walker, W.A. (2010). Probiotics and the gut microbiota in intestinal health and disease. *Nature Reviews Gastroenterology and Hepatology, 7*(9), 503–514. https://doi.org/10.1038/nrgastro.2010.117

Gaspar, C., Donders, G.G., Palmeira-de-Oliveira, R., Queiroz, J.A., Tomaz, C., Martinez-de-Oliveira, J., & Palmeira-de-Oliveira, A. (2018). Bacteriocin production of the probiotic *Lactobacillus acidophilus* KS400. *AMB Express, 8*(1), 153. https://doi.org/10.1186/s13568-018-0679-z

Ghavami, S.B., Rostami, E., Sephay, A.A., Shahrokh, S., Balaii, H., Aghdaei, H.A., & Zali, M.R. (2018). Alterations of the human gut *Methanobrevibacter smithii* as a biomarker for inflammatory bowel diseases. *Microbial Pathogenesis, 117*, 285–289. https://doi.org/10.1016/j.micpath.2018.01.029

Gibson, G.R., & Roberfroid, M.B. (1995). Dietary modulation of the human colonic microbiota: Introducing the concept of prebiotics. *Journal of Nutrition, 125*(6), 1401–1412. https://doi.org/10.1093/jn/125.6.1401

Goderska, K., Agudo Pena, So., & Alarcon, T. (2018). *Helicobacter pylori* treatment: Antibiotics or probiotics. *Applied Microbiology and Biotechnology, 102*, 1–7. https://doi:10.1007/s00253-017-8535-7

Guarner, F., & Schaafsma, G.J. (1998). Probiotics. *International Journal of Food Microbiology, 39*(3), 237–238. https://doi.org/10.1016/S0168-1605(97)00136-0

Heijtz, R.D., Wang, S., Anuar, F., Qian, Y., Björkholm, B., Samuelsson, A., Hibberd, M.L., Forssberg, H., & Pettersson, S. (2011). Normal gut microbiota modulates brain development and behavior. *Proceedings of the National Academy of Sciences of the United States of America, 108*(7), 3047–3052. https://doi.org/10.1073/pnas.1010529108

Hermiston, M.L., & Gordon, J.I. (1995). Inflammatory bowel disease and adenomas in mice expressing a dominant negative N-cadherin. *Science, 270*(5239), 1203–1207. https://doi.org/10.1126/science.270.5239.1203

Hill, C., Guarner, F., Reid, G., Gibson, G.R., Merenstein, D.J., Pot, B., Morelli, L., Canani, R.B., Flint, H.J., Salminen, S., Calder, P.C., & Sanders, M.E. (2014). Expert consensus document: The International Scientific Association for probiotics and prebiotics consensus statement on the scope and appropriate use of the term probiotic. *Nature Reviews Gastroenterology and Hepatology, 11*(8), 506–514. https://doi.org/10.1038/nrgastro.2014.66

Hoyles, L., Jiménez-Pranteda, M.L., Chilloux, J., Brial, F., Myridakis, A., Aranias, T., Magnan, C., Gibson, G.R., Sanderson, J.D., Nicholson, J.K., Gauguier, D., McCartney, A.L., & Dumas, M.E. (2018). Metabolic retroconversion of trimethylamine N-oxide and the gut microbiota. *Microbiome, 6*(1), 73. https://doi.org/10.1186/s40168-018-0461-0

Hoyles, L., Pontifex, M.G., Rodriguez-Ramiro, I., Anis-Alavi, M.A., Jelane, K.S., Snelling, T., Solito, E., Fonseca, S., Carvalho, A.L., Carding, S.R., Müller, M., Glen, R.C., Vauzour, D., & McArthur, S. (2021). Regulation of blood–brain barrier integrity by microbiome-associated methylamines and cognition by trimethylamine N-oxide. *Microbiome, 9*(1), 235. https://doi.org/10.1186/s40168-021-01181-z

Ibrahim, G.A., El-Sayed, H.S., El-Shafei, K., & Sharaf, O. (2015). Riboflavin and folate production in different media using encapsulated *Streptococcus Thermophilus* and *Lactobacillus Plantarum*. *Middle East Journal of Applied Sciences, 5*(3), 663–669.

Jakobsson, H.E., Rodríguez-Piñeiro, A.M., Schütte, A., Ermund, A., Boysen, P., Bemark, M., Sommer, F., Bäckhed, F., Hansson, G.C., & Johansson, M.E. (2015). The composition of the gut microbiota shapes the colon mucus barrier. *EMBO Reports, 16*(2), 164–177. https://doi.org/10.15252/embr.201439263

Jeon, S.G., Kayama, H., Ueda, Y., Takahashi, T., Asahara, T., Tsuji, H., Tsuji, N.M., Kiyono, H., Ma, J.S., Kusu, T., Okumura, R., Hara, H., Yoshida, H., Yamamoto, M., Nomoto, K., & Takeda, K. (2012). Probiotic *Bifidobacterium breve* induces IL-10-producing Tr1 cells in the colon. *PLoS Pathogens, 8*(5), 1–15. https://doi.org/10.1371/journal.ppat.1002714

Karczewski, J., Troost, F.J., Konings, I., Dekker, J., Kleerebezem, M., Brummer, R.J.M., & Wells, J.M. (2010). Regulation of human epithelial tight junction proteins by *Lactobacillus plantarum in vivo* and protective effects on the epithelial barrier. *American Journal of Physiology-Gastrointestinal and Liver Physiology, 298*(6), G851–G859. https://doi.org/10.1152/ajpgi.00327.2009

Kim, H.J., & White, P.J. (2009). *In vitro* Fermentation of oat flours from typical and high β-glucan oat lines. *Journal of Agricultural and Food Chemistry, 57*, 7529–7536. https://doi.org/10.1021/jf900788c

Koh, A., De Vadder, F., Kovatcheva-Datchary, P., & Bäckhed, F. (2016). From dietary fiber to host physiology: Short-chain fatty acids as key bacterial metabolites. *Cell, 165*(6), 1332–1345. https://doi.org/10.1016/j.cell.2016.05.041

Laitinen, K., Mokkala, K., & Kalliomäki, M. (2022). Impact of aerly nutrition on gut microbiota: Effects on immunity and long-term. *In Health, 9*, 229–256. https://doi.org/10.1016/B978-0-12-824389-3.00005-2.

LeBlanc, J.G., Chain, F., Martín, R., Bermúdez-Humarán, L.G., Courau, S., & Langella, P. (2017). Beneficial effects on host energy metabolism of short-chain fatty acids and vitamins produced by commensal and probiotic bacteria. *Microbial Cell Factories, 16*, 79. https://doi.org/10.1186/s12934-017-0691-z

Li, X., Yang, L., Xu, M., Qiao, G., Li, C.J., Lin, L., & Zheng, G. (2021). Smilax china L. polyphenols alleviates obesity and inflammation by modulating gut microbiota in high fat/high sucrose diet-fed C57BL/6J mice. *Journal of Functional Foods, 77*, 104332. https://doi.org/10.1016/j.jff.2020.104332

Liang, H., Dai, Z., Kou, J., Sun, K., Chen, J., Yang, Y., Wu, G., & Wu, Z. (2018). Dietary l-tryptophan supplementation enhances the intestinal mucosal barrier function in weaned piglets: Implication of tryptophan-metabolizing microbiota. *International Journal of Molecular Sciences, 20*(1), 20. https://doi.org/10.3390/ijms20010020

Lilly, D.M., & Stillwell, R.H. (1965). Probiotics: Growth-promoting factors produced by microorganisms. *Science, 147*(3659), 747–748. https://doi.org/10.1126/science.147.3659.747

Liu, Y., Alookaran, J.J., & Rhoads, J.M. (2018). Probiotics in autoimmune and inflammatory disorders. *Nutrients, 10*(10). https://doi.org/10.3390/nu10101537

Lyte, M. (2011). Probiotics function mechanistically as delivery vehicles for neuroactive compounds: Microbial endocrinology in the design and use of probiotics. *BioEssays, 33*(8), 574–581. https://doi.org/10.1002/bies.201100024

Ma, T., Jin, H., Kwok, L.Y., Sun, Z., Liong, M.T., & Zhang, H. (2021). Probiotic consumption relieved human stress and anxiety symptoms possibly via modulating the neuroactive potential of the gut microbiota. *Neurobiology of Stress, 14,* 100294. https://doi.org/10.1016/j.ynstr.2021.100294

Ma, T.Y., Iwamoto, G.K., Hoa, N.T., Akotia, V., Pedram, A., Boivin, M.A., & Said, H.M. (2004). TNF-α-induced increase in intestinal epithelial tight junction permeability requires NF-κB activation. *American Journal of Physiology-Gastrointestinal and Liver Physiology, 286*(3), G367–G376. (

Mafra, D., Ribeiro, M., Fonseca, L., Regis, B., Cardozo, L.F.M.F., Fragoso dos Santos, H., Emiliano de Jesus, H., Schultz, J., Shiels, P.G., Stenvinkel, P., & Rosado, A. (2022). Archaea from the gut microbiota of humans: Could be linked to chronic diseases?. *Anaerobe, 77,* 102629. https://doi.org/10.1016/j.anaerobe.2022.102629

Maslowski, K.M., Vieira, A.T., Ng, A., Kranich, J., Sierro, F., Di Yu, Schilter, H.C., Rolph, M.S., Mackay, F., Artis, D., Xavier, R.J., Teixeira, M.M., & Mackay, C.R. (2009). Regulation of inflammatory responses by gut microbiota and chemoattractant receptor GPR43. *Nature, 461*(7268), 1282–1286. https://doi.org/10.1038/nature08530

Mayer, E.A. (2011). Gut feelings: The emerging biology of gut–brain communication. *Nature Reviews Neuroscience, 12*(8), 453–466. https://doi.org/10.1038/nrn3071

McCann, K.B., Shiell, B.J., Michalski, W.P., Lee, A., Wan, J., Roginski, H., & Coventry, M.J. (2006). Isolation and characterization of a novel antibacterial peptide from bovine αS1-casein. *International Dairy Journal, 16*(4), 316–323. https://doi.org/10.1016/j.idairyj.2005.05.005

Metchnikoff, É. (1907). The prolongation of life; optimistic studies. *Wellcome Collection, 44*(0), 1864–1945.

Michel, L., & Prat, A. (2016). One more role for the gut: Microbiota and blood brain barrier. *Annals of Translational Medicine, 4*(1), 4–6. https://doi.org/10.3978/j.issn.2305-5839.2015.10.16

Min, M., Bunt, C.R., Mason, S L., & Hussain, M.A. (2019). Non-dairy probiotic food products: An emerging group of functional foods. *Critical Reviews in Food Science and Nutrition, 59*(16), 2626–2641. https://doi.org/10.1080/10408398.2018.1462760

Mitsuoka, T. (2009). Intestinal flora and aging. Nutrition *Reviews, 50*(12), 438–446. https://doi.org/10.1111/j.1753-4887.1992.tb02499.x

Montassier, E., Valdés-Mas, R., Batard, E., Zmora, N., Dori-Bachash, M., Suez, J., & Elinav, E. (2021). Probiotics impact the antibiotic resistance gene reservoir along the human GI tract in a person-specific and antibiotic-dependent manner. *Nature Microbiology, 6*(8), 1043–1054. https://doi.org/10.1038/s41564-021-00920-0

Mora, F., Segovia, G., del Arco, A., de Blas, M., & Garrido, P. (2012). Stress, neurotransmitters, corticosterone and body-brain integration. *Brain Research, 1476,* 71–85. https://doi.org/10.1016/j.brainres.2011.12.049

Mu, C., Nikpoor, N., Tompkins, T.A., Rho, J.M., Scantlebury, M.H., & Shearer, J. (2022). Probiotics counteract hepatic steatosis caused by ketogenic diet and upregulate AMPK signaling in a model of infantile epilepsy. *EBioMedicine, 76,* 103838. https://doi.org/10.1016/j.ebiom.2022.103838

Musumeci, G., Castrogiovanni, P., Castorina, S., Imbesi, R., Szychlinska, M.A., Scuderi, S., Loreto, C., & Giunta, S. (2015). Changes in serotonin (5-HT) and brain-derived neurotrophic factor (BDFN) expression in frontal cortex and hippocampus of aged rat treated with high tryptophan diet. *Brain Research Bulletin, 119,* 12–18. https://doi.org/10.1016/j.brainresbull.2015.09.010

Núñez, I.N., Galdeano, C.M., de Moreno de LeBlanc, A., & Perdigón, G. (2014). Evaluation of immune response, microbiota, and blood markers after probiotic bacteria administration in obese mice induced by a high-fat diet. *Nutrition, 30*(11–12), 1423–1432. https://doi.org/10.1016/j.nut.2014.03.025

O'Toole, P.W., & Cooney, J.C. (2008). Probiotic bacteria influence the composition and function of the intestinal microbiota. *Interdisciplinary Perspectives on Infectious Diseases,* 175285. https://doi.org/10.1155/2008/175285

Papadimitriou, K., Alegría, Á., Bron, P.A., de Angelis, M., Gobbetti, M., Kleerebezem, M., Lemos, J.A., Linares, D.M., Ross, P., Stanton, C., Turroni, F., van Sinderen, D., Varmanen, P., Ventura, M., Zúñiga, M., Tsakalidou, E., & Kok, J. (2016). Stress physiology of lactic acid bacteria. *Microbiology and Molecular Biology Reviews, 80*(3), 837–890. https://doi.org/10.1128/mmbr.00076-15

Park, J.S., Lee, E.J., Lee, J.C., Kim, W.K., & Kim, H.S. (2007). Anti-inflammatory effects of short chain fatty acids in IFN-gamma-stimulated RAW 264.7 murine macrophage cells: Involvement of NF-kappaB and ERK signaling pathways. *International Immunopharmacology, 7*, 70–77. https://doi.org/10.1016/j.intimp.2006.08.015

Parker, A., Fonseca, S., & Carding, S.R. (2020). Gut microbes and metabolites as modulators of blood-brain barrier integrity and brain health. *Gut Microbes, 11*(2), 135–157. https://doi.org/10.1080/19490976.2019.1638722

Patterson, E., Ryan, P.M., Wiley, N., Carafa, I., Sherwin, E., Moloney, G., Franciosi, E., Mandal, R., Wishart, D.S., Tuohy, K., Ross, R.P., Cryan, J.F., Dinan, T.G., & Stanton, C. (2019). Gamma-aminobutyric acid-producing lactobacilli positively affect metabolism and depressive-like behaviour in a mouse model of metabolic syndrome. *Scientific Reports, 9*(1), 16323. https://doi.org/10.1038/s41598-019-51781-x

Pretzer, G., Snel, J., Molenaar, D., Wiersma, A., Bron, P.A., Lambert, J., de Vos, W.M., van der Meer, R., Smits, M.A., & Kleerebezem, M. (2005). Biodiversity-based identification and functional characterization of the mannose-specific adhesin of *Lactobacillus plantarum*. *Journal of Bacteriology, 187*(17), 6128–6136. https://doi.org/10.1128/JB.187.17.6128-6136.2005

Rastogi, Y.R., Thakur, R., Thakur, P., Mittal, A., Chakrabarti, S., Siwal, S.S., Thakur, V.K., Saini, R.V., & Saini, A.K. (2022). Food fermentation – Significance to public health and sustainability challenges of modern diet and food systems. *International Journal of Food Microbiology, 371*. https://doi.org/10.1016/j.ijfoodmicro.2022.109666

Rinninella, E., Raoul, P., Cintoni, M., Franceschi, F., Miggiano, G.A.D., Gasbarrini, A., & Mele, M.C. (2019). What is the healthy gut microbiota composition? A changing ecosystem across age, environment, diet, and diseases. *Microorganisms, 7*(1), 14. https://doi.org/10.3390/microorganisms7010014

Sánchez, B., & Urdaci, M.C. (2012). Extracelular proteins from *Lactobacillus plantarum* BMCM12 prevent adhesion of enteropathogens to mucin. *Current Microbiology, 64*, 592–596. https://doi.org/10.1007/s00284-012-0115-6

Shamsipour, S., Sharifi, G., & Taghian, F. (2021). Impact of interval training with probiotic (*L. plantarum / Bifidobacterium bifidum*) on passive avoidance test, ChAT and BDNF in the hippocampus of rats with Alzheimer's disease. *Neuroscience Letters, 756*, 135949. https://doi.org/10.1016/j.neulet.2021.135949

Shandilya, S., Kumar, S., Kumar Jha, N., Kumar Kesari, K., & Ruokolainen, J. (2022). Interplay of gut microbiota and oxidative stress: Perspective on neurodegeneration and neuroprotection. *Journal of Advanced Research, 38*, 223–244. https://doi.org/10.1016/j.jare.2021.09.005

Shi, H., ter Horst, R., Nielen, S., Bloemendaal, M., Jaeger, M., Joosten, I., Koenen, H., Joosten, L.A.B., Schweren, L.J.S., Vasquez, A.A., Netea, M.G., & Buitelaar, J. (2022). The gut microbiome as mediator between diet and its impact on immune function. *Scientific Reports, 12*(1), 1–10. https://doi.org/10.1038/s41598-022-08544-y

Sigel, E., & Steinmann, M.E. (2012). Structure, function, and modulation of GABAA receptors. *Journal of Biological Chemistry, 287*(48), 40224–40231. https://doi.org/10.1074/jbc.R112.386664

Slavin, J. (2013). Fiber and prebiotics: Mechanisms and health benefits. *Nutrients, 5*, 1417–1435. https://doi.org/10.3390/nu5041417

Taverniti, V., Del Bo', C., Fiore, W., Gargari, G., Arioli, S., Riso, P., Guglielmetti, S., & Frøkiær, H. (2022). Combination of different probiotics and berry-derived (poly)phenols can modulate immune response in dendritic cells. *Journal of Functional Foods, 94*, 105121. https://doi.org/10.1016/j.jff.2022.105121

Tejada-Simon, M.V., Lee, J.H., Ustunol, Z., & Pestka, J.J. (1999). Ingestion of yogurt Containing *Lactobacillus acidophilus* and *Bifidobacterium* to potentiate immunoglobulin a response to cholera toxin in mice. *Journal of Dairy Science, 82*(4), 649–660. https://doi.org/10.3168/jds.S0022-0302(99)75281-1

Tette, F.M., Kwofie, S.K., & Wilson, M.D. (2022). Therapeutic anti-depressant potential of microbial GABA produced by *Lactobacillus rhamnosus* strains for GABAergic signaling restoration and inhibition of addiction-induced HPA axis hyperactivity. *Current Issues in Molecular Biology, 44*(4), 1434–1451. https://doi.org/10.3390/cimb44040096

Thakur, K., Tomar, S.K., & De, S. (2016). Lactic acid bacteria as a cell factory for riboflavin production. *Microbial Biotechnology, 9*(4), 441–451. https://doi.org/10.1111/1751-7915.12335

Visser, L., Jan de Heer, H., Boven, L.A., van Riel, D., van Meurs, M., Melief, M.J., Zähringer, U., van Strijp, J., Lambrecht, B.N., Nieuwenhuis, E.E., & Laman, J.D. (2005). Proinflammatory bacterial peptidoglycan as a cofactor for the development of central nervous system autoimmune disease. *The Journal of Immunology, 174*(2), 808–816. https://doi.org/10.4049/jimmunol.174.2.808

Vodolazov, I.R., Dbar, S.D., Oleskin, A.V., & Stoyanova, L.G. (2018). Exogenous and endogenous neuroactive biogenic amines: Studies with *Lactococcus lactis* subsp. lactis. *Applied Biochemistry and Microbiology, 54*(6), 603–610. https://doi.org/10.1134/S0003683818060157

Vorbrodt, A.W., & Dobrogowska, D.H. (2003). Molecular anatomy of intercellular junctions in brain endothelial and epithelial barriers: Electron microscopist's view. *Brain Research Reviews, 42*(3), 221–242. https://doi.org/10.1016/S0165-0173(03)00177-2

Wang, X., Ba, T., Cheng, Y., Zhang, P., & Chang, X. (2021). Probiotics alleviate adipose inflammation in high-fat diet-induced obesity by restoring adipose invariant natural killer T cells. *Nutrition, 89*, 111285. https://doi.org/10.1016/j.nut.2021.111285

Westfall, S., Lomis, N., & Prakash, S. (2018). A polyphenol rich prebiotic in combination with a novel probiotic formulation alleviates markers of obesity and diabetes in Drosophila. *Journal of Functional Foods, 48*, 374–386. https://doi.org/10.1016/j.jff.2018.07.012

Wikoff, W.R., Anfora, A.T., Liu, J., Schultz, P.G., Lesley, S.A., Peters, E.C., & Siuzdak, G. (2009). Metabolomics analysis reveals large effects of gut microflora on mammalian blood metabolites. *Proceedings of the National Academy of Sciences of the United States of America, 106*(10), 3698–3703. https://doi.org/10.1073/pnas.0812874106

Wu, Q., Chen, T., El-Nezami, H., & Savidge, T.C. (2020). Food ingredients in human health: Ecological and metabolic perspectives implicating gut microbiota function. *Trends in Food Science and Technology, 100*, 103–117. https://doi.org/10.1016/j.tifs.2020.04.007

Wu, Y., Zhu, C., Chen, Z., Chen, Z., Zhang, W., Ma, X., Wang, L., Yang, X., & Jiang, Z. (2016). Protective effects of Lactobacillus plantarum on epithelial barrier disruption caused by enterotoxigenic Escherichia coli in intestinal porcine epithelial cells. *Veterinary Immunology and Immunopathology, 172*, 55–63. https://doi.org/10.1016/j.vetimm.2016.03.005

Yazdani, B., Shidfar, F., Salehi, E., Baghbani-arani, F., Razmpoosh, E., Asemi, Z., Cheshmazar, E., & Zarrati, M. (2018). Probiotic plus low-calorie diet increase gene expression of Toll-like receptor 2 and FOXP3 in overweight and obese participants. *Journal of Functional Foods, 43*, 180–185. https://doi.org/10.1016/j.jff.2018.02.013

Ye, S., Shah, B.R., Li, J., Liang, H., Zhan, F., Geng, F., & Li, B. (2022). A critical review on interplay between dietary fibers and gut microbiota. *Trends in Food Science and Technology, 124*, 237–249. https://doi.org/10.1016/j.tifs.2022.04.010

Ye, X., & Sun, M. (2017). AGR2 ameliorates tumor necrosis factor-α-induced epithelial barrier dysfunction via suppression of NF-κB p65-mediatedMLCK/p-MLC pathway activation. *International Journal of Molecular Medicine, 39*(5), 1206–1214. https://doi.org/10.3892/ijmm.2017.2928

Yin, J., Zhang, B., Yu, Z., Hu, Y., Lv, H., Ji, X., Wang, J., Peng, B., & Wang, S. (2021). Ameliorative effect of dietary tryptophan on neurodegeneration and inflammation in D-galactose-induced aging mice with the potential mechanism relying on AMPK/SIRT1/PGC-1α pathway and gut microbiota. *Journal of Agricultural and Food Chemistry, 69*(16), 4732–4744. https://doi.org/10.1021/acs.jafc.1c00706

Yu, Y., & Zhao, F. (2021). Microbiota-gut-brain axis in autism spectrum disorder. *Journal of Genetics and Genomics, 48*(9), 755–762. https://doi.org/10.1016/j.jgg.2021.07.001

Yusufu, I., Ding, K., Smith, K., Wankhade, U.D., Sahay, B., Patterson, G.T., Pacholczyk, R., Adusumilli, S., Hamrick, M.W., Hill, W.D., Isales, C.M., & Fulzele, S. (2021). A tryptophan-deficient diet induces gut microbiota dysbiosis and increases systemic inflammation in aged mice. *International Journal of Molecular Sciences, 22*(9), 5005. https://doi.org/10.3390/ijms22095005

Zeinali, T., Faraji, N., Joukar, F., Khan Mirzaei, M., Kafshdar Jalali, H., Shenagari, M., & Mansour-Ghanaei, F. (2022). Gut bacteria, bacteriophages, and probiotics: Tripartite mutualism to quench the SARS-CoV2 storm. *Microbial Pathogenesis, 170,* 105704. https://doi.org/10.1016/j.micpath.2022.105704

Zelante, T., Iannitti, R.G., Cunha, C., De Luca, A., Giovannini, G., Pieraccini, G., Zecchi, R., D'Angelo, C., Massi-Benedetti, C., Fallarino, F., Carvalho, A., Puccetti, P., & Romani, L. (2013). Tryptophan catabolites from microbiota engage aryl hydrocarbon receptor and balance mucosal reactivity via interleukin-22. *Immunity, 39*(2), 372–385. https://doi.org/10.1016/j.immuni.2013.08.003

Zhang, H., & Mo Y. 2023 The gut-retina axis: A new perspective in the prevention and treatment of diabetic retinopathy. *Frontiers in Endocrinology, 14,* 1205846 https://doi.org/10.3389/fendo.2023.1205846

Zhang, Z., Huang, Z., Tong, J., Wu, Q., Pan, Y., Malakar, P.K., & Zhao, Y. (2021). An outlook for food sterilization technology: Targeting the outer membrane of foodborne gram-negative pathogenic bacteria. *Current Opinion in Food Science, 42,* 15–22. https://doi.org/10.1016/j.cofs.2021.02.013

Zheng, J., Wittouck, S., Salvetti, E., Franz, C.M.A.P., Harris, H.M.B., Mattarelli, P., O'toole, P.W., Pot, B., Vandamme, P., Walter, J., Watanabe, K., Wuyts, S., Felis, G.E., Gänzle, M.G., & Lebeer, S. (2020). A taxonomic note on the genus Lactobacillus: Description of 23 novel genera, emended description of the genus *Lactobacillus beijerinck* 1901, and union of Lactobacillaceae and Leuconostocaceae. *International Journal of Systematic and Evolutionary Microbiology, 70*(4), 2782–2858. https://doi.org/10.1099/ijsem.0.004107

Chapter 13

Combined Effect of Nanoparticles Loaded with α-Lipoic Acid, Vitamins E and C for Treating Oxidative Stress Related to Physiological Disorders

Patricia Rosales-Martínez
Gustavo F. Gutiérrez-López
Humberto Hernández-Sánchez
and Maribel Cornejo-Mazón

CONTENTS

 DOI: 10.1201/9781003332558-15

13.1 INTRODUCTION

Oxidative damage is the effect observed when the antioxidant systems of the cells are insufficient to neutralize the elevation of reactive oxygen species (ROS), which are products of aerobic metabolic pathways. They contain oxygen in the form of peroxides and superoxide hydroxyl radicals, singlet oxygen, or hydrogen peroxide. ROS can be produced in high quantities under pathophysiological conditions, which can result in the development and aggravation of many health disorders such as diabetes, Parkinson's and Alzheimer's diseases, acute kidney and lung failures, and cancer (Jomova et al. 2023). In addition, there are various factors such as environmental stress and exposure to ionizing radiation or excess ultraviolet (UV) radiation that can increase the production of ROS. ROS can damage cell membranes, lipids, proteins, and DNA, causing serious damage and impairment of normal physiological functions. Oxidative stress, the imbalance between ROS production and antioxidant protection mechanisms, originates the inability of endogenous antioxidant defense mechanisms to protect against oxidative stress. Antioxidants generally exert their effects primarily preventing the production of ROS or by eliminating or degrading them (Anwar, 2022).

Antioxidant capacity of foods and beverages has attracted considerable interest since antioxidants are believed to possess protective properties against numerous pathologies. In this context, the antioxidant capacity is the primary indicator to evaluate the state and potential of oxidative stress in age-related diseases. Since imbalance between antioxidant and oxidative agents generates the condition of oxidative stress, the estimation of the reducing power/antioxidant capacity of systems such as cells, tissues, or organisms is the first step in the prediction of oxidative stress aiming to counteract the effects of aging-related stresses (Luo et al. 2020).

Currently, nanotechnological techniques are used in food, pharmaceutical, and cosmetics applications to obtain different types of products to encapsulate bioactive compounds to protect them from external agents, improve their bioavailability, or modify their release (Alu'datt et al. 2022). Consumption of antioxidants has been recommended for neutralizing adverse effects of ROS and various compounds, including α-lipoic acid, and vitamins E and C have been recognized to display synergic effects against free radicals and thus to treat physiological disorders such as diabetic neuropathy, Alzheimer's disease, rheumatoid arthritis, lupus erythematosus, and cardiac ischemia to name a few (Kruk et al. 2022). This chapter presents firstly a description of the main physicochemical characteristics of lipoic acid and vitamins E and C that contribute to their bioactivity and describes the preparation and potential therapeutic effects of nanoparticles loaded with these three compounds by taking advantage of synergistic effects among them and by using chitosan as wall material.

13.2 NANOTECHNOLOGY

Nanotechnology is a field of applied sciences, focused on the design, synthesis, characterization, and application of materials and devices on a nanometric scale. It deals with measuring, seeing, manipulating, and preparation of materials between 1 and 100 nanometers (Taha et al. 2022) and is a multi- and inter-disciplinary science in which areas such as chemical engineering, food and materials sciences, biotechnology, mechanics, robotics, and medicine among others converge (Ashraf et al. 2021).

The use of nanoparticles and nanoencapsulation compounds has been given a great deal of attention since these materials give rise to a controlled absorption of nutrients, enhanced flavors, color modification, improved food safety, and increased shelf life in food materials, which may positively impact bioactivity against physiological malfunctions (Mohammad et al. 2022). This applies to the encapsulation of additives (vitamins, antioxidants, dyes), representing a viable and efficient alternative to improve the physical stability and therapeutic effects of active substances, protect them from interactions with the ingredients of foods, and improve their bioactivity due to their subcellular size (Alu'datt et al. 2022). Many of these bioactive compounds, in addition to presenting antioxidant activity (e.g., carotenoids, omega-3 fatty acids, fatty acids, and phytosterols), are highly lipophilic and have poor absorption and limited bioavailability. For this reason, the total antioxidant content of a food as well as its bioavailability are of key metabolic importance (Chen et al. 2022). To exert these effects, the bioactive molecules must reach the site of action and therefore unless the target is the digestive tract, they must be absorbed, pass into the bloodstream, and distributed in the body (Vilas-Boas et al. 2021). In addition, by reducing the particle to nanometric sizes, the properties of food compounds can be improved (Goméz-Gómez et al. 2023). The use of transport systems for bioactive compounds with high specificity and activity to the site of bioactivity without toxic effects is a good strategy being investigated (Zhao et al. 2021; Kurek et al. 2022). There are various types of polymeric particles used in the administration of these compounds and depending on their size, they can be classified into micro- and nanoparticles (Elmowafy et al. 2023).

13.2.1 Nanoparticles

When adding nanoparticles to liquids, a dispersed system is formed, in which the active compound can be dissolved, trapped, encapsulated, or adsorbed within the matrix (El-Khawaga et al. 2023). Organic nanoparticles are mainly applied to improve the nutritional value of foods and are used, as in the case of nanocapsules, as a vehicle for the release of vitamins and other nutrients (Pateiro et al. 2021). For this to take place, it is important to consider some specific factors in the characterization of these systems as particle size and its distribution, charge, and surface properties.

13.2.2 Characteristics of Nanoparticles

13.2.2.1 Size of Nanoparticles

During the preparation of nanoparticles, it is important to control the size and distribution of the product, since these parameters determine their stability, charge, toxicity, and release of the active ingredient (Frank et al. 2020).

In small particles, most of the associated bioactive is closer to the particle surface, leading to a fast release. On the contrary, larger particles have larger nuclei, allowing the encapsulation of more bioactive which diffusion is slower owing to larger transport distances to the surface of the particle (Frank et al. 2020).

13.2.2.2 Surface Properties of Nanoparticles

A charged particle alters the distribution of ions in the medium that surrounds it. According to the Gouy-Chapman-Stern theory (Demir & Searles, 2020), the electric field produced by this charge attracts ions of opposite sign (counter ions) from the medium, which forms a thin compact layer near the surface of the particle. Then an electrical double layer is formed, and the potential in the plane that separates the fixed layer from the diffuse one is the Stern potential (Demir & Searles,

2020; Chu et al. 2023). In the diffuse layer, there is a hypothetical limit, close to the Stern potential, where any ion within that limit will move with the particle when it moves through the liquid, but any ion outside this limit will stay where it was originally found. This limit that separates the mobile area from the immobile area is known as the slipping plane. The potential in the slip plane is called Z (ζ)-potential or electrokinetic potential as shown in **Figure 13.1**.

ζ-potential is the electrostatic power that exists between the separations of the layers that surround the particles and the media in which it is dispersed (Muneer et al. 2023; Leon et al. 2023). Nanoparticles with a ζ-potential of +/- 30 mV have proven to be stable in suspension since the surface charge prevents the aggregation of the particles. The ζ-potential can also be used to determine whether an active compound is encapsulated within the nanoparticle or adsorbed on the

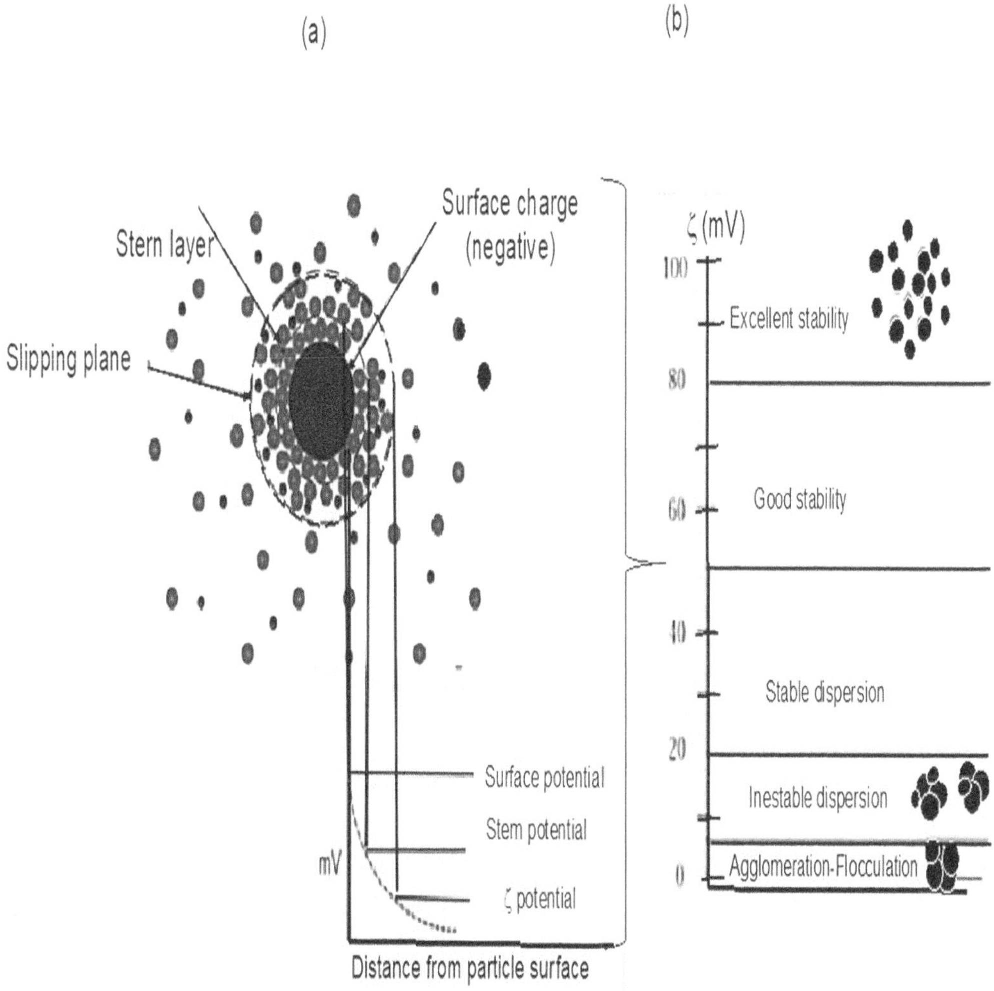

Figure 13.1 (a) Schematic representation of the ζ-potential. (b) Steady state of the fluid at different values of the ζ-potential.

hydrophilic surface (Sánchez *et al.* 2020). Virtually all colloidal particles acquire a surface charge when they are suspended in a polar medium such as water, and the associated potential between particle and medium is called surface potential (Ong et al. 2020; Delon et al. 2021).

When the power determining ion concentration is such that the positive and negative surface activities are equal, this concentration defines the point of zero charge or isoelectric point, and there is no electric double layer (Finney & Salvalaglio, 2023). Thus, for molecules and particles that are small enough, a high ζ-potential confers stability; that is, the solution or dispersion will resist aggregation. Therefore, dispersed systems with high ζ-potentials are electrically stabilized, whereas colloids with low ζ-potentials tend to coagulate or flocculate (Pochapski et al. 2021; Serrano-Lotina et al. 2023). Also, the stability of the dispersions depends on the balance between the repulsive electrostatic charges in the colloidal particle and the van der Waals forces of attraction (Benselfelt et al. 2023). As this property chiefly controls the stability of the dispersed system formed with the nanoparticles, it is necessary to elucidate the actual role of ζ-potential in real physiological fluid systems directed to therapeutic schemes.

13.3 MATERIALS USED IN NANOENCAPSULATION

In recent years, a special interest has been given to the development of biodegradable nanoparticles using polymers, as effective delivery systems for bioactive lipophilic components. Natural polymers have acquired importance in the biomedical area, especially in the development of new cell therapies (Zhu *et al.* 2021), since they are used to protect viable cells and/or tissues in semipermeable membranes or as encapsulating agents and carriers of various drugs and active ingredients for specific medical treatments.

A polymer can fulfill these functions when it meets certain characteristics such as biocompatibility, biodegradability, mechanical resistance, chemical stability, low toxicity, etc. (Reddy et al. 2021). Due to these physicochemical properties, the Food and Drug Administration (FDA) allowed its usage.

Due to their stability and ease of surface modification, natural and synthetic polymers can be used to prepare polymeric nanoparticles that provide additional advantages such as biodegradability, biocompatibility, and reduced toxicity. Furthermore, these nanoparticles are readily available from natural sources (e.g., cellulose, pectin, starch, collagen, silk fibroins, chitosan, and chitin). Among the wide range of existing natural polymers, chitosan stands out for its characteristics, uses, and potential to protect bioactive compounds and is extensively used in agriculture, food industry, cosmetics, and medical formulations for specific medical schemes (Zabot et al. 2022; Huesca-Urióstegui et al. 2022).

13.3.1 Chitosan

Chitosan is the N-deacetylated form of chitin and has good reactivity and solubility properties. Chitin is part of the support structure of many living organisms, such as arthropods, mollusks, and fungi (Iber et al. 2022). Chitosan is obtained by substituting its acetamido groups with amino ones and treating chitin with strong alkalis. It has been described as a linear, biodegradable, high-molecular-weight chain, easy to use, and environmentally friendly cationic polymer (Aranaz et al. 2021).

Chitosan dissolves easily in dilute solutions of most organic acids, such as formic, acetic, citric and tartaric, and in dilute mineral acids, except sulfuric acid (*Li* et al. 2023). Its degree of deacetylation (DD) varies from 60 to 90%, and the molecular weights (MW) are reported from 50 to 2000 KDa; this heterogeneity is attributed to the lack of control during processing (Triunfo et al. 2022).

Chitosan also has important biological, physiological, and pharmacological properties, mainly related to its cationic behavior (Yadav et al. 2022) such as being non-toxic, biodegradable, and biocompatible to plants and animals. This polymer has an important biological activity in relation to its analgesic, antimicrobial (Abd El-Hack et al. 2020), hemostatic (Wang et al. 2022), antioxidant, mucoadhesive, hemostatic, fungistatic, antitumoral (Silva et al. 2023), hypocholesterolemic, and immunological activity effects and possesses tissue regenerative properties (Gorantla et al. 2021). In addition, its polycationic nature gives it a high affinity to associated therapeutic effects (e.g., insulin, DNA, RNA, and heparin, among others) and antigenic macromolecules (Rial-Hermida et al. 2021).

Due to the abovementioned characteristics, chitosan has been used as a wall material to encapsulate bioactive substances and in the development of substance delivery systems (Maleki et al. 2022). Chitosan oligosaccharides can be obtained mainly by two methods: chemical and enzymatic. The enzymatic degradation of chitosan polymers is more effective, because the course of the hydrolysis reaction and the distribution of its products are subjected to better control, in addition to a higher yield (Shokri et al. 2022).

13.3.1.1 Enzymatic Degradation of Chitosan

Chitosan undergoes enzymatic transformation into non-toxic building blocks and is degraded *in vivo* by various enzymes, primarily by lysozyme (a non-specific protease present in all mammalian tissues). Its degradation products are oligosaccharides or monosaccharides, natural metabolites that, once absorbed, can be incorporated into the metabolic pathways of glycosaminoglycans or else excreted (Tao et al. 2021). *In vitro* degradations of chitosan by oxidation, chemical or enzymatic hydrolysis reactions are commonly used methods for the preparation of low-molecular-weight chitosan under controlled conditions (Allison et al. 2020).

Enzymatic hydrolytic degradation by lysozyme is the main mechanism of chitosan degradation within the human body and is easily modeled with *in vitro* testing (Khan & Alamry, 2021). Lysozyme has been measured in various body tissues, including serum, present at concentrations ranging from 1-14 μg/mL Lysozyme breaks up chitosan by cleavage of glycosidic bonds between polysaccharide units in the polymer (**Figure 13.2**). In addition, lysozyme degrades chitosan into saccharide and glucosamine byproducts that are incorporated into proteoglycans or metabolized by the body (Khalaf *et al.* 2023). Several enzymatic degradation studies have found evidence that the sequence of glucosamine and N-acetyl glucosamine units, that is, the distribution of block or random polymer units, affects lysozyme degradation (Sreekumar *et al.* 2022).

13.4 PREPARATION OF NANOPARTICLES

Special attention has been given to the development of biodegradable nanoparticles using polymers, as effective delivery systems for bioactive lipophilic components. In recent years, natural polymers or biopolymers have acquired great importance in the biomedical area, especially in the development of new cell therapies (Zhu *et al.* 2021). A polymer is suitable to fulfill these

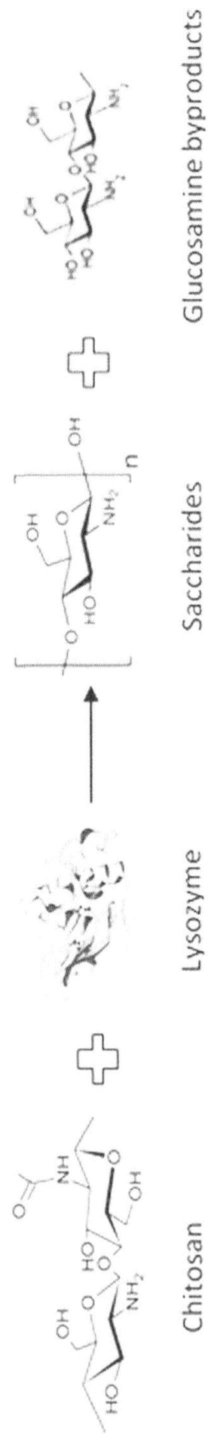

Figure 13.2 Reaction mechanism of the cleavage of β (1→4) glycosidic bonds by lysozyme in polymeric chitosan.

functions when it meets certain characteristics such as biocompatibility, biodegradability, mechanical resistance, chemical stability, low toxicity, etc. (Baranwal *et al.* 2022). Among the wide range of existing and commonly used natural polymers, chitosan stands out for its characteristics, uses, and as a potential carrier of bioactive molecules directed to specific medical treatments (Di Santo et al. 2021; Frigaard et al. 2022).

Chitosan nanoparticles were first described in 1994 when Ohya et al. (1994) proposed the intravenous administration of 5-fluorouracil, an anticancer drug, and using chitosan nanoparticles as a carrier system, obtained by emulsification and gelation. Since then, these systems have been extensively studied for drug delivery purposes and the original formulation was used for other applications, such as the incorporation of active substances in toothpaste (Liu et al. 2022), or modified by the application of different preparation methods (Bashir et al. 2022; Wang & Zhuang, 2022). In addition, many researchers have developed new chitosan nanoparticle formulations that include secondary matrix-forming materials (Tudu & Samanta, 2023; Strnad & Zemljič, 2023).

Polymers such as chitosan have been used for the preparation of nanoparticles, and for this purpose, there are different methodologies such as co-precipitation, chemical crosslinking, coacervation, emulsification, and ionotropic gelation, the latter of considerable interest since it is mostly used for the preparation of nanoparticles of polysaccharides (Bashir et al. 2022).

13.4.1 Ionotropic Gelation

This technique allows the crosslinking of polymeric chains that are arranged in nanostructures by interactions that can be covalent or non-covalent intermolecular interactions and that can be achieved with crosslinking agents that facilitate this bridge (Sacco et al. 2021). IG is based on the electrostatic interaction between the positively charged primary amino groups of chitosan and the negatively charged groups of a polyanion, such as sodium tripolyphosphate (TPP) or sodium pyrophosphate (SPP) (Kim et al. 2021), and a surfactant as Polysorbate 80 or Poloxamer 188 (Tosi et al. 2020; Kempe & Nicolazzo, 2021), which is illustrated in **Figure 13.3**. The use of this methodology offers a simple and gentle preparation method in an aqueous environment (Torrenegra-Alarcón et al. 2020; Lotfy et al. 2021) and avoids the use of high temperatures processes and the use of toxic organic solvents (Shoueir et al. 2021;Yammine et al. 2023). Chitosan-sodium tripolyphosphate nanoparticles have demonstrated their capacity for the encapsulation and release of proteins, genes, hydrophilic and hydrophobic drugs, vitamins, and polyphenolic compounds (Klojdová et al. 2023). In addition, the use of crosslinking agents is required for the effective release of bioactive compounds from the formed matrix.

13.4.2 Crosslinking Agents

Biodegradable polymer-based delivery systems need to be cross-linked to modulate their properties and maintain the stability of the matrix thus improving the release of the active compound (Calori et al. 2020). Glutaraldehyde is a widely used crosslinking agent, but its ability to transform into toxic reactive species has promoted the search for other safer crosslinking agents and procedures, such as sodium tripolyphosphate and genipin (Alavarse et al. 2022).

Sodium tripolyphosphate (TPP) is an inorganic compound used in the preparation of chitosan nanoparticles given that it is non-toxic and can gel fast. It is a sodium salt of polyphosphate pentaanion, which is a conjugated base of triphosphoric acid. It is used as a cross-linking agent for the formation of nanoparticles, as it has five negative charges that interact with the positive charges of the counterion (Silvestro et al. 2020).

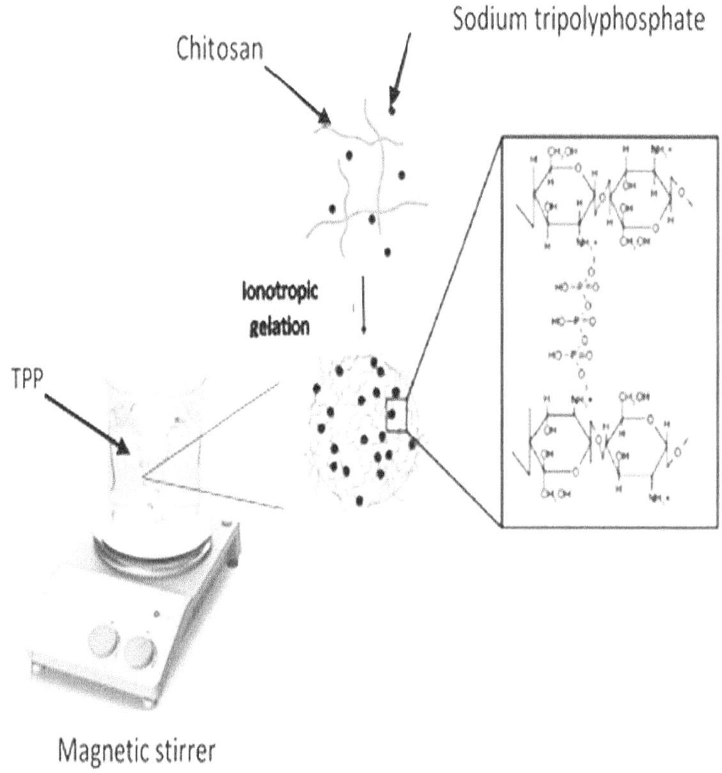

Figure 13.3 Schematic illustration of the formation of the chitosan–sodium tripolyphosphate (TPP) complex by ionotropic gelation.

13.4.3 Surfactants

For the preparation of nanoparticles, surfactants such as soy lecithin, egg lecithin and phosphatidylcholine, ethylene, oxide-propylene, oxide poloxamer 188 (Lutrol F-68), poloxamer 182, poloxamer 407, poloxamine 908, ethylene oxide, sorbitan/propylene oxide, polysorbate 20, polysorbate 60, polysorbate 80 (Tween 80), alkyl aryl alcohol polyether polymers such as tyloxapol, bile salts like sodium cholate, sodium glycolate, sodium taurocholate, and sodium taurodeoxycholate, and alcohols such as ethanol and butanol can be used (Miyazawa et al. 2021). Given the above, polymeric nanoparticles can be incorporated into the body by various routes of administration, which are described in section 13.6.

13.5 BIOACTIVE COMPOUNDS

13.5.1 α-Lipoic Acid (ALA)

ALA (**Figure 13.4**) can reduce reactive oxygen species, regenerate endogenous antioxidants, repair oxidative tissue damage, and chelating capacity (Tripathi *et al.* 2023). After oral intake, ALA

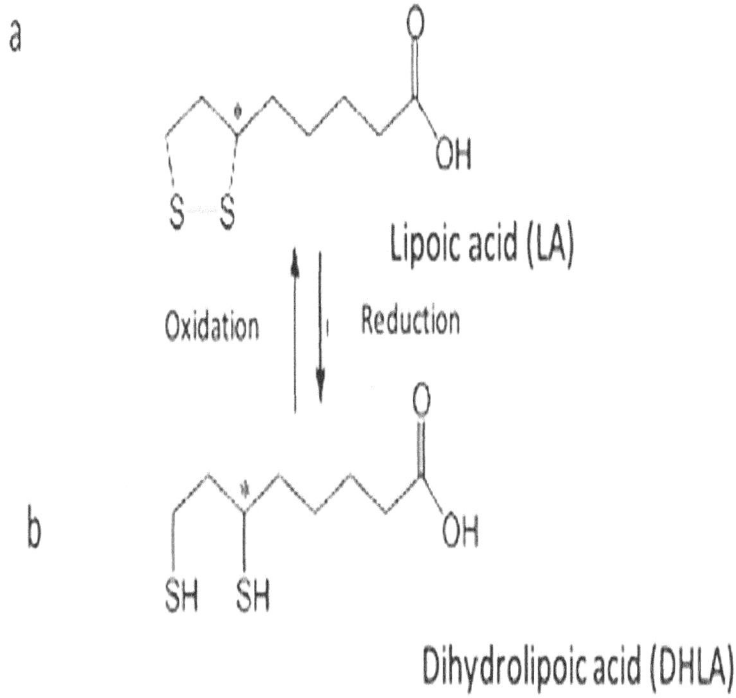

Figure 13.4 Biological forms of α-lipoic acid.

is easily absorbed in the intestinal tract and distributed throughout the body and can cross the blood-brain barrier under normal conditions. Intake and endogenous synthesis covers the basic needs of the organism (Balhaj et al. 2022). These properties vary depending on the compound and have been attributed to the pro-oxidant effects of α-dihydrolipoic acid. It has been noted that α-lipoic acid can block the pro-oxidant effect of α-dihydrolipoic acid by following the redox process shown in **Figure 13.4** (Farah et al. 2020). On the other hand, ALA is unstable under light and heat and gradually decomposes at room temperature. Temperatures higher than the boiling point of lipoic acid (59–62°C) cause its immediate polymerization and make it unusable because this polymer is insoluble in practically all solvents (Banik et al. 2021). ALA is abundant in red meat, brewer's yeast, wheat germ, spinach, broccoli, peas, and cabbage. In these foods, it remains covalently bound to the lysine residues of proteins (lipolysine), and the main substrate for the *in vivo* synthesis of ALA is monosulfurized octanoic acids (Gavazza et al. 2021).

13.5.2 Vitamin E (α-Tocopherol)

The term vitamin E is used for a group of fat-soluble compounds derived from tocopherol and tocotrienol. Four tocopherols and four tocotrienols (each in their α, β, δ, and γ forms) occur in nature and α -tocopherol has the highest biological activity and is the most widely available form of vitamin E present in foods (Divyadharsini et al. 2023). Tocopherols and tocotrienols inhibit lipid peroxidation mainly due to their ability to trap peroxyl radicals before they can react with fatty acids in cell membranes (Barouh et al. 2022). Beyond its antioxidant function, vitamin E has also been involved in the regulation of DNA synthesis, gene expression, mitochondrial metabolism, and cell proliferation and differentiation (Brabson et al. 2021). Under physiological conditions, after being ingested, vitamin E is rapidly absorbed in the small intestine and then is transported to plasmatic lipoproteins (Violet et al. 2020). The average percentage of absorption of vitamin E consumed in the diet (0.4–1 mg) is between 50 and 70%. However, the absorption efficiency can decrease when greater amounts of tocopherol are consumed. An increase in dietary polyunsaturated fatty acids, or co-administration of vitamin E with ferrous salts or bile acid-binding resins, interferes with the digestion and absorption of vitamin E (Kamath et al. 2022).

13.5.3 Vitamin C (Ascorbic Acid)

L-ascorbic acid, commonly called vitamin C, is considered one of the most potent antioxidant agents acting in the body (Boo, 2022). Ascorbic acid has a lactone structure, in which the hydroxyl groups associated with the double bond function as agents with high reducing potential, which allows it to participate in the direct reduction of oxygen, thus functioning as a donor substrate in reactions involving peroxidases (Parcheta et al. 2021). Most mammals and plants synthesize vitamin C endogenously from glucose and galactose. However, humans lack this ability (Murererehe et al. 2022).

Its two main dietary versions are L-ascorbic acid and dehydro-L-ascorbic acid. The latter is the reversible oxidized version, which also has vitamin activity (Tram et al. 2021) and reduces the phenoxyl radicals formed during vitamin E activity, restoring it and exerting a synergistic effect with this vitamin. Vitamins C and E, as well as A, are classified as interrupting antioxidants, because they work by stopping the chain reaction of free radical formation (Chaudhary et al. 2023).

The absorption of vitamin C varies depending on the dose. In the small intestine, between 80–90% is absorbed by a Na^+-dependent transport mechanism when the intake is up to 100 mg/day, while this percentage decreases rapidly with a dose of 500 mg/day (Goncharova et al. 2021). Likewise, some of the main sources of vitamin C are citrus fruits, broccoli, spinach, green peppers, cabbage, and tomatoes (Mieszczakowska-Frąc et al. 2021).

In this sense, antioxidants are enhanced when other compounds are called synergists of antioxidants, which are substances that regenerate the active form of the antioxidant or eliminate pro-oxidant substances that favor oxidation from the environment (Liang et al. 2021).

13.5.4 Synergism among Antioxidants (α-Lipoic Acid, α-Tocopherol and Ascorbic Acid)

In addition, to antioxidant activity, α-lipoic acid and its endogenous counterpart dihydrolipoic acid, which is formed rapidly after absorption into the body's cells, contribute to the non-enzymatic regeneration of reduced glutathione, as well as vitamin C, vitamin E, and coenzyme Q10, acting *in vivo* (Tripathi et al. 2023, Rodella, 2023). When meeting a free radical, vitamin E stabilizes it by donating one of its electrons, this makes it a free radical. The difference with the rest of the free radicals is that vitamin E becomes less reactive than oxygen radicals in a way that the body's antioxidant systems can neutralize it.

Vitamin E acts in conjunction with vitamin C by reducing it to its normal state, thus vitamin E ceases to be a free radical and can once again carry out antioxidant tasks. Thiols such as glutathione and α-lipoic acid act mainly by regenerating vitamin C (Boo, 2022).

Specifically, the mixture of antioxidants, such as α-lipoic acid and vitamin E/vitamin C, have been extensively studied in experimental and clinical conditions due to their nutritional and chemical properties (Tripathi et al. 2023). During the last two decades, various studies suggested that the combination of α-lipoic acid and vitamin E might be beneficial in limiting pathological processes, in which the excessive production of reactive oxygen species is involved in the mechanisms of initiation and damage progression. A combination of ALA and vitamin E/C provides synergistic protection against the pathological effects of ROS. *In vitro* studies have shown that antioxidant vitamins can prevent lipid peroxidation and that vitamins C and E have synergistic effects, but support from *in vivo* evidence is still poor (Tuell et al. 2023).

Ingested antioxidants also contribute to cellular protection against free radicals. Important dietary antioxidants include among others, vitamin E (tocopherols and tocotrienols), vitamin C, carotenoids (e.g., β-carotene), flavonoids, α-lipoic acid, and several trace elements in various pathologies such as diabetic neuropathy, Alzheimer's disease and other cognitive disorders, rheumatoid arthritis, lupus erythematosus and cardiac ischemia to name a few. In this respect in **Table 13.1**, a description of selected works on the positive physiological effects of vitamins E and

TABLE 13.1 STUDIES ON THE COMBINATION OF ALA/VITAMIN E/VITAMIN C, ON THERAPEUTIC EFFECTS AND PHYSIOLOGICAL DISORDERS

Involved Bioactives and Dosage when Available	Main Findings	Reference
Vitamin E (10,000 IU/kg), α-lipoic acid (1.65 g/kg)	Data reported that dietary supplementation of both vitamin E and α-lipoic acid reduced ischemia-reperfusion activity induced by myocardial injury evidenced by lower levels of lipid peroxidation during *in vivo* tests with adult rats.	Coombes et al. (2000)
Vitamins C (500 mg/day) and E (800 IU/day)	Combined administration of vitamins C and E to patients was associated with lipid peroxidation but did not affect endothelial function in patients with systemic lupus erythematosus after 3 months of therapy.	Tam et al. (2005)
Vitamins E, (800 IU/day), vitamin C (500 mg/day) and α-lipoic acid (900 mg/day)	It was shown that neuroprostane levels in the brain, did not affected sizes of Aβ plaques, or influenced neural changes, depicting an improvement of performance on spatial memory testing in subjects with mild-to-moderate Alzheimer disease.	Galasko et al. (2012)
Vitamins A, E and C	Rheumatoid arthritis patients treated with antioxidant vitamins A, E, and C for 12 weeks showed levels of MDA and increased concentrations of thiols and reduced glutathione (GSH) and vitamin C concentrations in blood samples.	Fonseca et al. (2019)
α-Lipoic acid. (600 mg/day)	It was found that α-lipoic acid supplementation is associated with amelioration in nerve conduction velocity scores, clinically significant of reported neuropathic pain, burning and paresthesia, as well as in serum triglycerides, improved insulin sensitivity, and quality of life.	Sementina et al. (2022)

(continued)

TABLE 13.1 (CONTINUED)

Involved Bioactives and Dosage when Available	Main Findings	Reference
Naturally occurring antioxidant therapy in Alzheimer's disease	The therapeutic potential of natural antioxidants as preventatives and/or treatment for Alzheimer's Disease is examined, with special attention paid to natural antioxidant combinations and conjugates that are currently being investigated in human clinical trials.	Collins et al. (2022)
Vitamins E (400 mg/day) and C (500 mg/day)	This study showed that the 60 patients included in this study completed consumed vitamin C and vitamin E. After 8 weeks of treatment in the study group, there was statistically significant improvement in pain score, tender joint score, swollen joint score, and DAS score.	Gomathi et al. (2022)
Vitamins E (20 0 IU/day) C (250 mg/day) and α-lipoic acid (90 mg/day)	The report suggested that triple antioxidant therapy: vitamin E, vitamin C and α-lipoic acid for a period of 6 weeks in diabetic volunteers attenuated the *in vitro* experimental oxidative (stress of methaemoglobin formation and hemoglobin glycation *in vivo*.	Coleman et al. (2003)

C, and α-lipoic acid among other antioxidants, is presented, based on the potential synergistic effects of combinations of these nutrients and positive effects on the free radicals neutralizing agents, properties, sources, and targets (Martemucci et al. 2022).

In summary, oxidative stress is involved in the pathogenesis of a wide spectrum of conditions. Therefore, the α-lipoic acid, vitamin E and vitamin C combination is potentially useful for preventing and treating pathological conditions that have an oxidative process as a major source of damage. Consequently, strategies directed at counteracting oxidative processes will certainly be important in clinical medicine. This is of great importance because the scientific evidence depicts that ascorbate recycles α-tocopherol after semidehydroascorbyl radicals are regenerated by ALA. In addition, lipoate can reduce glutathione disulfide, ubiquinone, and dehydroascorbate radicals, contributing to vitamin E regeneration. It seems, however, that more studies on the combined effects of ALA, vitamins E, and C administered via a nanoparticle carrier are needed to widen the positive effects of such combination and the basic knowledge on the synergic effects of these compounds and more *in vivo* studies need to be carried out in this respect.

13.6 ROUTES OF ADMINISTRATION

Historically, delivery systems have been developed to modify the route of administration for the benefit of the patient, improve bioavailability, change the release profile, or improve a formulation for a new presentation. Also, delivery systems must be developed to identify the target site, which can be an organ, a cell, a cellular compartment (nucleus, cytoplasm) or even within an organelle (Wang et al. 2021).

The release of the bioactive compounds must be controlled in response to an internal stimulus such as pH, presence of enzymes or a change in redox potential or an external one such as light, temperature, and magnetic field (Zhang et al. 2022). When these compounds are administered intravenously, they must have the ability to circulate freely through the blood capillaries and not be recognized as a foreign agent (Nowak-Jary & Machnicka, 2022).

Due to the subcellular size that nanoparticles present, they can be administered through different routes: intravenous, nasal, dermal application, and gastrointestinal ingestion (Zuzarte et al. 2022). Similarly, it is of great interest to point out the importance of the active principle in the production of nanoparticles. Some of the compounds that have been recently studied and have the capacity of being encapsulated are α-lipoic acid, α-tocopherol, and ascorbic acid.

In general, the challenges for the application of nanotechnology of bioactive compounds have become more important over the years. In this way, encapsulation carried out with nanotechnological tools is favorable since most studies show positive results regarding the ability of the nanocomposites to act in the body. The potential of their application as food additives, or to encapsulate vitamins and antioxidants, is promising; and in this respect, therapeutic effects of a nanocomposite system prepared by using a wall material as chitosan loaded with ALA, can contribute, when administered throughout various routes, to take advantage of the synergic effect of these antioxidants.

13.7 CONCLUSION

The nanoencapsulation of α-lipoic acid, α-tocopherol and ascorbic acid, in a chitosan matrix represents an efficient alternative to preserve the antioxidant properties of these bioactive compounds. Chitosan nanoparticles can be formed by ionotropic gelation by using two polyanions as sodium pyrophosphate (SPP) and sodium tripolyphosphate (TPP), incorporating ALA, vitamin E and vitamin C, and taking advantage of their synergistic effect, as a strategy to enhance their therapeutic effects. This system can be used to aid treating several metabolic disorders such as diabetic neuropathy, Alzheimer's disease, rheumatoid arthritis, lupus erythematosus as well as others. The triple antioxidant therapy (ALA/vitamin E/vitamin C) in diabetic volunteers attenuated oxidative stress which was possibly due to ascorbate recycles α-tocopherol given that semidehydroascorbyl radicals are regenerated by ALA. Moreover, lipoate can reduce glutathione disulfide, ubiquinone, and dehydroascorbate radicals, which provokes vitamin E regeneration. Loaded nanocapsules can be administered to patients through various routes as intravenous, nasal, dermic application, and gastrointestinal. More *in vivo* studies of the discussed therapy are needed as to further understand the basis of the synergistic effects observed, the controlled release of bioactives from the nanocomposite and the possible use of other compounds to enhance the biological activity of the proposed combination of ALA, vitamin E and vitamin C.

REFERENCES

Abd El-Hack, M.E., El-Saadony, M.T., Shafi, M.E., Zabermawi, N.M., Arif, M., Batiha, G.E., & Al-Sagheer, A.A. (2020). Antimicrobial and antioxidant properties of chitosan and its derivatives and their applications: A review. *International Journal of Biological Macromolecules*, 164, 2726–2744. https://doi.org/10.1016/j.ijbiomac.2020.08.153.

Alavarse, A.C., Frachini, E.C.G., da Silva, R.L.C.G., Lima, V.H., Shavandi, A., & Petri, D.F.S. (2022). Crosslinkers for polysaccharides and proteins: Synthesis conditions, mechanisms, and crosslinking efficiency, a review. *International Journal of Biological Macromolecules*, 202, 558–596. https://doi.org/10.1016/j.ijbiomac.2022.01.029.

Allison, C.L., Lutzke, A., & Reynolds, M.M. (2020). Identification of low molecular weight degradation products from chitin and chitosan by electrospray ionization time-of-flight mass spectrometry. *Carbohydrate Research*, 493, 108046. https://doi.org/10.1016/j.carres.2020.108046.

Alu'datt, M.H., Alrosan, M., Gammoh, S., Tranchant, C.C., Alhamad, M.N., Rababah, T., & Tan, T.C. (2022). Encapsulation-based technologies for bioactive compounds and their application in the food industry: A roadmap for food-derived functional and health-promoting ingredients. *Food Bioscience*, 50, 101971. https://doi.org/10.1016/j.fbio.2022.101971.

Anwar, M.M. (2022). Oxidative stress-A direct bridge to central nervous system homeostatic dysfunction and Alzheimer's disease. *Cell Biochemistry and Function*, 40(1), 17–27. https://doi.org/10.1002/cbf.3673.

Aranaz, I., Alcántara, A.R., Civera, M.C., Arias, C., Elorza, B., Heras Caballero, A., & Acosta, N. (2021). Chitosan: An overview of its properties and applications. *Polymers*, 13(19), 3256. https://doi.org/10.3390/polym13193256.

Ashraf, S.A., Siddiqui, A.J., Abd Elmoneim, O.E., Khan, M.I., Patel, M., Alreshidi, M., & Adnan, M. (2021). Innovations in nanoscience for the sustainable development of food and agriculture with implications on health and environment. *Science of the Total Environment*, 768, 144990. https://doi.org/10.1016/j.scitotenv.2021.144990.

Balhaj, M.O., Ibrahim, A.E., Alsemeh, A.E., & Mohammed, H.O. (2022). The Neurotoxic effects of methotrxate (MTX) on rat hippocampus and to explore the neuroprotective role of alpha lipoic acid (ALA). *The Egyptian Journal of Hospital Medicine*, 86(1), 762–769. https://doi.org/10.21608/EJHM.2022.216041.

Banik, S., Halder, S., Sato, H., & Onoue, S. (2021). Self-emulsifying drug delivery system of (R)-α-lipoic acid to improve its stability and oral absorption. *Biopharmaceutics & Drug Disposition*, 42(5), 226–233. https://doi.org/10.1002/bdd.2277.

Baranwal, J., Barse, B., Fais, A., Delogu, G.L., & Kumar, A. (2022). Biopolymer: A sustainable material for food and medical applications. *Polymers*, 14(5), 983. https://doi.org/10.3390/polym14050983.

Barouh, N., Bourlieu-Lacanal, C., Figueroa-Espinoza, M.C., Durand, E., & Villeneuve, P. (2022). Tocopherols as antioxidants in lipid-based systems: The combination of chemical and physicochemical interactions determines their efficiency. *Comprehensive Reviews in Food Science and Food Safety*, 21(1), 642–688. https://doi.org/10.1111/1541-4337.12867.

Bashir, S.M., Ahmed Rather, G., Patrício, A., Haq, Z., Sheikh, A.A., Shah, M.Z.U.H., & Fonte, P. (2022). Chitosan nanoparticles: A versatile platform for biomedical applications. *Materials*, 15(19), 6521. https://doi.org/10.3390/ma15196521.

Benselfelt, T., Kummer, N., Nordenström, M., Fall, A.B., Nyström, G., & Wågberg, L. (2023). The colloidal properties of nanocellulose. *ChemSusChem*, 16(8), e202201955. https://doi.org/10.1002/cssc.202201955.

Boo, Y.C. (2022). Ascorbic acid (vitamin C) as a cosmeceutical to increase dermal collagen for skin antiaging purposes: Emerging combination therapies. Antioxidants, *11*(9), 1663. https://doi.org/10.3390/antiox11091663.

Brabson, J.P., Leesang, T., Mohammad, S., & Cimmino, L. (2021). Epigenetic regulation of genomic stability by vitamin C. *Frontiers in Genetics*, 12, 675780. https://doi.org/10.3389/fgene.2021.675780.

Calori, I.R., Braga, G., de Jesus, P.D.C.C., Bi, H., & Tedesco, A.C. (2020). Polymer scaffolds as drug delivery systems. *European Polymer Journal*, 129, 109621. https://doi.org/10.1016/j.eurpolymj.2020.109621.

Chaudhary, P., Janmeda, P., Docea, A.O., Yeskaliyeva, B., Abdull Razis, A.F., Modu, B., & Sharifi-Rad, J. (2023). Oxidative stress, free radicals and antioxidants: Potential crosstalk in the pathophysiology of human diseases. *Frontiers in Chemistry*, 11, 1158198. https://doi.org/10.3389/fchem.2023.1158198.

Chen, L., Cao, H., Huang, Q., Xiao, J., & Teng, H. (2022). Absorption, metabolism and bioavailability of flavonoids: A review. *Critical Reviews in Food Science and Nutrition*, 62(28), 7730–7742. https://doi.org/10.1080/10408398.2021.1917508.

Chu, B., Biriukov, D., Bischoff, M., Předota, M., Roke, S., & Marchioro, A. (2023). Evolution of the electrical double layer with electrolyte concentration probed by second harmonic scattering. *Faraday Discussions*. https://doi.org/10.1039/D3FD00036B.

Coleman, M.D., Fernandes, S., & Khanderia, L. (2003). A preliminary evaluation of a novel method to monitor a triple antioxidant combination (vitamins E, C and α-lipoic acid) in diabetic volunteers using *in vitro* methaemoglobin formation. *Environmental Toxicology and Pharmacology*, 14(1-2), 69–75. https://doi.org/10.1016/S1382-6689(03)00027-9.

Collins, A.E., Saleh, T.M., & Kalisch, B.E. (2022). Naturally occurring antioxidant therapy in Alzheimer's disease. *Antioxidants*, 11(2), 213. https://doi.org/10.3390/antiox11020213.

Coombes, Powers, Demirel, Jessup, Vincent, Naito, & Packer. (2000). Effect of combined supplementation with vitamin E and alpha-lipoic acid on myocardial performance during in vivo ischaemia-reperfusion. *Acta Physiologica Scandinavica*, 169(4), 261–269. https://doi.org/10.1046/j.1365-201x.2000.00740.x.

Delon, T., Parsai, T., Kilic, U., Schubert, M., Morin, S.A., & Li, Y. (2021). Impacts of particle surface heterogeneity on the deposition of colloids on flat surfaces. *Environmental Science: Nano*, 8(11), 3365–3375. https://doi.org/10.1039/d1en00049g.

Demir, B., & Searles, D.J. (2020). Investigation of the ionic liquid graphene electric double layer in supercapacitors using constant potential simulations. *Nanomaterials*, 10(11), 2181. https://doi.org/10.3390/nano10112181.

Di Santo, M.C., D'Antoni, C.L., Rubio, A.P.D., Alaimo, A., & Pérez, O.E. (2021). Chitosan-tripolyphosphate nanoparticles designed to encapsulate polyphenolic compounds for biomedical and pharmaceutical applications– A review. *Biomedicine & Pharmacotherapy*, 142, 111970. https://doi.org/10.1016/j.biopha.2021.111970.

Divyadharsini, V., Maheswari, T.U., Rajeshkumar, S., & Umamaheswari, T.N. (2023). Assessment of antimicrobial activity of lycopene, Vitamin E, and lycopene-Vitamin E combination against *Staphylococcus aureus, Streptococcus mutans, Enterococcus faecalis*, and *Candida albicans*: An *in vitro* Study. *Cureus*, 15(7): e42419. https://doi.org/7759/cureus.42419.

El-Khawaga, A.M., Zidan, A., & Abd El-Mageed, A.I. (2023). Preparation methods of different nanomaterials for various potential applications: A review. *Journal of Molecular Structure*, 1281, 135148. https://doi.org/10.1016/j.molstruc.2023.135148.

Elmowafy, M., Shalaby, K., Elkomy, M.H., Alsaidan, O.A., Gomaa, H.A., Abdelgawad, M.A., & Mostafa, E.M. (2023). Polymeric nanoparticles for delivery of natural bioactive agents: Recent advances and challenges. *Polymers*, 15(5), 1123. https://doi.org/10.3390/polym15051123.

Farah, A.I., Ahmad, M.N., & Al-Qirim, T.M. (2020). The antioxidant and pro-oxidant impacts of varying levels of alpha-lipoic acid on biomarkers of myoglobin oxidation *in vitro*. *Jordan Journal of Agricultural Sciences*, 16(4), 89–99. https://doi.org/10.35516/jjas.v16i4.63.

Finney, A.R., & Salvalaglio, M. (2023). Properties of aqueous electrolyte solutions at carbon electrodes: Effects of concentration and surface charge on solution structure, ion clustering and thermodynamics in the electric double layer. *Faraday Discussions*, 249, 334-362. https://doi.org/10.1039/D3FD00133D.

Fonseca, L.J.S.D., Nunes-Souza, V., Goulart, M.O.F., & Rabelo, L.A. (2019). Oxidative stress in rheumatoid arthritis: What the future might hold regarding novel biomarkers and add-on therapies. *Oxidative Medicine and Cellular* Longevity, 2019, 7536805. https://doi.org/10.1155/2019/7536805.

Frank, L.A., Onzi, G.R., Morawski, A.S., Pohlmann, A.R., Guterres, S.S., & Contri, R.V. (2020). Chitosan as a coating material for nanoparticles intended for biomedical applications. *Reactive and Functional Polymers*, 147, 104459. https://doi.org/10.1016/j.reactfunctpolym.2019.104459.

Frigaard, J., Jensen, J.L., Galtung, H.K., & Hiorth, M. (2022). The potential of chitosan in nanomedicine: An overview of the cytotoxicity of chitosan based nanoparticles. *Frontiers in Pharmacology*, 13, 880377. https://doi.org/10.3389/fphar.2022.880377.

Galasko, D.R., Peskind, E., Clark, C.M., Quinn, J.F., Ringman, J.M., Jicha, G.A., & Aisen, P. (2012). Antioxidants for Alzheimer disease: A randomized clinical trial with cerebrospinal fluid biomarker measures. *Archives of Neurology*, 69(7), 836–841. https://doi.org/10.1001/archneurol.2012.85.

Gavazza, M., Marmunti, M., Leaden, P., Zeinsteger, P., & Palacios, A. (2021). Alpha-lipoic acid: Antioxidant activity against non-enzymatic peroxidation of rat kidney and liver mitochondria. *Journal of Clinical and Biomedical Investigation*, 1, 1–4. https://doi.org/10.52916/jcbi214007.

Gomathi, A., Chenthamarai, G., Manvizhi, S., & Gowrithilagam, T.G. (2022). Effects of Vitamin C and Vitamin E in rheumatoid arthritis-A randomized, open label, and comparative study in a tertiary care hospital. *National Journal of Physiology, Pharmacy and Pharmacology*, 12(9), 1463–1463. https://doi.org/10.5455/njppp.2022.12.08383202207082022.

Gómez-Gómez, A.L., Martínez-Ayala, A.L., Moguel-Concha, D.D.R., Borges-Martínez, J.E., Perea-Flores, M.D.J., & Dávila-Ortiz, G. (2023). Relationship of nanomaterials' structure based on their application in the food industry: Physicochemical and techno-functional characteristics. *Applied Sciences*, 13(12), 7167. https://doi.org/10.3390/app13127167.

Goncharova, P.S., Davydova, T.K., & Zhukova, N.G. (2021). Prospects for studying the pharmacokinetics, pharmacodynamics and pharmacogenetics of vitamin C in patients with neurological diseases and mental disorders. *Personalized Psychiatry and Neurology*, 1(2), 63–82. https://doi.org/10.52667/2712-9179-2021-1-2-63-82.

Gorantla, S., Dabholkar, N., Sharma, S., Rapalli, V.K., Alexander, A., & Singhvi, G. (2021). Chitosan-based microneedles as a potential platform for drug delivery through the skin: Trends and regulatory aspects. *International Journal of Biological Macromolecules*, 184, 438–453. https://doi.org/10.1016/j.ijbiomac.2021.06.059.

Huesca-Urióstegui, K., García-Valderrama, E.J., Gutierrez-Uribe, J.A., Antunes-Ricardo, M., & Guajardo-Flores, D. (2022). Nanofiber systems as herbal bioactive compounds carriers. *Current applications in healthcare. Pharmaceutics*, 14(1), 191. https://doi.org/10.3390/pharmaceutics14010191.

Iber, B.T., Kasan, N.A., Torsabo, D., & Omuwa, J.W. (2022). A review of various sources of chitin and chitosan in nature. *Journal of Renewable Materials*, 10(4), 1097. https://doi.org/10.32604/jrm.2022.018142.

Jomova, K., Raptova, R., Alomar, S.Y., Alwasel, S.H., Nepovimova, E., Kuca, K., & Valko, M. (2023). Reactive oxygen species, toxicity, oxidative stress, and antioxidants: Chronic diseases and aging. *Archives of Toxicology*, 97(10), 2499–2574. https://doi.org/10.1007/s00204-023-03562-9.

Kamath, B.M., Alonso, E.M., Heubi, J.E., Karpen, S.J., Sundaram, S.S., Shneider, B.L., & Sokol, R.J. (2022). Fat soluble vitamin assessment and supplementation in cholestasis. *Clinics in Liver Disease*, 26(3), 537–553. https://doi.org/10.1016/j.cld.2022.03.011.

Kempe, K., & Nicolazzo, J.A. (2021). Biodegradable polymeric nanoparticles for brain-targeted drug delivery. *Nanomedicines for Brain Drug Delivery*, 1–27. https://doi.org/10.1007/978-1-0716-0838-8_1.

Khalaf, E.M., Abood, N.A., Atta, R.Z., Ramírez-Coronel, A.A., Alazragi, R., Parra, R.M.R., & Farhood, B. (2023). Recent progressions in biomedical and pharmaceutical applications of chitosan nanoparticles: A comprehensive review. *International Journal of Biological Macromolecules*, 231, 123354. https://doi.org/10.1016/j.ijbiomac.2023.123354.

Khan, A., & Alamry, K.A. (2021). Recent advances of emerging green chitosan-based biomaterials with potential biomedical applications: A review. *Carbohydrate Research*, 506, 108368. https://doi.org/10.1016/j.carres.2021.108368.

Kim, E.S., Lee, J.S., & Lee, H.G. (2021). Quercetin delivery characteristics of chitosan nanoparticles prepared with different molecular weight polyanion cross-linkers. *Carbohydrate Polymers*, 267, 118157. https://doi.org/10.1016/j.carbpol.2021.118157.

Klojdová, I., Milota, T., Smetanová, J., & Stathopoulos, C. (2023). Encapsulation: A strategy to deliver therapeutics and bioactive compounds? *Pharmaceuticals*, 16(3), 362. https://doi.org/10.3390/ph16030362.

Kruk, J., Aboul-Enein, B.H., Duchnik, E., & Marchlewicz, M. (2022). Antioxidative properties of phenolic compounds and their effect on oxidative stress induced by severe physical exercise. *The Journal of Physiological Sciences*, *72*(1), 19. https://doi.org/10.1186/s12576-022-00845-1.

Kurek, M., Benaida-Debbache, N., Elez Garofulić, I., Galić, K., Avallone, S., Voilley, A., & Waché, Y. (2022). Antioxidants and bioactive compounds in food: Critical review of issues and prospects. *Antioxidants*, 11(4), 742. https://doi.org/10.3390/antiox11040742.

Li, H., Wang, Y., Zhao, P., Guo, L., Huang, L., Li, X., & Gao, W. (2023). Naturally and chemically acetylated polysaccharides: Structural characteristics, synthesis, activities, and applications in the delivery system: A review. *Carbohydrate Polymers*, 313(3), 120746. https://doi.org/10.1016/j.carbpol.2023.120746.

Leon, V.J., Blanc, B., Sonnert, S.D., & Varanasi, K.K. (2023). Externally tunable, low power electrostatic control of cell adhesion with nanometric high-k dielectric films. *Advanced Functional Materials*, 33(27). 2300732. https://doi.org/10.1002/adfm.202300732.

Liang, Z., Han, D., Han, F., Wu, Z., Zhao, X., Fu, W., & Niu, L. (2021). Novel strategy of natural antioxidant nutrition quality evaluation in food: Oxidation resistance mechanism and synergistic effects investigation. *Food Chemistry*, 359, 129768. https://doi.org/10.1016/j.foodchem.2021.129768.

Liu, Z., Wang, K., Peng, X., & Zhang, L. (2022). Chitosan-based drug delivery systems: Current strategic design and potential application in human hard tissue repair. *European Polymer Journal*, 166, 110979. https://doi.org/10.1016/j.eurpolymj.2021.110979.

Lotfy, N.S., Borg, T.M., & Mohamed, E.A. (2021). The promising role of Chitosan–poloxamer 188 nanocrystals in improving diosmin dissolution and therapeutic efficacy against ferrous sulfate-induced hepatic injury in rats. *Pharmaceutics*, 13(12), 2087. https://doi.org/10.3390/pharmaceutics13122087.

Luo, J., Mills, K., le Cessie, S., Noordam, R., & van Heemst, D. (2020). Ageing, age-related diseases and oxidative stress: What to do next? Ageing *Research Reviews*, 57, 100982. https://doi.org/10.1016/j.arr.2019.100982.

Maleki, G., Woltering, E.J., & Mozafari, M.R. (2022). Applications of chitosan-based carrier as an encapsulating agent in food industry. *Trends in Food Science & Technology*, *120*, 88–99. https://doi.org/10.1016/j.tifs.2022.01.001.

Martemucci, G., Costagliola, C., Mariano, M., D'andrea, L., Napolitano, P., & D'Alessandro, A.G. (2022). Free radical properties, source and targets, antioxidant consumption and health. *Oxygen*, 2(2), 48–78. https://doi.org/10.3390/oxygen2020006.

Mieszczakowska-Frąc, M., Celejewska, K., & Płocharski, W. (2021). Impact of innovative technologies on the content of vitamin C and its bioavailability from processed fruit and vegetable products. *Antioxidants*, 10(1), 54. https://doi.org/10.3390/antiox10010054

Miyazawa, T., Itaya, M., Burdeos, G.C., Nakagawa, K., & Miyazawa, T. (2021). A critical review of the use of surfactant-coated nanoparticles in nanomedicine and food nanotechnology. *International Journal of Nanomedicine*, 16, 3937–3999. https://doi.org/10.2147/IJN.S298606.

Mohammad, Z.H., Ahmad, F., Ibrahim, S.A., & Zaidi, S. (2022). Application of nanotechnology in different aspects of the food industry. *Discover Food*, *2*(1), 12. https://doi.org/10.1007/s44187-022-00013-9.

Muneer, R., Hashmet, M.R., Pourafshary, P., & Shakeel, M. (2023). Unlocking the power of artificial intelligence: Accurate zeta potential prediction using machine learning. *Nanomaterials*, 13(7), 1209. https://doi.org/10.2196/16607.

Murererehe, J., Uwitonze, A.M., Nikuze, P., Patel, J., & Razzaque, M.S. (2022). Beneficial effects of vitamin C in maintaining optimal oral health. *Frontiers in Nutrition*, 8, 805809. https://doi.org/10.3389/fnut.2021.805809.

Nowak-Jary, J., & Machnicka, B. (2022). Pharmacokinetics of magnetic iron oxide nanoparticles for medical applications. *Journal of Nanobiotechnology*, 20(1), 1–30. https://doi.org/10.1186/s12951-022-01510.

Ohya, Y., Shiratani, M., Kobayashi, H., & Ouchi, T. (1994). Release behavior of 5- fluorouracil from chitosan-gel nanospheres immobilizing 5-fluorouracil coated with polysaccharides and their cell specific cytotoxicity. *Journal of Macromolecular Science-Pure and Applied Chemistry*, 31(5), 629–642. https://doi.org/10.1080/10601329409349743.

Ong, G.M., Gallegos, A., & Wu, J. (2020). Modeling surface charge regulation of colloidal particles in aqueous solutions. *Langmuir*, 36(40), 11918–11928. https://doi.org/10.1021/acs.langmuir.0c02000.

Parcheta, M., Świsłocka, R., Orzechowska, S., Akimowicz, M., Choińska, R., & Lewandowski, W. (2021). Recent developments in effective antioxidants: The structure and antioxidant properties. *Materials*, 14(8), 1984. https://doi.org/10.3390/ma14081984.

Pateiro, M., Gómez, B., Munekata, P.E., Barba, F.J., Putnik, P., Kovačević, D.B., & Lorenzo, J.M. (2021). Nanoencapsulation of promising bioactive compounds to improve their absorption, stability, functionality and the appearance of the final food products. *Molecules*, 26(6), 1547. https://doi.org/10.3390/molecules26061547.

Pochapski, D.J., Carvalho dos Santos, C., Leite, G.W., Pulcinelli, S.H., & Santilli, C.V. (2021). Zeta potential and colloidal stability predictions for inorganic nanoparticle dispersions: Effects of experimental conditions and electrokinetic models on the interpretation of results. *Langmuir*, 37(45), 13379–13389. https://doi.org/10.1021/acs.langmuir.1c02056.

Reddy, M.S.B., Ponnamma, D., Choudhary, R., & Sadasivuni, K.K. (2021). A comparative review of natural and synthetic biopolymer composite scaffolds. *Polymers*, 13(7), 1105. https://doi.org/10.3390/polym13071105.

Rial-Hermida, M.I., Rey-Rico, A., Blanco-Fernandez, B., Carballo-Pedrares, N., Byrne, E.M., & Mano, J.F. (2021). Recent progress on polysaccharide-based hydrogels for controlled delivery of therapeutic biomolecules. *ACS Biomaterials Science & Engineering*, 7(9), 4102–4127. https://doi.org/10.1021/acsbiomaterials.0c01784.

Rodella, U., Honisch, C., Gatto, C., Ruzza, P., & D'Amato Tóthová, J. 2023. Antioxidant nutraceutical strategies in the prevention of oxidative stress related eye diseases. *Nutrients*, 15(10), 2283. https://doi.org/10.3390/nu15102283.

Sacco, P., Pedroso-Santana, S., Kumar, Y., Joly, N., Martin, P., & Bocchetta, P. (2021). Ionotropic gelation of chitosan flat structures and potential applications. *Molecules*, 26(3), 660. https://doi.org/10.3390/molecules26030660.

Sánchez, A., Mejía, S.P., & Orozco, J. (2020). Recent advances in polymeric nanoparticle-encapsulated drugs against intracellular infections. *Molecules*, 25(16), 3760. https://doi.org/10.3390/molecules25163760.

Sementina, A., Cierzniakowski, M., Rogalska, J., Piechowiak, I., Spichalski, M., & Araszkiewicz, A. (2022). A novel approach to alpha-lipoic acid therapy in the treatment of diabetic peripheral neuropathy. *Journal of Medical Science*, 91(3), e714-e714. https://doi.org/10.20883/medical.e714.

Serrano-Lotina, A., Portela, R., Baeza, P., Alcolea-Rodriguez, V., Villarroel, M., & Ávila, P. (2023). Zeta potential as a tool for functional materials development. *Catalysis Today*, 423, 113862. https://doi.org/10.1016/j.cattod.2022.08.004.

Silva, S.M.L., Fook, M.V.L., Montazerian, M., Barbosa, F.C., & Silva, H N. (2023). Composites based on chitosan and inorganic materials for biomedical applications. In Avalos Belmontes, F., González, F.J., & López-Manchado, M.Á. (eds.), *Green-Based Nanocomposite Materials and Applications* (pp. 119–139). Cham: Springer International Publishing. https://doi.org/10.1007/978-3-031-18428-4_7.

Shokri, Z., Seidi, F., Saeb, M.R., Jin, Y., Li, C., & Xiao, H. (2022). Elucidating the impact of enzymatic modifications on the structure, properties, and applications of cellulose, chitosan, starch and their derivatives: A review. *Materials Today Chemistry*, 24, 100780. https://doi.org/10.1016/j.mtchem.2022.100780.

Shoueir, K.R., El-Desouky, N., Rashad, M.M., Ahmed, M.K., Janowska, I., & El-Kemary, M. (2021). Chitosan based-nanoparticles and nanocapsules: Overview, physicochemical features, applications of a nanofibrous scaffold, and bioprinting. *International Journal of Biological Macromolecules*, 167, 1176–1197. https://doi.org/10.1016/j.ijbiomac.2020.11.072.

Silvestro, I., Francolini, I., Di Lisio, V., Martinelli, A., Pietrelli, L., Scotto d'Abusco, A., & Piozzi, A. (2020). Preparation and characterization of TPP-chitosan crosslinked scaffolds for tissue engineering. *Materials*, 13(16), 3577. https://doi.org/10.3390/ma13163577.

Sreekumar, S., Wattjes, J., Niehues, A., Mengoni, T., Mendes, A.C., Morris, E.R., & Moerschbacher, B.M. (2022). Biotechnologically produced chitosans with nonrandom acetylation patterns differ from conventional chitosans in properties and activities. *Nature Communications*, 13(1), 7125. https://doi.org/10.1038/s41467-022-34483-3.

Strnad, S., & Zemljič, L.F. (2023). Cellulose–chitosan functional biocomposites. *Polymers*, 15(2), 425. https://doi.org/10.3390/polym15020425.

Taha, T.B., Barzinjy, A.A., Hussain, F.H.S., & Nurtayeva, T. (2022). Nanotechnology and computer science: Trends and advances. *Memories-Materials*, Devices, Circuits and Systems, 2, 100011. https://doi.org/10.1016/j.memori.2022.100011.

Tam, L.S., Li, E.K., Leung, V.Y., Griffith, J.F., Benzie, I.F., Lim, P.L., & Tomlinson, B. (2005). Effects of vitamins C and E on oxidative stress markers and endothelial function in patients with systemic lupus erythematosus: A double blind, placebo controlled pilot study. *The Journal of Rheumatology*, 32(2), 275–282.

Tao, W., Wang, G., & Wei, J. (2021). The role of chitosan oligosaccharide in metabolic syndrome: A review of possible mechanisms. *Marine Drugs*, 19(9), 501. https://doi.org/10.3390/md19090501.

Torrenegra-Alarcón, M., Ortega-Toro, R., & Herrera-Barros, A. (2020). Recent advances in the use of nanomaterials with antimicrobial capacity. *Archivos Venezolanos de Farmacología y Terapéutica*, 39(8), 986–992. https://doi.org/10.5281/zenodo.4544021.

Tram, N.K., McLean, R.M., & Swindle-Reilly, K.E. (2021). Glutathione improves the antioxidant activity of vitamin C in human lens and retinal epithelial cells: Implications for vitreous substitutes. *Current Eye Research*, 46(4), 470–481. https://doi.org/10.1080/02713683.2020.1809002.

Tripathi, A.K., Ray, A.K., Mishra, S.K., Bishen, S.M., Mishra, H., & Khurana, A. 2023. Molecular and therapeutic insights of alpha-lipoic acid as a potential molecule for disease prevention. *Revista Brasileira de Farmacognosia*, 33(2), 272–287. https://doi.org/10.1007/s43450-023-00370-1.

Triunfo, M., Tafi, E., Guarnieri, A., Salvia, R., Scieuzo, C., Hahn, T., & Falabella, P. (2022). Characterization of chitin and chitosan derived from Hermetia illucens, a further step in a circular economy process. *Scientific Reports*, 12(1), 6613. https://doi.org/10.1038/s41598-022-10423-5.

Tosi, G., Duskey, J.T., & Kreuter, J. (2020). Nanoparticles as carriers for drug delivery of macromolecules across the blood-brain barrier. *Expert Opinion on* Drug Delivery, 17(1), 23–32. https://doi.org/10.1080/17425247.2020.1698544.

Tudu, M., & Samanta, A. (2023). Natural polysaccharides: Chemical properties and application in pharmaceutical formulations. *European Polymer Journal*, 184, 111801. https://doi.org/10.1016/j.eurpolymj.2022.111801.

Tuell, D.S., Los, E.A., Ford, G.A., & Stone, W.L. (2023). The role of natural antioxidant products that optimize redox status in the prevention and management of Type 2 diabetes. *Antioxidants*, 12(6), 1139. https://doi.org/10.3390/antiox12061139.

Vilas-Boas, A.A., Pintado, M., & Oliveira, A.L.S. (2021). Natural bioactive compounds from food waste: Toxicity and safety concerns. *Foods*, 10 (7), 1564. https://doi.org/10.3390/foods10071564.

Violet, P.C., Ebenuwa, I.C., Wang, Y., Niyyati, M., Padayatty, S.J., Head, B., & Levine, M. (2020). Vitamin E sequestration by liver fat in humans. *The Journal of Chemical Investigation Insight*, 5(1). https://doi.org/10.1172/jci.insight.133309.

Wang, J., Ni, Q., Wang, Y., Zhang, Y., He, H., Gao, D., & Liang, X.J. (2021). Nanoscale drug delivery systems for controllable drug behaviors by multi-stage barrier penetration. *Journal of Controlled Release*, 331, 282–295. https://doi.org/10.1016/j.jconrel.2020.08.045.

Wang, L., Hao, F., Tian, S., Dong, H., Nie, J., & Ma, G. (2022). Targeting polysaccharides such as chitosan, cellulose, alginate and starch for designing hemostatic dressings. *Carbohydrate Polymers*, 291, 119574. https://doi.org/10.1016/j.carbpol.2022.119574.

Wang, J., & Zhuang, S. (2022). Chitosan-based materials: Preparation, modification and application. *Journal of Cleaner Production*, 355, 131825. https://doi.org/10.1016/j.jclepro.2022.131825.

Vilas-Boas, A.A., Pintado, M., & Oliveira, A.L. (2021). Natural bioactive compounds from food waste: Toxicity and safety concerns. *Foods*, 10(7), 1564. https://doi.org/10.3390/foods10071564.

Yadav, N., Mudgal, D., Anand, R., Jindal, S., & Mishra, V. (2022). Recent development in nanoencapsulation and delivery of natural bioactives through chitosan scaffolds for various biological applications. *International Journal of Biological Macromolecules*, 220, 537–572. https://doi.org/10.1016/j.ijbiomac.2022.08.098.

Yammine, J., Chihib, N.E., Gharsallaoui, A., Ismail, A., & Karam, L. (2023). Advances in essential oils encapsulation: Development, characterization and release mechanisms. *Polymer Bulletin*, 81, 3837–3882. https://doi.org/10.1007/s00289-023-04916-0.

Zabot, G.L., Schaefer Rodrigues, F., Polano Ody, L., Vinícius Tres, M., Herrera, E., Palacin, H. & Olivera-Montenegro, L. (2022). Encapsulation of bioactive compounds for food and agricultural applications. *Polymers*, 14(19), 4194. https://doi.org/10.3390/polym14194194.

Zhang, Y., Dong, L., Liu, L., Wu, Z., Pan, D., & Liu, L. (2022). Recent advances of stimuli-responsive polysaccharide hydrogels in delivery systems: A review. *Journal of Agricultural and Food Chemistry*, 70(21), 6300–6314. https://doi.org/10.1021/acs.jafc.2c01080.

Zhao, Y., Zhang, Z., Pan, Z., & Liu, Y. (2021). Advanced bioactive nanomaterials for biomedical applications. *Exploration*, 1(3), 20210089. https://doi.org/10.1002/EXP.20210089.

Zhu, S., Tang, Y., Lin, C., Liu, X.Y., & Lin, Y. (2021). Recent advances in patterning natural polymers: From nanofabrication techniques to applications. *Small Methods*, 5(3), 2001060. https://doi.org/10.1002/smtd.202001060.

Zuzarte, M., Vitorino, C., Salgueiro, L., & Girão, H. (2022). Plant nanovesicles for essential oil delivery. *Pharmaceutics*, 14(12), 2581. https://doi.org/10.3390/pharmaceutics14122581.

Section III

Use of Technologies for the Scaling Up and Obtention of Bioactive Compounds (Nutraceuticals)

Chapter 14

Application of Green Technologies for the Extraction of Health Promoting Food Ingredients

Incinur Hasbay

CONTENTS

DOI: 10.1201/9781003332558-17

14.1 INTRODUCTION

The increasing evidence that correlates diet and chronic diseases, in addition to studies showing the adverse health effects of synthetic food additives, have resulted in increasing demand for natural food ingredients and health-promoting food products. Consequently, utilization of natural compounds in the food industry has become popular in recent years (Silva et al. 2022), leading to an increase in studies for the recovery of bioactive compounds from various natural sources (Faiza et al. 2023). In addition, incorporation of bioactive compounds to obtain health-promoting food products is gaining increased interest.

Extraction is the most important step to isolate bioactive compounds and phytochemicals from plant materials. Two important factors that commonly affect the efficiency of extraction are the solvent and technique (Panja, 2018). Usually, low toxicity solvents with a suitable polarity with the target compounds are selected. The efficiency of extraction also depends on several other factors such as temperature, extraction time, particle size, and solvent type (da Silva et al. 2016; Soquetta et al. 2018; Jahongir et al. 2019; Rodríguez De Luna et al. 2020).

Conventional extraction techniques commonly used for the extraction of bioactive compounds include maceration, Soxhlet extraction, and hydrodistillation (Soquetta et al. 2018). Maceration simply involves mixing the sample with the solvent and is based on leaching of the desired compounds from the solid material, which has been traditionally used for the extraction of bioactive compounds and essential oils for a long time (Picot-Allain et al. 2021). Agitation and heating may be applied to increase the mass transfer and solubility of compounds (Picot-Allain, 2021). Soxhlet extraction involves the continuous cycling of the extraction solvent though the matrix containing the bioactive compounds. This technique is widely studied in the literature, including the optimization of process conditions for the extraction (Soquetta et al. 2018). Hydrodistillation is generally used for the extraction of volatile compounds, mostly essential oils (Soquetta et al. 2018; Picot-Allain, 2021). This process is performed with distilled water with no requirement of organic solvents. However, it has drawbacks such as consumption of high levels of energy and long extraction times (Soquetta et al. 2018).

Classical extraction techniques require large amounts of solvents, high energy consumption, and extended extraction times, which may have adverse effects on human health and/or the environment (Rodríguez De Luna et al. 2020; Carpentieri et al. 2021). Moreover, intensive heat treatment during these processes causes thermal degradation of bioactive compounds, which results in extracts of poor quality that render them unusable as ingredients in health-promoting foods (Rodríguez De Luna et al. 2020). To overcome these limitations, green extraction techniques are widely proposed and investigated.

Green extraction is based on the discovery and design of processes that will reduce energy consumption, allow use of alternative solvents and renewable natural products to reduce the use of hazardous substances, and ensure a safe and high-quality extract (Chemat et al. 2012; Carpentieri et al. 2021). The type of solvent is one of the key factors for development of sustainable extraction processes, to obtain bioactive compounds (Carpentieri et al. 2021). Bioactive compounds exhibit varying polarity and solubility, thus an appropriate solvent type should be selected for their effective extraction. Accordingly, solvent selection is important in terms of yield and quality of the extracts, as well as cost and environmental safety (Carpentieri et al. 2021). In this respect, non-toxic, recyclable, biodegradable solvents that have a low energy cost for synthesis have been proposed as green alternatives to petrochemical solvents (Carpentieri et al. 2021). In addition to solvent, the technique used is also critical for a sustainable process. Green extraction techniques should allow process intensification by providing higher extraction rate, less energy use, increased mass transfer, reduced equipment size, and decreased number of procesing steps (Jacotet-Navarro et al. 2016).

The aim of this chapter is to discuss the application of green techniques for the extraction of bioactive compounds to be used as health-promoting food ingredients. The techniques discussed are supercritical fluid, ultrasound-assisted (UAE), microwave-assisted (MAE), pressurized liquid (PLE), and pulsed electric field (PEF)-assisted extractions. The current state of knowledge about these techniques, main parameters that effect the treatment efficiency, and the main advantages and drawbacks presented in recent studies are summarized.

14.2 SUPERCRITICAL FLUID EXTRACTION (SFE)

14.2.1 General Description

A supercritical fluid is any substance at temperature and pressure conditions above its critical point that has properties between gas and liquid. Its viscosity and diffusivity are like a gas, which provides its diffusion through solid matrices. Its density is similar to a liquid and subsequently dissolves materials like normal liquids (Ahmad et al. 2019; Ahangari et al. 2021). Many SFEs are conducted at conditions close to the critical point of the fluid, and hence small changes in pressure or temperature result in large changes of the fluid's density, which provides density-dependent selectivity of the technique (Ahangari et al. 2021).

SFE basically occurs in two steps as solubilization of the compounds present in the sample and their separation into the supercritical solvent (da Silva et al. 2016). The mostly preferred solvent for supercritical extraction is CO_2 (Faiza et al. 2023). It has a low critical temperature (31.1°C), which allows for the extraction of thermally labile compounds at lower temperatures than those used for conventional techniques (Heng et al., 2013). It is also non-flammable, inexpensive as well as non-polluting and non-toxic (Barba et al. 2016; Ahangari et al. 2021).

Due to the low polarity of CO_2, most target compounds obtained by SFE are non-polar or mid-polar compounds such as lipids, essential oils, and carotenoids. Supercritical CO_2 can solubilize hydrocarbon and oxygenated mono- and sesquiterpenes, which are the main constituents of essential oils. Owing to this property, SFE has been described by some authors as providing high-quality essential oils and oleoresin (Santoyo et al. 2006). For the extraction of more polar compounds such as phospholipids or polyphenols, SFE processes can be aided by a co-solvent such as ethanol in addition to CO_2 (Montanari et al. 1999).

SFE can also be used to increase the efficiency of a subsequent extraction as a combined process. Gallo-Molina et al. (2019) applied SFE followed by a solid phase extraction process for the purification of tetrahydrocannabinol (THC) from the *Cannabis sativa* plant. Babova et al. (2016) also extracted anthocyanins and other phenolic compounds from bilberries (*Vaccinium myrtillus*) using SFE followed by pressurized liquid extraction (PLE), with aqueous ethanol as co-solvent. An anthocyanin-rich fraction was obtained selectively from juçara residues by PLE followed by SFE (Garcia-Mendoza et al. 2017). The best solvent was selected in PLE studies and used as co-solvent for SFE.

14.2.2 Main Parameters

Parameters that effect the efficiency of SFE are mainly the fluid flow rate, extraction temperature, pressure, particle size, modifier type (such as ethanol and methanol), and ratio. The optimized configuration of these parameters leads to maximum extraction yield and minimized extraction time, energy, and sample loss (Yousefi et al. 2019). Some examples of SFE conditions of bioactive compounds from various bioresources are presented in **Table 14.1**, together with their bioactivity.

The effect of flow rate is generally reported to be significant on extraction yield. Higher flow rates could overcome interface resistance and improve the transport of target compounds from the matrix to the fluid, and hence could provide faster extractions and higher recoveries (Yi et al. 2009). However, if diffusion from the inner cells of plant material is slow, increasing flow rate of CO_2 does not have a strong effect in extraction. In this case, increasing the temperature is more appropriate (Rodríguez De Luna et al. 2020).

Extraction temperature affects both the fluid density and mass transfer of the solutes. The fluid density is very sensitive to temperature at regions close to the critical pressure. An increase in temperature can decrease the fluid density to a great extent, which leads to a decrease in solute solubility (Wang et al. 2008; Yousefi et al. 2019). On the other hand, increase in temperature also accelerates mass transfer, which leads to an increase in extraction yield. This indicates that the predition of temperature effect is difficult and its optimization with other parameters is required, considering their interaction effects (Wang et al. 2008). Generally, for extraction of polyphenols, the effect of temperature is reported to be negative at low pressures (10–15 MPa) up to a certain value, beyond which the effect of temperature becomes positive. The pressure value is called the cross-over pressure, which arises from the solubility being controlled by a balance between the solvent density and the change in the solute vapors pressure with increase in temperature (Adil et al. 2008).

The effect of pressure is similar to the effect of temperature in terms of increasing the fluid density and solvating power, as the extraction pressure is increased. It is reported that high pressures lead to high recoveries due to stronger interactions between the fluid and sample matrix (Yi et al. 2009). However, the effect of temperature on extraction efficiency is larger than that of flow rate and pressure (Rodríguez De Luna et al. 2020).

The effect of particle size is also another highly studied parameter for its effect on SFE efficiency. Theoretically, small particle sizes increase the extraction efficiency through increasing the surface area (da Silva et al. 2016; Jahongir et al. 2019). However, extended reduction of particle size may cause agglomeration of the sample in the compact channels of the SFE equipment and produce beds that decrease the mass transfer due to poor solvent flow (Rodríguez De Luna et al. 2020). In this case, the extraction rate decreases, which requires the use of dispersive agents to improve extration (da Silva et al. 2016; Jahongir et al. 2019).

The effect of modifier type and ratio on the solubility of target compounds depend on their polarity. Therefore, different types and concentrations of modifiers should be studied and optimized for the compounds to be extracted (Adil et al. 2008). Ethanol is the mostly used co-solvent, since it is permitted in food industry and easy to eliminate from the extract by evaporation at room temperature (Barba et al. 2016). Methanol and propanol are also occasionally used. However, the use of such modifiers is not recommended, if organic solvent-free extracts are required. In this case, it is preferable to manipulate the pressure and temperature conditions of the process to modify the density of CO_2 (Rodríguez De Luna et al. 2020).

TABLE 14.1 EXAMPLES FOR THE SFE CONDITIONS OF BIOACTIVE COMPOUNDS FROM BIORESOURCES

Target Compound	Plant Material	Extraction Conditions	Bioactivity	Reference
Curcumin	Turmeric rhizomes (*Curcuma longa* L.)	Temperature: 50°C Pressure: 30 MPa Time: 270 min Flow rate: 5 ml/min Co-solvent: 10% ethanol	Antioxidant, anticancer, antiviral, antiinfectious, and antiamyloidogenic activities	Wakte et al. (2011)
Polyacetylenes	*Bidens pilosa*	Temperature: 40°C Pressure: 250 bar Time: 240 min Flow rate: 15g/min	Antitumor, antimicrobial activities	Kviecinski et al. (2011)
Thymol	Ajwain (*Carum copticum* L.) and thyme (*Thymus vulgaris* L.)	Temperature: 40°C Pressure: 104 bar and 167 bar Time: 90 min Flow rate: 8 ml/min	Antimicrobial, antitussive, antispasmodic, and antioxidant activities	Morsy, (2020)
Flavonoids	*Odontonema strictum* leaves	Temperature:65°C Pressure:200 bar Time:270 min Flow rate: 15 g/min Co-solvent: 95% ethanol	Antihypertensive activities	Ouédraogo et al. (2018)
Phenolic compounds	Carob *(Ceratonia siliqua L.)*	Temperature: 40°C Pressure: 220 bar Time: 450 min Flow rate: 0.29 kg/h Co-solvent: 10% ethanol/water (8:2)	Antitumor, antimicrobial activities	Roseiro et al. (2013)
Oregano essential oils	*Oregano vulgare* leaves	Temperature: 40°C Pressure: 150 bar Time: 1 hour Co-solvent: 7% ethanol	Anti-inflammatory, antimicrobial activities	Santoyo et al. (2006)
Carnosic acid and carnosol rich extract	Rosemary	Two-stage extraction: Temperature: 40°C Pressure: S1:150-300 bar, S2:300 bar Flow rate: 60 g/min Time: S1:60 min, S2:120 min Co-solvent: S2: 7% (w/w) ethanol	Anti-proliferative, anti-inflammatory, neuroprotective, antioxidant activities	del Pilar Sanchez-Camargo et al. (2014)

(continued)

TABLE 14.1 (CONTINUED)

Target Compound	Plant Material	Extraction Conditions	Bioactivity	Reference
Cannabinoids	*Cannabis Sativa* L.	Temperature: 55°C Pressure: 34 MPa Flow rate: 200 g/min Co-solvent: 0-10% (w/w) ethanol	Therapeutic potential of controlling chronic pain, alleviating nausea and vomiting associated with chemotherapy	Rovetto and Aieta, (2017)
Phenolic compounds	*Thymus munbyanus*	Temperature: 70°C Pressure: 45 MPa Time: 30 + 180 Flow rate: 2 ml/min Co-solvent:	Antioxidant, antimicrobial activities	Bendif et al. (2018)
Piperine	Black pepper (*Piper nigrum*)	Temperature:40°C Pressure: 20-30 MPa Time: 240 min Flow rate:8±2 g/min Co-solvent: Ethanol or methanol	Pharmacological impact on nervous and neuromuscular systems, sedative effect, helps digestion	Andrade et al. (2017)

SFE = Supercritical fluid extraction

14.2.3 Advantages and Disadvantages

SFE has emerged as a superior alternative technique for the extraction of bioactive compounds due to several advantages of the process. SFE provides a selective extraction and results in an extract enriched in the desirable bioactive compounds. In addition, because the process takes place at low temperatures and in the absence of light and air, loss of compounds due to degradation is minimized (Heng et al. 2013; Barba et al. 2016). The final extract obtained by SFE is free of organic solvents, because CO_2 is a gas at room temperature and evaporates when the extraction is completed, and the system decompressed (Heng et al. 2013; da Silva et al. 2016). When carbon dioxide consumption is high at industrial production, the operation can be designed to recycle it (da Silva et al. 2016). Also, the CO_2 used in some SFE processes is largely a by-product of industrial processes, such as beer production process, which reduces emissions to the environment (Rodríguez De Luna et al. 2020). It is also noteworthy that the extracts obtained by SFE were proven to have good functional and biological activities and antibacterial and antifungal properties (Geeta et al. 2020).

The main disadvantage of SFE is the high initial cost of equipment and expertise required to run SFE extraction procedures (Rodríguez De Luna et al. 2020). Another disadvantage is that it is less effective for the extraction of polar compounds such as polyphenols, since CO_2 is non-polar molecule. In this case, it is used in combination with a suitable ratio of organic co-solvents as ethanol or methanol, also called modifiers (Barba et al. 2016; da Silva et al. 2016; Ahangari et al. 2021).

14.3 ULTRASOUND-ASSISTED EXTRACTION (UAE)

14.3.1 General Description

Ultrasound has been defined as sound waves with frequencies higher than 20 kHz, which is the threshold for human auditory detection. The output source of ultrasound is generally a vibrating body, which also provides vibration of the surrounding medium. Ultrasound wave transfers energy to other neighboring particles (Wen et al. 2018).

These waves create cavitation bubbles, vibration, crushing, mixing, and other physical effects in the media at certain frequencies and amplitudes. These bubbles implode when they reach a non-stable point, which leads to the release of high temperature and high pressure. This phenomenon can break cell walls and accelerate the release of metabolites, which enhances the extraction mechanism (Wen et al. 2018; Lefebvre et al. 2021).

The main physical parameters that modulate ultrasonic waves include power, frequency, and amplitude. Ultrasound power (W) is the energy level at which ultrasound propagates through the medium, which can also be expressed as ultrasound intensity (W/cm^2) or acoustic energy density (W/cm^3 or W/mL). Acoustic energy density has been extensively used in the evaluation of ultrasound power levels (Lefebvre et al, 2021; Wen et al. 2018). Optimization of ultrasound variables according to a specific plant matrix is important for achieving high extraction yield (Ma et al. 2008; Rashid, 2023).

14.3.2 Main Parameters

The extraction temperature is usually critical for UAE of bioactive compounds, as for the other extraction techniques. The other parameters that effect the efficiency of UAE are ultrasound power, frequency, and amplitude. Other factors such as solvent type, solid: solvent ratio, and particle size also have similar effects on UAE as for all the other extraction techniques. The extraction conditions for the UAE of some health-promoting bioactive compounds are given in **Table 14.2.**

Temperature has a positive effect on the extraction of bioactive due to the increased solubility and diffusivity of the compounds into the solution, which is a result of acceleration of molecular movements at higher temperatures (Maran et al. 2013). Decrease in solvent viscosity and density also leads to increased mass transfer. In addition, decreased viscosity causes easier formation of cavitation bubbles within the fluid, because the cohesive force and hence the tensile strength of the liquid is reduced (Hemwimol et al. 2006). In most UAE studies, temperature had a significant positive effect on extraction efficiency. Maran et al. (2013) reported that yield of polysaccharide extraction from pumpkins increased linearly as the temperature was increased from 50 to 70°C and chose 70°C as the optimum extraction temperature. However, high increase in temperature may also have a negative effect on the extraction of heat-sensitive compounds. In their study for determining the effect of different extraction parameters on UAE of polyphenolic compounds from *Aronia melanocarpa*, Ramić et al. (2015) reported that extraction time and temperature exhibited a significant negative influence on monomeric anthocyanin content and suggested that prolonged ultrasonic treatment at higher temperature and power levels may induce chemical decomposition of anthocyanins.

Ultrasound power also has an important effect on extraction yield in terms of enhanced cavitation effects as number of cavities and more violent collapse of cavities at higher power dissipation levels (Purohit & Gogate, 2015). Maran et al. (2013) studied the effect of ultrasound power (50–70 W) on the extraction yield of polysaccharides from *Cucurbita moschata* and reported that

TABLE 14.2 EXAMPLES FOR THE UAE CONDITIONS OF BIOACTIVE COMPOUNDS FROM BIORESOURCES

Target Compound	Plant Material	Extraction Conditions	Bioactivity	Reference
Beta-glucan	Barley	Solvent: Water Solid/Liquid ratio: 1/10 (w/v) Temperature: 55°C Time: 2-17 min Amplitute: 60-100 %	Reduce plasma cholesterol and postprandial serum glucose levels	Benito-Román et al. (2013)
Pectin	Grapefruit	Solvent: Water (pH:1.5) Solid/Liquid ratio: 1/50 (w/v) Temperature: 50-70 °C Time: 4-30 min	Reduce heart disease and gallstones	Bagherian et al. (2011)
Rosmarinic, carnosic and ursolic acid	Rosemary	Temperature: 40 °C Time: 30 min Solid/Liquid ratio: 1/20 (w/v)	Antioxidant, antibacterial, antimutagenic, antitumor activities	Jacotet-Navarro et al. (2015)
Hesperidin	Penggan (*Citrus reticulata*) peel	Solvent: Methanol Frequency: 60kHz Temperature: 40°C Time: 60 min	Antioxidant activity	Ma et al. (2008)
Anthraquinones	*Morinda citrifolia* roots	Solvent:Acetone, acetonitrile, methanol, ethanol Solid/Liquid ratio: 1/100 (w/v) Temperature: 50-70 °C Time: 15-90 min Power: 3-9 W	Anti-bacterial, antiviral, and anticancer activity, analgesic effects	Hemwimol et al. (2006)
Polyphenolic compounds	*Aronia melanocarpa*	Solvent: 50 % ethanol Frequency: 40 kHz Temperature: 30-70°C Time: 30-90 min Power: 72-216 W	Protective on colon cancer and cardiovascular disease, hepatoprotective effect	Ramić et al. (2015)
Phenolic compounds	Carob *(Ceratonia siliqua L.)*	Solvent: Water and 70 % aqueous acetone Solid/Liquid ratio: 1/5 (w/v) Frequency: 35 kHz Power: 350 W Temperature: 30°C Time: 30 min	Antitumor, antimicrobial activities	Roseiro et al. (2013)

TABLE 14.2 (CONTINUED)

Target Compound	Plant Material	Extraction Conditions	Bioactivity	Reference
Curcumin	Turmeric rhizomes (*Curcuma longa* L.)	Solvent: Acetone Power: 150 W Solid/Liquid ratio: 1:3 w/v Temperature: 21°C Time: 5 min	Antioxidant, anticancer, antiviral, antiinfectious and antiamyloidogenic properties	Wakte et al. (2011)
Flavonoids	*Radix Astragali* (root of *Astragalus;* Huangqi)	Solvent: Methanol Solid/Liquid ratio: 1/20 w/v Temperature: 60°C Time: 30 min x 2 cycles	Antioxidant activity, pharmacological properties such as effect on impairment of barrier function induced by hypoxia	Xiao et al. (2008)
Biflavonoids	*Selaginella helvetica*	Solvent: 95% ethanol Solid/Liquid ratio: 1/12.7 w/v Temperature: 47°C Time: 40 min Power: 250 W	Anticancer, antiviral, anti-inflammatory, anti-Alzheimer's disease, antimyocardial ischemia, and antimicrobial effects	Jiang et al. (2018)

UAE = Ultrasound-assisted extraction

the yield increased linearly with increasing power. However, the extent of increase in extraction yield may not always be at the same proportion as the increase in power. In their study on UAE of beta-carotene from waste carrot, Purohit and Gogate (2015) showed that increasing the power from 20 W to 60 W increased the extraction yield by 50%, whereas further increase in power from 60 W to 100 W rendered only 14.8% increase in extraction yield. On the other hand, in a study of UAE of anthraquinones from roots of *Morinda citrifolia,* Hemwimol et al. (2006) evaluated different ultrasonic powers (15.7, 39.7, 56.14 W) and reported that power was not the key factor affecting extraction efficiency, as they predicted. This was because only a small fraction of the electric energy entered the extraction solvent in the ultrasonic bath system, and the remained energy was absorbed by the water in the bath and the sides of the metal bath.

Multiple effects of high ultrasound frequencies have been reported such as peeling, erosion, particle breakdown, and also interparticle collisions, possible fragmentation, sonoporation, capillarity, and detexturization when combined with the effect of other parameters (Rodríguez De Luna et al. 2020). It has been reported that ultrasound cavitation produces lower extraction yield as frequency increases because the cavitation bubbles tend to be smaller and less energetic, which leads to a reduction in yield (Kulkarni et al. 2014). The effective frequency of ultrasound has been reported to generally range from 20 kHz to 50 kHz (Wen et al. 2018). Kulkarni et al. (2014) evaluated the effect of frequency in their work for mapping of an ultrasonic bath for the maximum extraction of mangiferin from *Mangifera indica* leaves. The extraction was carried out at 25 and 40 kHz at 204 W, and they reported that the extraction yield was higher at 25 kHz. There are also studies that show effective frequencies out of the range of 20–50 kHz. In a study on the UAE of hesperidin from Penggan (*C. reticulata*) peel, Ma et al. (2008) showed that the yield was higher at 60 kHz than at 20 kHz and 100 kHz. Chen et al. (2009) studied the effect of ultrasound frequency at much lower levels of 2–8 kHz in the UAE of betulin from white birch bark and showed that the effect of frequency was less significant than those of the other process parameters such as ethanol concentration and solid:solvent ratio.

Ultrasound amplitute generally has a positive effect on the extraction of bioactive compounds. As the amplitude increases, the collapse of the cavitation bubbles becomes more violent, which leads to a more efficient release of bioactive compounds (Rodríguez De Luna et al. 2020). In a UAE optimization study of antioxidant compounds from marjoram (*Origanum majorana* L.), Hossain et al. (2012) determined that amplitude showed the highest effect on extraction efficiency, followed by temperature and time. Oroian et al. (2020) also reported that amplitute induces a direct effect in cavitation, and hence has a significant effect on the UAE of polyphenols from crude pollen.

14.3.3 Advantages and Disadvantages

The multiple advantages of UAE are due to the effect of ultrasounds on creating cavitation, vibration, crushing, and mixing, which can break cell walls and enhance the extraction rate (Wen et al. 2018; Lefebvre et al. 2021). Accordingly, UAE has been proven to significantly decrease the extraction time and increase extraction yields with high selectivity and less solvent consumption, which also makes the process energy saving and easy to operate (Panja, 2018; Rodríguez De Luna et al. 2020). These advantages of UAE allow its application to food processing, pharmaceutical, nutraceutical, and cosmetic fields as a new green technology (Ma et al. 2008; Lefebvre et al. 2021).

One of the disadvantages of UAE is that only small amounts of mass can be treated due to the probe system. However, this can be overcome by operating in series. In addition, the disadvantage of the bath system is its low reproducibility, because the water contained in the tank attenuates the energy that is not adequately transmitted to the solvent (Rodríguez De Luna et al. 2020).

14.4 MICROWAVE-ASSISTED EXTRACTION (MAE)

14.4.1 General Description

Microwaves are electromagnetic waves with frequency between 300 MHz to 300 GHz (Bagade & Patil, 2019). MAE involves the penetration of microwave irradiation into a material, interaction with polar compounds, and induction of voluminous heating via ionic conduction or dipole rotation, which leads to high yield of extraction (Ekezie et al. 2017). The extent of heating by microwave irradiation depends on the dielectric constant of the molecules. Polar molecules such as water and ethanol have high dielectric constants of 78.5 and 24.3 at 25°C, respectively. Therefore, these molecules are able to absorb microwave energy and re-emit it, resulting in heating of the system. Solvents with a low dielectric constant such as hexane (1.87 at 250°C) are insensitive to microwaves (Lefebvre et al. 2021). In this respect, the common idea is that polar solvents are better for MAE. However, another theory states that some solvents are microwave-transparent, while others are microwave-absorbed and if microwave-transparent solvent is used, the plant materials will absorb microwave energy without much hindrance (Hao et al. 2002).

A basic MAE system consists of a magnetron, an isolator, a wave guide, a cavity, and a mode stirrer. The magnetron generates microwave energy that propagates through the wave guide into the cavity, and the mode stirrer disperses the energy via several routes. The system also includes a turntable that ensures uniform energy distribution when a sample is loaded into the system, regardless of its position (Ekezie et al. 2017).

14.4.2 Main Parameters

Extraction parameters for the MAE of some bioactive compounds from bioresources are given in **Table 14.3**. As in most of the other extraction techniques, the main parameter for MAE is temperature. Although it is not possible to adjust the temperature directly in this technique,

TABLE 14.3 EXAMPLES FOR THE MAE CONDITIONS OF BIOACTIVE COMPOUNDS FROM BIORESOURCES

Target Compound	Plant Material	Extraction Conditions	Bioactivity	Reference
Pectin	Grapefruit	Power: 0.45-0.90 kW Solvent: Water (pH:1.5) Solid/liquid ratio: 1/50 (w/v) Time: 2-14 min	Reduce heart disease and gallstones	Bagherian et al. (2011)
Rosmarinic, carnosic and ursolic acid	Rosemary	Power: 210 W (1 W/g) Temperature = 78 °C Time = 30 min	Antioxidant, antibacterial, antimutagenic, antitumor activities	Jacotet-Navarro et al. (2015)
Liquiritin and glycyrrhizic acid	Liquorice root	Solvent: Ethanol 40-60%; Time: 5 min Temperature: 85-90 °C	Anti-viral, anti-inflammatory and antioxidative activities	Chen et al. (2007b)
Artemisinin	*Artemisia annua* L.	Power: 650 W Solvent: No. 6 extraction solvent oil Time: 12 min Solid/liquid ratio: 100/30 – 350/30 ml/g	Antimalaria, antiulcerogenic, antifibrotic and antitumoral activities	Hao et al. (2002)
Curcumin	Turmeric rhizomes (*Curcuma longa* L.)	Solvent: Acetone Power: 60 W Solid/Liquid ratio: 1/3 w/v Temperature: 50°C Time: 5 min	Antioxidant, anticancer, antiviral, antiinfectious and antiamyloidogenic properties	Wakte et al. (2011)

(continued)

TABLE 14.3 (CONTINUED)

Target Compound	Plant Material	Extraction Conditions	Bioactivity	Reference
Silybinin	*Silybum marianum* (L.)	Solvent: 80 % ethanol Power: 600 W Time:12 min (6 min x 2 cycles) Solid/Liquid ration: 1/25 w/v	Prevent lipid peroxide formation in liver cells	Dhobi et al. (2009)
Solanesol	Tobacco leaves	Solvent: Hexane:ethanol (1:3 v/v)+NaOH (0.05 mol/L) Power: 700 W Time: 40 min Solid/Liquid ration: 1/25 w/v	Potential anticancer drug, cardiac stimulant, lipid antioxidant, antibiotics, starting material for many high-value biochemicals	Zhou et al. (2006)
Flavonoids	*Radix Astragali* (root of *Astragalus;* Huangqi)	Solvent: 95% ethanol Power: 1000 W Solid/Liquid ratio: 1/20 w/v Temperature: 110°C Time: 25 min x 2 cycles	Strong antioxidant activity, pharmacological properties such as effect on impairment of barrier function induced by hypoxia	Xiao et al. (2008)
Isoflavones	Soy beans	Solvent: 50% ethanol Power: 500 W Solid/iquid ratio: 1/50 w/v Temperature: 50°C Time: 20 min	Estrogenic/ antiestrogenic activity, antiproliferation, induction of cell-cycle arrest and apoptosis, prevention of oxidation	Rostagno et al. (2007)
Polyphenols	Black rice (*Oryza sativa* cv. Poireton) husk	Solvent: 67 % ethanol Frequency: 2450 MHz Solid/Liquid ratio: 1/41 w/v Temperature: 50°C Time: 31 min	Eliminates reactive oxygen species and lipid peroxide that can lower cholesterol content	Jha et al. (2017)

MAE = Microwave-assisted extraction

temperature can be set by changing the irradiation duration and power (Lefebvre et al. 2021). Therefore, extraction time and power are two key parameters for the microwave extraction. Xiao et al. (2008) studied the effect of extraction parameters such as microwave power, extraction cycles, ethanol concentration, extraction temperature, irradiation time, and solvent to material ratio for the extraction of flavonoids from *Radix astragali*. They determined that the yield of extraction increased with the increase of extraction time up to 25 min and then decreased slightly. This behavior is also reported in other studies where the extraction yield increases up to a certain extraction time, but then decreases or insignificantly increases with further increase in extraction time (Chen et al. 2007a; Eskilsson et al. 2000). This may be due to the degradation of the target compounds with overheating of the solute/solvent system as a result of prolonged application of microwave (Routray & Orsat, 2012).

The intensity of microwave power controls the amount of energy supplied to the sample, which is converted to heat energy in the dielectric material and increases the temperature. The power affects interactions and equilibrium rate, and controls partition of analytes between sample and solvent (Routray & Orsat, 2012). Xiao et al. (2008) reported that the extraction efficiency was improved by increasing power from 200 to 1000 W during short irradiation times of 5 and 10 min. The reason for this increase is considered to be due to the improved solvent penetration into materials by molecular interaction within the electromagnetic field and a rapid transfer of energy between solvent and plant matrix (Liu et al. 2011). However, when the extraction solutions were heated for up to 15 min, the effect of power was insignificant (Xiao et al. 2008). In some studies, excessive increase in power was also determined to decrease the extraction yield probably due to internal over-heating, which leads to degradation of the desired compounds (Lefebvre et al. 2021). In a study for the extraction of resveratrol from tree peony seed oil-extracted residues, Chen et al. (2016) determined that the extraction yield increased by increasing the power from 120 to 385 W, but decreased when the power was further increased to 700 W. Similar behaviour was observed by Kaur et al. (2019) for the extraction of swertiamarin, amarogentin, and mangiferin from different species of *Swertia*. The yield increased until a microwave power around 500 W and then decreased with the further increase in power.

Solvent type, solid/solvent ratio, extraction time, and particle size are the other general parameters that effect the efficiency of MAE as in the other extraction techniques (Liu et al. 2016; Rodríguez De Luna et al. 2020). However, particle size is indicated to have the least significant effect on MAE. A decrease in particle size does not imply higher extraction yield for most bioactive compounds, and microwave effect in yield may be roughly independent of particle size (Rodríguez De Luna et al. 2020).

14.4.3 Advantages and Disadvantages

Among other green extraction technologies, MAE has been widely used for the extraction of bioactive compounds in recent years. Advantages such as extraction efficiency, shorter extraction time, and less solvent consumption make this technique attractive for extraction of especially sensitive bioactive compounds. It can be used with certain modifications in parameters such as system pressure, atmospheric pressure, and flow of oxygen (Bagade & Patil, 2019; Hao et al. 2002). It is even possible to perform microwave extractions without any solvent. In addition, MAE allows the design of different schemes for varying the selectivity for specific substances. MAE may also be combined with other techniques such as ultrasound and enzyme-assisted extractions to improve the extraction and achieve higher yields (Rodríguez De Luna et al. 2020).

The only disadvantage of MAE may be the possible degradation of some thermosensitive compounds, as reported by Liazid et al. (2007), who studied the stability of 22 phenolic compounds under different temperature conditions of MAE. The influence on the stability affected by the working temperature between 50 and 175 °C was evaluated, and it was reported that as the temperature was increased, there was a significant degradation of the phenolic compounds, especially epicatechin, resveratrol, and myricetin.

14.5 PRESSURIZED LIQUID EXTRACTION (PLE)

14.5.1 General Description

PLE is a fully automated sample preparation technique that involves high pressures and temperatures, which enhance the extraction performance as compared to those techniques carried out at atmospheric pressure and lower temperatures (Mustafa & Turner, 2011). Application of high pressures allows the use of temperatures above the boiling point of solvents, which offers advantage in terms of enhanced solubility, higher mass transfer, and increased extraction rate (Mustafa & Turner, 2011; Wijngaard et al. 2012).

Higher temperatures generally provide increased capacity of solvents to solubilize solutes, increased diffusion rates, better disruption of solute–matrix bonds, decreased solvent viscosity, and decreased surface tension (Wijngaard et al. 2012). High pressure enhances the extraction in two ways: It ensures that the solvent maintains liquid state at the applied temperature, which is the main purpose to apply high pressure. It also helps drive the solvent into the pores of the matrix and enhance analyte solubility (Wijngaard et al. 2012; Joana Gil-Chávez et al. 2013).

PLE is also known as accelerated solvent extraction, enhanced solvent extraction, pressurized solvent extraction, or subcritical solvent extraction (www.sciencedirect.com/science/article/pii/S0963996911005618Mustafa and Turner, 2011). In the case when water is used as the extraction solvent, the technique is referred to as pressurized hot water extraction, sub-critical water extraction, or superheated water extraction (Mustafa & Turner 2011; Wijngaard et al. 2012). In ambient conditions, water is not a suitable solvent for dissolving lipids and organics due to its high polarity. Working at elevated pressures allows increasing the temperature above the boiling point, resulting in increase in dielectric constant and decrease in polarity (Chang & Chang, 2007; Joana Gil-Chávez et al. 2013).

14.5.2 Main Parameters

As already mentioned for the other extraction techniques, optimization of PLE process also involves examining the effects of different factors on the extraction process. The most important ones to mention are temperature, pressure, and extraction time (static or dynamic). Other extraction parameters studied widely are given in **Table 14.4** as examples for the extraction of health-promoting bioactive compounds from bioresources.

Extraction temperature is one of the most critical parameters that affect the extraction efficiency and selectivity of PLE (Zaibunnisa et al. 2009; Mustafa & Turner, 2011; Herrero et al. 2013), especially for bioactive and marker compounds in botanicals and medicinal plants (Ong et al. 2006). Theoretically, high temperatures provide the high extraction efficiency. However, high temperatures may also have some drawbacks for the extraction, such as increased amount of

TABLE 14.4 EXAMPLES FOR THE PLE CONDITIONS OF BIOACTIVE COMPOUNDS FROM BIORESOURCES

Target Compound	Plant Material	Extraction Conditions	Bioactivity	Reference
Hesperidin, nobiletin, tangeretin	*Citrus reticulata* 'Chachi' (Guangchenpi)	Solvent: 70% methanol Temperature: 160 °C Pressure: 1500 psi, Number of cycles: 1 Static extraction time: 20 min	Anticancer, antibacterial, antioxidant, antileukaemia activities	Li et al. (2012)
Anthraquinones	*Morinda citrifolia*	Solvent: Water Temperature: 100,170, 220 °C Pressure: 7 MPa Dynamic extraction time: 180 min	Antibacterial, antiviral, and anticancer activities, as well as analgesic effects	Shotipruk et al. (2004)
Rosmarinic and carnosic acid	Rosemary (*Rosmarinus officinalis* L.	Solvent: Water Temperature: 60-100°C Pressure: 1500 psi Static extraction time: 25 min	Antioxidant activity	Herrero et al. (2005)
Isoflavones	Defatted soybean flakes	Solvent: 80% ethanol Temperature: 383 K Pressure: 551 kPa Dynamic extraction flow rate: 25 mL/min	Reducing the risk of cardiovascular disease, decreasing growth rates of prostate, breast, colon cancers	Chang et al. (2007)
Policasanols	Wheat straw, germ, bran	Solvent: n-hexane, ethanol, petroleum ether, chloroform Temperature: 80-125°C Pressure: 1500 psi Static extraction time: 3 x 15 min	Decreasing low-density lipoprotein (LDL), increasing high-density lipoprotein (HDL) cholesterol level	Dunford et al. (2010)
Mangiferin	*Mahkota Dewa (Phaleria macrocarpa)*	Solvent: Water Temperature: 323-423 K Pressure: 0.7-4.0 MPa Static extraction time: 1-7 h	Antidiabetic, anti-HIV, anticancer, immunomodulatory, and antioxidant activity	Kim et al. (2010)
Damnacanthal	*Morinda citrifolia* roots	Solvent: Water Temperature: 120-200°C Pressure: 4 MPa Dynamic extraction flow rate: 1.6, 2.4, 3.2 ml/min	Anticancer activity	Anekpankul et al. (2007)
Gallic acid, ellagic acid, corilagin	*Terminalia chebula* Retz. fruits	Solvent: Water Temperature: 120-220°C Pressure: 4 MPa Dynamic extraction flow rate: 2-4 ml/min	Anticancer, antimicrobial, anti-inflammatory activities	Rangsriwong et al. (2009)
Quercetin	Onion skin	Solvent: Water Temperature: 100-190°C Pressure: 90-131 bar Static extraction time: 5-30 min	Anti-inflammatory, anticancer and antivirus activities, antihistamine effect, allergy medication, reduction of blood pressure	Ko et al. (2011)

(continued)

TABLE 14.4 (CONTINUED)

Target Compound	Plant Material	Extraction Conditions	Bioactivity	Reference
Polyphenols (Mainly catechin, epicatechin, and procyanidins)	Cocoa	Solvent: Water Temperature: 125°C Pressure: 1500 psi Static extraction time: 3 min	Antioxidant, antiradical, antihypertensive, and anticarcinogenic properties	Plaza et al. (2017)
Rutin	*Amaranthus paniculatus* leaves	Solvent: 70/30 water/ethanol Temperature: 188°C Pressure: 10 MPa Static extraction time: 20 min	Cytoprotective, antiplatelet, anti-inflammatory, antitumor, and antibacterial activities	Kraujalis et al. (2015)

PLE = Pressurized liquid extraction

co-extracted analytes due to decreased selectivity of the extraction (Mustafa & Turner, 2011) and negative effects on the bioactivity of some thermo-labile compounds (Ong et al. 2006; Herrero et al. 2013). Therefore, the effect of different extraction temperatures on extraction efficiency, selectivity, and bioactivity should be examined and optimized. Li et al. (2012) examined the effect of temperature on PLE of three major flavones in *Citrus reticulata* 'Chachi' (Guangchenpi) and determined that increasing the temperature from 80 to 160°C at isocratic pressure significantly increased the extraction yield. However, above 160°C, the extraction yields slowly decreased, which was predicted to be due to degradation of flavones above that temperature. Decreased selectivity of PLE or possible formation of new components as a result of different reactions during the extraction process may also be an advantage in terms of the bioactivity of the co-extracted analytes (Herrero et al. 2013). For example, Cha et al. (2010) reported that some antioxidant chlorophyll derivatives such as pyropheophytins a and b were formed in significant amounts during the PLE of microalga *Chlorella vulgaris* at high temperatures (135–160°C). This compound, which is not naturally present in the sample, might have potent antimutagenic activity and can inhibit tumor cell growth. The authors reported that these derivatives were formed through the severe thermal processing associated with PLE, which generates chemical structure deformations and, in this way, properties of the extracts might be enhanced www.sciencedirect.com/science/article/pii/S016599361200372X (Cha et al. 2010).

The effect of pressure on PLE have been reported to be less critical. It has been shown in different studies that extraction pressure is required to be high enough to keep the solvent in liquid state, and beyond that pressure level, the effect of pressure is insignificant (Shotipruk et al. 2004; Zaibunnisa et al. 2009; Herrero et al. 2013). Therefore, most PLE processes to obtain bioactive natural products are applied at an extraction pressure high enough to maintain the solvent in the liquid state (Herrero et al. 2013).

The effect of extraction time depends on whether PLE is applied in static or dynamic mode. Static mode is the mostly commonly applied one in which a certain volume of solvent is maintained in contact with the sample for a given time under certain conditions of pressure and temperature.

At static mode, equilibria between the sample components still bound to the matrix and the ones that are already solubilized in the solvent might be reached, and the efficiency of the extraction procedure will not increase beyond this point (Herrero et al. 2013). Therefore, static time in combination with temperature have significant effect on most PLE studies, up to the equilibrium point (Zaibunnisa et al. 2009; Herrero et al. 2013).

Further increase in extraction efficiency may be provided by applying multiple extraction cycles (Herrero et al. 2013). Luthria et al. (2006) studied the effect of the number of extraction cycles on the extraction efficiency of phenolic compounds from parsley flakes and reported that four cycles of extraction were sufficient for optimum recovery of phenolic compounds. A different result was reported by Lee & Kim (2010) for the PLE of lignans from the fruits of *Schisandra chinensis*, where no significant change was observed in extraction efficiency as the number of cycles was increased. They concluded that one cycle was optimum. For static extraction and one extraction cycle, extraction time has a positive significant effect up to a certain point, and the efficiency decreases after that. The negative effect of increasing extraction time was reported as the possible degradation of target compounds exposed to high temperatures for prolonged times (Lee & Kim, 2010; Kim et al. 2010; Li et al. 2012).

14.5.3 Advantages and Disadvantages

The use of PLE is considered advantageous due to its short extraction time, simplicity, ability to be automated, low cost, low solvent use, and hence lower negative environmental impact (Zaibunnisa et al. 2009). Another advantage of PLE is that the extraction temperature can be adjusted to a value that would achieve the polarity suitable for dissolving the compounds of interest (Gil-Chávez et al. 2013). When water is heated to 473–523 K at elevated pressures, its polarity measured by dielectric constant decreases from 80 to 27, which allows dissolving organic compounds (Chang & Chang, 2007). In this respect, the use of PLE allows obtaining food-grade extracts with the use of water or other Generally Recognized As Safe (GRAS) solvents such as ethanol (Gil-Chávez et al. 2013).

The only disadvantage of this technique could be the PLE system high cost. However, owing to the advantages offered, PLE systems are commercially available. In addition, these extraction systems can be programmed and automatically run, which is also convenient for quality control (Rodríguez De Luna et al. 2020).

14.6 PULSED ELECTRIC FIELD (PEF)-ASSISTED EXTRACTION

14.6.1 General Description

An emerging technique of extraction, PEF has gained increasing interest in recent decades due to its economic efficiency in food and pharmaceutical industries (Yan et al. 2017). The earliest studies of PEF focused on non-thermal sterilization and enzyme inactivation (Yin et al. 2008; Yan et al. 2017). Later, it was discovered that PEF can be used to extract active ingredients from natural products, providing advantages such as being non-thermal and fast, providing higher efficiency with lower power and creating low pollution (Yin et al. 2008). Moreover, PEF-assisted extraction might reduce degradation of heat-sensitive compounds and facilitate the extract purification (Barbosa-Pereira et al. 2018).

A PEF device consists of a high-voltage pulse generator, a treatment chamber with a fluid managing assembly, and a monitoring and control system. In addition, PEF equipment has a charger to convert direct current (DC) to alternating current (AC), and a device that stores energy in the generator (Ranjha et al. 2021). For PEF treatment, the ingredient is placed into a container as the electrolyte and short duration pulses of moderate electric voltage are applied (Yin et al. 2008; Kumari et al. 2018), which generate instantaneous pressure between the two electrodes that acts on the ingredient (Yin et al. 2008). The biological membrane is permeabilized by electroporation by the effect of electric pulses with moderate electric strength, which enhances the mass transfer through the membrane. In this way, application of PEF accelerates release of intracellular compounds, increases mass transfer, and improves the extraction yields of bioactive ingredients significantly (Yan et al. 2017; Barbosa-Pereira et al. 2018; Rajha et al. 2019; Kumari et al. 2018). The most generally accepted mechanism of membrane electropermeabilization consists of several steps (Savova et al. 2005): (1) charge separation across the non-conductive membrane results with polarization of the membrane; (2) the induced intracellular stress results in the temporal destabilization and reduction in thickness of the membrane; (3) reversible breakdown of the membrane near the critical potential results in formation of small conductive channels, leading to an immediate discharge at the membrane, re-sealing of pores, followed by a repeated charging process; and (4) increasing the intensity of the electric field, pulse duration, and number of pulses results in irreversible pore formation (Savova et al. 2005; Yin et al. 2008).

14.6.2 Main Parameters

Parameters that mainly affect the efficiency of PEF-assisted extraction are electric voltage, pulse duration, pulse number, and temperature. However, the effect of temperature on PEF-assisted extraction efficiency is not as significant as that on the efficiency of other extraction techniques. Therefore, most extraction studies with PEF treatment are carried out at room temperature (**Table 14.5**).

Typically, the electric voltage for PEF-assisted extraction is moderate up to 10 or 20 kV/cm (Yan et al. 2017; Kumari et al. 2018). Higher electric voltages are usually applied for enzyme or microbial inactivation purposes (Kumari et al. 2018), but not preferred for extraction of most bioactive compounds because higher intensity may lead to the generation of strong hydrogen radicals that might attack and decompose the bioactives (Yin et al. 2008; Yongguang et al. 2006; Yan et al. 2017). However, applications of high intensity PEF (typically between 20 and 80 kV/cm) also take

TABLE 14.5 EXAMPLES FOR THE PEF CONDITIONS OF BIOACTIVE COMPOUNDS FROM BIORESOURCES

Target Compound	Plant Material	Extraction Conditions	Bioactivity	Reference
Anthocyanins	Grape by-products	EF intensity: 3 kV/cm Number of pulses: 30 Frequency: 2 Hz PEF duration: 15 s Extractiontime:60min Solvent: 50% ethanol Solid/Liquid: 1/ 4.5 w/v Temperature: 70 °C	Antibacterial, antiviral, antioxidant, antiinflammatory, anticancerogenic properties, and can prevent cardiovascular diseases	Corrales et al. (2008)

TABLE 14.5 (CONTINUED)

Target Compound	Plant Material	Extraction Conditions	Bioactivity	Reference
Betulin	*Inonotus Obliquus*	EF intensity: 40 kV/cm Number of pulses: 2 Frequency: 600 Hz PEF duration: 0.4 μs Solvent: %75 ethanol Solid/Liquid: 1/25 w/v Temperature: RT	Protective effects against Cadmium (Cd)-induced cytotoxicity. Converted to betulinic acid with antimalarial, anti-inflammatory, anti-HIV, and cytotoxicity activities	Yin et al. (2008)
Chondroitin sulphate	Fish bones	EF intensity: 16.88 kV/cm Number of pulses: 9 Frequency: 40-300 Hz PEF duration: 18 μs Extraction time: 5 min Solvent: 3.24 % NaOH Solid/Liquid: 1/15 w/v Temperature: RT	Anticoagulant, antitumor, preventing artery hardening, and relieving arthritis	He et al. (2014)
Polysaccharides	Corn silk	EF intensity: 30kV/cm Number of pulses: 3 Frequency: 1-3000 Hz PEF duration: 6 μs Solvent: Water Solid/Liquid: 1/50 w/v Temperature: RT	Improve gastrointestinal movement, help weight loss	Zhao et al. (2011)
Polysaccharides	*Rana temporaria chensinensis* David	EF intensity: 20 kV/cm Number of pulses: 3 Frequency: 1200 Hz PEF duration: 6 μs Solvent: 0.5 % KOH Temperature: RT	Anticoagulant, anticancer, antioxidant, and prevents diabetes	Yongguang et al. (2006)
Polyphenols	Pomegranate peel	EF intensity: 10 kV/cm Extraction time: 7 min Solvent: Water Solid/Liquid: 1/10 w/v Temperature: 50°C	Antibacterial, anti-inflammatory, hepatoprotectiv,e and antigenotoxic activities	Rajha et al. (2019)

(continued)

TABLE 14.5 (CONTINUED)

Target Compound	Plant Material	Extraction Conditions	Bioactivity	Reference
Steroidal alkaloids	Potato peel	EF intensity: 0.75 kV/cm Pulse number: 200 PEF duration: 3 μs Extraction time: 60 min Frequency: 10 Hz Solvent: Methanol Solid/Liquid: 1/5 w/v Temperature: RT	Anti-inflammatory, antimicrobial, and anticarcinogenic activities	Hossain et al. (2015)
Phytosterol, germ oil	Maize germ and hull	EF intensity: 0.6 kV/cm Pulse number: 120 Extraction time: 60 min Solvent: Hexane Solid/Liquid: 1/40 w/v Temperature: RT	Reduce the serum level of low-density lipoprotein (LDL) cholesterol associated with atherosclerosis	Guderjan et al. (2005)

PEF = Pulsed electric field

place especially for the extraction of some intact compounds such as bovine serum albumin. Also, there are advanced continuous PEF process systems designed to work at high intensities that have shown to efficiently extract several kinds of bioactive compounds (Yan et al. 2017). Yin et al. (2008) showed that the extraction yield of betulin increased from 2.5 to 6.5 g/kg when the PEF strength increased from 10 to 60 kV/cm. However, they chose 40 kV/cm as the maximum strength, because the higher the PEF strength, the more expensive the equipment will be. Additionally, the PEF system is likely to be damaged at high intensities. In their extraction study of polysaccharides from *Rana temporaria chensinensis* David, Yongguang et al. (2006) reported a significant increase in the extraction ratio of polysaccharides with the increase in electric field intensity up to 20 kV/cm. When the electric field intensity was increased from 20 to 40 kV/cm, the extraction ratio of polysaccharides decreased from 26.87 to 21.41%, which indicates that higher PEF intensities caused decomposition of the polysaccharide. He et al. (2014) showed that the amount of chondroitin sulphate extracted from fish bone increased when the electric field strength was incresed up to 15 kV/cm and leveled off at that value. Further increase to 25 kV/cm maximized the chondroitin sulphate content, but the researchers indicated that higher electric field strengths may cause irreparable damage to the PEF system.

The pulse duration used in the application of PEF is usually as short as milliseconds or microseconds (Yan et al. 2017). Long pulse durations may lead to permeation of the compounds of interest due to enlargement of pores in the cell membranes and also decomposition of the bioactive compounds of interest (Yongguang et al. 2006; Zhao et al. 2011). For the extraction of polysaccharides from *Rana temporaria chensinensis* David, Yongguang et al. (2006) showed that the largest extraction rate of polysaccharides was obtained at a pulse duration of 6 μs. Zhao et al. (2011) showed similar results for the extraction of polysaccharides from corn silk, as they also obtained the largest yield of polysaccharides at 6 μs. When the electric pulse duration was increased from 6 to 10 μs, the yield of polysaccharides decreased significantly.

The number of pulses in combination with the duration of each pulse was also reported to be one of the parameters that significantly affects the PEF-assisted extraction (Yin et al. 2008; He et al. 2014). The use of two pulses, each with a pulse duration of 2 μs, yielded maximum extraction of betulin from *Inonotus Obliquus* (Yin et al. 2008). Whereas nine pulses were determined to be the optimum for the extraction of chondroitin sulfate from fish bone (He et al. 2014). When the pulse number was increased further than the optimum number, the extraction yield did not change significantly (Yin et al. 2008; He et al. 2014), because the equilibrium of potential between the inside and outside of the cell was achieved (Yin et al. 2008).

It has been shown that PEF pre-processing accelerated the extraction even at low room temperature (Savova et al. 2005). For the extraction of water-soluble polysaccharide fractions from linseeds, Savova et al. (2005) showed that 15 min of PEF-assisted extraction gave similar yields with 1 hour extraction without PEF treatment. Yin et al. (2008) also reported that the yield of betulin extraction-assisted with PEF for 4 μs was higher than those with maceration extraction for 24 h under the same experimental conditions, which indicates a significant reduction in extracting time. PEF has been shown to be a promising method for the extraction of polysaccharides, as it has been reported that the extraction ratio and polysaccharide content of PEF-assisted extraction are higher than those of conventional extraction methods, such as alkaline extraction, enzyme extraction, and compound extraction (Yongguang et al. 2006). For the extraction of chondroitin sulfate from fish bone, the use of PEF for 5 min and pulse number = 9, 18 μs produced higher content of chondroitin sulfate than the other methods studied. The content of chondroitin sulfate by PEF extraction was 2.02 times higher than that obtained by enzyme extraction for 4 h, 1.84 times higher than alkali extraction for 2 h, and 1.42 times higher than ultrasonic extraction for 23 min (He et al. 2014).

14.6.3 Advantages and Disadvantages

One of the main advantages of PEF-assisted extraction is its low energy demand, which results in lower processing cost (Bocker & Silva, 2022). PEF also provides non-thermal processing in shorter time with less solvent consumption (Panja, 2018; Ranjha et al. 2021). Owing to these advantages, PEF-assisted extraction is a promising method for industrial applications due to its energy-efficient nature (Ranjha et al. 2021), which compensates for the high initial investment costs (Bocker & Silva, 2022).

Besides the indicated advantages, it should be considered that PEF intensities may influence secondary metabolism activities due to degradation of phenolic compounds. More studies are needed in order to understand the mechanism of interaction of PEF with plant matrices and to evaluate the possible degradation of phenolic compounds as a consequence of high temperature and chemical medium (Rodríguez De Luna et al. 2020).

14.7 CONCLUSION

Extraction of bioactive compounds involves complex mechanisms and can be accomplished by various techniques. It has been proven that replacement of conventional techniques by green technologies is promising for improving extraction yields, decreasing process time, and reducing environmental damage caused by toxic solvents. In addition, degradation of bioactive compounds is almost eliminated or minimized by the application of green technologies, and high-quality extracts are obtained, which allows them to be used as ingredients for developing health-promoting food products.

Studies have also suggested that combination of several extraction techniques may improve the extraction efficiency of biocompounds. Although important advances have been developed in the application of green technologies for the obtention of health ingredients, a complete understanding of the corresponding process for each novel green technology is required in order to determine the thermodynamic and kinetic mechanisms that govern each of the techniques.

REFERENCES

Adil, I.H., Yener, M.E., & Bayındırlı, A. (2008). Extraction of total phenolics of sour cherry pomace by high pressure solvent and subcritical fluid and determination of the antioxidant activities of the extracts. *Separation Science and Technology, 43*(5), 1091–1110. DOI: 10.1080/01496390801888243

Ahangari, H., King, J.W., Ehsani, A., & Yousefi, M. (2021). Supercritical fluid extraction of seed oils–A short review of current trends. *Trends in Food Science & Technology, 111*, 249–260. DOI: 10.1016/j.tifs.2021.02.066

Ahmad, T., Masoodi, F.A., Rather, S.A., Wani, S.M., & Gull, A. (2019). Supercritical fluid extraction: A Review. *Journal of Biological and Chemical. Chronicles, 5*, 114–122. ISSN (Print):2454–7468. ISSN (Online): 2454–7476

Andrade, K.S., Trivellin, G., & Ferreira, S.R. (2017). Piperine-rich extracts obtained by high pressure methods. *The Journal of Supercritical Fluids, 128*, 370–377. DOI: 10.1016/j.supflu.2017.05.001

Anekpankul, T., Goto, M., Sasaki, M., Pavasant, P., & Shotipruk, A. (2007). Extraction of anti-cancer damnacanthal from roots of morinda citrifolia by subcritical water. *Separation and Purification Technology, 55*(3), 343–349. DOI: 10.1016/j.seppur.2007.01.004

Babova, O., Occhipinti, A., Capuzzo, A., & Maffei, M.E. (2016). Extraction of bilberry (*Vaccinium myrtillus* antioxidants using supercritical/subcritical CO2 and ethanol as co-solvent. *The Journal of Supercritical Fluids, 107*, 358–363. DOI: 10.1016/j.supflu.2015.09.029

Bagade, S.B., & Patil, M. (2019). Recent advances in microwave assisted extraction of bioactive compounds from complex herbal samples: A review. *Critical Reviews in Analytical Chemistry, 51*(2), 138–149. DOI: 10.1080/10408347.2019.1686966

Bagherian, H., Ashtiani, F.Z., Fouladitajar, A., & Mohtashamy, M. (2011). Comparisons between conventional, microwave-and ultrasound-assisted methods for extraction of pectin from grapefruit. *Chemical Engineering and Processing: Process Intensification, 50(*11-12), 1237–1243. DOI: 10.1016/j.cep.2011.08.002

Barba, F.J., Zhu, Z., Koubaa, M., Sant'Ana, A.S., & Orlien, V. (2016). Green alternative methods for the extraction of antioxidant bioactive compounds from winery wastes and by-products: A review. *Trends in Food Science & Technology, 49*, 96–109. DOI: 10.1016/j.tifs.2016.01.006

Barbosa-Pereira, L., Guglielmetti, A., & Zeppa, G. (2018). Pulsed electric field assisted extraction of bioactive compounds from cocoa bean shell and coffee silverskin. *Food and Bioprocess Technology, 11*(4), 818–835. DOI: 10.1007/s11947-017-2045-6

Bendif, H., Adouni, K., Miara, M.D., Baranauskienė, R., Kraujalis, P., Venskutonis, P.R., & Maggi, F. (2018). Essential oils (EOs, pressurized liquid extracts (PLE) and carbon dioxide supercritical fluid extracts (SFE-CO2) from Algerian Thymus munbyanus as valuable sources of antioxidants to be used on an industrial level. *Food Chemistry, 260*, 289–298. DOI: 10.1016/j.foodchem.2018.03.108

Benito-Román, Ó., Alonso, E., & Cocero, M.J. (2013). Ultrasound-assisted extraction of β-glucans from barley. *LWT-Food Science and Technology, 50*(1), 57–63. DOI: 10.1016/j.lwt.2012.07.006

Bocker, R., & Silva, E.K. (2022). Pulsed electric field assisted extraction of natural food pigments and colorings from plant matrices. *Food Chemistry: X, 15*, 100398. DOI: 10.1016/j.fochx.2022.100398

Carpentieri, S., Soltanipour, F., Ferrari, G., Pataro, G., & Donsì, F. (2021). Emerging green techniques for the extraction of antioxidants from agri-food by-products as promising ingredients for the food industry. *Antioxidants, 10*(9), 1417. DOI: 10.3390/antiox10091417

Cha, K.H., Kang, S.W., Kim, C.Y., Um, B.H., Na, Y.R., & Pan, C.H. (2010). Effect of pressurized liquids on extraction of antioxidants from *Chlorella vulgaris. Journal of Agricultural and Food Chemistry, 58*(8), 4756–4761. DOI: 10.1021/jf100062m

Chang, L.H., & Chang, C.M.J. (2007). Continuous hot pressurized fluids extraction of isoflavones and soyasaponins from defatted soybean flakes. *Journal of the Chinese Institute of Chemical Engineers, 38*(3-4), 313–319. DOI: 10.1016/j.jcice.2007.03.003

Chemat, F., Vian, M.A., & Cravotto, G. (2012). Green extraction of natural products: Concept and principles. *International Journal of Molecular Sciences, 13*(7), 8615–8627. DOI: 10.3390%2Fijms13078615

Chen, F., Zhang, X., Zhang, Q., Du, X., Yang, L., Zu, Y., & Yang, F. (2016). Simultaneous synergistic microwave–ultrasonic extraction and hydrolysis for preparation of trans-resveratrol in tree peony seed oil-extracted residues using imidazolium-based ionic liquid. *Industrial Crops and Products, 94*, 266–280. DOI: 10.1016/j.indcrop.2016.08.048

Chen, Q.H., Fu, M.L., Liu, J., Zhang, H.F., He, G.Q., & Ruan, H. (2009). Optimization of ultrasonic-assisted extraction (UAE) of betulin from white birch bark using response surface methodology. *Ultrasonics Sonochemistry, 16*(5), 599–604. DOI: 10.1016/j.ultsonch.2008.11.009

Chen, Y., Xie, M.Y., & Gong, X.F.(2007a). Microwave-assisted extraction used for the isolation of total triterpenoid saponins from Ganoderma atrum. *Journal of Food Engineering, 81*(1), 162–170. DOI: 10.1016/j.jfoodeng.2006.10.018

Chen, S.U.N., Yuchun, X.I.E., & Huizhou, L.I.U. (2007b). Microwave-assisted micellar extraction and determination of glycyrrhizic acid and liquiritin in licorice root by HPLC. *Chinese Journal of Chemical Engineering, 15*(4), 474–477. DOI: 10.1016/S1004-9541(07)60111-6

Corrales, M., Toepfl, S., Butz, P., Knorr, D., & Tauscher, B. (2008). Extraction of anthocyanins from grape by-products assisted by ultrasonics, high hydrostatic pressure or pulsed electric fields: A comparison. *Innovative Food Science & Emerging Technologies, 9*(1), 85–91. DOI: 10.1016/j.ifset.2007.06.002

Da Silva, R.P.F.F., Rocha-Santos, T.A., & Duarte, A.C. (2016). Supercritical fluid extraction of bioactive compounds. *TrAC Trends in Analytical Chemistry, 76*, 40–51. DOI: 10.1016/j.trac.2015.11.013

Dhobi, M., Mandal, V., & Hemalatha, S. (2009). Optimization of microwave assisted extraction of bioactive flavonolignan-silybinin. *Journal of Chemical Metrology, 3*(1), 13. ISSN:1307-6183.

Dunford, N.T., Irmak, S., & Jonnala, R. (2010). Pressurised solvent extraction of policosanol from wheat straw, germ and bran. *Food Chemistry, 119*(3), 1246–1249. DOI: 10.1016/j.foodchem.2009.07.039

Ekezie, F.G.C., Sun, D.W., & Cheng, J.H. (2017). Acceleration of microwave-assisted extraction processes of food components by integrating technologies and applying emerging solvents: A review of latest developments. *Trends in Food Science & Technology, 67*, 160–172. DOI: 10.1016/j.tifs.2017.06.006

Eskilsson, C.S., & Björklund, E. (2000). Analytical-scale microwave-assisted extraction. *Journal of Chromatography A, 902*(1), 227–250. DOI:10.1016/S0021-9673(00)00921-3

Faiza, N., Imran, A., Arshad, M.U., Arshad, M.S., & Shah, M.A. (2023). Valorization and characterization of corn by-product polyphenols through green extraction technologies. *Frontiers in Nutrition, 10*, 1107067. https://doi.org/10.3389/fnut.2023.1107067

Gallo-Molina, A.C., Castro-Vargas, H.I., Garzón-Méndez, W.F., Ramírez, J.A.M., Monroy, Z.J.R., King, J.W., & Parada-Alfonso, F. (2019). Extraction, isolation and purification of tetrahydrocannabinol from the Cannabis sativa L. plant using supercritical fluid extraction and solid phase extraction. *The Journal of Supercritical Fluids, 146*, 208–216. DOI: 10.1016/j.supflu.2019.01.020

Garcia-Mendoza, M.P, Espinosa-Pardo, F.A., Baseggio, A.M., Barbero, G.F., Junior, M.R.M., Rostagno, M.A., & Martínez, J. (2017). Extraction of phenolic compounds and anthocyanins from juçara (*Euterpe edulis* Mart.) residues using pressurized liquids and supercritical fluids. *The Journal of Supercritical Fluids, 119*, 9–16. DOI: 10.1016/j.supflu.2016.08.014

Geeta, H.P., Srinivas, G., & Champawat, P.S. (2020). Supercritical fluid extraction of bioactive compounds from bioresource: A review. *International Journal of Current Microbiology Applied Science, 9*(4), 559–566. DOI: https://doi.org/10.20546/ijcmas.2020.904.068

Guderjan, M., Töpfl, S., Angersbach, A., & Knorr, D. (2005). Impact of pulsed electric field treatment on the recovery and quality of plant oils. *Journal of Food Engineering, 67*(3), 281–287. DOI: 10.1016/j.jfoodeng.2004.04.029

Hao, J.Y., Han, W., Xue, B.Y., & Deng, X. (2002). Microwave-assisted extraction of artemisinin from Artemisia annua L. *Separation and Purification Technology, 28*(3), 191–196. DOI: 10.1016/S1383-5866(02)00043-6

He, G., Yin, Y., Yan, X., & Yu, Q. (2014). Optimisation extraction of chondroitin sulfate from fish bone by high intensity pulsed electric fields. *Food Chemistry, 164*, 205–210. DOI: 10.1016/j.foodchem.2014.05.032

Hemwimol, S., Pavasant, P., & Shotipruk, A. (2006). Ultrasound-assisted extraction of anthraquinones from roots of Morinda citrifolia. *Ultrasonics Sonochemistry, 13*(6), 543–548. DOI: 10.1016/j.ultsonch.2005.09.009

Heng, M.Y., Tan, S.N., Yong, J.W.H., & Ong, E.S. (2013). Emerging green technologies for the chemical standardization of botanicals and herbal preparations. *TrAC Trends in Analytical Chemistry, 50*, 1–10. DOI: 10.1016/j.trac.2013.03.012

Herrero, M., Arráez-Román, D., Segura, A., Kenndler, E., Gius, B., Raggi, M.A., ... & Cifuentes, A. (2005). Pressurized liquid extraction–capillary electrophoresis–mass spectrometry for the analysis of polar antioxidants in rosemary extracts. *Journal of Chromatography A, 1084*(1-2), 54–62. DOI: 10.1016/j.chroma.2004.10.003

Herrero, M., Castro-Puyana, M., Mendiola, J.A., & Ibañez, E. (2013). Compressed fluids for the extraction of bioactive compounds. *TrAC Trends in nalytical Chemistry, 43*, 67–83. DOI: 10.1016/j.trac.2012.12.008

Hossain, M.B., Aguiló-Aguayo, I., Lyng, J.G., Brunton, N.P., & Rai, D.K. (2015). Effect of pulsed electric field and pulsed light pre-treatment on the extraction of steroidal alkaloids from potato peels. *Innovative Food Science & Emerging Technologies, 29*, 9–14. DOI: 10.1016/j.ifset.2014.10.014

Hossain, M.B., Brunton, N.P., Patras, A., Tiwari, B., O'donnell, C.P., Martin-Diana, A.B., & Barry-Ryan, C. (2012). Optimization of ultrasound assisted extraction of antioxidant compounds from marjoram (*Origanum majorana* L.) using response surface methodology. *Ultrasonics Sonochemistry, 19*(3), 582–590. DOI: 10.1016/j.ultsonch.2011.11.001

Jacotet-Navarro, M., Rombaut, N., Deslis, S., Fabiano-Tixier, A.S., Pierre, F.X., Bily, A., & Chemat, F. (2016). Towards a "dry" bio-refinery without solvents or added water using microwaves and ultrasound for total valorization of fruit and vegetable by-products. *Green Chemistry, 18*(10), 3106–3115. DOI: 10.1039/C5GC02542G

Jacotet-Navarro, M., Rombaut, N., Fabiano-Tixier, A.S., Danguien, M., Bily, A., & Chemat, F. (2015). Ultrasound versus microwave as green processes for extraction of rosmarinic, carnosic and ursolic acids from rosemary. *Ultrasonics Sonochemistry, 27*, 102–109. DOI: 10.1016/j.ultsonch.2015.05.006

Jahongir, H., Miansong, Z., Amankeldi, I., Yu, Z., & Changheng, L. (2019). The influence of particle size on supercritical extraction of dog rose (Rosa canina) seed oil. *Journal of King Saud University-Engineering Sciences, 31*(2), 140–143. DOI: 10.1016/j.jksues.2018.04.004

Jha, P., Das, A.J., & Deka, S.C. (2017). Optimization of ultrasound and microwave assisted extractions of polyphenols from black rice (*Oryza sativa* cv. Poireton) husk. *Journal of Food Science and Technology, 54*(12), 3847–3858. DOI: 10.1007/s13197-017-2832-0

Jiang, Y., Li, D., Ma, X., Jiang, F., He, Q., Qiu, S., Li, Y.,& Wang, G. (2018). Ionic liquid–ultrasound-based extraction of biflavonoids from selaginella helvetica and investigation of their antioxidant activity. *Molecules, 23*(12), 3284. DOI: 10.3390/molecules23123284

Gil-Chávez, G.J., Villa, J.A., Ayala-Zavala, F., Basilio Heredia, J., Sepulveda, D., Yahia, E.M., & González-Aguilar, G.A. (2013). Technologies for extraction and production of bioactive compounds to be used as nutraceuticals and food ingredients: An overview. Comprehensive *Reviews in Food Science and Food Safety, 12*(1), 5–23. DOI: 10.1111/1541-4337.12005

Kaur, P., Pandey, D.K., Gupta, R.C., & Dey, A. (2019). Simultaneous microwave assisted extraction and HPTLC quantification of mangiferin, amarogentin, and swertiamarin in Swertia species from Western Himalayas. *Industrial Crops and Products, 132*, 449–459. DOI: 10.1016/j.indcrop.2019.02.055

Kim, W.J., Veriansyah, B., Lee, Y.W., Kim, J., & Kim, J.D. (2010). Extraction of mangiferin from Mahkota Dewa (*Phaleria macrocarpa*) using subcritical water. *Journal of Industrial and Engineering Chemistry, 16*(3), 425–430. DOI: 10.1016/j.jiec.2009.08.008

Ko, M.J., Cheigh, C.I., Cho, S.W., & Chung, M.S. (2011). Subcritical water extraction of flavonol quercetin from onion skin. *Journal of Food Engineering, 102*(4), 327–333. DOI: 10.1016/j.jfoodeng.2010.09.008

Kraujalis, P., Venskutonis, P R., Ibanez, E., & Herrero, M. (2015). Optimization of rutin isolation from Amaranthus paniculatus leaves by high pressure extraction and fractionation techniques. *The Journal of Supercritical Fluids, 104*, 234–242. DOI: 10.1016/j.supflu.2015.06.022

Kulkarni, V.M., & Rathod, V.K. (2014). Mapping of an ultrasonic bath for ultrasound assisted extraction of mangiferin from Mangifera indica leaves. Ultrasonics sonochemistry, *21*(2), 606–611. DOI: 10.1016/j.ultsonch.2013.08.021

Kumari, B., Tiwari, B.K., Hossain, M.B., Brunton, N.P., & Rai, D.K. (2018). Recent advances on application of ultrasound and pulsed electric field technologies in the extraction of bioactives from agro-industrial by-products. *Food and Bioprocess Technology, 11*(2), 223–241. DOI: 10.1007/s11947-017-1961-9

Kviecinski, M.R., Benelli, P., Felipe, K.B., Correia, J.F.G., Pich, C.T., Ferreira, S.R.S., & Pedrosa, R.C. (2011). SFE from Bidens pilosa Linné to obtain extracts rich in cytotoxic polyacetylenes with antitumor activity. *The Journal of Supercritical Fluids, 56*(3), 243–248. DOI: 10.1016/j.supflu.2010.12.011

Lee, H.J., & Kim, C.Y. (2010). Simultaneous determination of nine lignans using pressurized liquid extraction and HPLC-DAD in the fruits of Schisandra chinensis. *Food Chemistry, 120*(4), 1224–1228. DOI: 10.1016/j.foodchem.2009.11.068

Lefebvre, T., Destandau, E., & Lesellier, E. (2021). Selective extraction of bioactive compounds from plants using recent extraction techniques: A review. *Journal of Chromatography A, 1635*, 461770. DOI: 10.1016/j.chroma.2020.461770

Li, W., Wang, Z., Wang, Y.P., Jiang, C., Liu, Q., Sun, Y.S., & Zheng, Y.N. (2012). Pressurised liquid extraction combining LC–DAD–ESI/MS analysis as an alternative method to extract three major flavones in Citrus reticulata 'Chachi'(Guangchenpi). *Food Chemistry, 130*(4), 1044–1049. DOI: 10.1016/j.foodchem.2011.07.129

Liazid, A., Palma, M., Brigui, J., & Barroso, C.G. (2007). Investigation on phenolic compounds stability during microwave-assisted extraction. *Journal of Chromatography A, 1140*(1-2), 29–34. DOI: 10.1016/j.chroma.2006.11.040

Liu, J.L., Li, L.Y., & He, G.H. (2016). Optimization of microwave-assisted extraction conditions for five major bioactive compounds from Flos Sophorae immaturus (cultivars of Sophora japonica L.) using response surface methodology. *Molecules, 21*(3), 296. DOI: 10.3390/molecules21030296

Liu, T., Sui, X., Zhang, R., Yang, L., Zu, Y., Zhang, L., ... & Zhang, Z. (2011). Application of ionic liquids based microwave-assisted simultaneous extraction of carnosic acid, rosmarinic acid and essential oil from Rosmarinus officinalis. *Journal of Chromatography A, 1218*(47), 8480–8489. DOI: 10.1016/j.chroma.2011.09.073

Luthria, D.L., Mukhopadhyay, S., & Kwansa, A.L. (2006). A systematic approach for extraction of phenolic compounds using parsley (Petroselinum crispum) flakes as a model substrate. *Journal of the Science of Food and Agriculture, 86*(9), 1350–1358. DOI: 10.1002/jsfa.2521

Ma, Y., Ye, X., Hao, Y., Xu, G., Xu, G., & Liu, D. (2008). Ultrasound-assisted extraction of hesperidin from Penggan (Citrus reticulata) peel. *Ultrasonics Sonochemistry, 15*(3), 227–232. DOI: 10.1016/j.ultsonch.2007.03.006

Maran, J.P., Mekala, V., & Manikandan, S. (2013). Modeling and optimization of ultrasound-assisted extraction of polysaccharide from Cucurbita moschata. *Carbohydrate Polymers, 92*(2), 2018–2026. DOI: 10.1016/j.carbpol.2012.11.086

Montanari, L., Fantozzi, P., Snyder, J.M., & King, J.W. (1999). Selective extraction of phospholipids from soybeans with supercritical carbon dioxide and ethanol. *The Journal of Supercritical Fluids, 14*(2), 87–93. DOI: 10.1016/S0896-8446(98)00110-7

Morsy, N.F. (2020). Production of thymol rich extracts from ajwain (Carum copticum L.) and thyme (*Thymus vulgaris* L.) using supercritical CO2. *Industrial Crops and Products, 145*, 112072. DOI: 10.1016/j.indcrop.2019.112072

Mustafa, A., & Turner, C. (2011). Pressurized liquid extraction as a green approach in food and herbal plants extraction: A review. *Analytica Chimica Acta, 703*(1), 8–18. DOI: 10.1016/j.aca.2011.07.018

Ong, E.S., Cheong, J.S.H., & Goh, D. (2006). Pressurized hot water extraction of bioactive or marker compounds in botanicals and medicinal plant materials. *Journal of Chromatography A, 1112*(1-2), 92–102. DOI: 10.1016/j.chroma.2005.12.052

Oroian, M., Ursachi, F., & Dranca, F. (2020). Ultrasound-assisted extraction of polyphenols from crude pollen. *Antioxidants, 9*(4), 322. DOI: 10.3390/antiox9040322

Ouédraogo, J.C.W., Dicko, C., Kini, F.B., Bonzi-Coulibaly, Y.L., & Dey, E.S. (2018). Enhanced extraction of flavonoids from Odontonema strictum leaves with antioxidant activity using supercritical carbon dioxide fluid combined with ethanol. *The Journal of Supercritical Fluids, 131*, 66–71. DOI: 10.1016/j.supflu.2017.08.017

Panja, P. (2018). Green extraction methods of food polyphenols from vegetable materials. *Current Opinion in Food Science, 23*, 173–182. DOI: 10.1016/j.cofs.2017.11.012

Picot-Allain, C., Mahomoodally, M.F., Ak, G., & Zengin, G. (2021). Conventional versus green extraction techniques-A comparative perspective. *Current Opinion in Food Science, 40*, 144–156. DOI: 10.1016/j.cofs.2021.02.009

Plaza, M., Oliveira, D., Nilsson, A., & Turner, C. (2017). Green and efficient extraction method to determine polyphenols in cocoa and cocoa products. *Food Analytical Methods, 10*(8), 2677–2691. DOI: 10.1007/s12161-017-0830-5

Purohit, A.J., & Gogate, P.R. (2015). Ultrasound-assisted extraction of β-carotene from waste carrot residue: Effect of operating parameters and type of ultrasonic irradiation. *Separation Science and Technology, 50*(10), 1507–1517. DOI: 10.1080/01496395.2014.978472

Rajha, H.N., Abi-Khattar, A.M., El Kantar, S., Boussetta, N., Lebovka, N., Maroun, R.G., & Vorobiev, E. (2019). Comparison of aqueous extraction efficiency and biological activities of polyphenols from pomegranate peels assisted by infrared, ultrasound, pulsed electric fields and high-voltage electrical discharges. *Innovative Food Science & Emerging Technologies, 58*, 102212. DOI: 10.1016/j.ifset.2019.102212

Ramić, M., Vidović, S., Zeković, Z., Vladić, J., Cvejin, A., & Pavlić, B. (2015). Modeling and optimization of ultrasound-assisted extraction of polyphenolic compounds from Aronia melanocarpa by-products from filter-tea factory. *Ultrasonics Sonochemistry, 23*, 360–368. DOI: 10.1016/j.ultsonch.2014.10.002

Rangsriwong, P., Rangkadilok, N., Satayavivad, J., Goto, M., & Shotipruk, A. (2009). Subcritical water extraction of polyphenolic compounds from Terminalia chebula Retz. fruits. *Separation and Purification Technology, 66*(1), 51–56. DOI: 10.1016/j.seppur.2008.11.023

Ranjha, M.M A., Kanwal, R., Shafique, B., Arshad, R N., Irfan, S., Kieliszek, M., & Aadil, R.M. (2021). A critical review on pulsed electric field: A novel technology for the extraction of phytoconstituents. *Molecules, 26*(16), 4893. DOI: 10.1016/j.seppur.2008.11.023

Rashid, R., Mohd Wani, S., Manzoor, S., Masoodi, F.A., & Masarat Dar, M. (2023). Green extraction of bioactive compounds from apple pomace by ultrasound assisted natural deep eutectic solvent extraction: Optimisation, comparison and bioactivity. *Food Chemistry, 398*, 133871. https://doi.org/10.1016/j.foodchem.2022.133871

Rodríguez De Luna, S.L., Ramírez-Garza, R.E., & Serna Saldívar, S.O. (2020). Environmentally friendly methods for flavonoid extraction from plant material: Impact of their operating conditions on yield and antioxidant properties. *The Scientific World Journal, 2020*, 6792069. DOI: 10.1155/2020/6792069

Roseiro, L.B., Duarte, L.C., Oliveira, D L., Roque, R., Bernardo-Gil, M.G., Martins, A.I., ... & Rauter, A.P. (2013). Supercritical, ultrasound and conventional extracts from carob (Ceratonia siliqua L.) biomass: Effect on the phenolic profile and antiproliferative activity. *Industrial Crops and Products, 47*, 132–138. DOI: 10.1016/j.indcrop.2013.02.026

Rostagno, M.A., Palma, M., & Barroso, C.G. (2007). Microwave assisted extraction of soy isoflavones. *Analytica Chimica Acta*, 588(2), 274–282. DOI: 10.1016/j.aca.2007.02.010

Routray, W., & Orsat, V. (2012). Microwave-assisted extraction of flavonoids: A review. *Food and Bioprocess Technology, 5*(2), 409–424. DOI: 10.1007/s11947-011-0573-z

Rovetto, L.J., & Aieta, N.V. (2017). Supercritical carbon dioxide extraction of cannabinoids from Cannabis sativa L. *The Journal of Supercritical Fluids, 129*, 16–27. DOI: 10.1016/j.supflu.2017.03.014

Santoyo, S., Cavero, S., Jaime, L., Ibanez, E., Senorans, F.J., & Reglero, G. (2006). Supercritical carbon dioxide extraction of compounds with antimicrobial activity from Origanum vulgare L.: Determination of optimal extraction parameters. *Journal of Food Protection, 69*(2), 369–375. DOI: 10.4315/0362-028x-69.2.369

del Pilar Sánchez-Camargo, A.P., Valdés, A., Sullini, G., García-Cañas, V., Cifuentes, A., Ibáñez, E., & Herrero, M. (2014). Two-step sequential supercritical fluid extracts from rosemary with enhanced anti-proliferative activity. *Journal of Functional Foods, 11*, 293–303. DOI: 10.1016/j.jff.2014.10.014

Savova, M., Bart, H., & Seikova, I. (2005). Enhancement of mass transfer in solid-liquid extraction by pulsed electric field. *Journal of the University of Chemical Technology and Metallurgy, 40*(4), 329–334.

Silva, M.M., Reboredo, F.H., & Lidon, F.C. (2022). Food colour and Additives: A synoptical overview on their chemical properties, applications in food products and health side effects. *Foods, 11*, 379. DOI: 10.3390/foods11030379

Shotipruk, A., Kiatsongserm, J., Pavasant, P., Goto, M., & Sasaki, M.(2004). Pressurized hot water extraction of anthraquinones from the roots of Morinda citrifolia. *Biotechnology Progress, 20*(6), 1872–1875. DOI: 10.1021/bp049779x

Soquetta, M.B., Terra, L.D.M., & Bastos, C.P. (2018). Green technologies for the extraction of bioactive compounds in fruits and vegetables. *CyTA-Journal of Food, 16*(1), 400–412. DOI: 10.1080/19476337.2017.1411978

Wakte, P.S., Sachin, B.S., Patil, A.A., Mohato, D.M., Band, T.H., & Shinde, D.B. (2011). Optimization of microwave, ultra-sonic and supercritical carbon dioxide assisted extraction techniques for curcumin from Curcuma longa. *Separation and Purification Technology, 79*(1), 50–55. DOI: 10.1016/j.seppur.2011.03.010

Wang, L., Yang, B., Du, X., & Yi, C. (2008). Optimisation of supercritical fluid extraction of flavonoids from Pueraria lobata. *Food Chemistry, 108*(2), 737–741. DOI: 10.1016/j.foodchem.2007.11.031

Wen, C., Zhang, J., Zhang, H., Dzah, C.S., Zandile, M., Duan, Y., & Luo, X. (2018). Advances in ultrasound assisted extraction of bioactive compounds from cash crops–A review. *Ultrasonics Sonochemistry, 48*, 538–549. DOI: 10.1016/j.ultsonch.2018.07.018

Wijngaard, H., Hossain, M.B., Rai, D.K., & Brunton, N. (2012). Techniques to extract bioactive compounds from food by-products of plant origin. *Food Research International, 46*(2), 505–513. DOI: 10.1016/j.foodres.2011.09.027

Xiao, W., Han, L., & Shi, B.(2008). Microwave-assisted extraction of flavonoids from Radix astragali. *Separation and Purification Technology, 62*(3), 614–618. DOI: 10.1016/j.seppur.2008.03.025

Yan, L.G., He, L., & Xi, J. (2017). High intensity pulsed electric field as an innovative technique for extraction of bioactive compounds—A review. *Critical Reviews in Food Science and Nutrition, 57*(13), 2877–2888. DOI: 10.1080/10408398.2015.1077193

Yi, C., Shi, J., Xue, S.J., Jiang, Y., & Li, D. (2009). Effects of supercritical fluid extraction parameters on lycopene yield and antioxidant activity. *Food Chemistry, 113*(4), 1088–1094. DOI: 10.1016/j.foodchem.2008.08.083

Yin, Y., Cui, Y., & Ding, H. (2008). Optimization of betulin extraction process from Inonotus obliquus with pulsed electric fields. *Innovative Food Science & Emerging Technologies, 9*(3), 306–310. DOI: 10.1016/j.ifset.2007.07.010

Yongguang, Y., Yuzhu, H., & Yong, H. (2006). Pulsed electric field extraction of polysaccharide from Rana temporaria chensinensis David. *International Journal of Pharmaceutics, 3121-3122*, 33–36. DOI: 10.1016/j.ijpharm.2005.12.021

Yousefi, M., Rahimi-Nasrabadi, M., Pourmortazavi, S.M., Wysokowski, M., Jesionowski, T., Ehrlich, H., & Mirsadeghi, S. (2019). Supercritical fluid extraction of essential oils. *TrAC Trends in Analytical Chemistry, 118*, 182–193. DOI: 10.1016/j.trac.2019.05.038

Zaibunnisa, A.H., Norashikin, S., Mamot, S., & Osman, H. (2009). An experimental design approach for the extraction of volatile compounds from turmeric leaves (*Curcuma domestica*) using pressurised liquid extraction (PLE). *LWT-Food Science and Technology, 42*(1), 233–238. DOI: 10.1016/j.lwt.2008.03.015

Zhao, W., Yu, Z., Liu, J., Yu, Y., Yin, Y., Lin, S., & Chen, F. (2011). Optimized extraction of polysaccharides from corn silk by pulsed electric field and response surface quadratic design. *Journal of the Science of Food and Agriculture, 91*(12), 2201–2209. DOI: 10.1002/jsfa.4440

Zhou, H.Y., & Liu, C.Z. (2006). Microwave-assisted extraction of solanesol from tobacco leaves. *Journal of Chromatography A, 1129*(1), 135–139. DOI: 10.1016/j.chroma.2006.07.083

Chapter 15

Nutraceuticals Loaded in Polyelectrolyte Complexes (PECs) for the Treatment of Human Diseases

Evangelina García-Armenta
Lorenzo A. Picos-Corrales
Julieta del Carmen Villalobos-Espinosa
and María del Rocío Espinosa-Velázquez

CONTENTS

15.1 INTRODUCTION

In recent years, human diseases such as cancer and diabetes have gained prevalence as a worldwide epidemic, mainly due to a sedentary lifestyle, inappropriate nutrition habits and environmental factors (Wolin et al. 2010), given that these conditions promote cell inflammation. Chronic inflammation triggers systemic complications, commonly derived from obesity, and it is the main cause of tumor development and metastasis progression in cancer (Deng et al. 2016). Therefore, inflammation must be treated to prevent and handle these diseases. Conventional treatments involve the use of synthetic drugs in long, toxic, and hard emotional

 DOI: 10.1201/9781003332558-18

therapies (Tahrani et al. 2016). Alternatively, natural products such as food ingredients extracted from plants and animals can be a suitable option to develop friendlier treatments. It has been reported that nutraceuticals such as curcumin, flavonoids, alkaloids, carotenoids, and essential fatty acids can help to prevent and treat human diseases, such as different types of cancer and obesity. These substances act as antioxidants that decrease the inflammatory episode that activates the illness condition (Kuppusamy et al. 2014). However, these nutraceutical compounds are characterized by their susceptibility to degradation or loss as a consequence of differences in pH (during the gastrointestinal pathway), temperature conditions, other chemical interactions, non-specific release targets, and other factors in their microenvironment (Akhavan et al. 2018). Hence, several encapsulation systems have been developed to protect these bioactive compounds, such as emulsions, capsule powders, liposomes, micelles, vesicles, nanogels, and polyelectrolyte complexes (PECs) (Picos-Corrales et al. 2019). PECs in particular present potential advantages such as non-expensive and easy preparation, and their formulations can be based on biopolymers resulting in biocompatible and eco-friendly carriers, avoiding subsequent environmental contamination due to excreted nanocarriers (Delair, 2011). PECs involve colloidal systems formed by two macromolecules having different surface charge, namely cationic and anionic nature, resulting in an aggregation system governed by electrostatic interactions (Kulkarni et al. 2016). Recently, PECs have been developed for aiding in different human diseases. According to the Scopus database, in the last nine years (from 2013 to 2022), PEC publications increased. However, research papers on PECs based on food ingredients are fewer than the total reports on PECs prepared for biomedical purposes (**Figure 15.1**). In most of the studies, PECs have been prepared from biopolymers, mainly complexes constituted by chitosan and alginate. The formation of this system is spontaneous by a titration process and adjusting different parameters (related to the nature of the materials, molar ratios, and conditions of the aqueous medium) leads to systems at the macro-, micro-, or nano-scales. Regarding drug incorporation, the goal can be reached as follows: entrapment of the food ingredient from a solution by van der Waals forces during the formation of PECs or using preformed PECs; incorporation by chemical coupling of the active compound to either

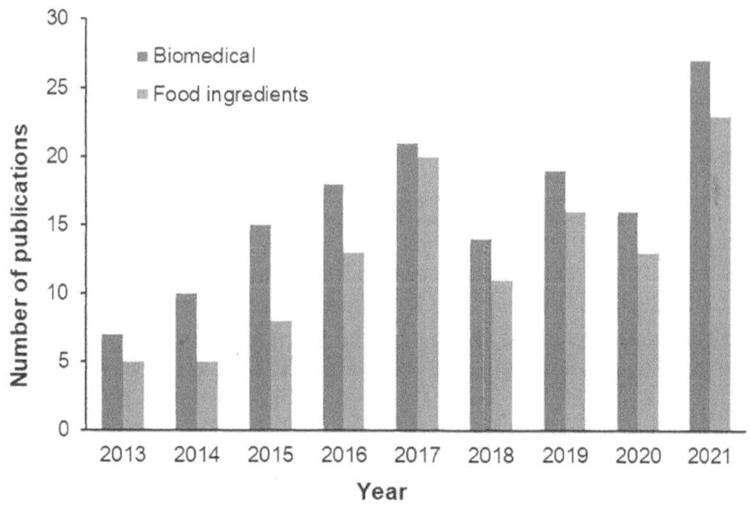

Figure 15.1 Number of publications regarding polyelectrolyte complexes prepared from food ingredients in the 2013–2021 years.

the polyanion or the polycation before PEC formation; and PEC formation using the food ingredient as partner by replacing one of the two polyelectrolytes (Meka et al. 2017).

Also, to guarantee suitable properties of PECs, it is necessary to perform characterization techniques according to specific purposes. As examples, electrostatic properties can be measured by potentiometric and conductimetric titrations (Silva et al. 2008); thermal properties can be evaluated by Differential Scanning Calorimetry (DSC) and Thermogravimetric Analysis (TGA) (Minkal et al. 2018); chemical composition and detection of the bioactive compound can be assessed by Fourier-transform Infrared Spectroscopy (FTIR), proton Nuclear Magnetic Resonance (H NMR), and X-ray diffractometry (Brar & Kaur, 2018); particle size (hydrodynamic diameter) and zeta potential can be determined by Dynamic Light Scattering (DLS) and Laser Doppler Electrophoresis, respectively (Yan et al. 2018); the morphology has been usually observed by Scanning Electron Microscopy (SEM) and Transmission Electron Microscopy (Sankalia et al. 2007); the entrapment efficiency is calculated from different analytical techniques, usually UV-Vis Spectroscopy (Sankalia et al. 2007); and *in vitro* release assays have been commonly carried out to determine the values of constants associated with kinetic models (Al-Kurdi et al. 2015).

Regarding human disease treatment, most reports mentioned that PECs loaded with curcumin have been reported to have properties against cancer and diabetes. Curcumin is a phenolic compound named diarylheptanoid obtained from turmeric (*Curcuma longa*) and numerous studies have supported several benefits to human health (Pivari et al. 2019). Curcumin loaded in PECs have decreased proliferation and increased apoptosis (Ahmadi et al. 2017) of cancer cells, possessed appropriate cytotoxicity (Ahmadi Nasab et al. 2018), and increased the activity of enzymes responsible for the protection of cellular damage (Khatun et al. 2018). For diabetes treatment, nutraceuticals such as curcumin and resveratrol can be loaded into these colloidal systems and food ingredients such as peptides derived from egg or nuts can also be contained in PECs and additionally loaded with insulin (Papagiannopoulos, 2021). Furthermore, several reports provide studies regarding the effectiveness of PECs in other human diseases, such as cardiovascular conditions, osteoporosis, wound dressings, and antibacterial and antifungal properties.

Therefore, the aim of this chapter is to revise and analyze the properties of food ingredients as a part of PECs used for the treatment of human diseases such as cancer and diabetes.

15.2 PREPARATION OF POLYELECTROLYTE COMPLEXES (PECS) CONTAINING FOOD INGREDIENTS

It is known that PEC formation is driven by proper electrostatic interactions between oppositely charged polyelectrolytes, and therefore the chemical structure of the polymer chains used plays a key role, given that each material presents different characteristics; as described by Neitzel et al. (2021), the PEC phase behavior can be influenced by some relevant parameters related to the polyelectrolyte nature, such as chain length and concentration, hydrophobicity and charge density, cation–π and π–π interactions, hydrogen bonding, sequence in its chemical structure, upper and lower critical solution temperatures (Neitzel et al. 2021). In order to manipulate the PEC phase, this system is usually designed from polymers classified as natural [e.g. hyaluronic acid and poly(L-lysine)], semi-synthetics (e.g. chitosan and carboxymethyl cellulose); and synthetics [e.g. poly(acrylic acid), poly(ethyleneimine) and poly(4-(N,N dimethylaminomethylstyrene))], showing different complexation capacity. Namely, PECs can be prepared from polymers with a single repeating unit (homopolymer) or by combining two (or more) different types of monomer (copolymers) (Meka et al. 2017). Regarding the medium and method of preparation, pH, ionic strength, temperature, mixing order, molar ratios, the use of cosolvents, and interaction time are processing factors affecting PEC production (Siyawamwaya et al. 2015). According to the survey by Kulkarni et al. (2016), PEC formation can be performed either by a stoichiometric feed (prepared using polymers in equimolar ratio) or by a non-stoichiometric feed where one type of polymer is in excess compared to the other one. As a result of the polycation-polyanion interactions, water-soluble, colloidally stable, or two-phase systems can be obtained. Water-soluble aggregates can be attained from polyelectrolytes bearing weak ionizable groups and considerably different molar mass by mixing at a non-stoichiometric ratio; colloidal stable PECs can be formed by adjusting the concentration and charge ratio of both polyelectrolytes at non-stoichiometric conditions, as well as the salts content; macroscopically heterogeneous systems can be prepared from high or similar molecular weight polyelectrolytes, and highly concentrated solutions near to a stoichiometric ratio yield two-phase systems, involving a liquid phase and a PECs-rich phase (Kulkarni et al. 2016). In some cases, variations of pH can strongly affect PEC formation and the stability of these systems because of changes in the number of dissociated groups onto the polyelectrolyte (Meka et al. 2017). Thus, the control of some parameters allows for obtaining PEC-based systems at the macro-, micro-, or nano-scale (Marciel et al. 2017). For the encapsulation of active compounds (e.g., food ingredients), the incorporation of the guest molecule into a PEC can be achieved in different ways, such as (a) adding the active compound during PEC formation, (b) adding the active compound into the PEC that has been previously formed, and (c) mixing/coupling the active compound with one of the polyelectrolytes before the PEC is formed; also, charged active compounds could act as one of the polyelectrolytes (Siyawamwaya et al. 2015). The chemical association or complexation between polymer chains and guest molecules can occur via ionic and hydrogen bonding, van der Waals forces, and hydrophobic interaction (Tonelli et al. 2018). For example, a PEC system formed by sodium alginate and chitosan and loaded with an active agent by mixing is schematized in **Figure 15.2**. In the case of chitosan, it is an outstanding polysaccharide having advantageous properties, such as biocompatibility, biodegradability, antimicrobial effect, mucoadhesive properties, and film-forming capacity, which in part is attributed to the presence of ionizable and reactive primary amino groups that provide strong chemical interactions with other substances (Picos-Corrales et al. 2023).

With the aim of presenting specific details of the preparation methods for the development of PECs used for encapsulation of food ingredients (at macro-, micro-, and nano-scales), a series of literature reports addressing this type of formulation is described below.

15.2.1 PECs and Food Ingredients – Macro-Scale System

The macro-scale system based on PECs is a macroscopically visible and stable structure and its preparation is carried out by manipulating the parameters mentioned above. For instance, researchers studied the effect of molar ratio, pH, and temperature on the formation and behavior of the supramolecular structure of macro-PECs obtained from chitosan (cationic polyelectrolyte) and carboxymethylcellulose (anionic polyelectrolyte); for that, prescribed volumes of CM-cellulose dispersions were poured dropwise under magnetic stirrer into the chitosan dispersions and then the mixtures were stirred for 10 min. The authors stated that a 1:2 mixing ratio (chitosan: CM-cellulose), at 25 °C and pH around 4, maximized the interactions yielding macro-PECs (Ferreira et al. 2022).

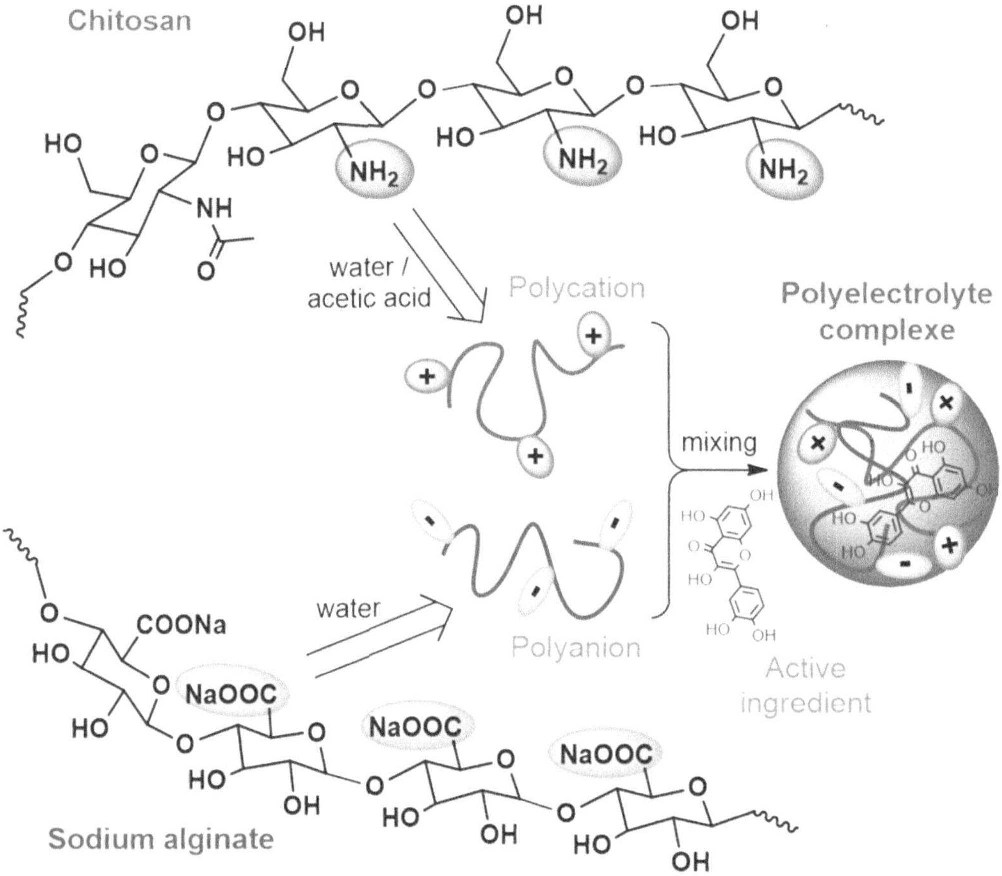

Figure 15.2 Representation of the formation of a polyelectrolyte complexes system (PECs) prepared from chitosan and sodium alginate loaded with an active ingredient.

In another work, networks from sodium hyaluronate (HA) and phosphonium polyelectrolytes (PPE) were prepared for the encapsulation of anionic molecules (e.g., adenosine-5′-triphosphate). For the preparation of the PEC networks, HA and PPE were dissolved individually in deionized water at a concentration of 0.01 M; after that, the solutions were poured simultaneously into a beaker under vigorous stirring and then stirred for 30 min. Next, the liquid was decanted, and the polymer network was placed into proper tubes to be centrifuged at 187000 g for 1 h. Finally, the product was removed from the tubes and cut into narrow disks, which were dried under vacuum until the weight was constant. The authors indicated that network properties were influenced by the structure of the PPE and the presence of salts (Harrison et al. 2020).

In the same way, Hu et al. (2020a) developed self-assembled polyelectrolyte complex hydrogel for sustained delivery of green tea polyphenols. Briefly, 2 g of Salecan and 1.5 g of *N,N,N*-trimethyl chitosan (TMC) were dissolved separately in 100 mL of deionized water. The Salecan solution

(2%, w/v) was added dropwise into the TMC solution (1.5%, w/v) for prescribed volume ratios, applying sonication for 30 min. Later, the mixture was poured into glass moulds placed in a desiccator containing a suitable volume of acetic acid solution (3%) and stayed for 2 h until the PEC hydrogels were obtained. Afterwards, the PEC hydrogels were taken out and washed thoroughly with deionized water. Finally, the samples were frozen and then lyophilized in a freeze dryer for 24 h. The authors remarked on the importance of the acetic acid atmosphere that led to a gradual protonation of amino groups resulting in a homogeneous and stable structure (Hu et al. 2020a).

15.2.2 PECs and Food Ingredients – Micro-Scale System

Micro-PECs are versatile platforms having micrometer particle sizes, usually considered as larger than 100 nm to several micrometers. Naturally, a precise design of this system allows adjusting its physicochemical properties for potential application in the encapsulation of active substances. For instance, in the work by Tabandeh & Leon (2019), peptide-based PECs were designed with an increased hydrophobicity. Polycations and polyanions were intended with an alternating sequence of D- and L-chiral patterns of glutamic acid or lysine combined with glycine, alanine, or leucine. The authors reported that polypeptide complexation was carried out by sequential addition of equivalent quantities (stock solutions of 10 mM) of the polyanion (first) and polycation (second) into different concentrations of salt or urea solutions, vortexing after each step. As a result, the stability of the system and encapsulation of a model hydrophobic dye were higher with increasing polymer hydrophobicity, and peptides showing increased charge density were even more stable (Tabandeh & Leon, 2019).

Researchers have also studied the effect of different concentrations of assai pulp oil and calcium chloride in the PECs formation, starting from chitosan and alginate (Teixeira-Costa et al. 2020). For the preparation of microcapsules from chitosan and alginate in ratio 8:2 (stock solutions at 1% v/v, in water), assai pulp oil (0.0–0.75% w/w) was added to the alginate solution (15 g) and homogenized (8000 rpm) for 5 min using an Ultra-Turrax, obtaining an oil emulsion. Later, the chitosan solution (60 g) was incorporated by a peristaltic pump into the oil emulsion. Then, the resulting mixture was homogenized (8000 rpm) for 10 min and the pH of the medium was adjusted to 6.0 to promote the PEC formation. Next, calcium chloride (solution at 0.0–1.0% w/v) was added as a crosslinking agent under constant stirring at room temperature for 3 h. The products were separated by filtration and dried at room temperature. The PECs were maintained under vacuum in a desiccator containing anhydrous calcium chloride (0% RH) until a constant weight was measured. The final PECs were stored in a refrigerator. From these trials, it was found that assai pulp oil and calcium chloride affected considerably the particle size and polydispersity index of the PECs, but the encapsulation efficiency was maintained at >99% (Teixeira-Costa et al. 2020).

In the same sense, Qin et al. (2021) prepared anthocyanins-loaded complexes as follows: an anthocyanins solution (3 mL, 2 mg/mL) was mixed with 7.5 mL of O-carboxymethyl chitosan solution (pH 6.0, 5.75 mg/mL). Afterwards, 15 mL of chitosan hydrochloride solution (pH 4.9, 2 mg/mL) was poured dropwise into the mixture containing anthocyanins/O-carboxymethyl chitosan and stirred at 20 °C for 30 min. The final pH registered for the system was 5.62. The authors indicated that the ratio of chitosan polymer and anthocyanins is a key factor in the PEC formation. In this case, particles with a size of 242 nm, a zeta potential of +28.13 mV, and polydispersity index of 0.186 were obtained (Qin et al. 2021).

15.2.3 PECs and Food Ingredients – Nano-Scale System

According to several reports (Geoffrion et al., 2020), the nanoscale has been assigned to particles having sizes less than 100 nm. For example, chitosan-based PECs containing curcumin have been prepared yielding particle size <50 nm. For nanoparticle formation, a chitosan solution (3 mL, 0.15% w/v) was mixed with the curcumin solution (1 mL, 0.25% w/v), and the resulting mixture was incorporated dropwise into a solution containing tripolyphosphate (15 mL, 0.3%) and alginate (10 mL, 0.2% w/v) under agitation. From that, colloids were obtained and allowed to cure for 1 h. Then, the product was centrifuged applying 10,000 rpm for 15 min and separated from the supernatant for further analysis. The authors stated that the complexation between chitosan and alginate reduced the loss of curcumin and improved the control of the release from the nanoparticles (Akolade et al. 2017).

Active compounds can also be covalently attached to a polyelectrolyte prior to the formation as reported by Hu et al. (2016), where the coupling of gallic acid to hydroxyl or amino group of chitosan was accomplished using ascorbic acid/hydrogen peroxide as redox pair. Afterwards, PECs were formed by ionic gelation between the polycation from gallic acid-chitosan conjugate (GA-CS) and gum arabic as polyanion. For that, aqueous solutions of gum arabic (at 2 mg/mL) and polyethylene glycol (100 mg/mL) were prepared. Then, gum arabic solution (5 mL) was added dropwise to GA-CS aqueous solution (10 mL) at different pH, under stirring (600 rpm) for 30 min at room temperature. In this case, 3:1 was found to be the optimal mass ratio of GA-CS to gum arabic; the optimal pH of GA-CS solution producing nano-PECs was 5.5; when the mass ratios of GA-CS to polyethylene glycol was 1:0.5, uniform particles with smaller size were obtained. The mass ratio GA-CS/gum arabic and the pH of the GA-CS solution were key parameters to control the size of nanoparticles. Polyethylene glycol was essential in maintaining the structural properties and stability of the nanoparticles. As result, a formulation exhibiting a particle size of 122 nm and zeta potential of +26 mV was developed (Hu et al. 2016).

15.3 CHARACTERIZATION TECHNIQUES

In the development of new materials and encapsulates such as PECs, the interaction between the polymeric platforms used for their preparation is important, as well as the additional stability provided by the active compound or drug. The requirement to deliver guest molecules or active compounds in their original form at a minimum dose is the attractiveness of such a drug delivery method, ensuring the release of the active compound in a specific area of the human body over an appropriate time to prevent or decreased human diseases (Emmanuel et al. 2018). That is why one of the main issues for the development and application of PECs is their physical, chemical, and toxicological characterization. **Table 15.1** shows examples of characterization data for PECs containing food ingredients or synthetic drugs, where the nutraceutical substance can be part of the PEC or represent the guest molecule. As can be seen when comparing the data summarized, particle size, encapsulation efficiency, and stability of systems containing food ingredients are also well controlled similar to formulations using synthetic drugs.

PECs have been studied using a variety of characterization methods such as turbidity measurements, static and dynamic light scattering (SLS/DLS), Laser Doppler Electrophoresis, electron and X-ray absorption microscopy, Fourier-transform infrared spectroscopy, scanning electron microscopy, powder X-ray diffraction, and atomic force microscopy, among others (Shah & Leon, 2021; Hamed et al. 2022).

TABLE 15.1 COMPARISON OF CHARACTERIZATION DATA FOR POLYELECTROLYTE COMPLEXES (PECS) CONTAINING FOOD INGREDIENTS OR SYNTHETIC DRUGS

Polymers	Active Agent	Scale	Particle Size	Z potential (mV)	Drug Encapsulation Efficiency	Human Disease	Reference
Positively charged heat-denatured lactoferrin particles Negatively charged pectin	Curcumin	Nano	~208 nm	-32	~85.3%	Antioxidant activity of hydrophobic curcumin	Yan et al. (2017)
Alginate Chondroitin sulfate Dextran Hyaluronic acid Heparin Trimethyl chitosan (TMC)	Lipopeptide vaccine candidate (LCP-1)	Nano	182 – 237 nm	27.2 – 33.8	*	Nasal vaccines against Group A *Streptococcus*	Zhao et al. (2020)
Carboxymethyl chitosan Alginate	5-Fluorouracil combined with Bisdemethoxycurcumin	Nano	203 nm	42.42	89.4% (5-Fluorouracil) and 85.8% (Bisdemethoxycurcumin)	Colorectal cancer	Anirudhan et al. (2019)
Chitosan Alginate	Assai pulp oil (*Euterpe oleracea*)	Micro	367.7 ± 1.0 μm	-1.7 ± 0.6	99.95%	Antioxidant activity	Teixeira-Costa et al. 2020)
Salecan N, N, N-trimethyl chitosan	Green tea polyphenols	Macro	Hydrogel (pore size: 19.1 – 55.5 μm)	*	94.1%	Intestinal targeted nutrient delivery	Hu et al. (2020a)
Gellan gum Chitosan	Thyme essential oil nanoemulsion	Macro	Films	*	Concentration: 2 – 6%	Antimicrobial activity against *E. coli*	Xinhui Zhang et al. (2021)

(continued)

TABLE 15.1 (CONTINUED)

Polymers	Active Agent	Scale	Particle Size	Z potential (mV)	Drug Encapsulation Efficiency	Human Disease	Reference
Polymyxin B Poly (styrene sulfonate) (PSS)	Polymyxin B	Nano	174 ± 49 nm	-30	*	Antimicrobial activity against *P. aeruginosa*	Insua et al. (2017)
Poly (allylamine hydrochloride) (PAH) Poly (sodium 4-styrenesulfonate) (PSS) Polyacrylic acid (PAA) Bovine serum albumin (BSA)	Monoclonal antibody Trastuzumab	Micro	4.2 – 6.1 µm	−29.6 ± 0.8 mV (before conjugation)	QD-encoded microcapsules containing 30 to 300 µg of the anti-HER2 mAb	Cancer (human breast adenocarcinoma)	Nifontova et al. (2019)
Sodium alginate Pluronic F127 Eudragit E100	5-Fluorouracil	Micro	58 ± 22 µm	*	50%	Cancer (colon)	Dalmoro et al. (2017)
Oligochitosan Poloxamer 407 Poloxamer 188	Methotrexate	Micro	470 ± 166 nm	26 ± 2	74.3 ± 5.8%	Rheumatoid arthritis	Agostini et al. (2021)

***Data not reported.**

In this regard, Turbiscan® is widely reported in the literature as a standard tool to assess the stability of colloidal systems for several days (Bonaccorso et al. 2021). The techniques most extensively used for estimating the size or size distribution of PEC particles are SLS and DLS. SLS provides the weight average molecular weight (M_w) and the radius of gyration (R_g) from data treatment performed by the Zimm plot, while DLS is used to determine the hydrodynamic diameter (D_h) and polydispersity index (PDI), where D_h is obtained from the Stokes-Einstein equation for spheres (Kulkarni et al. 2016; Picos-Corrales et al. 2014). Also, zeta potential obtained by Laser Doppler Electrophoresis helps assess the stability of the colloidal dispersions since it is associated with the electrostatic interaction inside the PECs and with the aqueous medium. A high surface charge (positive or negative) suggests that the system is stable (Garcia-Carrasco et al. 2022).

On the other hand, for morphological characterization: Scanning, Transmission and Environmental Electron Microscopy (SEM/TEM/ESEM) and Atomic Force Microscopy (AFM) can be used to examine the size and morphology/topography of colloidal PEC particles. PECs are often polydisperse spherical objects, according to several researchers, and are sometimes impacted by vacuum drying during SEM investigation (Nedĕla et al. 2020). In addition, images of PECs obtained from microscopy techniques can be useful to determine morphometric parameters extracted from Digital Image Analysis (DIA) such as fractal dimension and lacunarity, which have been reported as good indicators of encapsulation efficiency and stability (García-Armenta et al. 2021; García-Armenta & Gutiérrez-López, 2022).

Furthermore, it is important to characterize the complex creation process of PECs, which has been assessed by FT-IR spectroscopy (solid-state characterization). The appearance of additional bands, a change in intensity of existing bands, and the displacement of bands from their original position indicate the interaction of materials yielding the complex.

Alternatively, X-ray powder diffraction (XRD) analysis can be used to evaluate the nature and degree of crystallinity of the PEC particles (Yang et al. 2020). Similarly, Nuclear Magnetic Resonance spectroscopy (NMR) can be applied to assess the chemical composition of PEC particles; the chemical shifts can give information on the physicochemical condition of PEC particles to be obtained (Kulkarni et al. 2016).

In this sense, the Differential Scanning Calorimetry and Thermogravimetric Analysis (DSC/TGA) can be used to investigate the thermal stability of PEC particles. The thermal characterization shows events associated with the glass transition temperature (T_g), the melting point (T_m), and the decomposition temperature of the PEC components (Garcia-Carrasco et al. 2022).

The amount of the active compound in chemical formulations at micro- and nanoscale can be calculated from measurements by some analytical methods such as fluorescence and UV-Vis spectroscopy (Manzanares-Guevara et al. 2020).

15.4 TREATMENTS FOR HUMAN DISEASES

15.4.1 Cancer

Despite various scientific efforts, cancer is one of the more severe threats to humankind and is one of the leading causes of death, accounting for nearly 10 million deaths in 2020 (WHO, 2022). There are many therapies for treating different tumors, where the microenvironment surrounding it results in the biggest challenge to confront, due to its complexity and heterogeneity.

A variety of therapeutic drugs have been developed and used, and most of them are effective for killing tumors, but not effective in targeting tumors, and some show low bioavailability and toxicity problems (Zhang et al. 2021). For these reasons, novel methods with higher therapeutic efficiency and better delivery techniques are needed.

Nanodrug delivery systems are functional drug carriers that can be applied in cancer therapy. For this purpose, oppositely charged materials like PECs are deposited on a template that is removed afterwards when the specific structure and thickness are reached (Meka et al. 2017). Due to the nonspecific electrostatic interactions, functional components, nutraceuticals, and other biological macromolecules can be incorporated into the films yielding functionalized composite membranes (Liu et al. 2013). Their main advantage as drug/functional compound delivery systems is their capacity for the controlled release of their functional components in response to physical, chemical, and biochemical triggers.

PECs loaded with drugs having antitumor and anti-inflammatory activities have been widely applied to treat cancer. However, the use of food ingredients in the preparation of PECs is more required due to there exist several reports detecting their benefits and advantages against this disease.

A group of authors studied the release kinetics of PECs towards combinatory therapy, exploring a system based on Bisdemethoxycurcumin and 5-Fluorouracil. According to the results, both active agents were released at >75% at pH 7.4, and in a percentage higher than 55% at pH 1.2. Namely, the designed formulation presented a release profile dependent on pH, demonstrating that it was possible to release both drugs for improving the therapeutic effect (**Figure 15.3**).

Kalenichenko et al. (2021) developed PECs (using calcium carbonate microsized) with different structures (core/shell, shell types) with microbeads containing doxorubicin. To label the microcapsules, they used quantum dots (fluorescent semiconductor nanocrystals), which also weakened their potential toxic effect on live cells. The results showed that the microcapsules developed can be used as efficient systems of delivery for low-molecular-weight antitumor agents, ensuring a prolonged release of doxorubicin, and could be the vehicle of many nutraceuticals used in cancer as curcumin.

Cardoso et al. (2015) made a drug delivery system to modulate the phenotype of macrophages, converting them from protumor M2 phenotype to antitumor M1 phenotype, with the incorporation of interferon-gamma (a pro-inflammatory cytokine) in chitosan/poly gamma-glutamic acid) complexes, whose components interact electrostatically due to opposite charges. The loaded cytokine into PEC could effectively be incorporated as maintaining its biological activity and did not affect the polymer's assembly as a part of the colloidal system. Also, *in vitro* proved sustained release of IFN-1-gamma and a prominent effect in reversing M2-like macrophages, demonstrated by the production of higher levels of IL-6 and IL-10.

Ahmadi et al. (2017) prepared curcumin-based chitosan-alginate sodium tripolyphosphate nanoparticles by an ultrasonic mediated synthesis. Nanoparticles were dispersed in an aqueous solution, with an ideal size to be internalized into Hela cancer (cervical epithelioid carcinoma) cell lines, which led to suppression of proliferation and an increase in the expression of apoptotic genes in cells treated.

Some bioactive compounds have been studied as chemo preventives, but they have low chemical stability and restricted bioavailability and limited therapeutic application. To avoid this limitation, resveratrol (a potent antioxidant), has been loaded with bovine serum albumin to elevate the cytotoxic performance, and chitosan was used to provide a positive charge on the surface of the nanoparticles (Ghorbani et al. 2018). Cell uptake was observed by fluorescence microscopy

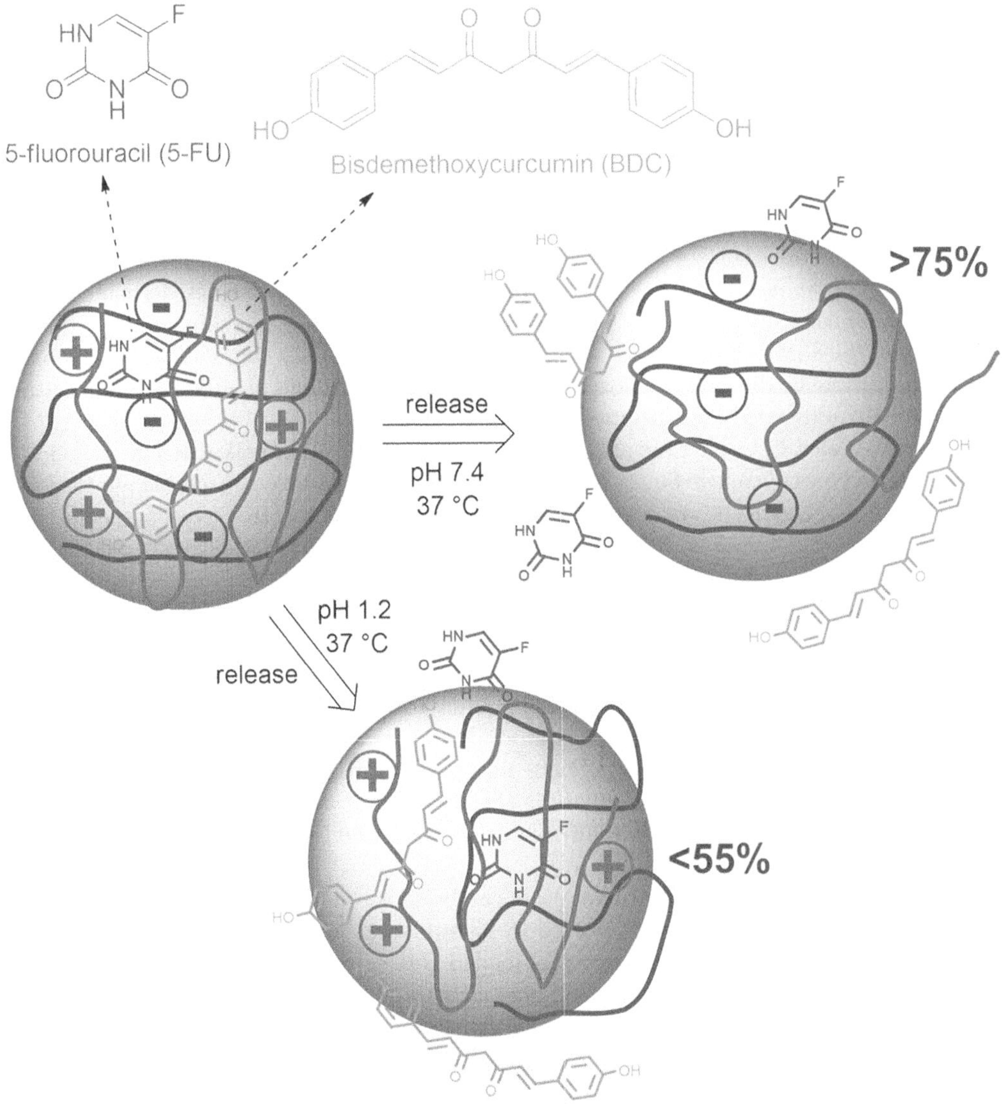

Figure 15.3 Example of the effect of pH on drug release percentage of bisdemethoxycurcumin (BDC) and 5-fluorouracil (5-FU) loaded in polyelectrolyte complexes system PECs for cancer therapy.

and flow cytometry, showing 1.5 times more cell internalization when chitosan was added, and 3 times higher cell cytotoxicity after 24 and 48 hours.

Newham et al. (2021) obtained a polyelectrolyte templated synthesis of L-Arginine/poly (acrylic acid) polyelectrolyte complexes in porous silica nanoparticles that allowed an efficient drug (Dox) loading when tested in primary patient-derived glioblastoma GBM1 and GBM2 cells. The strong interactions between Arginine and poly (acrylic acid) resulted in an optimum uniformity, and also enabled control of the diameter and lead to increased porosity, without the need for template removal. This type of molecule also resulted in a pH-responsive drug release.

The design of amphiphilic polyaspartamide PECs that solubilize hydrophobic drugs in an aqueous medium and enhance their cellular uptake was carried out by Fatima et al. (2016). They synthesized PECs through controlled derivatization of polysuccinimide precursor polymer with amines with various chain lengths and incorporated curcumin as the hydrophobic drug. The complexes were internalized by mammalian cells (HEK-293T, MDA-MB-231, and U2OS) and the uptake of curcumin was directly correlated with the length of the alkyl chain. The polyelectrolyte complex improved the nuclear transport of curcumin in cancer cells, resulting in death by apoptosis, and non-cancer cells were completely unaffected.

On the other hand, a solvent-antisolvent method has also been used to prepare Chitosan-Kheri gum PECs (Malviya et al. 2020), which exhibited significant anticancer activity against various cancer human cells: breast cancer (MCF-7), colon cancer (HT-29), and skin cancer (SK-MEL-2). In this study, the SEM images showed a cubic morphology of PECs, a characteristic that positively affects the uptake by the cells; the formation of cubic nanoparticles enables them for better cellular internalization and biodistribution.

In recent studies, PECs have been added with special features that make them more efficient in treating some types of cancer. Souza et al. (2021) incorporated a water-insoluble photosensitizer in a PEC. Photosensitizers are involved in the generation of reactive oxygen species and or singlet oxygen, which causes damage to the diseased cells. They worked with chitosan/chondroitin sulphate coated with polystyrene-block-poly (acrylic acid) nanoparticles loaded with chloroaluminium phthalocyanine, which has proven to be cytotoxic against colorectal cancer cells (HT-29 cells).

Dual stimuli-responsive PECs for chemo-photothermal synergistic therapy of colon cancer cells were developed by Anirudhan et al. (2019). They used dextran and cellulose and chemically modified graphene oxide. The bioactive compound incorporated was curcumin. With this composition, the authors proved to combine chemo-photothermal features and the cellulose act as pH-sensitive gatekeepers and biocompatible shell layers. Cytotoxicity was carried out in colon cancer cells (HCT-116), proving the safety of drug carriers and the activity of curcumin in killing cancerous cells.

Similarly, imaging agents and drug carriers can be targeted toward cancer cells with specific recognition molecules. In a study reported by Nifontova et al. (2019), prepared quantum dot-encoded polyelectrolyte microcapsules were loaded with antibodies (HER2) to ensure they had specific and selective interaction with the target cells, and that, indeed, were capable of a specific interaction with HER2 tumor cells. These types of complexes could be used as targeted systems for the diagnosis and therapy of a variety of tumors.

Antitumor therapy is also an application of PECs in cancer treatment. Lybaert et al. (2017) obtained microparticles by mixing mannitol and live cancer cells with oppositely charged polyelectrolytes, dextran sulphate, and poly-L-arginine, which *in vitro* experiments revealed were nontoxic. *In vivo* assessment revealed the potential of the polyelectrolytes-enrobed cancer cells as promotors of antigen cross-presentation. The PEC synthesis has emerged as a promising strategy to prepare nanoparticles suitable for drug or bioactive compound delivery in various types of cancer, and other therapeutic uses, but further research is needed to incorporate them into clinical practice.

15.4.2 Diabetes

PECs have been developed for the treatment of human diseases such as diabetes and hyperglycaemic conditions. These delivery systems have been prepared using food-based ingredients, focusing on two approaches: 1) the food ingredient as the active agent and 2) the food ingredient as a part of the material carrying a specific drug.

According to the first approach, several authors have proposed chitosan as the material to form complexes with food agents to treat diabetes (Bourganis et al. 2017). In this regard, nanoencapsulation in chitosan-alginate polyelectrolyte complexed with curcumin has been studied to evaluate antidiabetic activity. Curcumin is a phenolic compound present in a herbal spice plant (*Curcuma longa*) whose hepato-protective and hypoglycaemic properties have been widely proved (Pivari et al. 2019). Akolade et al. (2017) reported that chitosan/alginate loaded with 1 mg/mL curcumin inhibited the α-amylase enzyme activity after 30 min of a reaction in an *in vitro* assay. This enzyme is important for controlling diabetes since it metabolizes complex carbohydrates into simple carbohydrates, increasing the postprandial free glucose in the blood. Therefore, in diabetic conditions, it is remarkable to have low α-amylase enzyme activity. Also, in an *in vivo* experiment, Akolade et al. (2017) compared the effect of oral administration of non-encapsulated curcumin and its complexation with chitosan, finding that hyperglycaemia in diabetic-induced rats was significantly reduced using PECs at a concentration of 50 mg/kg b.wt. equivalent of curcumin. In this sense, PECs loaded with curcumin have exhibited suitable *in vitro* release percentages, showing a progressive release process of around 70% after 3 h of simulated gastrointestinal conditions (Tan et al. 2016) and nearly 80% after 120 h under the same conditions (Yan et al. 2017). In this regard, it has been reported (Sarika & James, 2016) that curcumin loaded in PECs was not easily metabolized by cells; otherwise this compound was protected from degradation and hydrolysis as shown in an intracellular uptake assay.

Furthermore, regarding the prevention of diabetes, vitamin D_3 has been reported as a good agent for blood glucose regulation and control of insulin levels (Seida et al. 2014). Therefore, the encapsulation of this vitamin in the form of PECs reflects an appropriate strategy to cover the daily supply of the human body. In this sense, vitamin D_3 has been loaded in a complex based on bovine protein β-lactoglobulin and lysozyme, finding that in an *in vivo* study all rats survived with weight gain; besides, after 1 h of simulated digestion, 75% of the vitamin was released (Diarrassouba et al. 2015). Additionally, Diarrassouba et al. (2015) carried out an *in vivo* assay in which PECs containing vitamin D_3 complexed with β-lactoglobulin and pancreatin were supplied to rats, where the serum level of the vitamin was higher in rats force-fed with the protected vitamin D_3 in comparison to the unprotected vitamin. Hence, this study showed the enhancement of the bioavailability of vitamin D_3 in the form of a polyelectrolyte complex. Similarly, other vitamins have been loaded in complex systems such as vitamin C, which was encapsulated in a Salecan/chitosan polyelectrolyte complex hydrogel, which presented a suitable cytocompatibility and an *in vivo* experiment showed that vitamin C was found in the blood of rats after 6h of oral administration since the hydrogels possessed a slow release during gastrointestinal conditions, giving adequate protection to the vitamin (Hu et al. 2020b). In this sense, it has been reported that the encapsulation of nutraceuticals by the preparation of PECs provides efficient protection and good release percentages of alginate/chitosan-coated zein loaded with resveratrol (Khan et al. 2019), mealworm protein/chitosan loaded with curcumin (Okagu et al. 2020), and xanthan-based polysaccharide/bovine serum albumin loaded with curcumin (Papagiannopoulos & Sklapani, 2021).

On the other hand, PECs have been prepared to treat diabetes, considering the second approach using the food ingredient as a part of the material nanocarrier of targeting drugs, such as insulin. For example, Al-Remawi et al. (2017) stated that the oral administration of nanoparticles of chitosan and lecithin (obtained from egg-yolk) complexed with insulin, in diabetic rats, enhanced the properties of insulin by itself by reducing fasting glucose by 20% and achieving the normal glucose levels after 8 h. Also, these authors performed an *in vitro* dissolution assay, observing that the insulin release from nanoparticles was carried out after 30 min of immersion in 0.1M HCl, indicating that stomach pH will not affect the delivery system during its transit through the stomach. It has been reported that the complexation of chitosan and insulin possess suitable release percentages concerning time in simulated gastric fluids (Bezerra et al. 2021).

15.4.3 Other Human Diseases

Besides the antidiabetic and anticancer properties of PECs based on food ingredients, these nanocarriers have been prepared for other purposes. In this sense, Blachman et al. (2020) analyzed the effect of dermatan sulfate/chitosan complexed with egg-derived tripeptide Ile-Arg-Trp (concentration: 10 μg/mL) on injured endothelial cells simulating damage by a cardiovascular human condition. These authors found that the interaction between the anti-mouse CD44 antibody and dermatan sulphate mediated the nanoparticle uptake, hence, the polyelectrolyte complex possessed anti-inflammatory properties by having the ability to distinguish between healthy and inflamed cells. Also, anti-inflammatory properties have been detected in other nutraceutical compounds such as catechins. These nutraceuticals have been used in formulations of PECs with biopolymers. In this regard, chitosan complexed with (−)-epigallocatechin gallate (EGCG), from green tea leaves, has shown higher radical scavenging than not-loaded EGCG and this carrier system has been reported as a good antibacterial agent (Dahiya et al. 2017). In addition, other PECs have exhibited antibacterial and antifungal activities such as whey protein isolated complexed with guar gum loaded with cumin seed oil with no apparent toxicity against healthy cells (Farshi et al. 2019).

Alternatively, food ingredient-based PECs have been applied in the development of tissue engineering for osteoporosis disease. Dash et al. (2018) prepared complexes with chitosan and ulvan (a polysaccharide extracted from edible algae) enzymatically treated with alkaline phosphatase and found that osteoblastic cells were capable to grow and form an extracellular matrix layer on the prepared scaffold containing the PECs, indicating that biomimetic mineralization was successful. Similarly, another application of PECs is the development of wound dressings such as the complex system composed of chitosan/alginate loaded with thyme oil nanoemulsion, which had no apparent cytotoxicity, possessed a suitable antibacterial activity against *E. coli* and *S. aureus*, and an appropriate release with 60% after the first 6 – 7 h (Hamedi et al. 2019).

15.5 CONCLUSION

PECs have been developed by several research groups for targeted therapy of human diseases such as cancer, diabetes, obesity, cardiovascular condition, and tissue engineering. Among the formulations used for these carrier systems, food ingredients (biopolymers and nutraceuticals) have been exploited due to their biocompatibility and non-toxic nature. In this sense, there have been two approaches for the preparation of PECs using food materials: 1) the food component as one of the charged chains composing the PEC loaded with a synthetic drug, and 2) the food component loaded into the PEC. Regarding the first approach, most PEC formulations were based on the system: chitosan-alginate. On the other hand, curcumin was detected as the most commonly used active agent loaded into PECs, showing suitable properties against human diseases, such as cytotoxicity, improvement of blood glucose regulation, and liver protection. For all cases, PECs exhibited protective and enhancing properties of the active substance.

According to the studies revised in this chapter, PECs can be a promising of drug administration via and an appropriate alternative for cancer and diabetes therapies. However, these carrier systems have been not fully explored with the use of other active agents and different preparation methods. Therefore, further investigation is needed concerning the development of PECs based on food ingredients.

ACKNOWLEDGMENTS

Dr. Evangelina García-Armenta and Dr. Lorenzo Antonio Picos Corrales thank Universidad Autónoma de Sinaloa for the support through projects PROFAPI UAS PRO_A7_016 and PRO_A2_005, respectively.

REFERENCES

Agostini, S.B.N., Malta, I.H.S., Rodrigues, R.F., Freitas, J.T.J., Lino, M.E.D.S., Dos Santos, R.S., Elisei, L.S., Moraes, T.R., Giusto, L.A.D.R., de Oliveira, M.K., Da Silva, J.B., Bruschi, M.L., Dos Santos, A.M., Nogueira, D.A., Novaes, R.D., Pereira, G.R., Galdino, G., & Carvalho, F.C. (2021). Preclinical evaluation of methotrexate-loaded polyelectrolyte complexes and thermosensitive hydrogels as treatment for rheumatoid arthritis. *European Journal of Pharmaceutical Sciences, 163*, 105856. https://doi.org/10.1016/j.ejps.2021.105856

Ahmadi, F., Ghasemi-Kasman, M., Ghasemi, S., Gholamitabar Tabari, M., Pourbagher, R., Kazemi, S., & Alinejad-Mir, A. (2017). Induction of apoptosis in HeLa cancer cells by an ultrasonic-mediated synthesis of curcumin-loaded chitosan-alginate-STPP nanoparticles. *International Journal of Nanomedicine, 12*, 8545–8556. https://doi.org/10.2147/IJN.S146516

Ahmadi Nasab, N., Hassani Kumleh, H., Beygzadeh, M., Teimourian, S., & Kazemzad, M. (2018). Delivery of curcumin by a pH-responsive chitosan mesoporous silica nanoparticles for cancer treatment. *Artificial Cells, Nanomedicine, and Biotechnology, 46*(1), 75–81. https://doi.org/10.1080/21691401.2017.1290648

Akhavan, S., Assadpour, E., Katouzian, I., & Jafari, S.M. (2018). Lipid nano scale cargos for the protection and delivery of food bioactive ingredients and nutraceuticals. *Trends in Food Science & Technology, 74*, 132–146. https://doi.org/10.1016/j.tifs.2018.02.001

Akolade, J.O., Oloyede, H.O.B., & Onyenekwe, P.C. (2017). Encapsulation in chitosan-based polyelectrolyte complexes enhances antidiabetic activity of curcumin. *Journal of Functional Foods, 35*, 584–594. https://doi.org/10.1016/j.jff.2017.06.023

Al-Kurdi, Z., Chowdhry, B., Leharne, S., Al Omari, M., & Badwan, A. (2015). Low molecular weight chitosan–insulin polyelectrolyte complex: characterization and stability studies. *Marine Drugs, 13*(4), 1765–1784. https://doi.org/10.3390/md13041765

Al-Remawi, M., Elsayed, A., Maghrabi, I., Hamaidi, M., & Jaber, N. (2017). Chitosan/lecithin liposomal nanovesicles as an oral insulin delivery system. *Pharmaceutical Development and Technology, 22*(3), 390–398. https://doi.org/10.1080/10837450.2016.1213745

Anirudhan, T.S., Chithra Sekhar, V., Shainy, F., & Thomas, J.P. (2019). Effect of dual stimuli responsive dextran/nanocellulose polyelectrolyte complexes for chemophotothermal synergistic cancer therapy. *International Journal of Biological Macromolecules, 135*, 776–789. https://doi.org/10.1016/j.ijbiomac.2019.05.218

Anirudhan, T.S., Sekhar V.C., & Nair, S.S. (2019). Polyelectrolyte complexes of carboxymethyl chitosan/alginate based drug carrier for targeted and controlled release of dual drug. *Journal of Drug Delivery Science and Technology, 51*, 569–582. https://doi.org/10.1016/j.jddst.2019.03.036

Bezerra, J.M.N.A., Oliveira, A.C.J., Silva-Filho, E.C., Severino, P., Souto, S.B., Souto, E.B., Soares, M.F.L.R., & Soares-Sobrinho, J.L. (2021). The potential role of polyelectrolyte complex nanoparticles based on cashew gum, tripolyphosphate and chitosan for the loading of insulin. *Diabetology, 2*(2), 107–116. https://doi.org/10.3390/diabetology2020009

Blachman, A., Funez, F., Birocco, A.M., Saavedra, S.L., Lázaro-Martinez, M., Camperi, S.A., Glisoni, R., Sosnik, A., & Calabrese, C. (2020). Targeted anti-inflammatory peptide delivery in injured endothelial cells using dermatan sulfate/chitosan nanomaterials. *Carbohydrate Polymers, 230,* 115610. https://doi.org/10.1016/j.carbpol.2019.115610

Bonaccorso, A., Carbone, C., Tomasello, B., Italiani, P., Musumeci, T., Puglisi, G., & Pignatello, R. (2021). Optimization of dextran sulfate/poly-l-lysine based nanogels polyelectrolyte complex for intranasal ovalbumin delivery. *Journal of Drug Delivery Science and Technology, 65,* 102678. https://doi.org/10.1016/j.jddst.2021.102678

Bourganis, V., Karamanidou, T., Kammona, O., & Kiparissides, C. (2017). Polyelectrolyte complexes as prospective carriers for the oral delivery of protein therapeutics. *European Journal of Pharmaceutics and Biopharmaceutics, 111,* 44–60. https://doi.org/10.1016/j.ejpb.2016.11.005

Brar, V., & Kaur, G. (2018). Preparation and characterization of polyelectrolyte complexes of Hibiscus esculentus (okra) gum and chitosan. *International Journal of Biomaterials, 2018,* 1–7. https://doi.org/10.1155/2018/4856287

Cardoso, A.P., Gonçalves, R.M., Antunes, J.C., Pinto, M.L., Pinto, A.T., Castro, F., Monteiro, C., Barbosa, M.A., & Oliveira, M.J. (2015). An interferon-γ-delivery system based on chitosan/poly(γ-glutamic acid) polyelectrolyte complexes modulates macrophage-derived stimulation of cancer cell invasion in vitro. *Acta Biomaterialia, 23,* 157–171. https://doi.org/10.1016/j.actbio.2015.05.022

Dahiya, S., Rani, R., Kumar, S., Dhingra, D., & Dilbaghi, N. (2017). Chitosan-gellan gum bipolymeric nanohydrogels—a potential nanocarrier for the delivery of epigallocatechin gallate. *BioNanoScience, 7*(3), 508–520. https://doi.org/10.1007/s12668-017-0416-0

Dalmoro, A., Sitenkov, A.Y., Cascone, S., Lamberti, G., Barba, A.A., & Moustafine, R.I. (2017). Hydrophilic drug encapsulation in shell-core microcarriers by two stage polyelectrolyte complexation method. *International Journal of Pharmaceutics, 518*(1–2), 50–58. https://doi.org/10.1016/j.ijpharm.2016.12.056

Dash, M., Samal, S.K., Morelli, A., Bartoli, C., Declercq, H.A., Douglas, T.E.L., Dubruel, P., & Chiellini, F. (2018). Ulvan-chitosan polyelectrolyte complexes as matrices for enzyme induced biomimetic mineralization. *Carbohydrate Polymers, 182,* 254–264. https://doi.org/10.1016/j.carbpol.2017.11.016

Delair, T. (2011). Colloidal polyelectrolyte complexes of chitosan and dextran sulfate towards versatile nanocarriers of bioactive molecules. *European Journal of Pharmaceutics and Biopharmaceutics, 78*(1), 10–18. https://doi.org/10.1016/j.ejpb.2010.12.001

Deng, T., Lyon, C.J., Bergin, S., Caligiuri, M.A., & Hsueh, W.A. (2016). Obesity, inflammation, and cancer. *Annual Review of Pathology: Mechanisms of Disease, 11*(1), 421–449. https://doi.org/10.1146/annurev-pathol-012615-044359

Diarrassouba, F., Garrait, G., Remondetto, G., Alvarez, P., Beyssac, E., & Subirade, M. (2015). Food protein-based microspheres for increased uptake of vitamin D3. *Food Chemistry, 173,* 1066–1072. https://doi.org/10.1016/j.foodchem.2014.10.112

Emmanuel, B.D., Abu-Thabit, N.Y., & Ngwuluka, N.C. (2018). Responsive polyelectrolyte complexes based on natural polysaccharides for drug delivery applications. In A. Hamdy Makhlouf & N. Abu-Thabit (Eds.), *Stimuli Responsive Polymeric Nanocarriers for Drug Delivery Applications, Volume 1* (pp. 267–287). Woodhead Publishing. https://doi.org/10.1016/B978-0-08-101997-9.00014-X

Farshi, P., Tabibiazar, M., Ghorbani, M., Mohammadifar, M., Amirkhiz, M.B., & Hamishehkar, H. (2019). Whey protein isolate-guar gum stabilized cumin seed oil nanoemulsion. *Food Bioscience, 28,* 49–56. https://doi.org/10.1016/j.fbio.2019.01.011

Fatima, M.T., Chanchal, A., Yavvari, P.S., Bhagat, S.D., Gujrati, M., Mishra, R.K., & Srivastava, A. (2016). Cell permeating nano-complexes of amphiphilic polyelectrolytes enhance solubility, stability, and anti-cancer efficacy of curcumin. *Biomacromolecules, 17*(7), 2375–2383. https://doi.org/10.1021/acs.biomac.6b00417

Ferreira, D.C.M., Ferreira, S.O., de Alvarenga, E.S., Soares, N. de F.F., dos Reis Coimbra Coimbra, J.S., & de Oliveira, E.B. (2022). Polyelectrolyte complexes (PECs) obtained from chitosan and carboxymethylcellulose: a physicochemical and microstructural study. *Carbohydrate Polymer Technologies and Applications, 3,* 100197. https://doi.org/10.1016/j.carpta.2022.100197

García-Armenta, E., & Gutiérrez-López, G.F. (2022). Fractal microstructure of foods. *Food Engineering Reviews, 14*(1), 1–19. https://doi.org/10.1007/s12393-021-09302-y

García-Armenta, E., Picos-Corrales, L.A., Gutiérrez-López, G.F., Gutiérrez-Dorado, R., Perales-Sánchez, J.X.K., García-Pinilla, S., Reynoso-García, F., Martínez-Audelo, J.M., & Armenta-Manjarrez, M.A. (2021). Preparation of surfactant-free emulsions using amaranth starch modified by reactive extrusion. *Colloids and Surfaces A: Physicochemical and Engineering Aspects, 608*, 125550. https://doi.org/10.1016/j.colsurfa.2020.125550

Garcia-Carrasco, M., Picos-Corrales, L.A., Gutiérrez-Grijalva, E.P., Angulo-Escalante, M.A., Licea-Claverie, A., & Heredia, J.B. (2022). Loading and release of phenolic compounds present in Mexican oregano (Lippia graveolens) in different chitosan bio-oolymeric cationic matrixes. *Polymers, 14*(17), 3609. https://doi.org/10.3390/polym14173609

Geoffrion, L.D., Hesabizadeh, T., Medina-Cruz, D., Kusper, M., Taylor, P., Vernet-Crua, A., Chen, J., Ajo, A., Webster, T.J., & Guisbiers, G. (2020). Naked selenium nanoparticles for antibacterial and anticancer treatments. *ACS Omega, 5*(6), 2660–2669. https://doi.org/10.1021/acsomega.9b03172

Ghorbani, M., Hamishehkar, H., & Tabibiazar, M. (2018). BSA/chitosan polyelectrolyte complex: a platform for enhancing the loading and cancer cell-uptake of resveratrol. *Macromolecular Research, 26*(9), 808–813. https://doi.org/10.1007/s13233-018-6112-2

Hamed, A.A., Saad, G.R., Abdelhamid, I.A., Elwahy, A.H.M., Abdel-Aziz, M.M., & Elsabee, M.Z. (2022). Chitosan Schiff bases-based polyelectrolyte complexes with graphene quantum dots and their prospective biomedical applications. *International Journal of Biological Macromolecules, 208*, 1029–1045. https://doi.org/10.1016/j.ijbiomac.2022.03.199

Hamedi, H., Moradi, S., Tonelli, A.E., & Hudson, S.M. (2019). Preparation and characterization of chitosan–alginate polyelectrolyte complexes loaded with antibacterial thyme oil nanoemulsions. *Applied Sciences, 9*(18), 3933. https://doi.org/10.3390/app9183933

Harrison, T.D., Yunyaeva, O., Borecki, A., Hopkins, C.C., de Bruyn, J.R., Ragogna, P.J., & Gillies, E.R. (2020). Phosphonium polyelectrolyte complexes for the encapsulation and slow release of ionic cargo. *Biomacromolecules, 21*(1), 152–162. https://doi.org/10.1021/acs.biomac.9b01115

Hu, Q., Wang, T., Zhou, M., Xue, J., & Luo, Y. (2016). Formation of redispersible polyelectrolyte complex nanoparticles from gallic acid-chitosan conjugate and gum arabic. *International Journal of Biological Macromolecules, 92*, 812–819. https://doi.org/10.1016/j.ijbiomac.2016.07.089

Hu, X., Wang, Y., Zhang, L., & Xu, M. (2020a). Construction of self-assembled polyelectrolyte complex hydrogel based on oppositely charged polysaccharides for sustained delivery of green tea polyphenols. *Food Chemistry, 306*, 125632. https://doi.org/10.1016/j.foodchem.2019.125632

Hu, X., Wang, Y., Zhang, L., & Xu, M. (2020b). Formation of self-assembled polyelectrolyte complex hydrogel derived from salecan and chitosan for sustained release of Vitamin C. *Carbohydrate Polymers, 234*, 115920. https://doi.org/10.1016/j.carbpol.2020.115920

Insua, I., Majok, S., Peacock, A.F.A., Krachler, A.M., & Fernandez-Trillo, F. (2017). Preparation and antimicrobial evaluation of polyion complex (PIC) nanoparticles loaded with polymyxin B. *European Polymer Journal, 87*, 478–486. https://doi.org/10.1016/j.eurpolymj.2016.08.023

Kalenichenko, D., Nifontova, G., Karaulov, A., Sukhanova, A., & Nabiev, I. (2021). Designing functionalized polyelectrolyte microcapsules for cancer treatment. *Nanomaterials, 11*(11), 3055. https://doi.org/10.3390/nano11113055

Khan, M.A., Yue, C., Fang, Z., Hu, S., Cheng, H., Bakry, A.M., & Liang, L. (2019). Alginate/chitosan-coated zein nanoparticles for the delivery of resveratrol. *Journal of Food Engineering, 258*, 45–53. https://doi.org/10.1016/j.jfoodeng.2019.04.010

Khatun, B., Banik, N., Hussain, A., Ramteke, A., & Maji, T. (2018). Genipin crosslinked curcumin loaded chitosan/montmorillonite K-10 (MMT) nanoparticles for controlled drug delivery applications. *Journal of Microencapsulation, 35*(5), 439–453. https://doi.org/10.1080/02652048.2018.1524524

Kulkarni, A.D., Vanjari, Y.H., Sancheti, K.H., Patel, H.M., Belgamwar, V.S., Surana, S.J., & Pardeshi, C.V. (2016). Polyelectrolyte complexes: mechanisms, critical experimental aspects, and applications. *Artificial Cells, Nanomedicine, and Biotechnology, 44*(7), 1615–1625. https://doi.org/10.3109/21691401.2015.1129624

Kuppusamy, P., Yusoff, M.M., Maniam, G.P., Ichwan, S.J.A., Soundharrajan, I., & Govindan, N. (2014). Nutraceuticals as potential therapeutic agents for colon cancer: a review. *Acta Pharmaceutica Sinica B*, *4*(3), 173–181. https://doi.org/10.1016/j.apsb.2014.04.002

Liu, S., Lin, T.P., Li, D., Leamer, L., Shan, H., Li, Z., Gabbaï, F.P., & Conti, P.S. (2013). Lewis acid-assisted isotopic 18 F-19 F exchange in BODIPY dyes: facile generation of positron emission tomography/fluorescence dual modality agents for tumor imaging. *Theranostics*, *3*(3), 181–189. https://doi.org/10.7150/thno.5984

Lybaert, L., Ryu, K.A., De Rycke, R., Chon, A.C., De Wever, O., Vermaelen, K.Y., Esser-Kahn, A., & De Geest, B.G. (2017). Polyelectrolyte-enrobed cancer cells in view of personalized immune-therapy. *Advanced Science*, *4*(6), 1700050. https://doi.org/10.1002/advs.201700050

Malviya, R., Sharma, P.K., & Dubey, S.K. (2020). Efficiency of self-assembled etoricoxib containing polyelectrolyte complex stabilized cubic nanoparticles against human cancer cells. *Precision Medical Sciences*, *9*(1), 9–22. https://doi.org/10.1002/prm2.12004

Manzanares-Guevara, L.A., Licea-Claverie, A., Oroz-Parra, I., Bernaldez-Sarabia, J., Diaz-Castillo, F., & Licea-Navarro, A.F. (2020). Smart nanoformulation based on stimuli-responsive nanogels and curcumin: promising therapy against colon cancer. *ACS Omega*, *5*(16), 9171–9184. https://doi.org/10.1021/acsomega.9b04390

Marciel, A.B., Chung, E.J., Brettmann, B.K., & Leon, L. (2017). Bulk and nanoscale polypeptide based polyelectrolyte complexes. *Advances in Colloid and Interface Science*, *239*, 187–198. https://doi.org/10.1016/j.cis.2016.06.012

Meka, V.S., Sing, M.K.G., Pichika, M.R., Nali, S.R., Kolapalli, V.R.M., & Kesharwani, P. (2017). A comprehensive review on polyelectrolyte complexes. *Drug Discovery Today*, *22*(11), 1697–1706. https://doi.org/10.1016/j.drudis.2017.06.008

Minkal, Ahuja, M., & Bhatt, D.C. (2018). Polyelectrolyte complex of carboxymethyl gum katira-chitosan: preparation and characterization. *International Journal of Biological Macromolecules*, *106*, 1184–1191. https://doi.org/10.1016/j.ijbiomac.2017.08.128

Neděla, V., Tihlaříková, E., Maxa, J., Imrichová, K., Bučko, M., & Gemeiner, P. (2020). Simulation-based optimisation of thermodynamic conditions in the ESEM for dynamical in-situ study of spherical polyelectrolyte complex particles in their native state. *Ultramicroscopy*, *211*, 112954. https://doi.org/10.1016/j.ultramic.2020.112954

Neitzel, A.E., De Hoe, G.X., & Tirrell, M.V. (2021). Expanding the structural diversity of polyelectrolyte complexes and polyzwitterions. *Current Opinion in Solid State and Materials Science*, *25*(2), 100897. https://doi.org/10.1016/j.cossms.2020.100897

Newham, G., Mathew, R.K., Wurdak, H., Evans, S.D., & Ong, Z.Y. (2021). Polyelectrolyte complex templated synthesis of monodisperse, sub-100 nm porous silica nanoparticles for cancer targeted and stimuli-responsive drug delivery. *Journal of Colloid and Interface Science*, *584*, 669–683. https://doi.org/10.1016/j.jcis.2020.10.133

Nifontova, G., Ramos-Gomes, F., Baryshnikova, M., Alves, F., Nabiev, I., & Sukhanova, A. (2019). Cancer cell targeting with functionalized quantum dot-encoded polyelectrolyte microcapsules. *Frontiers in Chemistry*, *7, 34*. https://doi.org/10.3389/fchem.2019.00034

Okagu, O.D., Verma, O., McClements, D.J., & Udenigwe, C.C. (2020). Utilization of insect proteins to formulate nutraceutical delivery systems: encapsulation and release of curcumin using mealworm protein-chitosan nano-complexes. *International Journal of Biological Macromolecules*, *151*, 333–343. https://doi.org/10.1016/j.ijbiomac.2020.02.198

Papagiannopoulos, A. (2021). Current research on polyelectrolyte nanostructures: from molecular interactions to biomedical applications. *Macromol*, *1*(2), 155–172. https://doi.org/10.3390/macromol1020012

Papagiannopoulos, A., & Sklapani, A. (2021). Xanthan-based polysaccharide/protein nanoparticles: preparation, characterization, encapsulation and stabilization of curcumin. *Carbohydrate Polymer Technologies and Applications*, *2*, 100075. https://doi.org/10.1016/j.carpta.2021.100075

Picos-Corrales, L.A., Garcia-Carrasco, M., Licea-Claverie, A., Chavez-Santoscoy, R.A., & Serna-Saldívar, S.O. (2019). NIPAAm-containing amphiphilic block copolymers with tailored LCST: aggregation behavior, cytotoxicity and evaluation as carriers of indomethacin, tetracycline and doxorubicin. *Journal of Macromolecular Science, Part A*, *56*(8), 759–772. https://doi.org/10.1080/10601325.2019.1586440

Picos-Corrales, L.A., Licea-Claveríe, A., & Arndt, K.-F. (2014). Bisensitive core–shell nanohydrogels by e-Beam irradiation of micelles. *Reactive and Functional Polymers, 75*, 31–40. https://doi.org/10.1016/j.reactfunctpolym.2013.11.010

Picos-Corrales, L.A., Morales-Burgos, A.M., Ruelas-Leyva, J.P., Crini, G., García Armenta, E., Jimenez-Lam, S.A., Ayón-Reyna, L.E., Rocha-Alonzo, F., Calderón-Zamora, L., Osuna-Martínez, U., Calderón-Castro, A., De-Paz-Arroyo, G., & Inzunza-Camacho, L.N. (2023). Chitosan as an outstanding polysaccharide improving health-commodities of humans and environmental protection. *Polymers, 15*(3), 526. https://doi.org/10.3390/polym15030526

Pivari, F., Mingione, A., Brasacchio, C., & Soldati, L. (2019). Curcumin and type 2 diabetes mellitus: prevention and treatment. *Nutrients, 11*(8), 1837. https://doi.org/10.3390/nu11081837

Qin, Y., Yun, D., Xu, F., Chen, D., Kan, J., & Liu, J. (2021). Smart packaging films based on starch/polyvinyl alcohol and Lycium ruthenicum anthocyanins-loaded nano-complexes: functionality, stability and application. *Food Hydrocolloids, 119*, 106850. https://doi.org/10.1016/j.foodhyd.2021.106850

Sankalia, M.G., Mashru, R.C., Sankalia, J.M., & Sutariya, V.B. (2007). Reversed chitosan–alginate polyelectrolyte complex for stability improvement of alpha-amylase: optimization and physicochemical characterization. *European Journal of Pharmaceutics and Biopharmaceutics, 65*(2), 215–232. https://doi.org/10.1016/j.ejpb.2006.07.014

Sarika, P.R., & James, N.R. (2016). Polyelectrolyte complex nanoparticles from cationised gelatin and sodium alginate for curcumin delivery. *Carbohydrate Polymers, 148*, 354–361. https://doi.org/10.1016/j.carbpol.2016.04.073

Seida, J.C., Mitri, J., Colmers, I.N., Majumdar, S.R., Davidson, M.B., Edwards, A.L., Hanley, D.A., Pittas, A.G., Tjosvold, L., & Johnson, J.A. (2014). Effect of vitamin D3 supplementation on improving glucose homeostasis and preventing diabetes: a systematic review and meta-analysis. *The Journal of Clinical Endocrinology & Metabolism, 99*(10), 3551–3560. https://doi.org/10.1210/jc.2014-2136

Shah, S., & Leon, L. (2021). Structural dynamics, phase behavior, and applications of polyelectrolyte complex micelles. *Current Opinion in Colloid & Interface Science, 53*, 101424. https://doi.org/10.1016/j.cocis.2021.101424

Silva, C.L., Pereira, J.C., Ramalho, A., Pais, A.A.C.C., & Sousa, J.J.S. (2008). Films based on chitosan polyelectrolyte complexes for skin drug delivery: development and characterization. *Journal of Membrane Science, 320*(1–2), 268–279. https://doi.org/10.1016/j.memsci.2008.04.011

Siyawamwaya, M., Choonara, Y.E., Bijukumar, D., Kumar, P., Du Toit, L.C., & Pillay, V. (2015). A review: overview of novel polyelectrolyte complexes as prospective drug bioavailability enhancers. *International Journal of Polymeric Materials and Polymeric Biomaterials, 64*(18), 955–968. https://doi.org/10.1080/00914037.2015.1038816

Souza, P.R., Vilsinski, B.H., Nunes, C.S., Bonkovoski, L.C., Garcia, F., Nakamura, C.V., Caetano, W., Valente, A.J.M., Martins, A.F., & Muniz, E.C. (2021). Application of a polyelectrolyte complex based on biocompatible polysaccharides for colorectal cancer inhibition. *Carbohydrate Research, 499*, 108194. https://doi.org/10.1016/j.carres.2020.108194

Tabandeh, S., & Leon, L. (2019). Engineering peptide-based polyelectrolyte complexes with increased hydrophobicity. *Molecules, 24*(5), 868. https://doi.org/10.3390/molecules24050868

Tahrani, A.A., Barnett, A.H., & Bailey, C.J. (2016). Pharmacology and therapeutic implications of current drugs for type 2 diabetes mellitus. *Nature Reviews Endocrinology, 12*(10), 566–592. https://doi.org/10.1038/nrendo.2016.86

Tan, C., Xie, J., Zhang, X., Cai, J., & Xia, S. (2016). Polysaccharide-based nanoparticles by chitosan and gum arabic polyelectrolyte complexation as carriers for curcumin. *Food Hydrocolloids, 57*, 236–245. https://doi.org/10.1016/j.foodhyd.2016.01.021

Teixeira-Costa, B.E., Silva Pereira, B.C., Lopes, G.K., & Tristão Andrade, C. (2020). Encapsulation and antioxidant activity of assai pulp oil (Euterpe oleracea) in chitosan/alginate polyelectrolyte complexes. *Food Hydrocolloids, 109*, 106097. https://doi.org/10.1016/j.foodhyd.2020.106097

Tonelli, A., Narayanan, G., & Gurarslan, A. (2018). Host–guest polymer complexes. *Polymers, 10*(8), 911. https://doi.org/10.3390/polym10080911

WHO. (2022). *Cancer.* World Health Organization. www.who.int/news-room/fact-sheets/detail/cancer

Wolin, K.Y., Carson, K., & Colditz, G.A. (2010). Obesity and cancer. *The Oncologist, 15*(6), 556–565. https://doi.org/10.1634/theoncologist.2009-0285

Yan, J.K., Qiu, W.Y., Wang, Y.Y., & Wu, J.Y. (2017). Biocompatible polyelectrolyte complex nanoparticles from lactoferrin and pectin as potential vehicles for antioxidative curcumin. *Journal of Agricultural and Food Chemistry, 65*(28), 5720–5730. https://doi.org/10.1021/acs.jafc.7b01848

Yan, J.K., Qiu, W.Y., Wang, Y.Y., Wu, L.X., & Cheung, P.C.K. (2018). Formation and characterization of polyelectrolyte complex synthesized by chitosan and carboxylic curdlan for 5-fluorouracil delivery. *International Journal of Biological Macromolecules, 107*, 397–405. https://doi.org/10.1016/j.ijbiomac.2017.09.004

Yang, J., Shen, M., Wu, T., Luo, Y., Li, M., Wen, H., & Xie, J. (2020). Role of salt ions and molecular weights on the formation of Mesona chinensis polysaccharide-chitosan polyelectrolyte complex hydrogel. *Food Chemistry, 333*, 127493. https://doi.org/10.1016/j.foodchem.2020.127493

Zhang, X., Liu, D., Jin, T.Z., Chen, W., He, Q., Zou, Z., Zhao, H., Ye, X., & Guo, M. (2021). Preparation and characterization of gellan gum-chitosan polyelectrolyte complex films with the incorporation of thyme essential oil nanoemulsion. *Food Hydrocolloids, 114*, 106570. https://doi.org/10.1016/j.foodhyd.2020.106570

Zhang, X., Liang, T., & Ma, Q. (2021). Layer-by-Layer assembled nano-drug delivery systems for cancer treatment. *Drug Delivery, 28*(1), 655–669. https://doi.org/10.1080/10717544.2021.1905748

Zhao, L., Jin, W., Cruz, J.G., Marasini, N., Khalil, Z.G., Capon, R.J., Hussein, W.M., Skwarczynski, M., & Toth, I. (2020). Development of polyelectrolyte complexes for the delivery of peptide-based subunit vaccines against group A Streptococcus. *Nanomaterials, 10*(5), 823. https://doi.org/10.3390/nano10050823

Chapter 16

Starch Nanocrystals as Pickering Emulsion Stabilizers

Raúl E. López-Hernández
Ilse Monroy-Rodríguez
Sandra E. García-Solís
Maribel Cornejo-Mazón
Georgina Calderón-Domínguez
and Gustavo F. Gutiérrez-López

CONTENTS

16.1 INTRODUCTION

Pickering emulsions are dispersed systems stabilized by solid particles that are wetted by both liquid phases, and, in recent years, these emulsions are progressively gaining attention in food science and technology due to their higher stability when compared to surfactant stabilized emulsions. This is further discussed in section 16.4.1. Starch and its derivatives are popular materials used as stabilizers of these emulsions and starch nanocrystals have been used for this purpose (López-Hernández et al. 2022). Starch is a storage polysaccharide present in plants; it is the second most abundant material of biological origin in nature after cellulose and is present in plant roots, stems, legumes, unripe fruits, and seeds of basic crops such as rice, corn, wheat, barley, potato, amaranth, and quinoa, among others (Jia et al. 2023). Due to being a cheap, natural, and biodegradable polymer, it is considered a compound of major interest in industry

DOI: 10.1201/9781003332558-19

(Dominguez-Ayala et al. 2022). In this respect, the use of starch nanocrystals as an ingredient of Pickering emulsions is promising and needs to be extensively studied aiming at obtaining the best properties of starch nanocrystals (SNCs) to be used in the preparation of these emulsions (López-Hernández et al. 2022). This will be further discussed in sections 16.4.1, 16.4.1.1. and 16.4.1.2.

Starch is a polymer composed of a mixture of two main components: amylose, a linear chain of glucose linked by α(1→4) bonds, and amylopectin, branched molecules linked by α(1→6) bonds. According to the proportion of amylose and amylopectin in the granule, starches are classified as standard or 'normal' starches: those with 25–30% amylose and 70–75% amylopectin content; starches composed almost exclusively of amylopectin (98–99%) are known as waxy starches; and a last group comprised by starches with higher amylose contents (50–70%) (Bello-Perez et al. 2020).

Starch can be isolated from many plant-based structures; due to its botanical origin, granules have a great variation in terms of size (0.5 to 100 μm), shape (round, lenticular, polygonal), size distribution (uni- or bimodal), and composition (protein content, lipid, amylose/amylopectin ratio), even among same species, the growing conditions of the source material have an influence on its characteristics. (Tester et al., 2004; Alcázar-Alay & Meireles, 2015). **Table 16.1** summarizes some properties of starches from different sources. The worldwide availability of starch and extensive studies on its applications and possibilities of obtaining derivatives as starch nanocrystals make

TABLE 16.1 MOST COMMON STARCH SOURCES AND THEIR MORPHOLOGICAL AND CRYSTALLINE PROPERTIES

Source	Size (μm)	Shape	XRD Pattern	References
Potato (*Imilla blanca*)	31.6 Unimodal distribution	Oval and ellipsoidal	B-type	Martínez et al. (2019)
Corn	3.55	Irregular polygonal	A-type	Sathyan & Nisha, (2022) Zhang et al. (2022b)
Wheat	1) 10–35 2) 10–10 Bimodal distribution	Disk shape Spherical	A-type B-type	Zhang et al. (2021)
Rice	5.87 (D_{50}) Unimodal	Polyhedral and Irregular	A-type	Zhang et al. (2023)
Amaranth	1–3 Unimodal distribution	Polyhedral	A-type	Gutiérrez-Cortez et al. (2021)
Barley	1) 20–30 2) 10 Bimodal distribution	Lenticular Spherical	A-type	Park et al. (2023)
Sorghum	14.5 Unimodal distribution	Irregular spherical or polygonal	A-type	Yan et al. (2023)

this polymer suitable for obtaining nanocrystals aimed at preparation of Pickering emulsions to be used in healthy food preparations.

16.2 USES OF STARCH

Starch can be used in its native form or undergo chemical modifications to confer its specific properties (He et al. 2023). Native starches are not always suitable for specific industrial applications due to functional limitations, such as low resistance to shear stress, poor thermal properties, and a high tendency to retrograde (He et al. 2023; Kumari & Sit, 2023).

The chemical modification of starches has been the subject of intensive research over the last century. Many types of chemically modified starches have been developed and applied for the food, paper, and textile industries; the most widely used method of chemical modification is acid hydrolysis (Compart et al. 2023).

This polymer can be used in a wide range of applications in food products, generally as a binding agent in formed meat, snacks, seasonings, battered and breaded foods; crisping in fried and baked foods; as dustings for chewing gum and bakery products; emulsifier in beverages and canned food; fat replacer in ice creams and salad dressings; flavor encapsulating agent; thickening for gravies and soups; moisture retainer in cakes, texturizing agent, among others (Omoregie Egharevba, 2020; Pavela et al. 2020; Ogori & Taofeek, 2022). A growing area of application of starch is the production of starch nanocrystals, which will be discussed in the next section.

16.3 STARCH NANOCRYSTALS

In nature, starch is bio-synthesized in the form of semi-crystalline granules, in which amylose and amylopectin are densely packed. It has been proposed that inside the granule, amylose and amylopectin are arranged in the form of rings that are composed of amorphous and crystalline lamellae, have 100 – 400 nm in thickness, and grow alternately surrounding the central point of the granule called hilum (Dominguez-Ayala et al. 2022; Šárka et al. 2023).

The crystalline regions of the starch granules can be isolated by acid hydrolysis using hydrochloric or sulfuric acid. It is believed that, at temperatures below the gelatinization temperature (<60°C), the acid preferentially hydrolyzes the amorphous regions of the granule; and at the end of the hydrolysis, solids of nanometric scale and enhanced crystallinity known as starch nanocrystals (SNCs) are obtained (Lin et al. 2011) and may be used as stabilizers of Pickering emulsions (López-Hernández et al. 2022) as described in section 16.4.1.

It has been stated that there are three types of crystalline structures: type A, present in cereal starches, type B characteristic of tubers, and type C found in legumes (Alcázar-Alay & Meireles, 2015). **Table 16.2** summarizes some works aimed to obtain starch nanocrystals and their main results.

Putaux et al. (2003) were the first to produce starch nanocrystals after treating waxy maize starch granules with 2.2 N HCl at 36°C for 2 weeks. A short time later, Angellier et al. (2004) reported that it was possible to obtain nanocrystals with similar characteristics. These authors hydrolyzed waxy corn starch by using sulfuric acid 3.16 M at 40°C; and under these conditions the reaction time to obtain nanocrystals was reduced to 5 days. Currently, these conditions are standard and

TABLE 16.2 STARCH NANOCRYSTALS (SNCS) OBTAINED FROM DIFFERENT SOURCES AND THEIR CHARACTERIZATION

Source	Process	Temperature	Time	Yield	Size	Reference
Amaranth	Acid hydrolysis (H_2SO_4 3.16 M)	40°C	3, 5 & 10 days	17.3% (3 days) 9.3% (5 days) 3.6% (10 days)	484 nm (3 days) 374 nm (5 days) 376 nm (10 days) (DLS)	Sanchez de la Concha et al. (2018)
Cassava	Acid hydrolysis (H_2SO_4 3.16 M)	37 °C	7 days	-	10–100 nm (FESEM)	Thangrongthong et al. (2020)
Corn	Acid hydrolysis (H_2SO_4 3.16 M)	40 °C	-	-	≈ 200 nm (FESEM)	Mittal et al. (2021)
Corn	Acid hydrolysis (H_2SO_4 3.16 M)	40 °C	6 days	-	10–100 nm (FESEM)	Priyan et al. (2021)
Corn	Acid hydrolysis (H_2SO_4 18% v/v)	40 °C	5 days	-	102 (DLS)	Kalateh-Seifari et al. (2021)
Rice	Acid hydrolysis (H_2SO_4 3.16 M)	40 °C	7 days	-	300 nm (DLS)	Martins et al. (2022)
Rice	Acid hydrolysis (H_2SO_4 3.16 M)	40 °C	7 days	9.7%	203.2 nm (DLS)	Xiong et al. (2021)
Quinoa	Acid hydrolysis (H_2SO_4 3.16 M)	40°C	5 days	-	~50–100 (AFM)	Velásquez-Castillo et al. (2023)
Waxy corn	Ball milling and acid hydrolysis (H_2SO_4 3.16 M)	0, 15, 30, 45, 60, 75, 90 min (milling) 1–5 days (acid hydrolysis)	40°C	13.5% (5 days of hydrolysis)	88.6 ± 5.7 (30-min milling time + 3 days of acid hydrolysis) 104.0 ± 7.5 (5 days of acid hydrolysis)	Dai et al. (2018)

TABLE 16.2 (CONTINUED)

Source	Process	Temperature	Time	Yield	Size	Reference
Waxy maize	Acid hydrolysis (1:1 mixture, H_2SO_4 & HCl 3.16 M, final solution concentration 6.32 M)	40 °C	1 hour	-	≈ 200	Li et al. (2023)

*DLS: Dynamic Light Scattering
**FESEM: Field emission gun scanning electron microscopy
***AFM: Atomic force microscopy

the most commonly used to obtain nanocrystals to be used as Pickering emulsions stabilizers (Peito et al. 2022).

Xiong et al. (2021) produced SNCs from rice starch and evaluated their morphological and rheological properties. They concluded that rice SNCs possessed excellent structure and colloidal properties for their use as a reinforcer for new biodegradable materials, food packages, fat replacers, thickeners, and Pickering emulsion stabilizers.

Yang et al. (2023a) evaluated the effect of rice SNCs and acetylated SNCs (0–25 %) on the properties of rice starch-based films blended with glycerol. They reported that by incorporating SNCs (both acetylated and non-acetylated) to the blend, up to 10% results in films with enhanced tensile strength; and the large number of hydroxyl groups present in SNCs caused their aggregation, thus leading to micro-phase separation of the structure. Likewise, Martins et al. (2022) incorporated potato SNCs along with glycerol and rice starch to elaborate films in order to analyze their physicochemical properties.

Similarly, Velásquez-Castillo et al. (2023) employed quinoa starch nanocrystals to enhance cassava starch films. They observed that by adding 5% of SNCs to formulation, the films' tensile strength improved from 6.5 to 16.5 MPa, and other properties such as roughness, water contact angle (important towards preparation of Pickering emulsions), and opacity also increased. Mittal et al., (2021) studied the effect of SNCs on the tensile strength of nanocomposite films made of polyvinyl alcohol, glycerol, and corn starch. The authors found that adding 5% of SNCs to the matrix significantly increased the tensile strength from 11.12 to 18.25. Also, SNCs improved the biodegradability of the films, which initially showed a 19.26% weight loss after 120 days of degradation, but when mixed with 7% of SNCs increased loss weight to 32.4%. This was attributed to the increased crystallinity of the SNCs and their good dispersion into the PVA/S matrix. In general, the rheological characteristics of SNCs may be used as indicators of their suitability for the preparation of Pickering emulsions (López-Hernández et al. 2022).

Thangrongthong et al. (2020) evaluated the effect of cassava SNCs as wall material reinforcement for the encapsulation of lactic-acid bacteria (LAB) and evaluated their survival under simulated gastrointestinal conditions. They found that LAB encapsulated in a matrix of SNCs-alginate had a survival higher survival rate (94.9%) under simulated gastrointestinal conditions and during storage (4°C) compared to uncovered cells and also discovered that LAB can survive up to 4 weeks at 25°C. In the case of Pickering emulsions, SNCs are used as protective wall material stabilizer (Li et al. 2022).

Priyan et al. (2021) investigated the potential damage of SNCs in HEK-293 cells. They found that SNCs did not affect the viability of HEK-293 cells at concentrations ranging from 0 to 25 μg/ml in a cell morphology assay. Morphological observations of the HEK-293 cells showed that SNC treatment did not cause any cytotoxicity, proving the safe use of SNCs.

Other studies have looked to optimize the obtention process of SNCs. Dai et al. (2018) evaluated the combination of ball milling with acid hydrolysis to obtain SNCs. After a combination of 30 min ball milling and 3 days of acid hydrolysis, it was possible to obtain SNCs of 88.6 nm with a yield of 19.3%. Although promising, mechanical methods like ball milling are unsuitable for the preparation of SNCs due to the destruction of the starch crystalline structure. Recently, Li et al. (2023), developed an innovator method to obtain SNCs, which consisted of using hydrochloric acid along with sulfuric acid, saving up to 83% of time. The resulting SNCs showed higher crystallinity and smaller size than SNCs obtained by using only sulfuric acid; no differences in morphology or structure were observed.

16.4 EMULSIONS

An emulsion is a system that consists of a mixture of liquids that are immiscible with each other (the most common in food science are oil and water), which combine to form a macroscopically homogeneous, but microscopically heterogeneous system, in which one of the liquids is immersed within the other in the form of small droplets (Zhou et al. 2021; Watts et al. 2023).

It is possible to increase their kinetic stability through the use of emulsifying agents. The substances used the most in the food industry are low molecular weight molecules, such as phospholipids, proteins, and some polysaccharides (Kumar et al. 2022).

Polysaccharides are an alternative to conventional proteins and phospholipid emulsifiers. Most polysaccharides are highly hydrophilic molecules but can be modified by joining non-polar groups to their structure, reducing their hydrophilic character, thus enhancing their emulsifying properties (Ozturk & McClements, 2016). The mechanism through which polysaccharides stabilize droplets is by steric hindrance; they have large hydrophilic groups that protect the droplets, avoiding their aggregation. Moreover, polysaccharides form a network-like structure, which reduces the movement of the droplets and increases the viscosity of the system, improving the stability of the emulsion. Another advantage of polysaccharides is that, unlike proteins or lecithins, they are stable against changes in pH, ionic strength, and temperature (Rayner et al. 2014; Ozturk & McClements, 2016). Furthermore, polysaccharides as starch have been used to prepare stabilizers of Pickering emulsions (Leal-Castañeda et al. 2018; López-Hernández et al. 2022).

16.4.1 Pickering Emulsions

Pickering emulsions are emulsions of both oil-in-water (O/W) and water-in-oil (W/O) types that have the basic properties of the emulsions stabilized by surfactants; yet the agent used to stabilize the interface of the droplets is not an amphiphilic molecule. In Pickering emulsions, the stabilization of the droplets is achieved by utilizing solid particles as SNCs (López-Hernández et al. 2022). The idea of using solid particles for emulsion stabilization was first reported more than a century ago by Pickering (1907). They observed that solid particles when adsorbed to an interface were wetted by both phases, and in this way get the droplets stabilized. The stabilization mechanism of Pickering emulsions is relatively simple; the dispersed particles adsorbed at the droplet oil–water interface to form a mechanical/steric barrier that prevents emulsion droplets against coalescence (Peito et al. 2022).

For solid particles to act as stabilizers in an emulsion, there are three criteria to take into account: 1) the particles must be partially wetted both by the continuous and dispersed phases, but they must not be soluble in any of them; 2) the particles must preserve adequate partial wettability to be adsorbed at the water–oil interface; and 3) the particle size must be smaller than the droplet size of the emulsion to be obtained (Peito et al. 2022; Sun et al. 2022). These criteria are met by a number of materials, including SNCs, which are gaining attention for the preparation of Pickering emulsions (López-Hernández et al. 2022).

Compared with conventional emulsions, Pickering emulsions have greater stability against phenomena such as coalescence and Ostwald ripening due to the adsorption practically irreversible of the particles in the water–oil interface, which means a decrease in droplet size variation and longer shelf life (Yang et al. 2021). In Pickering emulsions, it is also possible to incorporate thermolabile or oxygen-susceptible bioactive compounds such as vitamins, drugs, or even probiotics since the barrier formed by the particles around the droplets provides greater protection against temperature and oxidation compared to that provided for surfactants, which are molecules susceptible to factors such as pH, temperature, and ionic strength (Tavernier et al. 2016; Wu & Ma, 2016).

Another advantage of Pickering emulsion is their low toxicity; when biological materials are used to stabilize emulsions, the risk of toxicity is low (Wang et al. 2023c). By eliminating surfactants of emulsions, toxicity is further reduced since some of these compounds have triggered side effects, such as hypersensitivity reactions, neurotoxicity, and damage to the cell membrane (Wu & Ma, 2016; Xiao et al. 2016).

The physicochemical properties of Pickering emulsions are related to the particle used, such as wettability, size, concentration, and shape (Pan et al. 2023)

The composition of the oil phase also contributes to the properties of the emulsion. Liang et al. (2016) reported that a greater quantity of waxy cornstarch nanocrystals with a determined wettability was needed to stabilize an emulsion prepared with long-chain triglycerides compared to emulsions made with short- or medium-chain triglycerides.

The size of the particles that can be used to stabilize the emulsions varies from nanometers to several microns. For successful stabilization, the particles must be several orders of size smaller than the emulsified droplets for the particles to be properly located around the droplets; and in this respect, SNCs are suitable stabilizers of Pickering emulsions (Liang et al. 2016; López-Hernández et al. 2022). Additionally, the size of the emulsion droplets is, in part, determined by the number of particles dispersed within the continuous phase (Tavernier et al. 2016).

16.4.1.1 Wettability

During homogenization, the dispersed solid particles come into contact with the Pickering emulsion droplets containing the bioactive compounds, adsorbing practically irreversibly into the oil–water interface (Wang et al. 2023c), as shown in **Figure 16.1**.

Particle adsorption occurs depending on its affinity to oil or aqueous phase, which this is called particle wettability. Contact angle is a measure of the wettability of a solid surface by a liquid and is commonly used to characterize the interaction between a liquid and a solid. In Pickering emulsions, the contact angle wettability is characterized by three-phase contact angles (θ) (Ming et al. 2023; Muiz et al. 2023). Also, θ is related to the free energy balance on the surface between

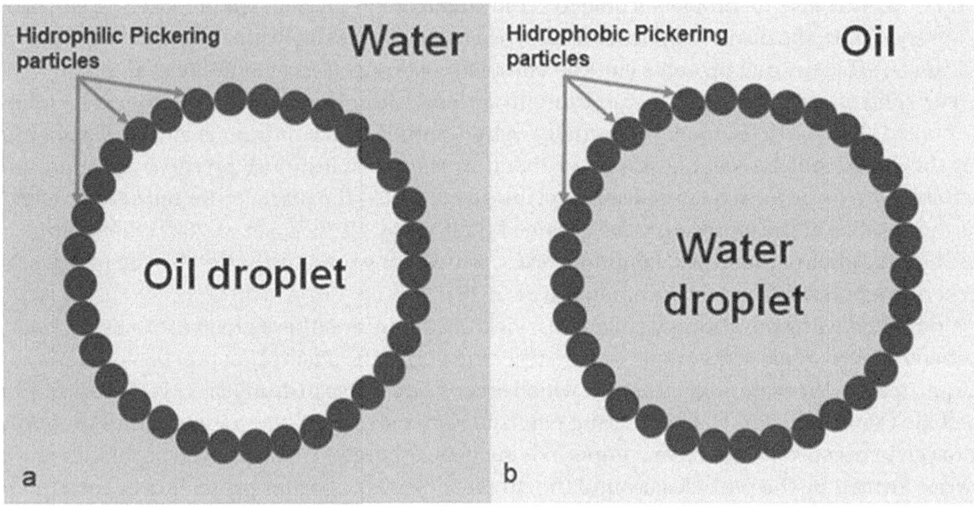

Figure 16.1 Representation of pickering particles covering droplets in a) O/W emulsions and b) W/O emulsions.

particle–oil, particle–water, and oil–water interface. Contact angle in Pickering emulsions is described by Young's equation as depicted in Equation 1(Wang et al. 2023c):

$$\cos\theta = \frac{\left(\gamma_{po} - \gamma_{pw}\right)}{\gamma_{ow}} \quad (1)$$

where γ_{po}, $\gamma_{pw,}$ and γ_{ow} represent the interfacial tension between the oil phase and the particles, the interfacial tension between particles and the water phase, and the interfacial tension between the water phase and oil phase, respectively (Wang et al. 2023c).

Also, the huge energy required to remove them from the oil–phase interface is described by Equation 2 (Ming et al., 2023):

$$E = \pi r^2 \gamma_{ow} (1 - \cos\theta)^2 \quad (2)$$

where E is the energy required to detach the particle from the interface; γ_{ow} is the oil–water interfacial tension, *r* is particle radius, and θ is contact angle.

In Pickering emulsions, the contact angle may be measured to help understand the interactions between the solid particles and the liquid phases. Pickering emulsions are stable when the particle's contact angle ranges from 30° to 150° and when the particle contact angle is below 90°, an oil-in-water emulsion (indicating a hydrophilic particle) is formed. On the contrary, if the contact angle exceeds 90°, it stabilizes a water-in-oil emulsion (characteristic of a hydrophobic particle). Ideally, the particle's contact angle should be close to 90°, so the particle is equally wetted in both the dispersed and continuous phases (Muiz et al. 2023). In this respect, barley SNCs are good stabilizers for Pickering emulsions that aid forming a stable system (López-Hernández et al.

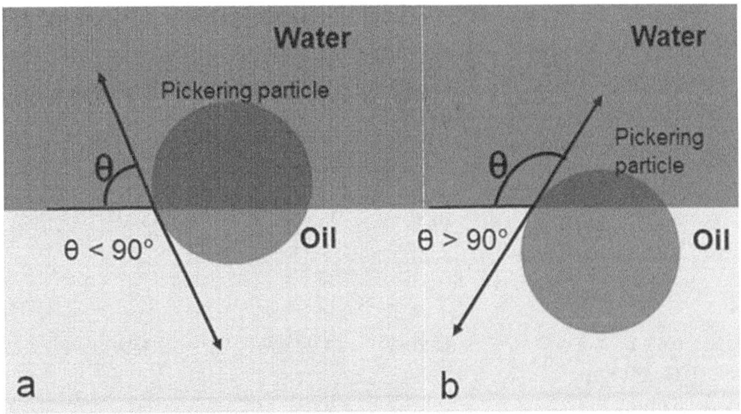

Figure 16.2 Representation of the contact angle in pickering emulsions: a) Contact angle (θ) when Particle is hydrophilic and b) contact angle (θ) when particle is hydrophobic.

2022). **Figure 16.2** shows the contact angle formed in Pickering emulsions by adsorbed Pickering particles.

16.4.1.2 Solid Particles as Stabilizers of Pickering Emulsions

The use of inorganic particles in Pickering emulsions has been widely studied; current knowledge about Pickering emulsions was obtained from systems stabilized by silica nanoparticles (silicon dioxide SiO_2) due to their shape and defined particle size (from a few nanometers to several microns), as well as the ease of modifying their surface (increased hydrophobicity) (Lam et al. 2014; Berton-Carabin & Schroën, 2015).

However, the applications of these particles in the food and pharmaceutical industries have been limited due to their inorganic origin (Rayner et al.,2014; Carabin & Schroën, 2015; Xiao et al. 2016). One of the current challenges in this field is to use particles obtained from natural polymers that are effective as stabilizers and that, in turn, are acceptable for their use in foods.

The particles used to stabilize food emulsions must be food-grade. In this regard, a variety of substances have been reported, including proteins (Deng et al. 2023), fat crystals (Tenorio-Garcia et al. 2023), flavonoids (Lu et al. 2023), cellulose and its derivatives (Cui et al. 2023), chitosan (Yang et al. 2023b), and native or modified starch granules (Leal-Castañeda et al. 2018; Lu et al. 2018; Matos et al. 2018). **Table 16.3** lists some works of Pickering emulsions stabilized by starch nanocrystals and other polysaccharides.

As seen in **Table 16.3**, many factors affect emulsion droplet size, such as the concentration and nature of Pickering particle, oil–water ratio, and emulsification conditions. The droplet size ranges between 1μm–100 μm..

It has been established that the size and form of starch nanocrystals are influenced by the botanical origin of starch; in this way, it is also possible that under the same processing conditions starch nanocrystals from different sources produce Pickering emulsions with distinct properties. Daniel et al. (2020) studied Pickering emulsions using nanocrystals from three different sources: cassava, corn, and bean. It was discovered that cassava nanocrystals produce emulsions with the smallest droplet size (14.0 μm), followed by bean (17.8 μm) and corn (25.4 μm). As for

TABLE 16.3 POLYSACCHARIDES EMPLOYED AND CONDITIONS FOR PREPARING PICKERING EMULSIONS

Particle	Particle Proportion (%) & Processing Conditions	Water/ Oil Ratio	Droplet Size	Emulsion Type	References
Modified (OSA) waxy corn starch nanocrystals	3	70:30	-	O/W	Sánchez de la Concha et al. (2020)
Cellulose nanocrystals	0.05-1 10,000 rpm 3 min	80:20	1.22 µm	O/W	Du Le et al. (2020)
Corn Starch Nanocrystals and Acetylated Corn Starch Nanocrystals	0.2, 0.5,0.8,1.0,1.5 & 2.0 10,000 rpm 2 min	80:20	1.23 µm (Normal nanocrystals) 780 nm (Acetylated nanocrystals)	O/W	Qian et al. (2020)
Cassava, bean, and corn starch nanocrystals	1,3 & 5 -	50:50	59.8 µm 44.0 µm 77.0 µm	O/W	Daniel et al. (2020)
Waxy and normal corn starch nanocrystals	3 20,000 rpm 5 min	50:50	2.66 µm (waxy starch) 8.33 µm (Normal starch)	O/W	Wang et al. (2020)
Sago starch nanocrystals	0.5 & 4.0 10,000 rpm 4 min	50:50	-	O/W	Azfaralariff et al. (2020)
Crosslinked corn starch	0.05, 0.1, 0.3, 0.5, 0.7 & 1	70:30	1.7 µm	O/W	Sufi-Maragheh et al. (2019)
OSA quinoa starch granules	3 22 000 rpm 30 s	93:7	32.6 µm	O/W	Marefati et al. (2017)
Chestnut Starch Nanocrystals	6,10, & 14. 13000 g 2 min	50:50	2.96, 1.84 & 1.47 µm	O/W	Zheng et al. (2022)
Waxy maize starch nanocrystals	0.15 8,000 rpm 1 min and then 11,000 rpm 1 min	95:5	538 nm	O/W	Jo et al. (2019)
Waxy corn starch nanoparticles and chitin nanofibers	Starch:1 & 2 Chitin: 0.2 & 0.4 20,000 rpm 2 min	90:10	-	O/W	Lee et al. (2021)

emulsifying index, corn-stabilized emulsions obtained the highest value (0.8), followed by cassava (0.75) and bean (0.7). Similarly, Sánchez de la Concha et al. (2020) studied Pickering emulsions by using corn and amaranth starch nanocrystals. Also, emulsions with OSA (octenyl succinic acid) corn and amaranth starch nanocrystals were prepared. Emulsions were stable for 10 days, and it was found that the OSA-treated nanocrystals possessed an enhanced emulsification capacity, which was attributed to an increase in the repulsive forces (ζ-potential values < -30mV). It was also found that amaranth starch nanocrystals exhibited superior emulsification ability; this was attributed to the lower crystal size in amaranth. In a similar way Qian et al. (2020) studied olive oil Pickering emulsions using cornstarch nanocrystals and acetylated cornstarch nanocrystals. Acetylation process of nanocrystals enhanced their emulsifying properties, producing emulsions with lower droplet size (780 nm) and unimodal size distribution in comparison with unacetylated emulsions (1.23 μm).

López-Hernández et al. (2022) studied and characterized Pickering emulsions stabilized successfully by barley starch nanocrystals. Emulsion droplet size (d_{32}) of freshly made emulsions ranged from 1.15–6.60 μm; and during 1 week of storage no phase separation was observed, and ζ-potential values were below -30 mV. **Figure 16.3** shows examples of macroscopic appearance of elaborated emulsions.

Regarding oil ratio Ahmad & Lazim (2018) employed different concentrations of sago starch nanocrystals to prepare Pickering emulsions at three water–oil ratios (1:3, 1:1, 3:1); when oil proportion was 25%, emulsions were stable at all starch nanocrystals concentrations but when oil content was 50 and 75% it was not possible to study emulsions. This could be attributed to the hydrophilic nature of starch nanocrystals. Liang et al. (2016) studied the influence of the oil type on the ability of waxy cornstarch nanocrystal to form Pickering emulsions by using oils with short-, medium-, and long-chain fatty acids (SCT, MCT, and LCT, respectively). All emulsions were stable against coalescence and Ostwald ripening for 90 days. They discovered that emulsion droplet size decreased in this order LCT > MCT > SCT. Furthermore, the creaming layers of MCT and LCT Pickering emulsions were at the top of the cell, whereas SCT emulsions were on the bottom, which was attributed to the higher density of tributyrin (ρ=1032 kg/m^3), the oil employed as SCT, in comparison to water (ρ=1000 kg/m^3).

It was already mentioned that Pickering emulsions could be used to encapsulate and deliver bioactive compounds into the human body that normally would be destroyed by processing

Figure 16.3 Macroscopic appearance of freshly made O/W emulsions elaborated with canola oil and water using different amounts (%) of barley SNCs.

and environmental conditions, storage, or hydrolyzed by the acid in the stomach. Curcumin is a polyphenolic compound with anticarcinogenic, anti-inflammatory, and antioxidant activities, among others (Lee et al. 2021). Zhou et al. (2023) encapsulated curcumin in Pickering emulsions stabilized by rice starch-fatty acid complexes, evaluating their behavior during simulated gastric and intestinal *in vitro* digestion. The release of curcumin reached 19.2 % after 2 h of simulated gastric digestion. In simulated intestinal digestion, curcumin release was 73.7%, demonstrating that Pickering emulsions exhibited good bioaccessibility of curcumin.

Likewise, in recent years, studies focused on using Pickering emulsions as delivery vehicle and protection for bioactive lipophilic compounds and drugs: curcumin (Zhang et al. 2022a; Jo et al. 2023; Wang et al. 2023b), ω-3 polyunsaturated fatty acids (Wang et al. 2023a; Huang et al. 2023), carotenoids (Chutia & Mahanta, 2021; Zhang et al. 2022c; Wei et al. 2021), vitamin D_3 (Winuprasith et al. 2018; Martakov et al. 2022; Zhao et al. 2023), essential oils (Harsanto et al. 2022; Lobato-Guarnido et al. 2023; Phosanam et al. 2023), flavonoids (Wang, et al. 2023d), pterostilbene (Cai et al. 2023), and immunosuppressive and anti-inflammatory agents (Dieng et al., 2020; Beladjine et al. 2023; Li & Yu, 2023).

From the above depicted evidence, it is possible to conclude that Pickering emulsions stabilized by starch nanocrystals have potential applications for the development of novel and functional food emulsion as vehicles for a wide variety of health-promoting compounds.

16.5 CONCLUSION

Starch is one of the most versatile polysaccharides in nature; it can be modified to be employed in food industry with diverse uses. Acid hydrolysis of starch eliminates most of amylose content, leaving at the end of the process particles of nanometric size and enhanced crystallinity, called starch nanocrystals. Starch nanocrystals usually are 200 nm in size; moreover, unlike other substances such as proteins, starch nanocrystals are not sensitive to environmental conditions such as pH and temperature, making them suitable for Pickering stabilization. Like native starch, starch nanocrystals can be modified and their stabilizing properties even more enhanced. Recent investigation on Pickering emulsions have proved that a low quantity of starch nanocrystals (1%) is capable of elaborating and keeping stable emulsions during long storage time (over a year). Also, studies have demonstrated that starch nanocrystals and other polysaccharides can be used to encapsulate several bioactive compounds in Pickering emulsion systems to deliver them into particular targets of the human body bringing beneficial effects to human health.

REFERENCES

Ahmad, A.A.A., & Lazim, A.M. (2018). Evaluation of emulsion emulsified by starch nanocrystal: A preliminary study. *AIP Conference Proceedings, 1940,* 020083. https://doi.org/10.1063/1.5027998

Alcázar-Alay, S.C., & Meireles, M.A.A. (2015). Physicochemical properties, modifications and applications of starches from different botanical sources. *Food Science and Technology (Campinas), 35*(2), 215–236. https://doi.org/10.1590/1678-457X.6749

Angellier, H., Choisnard, L., Molina-Boisseau, S., Ozil, P., & Dufresne, A. (2004). Optimization of the preparation of aqueous suspensions of waxy maize starch nanocrystals using a response surface methodology. *Biomacromolecules, 5*(4), 1545–1551. https://doi.org/10.1021/bm049914u

Azfaralariff, A., Fazial, F.F., Sontanosamy, R.S., Nazar, M.F., & Lazim, A.M. (2020). Food-grade particle stabilized pickering emulsion using modified sago (Metroxylon sagu) starch nanocrystal. *Journal of Food Engineering, 280*(February), 109974. https://doi.org/10.1016/j.jfoodeng.2020.109974

Beladjine, M., Albert, C., Sintès, M., Mekhloufi, G., Gueutin, C., Nicolas, V., Canette, A., Trichet, M., Tsapis, N., Michel, L., Agnely, F., & Huang, N. (2023). Pickering emulsions stabilized with biodegradable nanoparticles for the co-encapsulation of two active pharmaceutical ingredients. *International Journal of Pharmaceutics, 637*(March), 122870. https://doi.org/10.1016/j.ijpharm.2023.122870

Bello-Perez, L.A., Flores-Silva, P.C., Agama-Acevedo, E., & Tovar, J. (2020). Starch digestibility: Past, present, and future. *Journal of the Science of Food and Agriculture, 100*(14), 5009–5016. https://doi.org/10.1002/jsfa.8955

Berton-Carabin, C.C., & Schroën, K. (2015). Pickering emulsions for food applications: Background, trends, and challenges. *Annual Review of Food Science and Technology, 6*(1), 263–297. https://doi.org/10.1146/annurev-food-081114-110822

Cai, X., Du, X., Zhu, G., Shi, X., & Chen, Q. (2023). Fabrication of carboxymethyl starch/xanthan gum combinations Pickering emulsion for protection and sustained release of pterostilbene. *International Journal of Biological Macromolecules, 248*(130), 125963. https://doi.org/10.1016/j.ijbiomac.2023.125963

Chutia, H., & Mahanta, C.L. (2021). Properties of starch nanoparticle obtained by ultrasonication and high pressure homogenization for developing carotenoids-enriched powder and Pickering nanoemulsion. *Innovative Food Science & Emerging Technologies, 74*(July), 102822. https://doi.org/10.1016/j.ifset.2021.102822

Compart, J., Singh, A., Fettke, J., & Apriyanto, A. (2023). Customizing starch properties: A review of starch modifications and their applications. *Polymers, 15*(16), 3491. https://doi.org/10.3390/polym15163491

Cui, J., Hossain, M., Wang, Z., & Chang, C. (2023). Enzymatically produced nanocellulose as emulsifier for Pickering emulsion. *Food Hydrocolloids, 144*(June), 108997. https://doi.org/10.1016/j.foodhyd.2023.108997

Dai, L., Li, C., Zhang, J., & Cheng, F. (2018). Preparation and characterization of starch nanocrystals combining ball milling with acid hydrolysis. *Carbohydrate Polymers, 180*(July 2017), 122–127. https://doi.org/10.1016/j.carbpol.2017.10.015

Daniel, T.H.G., Bedin, A.C., Souza, É.C.F., Lacerda, L.G., & Demiate, I.M. (2020). Pickering emulsions produced with starch nanocrystals from Cassava (*Manihot esculenta* Crantz), Beans (*Phaseolus vulgaris* L.), and Corn (*Zea mays* L.). *Starch – Stärke, 72*(11–12), 1900326. https://doi.org/10.1002/star.201900326

Deng, F., Wang, Z., Wang, X., & He, R. (2023). Pickering emulsions stabilized by moringa seed protein: Regulating the emulsion properties by adjusting the Maillard reaction. *Industrial Crops and Products, 205*(September), 117574. https://doi.org/10.1016/j.indcrop.2023.117574

Dieng, S.M., Omran, Z., Anton, N., Thioune, O., Djiboune, A.R., Sy, P.M., Messaddeq, N., Ennahar, S., Diarra, M., & Vandamme, T. (2020). Pickering nano-emulsions stabilized by Eudragit RL100 nanoparticles as oral drug delivery system for poorly soluble drugs. *Colloids and Surfaces B: Biointerfaces, 191*(March), 111010. https://doi.org/10.1016/j.colsurfb.2020.111010

Dominguez-Ayala, J.E., Soler, A., Mendez-Montealvo, G., & Velazquez, G. (2022). Supramolecular structure and technofunctional properties of starch modified by high hydrostatic pressure (HHP): A review. *Carbohydrate Polymers, 291*(May), 119609. https://doi.org/10.1016/j.carbpol.2022.119609

Du Le, H., Loveday, S.M., Singh, H., & Sarkar, A. (2020). Pickering emulsions stabilised by hydrophobically modified cellulose nanocrystals: Responsiveness to pH and ionic strength. *Food Hydrocolloids, 99*(May 2019), 105344. https://doi.org/10.1016/j.foodhyd.2019.105344

Gutiérrez-Cortez, E., Hernadez-Becerra, E., Londoño-Restrepo, S.M., & Rodriguez-García, M.E. (2021). Physicochemical characterization of Amaranth starch insulated by mechanical separations. *International Journal of Biological Macromolecules, 177*, 430–436. https://doi.org/10.1016/j.ijbiomac.2021.02.138

Harsanto, B.W., Supriyanto, Kartini, I., & Pranoto, Y. (2022). The ability of breadfruit starch nanoparticle-stabilized pickering emulsion for encapsulating cinnamon essential oil. *Malaysian Applied Biology, 51*(1), 83–90. https://doi.org/10.55230/mabjournal.v51i1.2192

He, R., Li, S., Zhao, G., Zhai, L., Qin, P., & Yang, L. (2023). Starch modification with molecular transformation, physicochemical characteristics, and industrial usability: A state-of-the-art review. *Polymers, 15*(13), 2935. https://doi.org/10.3390/polym15132935

Huang, S., Peng, J., Zi, Y., Zheng, Y., Xu, J., Gong, H., Kan, G., Wang, X., & Zhong, J. (2023). Regenerated silk fibroin for the stabilization of fish oil-loaded Pickering emulsions. *Colloids and Surfaces A: Physicochemical and Engineering Aspects, 678*(October), 132523. https://doi.org/10.1016/j.colsurfa.2023.132523

Jia, R., Cui, C., Gao, L., Qin, Y., Ji, N., Dai, L., Wang, Y., Xiong, L., Shi, R., & Sun, Q. (2023). A review of starch swelling behavior: Its mechanism, determination methods, influencing factors, and influence on food quality. *Carbohydrate Polymers, 321*(April), 121260. https://doi.org/10.1016/j.carbpol.2023.121260

Jo, M., Ban, C., Goh, K.K.T., & Choi, Y.J. (2019). Influence of chitosan-coating on the stability and digestion of emulsions stabilized by waxy maize starch crystals. *Food Hydrocolloids, 94*(April), 603–612. https://doi.org/10.1016/j.foodhyd.2019.04.010

Jo, M., Ban, C., Goh, K.K.T., & Choi, Y.J. (2023). Enhancement of the intestinal permeability of curcumin using Pickering emulsions stabilized by starch crystals and chitosan. *Food Chemistry, 405*(PA), 134889. https://doi.org/10.1016/j.foodchem.2022.134889

Kalateh-Seifari, F., Yousefi, S., Ahari, H., & Hosseini, S.H. (2021). Corn starch-Chitosan nanocomposite film containing nettle essential oil nanoemulsions and starch nanocrystals: Optimization and characterization. *Polymers, 13*(13), 2113. https://doi.org/10.3390/polym13132113

Kumar, A., Kaur, R., Kumar, V., Kumar, S., Gehlot, R., & Aggarwal, P. (2022). New insights into water-in-oil-in-water (W/O/W) double emulsions: Properties, fabrication, instability mechanism, and food applications. *Trends in Food Science & Technology, 128*(March), 22–37. https://doi.org/10.1016/j.tifs.2022.07.016

Kumari, B., & Sit, N. (2023). Comprehensive review on single and dual modification of starch: Methods, properties and applications. *International Journal of Biological Macromolecules, 253*(P3), 126952. https://doi.org/10.1016/j.ijbiomac.2023.126952

Lam, S., Velikov, K.P., & Velev, O.D. (2014). Pickering stabilization of foams and emulsions with particles of biological origin. *Current Opinion in Colloid and Interface Science, 19*(5), 490–500. https://doi.org/10.1016/j.cocis.2014.07.003

Leal-Castañeda, E.J., García-Tejeda, Y., Hernández-Sánchez, H., Alamilla-Beltrán, L., Téllez-Medina, D.I., Calderón-Domínguez, G., García, H.S., & Gutiérrez-López, G.F. (2018). Pickering emulsions stabilized with native and lauroylated amaranth starch. *Food Hydrocolloids, 80*, 177–185. https://doi.org/10.1016/j.foodhyd.2018.01.043

Lee, Y.S., Tarté, R., & Acevedo, N.C. (2021). Curcumin encapsulation in Pickering emulsions co-stabilized by starch nanoparticles and chitin nanofibers. *RSC Advances, 11*(27), 16275–16284. https://doi.org/10.1039/D1RA01622A

Li, S., Zhou, W., Huang, C., Hu, Y., Gao, Q., & Chen, Y. (2023). Rapid preparation of starch nanocrystals by the mixed acid of sulfuric acid and hydrochloric acid. *International Journal of Biological Macromolecules, 232*(9), 123402. https://doi.org/10.1016/j.ijbiomac.2023.123402

Li, W., Jiao, B., Li, S., Faisal, S., Shi, A., Fu, W., Chen, Y., & Wang, Q. (2022). Recent advances on Pickering emulsions stabilized by diverse edible particles: Stability Mechanism and applications. *Frontiers in Nutrition, 9*(May), 1–17. https://doi.org/10.3389/fnut.2022.864943

Li, Z., & Yu, D. (2023). Controlled ibuprofen release from Pickering emulsions stabilized by pH-responsive cellulose-based nanofibrils. *International Journal of Biological Macromolecules, 242*(P3), 124942. https://doi.org/10.1016/j.ijbiomac.2023.124942

Liang, R., Jiang, Y., Yokoyama, W., Yang, C., Cao, G., & Zhong, F. (2016). Preparation of Pickering emulsions with short, medium and long chain triacylglycerols stabilized by starch nanocrystals and their: In vitro digestion properties. *RSC Advances, 6*(101), 99496–99508. https://doi.org/10.1039/c6ra18468e

Lin, N., Huang, J., Chang, P.R., Anderson, D.P., & Yu, J. (2011). Preparation, modification, and application of starch nanocrystals in nanomaterials: A review. *Journal of Nanomaterials, 2011*, 1–13. https://doi.org/10.1155/2011/573687

Lobato-Guarnido, I., Luzón, G., Ríos, F., & Fernández-Serrano, M. (2023). Synthesis and characterization of environmentally friendly Chitosan–Arabic gum nanoparticles for encapsulation of oregano essential oil in pickering emulsion. *Nanomaterials, 13*(19), 2651. https://doi.org/10.3390/nano13192651

López-Hernández, R.E., García-Solís, S.E., Monroy-Rodríguez, I., Cornejo-Mazón, M., Calderón-Domínguez, G., Alamilla-Beltrán, L., Hernández-Sánchez, H., & Gutiérrez-López, G.F. (2022). Preparation and characterization of canola oil-in-water Pickering emulsions stabilized by barley starch nanocrystals. *Journal of Food Engineering, 326*, 111037. https://doi.org/10.1016/j.jfoodeng.2022.111037

Lu, S., Li, X., Wei, X., Huang, C., Zheng, J., Ou, S., Yang, T., & Liu, F. (2023). Preparation and characterization of a novel natural Quercetin self-stabilizing pickering emulsion. *Foods, 12*(7), 1415. https://doi.org/10.3390/foods12071415

Lu, X., Xiao, J., & Huang, Q. (2018). Pickering emulsions stabilized by media-milled starch particles. *Food Research International, 105*(March 2017), 140–149. https://doi.org/10.1016/j.foodres.2017.11.006

Marefati, A., Bertrand, M., Sjöö, M., Dejmek, P., & Rayner, M. (2017). Storage and digestion stability of encapsulated curcumin in emulsions based on starch granule Pickering stabilization. *Food Hydrocolloids, 63*, 309–320. https://doi.org/10.1016/j.foodhyd.2016.08.043

Martakov, I.S., Vaseneva, I.N., Torlopov, M.A., Legki, P.V., Paderin, N.M., Patov, S.A., Mikhaylov, V.I., & Sitnikov, P.A. (2022). Biocompatible nanoparticle heteroaggregates as stabilizers of Pickering emulsions for Vitamin D3 efficient delivery. *ACS Applied Bio Materials, 5*(9), 4342–4353. https://doi.org/10.1021/acsabm.2c00520

Martínez, P., Peña, F., Bello-Pérez, L.A., Núñez-Santiago, C., Yee-Madeira, H., & Velezmoro, C. (2019). Physicochemical, functional and morphological characterization of starches isolated from three native potatoes of the Andean region. *Food Chemistry: X, 2*(May), 100030. https://doi.org/10.1016/j.fochx.2019.100030

Martins, P.C., Latorres, J.M., & Martins, V.G. (2022). Impact of starch nanocrystals on the physicochemical, thermal and structural characteristics of starch-based films. *LWT, 156*, 113041. https://doi.org/10.1016/j.lwt.2021.113041

Matos, M., Laca, A., Rea, F., Iglesias, O., Rayner, M., & Gutiérrez, G. (2018). O/W emulsions stabilized by OSA-modified starch granules versus non-ionic surfactant: Stability, rheological behaviour and resveratrol encapsulation. *Journal of Food Engineering, 222*, 207–217. https://doi.org/10.1016/j.jfoodeng.2017.11.009

Ming, L., Wu, H., Liu, A., Naeem, A., Dong, Z., Fan, Q., Zhang, G., Liu, H., & Li, Z. (2023). Evolution and critical roles of particle properties in Pickering emulsion: A review. *Journal of Molecular Liquids, 388*(April), 122775. https://doi.org/10.1016/j.molliq.2023.122775

Mittal, A., Garg, S., Premi, A., & Giri, A.S. (2021). Synthesis of polyvinyl alcohol/modified starch-based biodegradable nanocomposite films reinforced with starch nanocrystals for packaging applications. *Polymers and Polymer Composites, 29*(5), 405–416. https://doi.org/10.1177/0967391120922429

Muiz, A., Klojdová, I., & Stathopoulos, C. (2023). Utilization of by-products for preparation of Pickering particles. *European Food Research and Technology, 249*(12), 3069–3083. https://doi.org/10.1007/s00217-023-04349-z

Ogori, A.F., & Taofeek, A. (2022). Starch chemistry and application. *Journal of Food Technology and Preservation, 6*(6), 1–8. https://doi.org/10.35841/2591-796X-6.6.130

Omoregie Egharevba, H. (2020). Chemical Properties of Starch and Its Application in the Food Industry. In Martins Emeje (Ed.), *Chemical Properties of Starch* (pp. 1–26). IntechOpen. https://doi.org/10.5772/intechopen.87777

Ozturk, B., & McClements, D.J. (2016). Progress in natural emulsifiers for utilization in food emulsions. *Current Opinion in Food Science, 7*, 1–6. https://doi.org/10.1016/j.cofs.2015.07.008

Pan, J., Chen, J., Wang, X., Wang, Y., & Fan, J. (2023). Pickering emulsion: From controllable fabrication to biomedical application. *Interdisciplinary Medicine, 1*(3), e20230014. https://doi.org/10.1002/INMD.20230014

Park, J., Chung, H.J., Park, H.Y., Park, H.J., & Oh, S.K. (2023). Comparative analysis of malt quality and starch characteristics of three South Korean barley cultivars. *Food Science and Biotechnology, 33*(5), 1135–1145. https://doi.org/10.1007/s10068-023-01419-6

Pavela, G., Harman, T., Cardel, M.I., & Lee, A. (2020). Handbook of Eating and Drinking. In H.L. Meiselman (Ed.), *Handbook of Eating and Drinking*. Springer International Publishing. https://doi.org/10.1007/978-3-030-14504-0

Peito, S., Peixoto, D., Ferreira-Faria, I., Martins, A.M., Ribeiro, H.M., Veiga, F., Marto, J., & Paiva-Santos, A.C. (2022). Nano- and microparticle-stabilized Pickering emulsions designed for topical therapeutics and cosmetic applications. *International Journal of Pharmaceutics, 615*(October 2021), 121455. https://doi.org/10.1016/j.ijpharm.2022.121455

Phosanam, A., Moreira, J., Adhikari, B., Adhikari, A., & Losso, J.N. (2023). Stabilization of ginger essential oil Pickering emulsions by pineapple cellulose nanocrystals. *Current Research in Food Science, 7*(May), 100575. https://doi.org/10.1016/j.crfs.2023.100575

Pickering, S.U. (1907). CXCVI.—Emulsions. *Journal Chemical Society Transactions, 91,* 2001–2021. https://doi.org/10.1039/CT9079102001

Priyan, V.V., Shahnaz, T., Kunnumakkara, A.B., Rana, V., Saravanan, M., & Narayanasamy, S. (2021). Antioxidant, anti-inflammatory and biosorption properties of starch nanocrystals in vitro study: Cytotoxic and phytotoxic evaluation. *Journal of Cluster Science, 32*(5), 1419–1430. https://doi.org/10.1007/s10876-020-01905-5

Putaux, J.-L., Molina-Boisseau, S., Momaur, T., & Dufresne, A. (2003). Platelet nanocrystals resulting from the disruption of waxy maize starch granules by acid hydrolysis. *Biomacromolecules, 4*(5), 1198–1202. https://doi.org/10.1021/bm0340422

Qian, X., Lu, Y., Xie, W., & Wu, D. (2020). Viscoelasticity of olive oil/water Pickering emulsions stabilized with starch nanocrystals. *Carbohydrate Polymers, 230, 115575.* https://doi.org/10.1016/j.carbpol.2019.115575

Rayner, M., Marku, D., Eriksson, M., Sjöö, M., Dejmek, P., & Wahlgren, M. (2014). Biomass-based particles for the formulation of Pickering type emulsions in food and topical applications. *Colloids and Surfaces A: Physicochemical and Engineering Aspects, 458*(1), 48–62. https://doi.org/10.1016/j.colsurfa.2014.03.053

Sanchez de la Concha, B.B., Agama-Acevedo, E., Nuñez-Santiago, M.C., Bello-Perez, L.A., Garcia, H.S., & Alvarez-Ramirez, J. (2018). Acid hydrolysis of waxy starches with different granule size for nanocrystal production. *Journal of Cereal Science, 79,* 193–200. https://doi.org/doi.org/10.1016/j.jcs.2017.10.018

Sánchez de la Concha, B.B., Agama-Acevedo, E., Agurirre-Cruz, A., Bello-Pérez, L.A., & Alvarez-Ramírez, J. (2020). OSA esterification of amaranth and maize starch nanocrystals and their use in “Pickering” emulsions. *Starch – Stärke, 72*(7–8), 1900271. https://doi.org/10.1002/star.201900271

Šárka, E., Sinica, A., Smrčková, P., & Sluková, M. (2023). Non-traditional starches, their properties, and applications. *Foods, 12*(20), 3794. https://doi.org/10.3390/foods12203794

Sathyan, S., & Nisha, P. (2022). Optimization and characterization of porous starch from corn starch and application studies in emulsion stabilization. *Food and Bioprocess Technology, 15*(9), 2084–2099. https://doi.org/10.1007/s11947-022-02843-y

Sufi-Maragheh, P., Nikfarjam, N., Deng, Y., & Taheri-Qazvini, N. (2019). Pickering emulsion stabilized by amphiphilic pH-sensitive starch nanoparticles as therapeutic containers. *Colloids and Surfaces B: Biointerfaces, 181*(December 2018), 244–251. https://doi.org/10.1016/j.colsurfb.2019.05.046

Sun, Z., Yan, X., Xiao, Y., Hu, L., Eggersdorfer, M., Chen, D., Yang, Z., & Weitz, D.A. (2022). Pickering emulsions stabilized by colloidal surfactants: Role of solid particles. *Particuology, 64,* 153–163. https://doi.org/10.1016/j.partic.2021.06.004

Tavernier, I., Wijaya, W., Van der Meeren, P., Dewettinck, K., & Patel, A.R. (2016). Food-grade particles for emulsion stabilization. *Trends in Food Science & Technology, 50,* 159–174. https://doi.org/10.1016/j.tifs.2016.01.023

Tenorio-Garcia, E., Araiza-Calahorra, A., Rappolt, M., Simone, E., & Sarkar, A. (2023). Pickering water-in-oil emulsions stabilized solely by fat crystals. *Advanced Materials Interfaces, 10*(31), 1–16. https://doi.org/10.1002/admi.202300190

Tester, R.F., Karkalas, J., & Qi, X. (2004). Starch – Composition, fine structure and architecture. *Journal of Cereal Science, 39*(2), 151–165. https://doi.org/10.1016/j.jcs.2003.12.001

Thangrongthong, S., Puttarat, N., Ladda, B., Itthisoponkul, T., Pinket, W., Kasemwong, K., & Taweechotipatr, M. (2020). Microencapsulation of probiotic Lactobacillus brevis ST-69 producing GABA using alginate supplemented with nanocrystalline starch. *Food Science and Biotechnology, 29*(11), 1475–1482. https://doi.org/10.1007/s10068-020-00812-9

Velásquez-Castillo, L.E., Leite, M.A., Tisnado, V.J.A., Ditchfield, C., Sobral, P.J. do A., & Moraes, I.C.F. (2023). Cassava starch films containing quinoa starch nanocrystals: Physical and surface properties. *Foods, 12*(3), 1–18. https://doi.org/10.3390/foods12030576

Wang, K., Hong, Y., Gu, Z., Cheng, L., Li, Z., & Li, C. (2020). Stabilization of Pickering emulsions using starch nanocrystals treated with alkaline solution. *International Journal of Biological Macromolecules, 155*, 273–285. https://doi.org/10.1016/j.ijbiomac.2020.03.219

Wang, P., Wang, Y., Chen, C., & Fu, X. (2023a). The stability mechanism of Pickering emulsions fabricated by multi-functional amylose-based nanoparticles in a delivery system. *Food & Function, 14*(5), 2338–2348. https://doi.org/10.1039/D2FO02827A

Wang, R., Li, M., Liu, M., Wang, A., Strappe, P., Blanchard, C., & Zhou, Z. (2023b). Characterization of Pickering emulsion by SCFAs-modified debranched starch and a potent for delivering encapsulated bioactive compound. *International Journal of Biological Macromolecules, 231*(August 2022), 123164. https://doi.org/10.1016/j.ijbiomac.2023.123164

Wang, Y., Ai, C., Wang, H., Chen, C., Teng, H., Xiao, J., & Chen, L. (2023c). Emulsion and its application in the food field: An update review. *EFood, 4*(4), 1–18. https://doi.org/10.1002/efd2.102

Wang, Y., Kratzer, R., Murkovic, M., Eibinger, M., Machado Charry, E., Li, S., Zhang, T., Zhang, X., Zhang, M., & Chen, H. (2023d). Fabrication and characterization of a novel zein/pectin/pumpkin seed oil Pickering emulsion and the effects of myricetin on oxidation stability. *International Journal of Biological Macromolecules, 253*(P7), 127386. https://doi.org/10.1016/j.ijbiomac.2023.127386

Watts, S., Chow, C., & Lim, S. (2023). Protein nanocage engineering for pickering emulsions and potential food applications. *Current Opinion in Colloid & Interface Science*, 101761. https://doi.org/10.1016/j.cocis.2023.101761

Wei, Y., Zhou, D., Mackie, A., Yang, S., Dai, L., Zhang, L., Mao, L., & Gao, Y. (2021). Stability, interfacial structure, and gastrointestinal digestion of β-carotene-loaded pickering emulsions co-stabilized by particles, a biopolymer, and a surfactant. *Journal of Agricultural and Food Chemistry, 69*(5), 1619–1636. https://doi.org/10.1021/acs.jafc.0c06409

Winuprasith, T., Khomein, P., Mitbumrung, W., Suphantharika, M., Nitithamyong, A., & McClements, D.J. (2018). Encapsulation of Vitamin D3 in pickering emulsions stabilized by nanofibrillated mangosteen cellulose: Impact on in vitro digestion and bioaccessibility. *Food Hydrocolloids, 83*, 153–164. https://doi.org/10.1016/j.foodhyd.2018.04.047

Wu, J., & Ma, G.H. (2016). Recent studies of Pickering emulsions: Particles make the difference. *Small, 12*(34), 4633–4648. https://doi.org/10.1002/smll.201600877

Xiao, J., Li, Y., & Huang, Q. (2016). Recent advances on food-grade particles stabilized Pickering emulsions: Fabrication, characterization and research trends. *Trends in Food Science & Technology, 55*, 48–60. https://doi.org/10.1016/j.tifs.2016.05.010

Xiong, X., Huang, M., Zhou, X., Zhou, H., Zeng, C., Zhao, Z., & Xiao, Q. (2021). Physicochemical studies of nanocrystals of starches from two rice (Oryza sativa L.) types and their characteristics using various modern instrument techniques. *Journal of the Science of Food and Agriculture, 101*(3), 1038–1046. https://doi.org/10.1002/jsfa.10712

Yan, S., Li, Z., Wang, B., Li, T., Li, Z., Zhang, N., & Cui, B. (2023). Correlation analysis on physicochemical and structural properties of sorghum starch. *Frontiers in Nutrition, 9*(January), 1–16. https://doi.org/10.3389/fnut.2022.1101868

Yang, F., Wei, Y., Xiao, H., Zhang, Q., Li, J., Lin, Q., Zhu, D., Huang, Z., & Liu, G.Q. (2023a). Acetylated rice starch nanocrystals improved the physical, mechanical, and structural properties of native rice starch based films. *International Journal of Biological Macromolecules, 253*(P5), 127271. https://doi.org/10.1016/j.ijbiomac.2023.127271

Yang, W., Zhang, M., Wang, Q., Sun, J., & Song, A. (2021). Pickering emulsions stabilized by surfactant particles with smart responses to pH and metal-ligands. *Journal of Molecular Liquids, 324*, 114730. https://doi.org/10.1016/j.molliq.2020.114730

Yang, Y., Gupta, V.K., Amiri, H., Pan, J., Aghbashlo, M., Tabatabaei, M., & Rajaei, A. (2023b). Recent developments in improving the emulsifying properties of chitosan. *International Journal of Biological Macromolecules, 239*(March), 124210. https://doi.org/10.1016/j.ijbiomac.2023.124210

Zhang, C., Xue, W., Li, T., & Wang, L. (2023). Understanding the relationship between the molecular structure and physicochemical properties of soft rice starch. *Foods*, *12*(19), 3611. https://doi.org/10.3390/foods12193611

Zhang, K., Zhao, D., Huang, Q., Huang, J., & Wen, Q. (2021). Physicochemical, structural properties and in vitro digestibility of A- and B-type granules isolated from green wheat and mature wheat starch. *Starch – Stärke*, *73*(9–10), 1–10. https://doi.org/10.1002/star.202100065

Zhang, L., Chen, D.L., Wang, X.F., Xu, L., Qian, J.Y., & He, X.D. (2022a). Enzymatically modified quinoa starch based pickering emulsion as carrier for curcumin: Rheological properties, protection effect and in vitro digestion study. *Food Bioscience*, *49*(March), 101933. https://doi.org/10.1016/j.fbio.2022.101933

Zhang, S., Li, Q., Zhao, Y., Qin, Z., Zheng, M., Liu, H., & Liu, J. (2022b). Preparation and characterization of low oil absorption corn starch by ultrasonic combined with freeze–thaw treatment. *Food Chemistry: X*, *15*(July). https://doi.org/10.1016/j.fochx.2022.100410

Zhang, Z.H., Gu, Z.Y., Li, M.F., Liang, S., Huang, Z., Zong, M.H., Wu, X.L., & Lou, W.Y. (2022c). Oxidized high-amylose starch as pickering stabilizer for oil-in-water emulsion and delivery of bioactive compound. *Food Hydrocolloids for Health*, *2*(November), 100104. https://doi.org/10.1016/j.fhfh.2022.100104

Zhao, X., Yang, X., Bao, Y., Guo, Y., Luo, J., Jiang, S., & Zhang, W. (2023). Construction of vitamin D delivery system based on pine nut oil Pickering emulsion: Effect of phenols. *Journal of the Science of Food and Agriculture*, *103*(8), 4034–4046. https://doi.org/10.1002/jsfa.12363

Zheng, J., Zhao, L., Yi, J., Zhou, L., & Cai, S. (2022). Chestnut starch nanocrystal combined with macadamia protein isolate to stabilize pickering emulsions with different oils. *Foods*, *11*(21), 3320. https://doi.org/10.3390/foods11213320

Zhou, L., Zhang, J., Xing, L., & Zhang, W. (2021). Applications and effects of ultrasound assisted emulsification in the production of food emulsions: A review. *Trends in Food Science & Technology*, *110*(1), 493–512. https://doi.org/10.1016/j.tifs.2021.02.008

Zhou, Z., Liang, Z., Zhang, Y., Hu, H., Gan, T., & Huang, Z. (2023). Facile solid-phase synthesis of starch-fatty acid complexes via mechanical activation for stabilizing curcumin-loaded Pickering emulsions. *Food Research International*, *166*(August 2022), 112625. https://doi.org/10.1016/j.foodres.2023.112625

Chapter 17

Processing of Protein Microparticles and Their Application as Lipid Replacer for Low-Calorie Foods

Ilse Monroy-Rodríguez
Ismael Ramírez-Flores
Raúl E. López-Hernández
Judith Jaimez-Ordaz
Humberto Hernández-Sánchez
Maribel Cornejo-Mazón
and Gustavo F. Gutiérrez-López

CONTENTS

17.1 INTRODUCTION

In recent decades, the demand for healthy food has increased significantly (McClements & Grossmann, 2021; Derossi et al. 2020). One of the purposes of food is related to preserving the

DOI: 10.1201/9781003332558-20

health of consumers, by preventing the development of diseases, such as diabetes, heart disease, and hypertension (WHO 2003; Pereira et al. 2009). Due to the increase of overweight and obesity, government agencies, the food industry, and research organizations (Malik et al. 2013) have sought effective strategies to overcome them. Obesity is the result of an energy imbalance provoked by the consumption of more calories than those that are expended. High-calorie diets and sedentary lifestyles have been identified as two of the main factors leading to this health condition (Malik et al. 2013). One of the most popular solutions for obesity treatment is the creation of reduced-calorie foods with a healthier profile of ingredients (e.g., reduced lipids and carbohydrates and increased protein contents) (Buttriss, 2013). The consumption of reduced-calorie foods could lead to improved health and well-being of the general population by reducing obesity and related diseases (Malik et al. 2013; Chung et al. 2014; Yashini et al. 2019).

Since lipids have the highest number of calories per unit mass of the major food components (lipids, proteins, carbohydrates) (Ognean et al. 2006; Rawson, 2019), lipid reduction has been a key target for the development of low-calorie foods (Chung et al. 2014).

On the other hand, it is important to consider the role of lipids in foods for the improvement of the acceptability of consumers in terms of palatability, textural, aroma, and sensory properties (Yashini et al. 2019. Hence, the need to substitute lipids with compounds that have similar physical properties, and in this respect, lipid replacers provide lower caloric value than lipids. The protein-based lipid replacers can replace around 75 and up to 100% of total lipid content (Yashini et al. 2019). This chapter discusses the use of protein microparticles as lipid substitute in foods, their classification, and processing methods.

17.2 CLASSIFICATION OF LIPID REPLACERS

Conventional options to reduce the lipid content in food have been focused on removing lipids from the food by physical treatments as in light foods or using cooking methods that avoid the use of oil as in hot-air frying. Traditionally, lipids were replaced by water or air, using lean meat, skim milk instead of whole milk and baking instead of frying for making snack foods (Ognean et al. 2006).

The use of lipid replacers allows the possibility of better mimicking the sensory properties of foods, lowering caloric intake. Lipid replacers are categorized as lipid substitutes and lipid mimetics (Yashini et al. 2019). Lipid substitutes are macromolecules with chemical structures similar to lipids in which the triacylglycerols are modified by compounds as sucrose esters and cannot be digested or absorbed. They are either chemically synthesized or enzymatically derived from lipids and are suitable for cooking and frying. In addition, they can completely replace the lipid content in foods; however, there are no nutritional benefits derived from this replacement. They are known as lipid-based fat replacers and some commercial examples of these additives are Olestra™ and Salatrim® (Yashini, et al. 2019).

On the other hand, lipid mimetics are substances that imitate the function of lipids such as gums, modified starch, polydextrose, and other polysaccharides or protein-based compounds, which could be derived from animal and plant proteins to simulate the texture and sensory properties of lipids in food. Lipid mimetics are suitable for baking and heat processing and provide low energy densities ranging from 1 to 4 kcal/g (Ognean et al. 2006).

Research to enhance the texture properties of proteins used as lipid replacers resulted in obtaining protein microparticles (PMs), which are the product of the intentional aggregation of the protein. The protein aggregates are obtained by various methods, such as pH and ionic strength control, thermal and mechanical treatments that allow to obtain particles ranging between 1–10 μm (Ipsen, 2016; Schijven et al. 2023). Singer et al. (1988) patented the first method to obtain microparticles by using whey protein isolate (WPI) from the dairy waste industry. These first PMs were obtained by thermal aggregation at high shear and low pH in whey protein isolate resulting in spherical particles ranging in size from 0.1 to 2.5 μm. The initial patent was later extended to other proteins like bovine serum albumin, egg white albumin, and soy protein (Singer et al. 1990). The use of PMs as lipid substitute was approved by the US Food and Drug Administration (FDA) in 1990 and has been classified as a GRAS (Generally Recognized as Safe) additive.

Since then, various types of microparticles have been prepared from different protein sources and processing methodologies; the variety of PMs differ in size and shape, which allow them to provide a variety of sensory characteristics in the foods to which they are added (Ipsen, 2016). Some of the most important commercial PMs are:

- **Simplesse®**
  The NutraSweet Company, IL, manufactures Simplesse®, USA (Singer & Dunn, 1990). The microparticles are produced by simultaneously heating and homogenizing whey protein mainly and in some cases with egg white protein. These microparticles have spherical shape and a diameter around 0.1–10 μm and are insoluble due to the exposure of disulfide groups out the aggregates. This product has been used in dairy and bakery food products (Ipsen, 2016; Yashini et al. 2019).
- **Dairy-Lo®**
  Ault Foods in Canada manufactures these PMs (Yashini et al. 2021). This product is obtained by partial denaturation of whey protein by using ultrafiltered whey protein concentrate at pH 6.1 and subjected to a mild heat treatment at 80°C for 17 s. The partial denaturation of proteins starts with unfolding and self-aggregation of proteins. Since it is not fully denatured, many of the functionalities such as water binding and emulsification properties are preserved, compared to treatments with higher temperature and pressure (Miller, 1994).
- **Trailblazer®**
  This product is manufactured by Kraft General Foods in Glenview, IL. USA, based on microparticles made by xanthan gum, egg white, and whey protein. Microparticles are elaborated by electrostatic interaction between protein and polysaccharides to obtain particles of <7 μm in size (Yashini et al. 2019).
- **LITA®**
  This lipid substitute is manufactured by Opta Foods Ingredients, Inc. in Cambridge, Mass. USA. It is obtained by the microparticulation of zein protein by solubilizing the protein in organic solvents and then in aqueous solution leading to protein precipitation, forming microparticle suspension. Compared to the most common animal PMs, zein microparticles are heat-stable and non-denatured (Stark & Gross, 1990).

In addition to commercial brands, some other studies have focused on developing new methods for obtaining microparticles, considering different protein sources and emerging technologies, as presented in the following sections.

17.3 CLASSIFICATION OF PROTEIN MICROPARTICLES AND THEIR PROCESSING

As mentioned above, a crucial criterion in developing PMs is the choice of the type of protein and its animal or vegetable origin (Yashini et al. 2019). In this sense, most technological advances have been focused on proteins of animal origin, which is supported by the immense variety of dairy products where the use of PMs as a lipid replacer has been widely applied. Although animal proteins have a higher biological value (Millward et al. 2008) it is worth mentioning that the production of plant-based PMs has gained special interest given that it is considered environmentally sustainable and more affordable than animal origin products (Derossi et al. 2020).

Table 17.1 shows examples of microparticles obtained from animal proteins, their main characteristics, and processing conditions. It can be noted that whey protein has been used in most cases due to its biological value and compatibility with dairy products. In general, the microparticulation process consists of three stages:

1) Heat treatment that induces the formation of protein aggregates, decreasing protein solubility.
2) Reduction of particle size by mechanical processing, using techniques such as microfluidization, high-shear mixing, high-pressure homogenization, milling, and twin-screw extruder.
3) Microparticles drying by spray- or freeze-drying.

Suárez et al. (2016) conducted a WPM (whey protein microparticle) optimization and prediction of PMs properties by considering the following operating conditions: temperature (80–90°C), heating time (0–50 min), pH (4–7), homogenization pressure (10–30 MPa), and number of homogenization cycles (1–3) as variables. They determined that by increasing the amount of aggregation and size of the microparticles (1–17 μm) the viscosity of the preparation increased from 4.5 to 65 mPa, which indicated the dependence of the structure of the aggregates on their textural properties.

Figure 17.1 shows images obtained by atomic force microscopy (AFM) during soy protein microparticulation process by thermal treatment and microfluidization. In the micrographs it was possible to observe the decrement of the size of microparticles increasing the number of microfluidization cycles, which could improve the stability of the suspension of aggregates.

It is interesting to consider that the production of PMs from protein isolate involves the mix of proteins with different characteristics, such as molecular weight and isoelectric point, which are important for microparticle morphology, size, and solubility (Miller, 1994). These characteristics should be considered in the technological applications in the food matrices in which they are added due to the changes in their properties, such as water retention capacity, rheology, and ability to stabilize colloidal systems such as gels, foams, suspensions, and emulsions (Shi et al. 2021).

Table 17.2 presents a summary of the PMs of plant origin. One of the advantages of using this type of PM is related to the development of plant-based foods, which are an important category for vegan and vegetarian consumers (Derossi et al. 2020; McClements & Grossmann, 2021). Another interesting product was the application of a mix of proteins with polysaccharides, as in the case of Trailblazer®, a lipid replacer produced by the combination of xanthan gum, egg white, and whey protein (Yashini et al. 2019). The considerations and effects of processing on the techno-functional properties of PMs are presented as follows.

TABLE 17.1 PROCESSING OF ANIMAL PROTEIN MICROPARTICLES

Protein Suspension	Processing Conditions	Reference
EWP (diluted in water, pH 3.8)	Heat treatment 90°C for 30 min. Cold-set gel formation. Homogenization by rotor stator system and high-pressure homogenizer 30 MPa. Spray-dried (180/70 °C inlet/outlet temperature)	Shi et al. (2023)
WPI (microemulsification: water, oil and hydrophobic surfactant, pH 4.7–5.2)	Two-step emulsification and cold-set gel formation	Rouwhorst et al. (2021)
EWP (9% w/v, pH 3.6)	Heating 75°C for 13 min. Homogenization (10,000 rpm, 60 s).	Liu et al. (2018)
WPC (30% w/w, pH 6.7)	Heat treatment 90–120°C. Homogenization 100–800 rpm Twin screw extruder.	Wolz, et al. (2016)
EWP (2, 5, 10% w/w, pH 3.0 and 3.8)	Heating 90°C for 45 min. Homogenization 11,000 rpm for 2 min, 10 MPa, 3 times).	Chang et al. (2016)
WPC (12% w/w)	Superfine grinding. Heating 85°C for 30 min. Homogenization 10,000 rpm for 10 min.	Sun et al. (2016)
WPC (12% w/v, pH 4.5)	Superfine grinding. Heating 85°C for 35 min). Homogenization 6 min, 10,000 rpm and lyophilization	Sun et al. (2016)
WPI (25% w/w protein solution and sunflower oil at 1:9, pH 5.5 or 6.8)	Two-step emulsification. Heat treatment 80°C for 20 min. Mixing at 6500 rpm for 5 min.; centrifugation at 33,768 g for 1 h, 4 times.	Sağlam et al. (2014)
MWP, α-LA and β- LG	High shear mixer (10% w/w protein, pH 4.6, isolated α-LA at 90°C, 20 min for α-LA, 90°C, 5 min for β-LG, 95°C, 6 min for WP, 1,200 rpm) and spray- drying	Toro-Sierra et al. (2013)
WPI (40% w/w, pH 6.8)	Heating 90° C for 30 min. Mixing at 800 rpm for 50 min. Oven-drying (50–60°C) and milling (15,000 rpm, 120 s).	Purwanti et al. (2012)
WPC (7% w/w, pH 3.0 or 6.0)	Heat treatment 90°C for 20 min. Microfluidization (140 MPa, 5 passes) and spray-drying.	Dissanayake et al. (2012)
WPC (10% w/w, pH 7.0)	Heating 90° C for 20 min. Microfluidization (140 MPa, 1 or 5 passes) and spray-drying	Dissanayake & Vasiljevic (2009)
WPC (5% w/w, pH 3.8, 4.8, 5.8)	Heating 90°C for 10 min. Microfluidization (150 MPa, 3 passes), spray-drying or freeze-drying	Iordache & Jelen (2003)

(continued)

TABLE 17.1 (CONTINUED)

Protein Suspension	Processing Conditions	Reference
WPC (10% w/w, 0.05–0.33% w/w $CaCl_2$ pH 3.5–6.7)	Heat treatment 80–110°C. High shear mixer 1200 rpm.	Spiegel & Huss (2002)
WPI (20% w/w, pH 3.5–6.8)	Heat treatment 90–100°C. Twin screw extruder 100–200 rpm.	Queguiner et al. (1992)
WPC (21% w/w, pH 3.5–4.5)	Heat treatment 120°C for 5–15 min. High shear mixer 500 rpm.	Singer & Dunn (1990)

***WPC**; whey protein concentrate. **WPI**; whey protein isolate. α-**LA**; α lactoalbumin. β- **LG**; β lactoglobulin. **EWP**; egg white protein.

TABLE 17.2 PROCESSING OF PLANT PROTEIN MICROPARTICLES

Protein Suspension	Processing Conditions	Reference
Potato protein concentrate (75.5%, pH 6.9)	Twin-screw extruder with a screw diameter of 25mm, screw speed of 800 rpm, extruder barrel temperature 80°C	Tanger et al. (2022)
Pea protein concentrate (70.8%, pH 7.5)	Twin-screw extruder with a screw diameter of 25mm, screw speed of 800 rpm, extruder barrel temperature 130°C	Tanger et al. (2022) Tanger et al. (2021)
Soy protein isolate (3% w/w, pH 3, 5, 7, and 9).	Heat treatment (93°C for 15 min) Microfluidization (110MPa for 1–3 passes).	Monroy-Rodríguez et al. (2021)
Zein (1–3% w/w, 70% w/w ethanol)	Microfluidization (0–150 MPa, 2 passes)	Sun et al 2016; Sun et al. (2016)
Soy protein (7% w/w)	Microfluidization (0, 137 and 270 MPa for 0, 10, 20, and 30 passes)	Song et al. (2013)
Zein protein concentrate (20.67% w/w, pH 3.86, 6.0, 8.0, and 10.0)	Milling 20,000 rpm. Microfluidization (50 MPa followed by 125 MPa).	Öztürk & Mert, (2018)
Mung bean protein concentrate (5% w/w protein solution)	Heating 83°C, 5–15 min while homogenized at 17,000 rpm. Homogenization 23,000–27,000 rpm, 5 min and centrifugation 1000–4000g, 10 min.	Sirikulchayanont et al. (2007)

Figure 17.1 Atomic force microscopy images of microparticulation processing of soy protein. 1.1 Soy protein isolate (SPI) at pH 5: 1.2 SPI after heating treatment 93°C for 15 min; 1.3 SPI microparticles after 1 cycle of microfluidization at 110 MPa; 1.4 SPI microparticles after two cycles of microfluidization at 110 MPa; 1.5 SPI microparticles after three cycles of microfluidization at 110 MPa.

17.4 EFFECT OF PROCESSING ON FUNCTIONAL PROPERTIES OF PROTEIN MICROPARTICLES

In most cases, PMs are produced from protein isolates, concentrates, or protein mixtures. Prieto et al. (2022) evaluated the morphology and encapsulation efficiency of whey protein isolates as wall material with different percentages of purity (70, 80, and 90%) for PMs production. They found that it was possible to obtain less porous and more homogeneous PMs when the protein isolate purity was higher than 80%.

Furthermore, it is crucial to consider the functional properties of the proteins and their ability to interact with other components in food such as lipids, water, carbohydrates, and minerals (Schijven et al. 2023). **Figure 17.2** shows a diagram of the variables that determine the size, microstructure, and functional properties of PMs. During the process of microparticulation, the functionality of native proteins changes and there are reports in which the effect of microparticulation on the characteristics and functionality of PM were evaluated (Iordache & Jelen, 2003; Torres et al. 2016; Shi et al. 2021). Some of the most important properties of PMs are discussed below.

Particle size and microstructure: The particle size of PMs depends on the degree of denaturation and processing conditions. In general, spherical microparticles are produced by most procedures and their size ranges between 0.1–10 μm (Ipsen, 2016; Shi, et al. 2021). In addition, it is possible to relate the particle size with the functionality and mouthfeel. In this respect, small particles (<1 μm) can be better integrated into food, provide a creamy texture, and improve the water holding capacity (WHC). On the other hand, large particle sizes (>10 μm) could positively affect WHC in gels such as yogurt or cheeses (Torres et al. 2016).

Solubility: During aggregation, proteins are partially denatured, causing the exposure of hydrophobic groups and disulfide bonds that contributes to a decrement in solubility (Schokker & Singh, 2000). This property is relevant when using PMs in beverages to avoid phase separation (Dissanayake et al. 2009).

Water holding capacity: This property depends on the interaction between the protein and water molecules. WHC must be considered in the development of gel-type foods because it is related to their consistency, induction of syneresis, and yield. In general, small microparticles contribute to the formation of the gel network (Sun et al. 2016; Öztürk & Mert, 2018).

Gelation: The gelation of proteins depends on several factors, such as pH, thermal stability, ionic concentration, and mechanical stress (Dissanayake et al. 2010). Gelation can be favored by the increase in hydrophobic interactions between proteins. In the case of beverages processed by heat, the effect of pH and electrostatic repulsion forces on PMs should be considered to avoid gelation (Torres et al. 2016). Dissanayake et al. (2010) found that the addition of glucan-δ-lactone and/or NaCl controlled repulsive forces between the pre-denatured proteins at cold-set conditions and promoted intermolecular interactions thus determining gel strength.

Emulsifying: The microparticulation process causes an increase in the hydrophobic surface, which improves the protein's ability to interact with lipids (Shi et al. 2021). This effect has been demonstrated in whey (Dissanayake & Vasiljevic, 2009) and zein preparations (Öztürk & Mert, 2018). In addition, particle size is also relevant in emulsion stability. Besides PM concentration, particle size and electrostatic repulsion must be considered (Chang et al. 2016).

Foaming: Foaming properties could be determined by the ability of proteins to diffuse into the newly formed interface, unfold, reorient, and form viscoelastic films to stabilize the dispersed phase (Shi et al. 2021). The foaming properties can be favored by the denaturation of the protein because of thermal treatment and improve the formation of the protein crosslinking due to the decrease in particle size (Dissanayake & Vasiljevic, 2009; Sun et al. 2016). Monroy-Rodríguez et al. (2020)

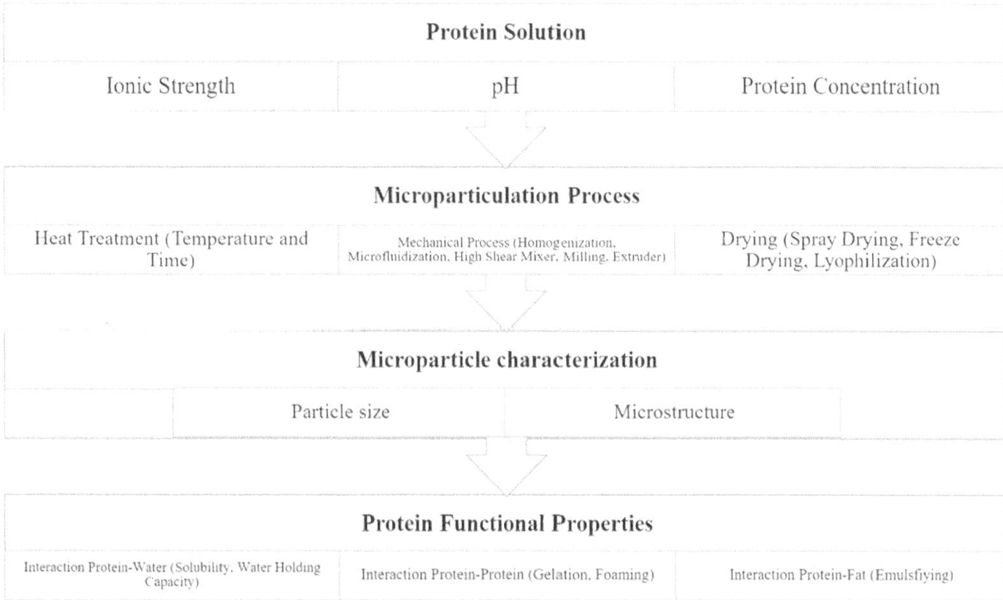

Figure 17.2 Conceptual scheme of variables in protein microparticulation.

found that protein partial denaturation by the effect of heating, pH control, and microfluidization provoked a decrease of protein solubility and might reduce the surface tension of the system, favoring the entry of air into the media, which leads to a greater volume of the foam. The improvement of this property could be relevant for the stability and quality in ice cream and desserts.

Some cases of low-lipid foods formulated with PM are presented in the following.

17.5 APPLICATIONS OF PROTEIN MICROPARTICLES FOR LOW-LIPID FOODS

17.5.1 Chocolate

Lipid properties are determinants of chocolate quality due to their emulsifying properties, palatability, and melting point. Godoi et al. (2021) produced chocolate-flavored milk with protein-based hydrocolloids like Simplesse® (microparticulated whey protein concentrate) and gelatin and collagen bovine hydrolyzate (GELITA Australia) as lipid replacers. They found that the flow consistency coefficient (K) increased in the reduction of protein hydrocolloids without significantly affecting the particle size distribution (most particles did not exceed 40 μm). In this formulation, coagulation was not observed despite the particle size being heterogeneous. The heating processing could cause a certain degree of denaturation, which could provoke larger aggregates. Despite this, their results indicated that the particles were smaller than the irregularities the human tongue can sense, which ranges between 42 and 95 μm. Therefore, since the use of PMs was not unfavorable to chocolate texture and palatability, they could represent a good option for the formulation of low-calorie chocolate (Godoi et al. 2021).

17.5.2 Dairy

In semi-skimmed dairy products PMs have found a considerable field of application as a lipid mimetic (Ipsen, 2016). This is because the microparticulation process has been focused mainly on the use of milk whey proteins. In this section, some advances in low-lipid dairy products formulated with PMs are discussed.

17.5.3 Beverages

In the case of dairy-based beverages, it is relevant to consider the emulsifying properties of proteins and their thermal stability to avoid phase separation, which can be considered a defect towards the appearance and palatability of the products (Torres et al. 2017). Emulsions prepared with microparticulated whey protein (MWP) have higher thermal stability compared to emulsions made with standard whey protein concentrate (WPC) (Cakir-Fuller, 2015). It is attributable to the fact that the MWP has fewer active sites, which increase by applying the secondary thermal process. Consequently, the main component determining the thermal equilibrium is the concentration of the undenatured protein. Banks (2004) found that dispersions having a total protein concentration of 18%, prepared with MWP at pH 5.5 and using a double emulsification method, exhibited exceptional thermal stability (90°C for 30 minutes at pH ~7). Thermal stability is a prerequisite during the production of high-protein beverages and therefore MWP is are a crucial component in such formulations (Sağlam et al. 2014).

Saglam et al. (2011) found good thermal stability at pH 5.5 in MWP due to non-covalent interactions (Saglam et al. 2011). Chung et al. (2014) proved that adding polysaccharides mixed with MWPs was a good strategy for the formulation of reduced-calorie food emulsions with appearance and consistency similar to full-lipid versions.

Mounsey et al. (2009) found that MWPs could replace 50 to 100% of lipids in reconstituted milk without any major adverse effects on viscosity or heat stability. Meza et al. (2016) investigated glaze materials (icing sugar, skim milk powder, water, and vegetable lipids) and the use of MWP as lipid mimetic and found that MWP was able to secure low-lipid glaze materials with rheological properties, microstructure, and film thickness similar to those of full-lipid formulations.

17.5.4 Cheese

The use of MWP in cheese is essential because for a long time whey represented a polluting residue of cheese production (Ipsen, 2016). It is possible to recycle cheese whey into cheese milk or curd matrix from thermal and shear processing to produce whey microparticles (Guyomarc'h, 2006; Mounsey et al. 2009). The addition of MWP to cheese milk can improve the texture of low-lipid cheese resulting in a smoother texture. This can provide better meltability and flowability in low-lipid cheese such as mozzarella (McMahon et al. 1996).

Desai & Nolting (1995) studied the microstructure of low-lipid cheddar cheese and found that MWP acted as a non-interacting bulking agent and that the particles appeared stable throughout cheese aging. Similarly, no influence of MWP aggregation on ripening was found for this type of cheese (Fenelon & Guinee, 1997). It has also been observed that the aggregation of carbohydrates or protein substitutes significantly reduce age-related proteolysis compared to a low-lipid cheese or full-lipid cheddar cheese (Haque et al. 2007). In addition, the higher moisture content in low-lipid, salted white cheese made with MWP was found to increase proteolytic and lipolytic activity during ripening (Romeih et al. 2002). MWP has also been applied in specialty cheeses such as Italian Cacciotta, for which no difference in cheese yield between standard and high-lipid cheeses was observed for MWP additions of up to 4% indicating its potential use as a lipid mimetic (Sturaro et al. 2015). MWPs remarkably help increase the moisture yield of low-lipid cheeses and are used in conjunction with starter cultures that produce exopolysaccharides and adjunct cultures to enhance flavor, which constitutes a possibility for making specialty low-lipid cheeses (Di Cagno et al. 2014).

Cheeses made with high levels of native whey proteins do not melt, given that the original whey proteins were incorporated into the cheese matrix, resulting in a denser composition due to the disulfide bonds as well as the non-covalent interactions created during their preparation. The degree of denaturation presented by the MWP could improve the melting properties of cheeses (Ipsen, 2016).

17.5.5 Ice Cream

The lipid content of ice cream is strongly associated with smoothness, creaminess, and mouth coverage, and low-lipid dairy firms cause an increase in mouth creaminess and coverage (Liou & Grun, 2007).

Yilsay, Yilmaz, & Bayizit (2006) and Liou & Grun (2007) reported the results of sensory evaluations of low-lipid ice cream formulated with PMs and found that this product presented lower viscosity but higher smoothness and similar creaminess to the full-lipid version. Liou & Grun (2007) also found synergy using Simplesse® added with a polydextrose lipids substitute (Litesse®). The obtained products exhibited a creamy texture similar to ice cream with 10% lipids, and although products with Simplesse® had flavor and aroma, their profile was more like full-lipid ice cream. A current proposal consists of incorporating vegetable proteins to produce plant-based dairy analogs. However, the challenge is to prevent the flavor of vegetable proteins from negatively affecting the sensory profile in products such as milk desserts and carry out the necessary modifications on proteins so that the techno-functional properties can stabilize dairy emulsions (McClements & Grossmann, 2021; Tanger et al. 2022).

17.5.6 Yogurt

Adding MWP to yogurt provides a creamy mouthfeel similar to low-lipid stirred yogurt (Janhoj, 2006) and improve the texture of settled low-lipid yogurt (Sandoval-Castilla et al. 2004). However, it should be noted that while creaminess in low-lipid yogurt is primarily a function of mouth viscosity and smoothness, this is not just a result of textural properties and the presence, appearance, and flavor of particles is also important (Janhoj et al. 2006). Tamime et al. (1995) noted that Simplesse® became an integral part of the yogurt microstructure but noted that the lipid replacer particles appeared to be larger than milk fat globules.

Through rheological measurements, it was also found that WPC provided a protein network that was more structurally similar to full-lipid yogurt than when MWP was added (Lobato-Calleros et al. 2004). Thus, in fermented dairy products as well as other dairy matrices, the type of added particulate protein component is relevant. Torres et al. (2011) examined several different MWP ingredients and found that a high ratio of natural to denatured whey protein resulted in a thicker and creamier low-lipid yogurt. They also found that the surface reactivity of the microparticles may be more significant than the initial particle size of the MWP aggregates in the final perception of particle size in yogurt.

The large particle size of the initial aggregates resulted in low particle size yogurt when the microparticles become part of the yogurt matrix during milk fermentation. Typically, MWP is added as a powder ingredient, so hydration is also crucial for functionality. Torres et al. (2016) investigated the hydration and water retention of two different MWP ingredients, which differed in particle size and content of native whey protein. They found that products with more natural whey protein had better water retention (less free water) and explained that this was the result of denaturation and further aggregation during heating of the milk prior to acidification.

It was found that MWP produced high levels of immobilized water in the protein network compared to WPC, and that 60 min of hydration of small whey particles (<1 μm) was sufficient to promote efficient interactions with other proteins and lead to high gel strengths to guide and improve the WHC. On the other hand, larger particles inhibited the gelation process and resulted in weaker yogurt and lower viscosity due to clog breakage, decreasing the need for native whey proteins to interact with other whey proteins and/or cheeses on the surface of proteins (Torres et al. 2016). In addition, MWP can be used to concentrate fermented products and to improve the organoleptic properties of strained yogurts for the enhancement of creaminess (Yazici & Akgun, 2004; Janhoj et al. 2009).

17.5.7 Mayonnaise

Mayonnaise is an oil-in-water emulsion traditionally prepared by mixing egg yolk, vinegar, oil, and salt. Whole traditional mayonnaise containing 70–80 g/100 g of lipids is one of the most commonly used dressings in the world (Chang et al. 2017). Microparticulate whey protein coats the mouth and taste buds, similar to lipids (Mirzanajafi-Zanjani, 2019). The use of MWP can help to mask some of the bitter and astringent flavors that seem more pronounced in low-lipid products (Mirzanajafi-Zanjani, 2019). The lipid mimetics based on MWP and pectin decrease the caloric value due to a lower proportion of oil in the formulation, and also maintain rheological properties and stability (Sun et al. 2016).

In a concentrated emulsion, flocculation leads to the formation of a three-dimensional network of small aggregate droplets. Increasing the shear rate causes progressive deformation and disruption of the aggregated particles, which in turn lowers the resistance of the emulsion to flow and reduces its apparent viscosity (Mirzanajafi-Zanjani, 2019).

Regarding its appearance, the color is one of the most important factors that can affect the acceptability of mayonnaise. Traditional whole mayonnaise is characterized by a bright and pale yellowish appearance, and sensory attributes (i.e., appearance, color, texture, smell, taste, and acceptability) of the low-lipid mayonnaises did not show a significant difference with original full-lipid products. Nevertheless, it was also observed that a high concentration of MWP caused flocculation and decreased the stability of the whole system (Chang et al. 2017).

Rheological attributes indicated that the addition of lipids mimetic influenced the steady-state dynamic flow characteristic and microrheological properties of low-lipid, lipid-free, and full-lipid mayonnaises. However, sensory evaluation results indicated that differences in rheology were too small for the human sense to distinguish such influence (Chang et al. 2017; Sun et al. 2016).

The decreasing ratio of the dispersed oil phase and the continuous aqueous phase resulted in the reduction of viscoelasticity of the oil-water emulsion. The MWP-pectin-based lipids mimetic could partially suppress the decrease in lipids substitution by up to 40% (Chang et al. 2017).

Long-term stability studies showed that the addition of up to 60% lipids mimetic could significantly improve the storage stability of mayonnaises by preventing droplet coalescence and flocculation. Both dynamic mechanical measurement and stability study results suggested that MWP and pectin could be a potential mimetic of the lipids used in mayonnaise preparation (Chang et al. 2017).

17.5.8 Meat

In recent years, it has become increasingly difficult not to associate the consumption of meat products with human health. The presence of excessive amounts of sodium, saturated lipids, cholesterol, and synthetic additives in meat products contributes to an increased risk of metabolic diseases, such as obesity, type 2 diabetes, and coronary heart disease (Domínguez et al. 2020). According to Domínguez et al. (2018) meat products can be reformulated by using healthy oils (vegetable and/or marine) such as lipid substitutes by incorporating them as emulsions or emulsion gels without affecting the characteristics of the final product. The lipid content can facilitate the release of moisture and is associated with the stability of emulsified systems in meat textural properties (McClements &Grossmann, 2021).

Teresinha et al. (2018) proposed a low-fat meat hamburger formulation with the substitution of 50% of pork lipids with microencapsulated chia oil. They concluded that reforming lipids with chia oil microparticles is a promising alternative to incorporating oils into cooked meat products when total lipids content is reduced.

The role of proteins in the structure and attributes of food is of paramount importance. Applications and formulations utilizing animal proteins have historically outperformed plant-based proteins due to more significant advancements in scientific and technological development. Many plant-based proteins are globular large proteins (McClements & Grossmann, 2021). Nevertheless, it is possible for this group of proteins to mimic the properties of animal proteins through chemical, physical, and enzymatic modifications, which can alter the structure and exposure of functional groups, thereby enhancing their integration and functional properties (McClements & Grossmann, 2021).

In most reported cases, the research and development of PMs contribute to the standardization of production processes, resulting in foods with improved nutritional and sensory properties and processing yields.

17.6 CONCLUSION

Currently, there are many protein alternatives suitable for meeting the needs of consumers such as plant-based diets or for patients with food allergies. Protein microparticles are a novel ingredient useful for the development of healthier foods. Low-lipid versions of dairy, dessert, dressing, and meat preparations using PMs have shown favorable results in mimicking the sensory properties of lipids and most applications have focused on dairy products. Developing meat analogs with a better nutritional profile than conventional meat would be an interesting and worthwhile endeavor through the application of PMs. The evaluation of the changes in the microstructure and functionality of the proteins subjected to the microparticulation should be considered as a control quality parameter and future research must be focused on the effect of microparticulation protein processing on determinant factors for human nutrition and health, such as biological value, digestibility, and allergenicity, among others.

REFERENCES

Banks, J.M. (2004). The technology of low-lipid cheese manufacture. *International Journal of Dairy Technology*, 57, 199–207. https://doi.org/10.1111/j.1471-0307.2004.00136.x

Buttriss, J.L. (2013). Food reformulation: The challenges to the food industry. *Proceedings of the Nutrition Society*, 72, 61–69. https://doi.org/10.1017/S0029665112002868

Cakir-Fuller, E. (2015). Enhanced heat stability of high protein emulsion systems provided by microparticulated whey proteins. *Food Hydrocolloids*, 47, 41–50. https://doi.org/10.1016/j.foodhyd.2015.01.003

Chang, C., Li, J., Li, X., Wang, C., Zhou, B., Su, Y., & Yang, Y. (2017). Effect of protein microparticle and pectin on properties of light mayonnaise. *LWT – Food Science and Technology*, 82, 8–14. https://doi.org/10.1016/j.lwt.2017.04.013

Chang, C., Niu, F., Gu, L., Li, X., Yang, H., Zhou, B., Wang, J., Su, Y., & Yanjun, Y. (2016). Formation of fibrous or granular egg white protein microparticles and properties of the integrated emulsions. *Food Hydrocolloids*, 61, 477–486. https://doi.org/10.1016/j.foodhyd.2016.06.002

Chung, C., Degner, B., & McClements, J.D. (2014). Development of reduced-calorie foods: Microparticulated whey proteins as lipid mimetics in semi-solid food emulsions. *Food Research International*, 56, 136–145. https://doi.org/10.1016/j.foodres.2013.11.034

Derossi, A., Husain, A., Caporizzi, R., & Severini, C. (2020). Manufacturing personalized food for people uniqueness. An overview from traditional to emerging technologies. *Critical Reviews in Food Science and Nutrition*, 60(7), 1141–1159. https://doi.org/10.1080/10408398.2018.1559796

Desai, N., & Nolting, J. (1995). Microstructure studies of reduced-lipid cheeses containing lipid substitute. In E.L. Malin & M.H. Tunick (Eds.), *Chemistry of structure-function relationships in cheese* (pp. 295–302). London, UK: Plenum Press.

Di Cagno, R., De Pasquale, I., De Angelis, M., Buchin, S., Rizzello, C.G., & Gobbetti, M. (2014). Use of microparticulated whey protein concentrate, exopolysaccharide-producing Streptococcus thermophilus, and adjunct cultures for making low-lipid Italian caciotta-type cheese. *Journal of Dairy Science*, 97, 72–84. https://doi.org/10.3168/jds.2013-7078

Dissanayake, M., Kelly, A., & Vasiljevic, T. (2010). Gelling properties of microparticulated whey proteins. *Journal of Agriculture and Food Chemistry*, 58(11), 6825–6832. https://doi.org/10.3168/jds.2008-1791

Dissanayake, M., Liyanaarachchi, S., & Vasiljevic, T. (2012). Functional properties of whey proteins microparticulated at low pH. *Journal of Dairy Science*, 95(4), 1667–1679. https://doi.org/10.3168/jds.2011-4823

Dissanayake, M., & Vasiljevic, T. (2009). Functional properties of whey proteins affected by heat treatment and hydrodynamic high-pressure shearing. *Journal of Dairy Science*, 92(4), 1387–1397. https://doi.org/10.3168/ jds.2008-1791

Domínguez R., Munekata P.E., Pateiro M., Lopez-Fernández O., & Lorenzo J.M. (2020). Immobilization of oils using hydrogels as strategy to replace animal lipids and improve the healthiness of meat products. *Current Opinion in Food Science*, 37, 135–144. https://doi.org/10.1016/j.cofs.2020.10.005

Fenelon, M A., & Guinee, T.P. (1997). The compositional, textural and maturation characteristics of reduced-lipid Cheddar made from milk containing added Dairy Lo™. *Milch Wissenschaft-Milk Science International*, 52, 385–389.

Godoi, F.C., Ningtyas, D.W., Geoffroy, Z., & Prakash S. (2021). Protein-based hydrocolloids: Effect on the particle size distribution, tribo-rheological behaviour and mouthfeel characteristics of low-lipid chocolate flavoured milk. *Food Hydrocolloids*, 115, 106628. https://doi.org/10.1016/j.foodhyd.2021.106628

Guyomarc'h, F. (2006). Formation of heat-induced protein aggregates in milk as a means to recover the whey protein fraction in cheese manufacture, and potential of heat-treating milk at alkaline pH values in order to keep its rennet coagulation properties. A review. *Lait*, 86, 1–20. https://doi.org/10.1051/lait:2005046

Haque, Z.U., Kucukoner, E., Aryana, K.J. (2007). Influence of lipid-replacing ingredients on process and age induced soluble nitrogen content and ultrastructure of low lipid Cheddar cheese. *Food Science and Technology Research*, 13, 338–344. https://doi.org/10.3136/fstr.13.338

Iordache, M., Jelen, P. (2003). High pressure microfluidization treatment of heat denatured whey proteins for improved functionality. *Innovative Food Science & Emerging Technologies*, 4(4), 367–376. https://doi.org/10.1016/s1466-8564(03)00061-4

Ipsen, R. (2016). Microparticulated whey proteins for improving dairy product texture. *International Dairy Journal*, 67, 73-79. https://doi.org/10.1016/j.idairyj.2016.08.009

Janhoj, T., Frost, M.B., Prinz, J., Ipsen, R. (2009). Sensory and instrumental characterization of low-lipid and non-lipid cream cheese. *International Journal of Food Properties*, 12, 211–227. https://doi.org/10.1080/10942910802252007

Janhoj, T., Petersen, C.B., Frost, M.B., & Ipsen, R. (2006). Sensory and rheological characterization of low lipid stirred yogurt. *Journal of Texture Studies*, 37, 276–299. https://doi.org/10.1111/j.1745-4603.2006.00052.x

Liou, B.K., & Grun, I.U. (2007). Effect of lipid level on the perception of five flavor chemicals in ice cream with or without lipid mimetics by using a descriptive test. *Journal of Food Science*, 72, S595–S604. https://doi.org/10.1111/j.1750-3841.2007.00494.x

Liu, R., Tian, Z., Song, Y., Wu, T., Sui, W., & Zhang, M. (2018). Optimization of the production of microparticulated egg white proteins as lipid mimetic in salad dressings using uniform design. *Food Science and Technology Re*search, 24(5), 817–827. https://doi.org/10.3136/fstr.24.817

Lobato-Calleros, C., Martinez-Torrijos, O., Sandoval-Castilla, O., Perez-Orozco, J.P., & Vernon-Carter, E.J. (2004). Flow and creep compliance properties of reduced-lipid yoghurts containing protein-based lipid replacers. *International Dairy Journal*, 14, 777–782. https://doi.org/10.1016/j.idairyj.2004.02.012

Malik, V.S., Willett, W.C., & Hu, F.B. (2013). Global obesity: Trends, risk factors and policy implications. *Nature Reviews Endocrinology*, 9, 13–27. https://doi.org/10.1038/nrendo.2012.199

McClements, D.J., & Grossmann, L. (2021). The science of plant-based foods: Constructing next-generation meat, fish, milk, and egg analogs. *Comprehensive Reviews in Food Science and Food Safety*, 20(4), 4049–4100. https://doi.org/10.1111/1541-4337.12771

McMahon, D.J., Alleyne, M.C., Fife, R.L., & Oberg, C.J. (1996). Use of lipid replacers in low lipid Mozzarella cheese. *Journal of Dairy Science*, 79, 1911–1921. https://doi.org/10.3168/jds.S0022-0302(96)76560-8

Meza, B.E., Peralta, J.M., & Zorrilla, S.E. (2016). Rheological characterization of full-lipid and low-lipid glaze materials for foods. *Journal of Food Engineering*, 171, 57–66. http://doi.org/10.1016/j.jfoodeng.2015.10.012

Miller, M.S. (1994). Proteins as lipid substitutes. In N. Hettiarachchy, & G. Ziegler (Eds.), *Protein functionality in food systems* (pp. 435–465). New York: Marcel Dekker Inc.

Millward, D.J., Layman, D.K., Tomé, D., & Schaafsma, G. (2008). Protein quality assessment: Impact of expanding understanding of protein and amino acid needs for optimal health. *American Journal of Clinical Nutrition*, 87(5), 1576S–81S. https://doi.org/10.1093/ajcn/87.5.1576S

Mirzanajafi-Zanjani, M., Yousefi M., & Ehsani A. (2019). Challenges and approaches for production of a healthy and functional mayonnaise sauce. *Food Science & Nutrition*, 7, 2471-2484. https://doi.org/10.1002/fsn3.1132

Monroy-Rodríguez, I., Monroy-Villagrana A., Cornejo-Mazón M., García-Pinilla S., Hernández-Sánchez H., & Gutiérrez-López G.F. (2020). Microfluidization in nano- food engineering. In: *Nano-food Engineering*, U. Hebbar, S. Ranjan, N. Dasgupta, R. Kumar Mishra (Eds.), Food Engineering Series. Ed. Springer. https://doi.org/10.1007/978-3-030-44552-2_6

Monroy-Rodríguez, I., Gutiérrez-López, G.F., Hernández-Sánchez, H., López-Hernández, R.E., Cornejo-Mazón, M., Dorantes-Álvarez, L., Alamilla-Beltrán, L. (2021). Surface roughness and textural image analysis, particle size and stability of microparticles obtained by microfluidization of soy protein isolate aggregates suspensions. *Revista Mexicana de Ingeniería Química*, 2, 787–805. http://rmiq.org/iqfvp/Numbers/V20/No2/Alim2311.pdf

Mounsey, J.S., O'Kennedy, B.T., Corrigan, B., Kelly, P.M., & O'Callaghan, D.J. (2009). Effect of microparticulated whey protein addition on the heat stability and rennet/acid gelling properties in model milk systems. *Milchwissenschaft-Milk Science International*, 64, 14–17.

Ognean, C.F., Darie, N., & Ognean, M. (2006). Lipid replacers-review. *Journal of Agroalimentary Processes and Technologies*, 12(2), 433–442. https://www.journal-of-agroalimentary.ro/admin/articole/74033L66_FAT_REPLACERS_final.pdf

Öztürk, O.K., & Mert, B. (2018). The effects of microfluidization on rheological and textural properties of gluten-free corn breads. *Food Research International*, 105,782–792. https://doi.org/10.1016/j.foodres.2017.12.008

Pereira, H.V.R., Saraiva, K.P., Carvalho, L.M.J., Andrade, L.R., Pedrosa, C., & Pierucci, A.P.T.R. (2009). Legumes seeds protein isolates in the production of ascorbic acid microparticles. *Food Research International*, 42(1), 115–21. https://doi.org/10.1016/j.foodres.2008.10.008

Prieto, C., Talón, E., Noreña, C.Z., & Lagaron, J.M. (2022). Effect of whey protein purity on the characteristics of algae oil-loaded encapsulates obtained by electrospraying assisted by pressurized gas. *Nanomaterials*, 12(18), 3096. https://doi.org/10.3390/nano12183096

Purwanti, N., Moerkens, A., Van Der Goot, A., & Boom, R. (2012). Reducing the stiffness of concentrated whey protein isolate (WPI) gels by using WPI microparticles. *Food Hydrocolloids*, 26(1), 240–248. https://doi.org/10.1016/j. foodhyd.2011.05.015

Queguiner, C., Dumay, E., Salou-Cavalier, C., & Cheftel, J.C. (1992). Microcoagulation of a whey protein isolate by extrusion cooking at acid pH. *Journal of Food Science*, 57(3), 610–616. https://doi.org/10.1111/j.1365-2621.1992.tb08054.x

Rawson. (2019). Protein-based lipid replacers – A review of recent advances. *Food Reviews International*, 37(2), 197-223. https://doi.org/10.1080/87559129.2019.1701007

Romeih, E.A., Michaelidou, A., Biliaderis, C.G., & Zerfiridis, G.K. (2002). Low-lipid white-brined cheese made from bovine milk and two commercial lipid mimetics: Chemical, physical and sensory attributes. *International Dairy Journal*, 12, 525–540. https://doi.org/10.1016/S0958-6946(02)00043-2

Rouwhorst, J., Van Baalen, C., Velikov, K., Habibi, M., Van Der Linden, E., & Schall, P. (2021). Protein microparticles visualize the contact network and rigidity onset in the gelation of model proteins. *Npj Science of Food*, 5(1), 32. https://doi.org/10.1038/s41538-021-00111-5

Saglam, D., Venema, P., de Vries, R., Sagis, L.M.C., & Van der Linden, E. (2011). Preparation of high protein micro-particles using two-step emulsification. *Food Hydrocolloids*, 25, 1139–1148. https://doi.org/10.1016/j.foodhyd.2010.10.011

Sağlam, D., Venema, P., De Vries, R., & Van Der Linden, E. (2014). Exceptional heat stability of high protein content dispersions containing whey protein particles. *Food Hydrocolloids*, 34(1), 68–77. https://doi.org/10.1016/j. foodhyd.2012.12.016

Sağlam, D., Venema, P., Van der Linden, E., & de Vries, R. (2014). Design, properties, and applications of protein micro- and nanoparticles. *Current Opinion in Colloid and Interface Science*, 19, 428–437. https://doi.org/10.1016/j.cocis.2014.09.004

Sandoval-Castilla, O., Lobato-Calleros, C., Aguirre-Mandujano, E., & Vernon-Carter, E.J. (2004). Microstructure and texture of yogurt as influenced by lipid replacers. *International Dairy Journal*, 14, 151–159. https://doi.org/10.1016/S0958-6946(03)00166-3

Schijven, L.M.I., Saggiomo, V., Velders, A.H., Bitter, J.H., & Nikiforidis, C.V. (2023). On the influence of protein aggregate sizes for the formation of solid and hollow protein microparticles. *Journal of Colloid and Interface Science*, 631, 181–190. https://doi.org/10.1016/j.jcis.2022.11.007

Schokker, E., Singh, H., & Creamer, L. (2000). Heat-induced aggregation of β-lactoglobulin A and B with α-lactalbumin. *International Dairy Journal*, 10(12), 843–853. https://doi.org/10.1016/s0958-6946(01)00022-x

Shi, D., Li, C., Stone, A.K, Guldiken, B., & Nickerson, M.T. (2021). Recent developments in processing, functionality, and food applications of microparticulated proteins, *Food Reviews International*, 39 (3), 1309-1332. https://doi.org/10.1080/87559129.2021.1933515

Shi, Y., Li, J., Gu, L., Su, Y., Chen, W., Zhang, M., Chang, C., & Yang, Y. (2023). Synergistic effect of gum arabic and xanthan gum on improving rheological properties of low-lipid mayonnaise with egg white protein microparticle as a lipid mimetic. *International Journal of Food Science & Technology*, 58(3), 1037–1048. https://doi.org/10.1111/ijfs.16237

Singer, N.S., & Dunn J.M. (1990). Protein microparticulation: The principle and the process. *Journal of the American College of Nutrition*, 9(4), 388-397. https://doi.org/10.1080/07315724.1990.10720397

Singer, N.S., Latella, J., & Shoji, Y. (1990). Lipid emulating protein products and process. Patent 4,961,953. John LaBatt Limited, London, ON, Canada.

Singer, N.S., Yamamoto, S., & Latella, J. (1988). Protein product base. Patent US 4,734,287. John LaBatt Limited, London, ON, Canada.

Sirikulchayanont, P., Jayanta, S., Pradipasena, P., & Miyawaki, O. (2007). Characteristics of microparticulated particles from mung bean protein. *International Journal of Food Properties*, 10(3), 621–630. https://doi.org/10.1080/10942910601051212

Song, X., Zhou, C., Fu, F., Chen, Z., & Wu, Q. (2013). Effect of high-pressure homogenization on particle size and film properties of soy protein isolate. *Industrial Crops and Products*, 43, 538–544. https://doi.org/10.1016/j.indcrop.2012.08.005

Spiegel, T., & Huss, M. (2002). Whey protein aggregation under shear conditions – effects of pH-value and removal of calcium. *International Journal Food Science & Technology*, 37(5), 559–568. https://doi.org/10.1046/j.1365-2621.2002.00612.x

Stark, L.E., & Gross, A.T. (1990). *Hydrophobic protein microparticles and preparation thereof* (patent no. WO 90/03123). World Intellectual Property Organization.

Sturaro, A., De Marchi, M., Zorzi, E., & Cassandro, M. (2015). Effect of microparticulated whey protein concentration and protein-to-lipid ratio on Caciotta cheese yield and composition. *International Dairy Journal*, 48, 46–52. https://doi.org/10.1016/j.idairyj.2015.02.003

Suárez, A., Fernández, L., Balbarie, P., Iglesias, J.R., & Riera, F.A. (2016). Predicting the properties of the whey protein microparticles produced by heat and mechanical treatments. *European Food Research and Technology*, 242(8), 1211–1220. https://doi.org/10.1007/s00217-015-2625-5

Sun, C., Lei, D., Fuguo, L., Yanxiang, G. (2016). Dynamic high pressure microfluidization treatment of zein in aqueous ethanol solution. *Food Chemistry*, 210, 388–95. https://doi.org/10.1016/j.foodchem.2016.04.138

Sun, C., Liu, R., Wu, T., Liang, B., Shi, C., Cong, X., Hou, T., & Zhang, M. (2016). Combined superfine grinding and heat-shearing treatment for the microparticulation of whey proteins. *Food and Bioprocessing and Technology*, 9(2), 378–386. https://doi.org/10.1007/s11947-015-1629-2

Tamime, A.Y., Kalab, M., Muir, D.D., & Barrantes, E. (1995). The microstructure of set-style, natural yogurt made by substituting microparticulate whey protein for milk lipid. *Journal of the Society of Dairy Technology*, 48, 107–111. https://doi.org/10.1111/j.1471-0307.1995.tb02478.x

Tanger, C., Schmidt, F., Utz, F., Kreissl, J., Dawid, C., & Kulozik, U. (2021). Pea protein microparticulation using extrusion cooking: Influence of extrusion parameters and drying on microparticle characteristics and sensory by application in a model milk dessert. Innovative *Food Science & Emerging Technologies*, 74, 102851. https://doi.org/10.1016/j.ifset.2021.102851

Tanger, C., Utz, F., Spaccasassi, A., Kreissl, J., Dombrowski, J., Dawid, C., & Kulozik, U. (2022). Influence of pea and potato protein microparticles on texture and sensory properties in a lipid-reduced model milk dessert. *ACS Food Science & Technology*, 2(1), 169–179. https://doi.org/10.1021/acsfoodscitech.1c00394

Teresinha, R.H., Bittencourt, M.F., José, A.C., Ragagnin de Menezes, C., Smanioto, J.B, Lorenzo, J.M., Wagner, R., & Bastianello Campagnol, P.C. (2018). Volatile compounds and sensory profile of burgers with 50% lipid replacement by microparticles of chia oil enriched with rosemary. *Meat Science*. https://doi.org/10.1016/j.meatsci.2018.10.017

Toro-Sierra, J., Schumann, J., & Kulozik, U. (2013). Impact of spray-drying conditions on the particle size of microparticulated whey protein fractions. *Dairy Science and Technology*, 93(4), 487–503. https://doi.org/10.1007/s13594-013- 0124-7

Torres, I.C., Janhoj, T., Mikkelsen, B.O., & Ipsen, R. (2011). Effect of microparticulated whey protein with varying content of denatured protein on the rheological and sensory characteristics of low-lipid yoghurt. *International Dairy Journal, 21*, 645–655. https://doi.org/10.1016/j.idairyj.2010.12.013

Torres, I.C., Mutaf, G., Larsen, F.H., & Ipsen, R. (2016). Effect of hydration of microparticulated whey protein ingredients on their gelling behaviour in a non-lipid milk system. *Journal of Food Engineering*, 184, 31–37. https://doi.org/10.1016/j.jfoodeng.2016.03.018

Torres, I.C., Nieto, G., Nylander, T., Simonsen, A.C., Tolkach, A., & Ipsen, R. (2017). Effect of homogenisation in formation of thermally induced aggregates in a non- and low- lipid milk model system with microparticulated whey proteins. *Journal of Dairy Research*, 84(2), 229–238. https://doi.org/10.1017/S002202991700019X

World Health Organization WHO. (2003). *Diet, nutrition, and the prevention of chronic diseases: Report of a joint WHO/FAO expert consultation* (Vol. 916). Geneva, Switzerland: World Health Organization.

Wolz, M., Kastenhuber, S., & Kulozik, U. (2016). High moisture extrusion for microparticulation of whey proteins – influence of process parameters. *Journal of Food Engineering*, 185, 56–61. https://doi.org/10.1016/j.jfoodeng.2016.04.002

Yashini, M., Sunil C.K., Sahana S., Hemanth S.D., Chidanand, D.V., & Ashish, R. (2019). Protein-based lipid replacers – A review of recent advances. *Food Reviews International.* 37(2), 197–223. https://doi.org/10.1080/87559129.2019.1701007

Yazici, F., & Akgun, A. (2004). Effect of some protein based lipid replacers on physical, chemical, textural, and sensory properties of strained yoghurt. *Journal of Food Engineering*, 62(3), 245–254. https://doi.org/10.1016/S0260-8774(03)00237-1

Yilsay, T.O., Yilmaz, L., & Bayizit, A.A. (2006). The effect of using a whey protein lipid replacer on textural and sensory characteristics of low-lipid vanilla ice cream. *European Food Research and Technology*, 222, 171–175. https://doi.org/10.1007/s00217-005-0018-x

Chapter 18

Electrohydrodynamic Encapsulation of Natural Food Additives in Breadmaking Using Chia Mucilage as Wall Material

A Kinetic Glucose Oxidase Release Study

Minerva Rentería-Ortega
Georgina Calderón-Domínguez
Ma de la Paz Salgado-Cruz
Eduardo Morales-Sánchez
Alitzel B. García-Hernández
and Fátima S. Serrano-Villa

CONTENTS

18.1 INTRODUCTION

Bread is one of the most common staple foods and represents a significant source of daily energy intake worldwide. However, due to its high demand in the market, in many countries its production has become highly mechanized, and technological alternatives and new formulations

DOI: 10.1201/9781003332558-21

are being sought, requiring the standardization of raw materials together with an increase in the use of additives, which are largely of chemical origin. On the other hand, consumers have become more demanding regarding nutritional, technological, and health-related specifications, being more aware of the harmful effects of chemicals in food products. They have started to ask for chemical-free foods and/or aggregated in smaller concentrations (Paranthaman et al. 2021; Compaore-Sereme et al. 2023).

The baking industry has adapted by introducing natural additives, such as enzymes, to improve the characteristics of both flour and final products (Dahiya et al. 2020; Shanmugavel et al. 2020; Xu et al. 2019). However, enzymatic additives have still not reached the same levels of performance as chemical additives; enzyme encapsulation has thus been considered as a potential way to control the release of enzymes during the different stages of the breadmaking process, and many researchers have focused their efforts on developing and improving such methods.

18.2 BREADMAKING ENZYMES

Within the baking industry, a wide range of additives has been used to improve flour and dough performance, as well as the quality and shelf-life of bread. As a response to rising consumer demand for healthier food options, the usage of natural additives in the food industry and specifically in breadmaking has risen in turn, reaching a world level production of more than 2.5 billion dollars by 2020, with bakery enzymes accounting for more than 35% of this total (Melim-Miguel et al. 2013). This demand is expected to continue increasing and there are many enzymes on the market that can be used as alternatives to chemical enhancement agents (Stemler & Scherf, 2022).

In the breadmaking industry, oxidizing enzymes are one of the most commonly used. These enzymes can be added alone or in commercial preparations as flour improvers, having beneficial effects on the characteristics of dough and bread: improving dough machinability, softening breadcrumb, increasing loaf volume, and improving products useful life; examples of these enzymes include hexose oxidase, pyranose oxidase, sulfhydryl oxidase, transglutaminase, peptidases, laccases, and glucose oxidase (**Table 18.1**) (Koksel et al. 2018). These beneficial effects have been commonly ascribed to the formation of disulfide or dityrosine bonds within the glutenin polymers, or by improving dough mechanical properties (Pourmohammadi & Abedi, 2021) through the production of hydrogen peroxide. In the case of transglutaminase its usefulness has been proven in the making of gluten-free breads, by inducing the formation of new covalent bonds in glutamine residues by the introduction of isopeptide bonds (ε-γ glutamyl-lysine) (Renzetti & Rosell, 2016).

In general, the primary objective of oxidative enzymes is to reinforce weak flours, to restore the gluten network of damaged wheat flour, mitigating the economic impact of product losses. However, these natural additives have a great disadvantage over other chemical additives, as their activity is affected by many factors, including temperature, pH, water activity, and enzyme concentration, leading researchers to seek ways to protect them and increase their efficiency. One such proposed method is enzyme encapsulation.

18.3 ENCAPSULATION OF ENZYMES

Encapsulation can be defined as a compartmentalization process that generates a unique environment by isolating active components from their surroundings through the use of wall materials

TABLE 18.1 BREADMAKING ENZYMES

Enzyme	Reactions Involved	Effect
Glucose oxidase [EC 1.1.3.4]	Catalyzes the oxidation of C1 of b-D-glucose, producing hydrogen peroxide and D-glucono-d-lactone. H_2O_2 may indirectly oxidize the gluten thiol groups crosslinking the gluten proteins.	Improves crumb texture, making a whiter and moister crumb with larger cells. Increased dough water absorption, and enhanced dough tolerance to mix.
Hexose oxidase [EC 1.1.3.5]	Catalyzes the oxidation of several mono and oligosaccharides, including lactose, to lactones and hydrogen peroxide, inducing the formation of disulfide and dityrosine bonds, this last one when enzyme is used at higher levels (30–40 U/kg).	Increases dough strength, improves specific volume and crumb structure, decreasing common crust defects.
Pyranose oxidase [EC 1.1.3.10]	Catalyzes the oxidation of C_2 or C_3 of monosaccharides and disaccharides by molecular oxygen to the corresponding dicarbonyl derivatives and hydrogen peroxide.	Increases the strength of the dough and the volume of the bread.
Sulfhydryl oxidase [1.8.3.2]	Catalyzes the formation of disulfide bonds from a variety of thiol groups. However, its affinity for thiol groups in protein chains seems limited.	Increases dough stability.
Transglutaminase [EC 2.3.2.13]	Introduces covalent isopeptide bonds through catalyzation by acyl-transfer reactions between the g-carboxamide group of peptide-bound glutamine residues and primary residues (acyl acceptors). Also between free amino group and the g-carboxamide group of glutamine. The formation of glutamic acid residues is also reported.	Promotes a more elastic and resistant crumb structure, improves loaf volume, being the change dependent of the enzyme concentration; low levels recommended.
Peptidases	Hydrolyze peptide bonds and have affinities for different amino-acid sequences.	Decreases dough strength/ viscosity, increases loaf volume, softens crumb, makes it coarser.
Laccases [EC 1.10.3.2]	Generates free radicals and induce cross-linking among proteins containing tyrosine, tryptophan, and cysteine.	Acts over gliadins, improving mechanical properties and water stability of this protein fraction.

Adapted from: Labat et al. (2000); De Jong et al. (2003); Gujral et al. (2003); Rasiah et al. (2005); Autio et al. (2005); Cauvain & Young, (2006); Beck et al. (2011); Boussard et al. (2012); McSweeney, (2011); Tebben & Li, (2018); Isaschar-Ovdat et al. (2018); Ooms & Delcour, (2019); Cui et al. (2020); Dahiya et al. (2020); Diowksz & Sadowska, (2021); Pourmohammadi & Abedi, (2021); Zhao et al. (2021); Gu et al. (2022).

that are selected based on process conditions, allowing for a controlled release of the isolated compound (Ghosh et al. 2023). Through this technique, the material to be protected from external deleterious influences is covered, generating a microenvironment capable of controlling the interactions between interior and exterior through the capsule wall (Borgogna et al. 2010). This technique provides resistance in processing and packaging conditions, and improves the flavor, aroma, stability, nutritional value, and appearance of the products (Yáñez et al. 2002; Đorđević et al. 2015).

18.3.1 Wall Materials

A wide variety of wall materials can be used to form capsules for food applications, with lipids, proteins, carbohydrates, and their combinations being the most common (del Río et al. 2013; Pedroza-Islas, 2002; Nesterenko et al. 2013; do Nascimento et al. 2022). Among the studied wall materials, gums and mucilages (16.9%), alginate (10.8%), maltodextrins (10.8%), gelatin (9.6%), starch (7.2%), chitosan (7.2%), pectin (3.6%), and cyclodextrins (3.6%) have been some of the most reported **(Table 18.2)**. Currently, research into new sources of natural wall materials continues, such as those extracted from industrial, regional, and sea or agricultural by-products, for example, chia mucilage (Us-Medina et al. 2017; Hernández-Nava et al. 2019), beeswax (Haghighat-Kharazi et al. 2020a), legume proteins (Sridhar et al. 2022), or phycocolloids from red seaweed (Cian et al. 2014), have also been proposed, with only a few being suggested for use in enzyme encapsulation (Rentería-Ortega et al. 2021; Haghighat-Kharazi et al. 2020b).

18.3.2 Core Materials

Many different types of core materials have been encapsulated, most of them due to their nutraceutical properties, such as hydrolyzed protein (Lima et al. 2021), peptides (Latorres et al. 2021), polyphenols (Zyaitdinov et al.2020), bioactive compounds (Zyaitdinov et al. 2020), essential oils (Xiao et al. 2014), and probiotics (Noor et al. 2022). Moreover, food juices (Wang et al. 2021), enzymes (Kawakita et al. 2021), and some pharmaceutical active compounds or cells (Sun et al. 2020) are also frequently encapsulated. According to the information in **Table 18.2**, it can be seen that for food applications enzymatic and non-enzymatic core materials have been studied at a very similar rate (48.3% and 51.7%), with glucose oxidase being one of the most reported, but not all of them being used for breadmaking.

Most encapsulation studies for enzymes used specifically for breadmaking have been carried out with glucose oxidase using a variety of compounds as wall materials including chia mucilage and sodium alginate (Renteria-Ortega et al. 2020), poly vinyl alcohol (PVA) (Alonso-González et al. 2020), agarose hydrogel (Zhang et al. 2016), and polyethyleneimine (PEI), along with tannic acid and a polyphenol coating (Zhang et al. 2015) and amylases employing beeswax maltodextrin and alginate (Haghighat-Kharazi et al. 2020a; Haghighat-Kharazi et al. 2019). However, these studies are still few in number, and those regarding the encapsulation of other breadmaking enzymes, such as hexose oxidases, pyranose oxidases, sulfhydryl oxidase, laccases, or peptidases, are even fewer. Hence, there is an open field to test different wall materials that could provide enzymes with different release properties.

TABLE 18.2 WALL AND CORE MATERIALS USED IN ENCAPSULATION TECHNOLOGY

Wall Material	Core materials		References
	Enzymes	Other Materials	
Gums (locust bean, xanthan, Arabic, Guar) & mucilages (chia)	GOX; transglutaminase; lactase	Melon juice; Cherokee blackberry pulp; Zn enzymatically hydrolyzed amaranth puree; ACE inhibitory peptides from brewer's spent grain; Bioactives from oat bran, polyphenol; lavender oil; mint oil; orange oil and lemon oil	Kshirsagar et al. (2008); Sarkar et al. (2012); Xiao et al. (2014); Traffano-Schiffo et al. (2017); de Souza et al. (2018); Isleroglu et al. (2019); Hernández-Nava et al. (2019); Hong et al. (2019); Zyaitdinov et al. (2020); Rentería-Ortega et al. (2021); Oro et al. (2023); Amin et al. (2021)
Alginate	Protease; naranginase; lactase; cellulase; α-amylase	Collagen-calcium chelates NIH/3T3 fibroblast cells	Freitas et al. (2012); Nguyen et al. (2016); Traffano-Schiffo et al. (2017); Mi et al. (2018); Housseiny & Aboelmagd, (2019); Kawakita et al. (2021); Noor et al. (2022)
Maltodextrin	Maltogenic amylase	Bacteria; melon juice; fish protein enzymatic hydrolysate; Cherokee blackberry pulp; Zn enzymatically hydrolyzed amaranth puree; bioactives from oat bran, polyphenols; goose blood hydrolysate	Wang et al. (2017); Haghighat-Kharazi et al. (2019); Haghighat-Kharazi et al. (2020b); Zyaitdinov et al. (2020); Wang et al. (2021); Lima et al. (2021); Oro et al. (2023); Amin et al. (2021); Noor et al. (2022)
Gelatin	Lactase	CoQ10; Isoquercetin; carcinoma cells (HeLa); mouse insulinoma; lavender oil; shrimp *Penaeus japonica* diet	Kshirsagar et al. (2008); Xie et al. (2011); Freitas et al. (2012); Xiao et al. (2014); de Souza et al. (2018); Hernández-Nava et al. (2019); Sun et al. (2020); Cao et al. (2014)
Starch	-------	Ascorbic acid; Zn enzymatically hydrolyzed amaranth puree; tea polyphenols; Lactobacillus rhamnosus; orange oil and lemon oil	Avila-Reyes et al. (2014); Hong et al. (2019); Chen et al. (2022); Amin et al. (2021)

(continued)

TABLE 18.2 (CONTINUED)

Wall Material	Core materials		References
	Enzymes	**Other Materials**	
Chitosan	Naranginase; horseradish peroxidase; urease	Red ginseng extract; collagen-calcium chelates; thymosin (T4)	Miguezb et al. (1997); Gavory et al. (2012); Bilal et al. (2017); Mi et al. (2018); Housseiny & Aboelmagd, (2019); Li et al. (2020); Kim et al. (2021)
Pectin	-------	Melon juice; Cherokee blackberry pulp; Isoquercetin	Wang et al. (2021); Oro et al. (2023)
Cyclodextrins	-------	Carvacrol-Tween 80; Caffeic acid; lavender flavor	Pinho et al. (2015); Zhu et al. (2015); Cheng et al. (2019)
Polyelectroytes (PAH, PSS)	Glucose oxidase; horseradish peroxidase, α-chymotrypsin	Bakery yeast cells	Krol et al. (2003); Pretrov et al. (2005); Stein et al. (2006)
Tannic acid polyphenol coating/PEI	Glucose oxidase	Imidazole curing agent, EMI-g-BGE	Fang et al. (2012); Zhang et al. (2015)
Casein	Transglutaminase	Goose blood hydrolysate	Miguezb et al. (1997); Wang et al. (2017); Isleroglu et al. (2019)
PVA	Glucose oxidase, urease	-------	Alonso-González et. (2020)
Carrageenan	-------	ACE inhibitory peptides from brewer's spent grain	
Agarose	Glucose oxidase	-------	Avila-Reyes et al. (2014); Zhang et al. (2016)
Inuline	-------	*Lactobacillus rhamnosus*	
Soy protein	-------	Melon juice	
Collagen	-------	Thymosin (T4)	Chiu et al. (2012)
Fucoidan	-------	Red ginseng extract	Kim et al. (2021); Wang et al. (2021)
Beewax	Maltogenic amylase	-------	Haghighat-Kharazi et al. (2020a)
Whey protein concentrate	-------	Bioactives, polyphenols	Zyaitdinov et al. (2020)
Mannitol	Transglutaminase	-------	Wang et al. (2017); Isleroglu et al. (2019)
CMC	-------	Goose blood hydrolysate	

TABLE 18.2 (CONTINUED)

Wall Material	Core materials		References
	Enzymes	**Other Materials**	
Ethyl cellulose	-------	Shrimp *Penaeus japonica* diet	Xie et al. (2011)
Trehalose	Alcohol dehydrogenase (ADH)	-------	Yoshii et al. (2008)
PLCL, PEOz,	-------	Vascular endothelial growth factor (VEGF) Basic fibroblast growth factor (bFGF)	Lakshmanan et al. (2016)
Lecithin/ colesterol	Flavourzyme	-------	Jahadi et al. (2020)
Polyacryalmide gel	Pectinase	-------	Rehman et al. (2020)
BSA	Catalase	-------	Schakowski et al. (2021)

18.3.3 Encapsulation Techniques

Encapsulation techniques are broadly categorized into two different classification systems. The first one depends on the specific reactions involved to produce capsules, separating them into: 1) Chemical; 2) Physicochemical; and 3) Physical or mechanical encapsulations (Mishra, 2015). This first classification relates to chemical interactions. The second classification separates these techniques based on the actual practical methods followed to produce the capsule, having two general categories: atomization and emulsion (Oxley, 2016; De & Godesmi, 2022). Within the first classification, the capsule formed by chemical processes is related to the chemical interactions between the wall materials and the encapsulated compound. The methods within these processes are interfacial polycondensation (Dhumal et al. 2008) and molecular inclusion (Yañez et al. 2002). Physicochemical encapsulation involves a mix of chemical interactions and drying processes. Some examples include simple or complex coacervation (Madene et al. 2006) and entrapment in liposomes (Kim et al. 1995). Physical or mechanical encapsulation involves the formation of the capsule or coating without chemical interaction between the constituent materials or those that participate in its formation. They are based on combining materials where the coating material is forced to form a deposit on the core material. Most of these methodologies share common steps (**Figure 18.1**), such as mixing, emulsification, filtration, and centrifugation, as well as a final step to set the capsule structure, mainly freeze drying (de Vos et al. 2012; López Hernández, 2010), fluidized bed coating (Madene et al.2006), chill or freeze spray (Gibbs et al. 2004; Gouin, 2004), and more recently electrohydrodynamic atomization.

Table 18.3 shows some of the most common encapsulation techniques related to the folding of enzymes or food materials using biopolymers as wall materials. Of these techniques, spray drying has been the most commonly reported encapsulation methodology (21.6%), followed by lyophilization (13.7%), coacervation (11.8%), gelation (9.8%), and coprecipitation (9.8%), representing

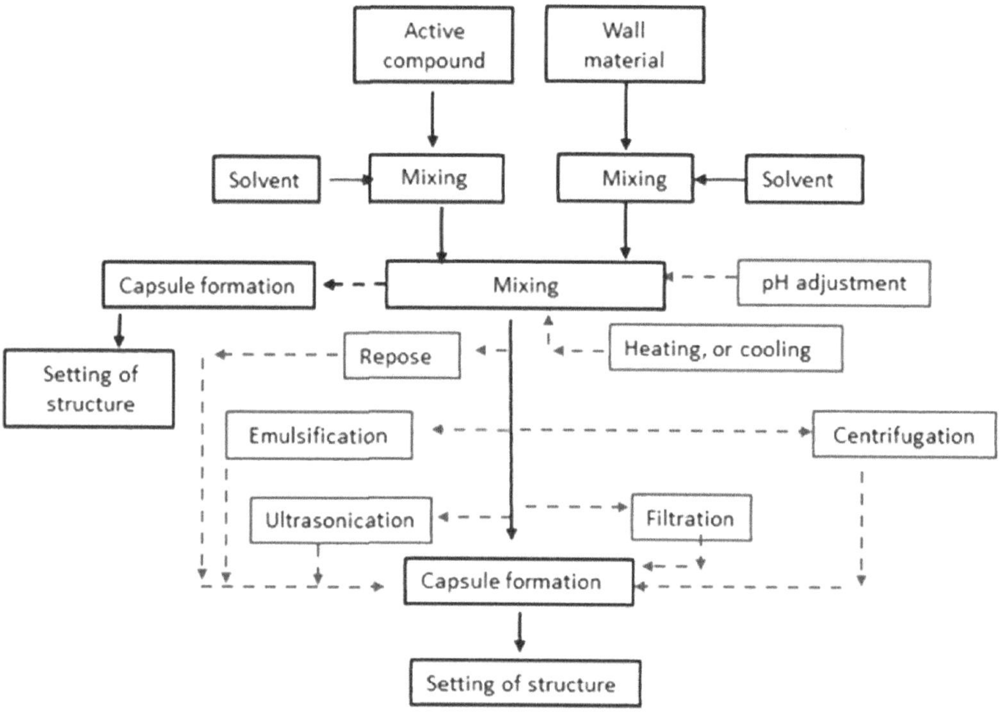

Figure 18.1 General method used to produce micro- and nanocapsules. Dotted lines indicate alternative methodologies. Setting of structure involves different methods, such as spray drying, freeze drying, etc.

TABLE 18.3 ENCAPSULATION METHODOLOGIES FOR FOOD MATERIALS AND ENZYMES

Encapsulation Techniques	Core material		References
	Enzymes	Other Materials	
Spray drying	Alcohol dehydrogenase (ADH)	Bacteria, melon juice, enzymatic hydrolysate of fish protein, Co-enzyme Q10, hydrolyzed amaranth puree, ACE inhibitory peptides, tea polyphenols, *Lactobacillus rhamnosus*, mint oil, orange oil, lemon oil	Yoshii et al. (2008) Kshirsagar et al. (2008) Sarkar et al (2012); Avila-Reyes et al. (2014); Hong et al. (2019); Wang et al. (2021); Lima et al. (2021); Cao et al. (2014); Amin et al. (2021); Noor et al. (2022)
Freeze-drying	Transglutaminase, maltogenic amylase	Ascorbic acid, blackberry pulp, hydrolyzed amaranth puree, Goose blood hydrolysate	Wang et al. (2017); Isleroglu et al. (2019); Haghighat-Kharazi et al. (2019); Chen et al. (2022); Oro et al. (2023); Amin et al. (2021)

TABLE 18.3 (CONTINUED)

Encapsulation Techniques	Core material		References
	Enzymes	Other Materials	
Coacervation	Lactase, urease	Isoquercetin, bioactives, polyphenols, lavender oil	de Miguezb et al.(1997); Xiao et al. (2014); Souza et al. (2018); Hernández-Nava et al. (2019); Zyaitdinov et al. (2020)
Gelation (ionic, alkaline)	Peroxidase, cellulase, lactase, amylase	Ginseng extract	Kim et al. (2021); Bilal et al. (2017); Nguyen et al. (2016); Freitas et al. (2012)
Coprecipitation/ adhesion	Catalase, glucose oxidase, peroxidase, α-chymotrypsin	Baker's yeast cells	Krol et al. (2003); Pretrov et al. (2005); Stein et al. (2006); Zhang et al. 2016; Schakowski et al. (2021)
Electrohydrodynamic atomization	Glucose oxidase,	Vascular endothelial growth factor, basic fibroblast growth factor	Lakshmanan et al. (2016); Alonso-González et al. (2020); Rentería-Ortega et al. (2021)
Congealing	Maltogenic amylase	Thymosin (T4)	Haghighat-Kharazi et al. (2020a); Haghighat-Kharazi et al. (2020b)
Fluidized bed spray coating	Protease	Shrimp Penaeus japonica diet	Xie et al. (2011); Kawakita et al. (2021)
Electrostatic droplet generator	-------	Collagen calcium, chelates	Mi et al. (2018)
Polymerization	Pectinase	-------	Rehman et al. (2020)
LbL inmersion	-------	HeLa cells, mouse insulinoma cell;	Sun et al. (2020)
Mixing/ homogenization	Naranginase	-------	Housseiny and Aboelmagd, (2019)
Mixing and oven drying	-------	Carvacrol-Tween 80	Cheng et al. (2019)
Inclusion complex	-------	Caffeic acid, lavender flavoring	Pinho et al. (2015); Zhu et al. (2015)
Coaxial air flow bead generator	-------	fibroblast cells	Kim et al. (2013)
Emulsion solvent evaporation	-------	Imidazole curing agent, EMI-g-BGE	Fang et al. (2012)

all hem almost 70% of the mentioned methodologies. In this sense, various enzymes have been encapsulated by some of these methods, including enzymes of interest in the baking industry, such as glucose oxidase (GOX), transglutaminase and microbial pentosanases, or endoxylanases. Other methodologies such as electrohydrodynamic atomization (EHDA) (Alonso-González et. 2020), electrostatic droplet generation (Mi et al. 2018), and layer by layer deposition (Lbl) (Sun et al. 2020) have been investigated. Among these, only EHDA has been applied for enzyme encapsulation purposes (Rentería-Ortega et al. 2021; Alonso-González et. 2020).

18.4 ELECTROHYDRODYNAMIC ATOMIZATION

Electrohydrodynamic atomization or spraying (EHDA) is based on the extrusion of a polymer solution under the influence of an electric field, resulting in the production of fibers (electrospinning) or particles (electrospray). The main parts of an EHDA setup are shown in **Figure 18.2** (Rentería-Ortega et al. 2020).

The electric charges employed in this technique allow for the generation of charged droplets, favoring their fission and reducing their size. This method also has the advantage of applying low temperatures and providing better control of the morphology of the particles (Liu et al. 2018; Niu et al. 2020). This effect is dependent on the physical properties of the feed (conductivity, viscosity, surface tension); the type of nozzle (axial, coaxial); and the processing parameters (voltage, working distance, flow rate, nozzle diameters). Currently, this technique has been used for the encapsulation of various food interesting compounds, such as proteins (Moreira et al. 2022; Mahalakshmi et al. 2023), enzymes (Vocadlo et al. 2001), stevia glycosides (Oikonomopoulou et al. 2022), bioactive compounds (Vázquez-González et al.2021; Jayaprakash et al. 2023), phytochemical compounds (Machmudah et al. 2022), fish oil (Cetinkaya et al. 2021), and catechin (Paximada et al. 2021), among others. The encapsulation by this technique aims to protect them from degradation due to the exposure to environmental conditions (oxygen, humidity, high temperatures or light); gastrointestinal conditions (enzymes, acids and bile salts); and also to maintain bioaccessibility or bioavailability (Radünz et al. 2021), during processing and/or storage (Rostamabadi, Falsafi

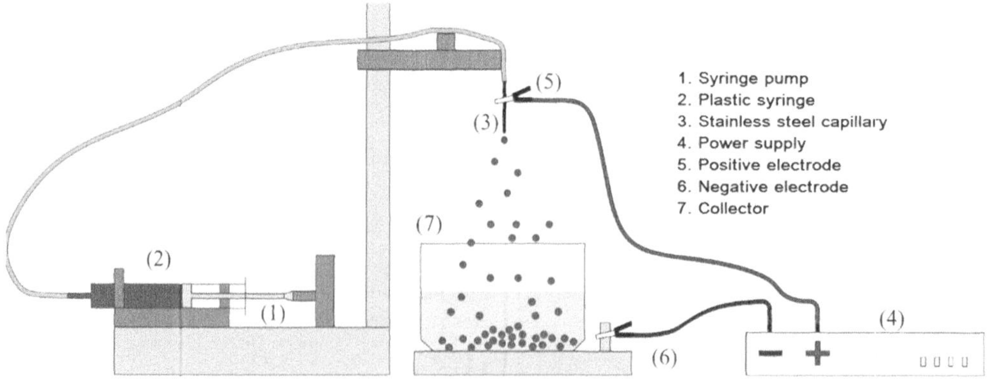

Figure 18.2 Electrospraying equipment in a vertical arrangement. (Reprinted with Permission from Rentería Ortega et al.2020. *Cyta-Journal of Food,* 18(1), 461–471.)

and Jafari, 2019), or to favor the organoleptic characteristics of the food (mask unpleasant flavors, appearance, or color) (Wang et al. 2020).

Once the capsule is formed, the active ingredient must reach the site of action at a suitable time and at a defined rate (Madene et al. 2006), which will be dependent on the wall material, specifically its chemical nature, transition temperature, the degree of swelling, diffusion capacity, and morphology of the membrane (de Vos et al. 2012; Fuchs et al. 2006; Yáñez et al. 2002). This action is defined as the release of core material and is analyzed through kinetic studies.

18.5 RELEASE MECHANISMS

The release of the core materials can be mediated by different mechanisms, including diffusion, swelling, and erosion, which are some of the most commonly studied. Diffusion is governed by the entry of the aqueous medium into the system where it dissolves the active compound and diffuses through the polymeric material, creating pores through which the rest of the core material is released (Vesely, 2008). In erosion, the constant change in the geometry of the capsule due to shear forces allows for the release of the active ingredient (Blandino et al. 2001). Many of the studies on this mechanism are focused on the generation of mathematical models that describe the kinetic mechanisms and the control of the release rate of the active compound from hydrophilic matrices (Siepmann & Peppas, 2012).

18.5.1 Release Kinetics Models Applied to Enzymes Encapsulation

Different mathematical equations (De Kee et al. 2005) have been applied to study the release of active principles. These models were developed to explain the release mechanism of drugs included in polymeric matrixes, with the Korsmeyer model (Korsmeyer et al.1983) being one of the most reported (Eq. 18.1). It describes the release kinetics in systems where release occurs mainly by diffusion (Kalam et al. 2007). In this model, Mt/Mα is the amount of material released in time t (h), k (s^{-n}) is a kinetic constant, and "n" is the diffusional exponent. The value of n describes the system, and when a system presents values of n <0.43, the diffusion is of the Fickian type, and if it is larger than 0.43 but less than one (0.43 < n <1), it is said to be anomalous or non-Fickian (Jurić et al. 2019). In Fickian diffusion, the active compound is transported through the polymer matrix and the rate of relaxation of the polymer is less than the rate of penetration of the medium into the capsule. When n has values between 0.43 and 0.85, an anomalous transport kinetic is presented, which combines diffusion and erosion of the polymer that actively participates in the release of the active compound (Lao et al. 2011; Sandoval et al. 2008; Marras-Marquez et al. 2015).

$$\left(\frac{M_t}{M_\pm}\right) = kt^n \tag{18.1}$$

Some of these mathematical models have been used to evaluate the controlled release of enzymes of food importance; however, there are few studies on the enzymatic stability or release of food enzymes and almost none on breadmaking enzymes such as glucose oxidase. In this regard, Vikartovská et al. (2007) immobilized glucose oxidase in capsules made of sodium alginate, cellulose sulfate, poly-methylene, and NaCl, while Liu et al. (2007) developed glucose

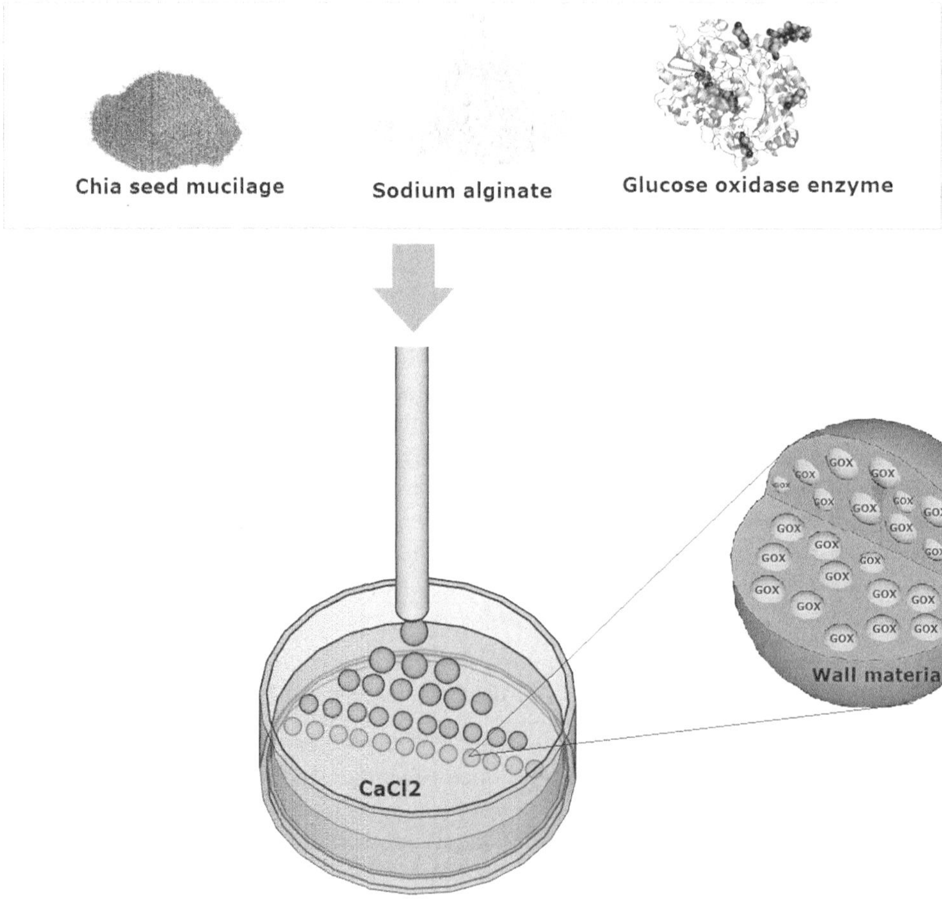

Figure 18.3 Gox encapsulation process.

oxidase microspheres in alginate/chitosan using internal emulsification-gelation. Rentería-Ortega et al. (2021) studied glucose oxidase release from chia mucilage capsules, prepared at different enzyme concentrations by two techniques (electrospraying, dripping), subjected to different stress conditions (mixing, pH, temperature) emulating the bread-making process.

To study the release of glucose oxidase, an experiment was carried out following the methodology reported by Rentería-Ortega et al. (2021), using the electrospraying method and collecting the drops in a calcium chloride solution as shown in **Figure 18.3**, followed by performing the enzyme release study. The results of this study allow to develop the glucose oxidase release profile from chia mucilage-alginate electrospun capsules, in which the effect of enzyme concentration is notable, releasing a higher proportion of enzymatic material at higher concentration values (**Figure 18.4**).

The glucose oxidase behavior is the result of rapid release during the first two weeks, followed by steady state. In this regard, other authors reported the same phenomenon for β-carotene encapsulated in Casein/Gum Tragacanth, attributing the fast core material liberation to more

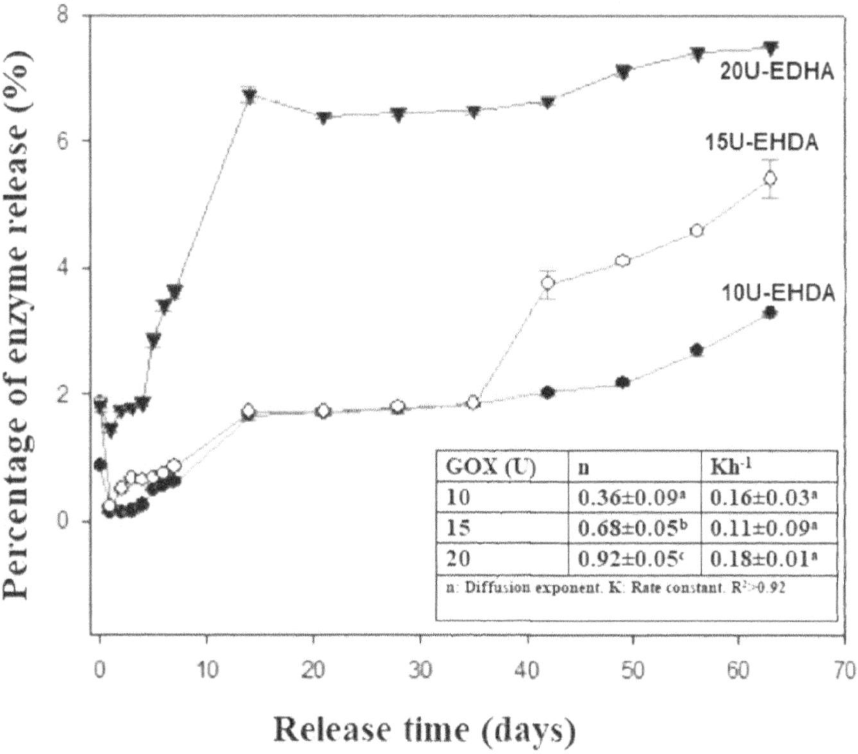

Figure 18.4 Enzyme release profiles of GOX-MC-AlgNa capsules stored at 8°C, pH 5.1 in acetate buffer by electrohydrodinamic atomization (Rentería-Ortega. 2020). GOX: glucose oxidase; GOX-MC-AlgNa: glucose oxidase chia mucilage sodium alginate capsules; EHDA: electrohydrodynamic atomization.

open structures as a result of matrix hydration, as well as to a higher enzyme concentration closer to the capsule surface (Jain et el. 2016).

Figure 18.4 data was adjusted to the Korsmeyer-Peppas model, observing again the effect of enzyme concentrations, since the diffusion exponent (n) notably changes with this variable, which means that GOX diffusion is influenced by the incoming fluid and the swelling characteristics of the polymer. Specifically, when n=0.36 (at the lowest concentration of the enzyme) a Fickian diffusion related to a marked influence of the matrix on transport is considered (Wu et al. 2019), while as the n value increases (n=0.68), the release process is considered a combination of non-Fickian and anomalous transport case II, more related to polymer relaxation (Maghsoodi & Barghi, 2011). For the highest value of n (n=0.92) at the highest enzyme concentration (20 IU/mL), it is considered to be related to a near-zero order, indicative of time-independent release, since the largest enzyme release process took place in the first 2 weeks, possibly due to relatively large amounts of enzyme on the surface of these capsules at this high enzyme concentration (Nayak et al. 2013). It has been reported that for matrix capsules, the bioactive can be present both on the surface and within the polymer wall and its relative amounts depend on the preparation method, concentration, and size (Bock et al. 2012).

K values, as in the case of n, are determined by the structure and size of the capsules by the way the enzyme is trapped (Ozaltin et al. 2019), and by the interactions between the wall material and the surrounding medium. From **Figure 18.4**, the K values were low due to small capsule sizes as observed by their Feret diameters (0.72±0.21 - 0.75±0.21), as well as by their hard and compact structures due to the large number of Ca^{+2} ions that hold the alginate chains together, making diffusion difficult (Deze et al. 2012). Additionally, it is also possible that there was no electrostatic interaction between the enzyme glucose oxidase and the polymer matrix for all encapsulation formulations, behaving similarly in all cases (Wu et al.2019).

18.6 ELECTROSPUN BREADMAKING ENZYMES

Very few studies have been carried out on the use of electrospun breadmaking enzymes and particularly those including glucose oxidase. Renteria-Ortega et al. (2021) reported that GOX encapsulated by electrospraying is released at a slower rate than when encapsulated by dripping, and the authors explained that this effect was related to a change in the capsule structure, as well as an effect of swelling and disaggregation. They also reported that when subjecting the encapsulated enzymes to different stages emulating the breadmaking process (mixing, fermentation, and baking), the stresses generated on this process affect the GOX release rate, with the highest values being observed at the fermentation stage, and with a protective effect on the breakdown of capsules at high GOX concentrations. Moreover, the authors concluded that the electrospraying methodology promotes a change in the capsule structure, modifying its release behavior, and the electrospraying methodology could be recommended when a slow enzymatic release is required, and the dripping mode when a faster release is needed.

18.7 CONCLUSION

Consumers are becoming increasingly health conscious and looking for foods free of chemical additives. In response to this, the food industry seeks to use components that are of natural origin, enzyme-based foods, food hydrocolloids, and some functional food ingredients; with this in mind, it is expected that in the future, there will be a strong inclination of consumers to spend more on foods and beverages of natural origin without additives. This tendency has led to the search for more natural and healthy alternatives than for chemical additives. In this context, enzyme encapsulation is an outstanding possibility to control the release of enzymes during the different stages of food processing by varying the wall materials. However, there are still few studies in this area. Regarding encapsulation techniques, electrospraying has the advantage of applying low temperatures retaining enzyme activity and better control of the capsule's morphology. However, to date, there are still just a few studies relating breadmaking enzymes, wall materials, the electrospraying technique, and the effect of the preparation process on enzyme activity and bread quality.

ACKNOWLEDGMENTS

Fátima Sarahí Serrano Villa thanks Instituto Politécnico Nacional (BEIFI-IPN) and CONACYT, and Alitzel Belem García Hernández thanks COMECyT (CAT2022-0011) for the scholarships provided.

This research was funded through projects: SIP-20221471; 20221510; 20230985; 20230993 from the Instituto Politécnico Nacional (IPN, Mexico).

REFERENCES

Alonso-González, M., Corral-González, A., Felix, M., Romero, A., & Martin-Alfonso, J.E. (2020). Developing active poly (vinyl alcohol)-based membranes with encapsulated antimicrobial enzymes via electrospinning for food packaging. *International Journal of Biological Macromolecules*, 162, 913–921. doi:10.1016/j.ijbiomac.2020.06.217.

Amin, S.F.M., Karim, R., Yusof, Y.A., & Muhammad, K. (2021). Effects of enzymatic liquefaction, drying techniques, and wall materials on the physicochemical p-roperties, bioactivities, and morphologies of Zinc-amaranth (Amaranthus viridis L.) powders. *International Journal of Food Science*, 2021 (1), 1819104. doi:10.1155/2021/1819104

Autio, K., Kruus, K., Knaapila, A., Gerber, N., Flander, L., & Buchert, J. (2005). Kinetics of transglutaminase-induced cross-linking of wheat proteins in dough. *Journal of Agricultural and Food Chemistry*, 53(4), 1039–1045. doi:10.1021/jf0485032

Avila-Reyes, S.V., Garcia-Suarez, F.J., Jiménez, M.T., San Martín-Gonzalez, M.F., & Bello-Perez, L.A. (2014). Protection of L. rhamnosus by spray-drying using two prebiotics colloids to enhance the viability. *Carbohydrate Polymers*, 102, 423–430. doi:10.1016/j.carbpol.2013.11.033

Beck, M., Jekle, M., Selmair, P.L., Koehler, P., & Becker, T. (2011). Rheological properties and baking performance of rye dough as affected by transglutaminase. *Journal of Cereal Science*, 54(1), 29–36. doi:10.1016/j.jcs.2011.01.012

Bilal, M., Iqbal, H.M., Hu, H., Wang, W., & Zhang, X. (2017). Enhanced bio-catalytic performance and dye degradation potential of chitosan-encapsulated horseradish peroxidase in a packed bed reactor system. *Science of the Total Environment*, 575, 1352–1360. doi:10.1016/j.scitotenv.2016.09.215

Blandino, A., Macıas, M., & Cantero, D. (2001). Immobilization of glucose oxidase within calcium alginate gel capsules. *Process Biochemistry*, 36(7), 601–606. doi:10.1016/s0032-9592(00)00240-5

Bock, N., Dargaville, T.R., & Woodruff, M.A. (2012). Electrospraying of polymers with therapeutic molecules: State of the art. *Progress in Polymer Science*, 37(11), 1510–1551. doi:10.1016/j.progpolymsci.2012.03.002

Borgogna, M., Bellich, B., Zorzin, L., Lapasin, R., & Cesàro, A. (2010). Food microencapsulation of bioactive compounds: Rheological and thermal characterisation of non-conventional gelling system. *Food Chemistry*, 122(2), 416–423. doi:10.1016/j.foodchem.2009.07.043

Boussard, A., Cordella, C.B.Y., Rakotozafy, L., Moulin, G., Buche, F., Potus, J., & Nicolas, J. (2012). Use of chemometric tools to estimate the effects of the addition of yeast, glucose-oxidase, soybean or horse bean flours to wheat flour on biochemical bread dough characteristics. *Chemometrics and Intelligent Laboratory Systems*, 113, 68–77. doi:10.1016/j.chemolab.2012.01.006

Cao, M., Che, L., Wang, J., Yang, M., Su, G., Fang, Z., ... & Wu, D. (2014). Effects of maternal over-and undernutrition on intestinal morphology, enzyme activity, and gene expression of nutrient transporters in newborn and weaned pigs. *Nutrition*, 30(11-12), 1442–1447. doi:10.1016/j.nut.2014.04.016

Cauvain, S.P., & Young, L.S. (2006). Ingredient qualities and quantities in the Chorleywood bread process. *The Chorleywood Bread Process*, first edition. Woodhead Publishing. Sawston, United Kingdom. p. 192. eBook ISBN: 9781845691431

Cetinkaya, T., Mendes, A.C., Jacobsen, C., Ceylan, Z., Chronakis, I.S., Bean, S.R., & García-Moreno, P.J. (2021). Development of kafirin-based nanocapsules by electrospraying for encapsulation of fish oil. *LWT – Food Science and Technology*, 136, 110297. doi:10.1016/j.lwt.2020.110297

Chen, G., Yu, Y., Fu, X., Wang, G., Wang, Z., Wu, X., & Zhao, Y. (2022). Microfluidic encapsulated manganese organic frameworks as enzyme mimetics for inflammatory bowel disease treatment. *Journal of Colloid and Interface Science*, 607, 1382–1390. doi:10.1016/j.jcis.2021.09.016

Cheng, X., Zhou, J., Chen, J., Xie, Z., Kuang, Q., & Zheng, L. (2019). One-step synthesis of thermally stable artificial multienzyme cascade system for efficient enzymatic electrochemical detection. *Nano Research*, 12, 3031–3036. doi:10.1007/s12274-019-2548-8

Chiu, L.L., Reis, L.A., Momen, A., & Radisic, M. (2012). Controlled release of thymosin β4 from injected collagen–chitosan hydrogels promote angiogenesis and prevents tissue loss after myocardial infarction. *Regenerative Medicine*, 7(4), 523–533. doi:10.2217/rme.12.35

Cian, R.E., Salgado, P.R., Drago, S.R., González, R.J., & Mauri, A.N. (2014). Development of naturally activated edible films with antioxidant properties prepared from red seaweed Porphyra columbina biopolymers. *Food* Chemistry, 146, 6–14. doi:10.1016/j.foodchem.2013.08.133

Compaore-Sereme, D., Hama-Ba, F., Tapsoba, F.W.B., Manner, H., Maina, N.H., Dicko, M.H., & Sawadogo-Lingani, H. (2023). Production and sensory evaluation of composite breads based on wheat and whole millet or sorghum in the presence of Weissella confusa A16 exopolysaccharides. *Heliyon*, 9(3). ISSN 2405-8440. doi:10.1016/j.heliyon.2023.e13837.

Cui, L., Wang, P., Fan, X., Wang, Q., & Yan, R. (2020). Laccase-mediated enhancement of the properties of regenerated fibers from wheat gliadin. *Journal of Cereal Science*, 96, 103129. doi:10.1016/j.jcs.2020.103129

Dahiya, S., Bajaj, B.K., Kumar, A., Tiwari, S.K., & Singh, B. (2020). A review on biotechnological potential of multifarious enzymes in bread making. *Process Biochemistry*, 99, 290–306. doi:10.1016/j.procbio.2020.09.002

De Jong, G.A.H., Wijngaards, G., & Koppelman, S.J. (2003). Transglutaminase Inhibitor from Milk. *Journal of Food Science*, 68(3), 820–825. doi:10.1111/j.1365-2621.2003.tb08249.x

De, B., & Goswami, T.K. (2022). Chemometrics optimized microencapsulation of ergocalciferol using emulsification technique and its physico-chemical characterization. *Applied Food Research*, 2(2), 100157. doi:10.1016/j.afres.2022.100157

De Kee, D., Liu, Q., & Hinestroza, J. (2005). Viscoelastic (non-Fickian) diffusion. *The Canadian Journal of Chemical Engineering*, 83(6), 913–929. doi:10.1002/cjce.5450830601

de Souza, V.B., Thomazini, M., Barrientos, M.A.E., Nalin, C.M., Ferro-Furtado, R., Genovese, M.I., & Favaro-Trindade, C.S. (2018). Functional properties and encapsulation of a proanthocyanidin-rich cinnamon extract (Cinnamomum zeylanicum) by complex coacervation using gelatin and different polysaccharides. *Food Hydrocolloids*, 77, 297–306. doi:10.1016/j.foodhyd.2017.09.040

de Vos, P., Spasojevic, M., de Haan, B.J., & Faas, M.M. (2012). The association between in vivo physicochemical changes and inflammatory responses against alginate based microcapsules. *Biomaterials*, 33(22), 5552–5559. doi:10.1016/j.biomaterials.2012.04.039

del Río, J.C., Prinsen, P., & Gutiérrez, A. (2013). Chemical composition of lipids in brewer's spent grain: A promising source of valuable phytochemicals. *Journal of Cereal Science*, 58(2), 248–254. doi:10.1016/j.jcs.2013.07.001

Deze, E.G., Papageorgiou, S.K., Favvas, E.P., & Katsaros, F.K. (2012). Porous alginate aerogel beads for effective and rapid heavy metal sorption from aqueous solutions: Effect of porosity in Cu^{2+} and Cd^{2+} ion sorption. *Chemical Engineering Journal*, 209, 537–546. doi:10.1016/j.cej.2012.07.133

Dhumal, S.S., Wagh, S.J., & Suresh, A.K. (2008). Interfacial polycondensation—Modeling of kinetics and film properties. *Journal of Membrane Science*, 325(2), 758–771. doi:10.1016/j.memsci.2008.09.002

Diowksz, A., & Sadowska, A. (2021). Impact of sourdough and transglutaminase on gluten-free buckwheat bread quality. *Food Bioscience*, 43, 101309. doi:10.1016/j.fbio.2021.101309

Đorđević, V., Balanč, B., Belščak-Cvitanović, A., Lević, S., Trifković, K., Kalušević, A., & Nedović, V. (2015). Trends in encapsulation technologies for delivery of food bioactive compounds. *Food Engineering Reviews*, 7, 452–490. doi:10.1007/s12393-014-9106-7

do Nascimento Filho, E., Silva, N.N.B., Converti, A., Grosso, C.R.F., Santos, A.M.P., Ribeiro, D.S., & Maciel, M.I.S. (2022). Microencapsulation of acerola (Malpighia emarginata DC) and ciriguela (Spondias purpurea L) mixed juice with different wall materials. *Food Chemistry Advances*, 1, 100046. https://doi:10.1016/j.focha.2022.100046.

Fang, Z., & Bhandari, B. (2012). Spray drying, freeze drying and related processes for food ingredient and nutraceutical encapsulation. Chapeter 4, In Encapsulation technologies and delivery systems for food ingredients and nutraceuticals. Garti, N., & McClements, D. J. (Eds.). Elsevier. (pp. 71–102). ISBN: 9780857091246 doi:10.1533/9780857095909.2.73

Freitas, F.F., Marquez, L.D.S., Ribeiro, G.P., Brandão, G.C., Cardoso, V.L., & Ribeiro, E.J. (2012). Optimization of the immobilization process of β-galatosidade by combined entrapment-cross-linking and the kinetics of lactose hydrolysis. *Brazilian Journal of Chemical Engineering*, 29, 15–24. doi:10.1590/S0104-66322012000100002

Fuchs, M., Turchiuli, C., Bohin, M., Cuvelier, M.E., Ordonnaud, C., Peyrat-Maillard, M.N., & Dumoulin, E. (2006). Encapsulation of oil in powder using spray drying and fluidised bed agglomeration. *Journal of Food Engineering*, 75(1), 27–35. doi:10.1016/j.jfoodeng.2005.03.047

Gavory, C., Abderrahmen, R., Valour, J.P., Chaussy, D., Belgacem, M.N., Fessi, H., & Briançon, S. (2012). Encapsulation of a pressure-sensitive adhesive by spray-drying: Microparticles preparation and evaluation of their crushing strength. *Journal of Microencapsulation*, 29(2), 185–193. doi:10.3109/02652048.2011.642014

Ghosh, V., Ranjha, R., & Gupta, A.K. (2023). Polymeric encapsulation of anti-larval essential oil nanoemulsion for controlled release of bioactive compounds. *Inorganic Chemistry Communications*, 150, 110507. doi:10.1016/j.inoche.2023.110507

Gibb, C.L., & Gibb, B.C. (2004). Well-defined, organic nanoenvironments in water: The hydrophobic effect drives a capsular assembly. *Journal of the American Chemical Society*, 126(37), 11408–11409. doi:10.1021/ja0475611

Gouin, S. (2004). Microencapsulation: Industrial appraisal of existing technologies and trends. *Trends in Food Science & Technology*, 15(7-8), 330–347. doi:10.1016/j.tifs.2003.10.005

Gu, M., Hong, T., Ma, Y., Xi, J., Zhao, Q., Xu, D., & Xu, X. (2022). Effects of a commercial peptidase on rheology, microstructure, gluten properties of wheat dough and bread quality. *LWT – Food Science and Technology*, 160, 113266. doi:10.1016/j.lwt.2022.113266

Gujral, H.S., Guardiola, I., Carbonell, J.V., & Rosell, C.M. (2003). Effect of cyclodextrinase on dough rheology and bread quality from rice flour. *Journal of Agricultural and Food Chemistry*, 51(13), 3814–3818. doi:10.1021/jf034112w

Haghighat-Kharazi, S., Milani, J.M., Kasaai, M.R., & Khajeh, K. (2019). Use of encapsulated maltogenic amylase in malotodextrins with different formulations in making gluten-free breads. *LWT – Food Science and Technology*, 110, 182–189. doi:10.1016/j.lwt.2019.04.076

Haghighat-Kharazi, S., Reza Kasaai, M., Milani, J.M, & Khajeh, K. (2020a). Antistaling properties of encapsulated maltogenic amylase in gluten-free bread. *Food Science & Nutrition*, 8(11), 5888–5897. doi:10.1002/fsn3.1865

Haghighat-Kharazi, S., Kasaai, M.R., Milani, J.M., & Khajeh, K. (2020b). Optimization of encapsulation of maltogenic amylase into a mixture of maltodextrin and beeswax and its application in gluten-free bread. *Journal of Texture Studies*, 51(4), 631–641. doi:10.1111/jtxs.12516

Hernández-Nava, R., López-Malo, A., Palou, E., Ramírez-Corona, N., & Jiménez-Munguía, M.T. (2019). Complex coacervation between gelatin and chia mucilage as an alternative of encapsulating agents. *Journal of Food* Science, 84(6), 1281–1287. doi:10.1111/1750-3841.14605

Hong, Y., Yang, J., Liu, W., Gu, Z., Li, Z., Cheng, L., & Duan, X. (2019). Sustained release of tea polyphenols from a debranched corn starch–xanthan gum complex carrier, *LWT – Food Science and Technology*, 103, 325–332. doi:10.1016/j.lwt.2019.01.014

Housseiny, M.M., & Aboelmagd, H.I. (2019). Nano-encapsulation of naringinase produced by Trichoderma longibrachiatum ATCC18648 on thermally stable biopolymers for citrus juice debittering. *Journal of Microbiology*, 57, 521–531. doi:10.1007/s12275-019-8528-6

Jahadi, M., Khosravi-Darani, K., Ehsani, M.R., Pimentel, T., Gomes, A., & Mozafari, M.R. (2020). Accelerating ripening of Iranian white brined cheesesusing liposome encapsulated and free proteinases. Biointerface *Research in Applied Chemistry*, 10(1), 4966-4971. doi:10.33263/BRIAC101.966971

Jayaprakash, P., Maudhuit, A., Gaiani, C., & Desobry, S. (2023). Encapsulation of bioactive compounds using competitive emerging techniques: Electrospraying, nano spray drying, and electrostatic spray drying. *Journal of Food Engineering*, 339, 111260. doi:10.1016/j.jfoodeng.2022.111260.

Isaschar-Ovdat, S., & Fishman, A. (2018). Crosslinking of food proteins mediated by oxidative enzymes–A review. *Trends in Food Science & Technology*, 72, 134–143. doi:10.1016/j.tifs.2017.12.011

Isleroglu, H., & Turker, I. (2019). Thermal inactivation kinetics of microencapsulated microbial transglutaminase by ultrasonic spray-freeze drying. *LWT – Food Science and Technology*, 101, 653–662. doi:10.1016/j.lwt.2018.11.091

Jain, A., Thakur, D., Ghoshal, G., Katare, O.P., & Shivhare, U.S. (2016). Characterization of microcapsulated β-carotene formed by complex coacervation using casein and gum tragacanth. *International Journal of Biological Macromolecules*, 87, 101–113. doi:10.1016/j.ijbiomac.2016.01.117

Jurić, S., Đermić, E., Topolovec-Pintarić, S., Bedek, M., & Vinceković, M. (2019). Physicochemical properties and release characteristics of calcium alginate microspheres loaded with Trichoderma viride spores. *Journal of Integrative Agriculture*, 18(11), 2534–2548. doi:10.1016/S2095-3119(19)62634-1

Kalam, M.A., Humayun, M., Parvez, N., Yadav, S., Garg, A., Amin, S., & Ali, A. (2007). Release kinetics of modified pharmaceutical dosage forms: A review. *Continental Journal Pharmaceutical Sciences*, 1, 30–35. https://www.researchgate.net/publication/222711448.

Kawakita, R., Strobel, S., Soares, B., Scher, H.B., Becker, T., Dale, D., & Jeoh, T. (2021). Fluidized bed spray-coating of enzyme in a cross-linked alginate matrix shell (CLAMshell). *Powder Technology*, 386, 372–381. doi:10.1016/j.powtec.2021.03.045

Kim, C.K., & Han, J.H. (1995). Lymphatic delivery and pharmacokinetics of methotrexate after intramuscular injection of differently charged liposome-entrapped methotrexate to rats. *Journal of Microencapsulation*, 12(4), 437–446. doi:10.3109/02652049509087256

Kim, H.R., Kim, D.J., Kang, D.R., Lee, J.G., Lim, S.M., Lee, C.Y., ... & Cho, B.C. (2013). Fibroblast growth factor receptor 1 gene amplification is associated with poor survival and cigarette smoking dosage in patients with resected squamous cell lung cancer. *Journal of Clinical Oncology*, 31(6), 731–737. doi:10.1200/JCO.2012.43.8622

Kim, E.S., Lee, J.S., & Lee, H.G. (2021). Improvement of antithrombotic activity of red ginseng extract by nanoencapsulation using chitosan and antithrombotic cross-linkers: Polyglutamic acid and fucoidan. *Journal of Ginseng Research*, 45(2), 236–245. doi:10.1016/j.jgr.2020.04.001

Koksel, F., & Scanlon, M.G. (2018). Investigation of the influence of bakery enzymes on non-yeasted dough properties during mixing. *Journal of Cereal* Science, 79, 86–92. doi:10.1016/j.jcs.2017.10.002

Korsmeyer, R.W., Gurny, R., Doelker, E., Buri, P., & Peppas, N.A. (1983). Mechanisms of solute release from porous hydrophilic polymers. *International Journal of Pharmaceutics*, 15(1), 25–35. doi:10.1016/0378-5173(83)90064-9

Krol, S., Cavalleri, O., Ramoino, P., Gliozzi, A., & Diaspro, A. (2003). Encapsulated yeast cells inside Paramecium primaurelia: A model system for protection capability of polyelectrolyte shells. *Journal of Microscopy*, 212(3), 239–243. doi:10.1111/j.1365-2818.2003.01251.x

Kshirsagar, A.C., & Singhal, R.S. (2008). Preparation of hydroxypropyl corn and amaranth starch hydrolyzate and its evaluation as wall material in microencapsulation. *Food Chemistry*, 108(3), 958–964. doi:10.1016/j.foodchem.2007.11.074.

Labat, E., Morel, M.H., & Rouau, X. (2000). Effects of laccase and ferulic acid on wheat flour doughs. *Cereal Chemistry*, 77(6), 823–828. doi:10.1094/CCHEM.2000.77.6.823

Lakshmanan, R., Kumaraswamy, P., Krishnan, U.M., & Sethuraman, S. (2016). Engineering a growth factor embedded nanofiber matrix niche to promote vascularization for functional cardiac regeneration. *Biomaterials*, 97, 176–195. doi:10.1016/j.biomaterials.2016.02.033

Lao, L.L., Peppas, N.A., Boey, F.Y.C., & Venkatraman, S.S. (2011). Modeling of drug release from bulk-degrading polymers. *International Journal of Pharmaceutics*, 418(1), 28–41. doi:10.1016/j.ijpharm.2010.12.020

Latorres, J.M., Aquino, S., da Rocha, M., Wasielesky Jr, W., Martins, V.G., & Prentice, C. (2021). Nanoencapsulation of white shrimp peptides in liposomes: Characterization, stability, and influence on bioactive properties. *Journal of Food Processing and Preservation*, 45(7), e15591. doi:10.1111/jfpp.15591

Lima, K.O., Alemán, A., López-Caballero, M.E., del Carmen Gómez-Guillén, M., Montero, M.P., Prentice, C., & Monserrat, J.M. (2021). Characterization, stability, and in vivo effects in Caenorhabditis elegans of microencapsulated protein hydrolysates from stripped weakfish (Cynoscion guatucupa) industrial byproducts. *Food Chemistry*, 364, 130380. doi:10.1016/j.foodchem.2021.13038

Liu, Q., Rauth, A.M., & Wu, X.Y. (2007). Immobilization and bioactivity of glucose oxidase in hydrogel microspheres formulated by an emulsification–internal gelation–adsorption–polyelectrolyte coating method. *International Journal of Pharmaceutics*, 339(1-2), 148–156. doi:10.1016/j.ijpharm.2007.02.027

Liu, Z.P., Zhang, L.L., Yang, Y.Y., Wu, D., Jiang, G., & Yu, D.G. (2018). Preparing composite nanoparticles for immediate drug release by modifying electrohydrodynamic interfaces during electrospraying. *Powder Technology*, 327, 179–187. doi:10.1016/j.powtec.2017.12.066

López Hernández, O.D. (2010). Microencapsulación de sustancias oleosas mediante secado por aspersión. *Revista Cubana de Farmacia*, 44(3), 381–389. ISSN 1561-2988.

Machmudah, S., Fitriana, M.W., Fatbamayani, N., Kanda, H., Winardi, S., & Goto, M. (2022). Phytochemical compounds extraction from medicinal plants by subcritical water and its encapsulation via electrospraying. *Alexandria Engineering Journal*, 61(3), 2116–2128. doi:10.1016/j.aej.2021.07.033

Madene, A., Jacquot, M., Scher, J., & Desobry, S. (2006). Flavour encapsulation and controlled release–A review. *International Journal of Food Science & Technology*, 41(1), 1–21. doi:10.1111/j.1365-2621.2005.00980.x

Maghsoodi, M., & Barghi, L. (2011). Polymer percolation threshold in multi-component HPMC matrices tablets. *Advanced Pharmaceutical Bulletin*, 1(1), 27–33. doi: 10.5681/apb.2011.004

Mahalakshmi, L., Choudhary, P., Moses, J.A., & Anandharamakrishnan, C. (2023) Emulsion electrospraying and spray drying of whey protein nano and microparticles with curcumin. *Food Hydrocolloids for Health*, 3, 100122. doi:10.1016/j.fhfh.2023.100122.

Marras-Marquez, T., Peña, J., & Veiga-Ochoa, M.D. (2015). Robust and versatile pectin-based drug delivery systems. *International Journal of Pharmaceutics*, 479(2), 265–276. doi:10.1016/j.ijpharm.2014.12.045

McSweeney, P.L.H. (2011). Enzymes exogenous to milk in dairy technology: Catalase, glucose oxidase, glucose isomerase and hexose oxidase. In: Fuquay J.W. (Ed.). Encyclopedia of Dairy Sciences. Second edition. Academic Press. Cambridge, USA. P.301-303. ISBN 9780123744074.

Melim-Miguel, A., Martins-Meyer, T.S., Figueiredo, E.V.D.C., Lobo, B.W.P., & Dellamora-Ortiz, G.M. (2013). Enzymes in bakery: Current and future trends. Chapater 14. Innocenzo Muzzalupo (Ed.) In: *Food Industry*, IntechOpen. London, UK. P: 287–321. doi:10.5772/53168.

Mi, Y., Liu, Z., Deng, J., Lei, H., Zhu, C., Fan, D., & Lv, X. (2018). Microencapsulation of phosphorylated human-like collagen-calcium chelates for controlled delivery and improved bioavailability. *Polymers*, 10(2), 185. doi:doi:10.3390/polym10020185.

Miguezb, M.J.B., Rodrigues, C., Sanchez, M.D.N., & Lafunjeira, M.C.M. (1997). Preparation and scanning electronic microscopy study of chitosan/polyvinyl (alcohol)-encapsulated crude urease extract. *Journal of Microencapsulation*, 14(5), 639–646. doi:10.3109/02652049709006816

Mishra, M. (Ed.). (2015). *Handbook of encapsulation and controlled release*. CRC Press, Florida, USA. p. 3–19. doi:doi:10.1201/b19038

Moreira, J.B., Vaz, B.D.S., Cardias, B.B., Cruz, C.G., Almeida, A.C.A.D., Costa, J.A.V., & Morais, M.G.D. (2022). Microalgae polysaccharides: An alternative source for food production and sustainable agriculture. *Polysaccharides*, 3(2), 441–457. doi:10.3390/polysaccharides3020027

Nayak, A., Pal, D., & Das, S. (2013). Calcium pectinate-fenugreek seed mucilage mucoadhesive beads for controlled delivery of metformin HCl. *Carbohydrate Polymers*, 96(1), 349–357. doi:10.1016/j.carbpol.2013.03.088

Nesterenko, A., Alric, I., Silvestre, F., & Durrieu, V. (2013). Vegetable proteins in microencapsulation: A review of recent interventions and their effectiveness. *Industrial Crops and Products*, 42(1), 469–479. doi:doi:10.1016/j.indcrop.2012.06.035

Nguyen, T.X., Huang, L., Gauthier, M., Yang, G., & Wang, Q. (2016). Recent advances in liposome surface modification for oral drug delivery. *Nanomedicine*, 11(9), 1169–1185. doi:10.2217/nnm.16.9

Niu, B., Shao, P., Luo, Y., & Sun, P. (2020). Recent advances of electrosprayed particles as encapsulation systems of bioactives for food application. *Food Hydrocolloids*, 99, 105376. doi:10.1016/j.foodhyd.2019.105376.

Noor, E., Yusron, M., & Sitorus, R.D.M.C. (2022). *Proteus penneri* encapsulation with maltodextrin and sodium alginate using a spray-drying method. *Food Science and Technology*, 42, e49621. doi:doi:10.1590/fst.49621

Oikonomopoulou, V., Stramarkou, M., Plakida, A., & Krokida, M. (2022). Optimization of encapsulation of stevia glycosides through electrospraying and spray drying. *Food Hydrocolloids*. 131, 107854. doi:10.1016/j.foodhyd.2022.107854

Ooms, N., & Delcour, J.A. (2019). How to impact gluten protein network formation during wheat flour dough making. *Current Opinion in Food Science*, 25, 88–97. doi:10.1016/j.cofs.2019.04.001

Oro, C.E.D., Paroul, N., Mignoni, M.L., Zabot, G.L., Backes, G.T., Dallago, R.M., & Tres, M.V. (2023). Microencapsulation of Brazilian Cherokee blackberry extract by freeze-drying using maltodextrin, gum Arabic, and pectin as carrier materials. *Food Science and Technology International*, 29(3), 255–265. doi:10.1177/10820132211068979

Oxley, J. (2016). Process-selection criteria. In *Handbook of encapsulation and controlled release. Mishra*, M. (Ed.). CRC Press, Florida, USA. Chapter 2. p.23–34. ISBN 9780429183478.

Ozaltin, K., Postnikov, P.S., Trusova, M.E., Sedlarik, V., & Di Martino, A. (2019). Polysaccharides based microspheres for multiple encapsulations and simultaneous release of proteases. *International Journal of Biological Macromolecules.* 132, (24-31) doi:10.1016/j.ijbiomac.2019.03.189

Paranthaman, R., Moses, J.A., & Anandharamakrishnan, C. (2021). Development and validation of a screening method for simultaneous detection of KBrO3 and KIO3 in baking ingredients and additives using powder XRD. *Journal of Food Composition and Analysis,* 102, 104007. doi:10.1016/j.jfca.2021.104007.

Paximada, P., Kanavou, E., Apostolidis, E., & Mandala, I.G. (2021). Encapsulation of catechin in electrosprayed food-grade particles. *Food Hydrocolloids for Health,* 1, 100021. doi:doi:10.1016/j.fhfh.2021.100021

Pedroza-Islas, R. (2002, 3–6 September). Alimentos Microencapsulados: Particularidades de los Procesos para la Microencapsulación de Alimentos para Larvas de Especies Acuícolas. In: *Avances En Nutrición Acuicola.. Memorias Del VI Simposium Internacional de Nutrición Acuicola* (pp. 438–447), Cruz-Suárez, L. E., Ricque-Marie, D., Tapia-Salazar, M., Gaxiola-Cortés, M. G., Simoes, N. (Eds.). Cancún, Quintana Roo, México. https://nutricionacuicola.uanl.mx/index.php/acu/article/view/249/247

Petrov, A.I., Volodkin, D.V., & Sukhorukov, G.B. (2005). Protein—calcium carbonate coprecipitation: A tool for protein encapsulation. *Biotechnology Progress,* 21(3), 918–925. doi:10.1021/bp0495825

Pinho, E., Soares, G., & Henriques, M. (2015). Evaluation of antibacterial activity of caffeic acid encapsulated by Evaluation of antibacterial activity of caffeic acid encapsulated by β-s-cyclodextrins. *Journal of Microencapsulation,* 32(8), 804–810. doi:10.3109/02652048.2015.1094531

Pourmohammadi, K., & Abedi, E. (2021). Enzymatic modifications of gluten protein: Oxidative enzymes. *Food Chemistry,* 356, 129679. doi:doi:10.1016/j.foodchem.2021.129679

Radünz, M., Camargo, T.M., dos Santos Hackbart, H.C., Blank, J.P., Hoffmann, J.F., Stefanello, F.M., & da Rosa-Zavareze, E. (2021). Encapsulation of broccoli extract by electrospraying: Influence of in vitro simulated digestion on phenolic and glucosinolate contents, and on antioxidant and antihyperglycemic activities. *Food Chemistry,* 339, 128075. doi:10.1016/j.foodchem.2020.128075

Rasiah, I.A., Sutton, K.H., Low, F.L., Lin, H.M., & Gerrard, J.A. (2005). Crosslinking of wheat dough proteins by glucose oxidase and the resulting effects on bread and croissants. *Food Chemistry,* 89(3), 325–332. doi:10.1016/j.foodchem.2004.02.052

Rehman, H.U., Nawaz, M.A., Pervez, S., Jamal, M., Attaullah, M., Aman, A., & Qader, S.A.U. (2020). Encapsulation of pectinase within polyacrylamide gel: Characterization of its catalytic properties for continuous industrial uses. *Heliyon,* 6(8), e04578. doi:10.1016/j.heliyon.2020.e04578

Rentería-Ortega, M., Salgado-Cruz, M.D.L.P., Morales-Sánchez, E., Alamilla-Beltrán, L., Farrera-Rebollo, R.R., Valdespino León, M., & Calderón-Domínguez, G. (2020). Effect of electrohydrodynamic atomization conditions on morphometric characteristics and mechanical resistance of chia mucilage-alginate particles. CyTA – *Journal of Food,* 18(1), 461–471. doi:10.1080/19476337.2020.1775706

Rentería-Ortega, M., Salgado-Cruz, M. de la P., Morales-Sánchez, E., Alamilla-Beltrán, L., Valdespino-León, M., & Calderón-Domínguez, G. (2021). Glucose oxidase release of stressed chia mucilage-sodium alginate capsules prepared by electrospraying. *Journal of Food Processing and Preservation,* 45(5). e15484 doi:10.1111/jfpp.15484

Renzetti, S., & Rosell, C.M. (2016). Role of enzymes in improving the functionality of proteins in non-wheat dough systems. *Journal of Cereal Science,* 67, 35–45. doi:10.1016/j.jcs.2015.09.008

Rostamabadi, H., Falsafi, S.R., & Jafari, S.M. (2019). Nanoencapsulation of carotenoids within lipid-based nanocarriers. *Journal of Controlled Release,* 298, 38–67. doi:10.1016/j.jconrel.2019.02.005

Sandoval, P.A., Baena, Y., Aragón, M., & Rosas, J.E. (2008). Mecanismos generales de cesión de principios activos a partir de matrices monolíticas hidrofílicas preparadas con éteres de celulosa. *Revista Colombiana de Ciencias Químico-Farmacéuticas,* 37(2), 105–121.

Sarkar, S., Gupta, S., Variyar, P.S., Sharma, A., & Singhal, R.S. (2012). Irradiation depolymerized guar gum as partial replacement of gum Arabic for microencapsulation of mint oil. *Carbohydrate Polymers,* 90(4), 1685–1694. doi:10.1016/j.carbpol.2012.07.051

Schakowski, K.M., Elm, C., Linders, J., & Kirsch, M. (2021). Synthesis and characterization of enzymatically active micrometer protein-capsules. *Artificial Cells, Nanomedicine, and Biotechnology,* 49(1), 605–612. doi:10.1080/21691401.2021.1955698

Shanmugavel, V., Santhi, K.K., Kurup, A.H., Kalakandan, S., Anandharaj, A., & Rawson, A. (2020). Potassium bromate: Effects on bread components, health, environment and method of analysis: A review. *Food Chemistry,* 311, 125964. doi:10.1016/j.foodchem.2019.125964

Siepmann, J., & Peppas, N.A. (2012). Modeling of drug release from delivery systems based on hydroxypropyl methylcellulose (HPMC). *Advanced Drug Delivery Reviews,* 64, 163–174. doi:10.1016/j.addr.2012.09.028

Sridhar, K., Bouhallab, S., Croguennec, T., Renard, D., & Lechevalier, V. (2022). Application of high-pressure and ultrasound technologies for legume proteins as wall material in microencapsulation: New insights and advances. *Trends in Food Science & Technology,* 127, 49–62. ISSN 0924-2244. doi:10.1016/j.tifs.2022.07.006

Stein, E.W., Volodkin, D.V., McShane, M., & Sukhorukov, G.B. (2006). Real-time assessment of spatial and temporal coupled catalysis within polyelectrolyte *microcapsules containing coimmobilized glucose oxidase and peroxidase. Biomacromolecules,* 7(3), 710–719. doi:10.1021/bm050304j

Stemler, C.D., & Scherf, K.A. (2022). Improvement of cake baking properties by lipases compared to a traditional emulsifier. *Food Chemistry:* X, 15, 100442. doi:10.1016/j.fochx.2022.100442.

Sun, H., Magnuson, Z., He, W., Zhang, W., Vardhan, H., Han, X., & Ma, S. (2020). PEG@ ZIF-8/PVDF nanocomposite membrane for efficient pervaporation desulfurization via a layer-by-layer technology. *ACS Applied Materials & Interfaces,* 12(18), 20664–20671. doi:10.1021/acsami.0c02513

Tebben, L., Shen, Y., & Li, Y. (2018). Improvers and functional ingredients in whole wheat bread: A review of their effects on dough properties and bread quality. *Trends in Food Science & Technology.* 81, 10–24 doi:10.1016/j.tifs.2018.08.015

Traffano-Schiffo, M.V., Castro-Giraldez, M., Fito, P.J., & Santagapita, P.R. (2017). Encapsulation of lactase in Ca (II)-alginate beads: Effect of stabilizers and drying methods. *Food Research International,* 100, 296–303. doi:10.1016/j.foodres.2017.07.020

Us-Medina, U., Ruiz-Ruiz, J.C., Quintana-Owen, P., & Segura-Campos, M.R. (2017). Salvia hispanica mucilage-alginate properties and performance as an encapsulation matrix for chia seed oil. *Journal of Food Processing and Preservation,* 41(6), e13270. doi:10.1111/jfpp.13270

Vázquez-González, Y., Prieto, C., Filizoglu, M.F., Ragazzo-Sánchez, J.A., Calderón-Santoyo, M., Furtado, R.F., & Lagaron, J.M. (2021). Electrosprayed cashew gum microparticles for the encapsulation of highly sensitive bioactive materials. *Carbohydrate Polymers,* 264, 118060. doi:10.1016/j.carbpol.2021.118060

Vesely, D. (2008). Diffusion of liquids in polymers. *International Materials Reviews,* 53(5), 299–315. doi:10.1179/174328008x324602

Vikartovská, A., Bučko, M., Mislovičová, D., Pätoprstý, V., Lacík, I., & Gemeiner, P. (2007). Improvement of the stability of glucose oxidase via encapsulation in sodium alginate–cellulose sulfate–poly (methylene-co-guanidine) capsules. *Enzyme and Microbial Technology,* 41(6-7), 748–755. doi:10.1016/j.enzmictec.2007.06.010

Vocadlo, D.J., Davies, G., Laine, R., & Withers, S.G. (2001). Catalysis by hen egg-white lysozyme proceeds via a covalent intermediate. *Nature,* 412(6849), 835–838. doi:10.1038/35090602

Wang, B., Cheng, F., Gao, S., Ge, W., & Zhang, M. (2017). Double enzymatic hydrolysis preparation of heme from goose blood and microencapsulation to promote its stability and absorption. *Food Chemistry,* 217, 699–704. doi:10.1016/j.foodchem.2016.09.007

Wang, Y. (2020). Exploring the functionality of in situ produced dextran in high-protein or wholegrain enriched wheat bread. Ph.D. thesis. University of Helsinki, Department of Food and Nutrition, Faculty of Agriculture and Forestry. Finland. ~~132, 15-17~~ https://helda.helsinki.fi/items/c1bd1258-c7a2-403b-8a39-3faa0cbb619c

Wang, L., Clardy, A., Hui, D., & Wu, Y. (2021). Physiochemical properties of encapsulated bitter melon juice using spray drying. *Bioactive Carbohydrates and Dietary Fibre,* 26, 100278. doi:10.1016/j.bcdf.2021.100278

Wu, I.Y., Bala, S., Škalko-Basnet, N., & Di Cagno, M.P. (2019). Interpreting non-linear drug diffusion data: Utilizing Korsmeyer-Peppas model to study drug release from liposomes. *European Journal of Pharmaceutical Sciences,* 138, 105026. doi:10.1016/j.ejps.2019.105026

Xiao, Z., Liu, W., Zhu, G., Zhou, R., & Niu, Y. (2014). A review of the preparation and application of flavour and essential oils microcapsules based on complex coacervation technology. *Journal of the Science of Food and Agriculture,* 94(8), 1482–1494. doi:10.1002/jsfa.6491

Xie, B., Zhang, M., Yang, H., & Jiang, W. (2011). Effects of conventional versus organic production systems on amino acid profiles and heavy metal concentrations in the Chinese shrimp Penaeus chinensis. *Fisheries Science*, 77, 839–845. doi:10.1007/s12562-011-0395-y

Xu, J., Wang, W., & Li, Y. (2019). Dough properties, bread quality, and associated interactions with added phenolic compounds: A review. *Journal of Functional Foods*, 52, 629–639. doi:10.1016/j.jff.2018.11.052

Yañez, J., Salazar, J., Chaires, L., Jiménez, J., Márquez, M., & Ramos, E. (2002). Aplicaciones biotecnológicas de la microencapsulación. *Avance y perspectiva*, 21, 313–319. ISSN:0185-1411.

Yoshii, H., Buche, F., Takeuchi, N., Terrol, C., Ohgawara, M., & Furuta, T. (2008). Effects of protein on retention of ADH enzyme activity encapsulated in trehalose matrices by spray drying. *Journal of Food Engineering*, 87(1), 34–39. doi:10.1016/j.jfoodeng.2007.03.014

Zhang, S., Jiang, Z., Wang, X., Yang, C., & Shi, J. (2015). Facile method to prepare microcapsules inspired by polyphenol chemistry for efficient enzyme immobilization. *ACS Applied Materials & Interfaces*, 7(35), 19570–19578. doi:10.1021/acsami.5b03823

Zhang, Z., Zhang, R., Chen, L., & McClements, D.J. (2016). Encapsulation of lactase (β-galactosidase) into κ-carrageenan-based hydrogel beads: Impact of environmental conditions on enzyme activity. *Food Chemistry*, 200, 69–75. doi:10.1016/j.foodchem.2016.01.014

Zhao, Y., Huang, Z.-H., Zhou, H.-M., Zhu, K.-X., Guo, X.-N., & Peng, W. (2021). Inhibition of hexose oxidase on the dark spots in fresh wet noodle sheets: A feasible prevention of dark spots. *Food Chemistry*, 339, 128021. doi:10.1016/j.foodchem.2020.12802

Zhu, G., Xiao, Z., Zhou, R., & Feng, N. (2015). Production of a transparent lavender flavour nanocapsule aqueous solution and pyrolysis characteristics of flavour nanocapsule. *Journal of Food Science and Technology*, 52, 4607–4612. doi:10.1007/s13197-014-1465-9

Zyaitdinov, D.R., Ewteew, A.V., & Bannikova, A.V. (2020). Immobilization of oat bran polyphenols in complex coacervates of whey protein and malthodextrin. *Food Processing Techniques and Technology*, 50, 460–469. doi:doi:10.21603/2074-9414-2020-3-460-469 Food

Chapter 19

Common Bean Post-Harvest Storage Process

Microstructural Changes and Their Relationship with the Hardening

Mayra Nicolás-García
Madeleine Perucini-Avendaño
María de Jesús Perea-Flores
Brenda H. Camacho-Díaz
Daniel Tapia-Maruri
and Gloria Dávila-Ortiz

CONTENTS

19.1 INTRODUCTION

The genus *Phaseolus* includes more than 70 species of which the common bean (*P. vulgaris* L.) is the most widely consumed along with *P. acutifolius*, *P. lunatus*, *P. coccineus*, and *P. polyanthus* (Porch

DOI: 10.1201/9781003332558-22

et al. 2013; Bitocchi et al. 2017; Capistrán-Carabarin et al. 2019; Nicolás-García, Jiménez-Martinez, Perucini-Avendaño, et al. 2021). The relevance of the common bean is due to its chemical, nutritional, and functional properties because it is a source of protein, complex carbohydrates, dietary fiber, vitamins, minerals, and phytochemicals (Ramírez-Jiménez et al. 2015; Mojica et al. 2015; Yang et al. 2018; Giusti et al. 2019; Nicolás-García, Jiménez-Martinez, Perucini-Avendaño, et al. 2021; Perera et al. 2023). However, inadequate post-harvest seed storage affects the nutritional quality mainly by hardening, also known as the hard-to-cook defect (HTC) (Reyes-Moreno et al. 1993; Dos Santos-Siqueira et al. 2016; Chigwedere et al. 2018; Chu et al. 2020; Nakhungu et al. 2020; Nicolás-García et al. 2022; Abay & Tolesa, 2023). During storage, complex reactions are activated within the seed, starting the browning phenomenon. Several groups of phenolic compounds have been found to contribute to non-enzymatic and enzymatic browning in different foods, mainly through their participation in oxidative steps and changes in flavonoid skeleton (Dos Santos-Siqueira et al. 2016; Miano & Augusto, 2018). The seed undergoes several biochemical changes, including darkening and hardening. Nicolás-García et al. (2022) reported color saturation indicating bean darkening, and an increase in hardness from 5.42 to 19.96 N during storage for 9 months under conditions of relative humidity and temperature (70% and 30 °C). On the other hand, it has been reported that hardening of bean seeds is caused by physical and chemical changes that develop inside the seed at the intracellular level and structural modifications during storage (Alvares et al. 2020; Chu et al. 2020; Nakhungu et al. 2020; Nicolás-García et al. 2024).

19.2 COMMON BEAN STRUCTURE

Legumes are diverse in shape and size, containing large amounts and characteristics of polymeric reserves (Bewley et al. 2013). The species *P. vulgaris* L. has different bean types and their commercial value is influenced by characteristics, such as bean size, color, and uniformity. The bean is constituted by three tissues: the seed coat, the cotyledon, and the embryo (**Figure 19.1**). These represent 10, 89, and 1% of the total bean weight, respectively (Chávez-Mendoza et al. 2019; Nicolás-García, Jiménez-Martinez, Perucini-Avendaño, et al. 2021). The seed is a crucial stage in the angiosperm life cycle, as it represents the new sporophytic generation (Ramos-Ordoñez et al. 2016).

The tissues that support and participate in the micromechanical properties of the seed coat are: 1) the epidermal palisade cell (PAL), 2) the subepidermal cell (SE), and 3) the spongy parenchymal cell (SP). These structures can be partially or completely impermeable to water (Velasco-González et al. 2013; Konzen & Tsai, 2014; Miano & Augusto, 2018; Nicolás-García, Arzate-Vázquez, Perea-Flores, et al. 2021; Nicolás-García et al. 2022). PAL consists of one dense layers of palisade cells or macrosclereids composed of cellulose, hemicellulose, and lignin that confer rigidity to the tissue and regulate seed water diffusion. SE is a layer constituted of osteosclereid cells that present calcium oxalate (CaOx) crystals inside. CaOx crystals have been reported to regulate calcium and participate in the structural rigidity of the seed coat. SP are tissues that facilitate gas exchange and in addition to PAL present chromoplasts with condensed tannins (see **Figure 19.1 and 19.2.1**). The presence of these macrosclereid and osteosclereid cells confers the seeds' characteristic impermeability (Nicolás-Bermúdez et al., 2018; Hernandez-Aguilar et al. 2020; Bassett et al. 2021; Nicolás-García, Arzate-Vázquez, Perea-Flores, et al 2021; Acosta et al. 2023). The seed coat of the common bean contains phenolic compounds, such as condensed tannins, anthocyanins, and flavanols like kaempferol and quercetin glycoside; whereas the cotyledon is rich in phenolic acids

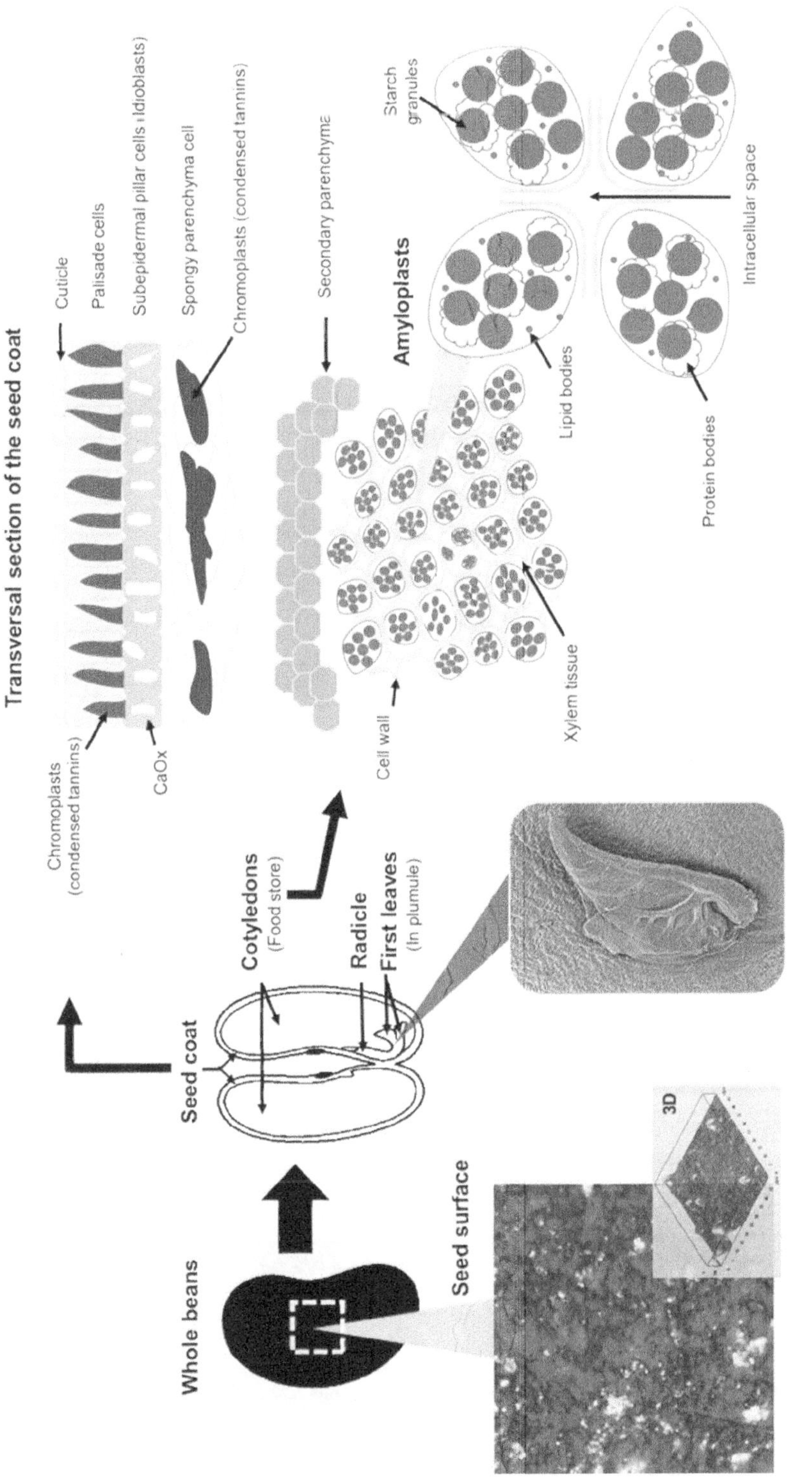

Figure 19.1 Schematic diagram of common bean seed structure.

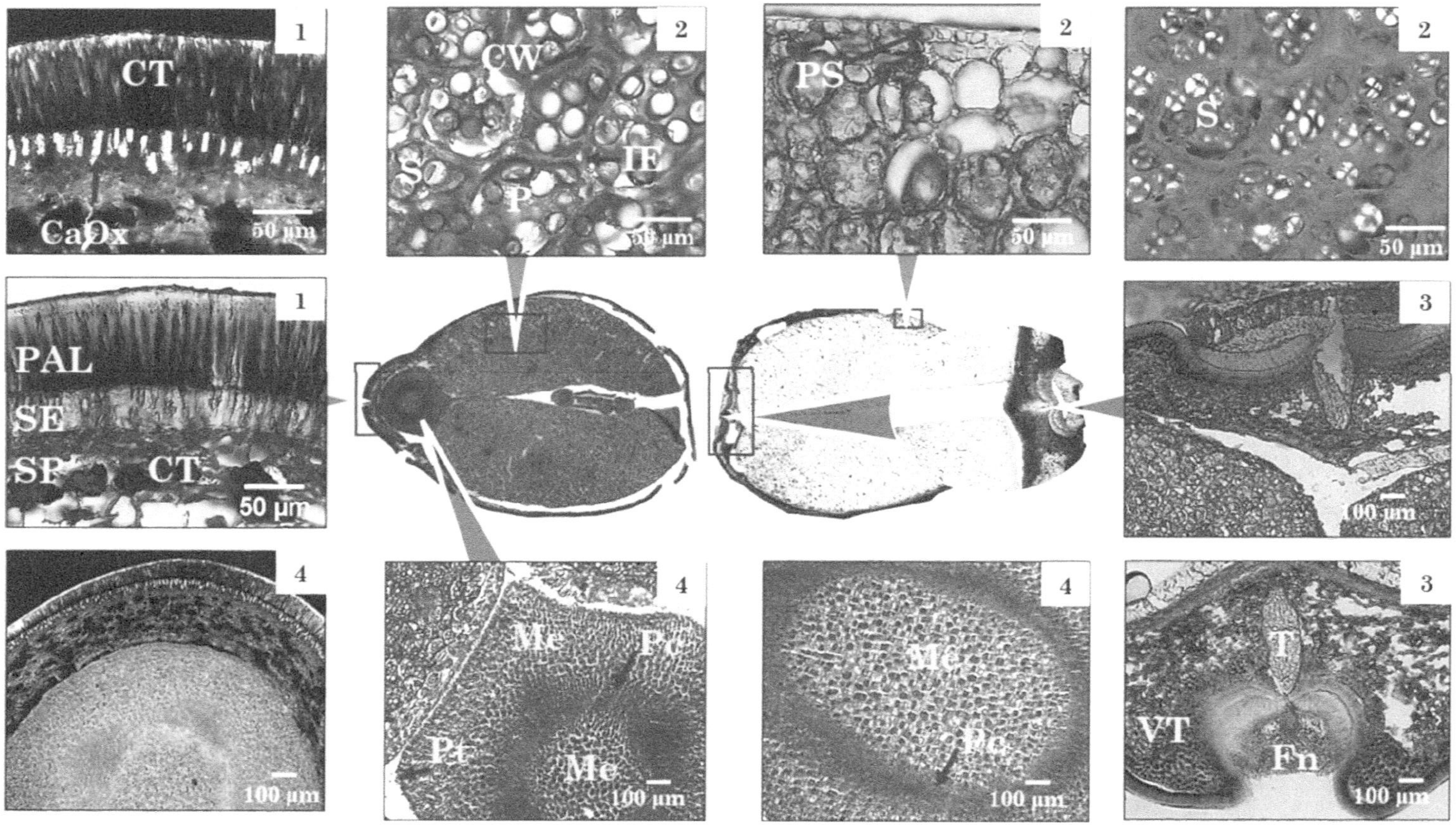

Figure 19.2 Bean seed morphology. Cross-section images obtained by optical microscopy of bean seed. 1: Seed coat (20×); 2: cotyledon (20×); 3: hilum double palisades (4×); 4: embryo (4×). PAL: epidermal palisade cell; SE: subepidermal cell; SP: spongy parenchymal cell; CT: condensed tannins; S: starch granule; P: protein matrix; CW: cell wall; IE: intracellular space; PS: parenchyma secondary; T: tracheid; VT: vascular tissue; Fn: funiculi; Me: meristems; Pt: protoderm; Pc: procambium; CaOx: calcium oxalate crystals.

like ferulic, sinapic, chlorogenic, and other hydroxycinnamic acids in smaller amounts (Ganesan & Xu, 2017; Los et al. 2018; Yang et al. 2018; Chávez-Mendoza et al. 2019; Giusti et al. 2019; Nurzyńska-Wierdak et al. 2019; Nicolás-García, Jiménez-Martinez, Perucini-Avendaño, et al. 2021; Nicolás-García et al. 2022).

The cotyledon contains the main food reserve substances, basically proteins (20–25%), carbohydrates (50–60%, mainly starch), and lipids (Velasco-González et al., 2013; De Oliveira et al. 2018; Los et al. 2018). **Figure 19.2.2** shows the amyloplasts and the secondary parenchyma cell. In the amyloplasts, the starch granules are protected by natural structural barriers or cell walls that prevent their complete swelling and gelatinization during cooking. The cell structure, as well as the protein matrix, act as physical barriers to the diffusion of enzymes that limit starch hydrolysis (Li et al. 2019; Do et al. 2019). Starch comprises 27–60% of carbohydrates, the rest being composed of non-starch polysaccharides and oligosaccharides (stachyose, raffinose, and verbascose) (Ramírez-Jiménez et al. 2014; Alcázar-Valle et al. 2020).

Moreover, the seed has structures of great relevance for the moisture absorption process, such as the raphe, hilum, and micropyle (Miano & Augusto, 2018). The seed presents a double palisade of the hilum and embryo (**Figure 19.2.3; Figure 19.2.4**). The latter participates as a reserve source for seed development and is composed of the embryonic axis and the cotyledons. The axis is composed of the embryonic root or radicle, the hypocotyl to which the cotyledons are attached, and the shoot apex with the first true leaf primordium (plumule) (Bewley et al. 2013). Embryonic cells can also be observed. In some cases, the embryo may be entirely absorbed so that most of the seed is occupied by the cotyledons. The hilum regulates the diffusion of water into the seed; it consists of a double layer of macrosclereid cells (palisade) that are seen in reverse, and below these layers of hilum cells are the tracheids (T). The tracheids are a pore tissue made up of elongated and tapered cells, relatively narrow and with thick and dotted walls without true perforations; these facilitate the exchange of gases and water vapor. They form the xylem tube system and transport water and solutes. In the subhilum region, vascular tissue (VT) as well as remnants of funiculi (Fn), cotyledons, and cotyledonary fissures are observed (Miano & Augusto, 2018; Zablatzká et al. 2021).

The structural properties, organization, and cell wall composition depict different mechanical behaviors, which depend on the seed's water absorption capacity. Several factors such as species, variety, genotype, structure, composition, shape, size, and color influence the hydration process (Miano, Saldaña, Campestrini, et al. 2018; Pramiu et al. 2019; Katuuramu et al., 2020). Miano, Sabadoti, Pereira, et al. (2018) demonstrated that color and size are important parameters in the hydration process, observing that the hydration time of dark-colored seeds is longer compared to light-colored seeds. Likewise, sigmoidal seeds hydrate at a higher rate than concave seeds and larger seeds hydrate more slowly. On the other hand, seed hydration and transpiration are influenced by the seed coat cuticle, which is made impermeable by the cutin, suberin, and waxes that are present (Mujica et al. 2015; Barthlott et al. 2017; Miano & Augusto, 2018; Nicolás-García, Arzate-Vázquez, Perea-Flores, et al. 2021, Nicolás-García et al. 2024). Several studies have evaluated the mechanical behavior of bean seed coat structure during processing, such as hydration and soaking (Miano, Saldaña, Campestrini, et al. 2018; Li et al. 2020), post-harvest storage (Ávila et al. 2020; Nicolás-García et al. 2022), germination, and cooking (Miano & Augusto, 2018). Mikac et al. (2015) observed that, during the bean soaking process, the greatest diffusion of water into the seed occurs in the micropyle region, and in the radicle and hypocotyl regions, as well as around the seed.

In addition, Li et al. (2020) observed, by magnetic resonance imaging (MRI), that water diffusion in the seed during the soaking process occurs first through the irregular wall and then through the inner region, which causes a texture modification. Nicolás-García, Arzate-Vázquez, Perea-Flores, et al. (2021) evaluated the micromechanical properties of the black seed coat and their relationship with the microstructure by observing the structural behavior of the seed multilayer. Morphophysiological properties, including dimension, shape, weight, volume, density, and surface area, of common bean seeds influence equipment for harvesting, sorting, transportation, and packaging or container design during storage (Wani et al. 2017; Kibar & Kibar, 2019). Kibar & Kibar (2019) observed changes in morpho-physiological properties of common bean seeds at different moisture contents (6.4, 11.7, and 17.3%) and stored at 4 °C and 65% relative humidity for one year. They observed that seeds with 11.7% and 17.3% moisture increased to 13.4% and 19.6%, respectively, indicating an increase in shape index and weight values.

19.3 BEAN POST-HARVEST STORAGE

P. vulgaris L. is one of the most consumed legumes due to its nutritional composition and functionality that provides benefits to the organism; its harvesting and post-harvest handling influence the seed's quality (García-Díaz et al. 2018; Mecha et al. 2019; Kibar & Kibar, 2019; Bulyaba et al. 2020; Perera et al. 2023). Cultivation is performed under different climatic conditions, soil types, and optimum temperature of 25 °C. After harvesting, the beans are stored for preservation until consumption. Post-harvest storage conditions are crucial to preserve quality and ensure seed stability, which depend on factors such as seed moisture content, relative humidity, the concentration of atmospheric conditions (O_2 and CO_2), temperature, and storage time (Kibar & Kibar, 2019; Li et al. 2019; Chu et al. 2020). Inadequate post-harvest storage management can cause changes in seed texture, chemical composition, and sensory properties when performed under adverse conditions of temperature and relative humidity above 25 °C and 65%, respectively. It has been reported that improper storage originates the hardening phenomenon during shelf life due to the physical, biological, and chemical changes the seed undergoes at the intercellular level favoring the hard-to-cook (HTC) defect in beans (Dos Santos-Siqueira et al. 2016; Chigwedere et al. 2018; Chu et al. 2020; Nakhungu et al. 2020; Nicolás-García et al. 2022; Abay & Tolesa, 2023).

19.3.1 Bean Hardening

The HTC defect results from physical and chemical changes at the intercellular level during storage, leading to an increase in middle lamina stability upon cooking and from intrinsic characteristics such as seed coat thickness, seed composition, and microstructure (Reyes-Moreno et al. 1993; Velasco-Gonzalez et al. 2013; Chu et al. 2020; Nakhungu et al. 2020; Perera et al. 2023). One of the main parameters affecting seed quality during the storage process is its moisture content (Kibar & Kibar, 2019). During storage, complex enzymatic and non-enzymatic reactions are activated inside the seed, causing the browning phenomenon. Several groups of phenolic compounds, such as flavonoids and condensed tannins, have been found to contribute directly to color changes during post-harvest storage (Dos Santos-Siqueira et al. 2016; Parmar et al. 2017; Chigwedere et al. 2018; Diaz-Contreras et al. 2018; Alvares et al. 2020; Nicolás-García et al. 2022). The behavior of the physical parameters or changes experienced by the bean seed during storage increases with increasing moisture content of the seeds exceeding 18%; therefore, the recommendation is to keep it at 15% to preserve their quality (Dos Santos-Siqueira et al. 2016; Kibar & Kibar, 2019; Abay & Tolesa, 2023). Nicolás-García et al. (2022) reported that the moisture content of black beans decreased from 9.7 to 7.8% after 9 months of storage. The color saturation observed during storage of the beans indicates that they experienced darkening as indicated by the significant increase in the chroma parameter (C*) from 1.10 to 1.72 observed in seeds with a decrease in moisture content after 150 days of storage.

Several mechanisms involved in this phenomenon have been reported, such as pectin insolubilization, enzymatic processes (pectin-methylesterases, polyphenol oxidases, and lipoxygenases), lipid oxidation, and pectin-phytate cation and protein-phenolic compound interactions. There are several types of interactions between phenolic compounds and proteins, which include non-covalent bonds, hydrogen bonds, hydrophobic, ionic and covalent bonds (Calderan-Rodrigues et al. 2017; Alves & Tavares, 2019; Nicolás-García, Jiménez-Martinez, Perucini-Avendaño, et al. 2021). The interactions or bonds responsible for the altered solubilization of the middle lamella during baking are believed to include divalent cationic cross-links between demethoxylated pectic substances and phenolic cross-links (Shahidi & Yeo 2016; Renard et al. 2017; Fernandes et al. 2019). Furthermore, changes in storage proteins, cell membrane degradation, and cell wall and middle lamella lignification of cotyledonary cells occur (Kinyanjui et al. 2015; Mujica et al. 2015; Koriyama et al. 2017; Parmar et al. 2017; Diaz-Contreras et al. 2018; Chigwedere et al. 2018; Jombo et al. 2018; Chigwedere et al. 2019; De Miranda et al. 2019; Chu et al. 2020; Coelho et al. 2020; Nakhungu et al. 2020; Nicolás-García, Jiménez-Martinez, Perucini-Avendaño, et al. 2021). In addition, the structure or cell wall determines the vegetable food texture; this is an important quality attribute of legume seeds. The composition, arrangement, and interactions of plant cell wall polymers, both physically and chemically, largely determine the mechanical properties of the food (Pallares, Rousseau, Chigwedere, et al. 2018; Nicolás-García, Arzate-Vázquez, Perea-Flores, et al. 2021).

The hardening process results in negative changes in the sensory and nutritional quality of beans (Dos Santos-Siqueira et al. 2016; Chigwedere et al. 2018; Chu et al. 2020; Perera et al. 2023). It has been reported to be related to enzyme activation and changes in lignin and phytates (Alves et al. 2021). Kibar & Kibar (2019) observed a decrease in mineral content, proteins, total phenols, flavonoids, and antioxidant activity, as well as in density and electrical conductivity, during storage (4 °C, 65% relative humidity, 12 months) in common bean with different moisture contents (6.4, 11.7, and 17.3%). Alves et al. (2021) indicated that storage for six months (22 °C and 65% relative humidity) was able to prevent physicochemical alterations and preserve important nutrients, phytochemicals and bioactive bioactive compounds, as well as the antioxidant capacity of the grains.

On the other hand, Nicolás-García et al. (2022) identified 37 compounds by UPLC-EDI-MS including caffeic acid, syringic acid, quercetin, and some glycosides, such as catechin-7-O-glucoside, that showed variations during storage. The increase in phenolic compounds, the color parameter "Chroma," seed hardness, and the saturation observed in the microstructure of the seed coat evidence the darkening and hardening of the black bean seed.

In Table 19.1, the principal changes in common bean composition during post-harvest storage across several primordial components as protein, lipids, minerals, phenolic, volatile compounds, and lignin, previously mentioned, are summarized.

TABLE 19.1 NUTRITIONAL AND FUNCTIONAL CHANGES IN COMMON BEAN DURING POSTHARVEST STORAGE

Compounds	Storage Conditions	Results	Reference
Protein	Fresh and hardened samples.	The protein content was reduced since 26.75 to 21.71% of fresh beans to hard to cook beans.	Ruiz-Ruiz et al. (2012)
	Three different seed moisture contents (6.4, 11.7, and 17.3%) and cold storage (4 ± 0.5 °C and 65 ± 5% r. h.) for 12 months.	Reported that the protein in beans ranged from 22.86 to 25.67%. However, this content decreased as seed moisture content increased and also decreased at the end of 12 months' storage compared to the fresh sample.	Kibar & Kibar (2019)
	Storage at room temperature (25 °C, 80% r. h.) and under refrigerated conditions (4°C, 80% r. h.) for 3, 6, 9, and 12 months.	The results indicated that variations in temperature and humidity (25 °C, 80% r. h.) exerted a negative effect on the proteomic profile during prolonged storage of beans for 12 months. Modifications in storage proteins such as phaseolin, legumins, and phytohemagglutinin were caused by physiological changes such as degradation or structural changes in amino acids and proteins.	Ávila et al. (2020)
Lipids	Storage at 20 °C for 225 days in the dark.	Reported an increase in lipid acidity in black bean seeds after 75 days in a period of 225 days of storage indicating an increase in free fatty acids resulting from triglycerides and phospholipids produced by the action of phospholipases and lipases present in the seed under elevated humidity and temperature conditions.	Elias et al. (2016)
Minerals	3 months at 35 °C and 83% r. h.	Bean aging caused a decrease in inositol hexaphosphate (InsP6) content resulting from calcium release. In addition, the bound calcium content of the cell wall increased, leading to an increase in cooking time.	Wainaina, Lugumira, Wafula, et al. (2022)
	Storage under ambient conditions (20.3 °C and 78.9% r. h.) during 36, 72, 108, 144, and 180 days.	They observed that Ca and Mg contents in the seed coat showed significant alterations after 180 days of storage. The presence of these minerals contributed to the hardness observed during the storage period due to their participation in the formation of pectate.	Bento et al. (2020)

TABLE 19.1 (CONTINUED)

Compounds	Storage Conditions	Results	Reference
Phenolics	6 months at 22 °C and 65% r. h.	Observed that the stored seeds showed higher antioxidant capacity (251.67 ± 11.40 mmol TE/g dry flour) compared to the fresh sample.	Alves et al. (2021)
Volatile compounds	6 months at different temperatures (25, 30, 35, and 42 °C) and seed moisture content (6.9%, 8.7%, 11.1%, 12.8%, and 14.5%).	The volatile compounds increased during storage at high temperature and high moisture content. They evidenced alcohols, aldehydes and ketones such as benzaldehyde, acetic acid, 1-propanol and 2-methyl, 2-butanone as the most abundant and with high concentrations, showing the existence of additional biochemical reactions and an increase in the reaction rate. These compounds are mainly related to the degradation of amino acids or proteins, as well as to the oxidation of lipids that affect seed quality.	Wainaina, Kyomugasho, Delbaere, et al. (2022)
Lignin	Storage at 12 °C and 85% r. h. for 22 days.	Reported an increase of lignin in common bean seed after storage for 22 days (12 °C and 85% r. h.).	Xie et al. (2020)

r. h.: relative humidity

19.4 MICROSCOPIC TECHNIQUES FOR BEAN MICROSTRUCTURAL CHARACTERIZATION DURING STORAGE: SEED COAT AND COTYLEDON

Food materials have different levels of structural organization, such as atomic, molecular, nanometer, microscopic and macroscopic levels, and their structure depends mainly on the interactions between atomic and molecular components. Structure can be related to the physical properties, quality, and acceptability of food materials (Perea-Flores et al. 2012; Verboven et al. 2018). Microstructural characteristics of natural foods provide important information on structural organization, as well as cell wall characteristics, starch granule arrangement, and the presence of proteins (Singh et al. 2014). Microstructural characterization systems, such as light and fluorescence microscopy, are available. Confocal laser scanning microscopes (CLSM) are used to study the internal microstructure of biological materials and to obtain topography and three-dimensional images. Likewise, scanning electron microscopy (SEM) (Perea-Flores et al. 2012; Phan et al. 2017) is a useful technique. Using these microscopic techniques, structural changes in bean seed have been evidenced (Konzen & Tsai, 2014; Mujica et al. 2015; Chigwedere et al. 2018; Miano & Augusto, 2018; Pallares, Rousseau, Chigwedere, et al. 2018; Bento et al. 2020; Chu et al. 2020;

Hernandez-Aguilar et al. 2020; Chen et al. 2021; Nicolás-García, Arzate-Vázquez, Perea-Flores, et al. 2021; Nicolás-García et al. 2022).

19.4.1 Optical Microscopy

Optical microscopy (OM) has been available for a long time and is used when microscopic structure and organization are important issues in food materials. Food materials can sometimes be examined whole, but studies of cell structure usually require sectioning (**Figure 19.3**). When using OM, a specific stain is applied to enhance visual contrast or differentiate tissues (Perea-Flores et al. 2012). Lugol's staining allows observing starch granules present in bean amyloplasts (Pallares, Miranda, Truong, et al. 2018) as shown in **Figure 19.3.1.** Also, Coomassie blue staining requires an acidic medium for electrostatic attraction between the dye molecules and the amino groups of proteins (Soto et al. 2013). Staining with Schiff's reagent is responsible for the pink coloration caused by the binding of leucofuchsin with aldehydes present in carbohydrates and polysaccharides, as shown in **Figure 19.3.2**, where cell wall components in seed coat and cotyledonary cells, as well as starch granules of amyloplasts, were stained (Gupta et al. 2009).

OM has been employed to study bean microstructure (Konzen & Tsai, 2014; Chigwedere et al. 2018; Pallares, Rousseau, Chigwedere, et al. 2018; Bento et al. 2020; Nicolás-García, Arzate-Vázquez, Perea-Flores, et al. 2021). Several studies of the structural changes undergone by bean kernels during different processes have implemented the use of these techniques (Do et al. 2019;

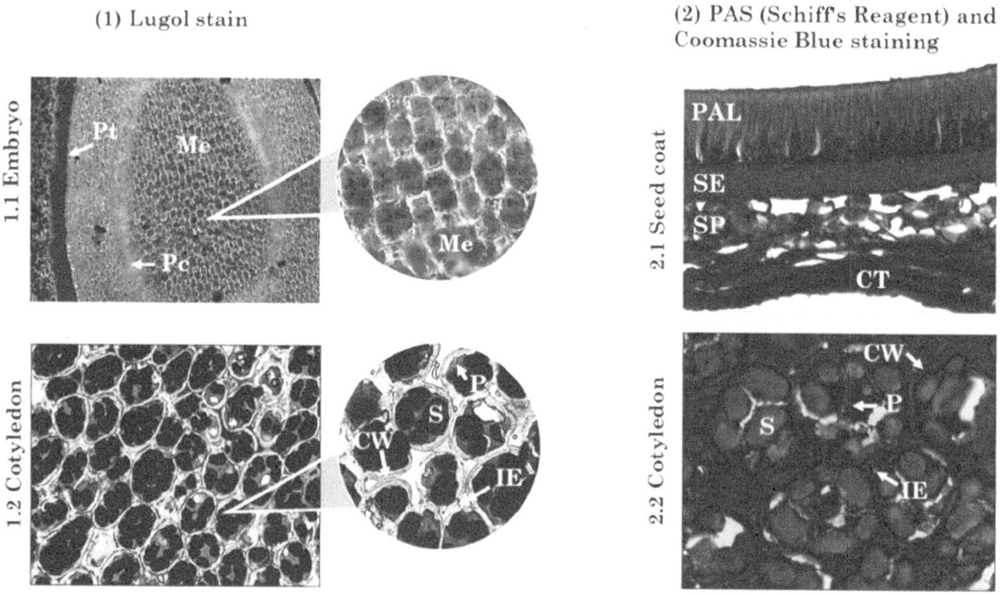

Figure 19.3 Bean seed cross-section images obtained by optical microscopy. 1: Lugol stain; 1.1: embryo (4×); 1.2: cotyledon (10×). 2: Schiff's reagent and Coomassie blue stain; 2.1: seed coat (60×); 2.2: cotyledon (60×). Me: meristems; Pt: protoderm; Pc: procambium; PAL: epidermal palisade cell; SE: subepidermal cell; SP: spongy parenchymal cell; CT: condensed tannins; S: starch granule; P: protein matrix; CW: cell wall; IE: intracellular space.

Kyomugasho et al. 2021). Pallares, Rousseau, Chigwedere, et al. (2018) evidenced the changes occuring in bean due to different heat treatment conditions through optical and polarized light microscopy. They observed the behavior of mechanically disintegrated cotyledons, even indicating the presence of starch that was not completely gelatinized. Rousseau et al. (2020) obtained optical images of common bean cotyledons after cooking for 60 min. They investigated the integrity of the cell wall and the participation of minerals in the mechanical disintegration by mixing and grinding, observing that the heat treatment did not affect the diffusion of minerals. Nicolás-García, Arzate-Vázquez, Perea-Flores, et al. (2021) evidenced calcium oxalate crystals through OM with polarized light. In regard to the HTC phenomenon, Cominelli et al. (2020) performed a histological analysis to identify the globulin protein fraction in a longitudinal section of bean cotyledon by staining with toluidine blue for the HTC study. In addition, Chen et al. (2021) by microscopic analysis evidenced the presence and increase of calcium-pectin and ferulic acid-pectin cross-links, accumulated mainly in the intracellular space of the cotyledon during the bean hardening process.

19.4.2 Scanning Electron Microscopy (SEM) and Environmental Scanning Electron Microscopy (ESEM)

SEM and ESEM allow for the observation and surface characterization of inorganic and organic materials, providing morphological information of the analyzed material (Salas-Cepeda, 2014). They consist of shining an electron beam on the sample and detecting the number of secondary electrons emitted from the material very close to the surface. This number depends on the surface topology and can therefore be used to create an image by scanning the entire surface (Verboven et al., 2018). Some works have applied these techniques to observe bean structure (Konzen & Tsai, 2014; Mujica et al. 2015; Do et al. 2019; Hernandez-Aguilar et al. 2020). Xiong et al. (2019), which revealed that hydrothermal treatment at different moisture contents (15%, 25%, and 35%) did not alter the round/oval shape and size of pinto bean starch granules by optical microscopy and SEM. In addition, polarized light revealed the ordered repeating structure of starch granules and the Maltese cross-centered at the hilum.

SEM and ESEM can be combined with X-ray techniques like energy-dispersive spectroscopy (EDS) (Jahn et al. 2012; Ando et al. 2018). EDS detector is used to separate the characteristic X-rays of different elements in the energy spectrum; an EDS system software is used to analyze the energy spectrum to determine the abundance of specific elements as well as their elemental composition (Salas-Cepeda, 2014). X-ray approaches allow combining the morphological investigation of a specimen with its chemical characterization. Currently, correlative microscopy has become increasingly important for the correlative analysis of the structure and function of cells and tissues (Jahn et al. 2012). Nicolás-García, Arzate-Vázquez, Perea-Flores, et al. (2021) corroborated the presence of calcium oxalate crystals by high-resolution scanning electron microscopy (HR-SEM) and EDS elemental analysis in the microstructure of the black bean seed coat (**Figure 19.4**).

The presence of minerals such as calcium (Ca), magnesium (Mg), zinc (Zn), silicon (Si), phosphorus (P), iron (Fe), and potassium (K) was evidenced by these techniques (SEM-EDS and ESEM-EDS). The role of minerals in post-harvest storage has been reported. Guzmán-Maldonado et al. (2003) indicate that both the environment and bean genotype affect Ca and Fe content. Kruger et al. (2014) mention that Ca and Mg are mobilized and concentrated around the cell wall in the middle lamella, where it probably binds to pectin, insolubilizing it. Furthermore, Cominelli et al.

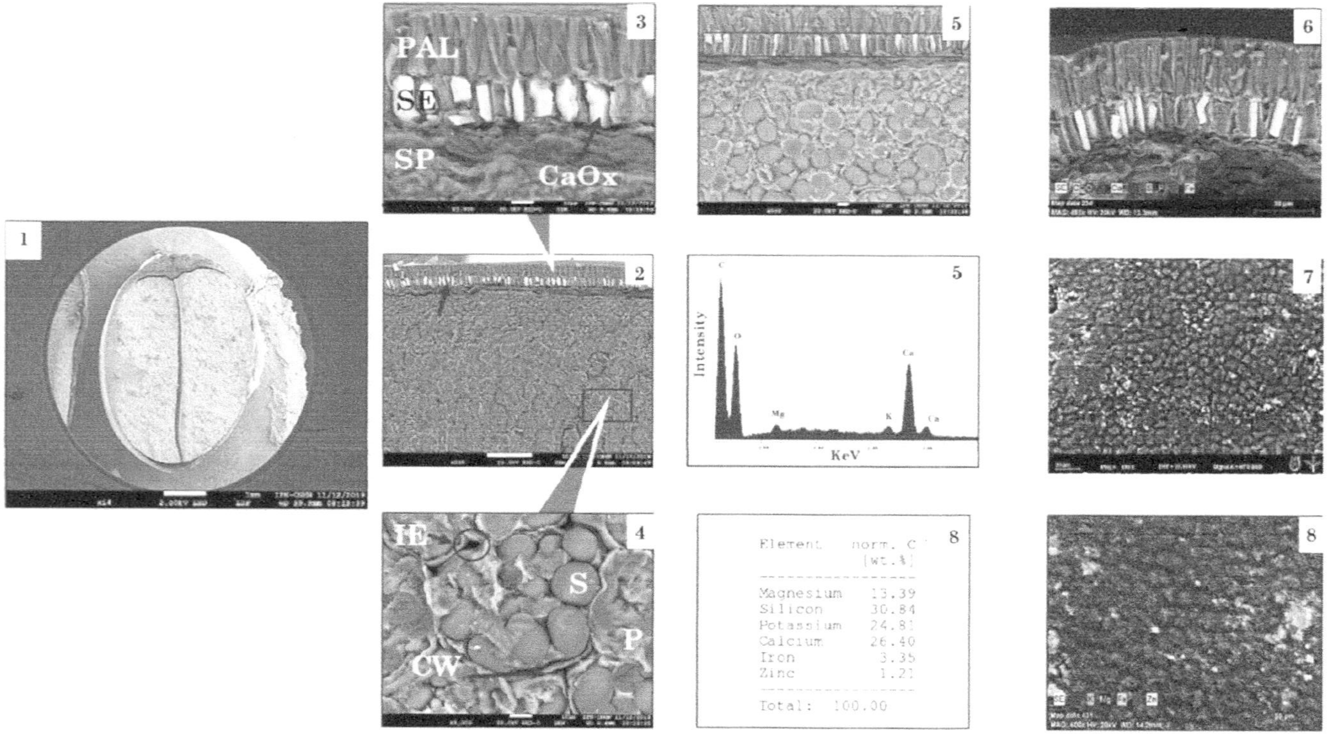

Figure 19.4 Whole bean cross-section images obtained by high-resolution scanning electron microscopy (HR-SEM) and environmental scanning electron microscopy (ESEM). 1: Cross-section of whole seed (seed coat and cotyledonary sac) (HR-SEM, 14×); 2: cross-section of seed (HR-SEM, 200×); 3: seed coat (HR-SEM, 1000×); 4: cotyledon (HR-SEM, 1000×); 5: energy-dispersive spectroscopy (EDS) elemental analysis of bean seed coat image by HR-SEM (500×); 6: EDS analysis of bean seed coat image by ESEM (495×); 7: seed surface image by ESEM (500×); 8: EDS analysis of seed surface by ESEM (500×). PAL: epidermal palisade cell; SE: subepidermal cell; SP: spongy parenchymal cell; CaOx: calcium oxalate crystals; S: starch granule; P: protein matrix; CW: cell wall; IE: intracellular space.

(2020) studied the effect of bean genotypes on HTC. They correlated the redistribution of calcium in cell walls, changes due to phytase-phytate-pectin interaction.

On the other hand, Mujica et al. (2015) observed an increase in the intercellular space of cotyledonary cells in hardened bean seeds, relating it to the reduction of cell wall thickness and stiffness. It has been mentioned that lignification of the middle lamella of the cell wall may participate in this behavior observed in bean hardening (Chigwedere et al. 2018; Bento et al. 2020; Xie et al. 2020).

19.4.3 Confocal Laser Scanning Microscopy

CLSM uses a focused beam of light in fluorescence mode microscopes (Verboven et al. 2018; Reilly & Obara, 2021). A laser scanning confocal microscope can be applied to gather this microstructure by dynamic and nearly non-invasive observation. An important feature of CLSM is the acquisition of a sample-specific focal section. This allows the visualization of interactions among structural elements without needing sample preparation by fixation and dehydration (Jekle & Becker, 2011). The spatial distribution of the components can be determined simultaneously and independently by labeling them with specific fluorochromes. This is one of the main advantages of CLSM. Therefore, it is desirable to have different fluorochromes with well-separated emission wavelengths so that the spatial distribution of food components can be determined simultaneously. This can be achieved by fluorochromes that bind non-specifically to a complete set of molecules such as proteins or lipids or by fluorochromes that bind specifically to a certain type of molecule, such as homogalacturonan in pectin. Examples of fluorochromes commonly used for food staining are acridine orange, Congo red, Nile red (fatty acids), Alexa Fluor (proteins), and Nile blue, which have been employed to study the microstructure of legumes like the cowpea (*Vigna unguiculata*) (Jombo et al. 2018; Coffigniez et al. 2019). **Figures 19.5.1** and **19.5.2** depict images acquired by CLSM of a cross-section of bean cotyledon, revealing the presence of carbohydrates (starch granules), proteins, cell wall components (cellulose and hemicellulose), and lipid bodies. Likewise, **Figure 20.5.3** shows the cuticle and cell wall components of bean seed coat sclereids. Xiong et al. (2019) evaluated starch granule porosity, starch size, and cell wall, evidencing the diffusion of α-amylase during their digestion in pinto bean samples using CLSM. Fluorescein isothiocyanate (FITC) staining showed that the higher the moisture content in the hydrothermal process, the greater the damage at the cellular level.

19.5 PERSPECTIVES ABOUT HARDENED SEEDS

The hardened common bean's physicochemical and sensory properties affect consumer preferences; this includes the required increased cooking times of these seeds. However, it has been evidenced that structures like the bean seed coat change during processing, among them color saturation, with phenolic compounds being responsible. The presence of these bioactive compounds is related to various health benefits, such as antioxidant and anticancer, among others. In addition, the presence of dietary fiber and minerals, among others, in this legume could be essential for its use as a food ingredient due to its nutritional, functional, and possible techno-functional properties. This legume could also be used for the development of new products and thus provide an added value to the hardened bean.

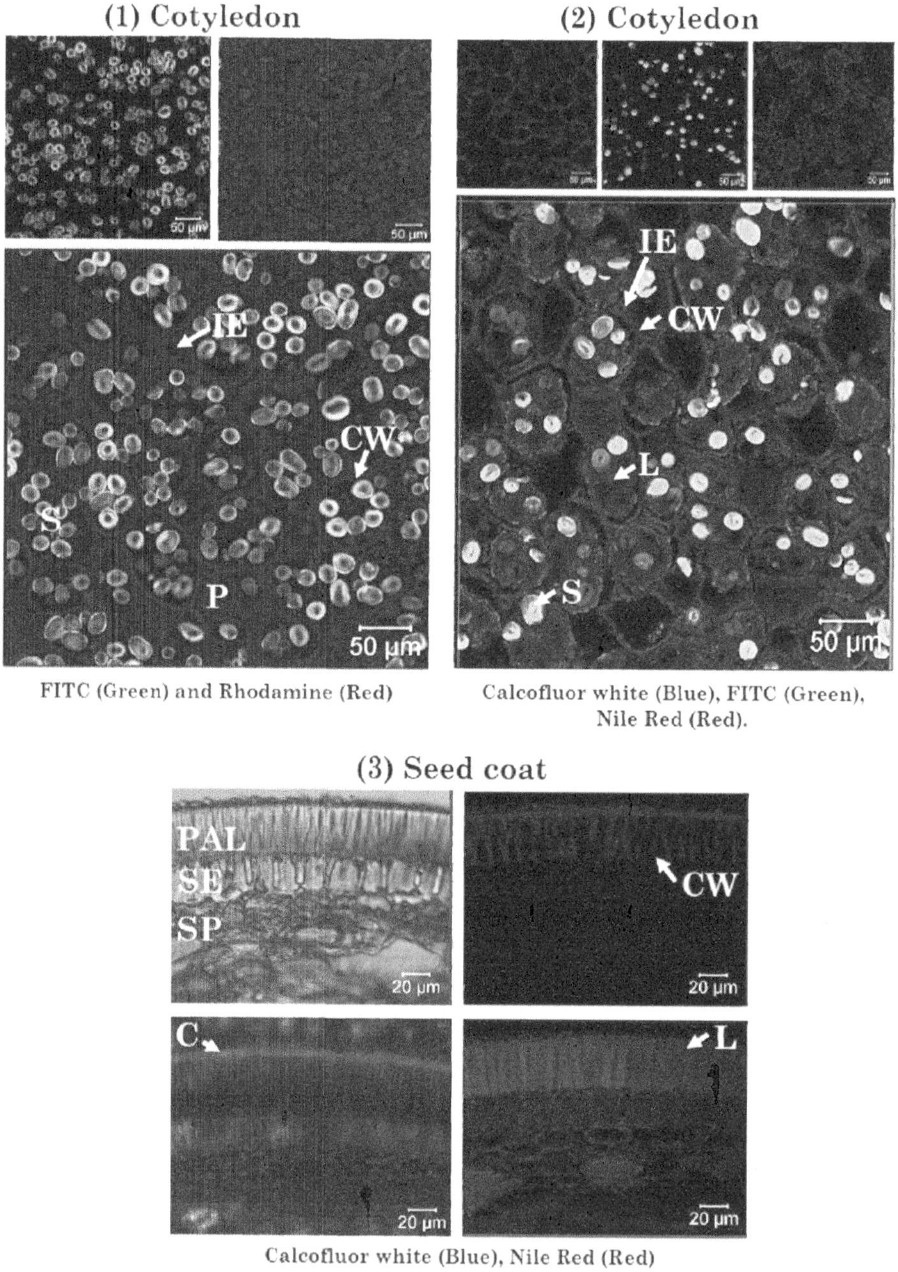

Figure 19.5 Bean seed cross-section images obtained by confocal laser scanning microscopy (CLSM). 1: Cotyledon staining with fluorescein isothiocyanate (FITC) and rhodamine; 2: cotyledon staining with calcofluor-white, FITC, and Nile red; 3: seed coat staining with calcofluor-white and Nile red. S: starch granule; P: protein matrix; CW: cell wall; IE: intracellular space; C: cuticle; L: lipids; PAL: epidermal palisade cell; SE: subepidermal cell; SP: spongy parenchymal cell.

19.6 CONCLUSION

The importance of the common bean resides in its nutritional and functional quality. However, this depends on its adequate handling and storage. Beans undergo a hardening process related to chemical, physical, and microstructural changes. It is important to note that the set of phenomena that include alteration of the cell components of the cotyledons, seed coat impermeability, the presence of phenolic compounds, and the seed coat color saturation contribute to the seed's hardening. Nevertheless, the study of these changes in beans during post-harvest storage is still in progress due to the diverse mechanisms they undergo and the relationship with their chemical composition. The hardened seeds could potentially be used as supplements due to their composition, such as phenolic compounds, and their antioxidant capacity.

ACKNOWLEDGEMENTS

The authors are grateful to the *Instituto Politécnico Nacional*, SIP projects (20181560, 20196640, and 20200291), the CONAHCyT-Mexico project (241756), and the *Instituto Nacional de Investigaciones Forestales, Agrícolas y Pecuarias* of Cotaxtla, Veracruz, MX, for providing the seeds for the structural analysis.

REFERENCES

Abay, M. M., & Tolesa, G. N. (2023). Effects of storage temperature and relative humidity on cooking time and moisture uptake of selected common bean (*Phaseolus vulgaris* L.) varieties grown in Ethiopia. *Cogent Food & Agriculture, 9* (1), 2163577. https://doi.org/10.1080/23311932.2022.2163577

Acosta, Y., Escobar-Gutiérrez, A., Ahmed, L. Q., Cejas, I., Martínez-Montero, M. E., Sánchez, J., Hajari, E., Höfer, M., Lorenzo, J. C., & Fontes, D. (2023). Morpho-anatomical evaluation of *Teramnus labialis* seeds: Strategies to overcome physical dormancy. *Biologia, 78* (8), 2003–2011. https://doi.org/10.1007/s11756-023-01341-6

Alcázar-Valle, M., Lugo-Cervantes, E., Mojica, L., Morales-Hernández, N., Reyes- Ramírez, H., Enríquez-Vara, J. N., & García-Morales, S. (2020). Bioactive compounds, antioxidant activity, and antinutritional content of legumes: A comparison between four *Phaseolus* species. *Molecules, 25* (15), 3528. https://doi.org/10.3390/molecules25153528

Alvares, R. C., Pereira, H. S., Melo, L. C., Melo, P., & Miklas, P. (2020). Induction of seed coat darkening in common beans (*Phaseolus vulgaris* L.) and the association with cooking time after storage. *Australian Journal of Crop Science, 14* (1), 21–27. https://doi.org/10.21475/ajcs.20.14.01.p1500

Alves, A. C., & Tavares, G. M. (2019). Mixing animal and plant proteins: Is this a way to improve protein techno-functionalities? *Food Hydrocolloids, 97* (105171), 1–10. https://doi.org/10.1016/j.foodhyd.2019.06.016

Alves, N. E., Gomes, M. J., Vasconcelos, C. M., Lima, A. C., de Lima, S. L., Brito, E. S., Bassinello, P. Z., Lima, A. C., & Martino, H. S. (2021). Six months under uncontrolled relative humidity and room temperature changes technological characteristics and maintains the physicochemical and functional properties of carioca beans (*Phaseolus vulgaris* L.). *Food Chemistry, 342*, 128390. https://doi.org/10.1016/j.foodchem.2020.128390

Ando, T., Bhamidimarri, S. P., Brending, N., Colin-York, H., Collinson, L., De Jonge, N., & Zifarelli, G. (2018). The 2018 correlative microscopy techniques roadmap. *Journal of Physics D: Applied Physics, 51* (44), 443001. https://doi.org/10.1088/1361-6463/aad055

Ávila, B. P., Nora, F. R., Neto, A. C. P. S., Rombaldi, C. V., da Silva Pinto, L., Gularte, M. A., & Elias, M. C. (2020). Proteomic and physicochemical characteristics: The search for a quality profile of beans (*Phaseolus vulgaris* L.) during long-term storage. *LWT-Food Science and Technology, 133*, 110057. https://doi.org/10.1016/j.lwt.2020.110057

Barthlott, W., Mail, M., Bhushan, B., & Koch, K. (2017). Plant surfaces: Structures and functions for biomimetic innovations. *Nano-Micro Letters, 9* (2), 23. https://doi.org/10.1007/s40820-016-0125-1

Bassett, A., Hooper, S., & Cichy, K. (2021). Genetic variability of cooking time in dry beans (*Phaseolus vulgaris* L.) related to seed coat thickness and the cotyledon cell wall. *Food Research International, 141*, 109886. https://doi.org/10.1016/j.foodres.2020.109886

Bento, J. A. C., Lanna, A. C., Bassinello, P. Z., Oomah, B. D., Pimenta, M. E. B., Carvalho, R. N., & Moreira, A. S. (2020). Aging indicators for stored carioca beans. *Food Research International, 134*, 109249. https://doi.org/10.1016/j.foodres.2020.109249

Bewley, J. D., Bradford, K. J., Hilhorst, H. W M., & Nonogaki, H. (2013). *Seeds. Physiology of development, germination and dormancy* (3rd Edition, pp. 1-24), Springer Science & Business Media. https://doi.org/10.1007/978-1-4614-4693-4

Bitocchi, E., Rau, D., Bellucci, E., Rodriguez, M., Murgia, M. L., Gioia, T., & Papa, R. (2017). Beans (*Phaseolus ssp.*) as a model for understanding crop evolution. *Frontiers in Plant Science, 8*, 722. https://doi.org/10.3389/fpls.2017.00722

Bulyaba, R., Winham, D. M., Lenssen, A. W., Moore, K. J., Kelly, J. D., Brick, M. A., & Ogg, J. B. (2020). Genotype by location effects on yield and seed nutrient composition of common bean. *Agronomy, 10* (3), 347. https://doi.org/10.3390/agronomy10030347

Calderan-Rodrigues, M. J., Fonseca, J. G., Labate, C. A., & Jamet, E. (2017). Cell wall proteomics as a means to identify target genes to improve second-generation biofuel production in E. Jacob.Lopes and L. Queiroz Zepka (Eds.), *Frontiers in Bioenergy and Biofuels* (1st Edition, pp. 5-24), IntechOpen. http://dx.doi.org/10.5772/66189

Capistrán-Carabarin, A., Aquino-Bolaños, E. N., García-Díaz, Y. D., Chávez-Servia, J. L., Vera-Guzmán, A. M., & Carrillo-Rodríguez, J. C. (2019). Complementarity in phenolic compounds and the antioxidant activities of *Phaseolus coccineus* L. and *P. vulgaris* L. Landraces. *Foods, 8* (8), 295. https://doi.org/10.3390/foods8080295

Chávez-Mendoza, C., Hernández-Figueroa, K. I., & Sánchez, E. (2019). Antioxidant capacity and phytonutrient content in the seed coat and cotyledon of common beans (*Phaseolus vulgaris* L.) from various regions in Mexico. *Antioxidants, 8* (1), 5. https://doi.org/10.3389/fpls.2017.00722

Chen, D., Pham, U. T. T., Van Loey, A., Grauwet, T., Hendrickx, M., & Kyomugasho, C. (2021). Microscopic evidence for pectin changes in hard-to-cook development of common beans during storage. *Food Research International, 141*, 110115. https://doi.org/10.1016/j.foodres.2021.110115

Chigwedere, C. M., Njoroge, D. M., Van Loey, A. M., & Hendrickx, M. E. (2019). Understanding the relations among the storage, soaking, and cooking behavior of pulses: A scientific basis for innovations in sustainable foods for the future. *Comprehensive Reviews in Food Science and Food Safety, 18* (4), 1135–1165. https://doi.org/10.1111/1541-4337.12461

Chigwedere, C. M., Olaoye, T. F., Kyomugasho, C., Kermani, Z. J., Pallares, A. P., Van Loey, A. M., & Hendrickx, M. E. (2018). Mechanistic insight into softening of Canadian wonder common beans (*Phaseolus vulgaris*) during cooking. *Food Research International*, 106, 522–531. https://doi.org/10.1016/j.foodres.2018.01.016

Chu, J., Ho, P., & Orfila, C. (2020). Growth region impacts cell wall properties and hard-to-cook phenotype of canned navy beans (*Phaseolus vulgaris*). *Food and Bioprocess Technology, 13* (5), 818–826. https://doi.org/10.1007/s11947-020-02436-7

Coelho, S. R. M., Alves Filho, E. G., Silva, L. M. A., Bischoff, T. Z., Ribeiro, P. R., Zocolo, G. J., Canuto, K. M., Bassinello, P. Z., & de Brito, E. S. (2020). NMR and LC-MS assessment of compound variability of common bean (*Phaseolus vulgaris*) stored under controlled atmosphere. *LWT-Food Science and Technology, 117* (108673), 1–8. https://doi.org/10.1016/j.lwt.2019.108673

Coffigniez, F., Briffaz, A., Mestres, C., Akissoé, L., Bohuon, P., & El Maâtaoui, M. (2019). Impact of soaking process on the microstructure of cowpea seeds in relation to solid losses and water absorption. *Food Research International, 119*, 268–275. https://doi.org/10.1016/j.foodres.2019.02.010

Cominelli, E., Galimberti, M., Pongrac, P., Landoni, M., Losa, A., Paolo, D., Daminati, M. G., Bollini, R., Cichy, K. A., Vogel-Mikuš, K., & Sparvoli, F. (2020). Calcium redistribution contributes to the hard-to-cook phenotype and increases PHA-L lectin thermal stability in common bean low phytic acid 1 mutant seeds. *Food Chemistry, 321*, 126680. https://doi.org/10.1016/j.foodchem.2020.126680

De Miranda, J. A. T., de Carvalho, L. M. J., de Castro, I. M., de Carvalho, J. L. V., de Alcântara Guimarães, A. L., & de Macêdo Vieira, A. C. (2019). Starch granules from cowpea, black, and carioca beans in raw and cooked forms. In *Legume Crops*. IntechOpen. http://dx.doi.org/10.5772/intechopen.85656

De Oliveira, A. P., Mateó, B. D. S. O., Fioroto, A. M., de Oliveira, P. V., & Naozuka, J. (2018). Effect of cooking on the bioaccessibility of essential elements in different varieties of beans (*Phaseolus vulgaris* L.). *Journal of Food Composition and Analysis, 67*, 135–140. https://doi.org/10.1016/j.jfca.2018.01.012

Diaz-Contreras, L. M., Erkinbaev, C., & Paliwal. J. (2018). Non-destructive and rapid discrimination of hard-to-cook beans using hyperspectral imaging. *Canadian Biosystems Engineering, 60*, 7.1–7.8. https://doi.org/10.7451/CBE.2018.60.7.1

Do, D. T., Singh, J., Oey, I., & Singh, H. (2019). Modulating effect of cotyledon cell microstructure on *in vitro* digestion of starch in legumes. *Food Hydrocolloids, 96*, 112–122. https://doi.org/10.1016/j.foodhyd.2019.04.063

Dos Santos Siqueira, B., Bassinello, P. Z., Malgaresi, G., Pereira, W. J., & Fernandes, K. F. (2016). Analyses of technological and biochemical parameters related to the HTC phenomenon in carioca bean genotypes by the use of PCA. *LWT-Food Science and Technology, 65*, 939–945. http://dx.doi.org/10.1016/j.lwt.2015.09.030

Elias, M. C., Ziegler, V., Romano, C. M., Alves, G. H., Paraginski, R. T., & de Oliveira, M. (2016). Physicochemical properties and enzymatic bean grains dried at different temperatures and stored for 225 days. Semina: *Ciências Agrárias, 37* (3), 1295–1305. https://doi.org/10.5433/1679-0359.2016v37n3p1295

Fernandes, P. A., Le Bourvellec, C., Renard, C. M., Wessel, D. F., Cardoso, S. M., & Coimbra, M. A. (2019). Interactions of arabinan-rich pectic polysaccharides with polyphenols. *Carbohydrate Polymers, 230*, 115644. https://doi.org/10.1016/j.carbpol.2019.115644

Ganesan, K., & Xu, B. (2017). Polyphenol-rich dry common beans (*Phaseolus vulgaris* L.) and their health benefits. *International Journal of Molecular Sciences, 18* (11), 2331. https://doi.org/10.3390/ijms18112331

García-Díaz, Y. D., Aquino-Bolaños, E. N., Chávez-Servia, J. L., Vera-Guzmán, A. M., & Carrillo-Rodríguez, J. C. (2018). Bioactive compounds and antioxidant activity in the common bean are influenced by cropping season and genotype. *Chilean Journal of Agricultural Research, 78* (2), 255–265. http://dx.doi.org/10.4067/S0718-58392018000200255

Giusti, F., Capuano, E., Sagratini, G., & Pellegrini, N. (2019). A comprehensive investigation of the behaviour of phenolic compounds in legumes during domestic cooking and *in vitro* digestion. *Food Chemistry, 285*, 458–467. https://doi.org/10.1016/j.foodchem.2019.01.148

Gupta, E., Bhalla, P., Khurana, N., & Singh, T. (2009). Histopathology for the diagnosis of infectious diseases. *Indian Journal of Medical Microbiology, 27*, 100–106. https://doi.org/10.4103/0255-0857.49423

Guzmán-Maldonado, S. H., Martínez, O., Acosta-Gallegos, J. A., Guevara-Lara, F., & Paredes-López, O. (2003). Putative quantitative trait loci for physical and chemical components of common bean. *Crop Science, 43* (3), 1029–1035. https://doi.org/10.2135/cropsci2003.1029

Hernandez-Aguilar, C., Dominguez-Pacheco, A., Palma Tenango, M., Valderrama-Bravo, C., Soto Hernandez, M., Cruz-Orea, A., & Ordonez-Miranda, J. (2020). Characterization of bean seeds, germination, and phenolic compounds of seedlings by UV-C radiation. *Journal of Plant Growth Regulation, 40* (2), 642–655.https://doi.org/10.1007/s00344-020-10125-0

Jahn, K. A., Barton, D. A., Kobayashi, K., Ratinac, K. R., Overall, R. L., & Braet, F. (2012). Correlative microscopy: Providing new understanding in the biomedical and plant sciences. *Micron, 43* (5): 565–582. https://doi.org/10.1016/j.micron.2011.12.004

Jekle, M., & Becker, T. (2011). Dough microstructure: Novel analysis by quantification using confocal laser scanning microscopy. *Food Research International, 44* (4), 984–991. https://doi.org/10.1016/j.foodres.2011.02.036

Jombo, T. Z., Minnaar, A., & Taylor, J. R. (2018). Effects of gamma-irradiation on cotyledon cell separation and pectin solubilisation in hard-to-cook cowpeas. *Journal of the Science of Food and Agriculture, 98* (5), 1725–1733. https://doi.org/10.1002/jsfa.8645

Katuuramu, D. N., Luyima, G. B., Nkalubo, S. T., Wiesinger, J. A., Kelly, J. D., & Cichy, K. A. (2020). On-farm multi-location evaluation of genotype by environment interactions for seed yield and cooking time in common bean. *Scientific reports, 10* (1), 1–12. https://doi.org/10.1038/s41598-020-60087-2

Kibar, H., & Kibar, B. (2019). Changes in some nutritional, bioactive and morpho-physiological properties of common bean depending on cold storage and seed moisture contents. *Journal of Stored Products Research, 84*, 101531. https://doi.org/10.1016/j.jspr.2019.101531

Kinyanjui, P. K., Njoroge, D. M., Makokha, A. O., Christiaens, S., Ndaka, D. S., & Hendrickx, M. (2015). Hydration properties and texture fingerprints of easy-and hard-to-cook bean varieties. *Food Science & Nutrition, 3* (1), 39–47. https://doi.org/10.1002/fsn3.188

Konzen, E. R., & Tsai, S. M. (2014). Seed coat shininess in *Phaseolus vulgaris*: Rescuing a neglected trait by its screening on commercial lines and landraces. *Journal of Agricultural Science, 6* (8), 113. https://doi.org/10.5539/jas.v6n8p113

Koriyama, T., Sato, Y., Iijima, K., & Kasai, M. (2017). Influences of soaking temperature and storage conditions on hardening of soybeans (*Glycine max*) and red kidney beans (*Phaseolus vulgaris*). *Journal of Food Science, 82* (7): 1546–1556. https://doi.org/10.1111/1750-3841.13749

Kruger, J., Pineda-Vargas, C., Minnis-Ndimba, R., & Taylor, J. R. N. (2014). Visualisation of the distribution of minerals in red non-tannin finger millet using PIXE microanalysis. *Journal of Cereal Science, 60* (1): 1–8. https://doi.org/10.1016/j.jcs.2014.04.005

Kyomugasho, C., Kamau, P. G., Aravindakshan, S., & Hendrickx, M. E. (2021). Evaluation of storage stability of low moisture whole common beans and their fractions through the use of state diagrams. *Food Research International, 140*, 109794. https://doi.org/10.1016/j.foodres.2020.109794

Li, P., Li, Y., Wang, L., Zhang, H., Qi, X., & Qian, H. (2020). Study on water absorption kinetics of black beans during soaking. *Journal of Food Engineering, 283*, 110030. https://doi.org/10.1016/j.jfoodeng.2020.110030

Li, P., Zhang, B., & Dhital, S. (2019). Starch digestion in intact pulse cells depends on the processing induced permeability of cell walls. *Carbohydrate Polymers, 225*, 115204. https://doi.org/10.1016/j.carbpol.2019.115204

Los, F. G. B., Zielinski, A. A. F., Wojeicchowski, J. P., Nogueira, A., & Demiate, I. M. (2018). Beans (*Phaseolus vulgaris* L.): Whole seeds with complex chemical composition. *Current Opinion in Food Science, 19*, 63–71. https://doi.org/10.1016/j.cofs.2018.01.010

Mecha, E., Leitão, S. T., Carbas, B., Serra, A. T., Moreira, P. M., Veloso, M. M., & Bronze, M. R. (2019). Characterization of soaking process' impact in common beans phenolic composition: Contribute from the unexplored Portuguese germplasm. *Foods, 8* (8), 296. https://doi.org/10.3390/foods8080296

Miano, A. C., & Augusto, P. E. D. (2018). The hydration of grains: A critical review from description of phenomena to process improvements. *Comprehensive Reviews in Food Science and Food Safety, 17* (2), 352–370. https://doi.org/10.1111/1541-4337.12328

Miano, A. C., Sabadoti, V. D., Pereira, J. D. C., & Augusto, P. E. D. (2018). Hydration kinetics of cereal and pulses: New data and hypothesis evaluation. *Journal of Food Process Engineering, 41* (1), e12617. https://doi.org/10.1111/jfpe.12617

Miano, A. C., Saldaña, E., Campestrini, L. H., Chiorato, A. F., & Augusto, P. E. D. (2018). Correlating the properties of different carioca bean cultivars (*Phaseolus vulgaris*) with their hydration kinetics. *Food Research International, 107*, 182–194. https://doi.org/10.1016/j.foodres.2018.02.030

Mikac, U., Sepe, A., & Serša, I. (2015). MR microscopy for noninvasive detection of water distribution during soaking and cooking in the common bean. *Magnetic Resonance Imaging, 33* (3), 336–345. https://doi.org/10.1016/j.mri.2014.12.001

Mojica, L., Meyer, A., Berhow, M. A., & de Mejía, E. G. (2015). Bean cultivars (*Phaseolus vulgaris* L.) have similar high antioxidant capacity, *in vitro* inhibition of α-amylase and α-glucosidase while diverse phenolic composition and concentration. *Food Research International, 69*, 38–48. https://doi.org/10.1016/j.foodres.2014.12.007

Mujica, M. V., Granito, M., & Soto, N. (2015). Cambios microestructurales en los granos de *Phaseolus vulgaris* endurecidos. *Archivos Latinoamericanos de Nutrición, 65* (2), 110–118.

Nakhungu, W. E., Njoki, W. I., Carolien, B., Nghia-Do-Trong, N., Kahenya, K. P., Wouter, S., & Marc, H. (2020). Application of near-infrared spectroscopy to predict the cooking times of aged common beans (*Phaseolus vulgaris* L.). *Journal of Food Engineering, 284*, 110056. https://doi.org/10.1016/j.jfoodeng.2020.110056

Nicolás-Bermúdez, J., Arzate-Vázquez, I., Chanona-Pérez, J. J., Méndez-Méndez, J. V., Rodríguez-Castro, G. A., & Martínez-Gutiérrez, H. (2018). Morphological and micromechanical characterization of calcium oxalate (CaOx) crystals embedded in the pecan nutshell (*Carya illinoinensis*). *Plant Physiology and Biochemistry, 132*, 566–570. https://doi.org/10.1016/j.plaphy.2018.10.008

Nicolás-García, M., Perucini-Avendaño, M., Jiménez-Martínez, C., Perea-Flores, M. D. J., Gómez-Patiño, M. B., Arrieta-Báez, D., & Dávila-Ortiz, G. (2021). Bean phenolic compound changes during processing: Chemical interactions and identification. *Journal of Food Science, 86* (3), 643–655. https://doi.org/10.1111/1750-3841.15632

Nicolás-García, M., Jiménez-Martínez, C., Perucini-Avendaño, M., Camacho-Díaz, B. H., Jiménez-Aparicio, A. R., & Dávila-Ortiz, G. (2021). Phenolic compounds in legumes: Composition, processing and gut health. *In* J. C. Jimenez-Lopez and A. Clemente (Eds.), *Legumes Research* (2nd Edition, pp. 1-17). IntechOpen. ISBN 978-1-83969-491 2. http://dx.doi.org/10.5772/intechopen.98202.

Nicolás-García, M., Arzate-Vázquez, I., de Jesús Perea-Flores, M., Méndez-Méndez, J. V., Perucini-Avendaño, M., Gómez-Patiño, M. B., & Dávila-Ortiz, G. (2021). An overview of instrumented indentation technique for the study of micromechanical properties in food: A case study on bean seed coat. *Biosystem Engineering, 204*, 377–385. https://doi.org/10.1016/j.biosystemseng.2021.02.006

Nicolás-García, M., Perucini-Avendaño, M., Arrieta-Báez, D., de Jesús Perea-Flores, M., Jiménez-Martínez, C., Gómez-Patiño, M. B., & Dávila-Ortiz, G. (2022). Phenolic compounds profile by UPLC-ESI-MS in black beans and its distribution in the seed coat during storage. *Food Chemistry, 395*, 133638. https://doi.org/10.1016/j.foodchem.2022.133638

Nicolás-García, M., Perucini-Avendaño, M., de Jesús Perea-Flores, M., Camacho-Díaz, B. H., & Dávila-Ortiz, G. (2024). Effect of post-harvest storage on the chemical and microstructural characteristics of the common bean (Phaseolus vulgaris L.). *Food Chemistry*, 140680. https://doi.org/10.1016/j.foodchem.2024.140680

Nurzyńska-Wierdak, R., Łabuda, H., Buczkowska, H., & Sałata, A. (2019). Pericarp of colored-seeded common bean (*Phaseolus vulgaris* L.) varieties a potential source of polyphenolic compounds. *Agronomy Research, 17* (5), 2005–2015. https://doi.org/10.15159/ar.19.187

Pallares, A. P., Rousseau, S., Chigwedere, C. M., Kyomugasho, C., Hendrickx, M., & Grauwet, T. (2018). Temperature-pressure-time combinations for the generation of common bean microstructures with different starch susceptibilities to hydrolysis. *Food Research International, 106*, 105–115. https://doi.org/10.1016/j.foodres.2017.12.046

Pallares, A. P., Miranda, B. A., Truong, N. Q. A., Kyomugasho, C., Chigwedere, C. M., Hendrickx, M., & Grauwet, T. (2018). Process-induced cell wall permeability modulates the *in vitro* starch digestion kinetics of common bean cotyledon cells. *Food & Function, 9* (12), 6544–6554.

Parmar, N., Singh, N., Kaur, A., & Thakur, S. (2017). Comparison of color, anti-nutritional factors, minerals, phenolic profile and protein digestibility between hard-to-cook and easy-to-cook grains from different kidney bean (*Phaseolus vulgaris*) accessions. *Journal of Food Science and Technology, 54* (4), 1023–1034. https://doi.org/10.1007/s13197-017-2538-3

Perea-Flores, M. D. J., Mendoza-Madrigal, A. G., Chanona-Pérez, J. J., Alamilla-Beltrán, L., & Gutierrez-López, G. F. (2012). Microscopy techniques and image analysis for the quantitative evaluation of food microstructure. in J. G., & Grandison, A. S. (Eds.), *Food Processing Handbook* (pp. 667–692). Brennan. https://doi.org/10.1002/9783527634361.ch22

Perera, D., Devkota, L., Garnier, G., Panozzo, J., & Dhital, S. (2023). Hard-to-cook phenomenon in common legumes: Chemistry, mechanisms and utilisation. *Food Chemistry, 415*, 135743. https://doi.org/10.1016/j.foodchem.2023.135743

Phan, A. D. T., Flanagan, B. M., D'Arcy, B. R., & Gidley, M. J. (2017). Binding selectivity of dietary polyphenols to different plant cell wall components: Quantification and mechanism. *Food Chemistry, 233*, 216–227. https://doi.org/10.1016/j.foodchem.2017.04.115

Porch, T. G., Beaver, J. S., Debouck, D. G., Jackson, S. A., Kelly, J. D., & Dempewolf, H. (2013). Use of wild relatives and closely related species to adapt common bean to climate change. *Agronomy, 3* (2), 433–461. https://doi.org/10.3390/agronomy3020433

Pramiu, P. V., Rizzi, R. L., Galante, G., Rizzi, C. B., & Coelho, S. R. M. (2019). Modeling grains hydration considering volumetric variations of spherical and elliptical geometries. *Computers and Electronics in Agriculture, 166*, 105007. https://doi.org/10.1016/j.compag.2019.105007

Ramírez-Jiménez, A. K., Reynoso-Camacho, R., Mendoza-Díaz, S., & Loarca-Piña, G. (2014). Functional and technological potential of dehydrated *Phaseolus vulgaris* L. flours. *Food Chemistry, 161*, 254–260. https://doi.org/10.1016/j.foodchem.2014.04.008

Ramírez-Jiménez, A. K., Reynoso-Camacho, R., Tejero, M. E., León-Galván, F., & Loarca-Piña, G. (2015). Potential role of bioactive compounds of *Phaseolus vulgaris* L. on lipid-lowering mechanisms. *Food Research International, 76*, 92–104. https://doi.org/10.1016/j.foodres.2015.01.002

Ramos-Ordoñez, M. F., Arizmendi, M. D. C., Flores-Enríquez, V., & Márquez-Guzmán, J. (2016). Extracción y morfología del embrión de *Bursera bipinnata* (Burseraceae). *Acta Botánica Mexicana, 117*, 27–35. https://doi.org/10.21829/abm117.2016.1165

Reilly, W. M., & Obara, C. J. (2021). Advances in confocal microscopy and selected applications. In: J. Brzostowski, and H. Sohn (Eds.), *Confocal Microscopy: Methods and Protocols*, (Vol. 2304, pp. 1-35). Humana, New York, NY. https://doi.org/10.1007/978-1-0716-1402-0_1

Renard, C. M., Watrelot, A. A., & Le Bourvellec, C. (2017). Interactions between polyphenols and polysaccharides: Mechanisms and consequences in food processing and digestion. *Trends in Food Science & Technology, 60*, 43–51. https://doi.org/10.1016/j.tifs.2016.10.022

Reyes-Moreno C., Paredes-López O., & González E. (1993). Hard-to-cook phenomenon in common beans: A review. *Critical Reviews in Foods Science and Nutrition,* 33 (3), 227–286. https://doi.org/10.1080/10408399309527621

Rousseau, S., Pallares, A. P., Vancoillie, F., Hendrickx, M., & Grauwet, T. (2020). Pectin and phytic acid reduce mineral bioaccessibility in cooked common bean cotyledons regardless of cell wall integrity. *Food Research International, 137*, 109685. https://doi.org/10.1016/j.foodres.2020.109685

Ruiz-Ruiz, J. C., Dávila-Ortíz, G., Chel-Guerrero, L. A., & Betancur-Ancona, D. A. (2012). Wet fractionation of hard-to-cook bean (*Phaseolus vulgaris* L.) seeds and characterization of protein, starch and fibre fractions. *Food and Bioprocess Technology, 5*, 1531–1540. https://doi.org/10.1007/s11947-010-0451-0

Salas Cepeda, R. (2014). Estudio de materiales nanométricos con microscopio electrónico de barrido SEM. Tesis Maestría. Centro de Investigación en Materiales Avanzados, Chihuahua, Chihuahua, México.

Shahidi, F., & Yeo, J. (2016). Insoluble-bound phenolics in food. *Molecules, 21* (9), 1216. https://doi.org/10.3390/molecules21091216

Singh, J., Berg, T., Hardacre, A., & Boland, M. J. (2014). Cotyledon cell structure and *in vitro* starch digestion in navy beans in food structures, digestion and health. In Golding, M., & Singh, H. (Eds.). *Food Structures, Digestion and Health* (pp. 223–242). Academic Press. https://doi.org/10.1016/B978-0-12-404610-8.00008-6

Soto, C., Guillermo, J., Plata, C., Sepúlveda, L., Efraín, R., Eligio, E.O ., Rodríguez, N., & Antonio, J. (2013). Evaluación de procedimientos de tinción para el análisis de proteínas por electroforesis (SDS-PAGE*)*, *Invurnus, 1*, 19–26. http://hdl.handle.net/20.500.12984/2145

Velasco-González, O., San Martín-Martínez, E., Aguilar-Méndez, M., Pajarito-Ravelero, A., & Mora-Escobedo, R. (2013). Propiedades físicas y químicas del grano de diferentes variedades de frijol (*Phaseolus vulgaris* L.). *Bioagro, 25* (3), 161–166.

Verboven, P., Defraeye, T., & Nicolai, B. (2018). Measurement and visualization of food microstructure: Fundamentals and recent advances. in S. (Ed.), *Food microstructure and its relationship with quality and stability devahastin* (pp. 3–28). Woodhead Publishing. https://doi.org/10.1016/B978-0-08-100764-8.00001-0

Wainaina, I., Kyomugasho, C., Delbaere, S., Wafula, E., Van Loey, A., Sila, D., & Hendrickx, M. (2022). (Bio) chemical reactions associated with ageing of red kidney beans (*Phaseolus vulgaris*) during storage probed by volatile profiling: The role of glass transition temperature. *Food Research International, 162*, 112102. https://doi.org/10.1016/j.foodres.2022.112102

Wainaina, I., Lugumira, R., Wafula, E., Kyomugasho, C., Sila, D., & Hendrickx, M. (2022). Insight into pectin-cation-phytate theory of hardening in common bean varieties with different sensitivities to hard-to-cook. *Food Research International, 151*, 110862. https://doi.org/10.1016/j.foodres.2021.110862

Wani, I. A., Sogi, D. S., Wani, A. A., & Gill, B. S. (2017). Physical and cooking characteristics of some Indian kidney bean (*Phaseolus vulgaris* L.) cultivars. *Journal of the Saudi Society of Agricultural Sciences, 16* (1), 7–15. https://doi.org/10.1016/j.jssas.2014.12.002

Xie, G., Yang, C., Fei, Y., & Ma, L. (2020). Physiological and proteomic analyses of 1-MCP treatment on lignifi cation in fresh common bean (*Phaseolus vulgaris* L) during storage. *Postharvest Biology and Technology, 160,* 111041. https://doi.org/10.1016/j.postharvbio.2019.111041

Xiong, W., Zhang, B., Dhital, S., Huang, Q., & Fu, X. (2019). Structural features and starch digestion properties of intact pulse cotyledon cells modified by heat-moisture treatment. *Journal of Functional Foods, 61,* 103500. https://doi.org/10.1016/j.jff.2019.103500

Yang, Q. Q., Gan, R. Y., Ge, Y. Y., Zhang, D., & Corke, H. (2018). Polyphenols in common beans (*Phaseolus vulgaris* L.): Chemistry, analysis, and factors affecting composition. Comprehensive *Reviews in Food Science and Food Safety, 17* (6), 1518–1539. https://doi.org/10.1111/1541-4337.12391

Zablatzká, L., Balarynová, J., Klčová, B., Kopecký, P., & Smýkal, P. (2021). Anatomy and histochemistry of seed coat development of wild (*Pisum sativum* subsp. elatius (M. Bieb.) Asch. et Graebn. and domesticated pea (*Pisum sativum* subsp. sativum L.). *International Journal of Molecular Sciences, 22* (9), 4602. https://doi.org/10.3390/ijms22094602

Chapter 20

Liposomes Loaded as Active Carriers to Enhance the Bioavailability of Phenolic Compounds as Nutraceuticals

Luis E. Palomo-Martínez
Iram Cisneros-García
Diana E. Leyva-Daniel
Sandra Rivera-Gutiérrez
Miguel Ángel A. Ibáñez-Hernández
and Liliana Alamilla-Beltrán

CONTENTS

 DOI: 10.1201/9781003332558-23

20.1 INTRODUCTION

Complex non-transmissible diseases (CD) represent a group of multifactorial disorders, including cancer, diabetes, cardiovascular issues, and obesity. The etiology of CD is only partially known and involves various interconnected risk factors such as personal habits, environment, and multiple genetic ones (Ezzati et al. 2023).

Nutraceuticals have recently drawn attention because of their ability to provide physiological benefits with less toxic effects (Dagar et al. 2023).

The use of nutraceuticals may offer a practical alternative (Marini et al. 2023), as they demonstrate great potential to combat the toxic and oxidative stress effects produced in the body, and most of them have been shown to have no side effects or harmful effects (Ronis et al. 2018); however, many of these present low bioavailability. This chapter focuses on the importance of bioactive carrier systems and encapsulation methods, mainly liposomes, to improve bioavailability, bioaccessibility, digestion and metabolism, and their therapeutic activities.

20.1.1 Nutraceutical Advantages and Disadvantages

The term nutraceutical, derived from the words nutrition and pharmaceutical (Dima et al. 2020; Marini et al. 2023), was defined in 1995 as a food or its parts that provides medical or health benefits, including prevention and treatment of various diseases (Helal et al. 2019; Dima et al.,2020). In addition to micro- and macronutrients, according to chemical structure, nutraceuticals can be classified into proteins and peptides, antioxidants, probiotics, prebiotics, phytochemicals (bioactive), fiber, lipids, carbohydrates, vitamins, minerals, and herbs (Sachdeva et al. 2020; Chandra et al. 2022).

Antioxidants such as phytochemicals are considered nutraceuticals with promising preventive efficacy in delaying or preventing age-related pathological conditions (Vaiserman et al. 2020, Jain et al. 2022), and the most notable of them are the phenolic compounds (PCs), which are

TABLE 20.1 ADVANTAGES AND DISADVANTAGES OF THE GENERAL USE OF NUTRACEUTICALS

Advantages	Disadvantages
Prevent various diseases by controlling risk factors, such as cholesterol, diacylglycerides, and glucose (Helal et al. 2019).	No patent protection Helal et al. (2019).
Natural and biocompatible, derived from natural resources such as spices, fruits, vegetables, and plants (Dutta et al. 2018).	Few have received significant attention related to biosafety studies Ronis et al. (2018).
Fewer side effects than pharmaceuticals (Pandey et al. 2022).	The industry may face high costs in obtaining raw materials Sachdeva et al. (2020).
Great potential to combat the toxic effects of toxins and drugs (Pandey et al. 2023).	The therapeutic potential may be restricted and limited in most cases by their low bioavailability Vaiserman et al. (2020).

antioxidant molecules characterized by possession of at least one or two phenyl rings and one or more hydroxyl substituents.

The mechanistic actions of food nutraceuticals are involved in various biological processes, including the activation of signal transduction pathways, antioxidant activity, gene expression, cell proliferation, differentiation, and preservation of mitochondrial integrity (Salami et al. 2023). Nutraceuticals should be administered in an appropriate pharmaceutical form to ensure high bioavailability and efficacy and should be used in areas that go "beyond diet and before drugs" (Santini et al. 2017). The advantages and disadvantages of these nutraceuticals are listed in **Table 20.1**.

20.1.2 Carrier Systems Used to Protect the Bioactivity of Phytochemicals

Phytochemicals can be classified as hydroxybenzoic acids, hydroxycinnamic acids, flavonoids, and non-flavonoids, or subdivided into many subclasses depending on the number of phenolic units within their molecular structure, substituent groups, and bonding type between phenolic units. For instance, the structural diversity of flavonoid molecules arises from variations in their hydroxylation patterns and oxidation states, resulting in a wide range of compounds, including anthocyanins, chalcones, flavanones, flavones, flavonols, and isoflavonoids (Singla et al. 2019).

Furthermore, the bioactivity of nutraceuticals derived from plant tissues can be achieved by employing active carrier systems (technological systems) such as emulsifying, spray drying, spray cooling, fluid bed coating, melt injection or extrusion, co-extrusion, coacervation, preparatioin of microspheres, liposomes, niosomes, lipid nanocarriers, inorganic colloids, protein-based colloids, and polysaccharide-based colloids (**Figure 20.1**) (Zuidam & Velikov, 2018).

Liposomes are effective carriers of drugs and PCs (Wu et al. 2015) with applications in the fields of nanotechnology for cancer diagnosis and therapy, gene delivery, cosmetics, agriculture, and food technology, food safety and processing, taste, nutrition, delivery methods, pathogen detection, food functionality, environmental protection, cost-effective storage, and distribution as well as improving the ability of plants to absorb nutrients (Sheoran et al. 2019).

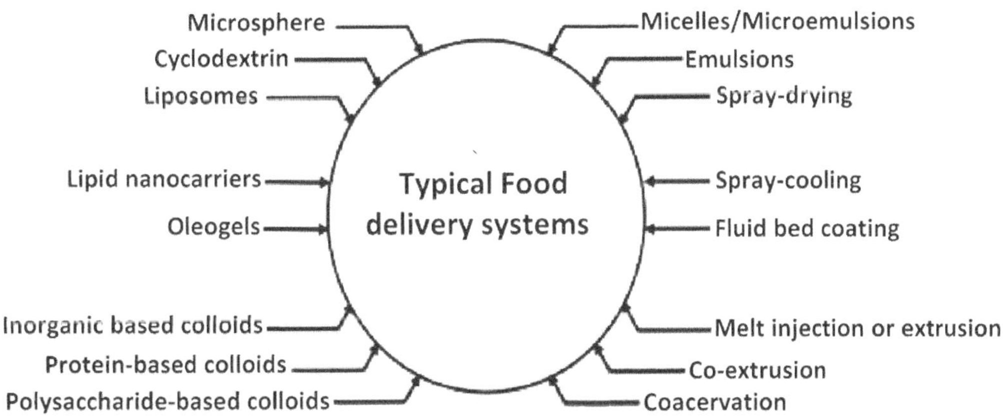

Figure 20.1 Carrier systems used to improve the bioavailability and bioaccessibility of nutraceuticals.

20.2 BIOAVAILABILITY AND BIOACCESSIBILITY OF PHENOLIC COMPOUNDS DURING GASTROINTESTINAL DIGESTION

The bioavailability of PCs is limited due to their transformation into inactive forms during digestion. They undergo chemical transformations, such as oxidation, reduction, and hydrolysis reactions, that alter their biological activities (McClements et al., 2015). Bioavailability refers to the fraction of the administered drug or nutrient that can reach the plasma and body tissues in the unchanged form, which has a major impact on the discovery and development of drugs intended for oral use, where gastrointestinal absorption is the main barrier between the active ingredient and systemic circulation (Vraníková & Gajdziok, 2015). Bioavailability is influenced by bioaccessibility, digestion and metabolism in the gastrointestinal tract (GIT), absorption into the bloodstream, tissue distribution, and bioactivity (Grgić et al. 2020).

The recommended daily dietary intake of PCs for humans is 1 g (Swallah et al. 2020). They are widely distributed in vegetal tissues (fruits, vegetables, and legumes), mainly as glycosides or aglycones (Singla et al. 2019). However, despite their reported (Gandhi et al. 2020) beneficial effects, such as antioxidant, neuroprotective, chemopreventive, and anti-inflammatory activities in treating various diseases, they have demonstrated restricted therapeutic potential due to their low bioavailability mainly attributed to their poor water solubility and intestinal permeability, which may limit their bioactivity and potential health benefits (Vaiserman et al. 2020). **Table 20.2** presents the oral bioavailability of several PCs.

20.2.1 Gastrointestinal Digestion of Phenolic Compounds

The fragmentation of ingested food occurs in the mouth and stomach, whereas absorption and enzymatic digestion occur predominantly in the small and large intestines (Jiao et al. 2017; Swallah et al. 2020). The digestion process begins in the mouth with chewing of food and releasing bioactive compounds from the food matrix (Swallah et al. 2020); this results in increased bioavailability of various nutrients. When an amount or fraction of a compound is released into the gastrointestinal

TABLE 20.2 BIOAVAILABILITY OF SOME PHENOLIC COMPOUNDS

Phenolic Compound Group	Subclass	Example	Oral Bioavailability in Humans (%)	Reference
Phenolic acids	Hydroxybenzoic acids	Gallic acid	36.4 ± 4.5	Bhuia et al. (2023)
	Hydroxycinnamic acids	Ferulic acid	~ 20	Kohno et al. (2020)
Flavonoids	Flavonol	Quercetin	~ 2	Li et al. (2016)
	Flavone	Apigenin	30	DeRango et al. (2021)
	Flavanol	Catechin	1.68	Cai et al. (2018)
	Flavanone	Naringenin	~ 15	Stabrauskiene et al. (2022)
	Isoflavone	Genistein	~ 23	Shete et al. (2023)
	Anthocyanin	Pelargonidin-3-O-rutinoside	1.13	Xu et al. (2021)
Non-flavonoid	Stilbenes	Resveratrol	12	Gowd et al. (2022)

tract and remains available for absorption, this phenomenon is known as bioaccessibility (Jiao et al. 2017; Tarko et al. 2020).

Digestive enzymes more easily degrade PCs in fruit extracts due to the lack of binding to a food matrix, making more compounds available in the gastrointestinal environment (Tarko et al. 2020). However, the biological membranes, including the gastrointestinal wall, act as lipid barriers or protective layers that limit the passage and absorption of nutrients, bioactive compounds, and drugs, reducing the bioaccessibility (McClements et al. 2015).

To achieve satisfactory bioaccessibility of nutrients during gastrointestinal digestion, hydrolytic enzymes, gastric acids, and the physicochemical properties of secretions act on the food bolus, releasing PCs (Swallah et al. 2020). The presence of a nutrient in its pure form is generally characterized by much higher bioaccessibility than the components associated with its food matrix (Tarko et al. 2020). Obtaining the pure form of PCs can be achieved by modifying food processing conditions (such as cooking, shearing, or homogenization), dietary habits (such as chewing duration), food matrix properties (such as composition and structure) (McClements et al. 2015), and the use of solvents such as ethanol (Tarko et al. 2020). In addition, the use of heat treatments exerts a positive effect on bioaccessibility, thus freeing them from their food matrix (Jiao et al. 2017).

20.2.2 Intestinal Absorption of Phenolic Compounds

Orally administered PCs reach the bloodstream through the intestinal wall after gastrointestinal digestion and are then transported to various parts of the body. However, numerous studies have revealed that they are unstable during this process (Jiao et al. 2017), are absorbed as xenobiotic substances, and their absorption is quite low compared to that of macro- and micronutrients. The transport of a bioactive substance through the mucus layer can also be delayed because of attractive interactions between nutraceuticals and mucus molecules (such as electrostatic or hydrophobic interactions), delaying intestinal absorption (McClements et al. 2015). Recent research has revealed that of the 100% total intake, only 5–10% is absorbed in the small intestine (monomeric, dimeric structure, and aglycones can be absorbed in the small intestine), and 90–95% is absorbed in the lumen of the large intestine along with other conjugates excreted in the bile (Swallah et al. 2020). Nutraceuticals that are relatively stable against molecular transformations within the GIT remain bioactive (> 75%) (McClements et al. 2015).

The absorption of dietary PCs is mainly determined by their chemical structure, which depends on factors such as the degree of glycosylation, acylation, basic structure (benzene or flavone derivatives), conjugation with other PCs, molecular size, polymerization, and solubility. PCs are then exposed to intestinal enzymes, and the colonic microbiota further breaks down the polyphenolic structures into smaller molecules to facilitate their absorption. In the large intestine, glycosides are hydrolyzed by gut microbiota, forming aglycones through heterocycle opening. This catabolic action reduces the complex structure of the phenolic hydroxyl group to low-molecular-weight phenolic metabolites that can be absorbed (Swallah et al. 2020). For example, highly hydrophilic anthocyanins are not readily absorbed in the small intestine but can be metabolized by enzymes present in the colon and converted into forms suitable for absorption, such as protocatechuic, syringic, vanillic, and 4-hydroxybenzoic acids (Mattioli et al. 2020). All these factors pose substantial challenges to the therapeutic use of PCs. Innovative biotechnological and nanotechnological approaches, such as encapsulation, are being actively developed to overcome the low bioavailability and improve the bioactivity of these polyphenols after oral administration, which increases their therapeutic efficacy (Vaiserman et al. 2020).

20.3 BIOACTIVE CARRIER SYSTEMS AND ENCAPSULATION METHODS

20.3.1 Carriers of Bioactive Compounds

20.3.1.1 Cyclodextrin Complex

The cyclodextrin system is based on molecular recognition, in which two or more molecules are joined by noncovalent intermolecular bonding forces to form complex entities or aggregates. Owing to the formation of a supramolecular host-guest system, they can be used to encapsulate PCs with a cylindrical structure, thereby improving water solubility and bioavailability. This system can be obtained via solvent evaporation, freeze-drying, and kneading other methods (Li et al. 2022).

20.3.1.2 Alginate Microspheres

Alginates are unbranched copolymers of D-manuronic acid (M-block) and L-guluronic acid (G-block) linked by β (1-4) glycosidic bonds. These blocks are arranged in an irregular pattern with different proportions of GG, MM, and MG blocks. Polymerization is based on the cross-linking of copolymers through ionic bonds between Ca^{2+} and alginate anions, forming biodegradable hydrogels. The most frequently used methods are spray drying and extrusion or emulsification/gelation (Letocha et al. 2022).

20.3.1.3 Niosomes

Niosomes (nonionic surfactant vesicles) obtained by hydration are microscopic lamellar structures formed by combining nonionic surfactants of alkyl ether or dialkyl polyglycerol class with cholesterol. Nonionic surfactants form closed bilayer vesicles in aqueous media, depending on their amphiphilic nature, using energy such as heat and physical agitation to form this structure. In the bilayer structure, the hydrophobic parts are separated from the aqueous solvent, whereas the hydrophilic heads remain in contact with the aqueous solvent. Niosomes can be mainly prepared using the traditional thin film hydration technique and the microfluidic mixing technique (Obeid et al. 2020).

20.3.1.4 Protein-Based Colloids

Micro- and nanoparticles can be produced by combining one or more types of proteins or by combining proteins and biopolymers, mainly polysaccharides. Nanoparticles based on proteins and polysaccharides can be produced by pH adjustment or heat treatment. Nanocomplexes of proteins and polysaccharides provide greater chemical and colloidal protection than the two polymers individually. The mixing of proteins and polysaccharides in an aqueous medium leads to two phenomena: repulsive and attractive interactions, depending on the charge of both biopolymers, protein/polysaccharide ratio, charge density, pH, ionic strength, molecular weight, and processing conditions, such as temperature, pressure, time, and shear rate (Ashfaq et al. 2022).

20.3.1.5 Polysaccharide-Based Colloids

These emulsion gels comprise polysaccharide food polymers with gelation capacity that depends on their source and structure. Alginate, agarose, modified starches (e.g., octenyl succinic anhydride), carrageenan, curdlan, inulin, konjac gum, and xanthan gum are the commonly employed biopolymers in the formation of emulsion gels. Various methods, including heating, cooling, high shear, freeze-thaw cycles, pH, ion-induced (Ca^{2+} and K^{+}), and salt addition ($CaCl_2$), can be applied to induce gelation, leading to the formation of emulsion gels. Polysaccharide-based emulsion gels are thermoreversible, and their derived colloidal particles are considered less effective in producing emulsion gels than protein-based emulsion gels. However, polysaccharide-based biopolymers are more effective at increasing viscosity and exhibit inherent resistance to digestive enzymes, rendering them particularly suitable as delivery vehicles for bioactive compounds that require a prolonged digestive process for the encapsulates (Abdullah et al. 2022).

20.3.1.6 Lipid Nanocarriers

Among all nanocarriers, lipid nanocarriers have great potential for drug encapsulation and favorable drug delivery in a predetermined manner to achieve bioavailability and avoid adverse effects. Phospholipids, cholesterol, and triglycerides are the most commonly used lipid monomers. Various types of lipid nanoparticles, such as liposomes, nanostructured lipid carriers, and lipid polymer hybrid nanoparticles, have been developed (Kumar et al. 2022).

20.3.1.7 Porous Microspheres

Traditional microspheres are microcapsules or monolithic particles; the charged molecule is encapsulated in the core, and the shell is composed of the matrix material, whereas in monolithic particles, the charged molecule is dispersed within the matrix. The density of traditional microspheres depends on the density of the matrix material, whereas their specific surface area is determined by the particle size. Different methods for synthesizing porous microspheres include solvent evaporation, polymerization, seed swelling, sintering, phase separation, and spray drying (Ghosh Dastidar et al. 2018).

20.3.1.8 Liposomes

Liposomes are spherical vesicles comprising one or more phospholipid bilayers. They have an aqueous core that can encapsulate hydrophilic molecules and a lipid bilayer that can trap lipophilic particles (Rakotondrabe et al. 2023). Therefore, various fractions can be loaded into the liposomes. Depending on the composition of the formulations, properties such as the size, zeta potential, and stiffness of the liposomes can be controlled. Among the novel drug delivery systems, liposomes are characterized by their advanced technology to deliver active molecules to their sites of action, and several formulations are currently in clinical use (Gbian et al. 2022).

20.3.2 Bioactive Compounds Encapsulation Methods

20.3.2.1 Emulsification

Emulsions are mixtures of two or more immiscible liquids. Emulsification occurs when mixing and heating produce sufficient energy to disperse oil droplets into the aqueous phase or vice versa. It is important to carefully define the mixing conditions during the development and scale-up (Jie & Chen, 2022).

20.3.2.2 Spray Drying

Spray drying is widely used in the food industry and is economical, rapid, and easy to apply. This technique transforms a material from a liquid to a powder. To achieve PC encapsulation by spray drying, a solution or suspension of PCs must first be obtained using a vehicle, followed by atomization in a stream of hot air. Therefore, it is important to consider the optimal conditions to avoid PC degradation (Buljeta et al. 2022).

20.3.2.3 Freeze Drying

Freeze drying, also known as lyophilization, is a sublimation-based drying technique. This technique allows for the long-term preservation of heat-sensitive and oxidation-prone compounds, foods, and other biological materials, as it is performed at low temperatures and under vacuum. This method has disadvantages, such as a higher unit cost and processing time. Lyophilization is widely used to obtain high-value food products and is considered the standard encapsulation method in most studies (Buljeta et al. 2022).

20.3.2.4 Spray Cooling

Spray cooling is a strategy that allows large heat fluxes to be dissipated and temperature to be precisely controlled. To achieve encapsulation, a mixture of the core and wall materials is atomized to form a mist inside a chamber into which cold air flows. The low temperature inside the chamber causes the microdroplets to solidify, forming a microencapsulated powder (Chen et al. 2022).

20.3.2.5 Extrusion

Extrusion technology can be used to produce dense microcapsules. It is based on the core and wall materials (immiscible) being passed through concentric nozzles, forming droplets containing the core surrounded by the wall material. Subsequently, solidification occurs by cooling or using a suitable gelling bath in which the droplets fall and solidify owing to complex formation. The formed encapsulates using this method are larger than those formed using other methods, and this technology is useful with limited wall materials (Choudhury et al. 2021).

20.3.2.6 Melt Injection

Melt injection is a vertical screwless extrusion process. In the melt injection extrusion process, the bioactive compound is dispersed in heat (80–140 °C) melted carbohydrate (composed of sucrose, maltodextrin, glucose syrup, polyols, and other mono- and disaccharides) and then pressed through one or more orifices (filter) into a bath of cold, dehydrating liquid (liquid nitrogen or isopropanol). The encapsulant or wall material solidifies with the liquid to form an encapsulating matrix that traps the bioactive compounds (Bamidele & Emmambux, 2021).

20.3.2.7 Hot-Melt Extrusion (HME)

HME technology has several applications, including solubility enhancement, taste masking, and sustained drug release. HME is another option for production in the pharmaceutical industry, mainly for producing amorphous solid dispersions. This process is solvent-free, easily scaled, and can be used for continuous production. Successful HME requires thermal and mechanical energy inputs. To convert a solid polymeric material into a fluid melt, a process temperature above the glass transition temperature of the respective polymer(s) is required. It is carried out through an orifice or die at a controlled temperature, pressure, and feed rate (Schmied et al. 2022).

20.3.2.8 Coacervation

Coacervation is a simple technique that involves the formation of a homogeneous layer of polymeric wall material around a core material. The system is obtained by altering the physicochemical properties of the wall material. In this process, the core and wall materials must be mixed to form an immiscible solution. Phase separation is achieved by changing the ionic strength, pH, or temperature to form coacervates, small liquid droplets composed of a dense polymer-rich phase. These coacervates surround the core material to form microcapsules (Choudhury et al. 2021).

20.4 LIPOSOME-LOADED ENCAPSULATION

Liposomes are spherical vesicles composed of one or more phospholipid bilayers that enclose a core aqueous phase (Jash et al. 2021). Their amphiphilic chemical composition allows for encapsulating hydrophilic biomolecules and/or drugs in the aqueous core and lipophilic biomolecules and/or drugs in the lipid bilayer (Jash et al. 2021; Ren et al. 2022) (**Figure 20.2**).

In 1964, Bangham and Horne published the first electron microscopy images of phospholipid vesicles. Since then, a notable succession of research advancements and discoveries has produced liposomes similar in composition and structural organization to cell membranes, which are the protagonists of numerous sectors of basic and applied research (De Leo et al. 2022). Liposomes are categorized according to their size, number of bilayers, and composition (**Figure 20.3**).

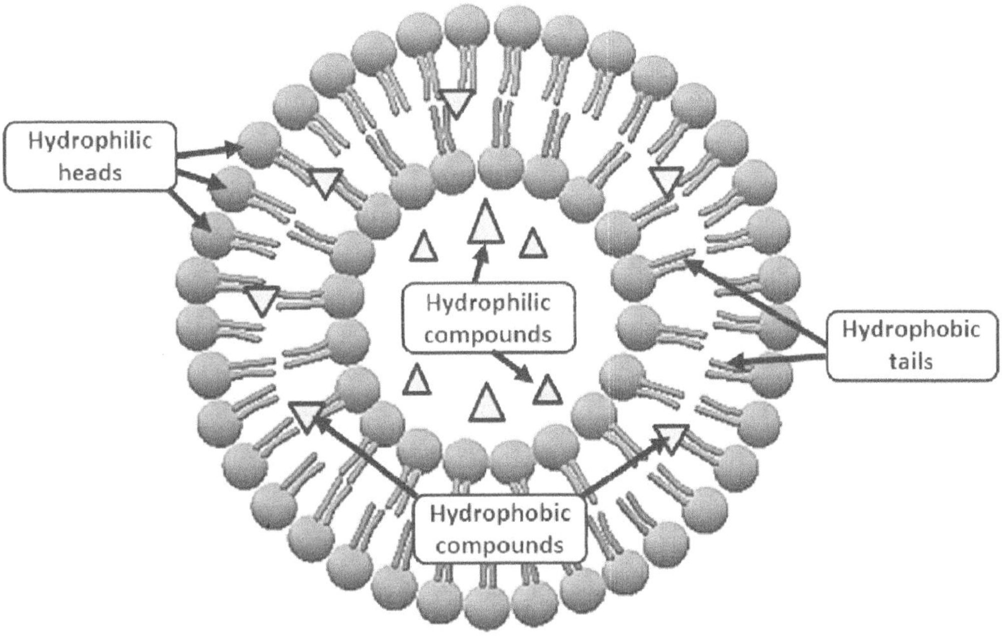

Figure 20.2 Structure of a liposome loaded with hydrophilic and hydrophobic compounds.

Based on the number of bilayers and their thicknesses, liposomes can be classified into two groups: unilamellar vesicles and multilamellar liposomes. The last one can also be subclassified into categories according to their size, such as giant unilamellar vesicles >1 μm, large ≥250 nm, intermediate 100–250 nm, and small ≤ 100 nm. Based on their surface charge, they are classified as anionic, cationic, or neutral (Zylberberg et al. 2016; Dymek et al. 2022)

20.4.1 Liposome Preparation Methods

The US Food and Drug Administration has approved liposome-based drug delivery systems for intravenous, intramuscular, dermatological, and oral administration. Research on liposome technology has progressed from conventional vesicles to "second generation liposomes," which include extended circulation liposomes, stimuli-responsive liposomes, nebulized liposomes, elastic liposomes for topical, oral, and transdermal delivery, and covalent lipid-drug complexes to enhance drug plasma membrane crossing and targeting to specific organelles (Zylberberg et al. 2016).

Liposomes are the most widely used nanocarriers for various potentially active hydrophobic and hydrophilic molecules, owing to their high biocompatibility, biodegradability, and low immunogenicity. In addition, they improve the solubility of PCs and their surface modification capability for targeted, prolonged, and sustained release. Based on their composition, liposomes have evolved from conventional, long-circulating, targeted, and immunoliposomes to stimuli-responsive and actively targeted liposomes. Many liposome-based drug delivery systems are

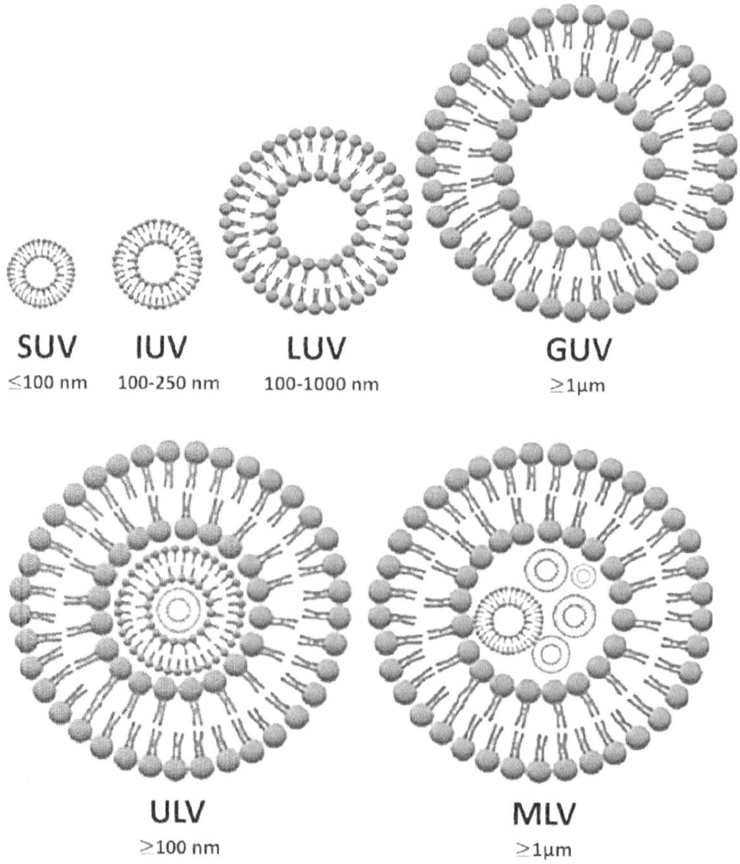

Figure 20.3 Liposomes categorization (Size and Number of Bilayers).

clinically approved for treating various diseases, such as cancer and fungal and viral infections; however, more liposomes have reached advanced stages in clinical trials (Nsairat et al. 2022). Some conventional methods of preparing liposomes are described below.

20.4.1.1 Thin Film Hydration (TFH)

This method consists of obtaining a thin film of lipids in a round bottom flask by removing an organic solvent. Subsequently, an aqueous solution (which may contain the bioactive compounds) is added, and the dispersion medium is stirred, and finally heterogeneous liposomes are formed (Jash et al. 2021).

20.4.1.2 Ethanol Injection Method

In this method, an ethanolic solution with lipids is rapidly injected into an aqueous medium through a needle, dispersing the phospholipids throughout the medium and promoting the formation of vesicles (liposomes) (Shaker et al. 2017).

20.4.1.3 Reverse Phase Evaporation

Bioactive compounds are dissolved in organic solvents, such as ether, chloroform, or methanol together with an aqueous solution, then the organic solvent is removed from the system, forming large unilamellar vesicles in the aqueous phase (Jash et al. 2021).

20.4.1.4 Others

Membrane extrusion, sonication, and freezing thawing are processes subsequent to the formation of liposomes; their purpose is to further reduce the size, lamellarity, heterogeneity, and polydispersity (Jash et al. 2021).

20.4.2 PCs-Loaded Liposomes

Polyphenol-loaded liposomes are promising, and flexible delivery systems are recognized in clinical and biological research for their role in human health, including liver protection, memory improvement, and reduction of cholesterol intake (Yang et al., 2020). Liposomes can carry drugs and nutraceuticals to deliver bioactive molecules to desired target tissues, decrease potential side effects, and protect the encapsulated molecules from enzymatic metabolic processes (Puri et al. 2022). In recent years, liposomes have received increasing attention in the field of food applications with specific dietary supplements that ensure high encapsulation efficiency and solubility of bioactive compounds such as polyphenolic compounds and food ingredients with lower toxicity owing to their phospholipid content and protection of sensitive ingredients against various degradation factors such as pH, light, digestive enzymes, and cell interactions (Subramani & Ganapathyswamy, 2020). Liposomal stability also depends on the PCs (the number of rings and hydroxyl groups), which influences its solubility and interaction with the liposome bilayer, including the partitioning of non-polar compounds into the hydrophobic bilayer of the membrane and the formation of hydrogen bonds between the more hydrophilic polyphenol and the polar head lipid groups. The chemical interaction between the lipid components and PCs could help understand and improve the encapsulation rate (Figueroa-Robles et al. 2021).

20.4.3 Liposomes for Oral Use

As presented in **Table 20.3**, liposomes can be used to promote effective applications in functional foods and nutraceuticals, and they are widely used in the development of nanomedicine because of their good biosafety and bioavailability (Ajeeshkumar et al. 2021; Jia et al. 2023). Oral drug delivery is the preferred route of drug administration clinically (Nguyen et al. 2017). Oral administration of liposomes can improve the bioavailability of nutraceuticals through mechanisms such as protection of the GIT, increased the cell contact and residence time, protection of pre-systemic metabolism and outflow, and increased diffusion through the epithelial and mucosal layers (Ren et al. 2022). Despite the growing body of research on oral delivery of liposomes, essential advances are still needed to develop and commercialize these products for clinical use. After the transit of liposomes from the stomach to the small intestine, they are gradually broken down, and the drug and/or nutrient payloads can be released immediately into the gastrointestinal lumen or transferred to secondary vehicles, such as mixed micelles and transported to the intestinal epithelium for absorption (He et al. 2019). Similarly, a challenge with conventional liposomes is that they are mainly composed of phospholipids and cholesterol (Wu et al. 2015; Lai et al. 2020), and

TABLE 20.3 EXAMPLES OF EFFECTS OF PHENOLIC COMPOUNDS (QUERCETIN, RESVERATROL, AND CURCUMIN) AND THEIR ENCAPSULATION IN LIPOSOMES

Polyphenol	Study and General Characterization	Effect	Reference
Quercetin	Soy lecithin dispersed in water High-pressure homogenization (50, 100, and 150 MPa, 1 pass; 150 MPa, 3 passes) PS 110.6 nm (by analytical ultracentrifugation) EE 31% to 42% *In vitro* model	The loaded nano-sized liposomes act as antitumor agent in colon cancer cells	Melchior et al. (2023)
Quercetin	Nanoliposomes with gallic acid, quercetin, and mix Thin-film hydration to synthetize nanocarriers PS >300nm (co-loaded liposomes) EE 84.7% (mean formula for the Mix (GA)) *In vitro* model	Nanoliposomes-loaded gallic acid, and mix showed an improvement in cytotoxicity against Lung cancer cells	Al-Samydai et al. (2023)
Quercetin	Film-sonication method Rice-bran phospholipids PS 157.33 78 nm EE 84.92% *In vitro* model	Higher radical scavenging activity compared with free quercetin	Rodriguez et al. (2018)
Quercetin	Sonication Phospholipid (P90G), stearyl amine, quercetin, phosphate-buffered saline, and Eudragit S100 PS 80 nm EE 84% *In vitro and in vivo* model	Increased the physical stability of fluids and reduced the production of reactive oxygen species	Caddeo et al. (2019)
Quercetin	Lipid hydration technique L-α-phosphatidylcholine, cholesterol, DSPE-PEG (2000), epigallocatechin gallate, quercetin, and doxorubicin Lipid hydration technique PS 342 nm EE 65.8–98% *In vitro* model	The loading of doxorubicin and quercetin in different compartments of liposome expose the cancerous cells in tandem, with anticancer and antimicrobial activity	Das et al. (2019)
Quercetin	Thin film evaporation method Phosphatidylcholine, phosphatidylserine, and cholesterol PS 142 ± 19 nm *In vivo* model	Increased bioactivity and bioavailability. Reduce liver injury and protect hepatocytes against damage.	Liu et al. (2020)

TABLE 20.3 (CONTINUED)

Polyphenol	Study and General Characterization	Effect	Reference
Resveratrol	Resveratrol-loaded liposomes Thin-lipid film evaporation method PS 75, 150, 300 nm (size-controlled) EE 80% *In vitro* release profile and in vivo oral bioavailability	The resveratrol-loaded liposomes (<100 nm) enhance the solubility, in vitro release profile, cellular properties, and in vivo oral availability	Baek et al. (2023)
Resveratrol	A thin-film hydration method Lecithin and cholesterol EE 85.32%. Liposomes integrated into the carbopol-based hydrogel matrix *In vitro* model	Showed a significant wound-healing effect.	Pandey et al. (2023)
Resveratrol	2-distearoyl-sn-glycero-3-phosphocholine ZP −17.99 mV PS 333 nm EE 84.69%	Better cell absorption efficiency and a stronger ability to reduce reactive oxygen species when compared with free resveratrol	Xu et al. (2023)
Resveratrol	Lipid hydration technique and sonication Lipoid S75 (soybean lecithin with phosphatidylcholine, phosphatidylethanolamine), quercetin, resveratrol, oleic acid, and chitosan PS 80.8–81.6 nm EE 70% *In vitro* model	Effective delivery systems for incorporating, protecting, and releasing bioactive polyphenols intended for oral delivery.	Caddeo et al. (2017)
Resveratrol	Film dispersion method followed by a membrane extrusion method. Phospholipids and cholesterol PS 110 nm EE 96% *In vitro* model	Higher physical and chemical stability. Induced uncoupling protein 1 (UCP1) mRNA expression and decreased white specific marker.	Zu et al. (2018)
Curcumin	Curcumin liposomes for intraarticular injection. Microfluidic method PS 132.9 nm EE 97.1% In vivo model	Curcumin liposomes have potential for the clinical treatment of inflammatory osteolysis.	Huang et al. (2023)

(continued)

TABLE 20.3 (CONTINUED)

Polyphenol	Study and General Characterization	Effect	Reference
Curcumin	Evaporation method Phospholipids and cholesterol PS 300 nm EE 45–74% *In vitro* model	Improved chemical stability	Shao et al. (2017)
Curcumin	Lipid-film hydration method Phosphatidylcholine *In vitro* model	Radioprotective activity in human blood cells, coupled with its well-established bioavailability.	Nguyen et al. (2017)
Curcumin	Reversed-phase evaporation method. Phosphatidylcholine, α-phosphatidic acid, and chitosan PS 129 nm *In vitro* model	Protection to improve the bioavailability of bioactive CU in the raw digesta and the bile salt micellar phase.	Cuomo et al. (2018)
Curcumin	Lipoid S100/cholesterol 10:1 w/w Sonication of a thin lipid film and encapsulated within Eudragit S100 using a pH-driven method PS 36–70 nm EE 16–98%	Suitable for colonic delivery of hydrophobic bioactive molecules such as curcumin	De Leo et al. (2018)

PS: Particle size (nm); EE: encapsulation efficiency; ZP: Potential zeta

their oral efficacy may be hampered by low stability in the GIT owing to their high susceptibility to physiological factors, such as gastric acid, bile salts, and phospholipases, whose destructive effects commonly result in loss of liposomal integrity and leakage of the payload. Another challenge is the poor permeability of liposomes through the gastrointestinal epithelium, which is the primary barrier to their absorption. However, controversy remains about whether the active ingredients are released firsthand before being absorbed or whether liposomes, together with their payload, can be absorbed intact by the gastrointestinal epithelium. The gastrointestinal epithelium blocks the entry of intact liposomes into the circulation because they are usually very large compared to easily absorbable small molecules. Oral administration of liposomes sometimes requires modifications to improve their stability under gastrointestinal conditions, such as using appropriate lipid compositions, coating with polymers, the addition of stabilizing lipids to liposome structures, preparation of double liposomes and pro-liposomes, and other innovative methods (Wu et al. 2015). **Table 20.3** presents some examples of the effects of PCs loaded into liposomes.

20.5 CONCLUSION

PCs have been widely studied for their various health benefits; however, most are poorly bioavailable in the GIT because of their low permeability and solubility; therefore, they remain difficult to use clinically. PC encapsulation in liposomes is an innovative technology with several applications to improve the potential health benefits of these compounds. Some PCs have been successfully encapsulated in liposomes. However, further studies are needed to determine the encapsulation efficiency, which varies according to the encapsulated polyphenol, the phospholipids and their surface characteristics, and the method employed. Finally, with improvements in liposome technologies, it is important to develop a protocol that can be used on a large scale and reduces liposome manufacturing costs, which are relatively high. In addition, the evolution of technology will lead to the emergence of new nanomaterials that can be used to manufacture new polyphenol delivery systems.

ACKNOWLEDGMENT

This study was supported by the Instituto Politécnico Nacional (SIP-IPN 20220528, SIP-IPN 20231809). The students LEPM and ICG are grateful for scholarships received from the Consejo Nacional de Ciencia y Tecnología (CONACyT), Mexico.

REFERENCES

Abdullah, L. L., Javed, H. U., & Xiao, J. (2022). Engineering emulsion gels asfunctional colloids emphasizing food applications: A review. *Frontiers in Nutrition*, 9, 890188. https://doi.org/10.3389/fnut.2022.890188

Ajeeshkumar, K. K., Aneesh, P. A., Raju, N., Suseela, M., Ravishankar, C. N., & Benjakul, S. (2021). Advancements in liposome technology: Preparation techniques and applications in food, functional foods, and bioactive delivery: A review. *Comprehensive Reviews in Food Science and Food Safety*, 20(2), 1280–1306. http://doi:10.1111/1541-4337.12725

Al-Samydai, A., Al Qaraleh, M., Al Azzam, K. M., Mayyas, A., Nsairat, H., Abu Hajleh, M. N., Al-Halaseh, L. K., Al-Karablieh, N., Akour, A., Alshaik, F., & Alshaer, W. (2023). Formulating co-loaded nanoliposomes with gallic acid and quercetin for enhanced cancer therapy. *Heliyon*, 9(6), 1–12. e17267. https://doi.org/10.1016/j.heliyon.2023.e17267.

Ashfaq, A., Jahan, K., Islam, R. U., & Younis, K. (2022). Protein-based functional colloids and their potential applications in food: A review. *Lebensmittel-Wissenschaft & Technologie*, 154, 112667. https://doi.org/10.1016/j.lwt.2021.112667

Baek, Y., Woo Jeong, E., & Gyu Lee, H. (2023). Encapsulation of resveratrol within size-controlled nanoliposomes: Impact on solubility, stability, cellular permeability, and oral bioavailability. *Colloids and Surfaces B: Biointerfaces*, 224. 13205. https://doi.org/10.1016/j.colsurfb.2023.113205.

Bamidele, O. P., & Emmambux, M. N. (2021). Encapsulation of bioactive compounds by extrusion technologies: A review. *Critical Reviews in Food Science and Nutrition*, 61(18), 3100–3118. https://doi.org/10.1080/10408398.2020.1793724

Buljeta, I., Pichler, A., Šimunović, J., & Kopjar, M. (2022). Polysaccharides as carriers of polyphenols: Comparison of freeze-drying and spray-drying as encapsulation techniques. *Molecules*, 27(16), 5069. https://doi.org/10.3390/molecules27165069

Bhuia, M. S., Rahaman, M. M., Islam, T., Bappi, M. H., Sikder, M. I., Hossain, K. N., Akter, F., Al Shamsh Prottay, A., Rokonuzzman, M., Gürer, E. S., Calina, D., Islam, M. T., & Sharifi-Rad, J. (2023). Neurobiological effects of gallic acid: Current perspectives. *Chinese Medicine*, 18(1), 27. https://doi.org/10.1186/s13020-023-00735-7

Caddeo, C., Gabriele, M., Fernàndez-Busquets, X., Valenti, D., Fadda, A. M., Pucci, L., & Manconi, M. (2019). Antioxidant activity of quercetin in Eudragit-coated liposomes for intestinal delivery. *International Journal of Pharmaceutics*, 565, 64–69. https://doi.org/10.1016/j.ijpharm.2019.05.007

Caddeo, C., Pons, R., Carbone, C., Fernàndez-Busquets, X., Cardia, M. C., Maccioni, A. M., & Manconi, M. (2017). Physico-chemical characterization of succinyl chitosan-stabilized liposomes for the oral co-delivery of quercetin and resveratrol. *Carbohydrate Polymers*, 157, 1853–1861. https://doi.org/10.1016/j.carbpol.2016.11.072

Cai, Z. Y., Li, X. M., Liang, J. P., Xiang, L. P., Wang, K. R., Shi, Y. L., Yang, R., Shi, M., Ye, J. H., Lu, J. L., Zheng, X. Q., & Liang, Y. R. (2018). Bioavailability of tea catechins and its improvement. *Molecules*, 23(9), 2346. https://doi.org/10.3390/molecules23092346

Chandra, S., Saklani, S., Kumar, P., Kim, B., & Coutinho, H. D. M. (2022). Nutraceuticals: Pharmacologically active potent dietary supplements. *BioMed Research International*, 2022, 2051017. https://doi.org/10.1155/2022/2051017

Chen, H., Ruan, X., Peng, Y., Wang, Y., & Changkun, Y. (2022). Application status and prospect of spray cooling in electronics and energy conversion industries. *Sustainable Energy Technologies and Assessments*, 52, 102181. https://doi.org/10.1016/j.seta.2022.102181

Choudhury, N., Meghwal, M., & Das, K. (2021). Microencapsulation: An overview on concepts, methods, properties and applications in foods. *Food Frontiers*, 2, 426–442. http://doi:10.1002/fft2.94

Cuomo, F., Cofelice, M., Venditti, F., Ceglie, A., Miguel, M., Lindman, B., & Lopez, F. (2018). In-vitro digestion of curcumin loaded chitosan-coated liposomes. *Colloids and Surfaces B: Biointerfaces*, 168, 29–34. https://doi.org/10.1016/j.colsurfb.2017.11.047

Dagar, N., Kale, A., Jadhav, H. R., & Gaikwad, A. B. (2023). Nutraceuticals and network pharmacology approach for acute kidney injury: A review from the drug discovery aspect. *Fitoterapia*, 168, 105563. https://doi.org/10.1016/j.fitote.2023.105563

Das, A., Konyak, P. M., Das, A., Dey, S. K., & Saha, C. (2019). Physicochemical characterization of dual action liposomal formulations: Anticancer and antimicrobial. *Heliyon*, 5(8), e02372. https://doi.org/10.1016/j.heliyon.2019.e02372

De Leo, V., Maurelli, A. M., Giotta, L., & Catucci, L. (2022). Liposomes containing nanoparticles: Preparation and applications. *Colloids and Burfaces. B, Biointerfaces*, 218, 112737. https://doi.org/10.1016/j.colsurfb.2022.112737

De Leo, V., Milano, F., Mancini, E., Comparelli, R., Giotta, L., Nacci, A., Longobardi, F., Garbetta, A., Agostiano, A., & Catucci, L. (2018). Encapsulation of curcumin-loaded liposomes for colonic drug delivery in a pH-responsive polymer cluster using a pH-driven and organic solvent-free process. *Molecules (Basel, Switzerland)*, 23(4), 739. https://doi.org/10.3390/molecules23040739

DeRango-Adem, E. F., & Blay, J. (2021). Does oral apigenin have real potential for a therapeutic effect in the context of human gastrointestinal and other cancers?. *Frontiers in Pharmacology*, 12, 681477. https://doi.org/10.3389/fphar.2021.681477

Dima, C., Assadpour, E., Dima, S., & Jafari, S. M. (2020). Bioavailability of nutraceuticals: Role of the food matrix, processing conditions, the gastrointestinal tract, and nanodelivery systems. *Comprehensive Reviews in Food Science and Food Safety*, 19(3), 954–994. https://doi.org/10.1111/1541-4337.12547

Dutta, S., Moses, J. A., & Anandharamakrishnan, C. (2018). Encapsulation of nutraceutical ingredients in liposomes and their potential for cancer treatment. *Nutrition and Cancer*, 70(8), 1184–1198. https://doi.org/10.1080/01635581.2018.1557212

Dymek, M., & Sikora, E. (2022). Liposomes as biocompatible and smart delivery systems – The current state. *Advances in Colloid and Interface Science*, 309, 102757. https://doi.org/10.1016/j.cis.2022.102757

Ezzati, A., Rosenkranz, S. K., Phelan, J., & Logan, C. (2023). The effects of isocaloric intermittent fasting vs daily caloric restriction on weight loss and metabolic risk factors for noncommunicable chronic diseases: A systematic review of randomized controlled or comparative trials. *Journal of the Academy of Nutrition and Dietetics*, 123(2), 318–329.e1. https://doi.org/10.1016/j.jand.2022.09.013

Figueroa-Robles, A., Antunes-Ricardo, M., & Guajardo-Flores, D. (2021). Encapsulation of phenolic compounds with liposomal improvement in the cosmetic industry. *International Journal of Pharmaceutics*, 593, 120125. https://doi.org/10.1016/j.ijpharm.2020.120125

Gandhi, G. R., Vasconcelos, A. B. S., Wu, D.-T., Li, H.-B., Antony, P. J., Li, H., & Gan, R.-Y. (2020). Citrus flavonoids as promising phytochemicals targeting diabetes and related complications: A systematic review of in vitro and in vivo studies. *Nutrients*, 12(10), 2907. http://doi:10.3390/nu12102907

Gbian, D. L., & Omri, A. (2022). Lipid-based drug delivery systems for diseases managements. *Biomedicines*, 10(9), 2137. https://doi.org/10.3390/biomedicines10092137

Ghosh Dastidar, D., Saha, S., & Chowdhury, M. (2018). Porous microspheres: Synthesis, characterization, and applications in pharmaceutical & medical fields. *International Journal of Pharmaceutics*, 548(1), 34–48. https://doi.org/10.1016/j.ijpharm.2018.06.015

Gowd, V., Kanika, Jori, C., Chaudhary, A. A., Rudayni, H. A., Rashid, S., & Khan, R. (2022). Resveratrol and resveratrol nano-delivery systems in the treatment of inflammatory bowel disease. *The Journal of Nutritional Biochemistry*, 109, 109101. https://doi.org/10.1016/j.jnutbio.2022.109101

Grgić, J., Šelo, G., Planinić, M., Tišma, M., & Bucić-Kojić, A. (2020). Role of the encapsulation in bioavailability of phenolic compounds. *Antioxidants*, 9(10), 923. http://doi:10.3390/antiox9100923

He, H., Lu, Y., Qi, J., Zhu, Q., Chen, Z., & Wu, W. (2019). Adapting liposomes for oral drug delivery. *Acta Pharmaceutica Sinica B*, 9(1), 36–48. http://doi.org/10.1016/j.apsb.2018.06.005

Helal, N. A., Eassa, H. A., Amer, A. M., Eltokhy, M. A., Edafiogho, I., & Nounou, M. I. (2019). Nutraceuticals' novel formulations: The good, the bad, the unknown and patents involved. *Recent Patents on Drug Delivery & Formulation*, 13(2), 105–156. http://doi:10.2174/1872211313666190503112040

Huang, S., Xu, D., Zhang, L., Hao, L., Jia, Y., Zhang, X., & Tao Cheng, Jian Chen. (2023). Therapeutic effects of curcumin liposomes and nanocrystals on inflammatory osteolysis: In vitro and in vivo comparative study. *Pharmacological Research*, 192. 106778. https://doi.org/10.1016/j.phrs.2023.106778

Jain, N., Radhakrishnan, A., & Kuppusamy, G. (2022). Review on nutraceuticals: Phase transition from preventive to protective care. *Journal of complementary & Integrative Medicine*, 19(3), 553–570. https://doi.org/10.1515/jcim-2022-0026

Jash, A., Ubeyitogullari, A., & Rizvi, S. S. H. (2021). Liposomes for oral delivery of protein and peptide-based therapeutics: Challenges, formulation strategies, and advances. *Journal of Materials Chemistry B*, 9(24), 4773–4792. https://doi.org/10.1039/d1tb00126d

Jia, W., Zhou, L., Li, L., Zhou, P., & Shen, Z. (2023). Nano-based drug delivery of polyphenolic compounds for cancer treatment: Progress, opportunities, and challenges. *Pharmaceuticals (Basel, Switzerland)*, 16(1), 101. https://doi.org/10.3390/ph16010101

Jiao, X., Li, B., Zhang, Q., Gao, N., Zhang, X., & Meng, X. (2017). Effect of in vitro -simulated gastrointestinal digestion on the stability and antioxidant activity of blueberry polyphenols and their cellular antioxidant activity towards HepG2 cells. *International Journal of Food Science & Technology*, 53(1), 61–71. http://doi:10.1111/ijfs.13516

Jie, Y., & Chen, F. (2022). Progress in the application of food-grade emulsions. *Foods*, 11(18), 2883. https://doi.org/10.3390/foods11182883

Kohno, M., Musashi, K., Ikeda, H. O., Horibe, T., Matsumoto, A., & Kawakami, K. (2020). Oral administration of ferulic acid or ethyl ferulate attenuates retinal damage in sodium iodate-induced retinal degeneration mice. *Scientific Reports*, 10(1), 8688. https://doi.org/10.1038/s41598-020-65673-y

Kumar, R., Dkhar, D., Kumari, R., Divya, N., Mahapatra, S., Dubey, V. K., & Chandra, P. (2022). Lipid based nanocarriers: Production techniques, concepts, and commercialization aspect. *Journal of Drug Delivery Science and Techn*ology, 74, 103526. https://doi.org/10.1016/j.jddst.2022.103526

Lai, W. F., Wong, W. T., & Rogach, A. L. (2020). Molecular design of layer-by-layer functionalized liposomes for oral drug delivery. *ACS Applied Materials & Interfaces*, 12(39), 43341–43351. https://doi.org/10.1021/acsami.0c13504

Łętocha, A., Miastkowska, M., & Sikora, E. (2022). Preparation and characteristics of alginate microparticles for food, pharmaceutical and cosmetic applications. *Polymers*, 14(18), 3834. https://doi.org/10.3390/polym14183834

Li, J., Xu, F., Dai, Y., Zhang, J., Shi, Y., Lai, D., Sriboonvorakul, N., & Hu, J. (2022). A review of cyclodextrin encapsulation and intelligent response for the release of curcumin. *Polymers*, 14(24), 5421. https://doi.org/10.3390/polym14245421

Li, Y., Yao, J., Han, C., Yang, J., Chaudhry, M., Wang, S., & Yin, Y. (2016). Quercetin, inflammation and immunity. *Nutrients*, 8(3), 167. http://doi:10.3390/nu8030167

Liu, X., Zhang, Y., Liu, L., Pan, Y., Hu, Y., Yang. P., & Liao M. (2020). Protective and therapeutic effects of nanoliposomal quercetin on acute liver injury in rats. *BMC Pharmacology & Toxicology*, 21(1), 11. http://doi:10.1186/s40360-020-0388-5

Marini, H. R., Bellone, F., Catalano, A., Squadrito, G., Micali, A., Puzzolo, D., Freni, J., Pallio, G., & Minutoli, L. (2023). Nutraceuticals as alternative approach against cadmium-induced kidney damage: A narrative review. *Metabolites*, 13(6), 722. https://doi.org/10.3390/metabo13060722

Mattioli, R., Francioso, A., Mosca, L., & Silva, P. (2020). Anthocyanins: A comprehensive review of their chemical properties and health effects on cardiovascular and neurodegenerative diseases. *Molecules*, 25(17), 3809. https://doi.org/10.3390/molecules25173809

McClements, D. J., Li, F., & Xiao, H. (2015). The nutraceutical bioavailability classification scheme: Classifying nutraceuticals according to factors limiting their oral bioavailability. *Annual Review of Food Science and Technology*, 6(1), 299–327. http://doi:10.1146/annurev-food-032814-014043.

Melchior, S., Codrich, M., Gorassini, A., Mehn, D., Ponti, J., Verardo, G., Tell, G., Calzolai, L., & Calligaris S. (2023). Design and advanced characterization of quercetin-loaded nano-liposomes prepared by high-pressure homogenization. *Food Chemistry*, 428, 136680. https://doi.org/10.1016/j.foodchem.2023.136680.

Nguyen, M.-H., Pham, N.-D., Dong, B., Nguyen, T.-H.-N., Bui, C.-B., & Hadinoto, K. (2017). Radioprotective activity of curcumin-encapsulated liposomes against genotoxicity caused by Gamma Cobalt-60 irradiation in human blood cells. *International Journal of Radiation Biology*, 93(11), 1267–1273. https://doi.org/10.1080/09553002.2017.1380329

Nsairat, H., Khater, D., Sayed, U., Odeh, F., Al Bawab, A., & Alshaer, W. (2022). Liposomes: Structure, composition, types, and clinical applications. *Heliyon*, 8(5), e09394. https://doi.org/10.1016/j.heliyon.2022.e09394

Obeid, M. A., Teeravatcharoenchai, T., Connell, D., Niwasabutra, K., Hussain, M., Carter, K., & Ferro, V. A. (2020). Examination of the effect of niosome preparation methods in encapsulating model antigens on the vesicle characteristics and their ability to induce immune responses. *Journal of Liposome Research*, 31(2), 195–202. http://doi:10.1080/08982104.2020.1768110

Pandey, S. N., Singh, G., Semwal, B. C., Gupta, G., Alharbi, K. S., Almalki, W. H., Albratty, M., Najmi, A., & Meraya, A. M. (2022). Therapeutic approaches of nutraceuticals in the prevention of Alzheimer's disease. *Journal of Food Biochemistry*, 46(12), e14426. https://doi.org/10.1111/jfbc.14426

Pandey, S., Shamim, A., Shaif, M., & Kushwaha, P. (2023). Development and evaluation of Resveratrol-loaded liposomes in hydrogel-based wound dressing for diabetic foot ulcer. *Naunyn-Schmiedeberg's Archives of Pharmacology*, 396(8), 1811–1825. https://doi.org/10.1007/s00210-023-02441-5

Puri, V., Nagpal, M., Singh, I., Singh, M., Dhingra, G. A., Huanbutta, K., Dheer, D., Sharma, A., & Sangnim, T. (2022). A comprehensive review on nutraceuticals: Therapy support and formulation challenges. *Nutrients*, 14(21), 4637. https://doi.org/10.3390/nu14214637

Rakotondrabe, T. F., Fan, M. X., Muema, F. W., & Guo, M. Q. (2023). Modulating inflammation-mediated diseases via natural phenolic compounds loaded in nanocarrier systems. *Pharmaceutics*, 15(2), 699. https://doi.org/10.3390/pharmaceutics15020699

Ren, Y., Nie, L., Zhu, S., & Zhang, X. (2022). Nanovesicles-mediated drug delivery for oral bioavailability enhancement. *International Journal of Nanomedicine*, 17, 4861–4877. https://doi.org/10.2147/IJN.S382192

Rodriguez, E. B., Almeda, R. A., Vidallon, M. L. P., & Reyes, C. T. (2018). Enhanced bioactivity and efficient delivery of quercetin through nanoliposomal encapsulation using rice bran phospholipids. *Journal of the Science of Food and Agriculture*, 99(4), 1980–1989. https://doi.org/10.1002/jsfa.9396

Ronis, M. J. J., Pedersen, K. B., & Watt, J. (2018). Adverse effects of nutraceuticals and dietary supplements. *Annual Review of Pharmacology and Toxicology*, 58, 583–601. https://doi.org/10.1146/annurev-pharmtox-010617-052844

Sachdeva, V., Roy, A., & Bharadvaja, N. (2020). Current prospects of nutraceuticals: A review. *Current Pharmaceutical Biotechnology*, 21(10), 884–896. https://doi.org/10.2174/1389201021666200130113441

Salami, M., Salami, R., Aarabi, M. H., Mafi, A., Ghorbanhosseini, S. S., Shafabakhsh, R., & Asemi, Z. (2023). Targeting glioma cells with nutraceuticals: Therapeutic effects based on molecular mechanisms, new evidence and perspectives. *Mini Reviews in Medicinal Chemistry*, 23(11), 1167–1192. https://doi.org/10.2174/1389557522666220531151137

Santini, A., & Novellino, E. (2017). To nutraceuticals and back: Rethinking a concept. *Foods*, 6(9), 74. http://doi.org/10.3390/foods6090074

Schmied, F. P., Bernhardt, A., & Klein, S. (2022). Preparation of Solid Self-Nanoemulsifying Drug Delivery Systems (S-SNEDDS) by co-extrusion of liquid SNEDDS and polymeric carriers-A new and promising formulation approach to improve the solubility of poorly water-soluble drugs. *Pharmaceuticals*, 15(9), 1135. https://doi.org/10.3390/ph15091135

Shaker, S., Gardouh, A. R., & Ghorab, M. M. (2017). Factors affecting liposomes particle size prepared by ethanol injection method. *Research in Pharmaceutical Sciences*, 12(5), 346–352. https://doi.org/10.4103/1735-5362.213979

Shao, X. R., Wei, X. Q., Zhang, S., Fu, N., Lin, Y. F., Cai, X. X., & Peng, Q. (2017). Effects of micro-environmental pH of liposome on chemical stability of loaded drug. *Nanoscale Research Letters*, 12, 504. https://doi.org/10.1186/s11671-017-2256-9

Sheoran, R., Khokra, S. L., Chawla, V., & Dureja, H. (2019). Recent patents, formulation techniques, classification and characterization of liposomes. Recent *Patents on Nanotechnology*, 13(1), 17–27. http://doi.org/10.2174/1872210513666181127110413

Shete, V. S., Telange, D. R., Mahajan, N. M., Pethe, A. M., & Mahapatra, D. K. (2023). Development of phospholipon®90H complex nanocarrier with enhanced oral bioavailability and anti-inflammatory potential of genistein. *Drug Delivery*, 30(1), 2162158. https://doi.org/10.1080/10717544.2022.2162158

Singla, R. K., Dubey, A. K., Garg, A., Sharma, R. K., Fiorino, M., Ameen, S. M., & Al-Hiary, M. (2019). Natural polyphenols: Chemical classification, definition of classes, subcategories, and structures. *Journal of AOAC International*, 102(5), 1397–1400. http://doi:10.5740/jaoacint.19-0133

Stabrauskiene, J., Kopustinskiene, D. M., Lazauskas, R., & Bernatoniene, J. (2022). Naringin and naringenin: Their mechanisms of action and the potential anticancer activities. *Biomedicines*, 10(7), 1686. https://doi.org/10.3390/biomedicines10071686

Subramani, T., & Ganapathyswamy, H. (2020). An overview of liposomal nano-encapsulation techniques and its applications in food and nutraceutical. *Journal of Food Science and Technology*, 57(10), 3545–3555. https://doi.org/10.1007/s13197-020-04360-2

Swallah, M. S., Fu, H., Sun, H., Affoh, R., & Yu, H. (2020). The impact of polyphenol on general nutrient metabolism in the monogastric gastrointestinal tract. *Journal of Food Quality*, 2020, 1–12. http://doi:10.1155/2020/5952834

Tarko, T., & Duda-Chodak, A. (2020). Influence of food matrix on the bioaccessibility of fruit polyphenolic compounds. *Journal of Agricultural and Food Chemistry*, 68(5), 1315–1325. https://doi.org/10.1021/acs.jafc.9b07680

Vaiserman, A., Koliada, A., Zayachkivska, A., & Lushchak, O. (2020). Nanodelivery of natural antioxidants: An anti-aging perspective. *Frontiers in Bioengineering and Biotechnology*, 7, 447. https://doi.org/10.3389/fbioe.2019.00447

Vraníková, B., & Gajdziok, J. (2015). Biologická dostupnost léčiva a možnosti jejího ovlivňování [Bioavailability and factors influencing its rate]. *Ceska a Slovenska Farmacie: Casopis Ceske Farmaceuticke Spolecnosti a Slovenske Farmaceuticke Spolecnosti*, 64(1-2), 7–13. https://pubmed.ncbi.nlm.nih.gov/26084643/

Wu, W., Lu, Y., & Qi, J. (2015). Oral delivery of liposomes. *Therapeutic Delivery*, 6(11), 1239–1241. http://doi:10.4155/tde.15.69

Xu, X., Tian, M., Deng, L., Jiang, H., Han, J., Zhen, C., Huang, L., & Liu, W. (2023). Structural degradation and uptake of resveratrol-encapsulated liposomes using an in vitro digestion combined with Caco-2 cell absorption model. *Food Chemistry*, 403, 133943. https://doi.org/10.1016/j.foodchem.2022.133943

Xu, Y., Li, Y., Xie, J., Xie, L., Mo, J., & Chen, W. (2021). Bioavailability, absorption, and metabolism of pelargonidin-based anthocyanins using Sprague-Dawley rats and Caco-2 cell monolayers. *Journal of Agricultural and Food Chemistry*, 69(28), 7841–7850. https://doi.org/10.1021/acs.jafc.1c00257

Yang, B., Dong, Y., Wang, F., & Zhang, Y. (2020). Nanoformulations to enhance the bioavailability and physiological functions of polyphenols. *Molecules*, 25(20), 4613. https://doi.org/10.3390/molecules25204613.

Zu, Y., Overby, H., Ren, G., Fan, Z., Zhao, L., & Wang, S. (2018). Resveratrol liposomes and lipid nanocarriers: Comparison of characteristics and inducing browning of white adipocytes. *Colloids and Surfaces B: Biointerfaces*, 164, 414–423. http://doi:10.1016/j.colsurfb.2017.12.044

Zuidam, N. J., & Velikov, K. P. (2018). Choosing the right delivery systems for functional ingredients in foods: An industrial perspective. *Current Opinion in Food Science*, 21, 15–25. http://doi:10.1016/j.cofs.2018.04.007

Zylberberg, C., & Matosevic, S. (2016). Pharmaceutical liposomal drug delivery: A review of new delivery systems and a look at the regulatory landscape. *Drug Delivery*, 23(9), 3319–3329. https://pubmed.ncbi.nlm.nih.gov/27145899/. http://doi:10.1080/10717544.2016.1177136

Chapter 21

Food Texture and Rheology

An Opportunity to Improve Wellness and Nutrition of Elderly Population

Gabriela Caez-Ramírez
Luz I. Sotelo-Díaz
and Katherine S. Torres-Ortiz

CONTENTS

21.1 INTRODUCTION

Global trends show that the elderly population is continuously growing, and, in this sense, by the year 2050, one in six people will be sixty-five, expecting 2.1 billion old adults and 426 million over 80 years old with increased levels of vulnerability (WHO, 2021). The growth of the elderly population directly impacts biomedical, social, and economic systems. The aging process is complex, presenting multiple biological and psychological changes, which can alter health, influence family dynamics, and generate considerable expenses for the health system (WHO, 2021). Furthermore, healthy aging depends on physical and mental abilities that enable well-being and is the intrinsic capacity of the person developed across a lifetime, a critical factor in overcoming elderly challenges, including adverse physical and social environments (Gomez et al. 2022).

DOI: 10.1201/9781003332558-24

TABLE 21.1 ESTIMATED DAILY LEVELS OF INTAKE OF DIETARY ENERGY AND MACRONUTRIENTS IN ELDERLY

	Males	Females
Energy (kcal)/Age Range		
51–70 years	2001 ±21.4	1571 ± 29.3
71–85 years	1863 ±37.4	1555 ± 63.2
Protein (g)		
51–70 years	76 ± 0.9	63 ± 1.3
71–85 years	71 ± 1.5	60 ± 2.7
Carbohydrate (g)		
51–70 years	210 ± 2.3	171 ± 3.6
71–85 years	202 ± 4.3	172 ± 7.6
Total Fat (g)		
51–70 years	87 ± 1.2	67 ± 2.1
71–85 years	79 ± 2.1	66 ± 3.3
Saturated Fat (g)		
51–70 years	30 ± 0.4	22 ± 0.6
71–85 years	27 ± 0.8}	22 ± 1.3
Dietary Fiber (g)		
51–70 years	16 ± 0.2	15 ± 0.4
71–85 years	17 ± 0.4	15 ± 0.7

Source: Foote et al. (2004); Rajabi et al. (2021).

Data presents mean ± standard error of the mean (SEM).

According to the World Health Organization (WHO, 2021), the most common issues that affect this population include hearing loss, cataracts, back pain, neck pain and osteoarthritis, chronic obstructive pulmonary disease, diabetes, depression, and dementia. Malnutrition is associated with other diseases, including cognitive impairment with a high risk of developing dementia. Moreover, malnourishment and weight loss are recognized as risk-critical factors for rapid cognitive decline, disease progression, degree of disability, and consequent lack of autonomy, so malnutrition and cognitive impairment have a two-way interrelationship (Takagi et al. 2017; Yu et al. 2021). The concept of hidden hunger was created by the WHO to refer to the chronic deficiency of micronutrients due to a low intake of vitamins and minerals, causing high mortality rates, and by 2020, more than two billion people had suffered from this condition (WHO, 2021). Malnutrition risk across all countries ranks from 8.5% in community settings to 28% in hospitals (Norman et al. 2021). In particular, the prevalence of malnutrition in Latin America varies between 40 and 60% and is detected when patients are admitted to hospitals (Gomez et al. 2022). **Table 21.1** shows the estimated daily dietary energy of free-living older adults by gender, and the estimated average daily intake of micronutrients is shown in **Table 21.2**.

The deficiency of these macro- and micronutrients leads to various problems in the elderly, as seen in **Table 21.3.** However, not only nutrient deficiency could be a problem. For instance,

TABLE 21.2 ESTIMATED DAILY LEVELS OF INTAKE OF DIETARY MICRONUTRIENTS IN THE ELDERLY

	Dietary Reference Intake	
Micronutrient / Age Range	**Males**	**Females**
Vitamin A (Retinol Equivalents)		
51–70 years	1459.90 ± 24.30	1237.10 ±31.60
71–85 years	1505.20 ± 47.20	1238.40 ± 72.50
Vitamin E (Tocopherol Equivalents)		
51–70 years	10.00 ± 0.14	8.30 ± 0.19
71–85 years	10.10 ± 0.37	8.50 ± 0.40
Thiamin (mg)		
51–70 years	1.30 ± 0.02	1.10 ± 0.02
71–85 years	1.30 ± 0.03	1.10 ± 0.04
Riboflavin (mg)		
51–70 years	2.00 ± 0.03	1.70 ± 0.04
71–85 years	1.90 ± 0.05	1.70 ± 0.09
Niacin (mg NE)		
51–70 years	20.20 ± 0.24	15.70 ± 0.32
71–85 years	18.90 ± 0.44	15.10 ± 0.59
Folate (μg)		
51–70 years	329.40 ± 4.66	282.10 ± 6.90
71–85 years	333.00 ± 9.98	284.30 ± 12.90
Vitamin B6 (mg)		
51–70 years	2.00 ± 0.02	1.90 ± 0.04
71–85 years	2.10 ± 0.06	1.90 ± 0.08
Vitamin B12 (μg)		
51–70 years	5.90 ± 0.12	5.50 ± 0.19
71–85 years	5.70 ± 0.27	5.10 ± 0.51
Vitamin C (mg)		
51–70 years	155.30 ± 3.28	155.20 ± 5.66
71–85 years	159.20± 4.71	162.60 ± 11.12

Adapted from Foote et al. (2000) Rajabi et al. (2021). Data presents mean ± standard error of the mean (SEM) per age range.

excessive calcium intake in the elderly promotes hypercalcemia, recognized as a trigger of weakening bones, forming kidney stones, and interfering with the functioning of the heart and brain (Shiizaki et al. 2021).

21.2 FOOD AND NUTRITION IN OLDER ADULTS WITH COGNITIVE IMPAIRMENT AND SWALLOWING LIMITATIONS

Changes in body weight in aging people can result in cognitive deterioration, emaciation, and loss of independence for daily activities (Bello-Corral et al. 2021). Dementia, a syndrome usually of a progressive nature and currently the seventh primary cause of death, leads to impairment in cognitive function. More than fifty-five million people live with dementia worldwide, with approximately ten million new cases yearly (WHO, 2021). Alzheimer's, one the most important causes of dementia, with 60–70% of cases, is a degenerative neuronal disease that worsens over time as neurons in different brain parts associated with the cognitive function of learning and memory decline. Older adults regularly live with Alzheimer's symptoms for years, which begin to interfere with daily activities such as performing essential bodily functions such as walking and swallowing until the person becomes completely dysfunctional (WHO, 2021; Lee et al. 2023). Swallowing is a process by which mechanical physiological sequences transport solid and liquid foods, voluntary and involuntary (Ortega et al. 2017), with a succession of central and peripheral neurological components responsible for transporting the already minced food from the cavity mouth to the stomach (Altman, 2018). Swallowing has four phases:

(1) Oral preparation stage with voluntary muscle involvement, in which food is chewed and mixed with saliva, forming the bolus.
(2) Oral phase with voluntary muscle involvement, where the tongue rises and pushes the bolus towards the back (Ortega et al. 2017).
(3) Pharyngeal phase with involuntary muscle involvement and in which there is a prominence of the soft palate, closure of the vocal cords, and movement of the pharynx towards the hypopharynx with the opening of the cricopharyngeal sphincter.
(4) Esophageal phase, involuntary muscle involvement, occurs with the relaxation of the cricopharyngeal sphincter, allowing the passage of food from the pharynx to the esophagus, entering the lower esophageal sphincter, which leads to the stomach (Gallegos et al. 2017). Dysphagia is a disorder in one or more stages of the swallowing process closely related to the severity of cognitive impairment (Takagi et al. 2017; Saitoh et al. 2018).

Energy intake should consider multiple aspects of food acceptance, such as sensory properties, mood, gastrointestinal health, and reaction speed. Typical recommendations for oral diet administration are varied menus and thickened or gelled water to adapt the texture of the food and enrich the diet with caloric and protein nutrients.

Some diets such as the Mediterranean diet, which mainly consists of fruits, vegetables, grains, beans, olive oil and fish, with considerable amounts of nutrients to modulate detoxifying enzymes and regulate cholesterol synthesis, may have neuroprotective effects (Gardener & Caunca, 2018). Antioxidants reduce the oxidative stress that causes cognitive deterioration. One of the most potent is vitamin E, which helps prevent the oxidation of unsaturated fatty acids, which is essential for cell membrane protection. Older adults are generally deficient in vitamin E (Śliwińska & Jeziorek, 2021) and are advised to consume dried fruits, seeds, nuts, and vegetable oils. **Figure 21.1** depicts dietary reference intakes for the elderly (Rodd et al. 2021).

Antioxidants such as selenium (Se); vitamins C, D, E, and B complex; transition metals such as zinc (Zn), iron (Fe), and copper (Cu); as well as omega-3 fatty acids have possible neuronal protective effects (Botchway et al. 2018). Older adults with cognitive impairments have low intakes

TABLE 21.3 MACRO AND MICRONUTRIENT DEFICIENCY AND SOURCES

Malnutrition Problem	Nutrient Deficiency	Main Sources	Reference
Osteoporosis; impaired immune function; anorexia; loss of taste; delayed wound healing; development of pressure ulcers; cardiovascular disease	Zinc	Meat, nuts, seeds, fruit, and vegetables	Bouvard et al.(2021) ; Kaur et al. (2019); Rodd et al. (2021)
Anemia; blood loss; decreased iron absorption; skin changes; nail changes	Iron	Organ meats, meats, poultry, legumes, nuts, and vegetables	Mohajeri et al. (2015); Reber et al. (2019)
Cardiovascular disease; Anemia; dermatitis; reduced growth; marker of risk of atherothrombosis, Alzheimer's, and Parkinson's disease.	Folate	Vegetables, legumes, liver	Reber et al. (2019); Rodd et al. (2021)
Reduce renal capacity; various skin's functions decrease **Osteoporosis**	Vitamin D	Fish such as salmon, sardines, herring, and mackerel; red meat; liver; egg yolks	Rodd et al. (2021)
Cancer and other chronic diseases; Malaise; weight loss; scurvy	Vitamin C	Fruits and vegetables (citrus, strawberries, tomatoes, peppers, and potatoes)	Mohajeri et al. (2015); Kaur et al. (2019); Reber et al. (2019)
Peripheral neuropathy; ataxia or cognitive alterations	Vitamin B_{12}	Animal origin	Reber et al. (2019); Rodd et al. (2021)
Decreased muscle mass (emaciation); slow metabolism; poor physical and intellectual performance; poor immune or defense system; fatigue, exhaustion; apathy; lack of concentration and mood	Protein	Egg, chicken and meat, fish, legumes or legumes, walnuts, or nuts	Mohajeri et al. (2015); Reber et al. (2019); Rodd et al. (2021).
Constipation	Fiber	Legumes, vegetables, whole grains, fruits	Reber et al. (2019); Rodd et al. (2021).
Dehydration decreased kidney function	Water	Water: Minimum fluid intake of 1,500 ml/day or 1 ml per calorie consumed	Reber et al. (2019); Rodd et al. (2021).

(continued)

TABLE 21.3 (CONTINUED)

Malnutrition Problem	Nutrient Deficiency	Main Sources	Reference
Hypernatremia, hyponatremia	Sodium	Salt: Recommend consumption of 1,500 to a maximum of 2,300 mg/day	Mohajeri et al. (2015); Reber et al. (2019); Rodd et al. (2021)

of these specific nutrients, which may also be associated with an increased risk of triggering the disease. Diets rich in saturated fatty acids and alcohol, low in antioxidants and vitamins, promote the onset of the disease, while diets rich in unsaturated fatty acids, vitamins, antioxidants, wine, turmeric, and certain spices inhibit its onset by sequestering free radicals and preventing oxidative damage (Hill et al. 2019).

According to Kuelzow et al. (2016), nutritional interventions and their outcomes in older adults with Alzheimer's improved nutritional status and oral intake. Keller et al. (2015) highlighted the importance of mealtime strategies, pointing out that nutritional status is related to health and well-being. Watkins et al. (2017) reported that there is a wide variety of strategies to promote food and fluid intake, addressing key factors that influence nutrition.

In recent years, various tools have been used to intervene in nutrition in this population, developing foods with modified textures to reduce the risk of aspiration. However, there is a problem around these foods and what old adults enjoy. Those changes in texture and visual appeal in foods tend to influence acceptability, resulting in dissatisfaction (Keller et al. 2015). The perception of unpalatable food decreases appetite in older adults, reducing food intake and, consequently, resulting in malnutrition. For this reason, pleasure during food consumption should be restored to increase appetite (Endo et al. 2017).

21.3 EUDAEMONIC AND PROPER FEEDING FOR THE ELDERLY

Research on well-being promoted by food consumption can be divided into two complementary terms. Firstly, hedonic is related to the pleasure achieved from elevated positive and low negative emotions, focusing on happiness (i.e., sweet food preference). Secondly, eudaimonia focuses on living life entirely and intensely enjoyable (i.e., the pleasure of eating with family or friends) (Delle Fave et al. 2011). Food is one of the highest sources of hedonic experiences in human life. Recent studies show that *indulgent food* relates to older adults considering food highly pleasurable, such as sweet desserts and delightful fatty meals (Tarrega et al. 2017). Indulgent food satisfies cravings, pursues hedonism, splurges on special occasions, enjoys premium products, experiments with food culture, and uses food as a status symbol.

Evidence suggests non-pharmacological therapies positively influence neuropsychiatric symptoms and quality of life in people with cognitive impairment (Savazzi et al. 2020). Activities of daily living such as mealtimes are enclosed in non-pharmacological therapies and are one of the few social activities of the day for institutionalized patients. Therefore, these moments should be as pleasant as possible, promoting the pleasure of eating (Beck et al. 2019).

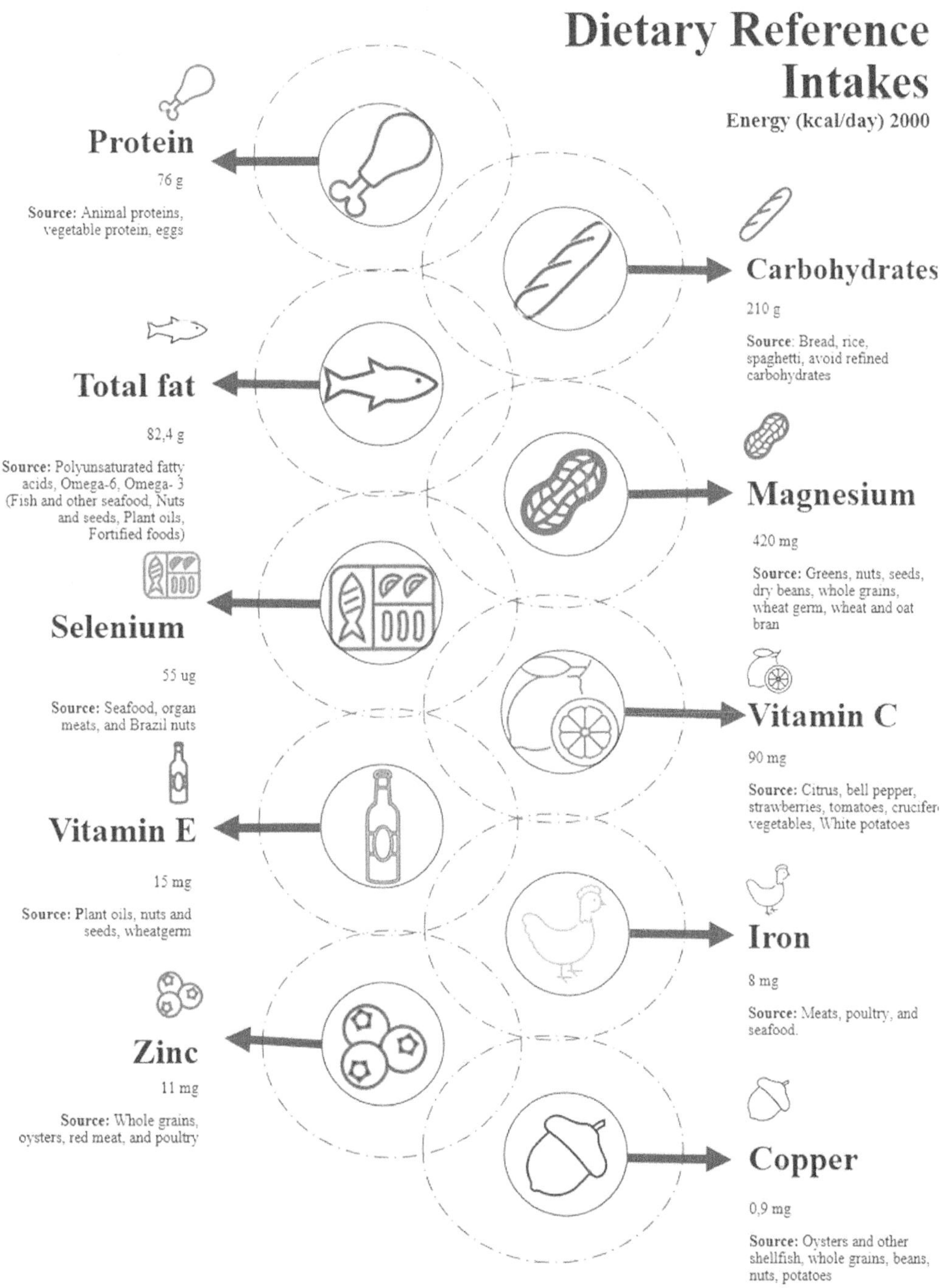

Figure 21.1 Dietary reference intakes for elderly.

Emotional states can cause alterations in the rhythm, frequency, and availability of food, depending on the characteristics of each person and emotional changes (Nifli, 2018). For instance, when older adults have sad days, their food intake decreases. For this reason, food should be used to overcome the feeling of sadness, making them consume food preparations that evoke a calm mood and well-being (Marcus, 2019).

Oral alterations usually go with aging, such as teeth deterioration, lower chewing performance (Fujita & Maki, 2018), or a decrease in salivary flow (Xu et al., 2019), which can affect the perception of food texture and therefore the emotions associated with eating (Farouk et al. 2018). These changes could affect the perception of foods' structural properties, such as tenderness, elasticity, and firmness (Vandenberghe-Descamps et al. 2017). A decrease in salivary flow could change oral relationships such as viscosity, tenderness, juiciness, and astringency (Kang et al., 2020). The mean particle size of the food bolus decreased in patients with dentures compared to the elderly without removable dental prostheses (Bourdiol et al. 2020). **Figure 21.2** shows frequent chemosensation disorders (Papazian et al. 2021).

Crispness and crunchiness are mechanical attributes related to the fracture properties of food during biting and chewing with audible aspects or sound emission (Tunick et al. 2013). Studies have shown that people consider crispy and crunchy food appealing and enjoyable (Szczesniak & Kahn, 1971, 1984; Fillion & Kilcast, 2002). Crunchiness is related to chewing sounds and influences food texture perception (Spence, 2015). Crispy foods generate highly inclined sounds that show an essential level of frequencies higher than 5 kHz, while crunchy foods generate low inclined sounds with a frequency range of 1.25 to 2 kHz (Wang et al. 2020). The effect of chewing sounds as a tool to intervene in elderly nutrition can improve food consumption from inner mouth sounds. For this reason, products that visually evoke crispy or crunchy attributes are more appetizing for older adults during mealtimes (Endo et al. 2017).

Zampini and Spence (2004) reported the first studies on the acceptability of foods by evaluating crispness and crunchiness perception. The research was replicated with commonly consumed food (Dematte et al. 2014) and food in the households of older adults (Endo et al. 2016). Furthermore,

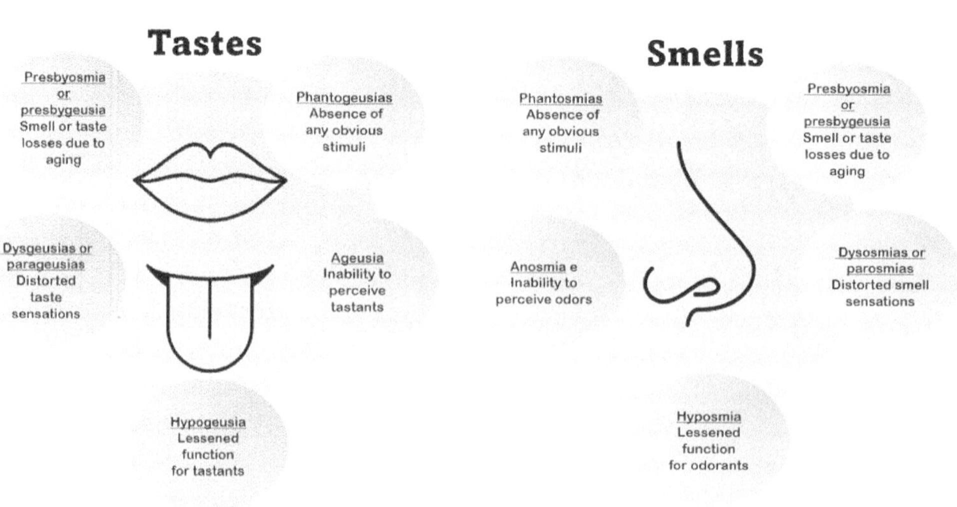

Figure 21.2 Chemo sensation disorders.

based on these investigations, it was possible to find that crunchy texture is visually related to improved food taste.

21.4 FOOD INTAKE AND COGNITIVE HEALTH

Food can promote elderly wellness. **Figure 21.3** shows the nutrients that should be present in elderly-prepared meals.

Fatty acids are classified into saturated fatty and unsaturated fatty acids, which can be grouped into mono- and polyunsaturated fatty acids. High consumption of saturated fatty acids has been related to an increased risk of metabolic disorders such as type 2 diabetes mellitus or cognitive impairment (Baranowski et al. 2020). Diets high in polyunsaturated fatty acids can enhance cognition and reduce impairment progression, endorsing fish as rich in long-chain omega-3 fatty acids, including α-linolenic, eicosapentaenoic, and docosahexaenoic acids (Avallone et al. 2019).

Carbohydrates are classified as complex and simple. The effect of simple sugars on cognition increases the risk of Alzheimer's (Baranowski et al. 2020). Carvalho et al. (2012) showed that sucrose intake related to Amyloid beta peptides (Aβ) peptide increase in cortical and hippocampal regions, promoting the risk of Alzheimer's.

Polyphenols are structurally characterized by various hydroxyl groups connected to aromatic rings and have been shown to enhance cognition, reduce neuroinflammation, and reduce Aβ pathologies. Fruits, vegetables, tea/coffee, red wine, olive oil, and dark chocolate are reliable sources of polyphenols (Godos et al. 2017), and high intake of polyphenols and flavonoids (Hagan et al. 2016) results in reduced cognitive decline.

Fruit juices, including orange, grape, and pomegranate juice, positively affect cognition health (Kean et al. 2015). Pomegranate juice has the highest concentration of phenolics and shows antioxidant activity, followed by red wine and blueberry juice (Harnly et al. 2006). Citrus fruit juices,

Figure 21.3 Nutrients that should be in prepared meals for the elderly.

Source: Kaur et al. (2019)

pomegranate, grape, and cherry juice, all with high quantities of polyphenols of different subtypes, have brain-protective effects that support berries consumption for improving cognitive well-being (Kent et al. 2017). Berries have high quantities of flavonoids, flavan-3-ols catechin, epicatechin, kaempferol, and myricetin, and anthocyanins such as delphinidin and petunidin, which are neuroprotective and can reduce brain aging (Gildawie et al. 2018; Spagnuolo et al. 2018; Potì et al. 2019).

Another polyphenol-rich food is cocoa, which flavan-3-ols have cognitive benefits (Fraga et al., 2019). Regular intake of cocoa flavanols also showed cognitive improvements in older adults with no cognitive impairments besides cardiovascular and metabolic enhancement, reducing blood pressure and insulin resistance (Zeli et al. 2022). Sources of caffeine, such as coffee and tea, are both excellent sources of polyphenols. Coffee has a significant quantity of phenolic acids, the hydroxycinnamic acids 4-caffeoylquinic acid, 5-caffeoylquinic acid, and catechol. Green tea has copious quantities of flavanols like catechins, epicatechins, and procyanidins; phenolic acids like hydroxybenzoic and hydroxycinnamic acids; and 5-caffeoylquinic acid (Burkholder-Cooley et al. 2016; Zamora-Ros et al. 2016). For these reasons, coffee and tea have cognitive benefits and are correlated with better cognitive effects, preventing the risk of cognitive disorders (Ma et al., 2016; Wu et al. 2017). The phenolics 2-hydroxy-4-methoxy benzaldehyde and 4-hydroxy-3-methoxy benzaldehyde can be extracted from *Hemidesmus indicus* and *Vanilla planifolia*, which have inhibitory potential against acetylcholinesterase used for treating Alzheimer's and other neurological impairments (Wilson et al. 2017).

Black chokeberry (*Aronia melanocarpa*), found in juices, purees, and jams, has prominent levels of polyphenols and flavonoids and helps alleviate chronic diseases such as diabetes and cardiovascular disorders (Jurikova et al. 2017). Red raspberries (*Rubus idaeus*) have various polyphenolic compounds, such as ellagitannins and anthocyanins, which may decrease the risk of non-communicable diseases and metabolically linked pathophysiology (Burton-Freeman et al. 2016)

Antioxidants protect from free radical-mediated destruction produced by toxic chain effects in neuronal cells and decrease the toxicity of Aβ in the brains of the elderly with Alzheimer´s (Feng & Wang, 2012). Fruits, vegetables, and olive oil have an essential quantity of antioxidants (Trichopoulou et al. 1995). Vitamin E is abundant in fruits and vegetables, with antioxidant properties to reduce Alzheimer´s progression (Dysken et al. 2014). The water-soluble antioxidant S-allyl mercapto cysteine, found in garlic, has biologically active compounds that help human health (Griffiths et al. 2016).

The role of antioxidants in balancing free radical creation is necessary for functional reasons against oxidative stress, which increases with age. Supervising the immune system in reaction to antioxidant dietary intake is another indicator of health or disease (Gonçalves et al. 2020). Dietary intake in the elderly has shown a vital role in antioxidants and micronutrients, essential in keeping the immune system healthy (Mocchegiani et al. 2014).

Spices contain essential oils with antidiabetic, antioxidant, anti-inflammatory, and hypocholesterolemic effects such as coriander, fennel, mustard, cinnamon, mace, clove, saffron, ginger, asafetida, bay leaf oil, cinnamon, cloves, cumin, fenugreek, turmeric, poppy seed, pomegranate, red chili, sesame seed, and so forth. Moreover, aromatic herbs like thyme, sage, oregano, parsley, dill, marjoram, chives, rosemary, mint, and lemongrass have significant antioxidant capacity and total phenolic content (Henning et al., 2011). Those spices hold two critical compounds used in neurodegenerative disease treatment: First, L-DOPA for Parkinson's disease, and second, galantamine for Alzheimer's disease. Both can help manage neurodegenerative conditions with their neuroprotective benefit, improving gastrointestinal functions and immunity (Gonçalves et al. 2020).

A further critical factor in promoting elderly wellness is the gut microbiome, involving bacteria, archaea, viruses, and other microbes in the gastrointestinal tract of humans (Kesika et al. 2021). They have an essential influence on the body's physiology in terms of health and disease due to those microbes' metabolic processes and defense mechanisms and their use in the digestion of food materials (Kesika et al., 2021). Also, these microorganisms can build an immune system in the intestinal mucosa that can protect the host from carcinogens by releasing short-chain fatty acids (Young, 2017). Metabolic diseases like diabetes (Karlsson et al. 2013), obesity (Sivamaruthi et al., 2019), colorectal cancer (Sivamaruthi et al. 2020), cardiovascular diseases, and inflammatory bowel diseases are related to changes in microbiota known as dysbiosis (Young, 2017).

The gut microbiome system influences the brain structure and involves an enteric nervous system, including an interrelated network of neurons in the spinal cord (Fernandez et al. 2015). The brain controls the gut's secretory and sensory role through signal communication eased by physiological networks like the neuroendocrine, autonomic nervous system, neuroimmune pathways, and signaling molecules generated by gut microbes (Wall et al. 2014). The production of γ-aminobutyric acid (GABA) through *Lactobacillus brevis* has an essential influence on the neuronal network of the body (Barrett et al. 2012). Neurotransmitters like monoamines and serotonin play a vital role in the signaling system of gut-brain microbes (O'Mahony et al. 2015). For this reason, a diet rich in probiotics could be a safe alternative treatment for oxidative diseases because probiotics enhance the body's gut microbiome. Probiotics should be consumed appropriately to deliver a health benefit to the body (Kesika et al. 2021). They could be supplied as food or through food ingredients, medical food, dietary supplements, and even medicines (Aguilar-Toalá et al. 2018).

However, regular exercise should also be adopted as another lifestyle factor in addition to nutritional interventions. Epidemiological studies point out that exercise reduces obesity, related metabolic diseases, and the risk of Alzheimer's (Ogino et al. 2019). However, these studies could not prove if there was a variation in type, duration, and intensity of exercise that can reduce the risk of Alzheimer's disease (Stephen et al. 2017). However, older adults may not have the energy or enthusiasm to do it. Hence, it is fundamental to educate this population on the importance of regular exercise for healthy aging. Possible strategies include creating an exercise program or suggesting community resources like walking groups and elderly fitness programs (Mumba et al. 2021). Fifteen years ago, the Age-Friendly Cities and Communities (AFCC) initiative was launched to create better living options for older adults. The initiative promotes autonomy, inclusion, dignity, health, and well-being. Through studying the experiences of communities in 33 cities, mostly in developed countries, the initiative identified eight areas where cities and communities can help promote healthy and active aging. These areas include outdoor spaces, transport and mobility, housing, social participation, social inclusion and non-discrimination, civic engagement, employment, communication and information, and community support and health services (WHO, 2023). Hence, long-term care implies person-centered strategies to foster community resources to respond to older people's needs.

21.5 FOOD STRUCTURE IMPLICATION IN THE ELDERLY NUTRITION

Swallowing problems have often been solved with thickened liquids and texture-modified foods. Thickened liquids refer to foods such as juices and milk with thickening agents, even pre-thickened water, specifically designed for people with dysphagia. Beverages can be designed to improve food and saliva transit along the oral and pharyngeal phase by slowing down, promoting better coordination, and making a cohesive bolus without breaking into drops moving safely to the esophagus; hence, rheological behavior characterization is essential (Vieira et al. 2020). Texture-modified foods refer to products with a solid or soft solid consistency, which, during oral processing, require low chewing, such as puree (Hadde & Chen, 2020).

Liquid food, appropriate for severe dysphagia cases, has traditionally been categorized using two indicators (Torres-Ortiz et al. 2022). The National Dysphagia Diet (NDD) classifies food viscosity responses at a shear rate of 50 s^{-1} as follows: liquids (0 – 50 mPa), nectar-like liquids (51 – 350 mPa), honey-like liquids (351 – 1750 mPa), and liquids thick or pudding (> 1750 mPa), having limitations for dysphagia disorders and showing non-Newtonian behavior (Hadde & Chen, 2020). In this respect, shear-thinning effects make swallowing more straightforward, as viscosity decreases with increasing shear rate during swallowing (Tobin et al. 2020).

For the second indicator, the International Dysphagia Diet Standardization Initiative (IDDSI) developed the globally standardized terminology and definitions for textured modified food and thickened liquids in all care settings and cultures for people with dysphagia (Torres-Ortiz et al. 2022). **Figure 21.4** shows the descriptors at five levels for the rheology of liquids and six levels for foods with a modified texture as proposed by the IDDSI (Cichero et al. 2017), together with the texture analysis profile conducted by Matsuo & Fujishima (2020).

On the other hand, recent studies reveal the importance of defining the rheological and mechanical properties of liquid, semi-solid, and solid foods according to the severity of dysphagia to support safe swallowing (Kim et al. 2020; Matsuo & Fujishima. 2020).

Gallego et al. (2022) proposed a relationship between the type of dysphagia, severity level, and food texture as an alternative to manage the effects of dysphagia by selecting proper food (Torres et al. 2022), as shown in **Figure 21.5.**

21.6 DESIGNING TEXTURE-MODIFIED FOODS AND THICK LIQUIDS

The essential ingredients for designing foods with modified textures and thickened liquids are proteins, lipids, and carbohydrates, which supply energy and vital nutrients for bodily functions. Moreover, these macronutrients adjust structurally and hierarchically, contributing to food's sensory and textural perception (Aguilera, 2016).

Proteins are the fundamental components of food, intervening as regulators in releasing nutrients or changing the texture (Ritzoulis & Karayannakidis, 2015).

Gelation is very important in achieving desired textures in food. It is induced by thermal denaturation, causing an increase in viscosity/texture with the gradual aggregation of network-denatured proteins (Nicolai, 2019). Due to the amphiphilic nature of proteins, emulsions, and foams can also be developed and stabilized. Their formation will depend on the air-water and oil-water interfaces in each protein's structure and composition (Yan et al. 2020). For the design of foods with modified textures and thick liquids, the structure of aerated foods is essential since it allows easy chewing and improves digestibility, flavor, and mouthfeel (Zúñiga et al., 2011). Examples of proteins that change the texture are eggs, milk, and gelatin of animal origin (Gallego et al. 2022).

Carbohydrates, such as fructose and saccharose, are sources of energy, and others, such as starch and fibers, release nutrients with a satiating effect, promoting the structure, texture, and sensory properties of foods. Polysaccharides function as gelling and thickening agents in aqueous dilutions and stabilizers for emulsions and foams (Gallego et al. 2017). Examples of carbohydrates

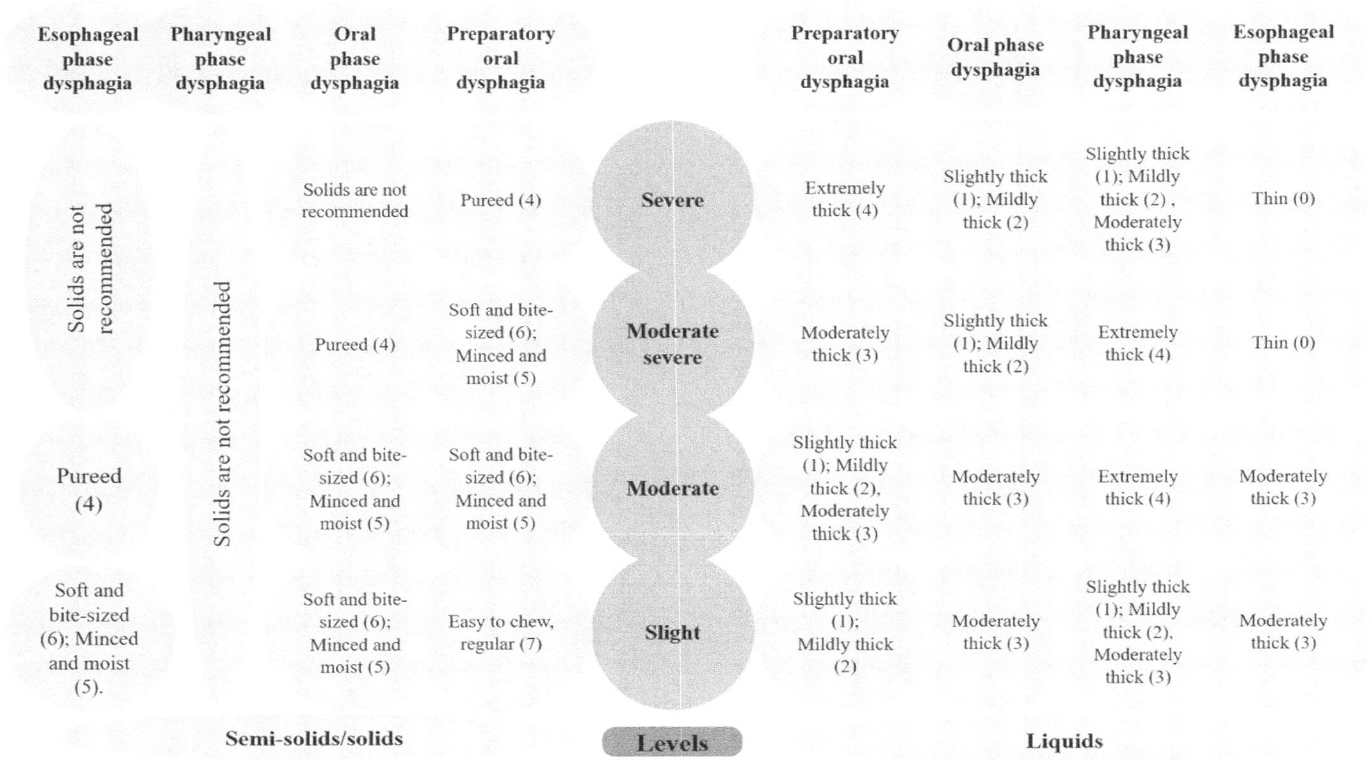

Figure 21.4 Level descriptors for liquid semi-solid and solid foods.

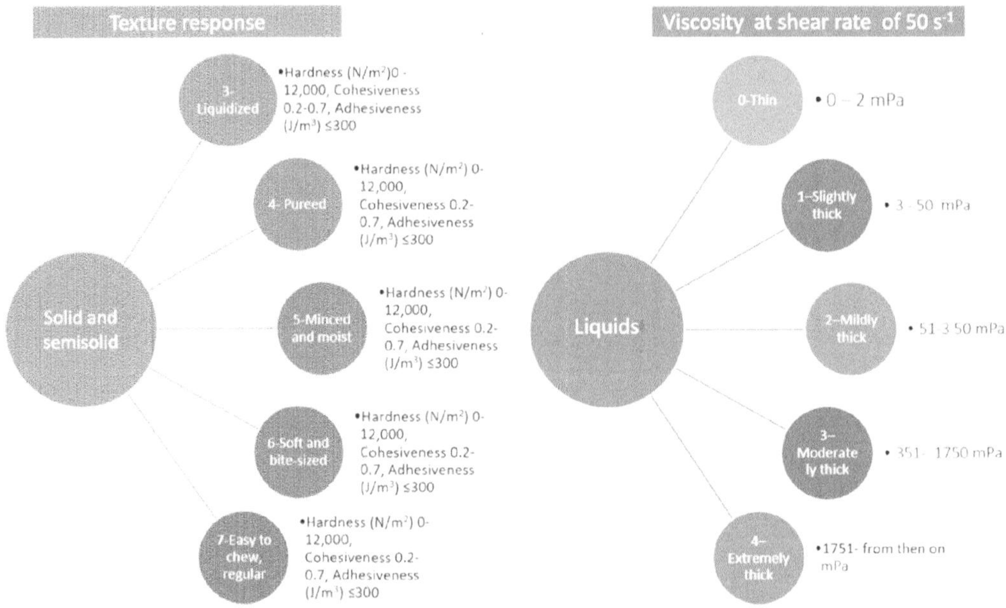

Figure 21.5 Type of dysphagia with its levels and consistency of semi-solid, solid, and liquid foods.

Note: Figures in brackets show the levels according to IDDSI (International Dysphagia Diet Standardization Initiative).

to increase viscosity in liquids are starch, xanthan gum, and cellulose derivatives to form gels, pectin, agar, and carrageenan. Some others could be emulsifying agents, such as gum Arabic and modified starches (Gallego et al. 2022). According to Wijaya et al. (2017), in the design of textured foods, polysaccharides, and proteins can be combined by covalent or non-covalent interactions, improving stability and mechanical properties. The behavior of the protein-polysaccharide structure will be determined by factors such as concentration, ionic strength, pH, heat, and application of mechanical procedures (Gentile, 2020).

Lipids, such as monoglycerides and phospholipids, are amphiphilic molecules commonly used as emulsifiers to create associations with nanometric structures, micelles, or vesicles (Aguilera et al. 2016). Triacylglycerols (triglycerides) crystallize and clump together, developing a fat network, resulting in texture and sensory attributes specific to foods (Marangoni & Wesdorp, 2019). Oleogels improve the texture of foods through an intrinsic lipid phase in a stable gel network (Marangoni, 2012). The functions of the lipids will depend on the physical properties, interfaces, emulsion stability, and droplet size (Farjami & Madadlou, 2019; Yan et al. 2020; Gao et al. 2021).

Texture-modified foods often enhance visually when accompanied by other evocative textures, acoustic sounds, and innovative technologies, such as 3D food printing, as a primary factor in food design for older adults with cognitive impairment (Yang et al. 2017).

Three-dimensional printing is a digital process that builds complex geometric structures layer by layer, assembling them through phase transitions or chemical reactions (Severini et al., 2016), which allows continuous production with environmental and economic benefits (Kim et al. 2017).

The basic concept of three-dimensional printing is a controlled process where a product is built from a digital design that is usually computer-aided (CAD) (Dankar et al. 2018). The software cuts the 3D model into layers. Then, it prints them individually, allowing the design and development of foods with various shapes, colors, flavors, and personalized nutritional needs, which have a different gastronomic experience for food consumption (Sun et al. 2015).

21.7 CONCLUSION

Promoting a healthy lifestyle has become a global priority as the world's population ages. Though aging is natural, simple changes can help to reduce the risk of an accelerated body and mental health decline. Malnutrition is a significant risk for rapid cognitive impairment. A strategy to reach a more extended lifespan with health and well-being should involve a better food choice, rich in antioxidants, neuroprotective, and essential nutrients, avoiding simple carbohydrates and indulgent foods. Moreover, food designed to enhance proper nutrient intake should boost older adults' preferences and consider present and future health conditions, including the advance and severity of swallowing limitations. Hence, textural and rheological properties will be essential in improving food intake under the customized condition of dysphagia. Nourishing for a longer time involves the whole human being's understanding of offering nutritious, sensory-pleasing food and recognizing the eating experience as an opportunity to elevate the emotional state. A deep understanding of the relationships among food hedonics, eudaimonia, appetite, and structure should prompt better food development, with a sense of balance among tasty and healthy food, improving nutritional status, longevity, and wellness. Finally, the rapidly growing older population around the world, expected to double by 2050, with the large majority living in low and middle-income countries, urges the promotion of whole strategies for elderly care in age-friendly environments to foster well-being and healthy lifestyles.

ACKNOWLEDGMENTS

Thanks to the Ministry of Sciences for the financial resources for the project at the National Program for Science, Technology, and Innovation in Health, ID 123084468143.

REFERENCES

Aguilar-Toalá, J.E., Garcia-Varela, R., Garcia, H.S., Mata-Haro, V., González-Córdova, A.F., Vallejo-Cordoba, B., & Hernández-Mendoza, A. (2018). Postbiotics: An evolving term within the functional foods field. *Trends in Food Science & Technology, 75*, 105–114. https://doi.org/10.1016/j.tifs.2018.03.009

Aguilera, J.M., & Park, D.J. (2016). Texture-modified foods for the elderly: Status, technology, and opportunities. *Trends in Food Science & Technology, 57*, 156–164. https://doi.org/10.1016/j.tifs.2016.10.001

Altman, K. W. 29 July;. . *Dysphagia Evaluation and Management in Otolaryngology* Chhetri, Dinesh, Dewan, Karuna. 2018 Neurologic dysphagia, (St. Louis, Missouri: Elsevier)107–114 9780323569361.

Avallone, R., Vitale, G., & Bertolotti, M. (2019). Omega-3 fatty acids and neurodegenerative diseases: New evidence in clinical trials. *International Journal of Molecular Sciences, 20*(17), 4256. https://doi.org/10.3390/ijms20174256

Baranowski, B.J., Marko, D.M., Fenech, R.K., Yang, A.J., & MacPherson, R.E. (2020). Healthy brain, healthy life: A review of diet and exercise interventions to promote brain health and reduce Alzheimer's disease risk. *Applied Physiology, Nutrition, and Metabolism, 45*(10), 1055–1065. https://doi.org/10.1139/apnm-2019-0910

Barrett, E., Ross, R.P., O'Toole, P.W., Fitzgerald, G.F., & Stanton, C. (2012). γ-Aminobutyric acid production by culturable bacteria from the human intestine. *Journal of Applied Microbiology, 113*(2), 411–417. https://doi.org/10.1111/j.1365-2672.2012.05344.x

Beck, M., Birkelund, R., Poulsen, I., & Martinsen, B. (2019). Hospital meals are existential asylums to hospitalized people with a neurological disease: A phenomenological–hermeneutical explorative study of the meaningfulness of mealtimes. *Nursing Open, 6*(2), 626-634. https://doi.org/10.1002/nop2.246

Bello-Corral, L., Sánchez-Valdeón, L., Casado-Verdejo, I., Seco-Calvo, J.Á., Fernández-Fernández, A.J., & Fernández-Martínez, N.M. (2021). The influence of nutrition in Alzheimer's disease: Neuroinflammation and the Microbiome vs. Transmissible Prion. *Frontiers in Neuroscience, 20*(15), 677777. https://doi.org/10.3389%2Ffnins.2021.677777

Botchway, B.O., Moore, M.K., Akinleye, F.O., Iyer, I.C., & Fang, M. (2018). Nutrition: Review on the possible treatment for Alzheimer's disease. *Journal of Alzheimer's Disease, 61*(3), 867–883. https://doi.org/10.3233/jad-170874

Bouvard, B., Annweiler, C., & Legrand, E. (2021). Osteoporosis in older adults. *Joint Bone Spine, 88*(3), 105135.

Bourdiol, P., Hennequin, M., Peyron, M.A., & Woda, A. (2020). Masticatory adaptation to occlusal changes. *Frontiers in Physiology,* 11:*263*. https://doi.org/10.3389%2Ffphys.2020.00263

Burkholder-Cooley, N., Rajaram, S., Haddad, E., Fraser, G.E., & Jaceldo-Siegl, K. (2016). Comparison of polyphenol intakes according to distinct dietary patterns and food sources in the Adventist Health Study-2 cohort. *British Journal of Nutrition, 115*(12), 2162–2169. https://doi.org/10.1017/s0007114516001331

Burton-Freeman, B.M., Sandhu, A.K., & Edirisinghe, I. (2016). Red raspberries and their bioactive polyphenols: Cardiometabolic and neuronal health links. *Advances in Nutrition, 7*(1), 44–65. https://doi.org/10.3945/an.115.009639

Carvalho, C., Cardoso, S., Correia, S.C., Santos, R.X., Santos, M.S., Baldeiras, I., Oliveira, C.R., & Moreira, P.I. (2012). Metabolic alterations induced by sucrose intake and Alzheimer's disease promote similar brain mitochondrial abnormalities. *Diabetes, 61*(5), 1234–1242. https://doi.org/10.2337/db11-1186

Cichero, J.A., Lam, P., Steele, C.M., Hanson, B., Chen, J., Dantas, R.O., Duivestein J., Kayashita, J., Lecko, C., Murray, J., Pillay, M., Riquelme, L., & Stanschus, S. (2017). Development of international terminology and definitions for texture-modified foods and thickened fluids used in dysphagia management: The IDDSI framework. *Dysphagia, 32*(2), 293–314. https://doi.org/10.1007/s00455-016-9758-y

Dankar, I., Haddarah, A., Omar, F.E., Sepulcre, F., & Pujolà, M. (2018). 3D printing technology: The new era for food customization and elaboration. *Trends in Food Science & Technology, 75*, 231–242, https://doi.org/10.1016/j.tifs.2018.03.018

Delle Fave, A., Brdar, I., Freire, T., Vella-Brodrick, D., & Wissing, M. (2011). The eudaimonic and hedonic components of happiness: Qualitative and quantitative findings. *Social Indicator Research, 100*, 185–207. https://doi.org/10.1007/s11205-010-9632-5

Dematte, M.L., Pojer, N., Endrizzi, I., Corollaro, M.L., Betta, E., Aprea, E., Charles, M., Biasioli, F., Zampini, M., & Gasperi, F. (2014). Effects of the sound of the bite on apple perceived crispness and hardness. *Food Quality and Preference, 38*, 58e64. http://dx.doi.org/10.1016/j.foodqual.2014.05.009.

Dysken, M.W., Sano, M., Asthana, S., Vertrees, J.E., Pallaki, M., Llorente, M., Love, S., Schellenberg, G.D., McCarte, J.R., Malphurs, J., Prieto, S., Chen P., Loreck, D.J., Trapp G., Bakshi, R.S., Mintzer, J.E., Heidebrink, J.L., Vidal-Cardona, A., Arroyo, L.M., Cruz, A.R., Zachariah, S., Kowall, N.W., Chopra, M.P., Craft, S., Thielke, S., Turvey, C.L., Woodman, C., Monnell, K.A., Gordon, K., Tomaska, J., Segal, Y., Peduzzi, P.N., & Guarino, P.D. (2014). Effect of vitamin E and memantine on functional decline in Alzheimer disease: The TEAM-AD VA cooperative randomized trial. *JAMA, 311*(1), 33–44. https://doi.org/10.1001/jama.2013.282834

Endo, H., Ino, S., & Fujisaki, W. (2016). The effect of a crunchy pseudo-chewing sound on perceived texture of softened foods. *Physiology and Behavior, 167*, 324e331. http://dx.doi.org/10.1016/j.physbeh.2016.10.001.

Endo, H., Ino, S., & Fujisaki, W. (2017). Texture-dependent effects of pseudo-chewing sound on perceived food texture and evoked feelings in response to nursing care foods. *Appetite, 116*, 493–501. https://doi.org/10.1016/j.appet.2017.05.051

Farjami, T., & Madadlou, A. (2019). An overview on preparation of emulsion-filled gels and emulsion particulate gels. *Trends in Food Science & Technology, 86*, 85–94. https://doi.org/10.1016/j.tifs.2019.02.043

Farouk, M.M., Yoo, M.J., Hamid, N.S., Staincliffe, M., Davies, B., & Knowles, S.O. (2018). Novel meat-enriched foods for older consumers. *Food Research International, 104*, 134–142. https://doi.org/10.1016/j.foodres.2017.10.033

Feng, Y., & Wang, X. (2012). Antioxidant therapies for Alzheimer's disease. *Oxidative Medicine and Cellular Longevity, 2012*, 472932. https://doi.org/10.1155%2F2012%2F472932

Fernandez-Real, J.M., Serino, M., Blasco, G., Puig, J., Daunis-i-Estadella, J., Ricart, W., Burcelin, R., Fernández-Arana, F., & Portero-Otin, M. (2015). Gut microbiota interacts with brain microstructure and function. *The Journal of Clinical Endocrinology & Metabolism, 100*(12), 4505–4513. https://doi.org/10.1210/jc.20153076

Fillion, L., & Kilcast, D. (2002). Consumer perception of crispness and crunchiness in fruits and vegetables. *Food Quality and Preference, 13*(1), 23–29. https://doi.org/10.1016/S0950-3293(01)00053-2

Foote, J. A., Murphy, S. P., Wilkens, L. R., Basiotis, P. P., & Carlson, A. (2004). Dietary variety increases the probability of nutrient adequacy among adults. *The Journal of nutrition, 134*(7), 1779-1785.

Fraga, C.G., Croft, K.D., Kennedy, D.O., & Tomás-Barberán, F.A. (2019). The effects of polyphenols and other bioactives on human health. *Food & Function, 10*(2), 514–528. https://doi.org/10.1039/c8fo01997e

Fujita, Y., & Maki, K. (2018). Association of feeding behavior with jawbone metabolism and tongue pressure. *Japanese Dental Science Review, 54*(4), 174–182. https://doi.org/10.1016/j.jdsr.2018.05.001

Gallego, M., Barat, J.M., Grau, R., & Talens, P. (2022). Compositional, structural design and nutritional aspects of texture-modified foods for the elderly. *Trends in Food Science & Technology, 119*, 152–163. https://doi.org/10.1016/j.tifs.2021.12.008

Gallegos, C., Brito-De La Fuente, E., Clavé, P., Costa, A., & Assegehegn, G. (2017). Nutritional aspects of dysphagia management. *Advances in Food and Nutrition Research, 81*, 271–318. https://doi.org/10.1016/bs.afnr.2016.11.008

Gao, H., Ma, L., Cheng, C., Liu, J., Liang, R., Zou, L., Wei, L., & McClements, D.J. (2021). Review of recent advances in the preparation, properties, and applications of high internal phase emulsions. *Trends in Food Science & Technology, 112*, 36–49. https://doi.org/ 10.1016/j.tifs.2021.03.041

Gardener, H., & Caunca, M.R. (2018). Mediterranean diet in preventing neurodegenerative diseases. *Current Nutrition Reports, 7*, 10–20. https://doi.org/10.1007%2Fs13668-018-0222-5

Gentile, L. (2020). Protein-polysaccharide interactions and aggregates in food formulations. *Current Opinion in Colloid & Interface Science, 48*, 18–27. https://doi.org/10.1016/j.cocis.2020.03.002

Gildawie, K.R., Galli, R.L., Shukitt-Hale, B., & Carey, A.N. (2018). Protective effects of foods containing flavonoids on age-related cognitive decline. *Current Nutrition Reports, 7*(2), 39–48. https://doi.org/10.1007/s13668-018-0227-0

Godos, J., Marventano, S., Mistretta, A., Galvano, F., & Grosso, G. (2017). Dietary sources of polyphenols in the mediterranean healthy eating, aging and lifestyle (MEAL) study cohort. *International Journal Food Science and Nutrition, 68*(6), 750–756. https://doi.org/10.1080/ 09637486.2017.1285870.

Gomez, G., Botero-Rodríguez, F., Misas, J.D., Garcia-Cifuentes, E., Sulo, S., Brunton, C., Venegas-Sanabria, L.C., Gracia, D., Cano-Gutiérrez, C.A., & Gutierrez, C.A.C. (2022). A nutritionally focused program for community-living older adults resulted in improved health and well-being. *Clinical Nutrition. 41*(7), 1549–1556. https://doi.org/10.1016/j.clnu.2022.05.003

Gonçalves, S., Mansinhos, I., & Romano, A. (2020). Neuroprotective compounds from plant sources and their modes of action: An update. In: Swamy, M. (Ed), *Plant-DERIVEd Bioactives*. Springer, Singapore. https://doi.org/10.1007/978-981-15-2361-8_19. http://dx.doi.org/10.1007/978-981-15-2361-8_19

Griffiths, K., Aggarwal, B.B., Singh, R.B., Buttar, H.S., Wilson, D., & De Meester, F. (2016). Food antioxidants and their anti-inflammatory properties: A potential role in cardiovascular diseases and cancer prevention. *Diseases, 4*(3), 28. https://doi.org/10.3390/diseases4030028

Hadde, E.K., & Chen, J. (2020). Texture and texture assessment of thickened fluids and texture-modified food for dysphagia management. *Journal of Texture Studies. 52*(1), 4–15. https://doi.org/10.1111/jtxs.12567

Hagan, K.A., Munger, K.L., Ascherio, A., & Grodstein, F. (2016). Epidemiology of major neurodegenerative diseases in women: Contribution of the Nurses' Health study. *American Journal Public Health, 106*(9), 1650–1655. https://doi.org/10.2105/AJPH.2016. 303324. PMID:27459462.

Harnly, J.M., Doherty, R.F., Beecher, G.R., Holden, J.M., Haytowitz, D.B., Bhagwat, S., & Gebhardt, S. (2006). Flavonoid content of US fruits, vegetables, and nuts. *Journal of Agricultural and Food Chemistry, 54*(26), 9966–9977. https://doi.org/10.1021/jf061478a

Henning, S.M., Zhang, Y., Seeram, N.P., Lee, R.P., Wang, P., Bowerman, S., & Heber, D. (2011). Antioxidant capacity and phytochemical content of herbs and spices in dry, fresh, and blended herb paste form. *International Journal of Food Sciences and Nutrition, 62*(3), 219–225. https://doi.org/10.3109/09637486.2010.530595

Hill, E., Goodwill, A.M., Gorelik, A., & Szoeke, C. (2019). Diet and biomarkers of Alzheimer's disease: A systematic review and meta-analysis. *Neurobiology of Aging, 76*, 45–52. https://doi.org/10.1016/j.neurobiolaging.2018.12.008

Jurikova, T., Mlcek, J., Skrovankova, S., Sumczynski, D., Sochor, J., Hlavacova, I., & Orsavova, J. (2017). Fruits of black chokeberry Aronia melanocarpa in the prevention of chronic diseases. *Molecules, 22*(6), 944. https://doi.org/10.3390/molecules22060944

Kang, M.G., Choi, J.H., & Kho, H.S. (2020). Relationships between gustatory function tests. *Oral Diseases, 26*(4), 830–837. https://doi.org/10.1111/odi.13291

Karlsson, F.H., Tremaroli, V., Nookaew, I., Bergström, G., Behre, C.J., Fagerberg, B. Nielsen, J., & Bäckhed, F. (2013). Gut metagenome in European women with normal, impaired, and diabetic glucose control. *Nature, 498*(7452), 99–103. https://doi.org/10.1038/nature12198

Kaur, D., Rasane, P., Singh, J., Kaur, S., Kumar, V., Mahato, D.K., Dey, A., Dhawan, K., & Kumar, S. (2019). Nutritional interventions for elderly and considerations for the development of geriatric foods. *Current Aging Science, 12*(1), 15–27. https://doi.org/10.2174/1874609812666190521110548

Kean, R.J., Lamport, D.J., Dodd, G.F., Freeman, J.E., Williams, C.M., Ellis, J.A., Butler, L., & Spencer, J.P. (2015). Chronic consumption of flavanone-rich orange juice is associated with cognitive benefits: An 8-wk, randomized, double-blind, placebo-controlled trial in healthy older adults. *The American Journal of Clinical Nutrition, 101*(3), 506–514. https://doi.org/10.3945/ajcn.114.088518

Keller, H., Beck, A.M., & Namasivayam, A. (2015). Improving food and fluid intake for older adults living in long-term care: A research agenda. *Journal of the American Medical Directors Association, 16*(2), 93–100. https://doi.org/10.1016/j.jamda.2014.10.017

Kent, K., Charlton, K.E., Netzel, M., & Fanning, K. (2017). Food-based anthocyanin intake and cognitive outcomes in human intervention trials: A systematic review. *Journal of Human Nutrition and Dietetics, 30*(3), 260–274. https://doi.org/10.1111/jhn.12431

Kesika, P., Suganthy, N., Sivamaruthi, B.S., & Chaiyasut, C. (2021). Role of gut-brain axis, gut microbial composition, and probiotic intervention in Alzheimer's disease. *Life Sciences, 264*, 118627. https://doi.org/10.1016/j.lfs.2020.118627

Kim, D.Y., Park, H.S., Park, S.W., & Kim, J.H. (2020). The impact of dysphagia on quality of life in stroke patients. *Medicine, 99*(34), e21795 https://doi.org/10.1097%2FMD.0000000000021795

Kim, H.W., Bae, H., & Park, H.J. (2017). Classification of the printability of selected food for 3D printing: Development of an assessment method using hydrocolloids as reference material. *Journal of Food Engineering, 215*, 23–32. https://doi.org/10.1016/j.jfoodeng.2017.07.017

Kuelzow, N., Witte, A.V., Kerti, L., Grittner, U., Schuchardt, J.P., Hahn, A., & Floeel, A. (2016). Impact of omega-3 fatty acid supplementation on memory functions in healthy older adults. *Journal of Alzheimer's Disease, 51*(3), 713–725. https://doi.org/10.3233/jad-150886

Lee, D., Clark, E.D., Antonsdottir, I.M., & Porsteinsson, A.P. (2023). A 2023 update on the advancements in the treatment of agitation in Alzheimer's disease. *Expert Opinion on Pharmacotherapy, 24*(6), 691–703, https://doi.org/10.1080/14656566.2023.2195539

Ma, Q.P., Huang, C., Cui, Q.Y., Yang, D.J., Sun, K., Chen, X., & Li, X.H. (2016). Meta-analysis of the association between tea intake and the risk of cognitive disorders. *PLOS ONE, 11*(11), e0165861. https://doi.org/10.1371%2Fjournal.pone.0165861

Marangoni, A.G. (2012). Organogels: An alternative edible oil-structuring method. *Journal of the American Oil Chemists' Society, 89*(5), 749–780. https://doi.org/10.1007/s11746-012-2049-3

Marangoni, A.G., & Leendert H.W. (2019). *Structure and properties of fat crystal networks.* CRC Press. (2nd. Edition), ISBN 13:978-1-4398-8764-6 https://doi.org/10.1201/b12883

Marcus, J.B. (2019). *Aging, nutrition and taste: Nutrition, food science and culinary perspectives for aging tastefully.* Academic Press, p. 526 . https://doi.org/10.1016/B978-0-12-813527-3.00009-0

Matsuo, K., & Fujishima, I. (2020). Textural changes by mastication and proper food texture for patients with oropharyngeal dysphagia. *Nutrients, 12*(6), 1613. https://doi.org/10.3390/nu12061613

Mohajeri, M. H., Troesch, B., & Weber, P. (2015). Inadequate supply of vitamins and DHA in the elderly: implications for brain aging and Alzheimer-type dementia. *Nutrition, 31*(2), 261–275.

Mocchegiani, E., Costarelli, L., Giacconi, R., Malavolta, M., Basso, A., Piacenza, F., Ostan R., Cevenini, E, Gonos, E.S., & Monti, D. (2014). Micronutrient–gene interactions related to inflammatory/immune response and antioxidant activity in ageing and inflammation. A systematic review. *Mechanisms of Ageing and Development, 136*, 29–49. https://doi.org/10.1016/j.mad.2013.12.007

Mumba, M.N., Nacarrow, A.F., Cody, S., Key, B.A., Wang, H., Robb, M., Jurczyk, A., Ford, C.K., & Allen, R.S., (2021). Intensity and type of physical activity predicts depression in older adults. *Aging & Mental Health, 25*(4), 664–671. https://doi.org/10.1080/13607863.2020.1711861

Nicolai, T. (2019). Gelation of food protein-protein mixtures. *Advances in Colloid and Interface Science, 270*, 147–164. https://doi.org/10.1016/j.cis.2019.06.006

Nifli, A.P. (2018). Appetite, metabolism and hormonal regulation in normal ageing and dementia. *Diseases, 6*(3), 66. https://doi.org/10.3390/diseases6030066

Norman, K., Haß, U., & Pirlich, M. (2021). Malnutrition in older adults—recent advances and remaining challenges. *Nutrients, 13*(8), 2764. https://doi.org/10.3390%2Fnu13082764

O'Mahony, S.M., Clarke, G., Borre, Y.E., Dinan, T.G., & Cryan, J.F. (2015). Serotonin, tryptophan metabolism and the brain-gut-microbiome axis. *Behavioural Brain Research, 277*, 32–48. https://doi.org/10.1016/j.bbr.2014.07.027

Ogino, E., Manly, J.J., Schupf, N., Mayeux, R., & Gu, Y. (2019). Current and past leisure time physical activity in relation to risk of Alzheimer's disease in older adults. *Alzheimer's & Dementia, 15*(12), 1603–1611. https://doi.org/10.1016/j.jalz.2019.07.013

Ortega, O., Martín, A., & Clavé, P. (2017). Diagnosis and management of oropharyngeal dysphagia among older persons, state of the art. *Journal of the American Medical Directors Association, 18*(7), 576–582. https://doi.org/10.1016/j.jamda.2017.02.015

Papazian, E.J., & Pinto, J.M. (2021). Olfactory loss and aging: Connections with health and well-being. *Chemical Senses, 46*, bjab045 https://doi.org/10.1093/chemse/bjab045

Potì, F., Santi, D., Spaggiari, G., Zimetti, F., & Zanotti, I. (2019). Polyphenol health effects on cardiovascular and neurodegenerative disorders: A review and meta-analysis. *International Journal of Molecular Sciences, 20*(2), 351. https://doi.org/10.3390%2Fijms20020351

Rajabi, H., Sabouri, M., & Hatami, E. (2021). Associations between physical activity levels with nutritional status, physical fitness, and biochemical indicators in older adults. *Clinical Nutrition ESPEN 45*, 389–398. https://doi.org/10.1016/j.clnesp.2021.07.014

Reber, E., Gomes, F., Vasiloglou, M.F., Schuetz, P., & Stanga, Z. (2019). Nutritional risk screening and assessment. *Journal of Clinical Medicine, 8*(7), 1065. https://doi.org/10.3390%2Fjcm8071065

Ritzoulis, C., & Karayannakidis, P.D. (2015). Proteins as texture modifiers. In: *Modifying food texture* (pp. 51–69). Woodhead Publishing. ISBN 9781782423331.

Rodd, B.G., Tas, A.A., & Taylor, K.D.A. (2021). Dysphagia, texture modification, the elderly and micronutrient deficiency: A review. *Critical Reviews in Food Science and Nutrition, 62*(26), 7354–7369. https://doi.org/10.1080/10408398.2021.1913571

Saitoh, E., Pongpipatpaiboon, K., Inamoto, Y., & Kagaya, H. (2018). *Dysphagia evaluation and treatment.* Springer. https://doi.org/10.1159/000266348

Savazzi, F., Isernia, S., Farina, E., Fioravanti, R., D'Amico, A., Saibene, F.L., & Baglio, F. (2020). "Art, Colors, and Emotions" Treatment (ACE-t): A Pilot study on the efficacy of an art-based intervention for people with Alzheimer's disease. *Frontiers in Psychology, 11*, 1467. https://doi.org/10.3389/fpsyg.2020.01467

Severini, C., Derossi, A., & Azzollini, D. (2016). Variables affecting the printability of foods: Preliminary tests on cereal-based products. *Innovative Food Science & Emerging Technologies*, 38, 281–291. https://doi.org/10.1016/j.ifset.2016.10.001

Shiizaki, K., Tsubouchi, A., Miura, Y., Seo, K., Kuchimaru, T., Hayashi, H., Iwazu, Y., ...& Kuro-O, M. (2021). Calcium phosphate microcrystals in the renal tubular fluid accelerate chronic kidney disease progression. *Journal Clinical Investigation, 131*(16), 1–16. https://doi.org/10.1172/jci145693

Sivamaruthi, B.S., Kesika, P., & Chaiyasut, C. (2020). The role of probiotics in colorectal cancer management. *Evidence-Based Complementary and Alternative Medicine*, 2020, 1–17. https://doi.org/10.1155/2020/3535982

Sivamaruthi, B.S., Kesika, P., Suganthy, N., & Chaiyasut, C. (2019). A review on role of microbiome in obesity and antiobesity properties of probiotic supplements. *BioMed Research International*, 2019, 1-21. https://doi.org/10.1155/2019/3291367

Spagnuolo, C., Moccia, S., & Russo, G.L. (2018). Anti-inflammatory effects of flavonoids in neurodegenerative disorders. *European Journal of Medicinal Chemistry, 153*, 105–115. https://doi.org/10.1016/j.ejmech.2017.09.001

Spence, C. (2015). Eating with our ears: Assessing the importance of the sounds of consumption on our perception and enjoyment of multisensory flavour experiences. *Flavour, 4*(3), 2–14. http://dx.doi.org/10.1186/2044-7248-4-3.

Stephen, R., Hongisto, K., Solomon, A., & Lönnroos, E. (2017). Physical activity and Alzheimer's disease: A systematic review. *The Journals of Gerontology: Series A, 72*(6), 733–739. https://doi.org/10.1093/gerona/glw251

Sun J., Peng, Z., Zhou, W., Fuh, J.Y.H., Hong, G.S., & Chiu, A. (2015). A review on 3D printing for customized food fabrication. *Procedia Manufacturing, 1*, 308–319. https://doi.org/10.1016/j.promfg.2015.09.057

Śliwińska S., & Jeziorek M. (2021). The role of nutrition in Alzheimer's disease. *National Institute of Public Health Roczniki Państwowego Zakładu Higieny, 72*(1), 29-39 https://doi.org/10.32394/rpzh.2021.0154

Szczesniak, A.S., & Kahn, E.L. (1971). Consumer awareness and attitudes to food texture. I: Adults. *Journal of Texture Studies, 2*, 280–295. https://doi.org/10.1111/j.1745-4603.1971.tb01005.x

Szczesniak, A.S., & Kahn, E.L. (1984). Texture contrasts and combinations: A valued consumer attribute. *Journal of Texture Studies, 15*, 285–301. https://doi.org/10.1111/j.1745-4603.1984.tb00385.x

Takagi, D., Hirano, H., Watanabe, Y., Edahiro, A., Ohara, Y., Yoshida, H., Kim, H., Murakami, K., & Hironaka, S. (2017). Relationship between skeletal muscle mass and swallowing function in patients with Alzheimer's disease. *Geriatrics & Gerontology International, 17*(3), 402–409. https://doi.org/10.1111/ggi.12728

Tarrega, A., Marcano, J., & Fiszman, S. (2017). Consumer perceptions of indulgence: A case study with cookies. *Food Quality and Preference, 62*, 80–89. https://doi.org/10.1016/j.foodqual.2017.07.001

Trichopoulou, A., Kouris-Blazos, A., Wahlqvist, M.L., Gnardellis, C., Lagiou, P., Polychronopoulos, E., Vassilacou, T., Lipworth L., & Trichopoulos, D. (1995). Diet and overall survival in elderly people. *BMJ, 311*(7018), 1457–1460. https://doi.org/10.1136/bmj.311.7018.1457

Tobin, A.B., Mihnea, M., Hildenbrand, M., Miljkovic, A., Garrido-Bañuelos, G., Xanthakis, E., & Lopez-Sanchez, P. (2020). Bolus rheology and ease of swallowing of particulated semi-solid foods as evaluated by an elderly panel. *Food & Function, 11*(10), 8648–8658. https://doi.org/10.1039/d0fo01728k

Torres-Ortiz K.S., Sotelo-Díaz, L.I., & Caez-Ramírez, G. (2022). Mechanical and rheological categorization of food patterns suitable for older adults with swallowing limitation. *International Journal of Food Properties, 25*(1), 2627–2660. https://doi.org/10.1080/10942912.2022.2140811

Tunick, M.H., Onwulata, C.I., Thomas, A.E., Phillips, J.G., Mukhopadhyay, S., Sheen, S., Liu Ch. K., Latona, N., Pimentel, M.R., & Cooke, P.H. (2013). Critical evaluation of crispy and crunchy textures: A review. *International Journal of Food Properties, 16*(5), 949–963. https://doi.org/10.1080/10942912.2011.573116

Vandenberghe-Descamps, M., Sulmont-Rossé, C., Septier, C., Feron, G., & Labouré, H. (2017). Using food comfortability to compare food's sensory characteristics expectations of elderly people with or without oral health problems. *Journal of Texture Studies, 48*(4), 280–287. https://doi.org/10.1111/jtxs.12250

Vieira, J.M., Oliveira Jr, F.D., Salvaro, D.B., Maffezzolli, G.P., de Mello, J.D.B., Vicente, A.A., & Cunha, R.L. (2020). Rheology and soft tribology of thickened dispersions aiming the development of oropharyngeal dysphagia-oriented products. *Current Research in Food Science, 3*, 19–29. https://doi.org/10.1016/j.crfs.2020.02.001

Wall, R., Cryan, J.F., Ross, R.P., Fitzgerald, G.F., Dinan, T.G., & Stanton, C. (2014). Bacterial neuroactive compounds produced by psychobiotics. *Advances in Experimental Medicine and Biology, 817*, 221–239. https://doi.org/10.1007/978-1-4939-0897-4_10.

Wang, X., Njehia, N.S., Katsuno, N., & Nishizu, T. (2020). An acoustic study on the texture of cellular brittle foods. *Reviews in Agricultural Science, 8*, 170–185. https://doi.org/10.7831/ras.8.0_170

Watkins, R., Goodwin, V.A., Abbott, R.A., Backhouse, A., Moore, D., & Tarrant, M. (2017). Attitudes, perceptions, and experiences of mealtimes among residents and staff in care homes for older adults: A systematic review of the qualitative literature. *Geriatric Nursing, 38*(4), 325–333. https://doi.org/10.1016/j.gerinurse.2016.12.002

Wijaya, W., Patel, A.R., Setiowati, A.D., & Van der Meeren, P. (2017). Functional colloids from proteins and polysaccharides for food applications. *Trends in Food Science & Technology, 68*, 56–69. https://doi.org/10.1016/j.tifs.2017.08.003

Wilson, D.W., Nash, P., Buttar, H.S., Griffiths, K., Singh, R., De Meester, F., & Takahashi, T. (2017). The role of food antioxidants, benefits of functional foods, and influence of feeding habits on the health of the older person: An overview. *Antioxidants, 6*(4), 81. https://doi.org/10.3390/antiox6040081

World Health Organization. (2023). National programmes for age-friendly cities and communities: A guide. https://iris.who.int/bitstream/handle/10665/366634/9789240068698-eng.pdf?sequence=1

World Health Organization. (2021). World Health Organization Ageing and health. Common health conditions associated with ageing (2021) www.who.int/news-room/fact-sheets/detail/ageing-and-health

Wu, L., Sun, D., & He, Y. (2017). Coffee intake and the incident risk of cognitive disorders: A dose–response meta-analysis of nine prospective cohort studies. *Clinical Nutrition, 36*(3), 730–736. https://doi.org/10.1016/j.clnu.2016.05.015

Xu, F., Laguna, L., & Sarkar, A. (2019). Aging-related changes in quantity and quality of saliva: Where do we stand in our understanding?. *Journal of Texture Studies, 50*(1), 27–35. https://doi.org/10.1111/jtxs.12356

Yan, X., Ma, C., Cui, F., McClements, D.J., Liu, X., & Liu, F. (2020). Protein-stabilized Pickering emulsions: Formation, stability, properties, and applications in foods. *Trends in Food Science & Technology, 103*, 293–303. https://doi.org/10.1016/j.tifs.2020.07.005

Yang, F., Zhang, M., & Bhandari, B. (2017). Recent development in 3D food printing. *Critical Reviews in Food Science and Nutrition, 57*(14), 3145–3153. https://doi.org/10.1080/10408398.2015.1094732

Young, V.B. (2017). The role of the microbiome in human health and disease: An introduction for clinicians. *The BMJ*, 356. https://doi.org/10.1136/bmj.j831

Yu, W., Yu, W., Liu, X., Wan, T., Chen, C., Xiong, L., Zhang, W., & Lü, Y. (2021). Associations between malnutrition and cognitive impairment in an elderly Chinese population: An analysis based on a 7-year database. *Psychogeriatrics, 21*(1), 80–88. https://doi.org/10.1111/psyg.12631

Zamora-Ros, R., Knaze, V., Rothwell, J.A., Hémon, B., Moskal, A., Overvad, K., Tjønneland, A., Kyrø, C., Fagherazzi, G., Boutron-Ruault, M.C., Touillaud, M., Katzke, V., Kühn, T., Boeing, H., Förster, J., Trichopoulou, A., Valanou, E., Peppa, E., Palli, D., et al. (2016). Dietary polyphenol intake in Europe: The European Prospective Investigation into cancer and nutrition (EPIC) study. *European Journal of Nutrition, 55*(4), 1359–1375. https://doi.org/10.1007/s00394-015-0950-x

Zampini, M., & Spence, C. (2004). The role of auditory cues in modulating potato chips' perceived crispness and staleness. *Journal of Sensory Studies, 19*, 347e363. http://dx.doi.org/10.1111/j.1745-459x.2004.080403.x.

Zeli, C., Lombardo, M., Storz, M.A., Ottaviani, M., & Rizzo, G. (2022). Chocolate and cocoa-derived biomolecules for brain cognition during aging. *Antioxidants, 11*(7), 1353. https://doi.org/10.3390/antiox11071353

Zúñiga, R.N., Kulozik, U., & Aguilera, J.M. (2011). Ultrasonic generation of aerated gelatin gels stabilized by whey protein β-lactoglobulin. *Food Hydrocolloids, 25*(5), 958–967. https://doi.org/10.1016/j.foodhyd.2010.09.010

Chapter 22

A Comparison of Traditional Oil and Hot-Air Frying with Respect to Composition, Morphological and Nutritional Characteristics of Fried Products

Ismael Ramírez-Flores
Ilse Monroy-Rodríguez
and Gustavo F. Gutiérrez-López

CONTENTS

22.1 INTRODUCTION

Frying of foods has become one of the most practiced cooking methods as well as one of the most profitable techniques for food preservation (Vandna et al. 2023). Traditionally, various types of vegetables are used as side dishes and prepared by conventional frying on a domestic

 DOI: 10.1201/9781003332558-25

and industrial scale. In this respect, frying is a dehydration process involving rapid heat and mass transfer when the food is immersed in hot oil (150–190°C), giving place to a crispy texture, pleasant odors, colors, and flavors (Xinli et al. 2023). However, these types of products have been related to public health problems such as obesity, hypertension, diabetes, as well as different types of cancer, due to their frequent consumption and high content of saturated lipids (Arash et al. 2018). Consequently, demand for hot-air fried foods has gradually increased due to consumer demand for healthier products, and in this respect, the challenge is to impact the sensory properties of the product as little as possible, since reduction of oil lowers palatability and flavor of fried foods. Hot-air frying was first developed aiming at reducing oil consumption but preserving, to an acceptable extent, the same sensorial characteristics of oil-fried products (Godswill et al. 2018).

22.2 HOT-AIR FRYING

Hot-air frying is, chiefly, a dehydration process involving a rapid exchange of heat and mass in the hot-air-food system producing several physical and chemical changes in the fried product (Arash et al. 2020), and consequently aiding in the preservation of materials by decreasing their moisture and water activity (a_w) (Xinli et al. 2023). During this process, the products can be covered by a small layer of oil before being introduced to the fryer aiming to develop the typical crust of fried foods which gradually forms during processing.

An essential characteristic of this process is that the oil content of foods is significantly lower (up to 90% less) than in conventionally oil-fried foods, while the sensory attributes are essentially similar (Vandna et al. 2023). Hot-air frying is projected as an alternative to the conventional oil immersion process to produce healthier foods that are used in the form of snacks, garnish for different dishes, or as a complement to salads (Yi et al. 2021).

The most relevant quality attributes of hot-air fried products are color, texture, flavor, nutrient retention, porosity, density, and specific volume and their overall quality substantially depends on the hot-air frying operating conditions (Vandna et al. 2023).

Forced hot-air frying is a relatively inexpensive method in which air is moved parallel or forced through a bed of food material, and there is a wide range of equipment, available worldwide for this process (Godswill et al. 2018). Under this condition, the fried food is obtained by direct contact of the air with the product and an external emulsion of oil droplets inside a processing chamber (Gao et al. 2022). The final product resembles the dehydration and crust formation typical of conventionally oil-fried goods. It has been shown (Mingchih et al. 2021) that the lipid content of products prepared by this method is significantly lower, acrylamide formation is reduced, and sensory attributes are similar to those processed by oil frying. Chong et al. (2019) evaluated the sensory profile of chips fried by hot-air frying and hot-oil immersion frying and found that, in general, the hot-air fried products showed the best profiles and similar results were obtained in French fries made with both methods (Teruel et al. 2018), which can be attributed to the oil content of the chip that increased the perception of the salty taste (Belkova et al. 2018). Chong et al. (2019) compared the effects of oil immersion frying and forced-air cooking methods, on the physicochemical and sensory properties of taro chips (*Colocasia esculenta*) and concluded that the hot-air frying method can be a healthier alternative that significantly reduces the oil content and presents acceptable sensory characteristics of fried products.

TABLE 22.1 COMPARISON BETWEEN CONVENTIONAL FRYING AND THE HOT-AIR FRYING PROCESS

Conventional Frying (Hot-Oil Immersion)	Hot-Air Frying
Temperatures from 140°C to 180°C depending on the nature and thickness of the food for 5 to 20 min.	Food is fried with a thin layer of oil at temperatures from 160°C to 180°C with times ranging from 8 to 12 min.
Conventionally fried products are mainly associated with the high oil content used.	The products are dehydrated, and the typical crust of fried foods gradually forms.
There are undesirable changes in the cooking media due to oil-water interactions and oxidation.	The oil content in these types of foods that are exposed to hot-air frying was significantly lower than in conventionally fried foods.
Consumers are prone to developing cardiovascular diseases, hypertension, obesity, and diabetes.	Are a better alternative to the conventional one to produce different healthy foods.

Source: Xinli et al. (2023)

León & Espejel, (2019) summarized the main advantages of hot-air frying in terms of a longer conservation, enhancement of sensory characteristics, reduced storage, and transportation costs.

On the other hand, according to Xinli, et al. (2023), the main differences between conventional oil and hot-air frying process are as presented in **Table 22.1.**

Hot-air frying not only brings about nutritional health benefits. This process also has environmental advantages, such as reducing fuel consumption and emissions. However, due to the limitations of the physical form of the oil used to promote heat transfer, some physical properties of the final products may be adversely affected, including texture, color, flavor, and moisture content (Yu et al. 2020) in a way that also influences the morphometry of final products.

22.3 THE OPERATING PRINCIPLES OF HOT-AIR FRYING PROCESS

The main parts of a small-scale air fryer are schematically presented in **Figure 22.1**. In fact, researchers have shown that oil absorption can be reduced by up to 90% with this method in which the product is moved constantly so that it comes in close contact with the hot air. Additionally, the food heats up, the water evaporates, and a crust gradually forms on its surface (Abd Rahman et al. 2018) and the high speed of the blowing air makes the entire process considerably shorter than conventional oil frying (Yi et al. 2021).

Hot-air frying involves different heat and mass transfer phenomena, which causes microscopic and macroscopic changes in the food's structure, which gives the product its characteristic textural features (Nguyen et al. 2023). A review of the engineering characteristics of this operation is presented next.

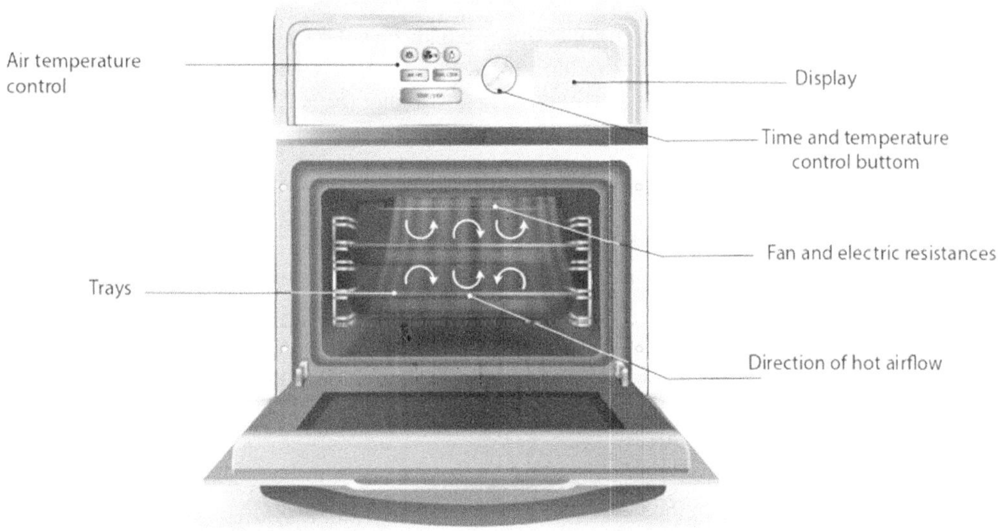

Figure 22.1 Illustration of a hot-air frying equipment, showing its main components.

22.3.1 Mass and Energy Transfer Phenomena during Hot-Air Frying in Foods

The rate of heat transfer to the surface of the food depends on the temperature difference between the food and the air as well as on the thermal conductivity of the surface of the solid and the rate at which heat penetrates into the material depends on its thermal conductivity (Yi et al. 2021) and thus on the structural characteristics of the food.

The limiting rates of heat and mass transfer phenomena makes frying one of the most difficult unit operations to understand, in which the convective heat transfer coefficient plays a predominant role (Arash et al. 2020). The most widely used models are presented next.

22.3.2 Empirical Models Used in Hot-Air Frying

These models will be useful for design purposes, especially for interpolation of operating conditions and, in this respect, Xinli et al. (2023) proposed an exponential model for the description of dehydration during the frying process:

$$W = ae^{-bt} \tag{22.1}$$

Where:
W = moisture content (kg/kg dried basis).
a = Drying constant.
b = Constant that is related to the thickness of the sample.
t = Frying time (s).

The same authors proposed a compartmental dynamic model, considering the mass transfer at the edges and in the center of the product. This model is based on a first-order general differential equation and considers that, when the food comes into contact with hot air, the effect does not start instantly due to the external resistance to heat transfer (Eq. 22.2):

$$\tau\frac{\partial y}{\partial x}+X=X_0e^{-\beta t} \tag{22.2}$$

Reordering and introducing a ratio constant K, the equation is:

$$\frac{\partial y}{\partial x}=-K\left(X-X_0e^{-\beta t}\right) \tag{22.3}$$

Equation 22.4 shows that the moisture gradient is time-dependent and that at time zero this gradient is zero. Upon integrating Eq. 22.3:

$$X=X_0e^{-Kt}+\frac{X_0K}{\beta-K}\left(e^{-Kt}-e^{-\beta t}\right) \tag{22.4}$$

These authors fitted the experimental data to Eq. 22.4 and found that the values of β were not statistically different for a given frying compartment, thus obtaining the following:

$$\frac{X}{X_0}=K\{e^{-Kct}+(1+K_ct)\}+(1-K)\left\{e^{-Ket}+(1+K_et)\right\} \tag{22.5}$$

Where:
X = Moisture of the solid (g/100g wet weight).
t = Time (s).
X_0 = Initial moisture (g H_2O).
K = Rate constant (min^{-1}).
c = Distance from the core to the crust (mm).
e = Crust thickness (mm).

Xinli et al. (2023) proposed using Fick's second law of diffusion for an infinite slab to describe the evolution of moisture in a solid and assuming that the material is homogeneous, and prevails a constant effective diffusion coefficient (D_{ef}):

$$\frac{\partial w(x,t)}{\partial t}=D_{ef}\{\frac{\partial w^2(x,t)}{\partial x^2}\} \tag{22.6}$$

Where:
W = moisture of the solid (dry basis)
t = Time (s)
X = Distance from surface (m)
D_{ef} = effective diffusivity (m^2s^{-1})

This type of mathematical modeling has helped to describe moisture loss during hot-air frying of different food matrices such as sweet potatoes, French fries, taro, and beet roots, among others; and authors mentioned that Fick's law derived models can help understanding the behavior of materials in relation to the microstructural and morphometric changes they have during the process (Xinli et al. 2023).

22.3.3 Effect of Air-Frying Processing Conditions on Food Morphology

In general, the shape of fried vegetables does not change much during processing, but their porosity increases in relation to the loss of water. The contact of the food with hot-air causes a rapid increase in the surface temperature of the food materials above the boiling temperature of water (T>100°C) (Yu et al. 2020). The water on the surface begins to boil immediately and, for the flow of steam to continue, enough water must be able to migrate from the center of the food to the crust and must remain permeable (Gouyo et al. 2021). At the end of the process, the fried products have three distinct regions: the peripheral crust, a highly porous zone surrounding the central region, and a core, the initial structure having an existing pre-crust that has formed as a result of pre-treatment in frozen fried foods; this peripheral pre-crust consists of a single layer of porous bubbles with a diameter of approximately 0.1 mm (Mohammad et al. 2022). **Figure 22.2** shows the most significant changes that occur in the microstructure of food during hot-air frying.

Deep frying, on the other hand, is a thermal process, in which heat and mass transfer occur simultaneously. During the frying process, heat is transferred by convection from the hot oil to the surface of the product, and subsequently heat transfer to the interior of the food occurs by conduction. The water that leaves the food generates a cooling in the evaporation front (Okonkwo et al. 2022). Mass transfer is characterized by the loss of water from the food as water vapor and the movement of oil into the food (Dehghannya & Ngadi, 2021).

The rate of heat transfer to the food is influenced by the thermal properties of the food, viscosity of the oil, and stirring conditions. Mass transfer is characterized by moisture loss and oil gain,

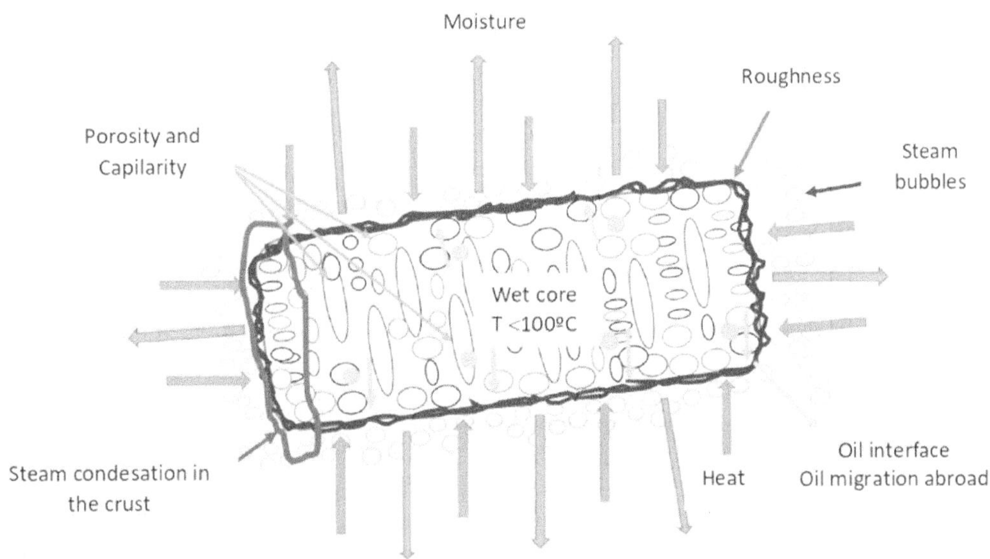

Figure 22.2 Representative illustration of changes at the microstructure level in plant tissues during the hot-air frying process.

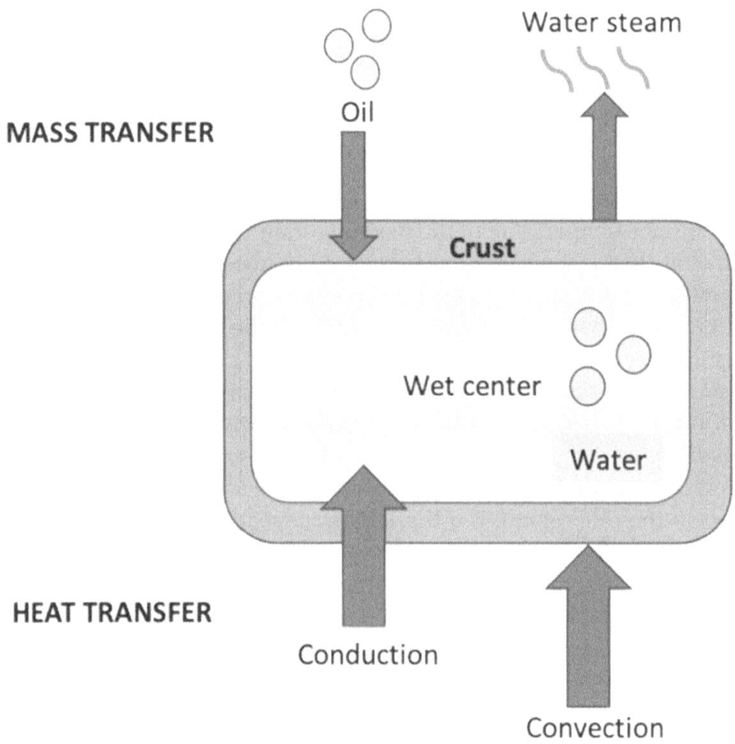

Figure 22.3 Heat and mass transfer during the oil-frying process.

phenomena that operate countercurrently (Okonkwo et al. 2022). A schematic representation of the main features of oil frying is shown in **Figure 22.3**.

Arash (2020) reported significant changes in the process conditions that affect the thickness of the crust, which is affected by the loss of water and the area of the frying region. For example, for potato chips with a water loss of about 60%, the thickness of the crust has average values between 0.5 and 1.5 mm.

In order to have a good crispness level in fried food, the frying process must generate heterogeneous pores with a high density of pores with diameters ranging from 0 to 0.15 mm and an average diameter of less than 0.2 mm in the first millimeter of the crust, so that it is not influenced by the different conditions of the process, especially the loss of moisture. Dehghannya & Ngadi (2021) studied the impact of hot-air frying of potato strips under similar conditions on oil absorption, moisture loss, color, texture, microstructural, calorimetric, and sensory properties. Hot-air frying had a longer processing time (21 min) compared to oil frying (9 min). The hot-air fried samples had considerably less oil absorption but similar moisture loss and color properties. In general, no differences were noted between the two frying techniques in the crust region with respect to the different frying times. In terms of structure, the deep-oil fried products had a thick, dry, crispy crust, which was attributed to high surface temperatures, leading to strong local evaporation of water that hinders starch gelatinization in the region (Verma et al. 2022).

However, in hot-air fried samples there is a slower evaporation of water, which results in the creation of a thinner and more homogeneous crust without irregularities. In addition, the hot-air fried potato strips, unlike the deep-oil fried products, greatly expanded and showed a regular distribution of pores in the central region. Furthermore, the hot-air fried samples, in contrast to the deep-oil fried products, demonstrated crust shrinkage during cooling, which, in fact, is a characteristic of the hot-air fried product (Mohammad et al. 2022). The slower rise in temperature during hot-air frying led to lower rates of moisture loss and color development reactions. In addition, the degree of starch gelatinization was lower during hot-air frying (Mohammad et al. 2022).

Previous studies have not reported differences between the two frying techniques in the crust region with respect to the different frying times. However, the core region in products such as hot-air potato fries showed higher hardness compared to those treated by oil-frying fries, and the variations in the texture of the central region were attributed to a lower degree of starch gelatinization in hot-air products, related to the appearance of lower temperatures within the samples (Verma et al. 2022).

In terms of structure, hot-air fried products resulted in a slower evaporation of water, which resulted in the creation of a thinner and more homogeneous crust without irregularities (Teruel et al 2018; Mohammad et al. 2022).

In another study, the impact of hot-air frying on the oil and moisture content, firmness, and color of sweet potato was investigated using a response surface methodology and the suggested frying conditions were a temperature of 150°C for 12 min. This process produced samples with significantly lower oil content (90% reduction); this was attributed to oil draining from the samples even though no oil was used during hot-air frying. In addition, the hot-air fried samples showed a lighter color and harder texture with better sensory attributes. It was noted that hot-air fried samples had a drier and tighter continuous network structure attributed to starch gelatinization on their surface (Abd Rahman et al. 2018).

Teruel et al. (2018) compared the effects of conventional and hot-air frying on potato chip quality and found that the oil content of air fried potatoes (0.4 to 1.1 g/100 g sample) was significantly lower than that of traditional French fries (5.6 to 13.8 g/100 g of sample). Similar results were obtained by Okonkwo et al. (2022), who treated potatoes with traditional oil and hot-air frying. Results showed that French fries produced by hot-air frying had an average oil content 70% lower than that of conventional French fries. Similarly, Arash et al. (2020) studied the changes in oil absorption during frying and confirmed that it occurred mainly at the end of the frying process, which was attributed to the condensation of water vapor within the food caused by the decrease in temperature and, as a result, some of the oil on the surfaces of the food product was drawn into the voids formed inside the product during frying. In this respect, water and steam were firstly discharged from the capillaries and then replaced by hot oil. According to Teruel et al. (2018), the differences in the final oil content of foods produced by hot-air and conventional frying were due to the frying medium that surrounds the products; hot oil in the case of deep frying and a mist of droplets of oil in the air in the case of hot-air frying. The frying temperature, the frying time, and the microstructure of the food crust also played an important role in the phenomena observed during processing (Gao et al. 2022).

The process conditions of the foods that are subjected to the hot-air frying process directly influence the attributes, morphology as well as the characterization of their microstructure. **Figure 22.4** presents the main micro- and macrostructural food characteristics.

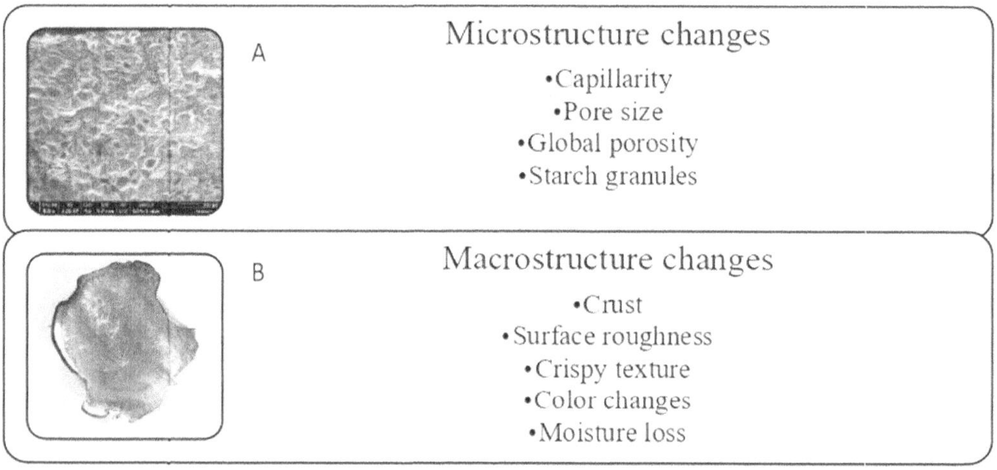

Figure 22.4 Main attributes involved in the morphology of hot-air fried foods. Images used as examples are: A) SEM image of a sweet potato slice hot-air fried. B) Image of a sweet potato slice hot-air fried.

22.3.4 Characterization of Food Microstructure during Hot-Air Frying

The literature review indicates that the majority of available research is related to the application of different frying methods for vegetables, evaluating different response variables and processing conditions; some of them are shown in **Table 22.2**.

Image texture analysis has shown to be an effective tool to quantify the microstructure of vegetables that have been subjected to hot-air frying processes. This has been reported in previous works (Abd Rahman et al. 2018; Tian et al. 2019; Gouyo et al. 2021). Also, Abd Rahman et al. (2018) used a scanning electron microscope to observe the surface cellular structure of the optimized air-fried product, finding that the air-fried sweet potato sample appeared to have a drier structure and a continuous closed network structure that could be attributed to the surface starch gelatinization. Tian et al. (2019) studied the microstructure and digestibility of potato strips produced by air-frying. In this respect, in a study by using scanning electron microscopy, the authors found differences in microstructural changes during conventional and hot-air frying; in both cases, microstructural characteristics could be correlated with the variation of digestibility of starch. Gouyo et al. (2021) reported the microstructure of French fries visualized by X-ray micro-computed tomography finding useful information on the pore size distribution, global porosity of the French fries, and local porosity of the crust region.

The authors studied microstructure analysis of hot-air frying to understand the texture of French fries. Optical scanning laser cutting is another useful technique that allows finding that the vapor inside the food samples condenses as the product cools. This technique is also useful for studying changes in the cell walls of fried foods, as the crust cells had a tendency to change their original shape as a result of starch swelling and water migration during frying, despite the high stresses to which cell walls are exposed during frying. It is worth mentioning that upon hot-air frying there were no signs of rupture of the cell wall and the changes in the shape of the cell wall were strongly related to the different frying times (Mohammad et al. 2022).

TABLE 22.2 RESEARCH STUDIES ON THE MICROSTRUCTURAL CHARACTERIZATION OF HOT-AIR FRIED FOODS

Sample	Frying Conditions	Findings	Reference
Potato fries	180°C to 200°C, the frying time was estimated between 5 and 10 min on average.	Pore size distribution, the global porosity of the potato chips, and the local porosity of the crust region. X-ray microcomputed tomography (XMT) (Bruker μCT, Kontich, Belgium).	Gouyo et al (2021)
Potato strips	4.5 g of oil per 100 g of potato strips, mixed well and fried at 185°C for 25 min.	The differences in the microstructural changes during conventional frying and hot-air frying are related to the variation of starch digestibility using a scanning electron microscope (TM-1000, HITACHI, Japan).	Tian et al (2019)
Sweet potato	Frying temperatures (140–160°C) and time (4–12 min).	They used a scanning electron microscope (Hitachi S-3400N, Japan), where they observed the cellular structure of the air-fried sweet potato surface. The sample had a drier structure and a closed network that could be attributed to surface gelatinization of starch.	Abd Rahman et al. (2018)

Scanning electron microscopy is a valuable tool for studying the topography of the surface of foods that are subjected to hot-air frying. Rurush et al. (2022) found that when these products were washed with petroleum ether, they lose a considerable amount of surface oil that allows the cells on the surface of fried foods to be clearly observed in the obtained images.

A study on the use of imaging magnetic resonance in oil frying aimed to determine the concentration gradients of oil and water and their interface within the fried samples. Also, Sharanagat et al. (2021) conducted a comparative study on the characteristics of French fries obtained by oil frying and hot air by using scanning electron microscopy fitted with a cryogenic stage. The results showed that the degree of starch gelatinization was also lower in the case of the hot-air fried product. According to the studies presented in **Table 22.4**, the different parameters extracted from the images generated by microscopy techniques in foods that were subjected to hot-air frying were useful to quantitatively describe the microstructure of the surface of the different food matrices.

A work by Yu et al. (2020) showed that image texture analysis in processes such as hot-air frying has become an efficient tool to quantitatively assess the microstructure of certain foods examined by different microscopy techniques such as scanning electron and confocal laser scanning microscopy.

The main barrier to obtain good images in frying processes is economic considerations (Belkova et al. 2018). An image processing system is not yet viable for many potential applications due to the unacceptable costs, since to meet the demand for cost-effective and multipurpose image processing systems, it is especially important to avoid bottlenecks in heavy-duty real-time applications that cannot easily handle large data streams (Mingchih et al. 2021). The development of a suitably efficient and accurate image processing algorithms can speed up the processing speed to meet modern manufacturing requirements. Also, the integration of image processing algorithms in specialized hardware can significantly reduce time for quality control of the products (Mohammad et al. 2022).

With the continuous decrease in costs associated and fast solutions in software and hardware, image processing techniques will certainly play an increasingly important role in the evaluation of food quality, including hot-air frying process (Teruel et al. 2018). Some applications of the use of digital image analysis in the evaluation of the micro- and macrostructure of fried foods are presented in the following.

22.3.5 Evaluation of the Morphometric Changes in Hot-Air Fried Foods with Micro and Macrostructure

The morphometric changes at the micro and macro levels of the product surface during hot-air frying are evaluated by obtaining the area, perimeter, entropy, contrast, and fractal dimension, among others (Gouyo et al. 2021; Arash et al. 2020; Montes de Oca, 2019). **Table 22.3** presents an overview of the morphometric evaluations of different products.

Image analysis is a useful tool to determine changes in the micro- and macrostructure of foods with quantitative parameters that allow characterizing and relating structural changes with the functionality and properties (Yi et al., 2021).

22.4 EFFECT OF HOT-AIR FRYING ON NUTRITIONAL AND SENSORY PROPERTIES

There are relatively few studies on the sensory attributes of hot-air frying foods, but one of the main drawbacks of these products is the depletion of sensory attributes, such as the sensation in

TABLE 22.3 BACKGROUND OF MORPHOMETRIC PARAMETERS ON VARIOUS FOOD MATRICES

Sample	Image Type	Software Used	Morphometric Parameters	Findings	Reference
French fries	Global thresholds were selected by the Otsu algorithm to segment the image to binary. Noise objects were removed from the binarized image by morphological despeckle operations.	Region of interest selection, segmentation to binary and morphometric analysis were all performed using SkyScan CT-Analyser ("CTAn") software (Version: 1.17.7.2, Bruker μCT, Kontich, Belgium).	The 3D morphometric parameters were calculated on the total volume of fries and on the volume of the crust. Porosity was determined as the percentage of voxels segmented as pores over the total number of voxels.	Principal component analysis (PCA) was performed between sensory, instrumental and morphometric parameters of the first millimeter of potato chips, correlating the microstructure of the crust developed during frying with the crisp texture of the crust.	Gouyo et al. (2021)

TABLE 22.3 (CONTINUED)

Sample	Image Type	Software Used	Morphometric Parameters	Findings	Reference
French fries	Image observations were recorded at different depths and treated to make them volumetric as well as 3D reconstructions at different product scales.	Microscopic observations of fried specimens were done with an inverted confocal laser scanning microscope (CLSM) TCS SP2 AOBS (Leica, Germany) equipped with UV–VIS sources and a motorized focal plane. Observations were made along the vertical plane to get images in (x; y) through the thickness of the sample. Resolution is expected to be higher than 0.3 lm in function of the pinhole opening. Observations in (x; z) along the horizontal plane were done for connectivity analysis. Resolution is expected to be lower in this case (0.6 lm) because of a higher light diffraction.	Some images show only cellular fluorescence except at the extreme surface due to cooling conditions. These cell images represent intact cells. These cells filled with starch constitute a barrier of the cell itself. On the contrary, some profiles are almost totally fluorescent in oil, which corresponds to cut or damage oil-filled cells.	The results showed that after frying, the cell structures were well preserved in terms of shape and size occupying about 50% of its volume partially damaged by cell detachment due to starch swelling and dehydration.	Arash et al. (2020)
Different apple varieties	The micrographs were captured in RGB format and stored in TIFF, then they were converted to grayscale and later to binary images by manual segmentation, managing a segmentation threshold.	The apple tissue is characterized through a morphometric analysis through image analysis with the ImageJ v1.49 program.	The morphometric parameters of the cells obtained from the different apple varieties were mainly the distribution of the pores by means of environmental scanning electron microscopy.	Scanning electron microscopy studies showed the distribution of porosity in the apple varieties studied, finding that the Red Delicious apple has a greater number of pores while the Golden Delicious variety has a higher pore density compared to all the varieties studied.	Montes de Oca (2019)

the mouth and altered flavor and aroma profiles, compared to foods produced by conventional oil frying (Yi et al. 2021).

In recent years, various studies have reported new technological alternatives for the processing of foods that reduce their oil content while maintaining their nutritional and sensory quality. And to a greater extent polycyclic aromatic hydrocarbon formation has been reported to occur as the degree of unsaturation of the added lipids increases (Kim et al. 2021). Also, Okonkwo et al. (2022) attributed this phenomenon to the fact that unsaturated fatty acids are more prone to oxidation during heating giving place to polycyclic aromatic hydrocarbons.

Mohammad et al. (2022) investigated the effects of hot-air and oil frying on the microstructure of potato chips. It was concluded that the starch molecules vibrate, breaking the intermolecular hydrogen bonds in amorphous zones of the granules, which causes swelling due to a progressive and irreversible absorption of water that finally joins the structure of the product, thus influencing the gelatinization and the digestibility of this polymer. In this aspect, starch gelatinization decreased from 59 to 86%, while its digestibility increased from 54 to 59% in hot-air fried samples. It is noteworthy that fried foods containing starch and protein such as French fries are the main source of acrylamide. The amount of acrylamide in a particular food commodity depends on factors such as the type of food, its moisture content, and frying conditions (Nguyen et al. 2023). Acrylamide is formed during a Maillard reaction between amino acids (e.g., asparagine) and reducing sugars (e.g., glucose and fructose) when the temperature of a food exceeds 120°C during an immersion frying process, and this compound is commonly produced during the frying of foods with high starch content (Yi et al. 2021).

Dehghannya & Ngadi (2021) found that the acrylamide content of hot-air French fries was significantly lower than that of high oil fries produced by traditional immersion processing. In this same study, acrylamide levels of traditional fries was 290 ppm, while in the hot-air products it was 78 ppm (73.11% decrement).

The reason why the hot-air fried samples had lower content of acrylamide and polycyclic aromatic hydrocarbons is due mainly to shorter processing times and lower oil content used with respect to oil-fried samples (Kim, et al., 2021).

In **Table 22.4**, a comparison of the quality attributes of selected materials treated with the conventional frying process and hot-air frying is shown. It is observed that the acrylamide content was higher in potato chips and potato strips in an oil-frying process.

The frying temperature and processing time and the consequent formation of crust also play an important role in the textural attributes of the foods fried by means of hot air. Another characteristic is that hot-air produced samples appeared swollen and dry, which has been related to an increase in the volume of gelatinized starch in their cells (Salamatullah et al. 2021). Hot-air fried samples also showed greater crust shrinkage during the cooling stage. However, hot-air fried samples retained higher levels of vitamins since chemical degradation reactions were inhibited (Rurush et al. 2022).

Color is another important quality attribute of fried foods. Yu et al. (2020) reported that the color of hot-air fried products such as potatoes or sweet potatoes changed progressively with increasing temperature and frying time, which can be attributed to moisture loss and formation of brown colors on surfaces due to Maillard mechanisms. In another study, the impact of hot-air frying on the oil and moisture content, firmness, and color of sweet potato was investigated using the response surface methodology (Abd Rahman et al. 2018). The suggested frying conditions were 150°C for 12 min, and compared to oil immersion processing, hot-air frying produced samples with significantly lower oil content (90 % reduction) (Sharanaga et al. 2021).

TABLE 22.4 EFFECTS OF HOT-AIR FRYING ON DIFFERENT SAMPLES

Materials (Samples)	Treatment	Acrylamide (ppm)	Polymer Content (%)	Moisture Content (g/100 g db)	Oil Content (g/100 g db)	Reference
Potato chips	CF (180 °C, 6 min) AF (180 °C, 10 min, 1400 W)	288 86	0.20 0.08	77.5 to 3.2 77.5 to 16.6	0 to 42 0 to 0.0024	Yi et al. (2021)
Sweet potato snacks	CF (150 °C, 8 min) AF (150 °C, 12 min, 1300 W)	- -	- -	- -	0 to 48.23 0 to 23.78	Godswill et al. (2018)
Potato strips	CF (180 °C, 6 min) AF (180 °C, 6 min, 1300 W)	232 118	0.16 0.10	- -	- -	Dehghannya & Ngadi (2021)

***CF, conventional oil-frying; AF, hot-air frying**

In addition, the hot-air fried samples showed a lighter color and harder texture with better sensory attributes and had a drier and tighter continuous network structure attributed to starch gelatinization on the surface of the samples. On the other hand, deep-oil fried products had gradual dehydration due to the development of a crispy surface and the brown color due to Maillard reactions (Sharanagat et al. 2021). Furthermore, Shoma et al. (2022) investigated the effects of hot-air frying and deep-oil frying on the microstructure of potato strips; hot-air frying produced samples with a compact and well-organized microstructure, while deep frying resulted in samples with a crispy, oil-soaked surface.

The morphology changes suffered by foods that have been subjected to hot-air frying processes are frequently studied by identifying the microstructural parameters of the crust that could explain the crispy texture of these foods. And in this respect, the microstructure of processed vegetables can be visualized through different techniques of microscopy such as optical, confocal, atomic force, environmental and scanning electronic, as well as by X-ray micro computed tomography, among others (Vandna et al. 2023).

Image analysis provides information on the pore size distribution, the global porosity of the products, and the local porosity of the crust region (Siqi et al. 2023). Gouyo et al. (2021) showed that foods such as pre-frozen French fries for the most part did not suffer any reduction in volume during frying; the total porosity created in these materials approximately corresponded to the volume of water lost during frying, and the difference between fries processed with hot air and fried oil can be related mainly to the diameter of the pores and to their size distribution in the crust. Also, these authors conducted a principal component analysis to relate the sensory and morphometric parameters showing that the volume of small pores and the width of the pore size distribution in the first few millimeters of the crust correlated well with the crispness of the product.

To achieve a good level of crispness in fried products, the frying process must generate a high density of heterogeneous pores with diameters ranging from 0 to 0.15 mm and an average diameter of less than 0.2 mm within a few millimeters of the cortex (Gouyo et al. 2021).

Therefore, foods that are subjected to hot-air frying will undergo changes in their microstructure, and rather frequently they will be influenced by the applied processing conditions (frying time and temperature) (Vandna et al. 2023).

22.5 CONCLUSION

Hot-air frying has gradually become a useful technology for obtaining products with better sensory characteristics, as well as healthier nutritional profiles than those obtained by immersion frying in hot oil; in the former case, lower lipid content is shown by products prepared by this technology. Also, the formation of acrylamide of hot-air fried products is approximately 70% lower than that of traditionally oil fried products; and the sensory attributes such as color, texture, flavor, nutrient retention, porosity, and other physical properties such as density and specific volume and overall quality are similar in those products elaborated by both treatments. The hot-air frying technology is giving rise to new foods that need to be further studied in their microstructure through the use of different microscopy techniques such as laser scanning confocal and electron scanning microscopies, as well as magnetic resonance and infrared spectroscopy. These techniques have proven to be useful for the characterization of changes in micropores and crust in fried plant tissues considering that microstructural changes are relevant for the description of nutritional, morphometric, and sensory properties in hot-air fried foods. Foods treated with the latter procedure possess cellular architecture with heterogeneous sizes and irregular and non-circular shapes with different levels of porosity. This process causes the cells to shrink, a phenomenon that is expected for this type of treatment, but that generates a product with a lower surface area while the perimeter increases significantly. Interestingly, the magnitude of the cited changes depends on the operating conditions of the corresponding process.

REFERENCES

Abd Rahman N.A., Razak S.Z.A., Lokmanalhakim L.A., Taip F.S., Kamal S.M.M. (2018). Response surface optimization for hot-air-frying technique and its effects on the quality of sweet potato snack. *Journal of Food Process Engineering*, 40(4), e12507. https://doi.org/10.1111/jfpe.12507

Arash G., Arash K., Mohebbat M., Michael O.N. (2018). Effect of deep oil and hot-air frying on doughnuts physical properties and kinetic of crust formation. *Journal of Cereal Science*, 83, 25–31. https://doi.org/10.1016/j.jcs.2018.07.006

Arash G., Mohebbat M., Arash K., Michael O.N. (2020). An agent-based coupled heat and water transfer model for air frying of doughnut as a heterogeneous multiscale porous material. *Innovative Food Science and Emerging Technologies*, 61, 102335. https://doi.org/10.1016/j.ifset.2020.102335

Belkova B., Hradecky J., Hurkova K., Forstova V., Vaclavick L., Hajslova J. (2018). Impact of vacuum frying on quality of potato crisps and frying oil. *Food Chemistry*, 241, 51–59. https://doi.org/10.1016/j.foodchem.2017.08.062

Chong M., Mazzitelli G., Quintero R. (2019). Efecto de los métodos de cocción por fritura en las propiedades fisicoquímicas y sensoriales de chips de taro (*Colocasia esculenta*). *Revista I+D Tecnológico*, 15(1), 30–37. https://https://doi.org/10.33412/idt.v15.1.2095

Dehghannya J., Ngadi M. (2021). Recent advances in microstructure characterization of fried foods: Different frying techniques and process modeling. *Trends in Food Science & Technology*, 116, 786–801. https://doi.org/10.1016/j.tifs.2021.03.033

Godswill A. Ch., Amagwula I.O., Igwe V.S., Gonzaga A.I. (2018). Effects of repeated deep frying on refractive index and peroxide value of selected vegetable oils. *International Journal of Advanced Academic Research*, 4(4), 106–119.

Gouyo T., Rondet E., Mestres C., Hofleitner C., Bohuon P. (2021). Microstructure analysis of crust during deep-oil or hot-air frying to understand French fry texture. *Journal of Food Engineering*, 298, 110484. https://doi.org/10.1016/j.jfoodeng.2021.110484

Kim H., Cho J., Jang A. (2021). Effect of charcoal type on the formation of polycyclic aromatic hydrocarbons in grilled meats. *Food Chemistry*, *343*:128453. https://doi.org/10.1016/j.foodchem.2020.128453

León-López L.P., Espejel-Blanco J.E. (2019). Dehydrated sweet potato as a new healthy consumption alternative for Sonorans. *Revista Inclusiones*, (Edited by Editorial Cuadernos de Sofía SPA de la República de Chile). Vol 6, ISSN 0719-4706

Mingchih F., Wen-Chieh S. (2021). Mass transfer and texture characteristics of fish skin during deep-oil-frying, electrostatic frying, air frying and vacuum frying. *Food Science and Technology*, 137, 110494. https://doi.org/10.1016/j.lwt.2020.110494

Mohammad F., Ibrahim K., Rokkaya S., Ebtihal K., Khadiga A.I., Mokhtar D. (2022). Optimization of the frying temperature and time for preparation of healthy falafel using air frying technology. *Foods*, 10, 2567. https://doi.org/10.3390/foods10112567

Montes de Oca Rosales L. (2019). Effect of frying treatment, addition of fresh oil and type of feeding on the oxidative stability of a vegetable oil. *MSc Tesis*, Autonomous University of the State of Mexico, Toluca, State of Mexico, Mexico.

Nguyen T.T., Rosselló C., Ratti C. (2023). Simple mathematical modelling to represent air-drying kinetics of potato peel. *Journal of Food Engineering*, 357, 111634. https://doi.org/10.1016/j.jfoodeng.2023.111634

Okonkwo C.E., Moses O.I., Nwonuma Ch., Abiola T., Benjamin B.O., Joshua O., Folorunsho J.O., Olaniran A.F., Pan Z. (2022). Infrared and Microwave as a dry blanching tool for Irish potato: Product quality, cell integrity, and artificial neural networks (ANNs) modeling of enzyme inactivation kinetic. *Innovative Food Science and Emerging Technologies*, 78, 103010. https://doi.org/10.1016/j.ifset.2022.103010

Rurush E., Alvarado M., Palacios P., Flores Y., Rojas M.L., Miano A.C. (2022). Drying kinetics of blueberry pulp and mass transfer parameters: effect of hot-air and refractance window drying at different temperatures. *Journal of Food Engineering*, 320, 110929. https://doi.org/10.1016/j.jfoodeng.2021.110929

Salamatullah A., Ahmed M., Alkaltham M., Hayat K., Aloumi N., Al-Dossari A., Al-Harbi L., Arzoo S. (2021). Effect of air-frying on the bioactive properties of eggplant (*Solanum melongena* L.). *Processes*, 9(3), 435, https://doi.org/10.3390/pr9030435

Sharanagat V.S., Nema P.K., Singh L., Kumar A. (2021). Formation of acrylamide in microwave-roasted sorghum and associated dietary risk. *International Journal of Food Science and Technology*, *57*, 1654–1665. https://doi.org/10.1111/ijfs.15528

Shoma D., Min Z., Ronghua J., Bhesh B. (2022). Recent development of innovative methods for efficient frying technology. *Critical Reviews in Food Science and Nutrition*, *61*(22),3709–3724. https://doi.org/10.1080/10408398.2020.1804319

Siqi Z., Shuyang W., Haizhen D., Zhenqi G., Mthokozisi B.C., Simelane Q.L., Tingting T., Liping G., Le Ch., Chao D. (2023). Heat transfer modelling for novel infrared peeling of potato. *Journal of Food Engineering*, 357, 111631. https://doi.org/10.1016/j.jfoodeng.2023.111631

Teruel M., Gordon M., Linares M., Garrido M., Ahromrit A., Niranjan K. A. (2018). Comparative study of the characteristics of french fries produced by deep oil-frying and air frying. *Food Engineering & Materials Science*, 80(2), E349–E358. https://doi.org/10.1111/1750-3841.12753

Tian J., Shiguo C., Join S., Jianchu C., Donghong L., Yidi., Yukiharu O., Xingqian Y. (2019). Microstructure and digestibility of potato strips produced by conventional frying and air-frying: An in vitro study. *Food Structure*, 14, 30–35. https://doi.org/10.1016/j.foostr.2017.06.001

Vandna V., Vijayta S., Om P. Ch., Neelam Y. (2023). Comparative evaluation of conventional and advanced frying methods on hydroxymethylfurfural and acrylamide formation in French fries. *Innovative Food Science and Emerging Technologies*, 83, 103233. https://doi.org/10.1016/j.ifset.2022.103233

Verma V., Singh Z., Yadav N. (2022). Effective method of minimizing the acrylamide and 5-hydroxymethylfurfural formation in French fries by lactic acid fermentation. *Journal of Food Measurement and Characterization*, 16, 5032–5043. https://doi.org/ 10.1007/s11694-022-01568-y

Xinli R., Dingsong L., Lingjie Z., Yifan L., Hongshun Y. (2023). Kinetic modelling of the mass and heat transfer of a plant-based fishball alternative during deep-oil-frying and air frying and the changes in physicochemical properties. *Journal of Food Engineering*, 350, 111457. https://doi.org/10.1016/j.jfoodeng.2023.111457

Yi W., Xianglei W., David J.M., Long Ch., Ming M., Zhengyu J. (2021). Effect of new frying technology on starchy food quality. *Foods*, 10, 1852. https://doi.org/ 10.3390/foods10081852

Yu X., Li L., Xue J., Wang J., Song G., Zhang Y., Shen Q. (2020). Effect of air-frying conditions on the quality attributes and lipidomic characteristics of surimi during processing. *Innovation Food Science Emerging Technology*, 60, 102305. https://doi.org/10.1016/j.ifset.2020.102305

Index

B

C

D

E

M

N

O

Q

R

S

Y

Z